Photovoltaic Systems Engineering for Students and Professionals

Photovoltaic Systems Engineering for Students and Professionals: Solved Examples and Applications examines photovoltaic (PV) power plants in a holistic way. PV installations of all types and sizes – from the smallest plant element to the largest system components – are approached from an electrical engineering perspective and further explained through worked examples. It presents the different forms of energy and the energy conversions between them in a clear and understandable way. This book is an essential resource for both students and practicing engineers working in the solar photovoltaic areas and critical work for all electrical engineers.

Features:

- Includes over 100 worked examples and more than 80 end-of-chapter problems
- Presents systematic techniques and approaches to problem solving
- Includes PowerPoint presentations and a solutions manual for instructors
- Considers the effects of environmental conditions on the performance of PV systems
- Presents step-by-step design of photovoltaic systems of all sizes from scratch

Photovoltaic Systems Engineering for Students and Professionals
Solved Examples and Applications

Mugdesem Tanrioven

CRC Press
Taylor & Francis Group
Boca Raton London New York

CRC Press is an imprint of the
Taylor & Francis Group, an **informa** business

MATLAB® and Simulink® are trademarks of The MathWorks, Inc. and are used with permission. The MathWorks does not warrant the accuracy of the text or exercises in this book. This book's use or discussion of MATLAB® and Simulink® software or related products does not constitute endorsement or sponsorship by The MathWorks of a particular pedagogical approach or particular use of the MATLAB® and Simulink® software.

Designed cover image: Drawings and design by Mugdesem Tanrioven

First edition published 2024
by CRC Press
6000 Broken Sound Parkway NW, Suite 300, Boca Raton, FL 33487-2742

and by CRC Press
4 Park Square, Milton Park, Abingdon, Oxon, OX14 4RN

CRC Press is an imprint of Taylor & Francis Group, LLC

© 2024 Mugdesem Tanrioven

Reasonable efforts have been made to publish reliable data and information, but the author and publisher cannot assume responsibility for the validity of all materials or the consequences of their use. The authors and publishers have attempted to trace the copyright holders of all material reproduced in this publication and apologize to copyright holders if permission to publish in this form has not been obtained. If any copyright material has not been acknowledged please write and let us know so we may rectify in any future reprint.

Except as permitted under U.S. Copyright Law, no part of this book may be reprinted, reproduced, transmitted, or utilized in any form by any electronic, mechanical, or other means, now known or hereafter invented, including photocopying, microfilming, and recording, or in any information storage or retrieval system, without written permission from the publishers.

For permission to photocopy or use material electronically from this work, access www.copyright.com or contact the Copyright Clearance Center, Inc. (CCC), 222 Rosewood Drive, Danvers, MA 01923, 978-750-8400. For works that are not available on CCC please contact mpkbookspermissions@tandf.co.uk

Trademark notice: Product or corporate names may be trademarks or registered trademarks and are used only for identification and explanation without intent to infringe.

ISBN: 978-1-032-54185-3 (hbk)
ISBN: 978-1-032-54187-7 (pbk)
ISBN: 978-1-003-41557-2 (ebk)

DOI: 10.1201/9781003415572

Typeset in Times
by codeMantra

Access the instructor and student resources: www.routledge.com/9781032541853

My Beloved Wife Fatma,
Children Yusuf and Zeynep.

Contents

Foreword xiii
Author xvii

Chapter 1 Energy and Energy Conversion 1

    1.1   Work, Energy, and Power 1
    1.2   Energy Conversion and Different Forms of Energy 1
            1.2.1   Electrical Energy 1
            1.2.2   Thermal Energy 2
            1.2.3   Light (Solar) Energy 3
            1.2.4   Chemical Energy 4
            1.2.5   Linear Kinetic Energy 4
            1.2.6   Electromagnetic Energy 5
            1.2.7   Mechanical Energy (Rotational Kinetic Energy) 5
            1.2.8   Nuclear Energy 6
            1.2.9   Potential Energy 7
            1.2.10 Sound Energy 8
    1.3   Energy Units and Conversion Factors 9
    1.4   Problems 10
    Bibliography 10

Chapter 2 The Sun and Solar Radiation 11

    2.1   The Sun as an Energy Source 11
    2.2   Solar Spectrum 11
    2.3   Solar Geometry 14
            2.3.1   The Orbit of Earth 15
            2.3.2   The Earth's Latitude and Longitude 18
            2.3.3   Solar Angles 20
    2.4   The Solar Incidence Angle 26
    2.5   Sun's Position, Solar Time, and Clock Time 29
    2.6   Solar Radiation Calculations 37
            2.6.1   Calculation of Clear-Sky Radiation 37
            2.6.2   Radiation Calculation Based on Atmospheric Transmission 40
            2.6.3   Calculation of Hourly, Daily, and Monthly Radiation 42
            2.6.4   Extraterrestrial Radiation on Horizontal Surfaces 43
    2.7   Insolation on Solar Tracking Surfaces 47
    2.8   Shading and Shadow Analysis 50
            2.8.1   Shadow Types 50
            2.8.2   Shadow Geometry 51
            2.8.3   Shadow Analysis 53
    2.9   Solar Radiation Measurements 56
            2.9.1   Pyranometer 57
            2.9.2   Pyrheliometer 59
            2.9.3   Albedometer 59
            2.9.4   Sunlight Recorder 59

	2.10 Problems	61
	Bibliography	63
Chapter 3	Introduction to Photovoltaic Systems and Photovoltaic Technologies	65
	3.1 Introduction	65
	3.2 Semiconductors and *p-n* Junction	65
	3.2.1 Energy Bands and Silicon Lattice	66
	3.2.2 The *p-n* Junction	70
	3.2.3 The *p-n* Junction Diode	72
	3.3 The PV Cell	75
	3.4 The PV Module and PV Array	76
	3.5 The PV Modeling and Electrical Characteristics of PV Systems	77
	3.5.1 Ideal PV Model	77
	3.5.2 The PV Cell with Shunt Resistance	79
	3.5.3 The PV Cell with Series Resistance	80
	3.5.4 The PV Cell with Series and Shunt Resistance	81
	3.5.5 Two-Diode Model	83
	3.6 Characteristics Curves of PV Cells	84
	3.7 Special Issues on *I–V* Curves	86
	3.8 Additional Performance Parameters of PV Cells	92
	3.9 Solution Methods of Photovoltaic Equations	94
	3.9.1 *I–V* Equations for Series-Parallel Connections	94
	3.9.2 Simple Approximation Method	96
	3.9.3 Parameter Estimation of Single-Diode Model	98
	3.9.4 Newton–Raphson Method	101
	3.9.5 Parameter Estimation of Double-Diode Model	106
	3.10 Photovoltaic Technologies	107
	3.10.1 Crystalline Silicon Technologies	107
	3.10.1.1 Monocrystalline Silicon	108
	3.10.1.2 Polycrystalline Silicon	108
	3.10.1.3 String Ribbon Silicon	108
	3.10.2 Thin-Film Technologies	108
	3.10.2.1 Amorphous Silicon	109
	3.10.2.2 Cadmium Telluride	109
	3.10.2.3 Copper Indium Gallium Selenide (CIGS-CIS)	109
	3.10.2.4 Gallium Arsenide Thin Film	109
	3.10.2.5 Thin-Film Microcrystalline (Polycrystalline) Silicon Cell	110
	3.10.2.6 Multi-Junction Concentrated PV Cells	110
	3.10.3 New Emerging PV Technologies	110
	3.10.3.1 Dye-Sensitized PV Cells	110
	3.10.3.2 Organic PV Cells	110
	3.10.3.3 Perovskite PV Cells	111
	3.10.3.4 Quantum-Dot PV Cells	111
	3.10.3.5 Graphene PV Cells	111
	3.10.3.6 Hybrid PV Cells	111
	3.10.3.7 Bifacial PV Cells	111
	3.11 Technical Specifications of Different PV Technologies	112
	3.12 Problems	125
	Bibliography	127

Chapter 4   Effect of Environmental Conditions on the Performance of Photovoltaic Systems ... 129

- 4.1 Introduction ... 129
- 4.2 Factors Affecting PV Module/PV Cell Operating Temperature ... 129
- 4.3 The Angle of Incidence Effect on PV Performance ... 138
- 4.4 Air Mass Effect on PV Performance ... 141
- 4.5 Wind Effect on PV Systems ... 154
- 4.6 Soiling/Dust Effect on PV Systems ... 155
- 4.7 Effect of Rain and Humidity on PV Systems ... 157
- 4.8 Snow and Icing Effect on PV Systems ... 157
- 4.9 Shading and Bypass Diodes Effect on PV System ... 158
  - 4.9.1 Partial Shading and Hotspot Phenomenon ... 159
  - 4.9.2 Estimation of Bypass Diodes Included in a Junction Box ... 159
  - 4.9.3 Mathematical Modeling of Partially Shaded PV System ... 164
  - 4.9.4 Blocking Diodes and Partial Shading ... 172
- 4.10 Tilt Angle Determination for Stand-Alone and Grid-Connected Systems ... 175
  - 4.10.1 Tilt Angle for Grid-Connected Systems ... 177
    - 4.10.1.1 Hourly Tilt Angle ... 177
    - 4.10.1.2 Daily Tilt Angle ... 183
    - 4.10.1.3 Monthly, Seasonally, and Yearly Tilt Angles ... 186
  - 4.10.2 Tilt Angle for Stand-Alone Systems ... 188
- 4.11 Problems ... 192
- Bibliography ... 195

Chapter 5   Photovoltaic Power Systems: Designing and Sizing ... 199

- 5.1 Introduction ... 199
- 5.2 Grid-Connected Photovoltaic Power Systems ... 199
  - 5.2.1 Photovoltaic Array Configurations ... 201
    - 5.2.1.1 Photovoltaic Array for Module-Inverter Concept ... 203
    - 5.2.1.2 Photovoltaic Array for String Inverter Concept ... 204
    - 5.2.1.3 Photovoltaic Array for Multi-String Inverter Concept ... 204
    - 5.2.1.4 Photovoltaic Array for Master-Slave Concept Inverters ... 206
    - 5.2.1.5 Photovoltaic Array for Team Concept Inverters ... 206
    - 5.2.1.6 Photovoltaic Array for Central Inverter Concept ... 206
    - 5.2.1.7 Alternative Photovoltaic Array Configurations ... 207
  - 5.2.2 DC Combiner Box and DC Distribution Box ... 208
  - 5.2.3 Photovoltaic Inverters ... 210
    - 5.2.3.1 Photovoltaic Grid Inverters ... 210
    - 5.2.3.2 Stand-Alone Inverters ... 216
    - 5.2.3.3 Bimodal (Hybrid) Inverters ... 219
  - 5.2.4 Photovoltaic Systems for Three-Phase Connection ... 219
  - 5.2.5 AC Distribution Box and Utility Connection ... 220
  - 5.2.6 Substation Layout for Utility-Connected PV Systems ... 222
  - 5.2.7 Protection Devices in PV Systems ... 223
    - 5.2.7.1 Surge Protection Devices ... 223
    - 5.2.7.2 Residual Current Devices ... 226
    - 5.2.7.3 Photovoltaic DC Fuses ... 227
    - 5.2.7.4 Disconnectors ... 229
    - 5.2.7.5 Load Switches ... 231
    - 5.2.7.6 Switch Disconnectors ... 231

		5.2.7.7	Circuit Breakers	231
		5.2.7.8	Contactors	232
	5.2.8	Other Balance of System Components		234
		5.2.8.1	Inverters	234
		5.2.8.2	PV Rapid Shutdown System	236
		5.2.8.3	Monitoring and Communication Systems	236
		5.2.8.4	Bidirectional Net Metering	237
		5.2.8.5	Photovoltaic Cables and Cable Systems	239
5.3	Grid-Connected PV Systems with Battery Storage			244
	5.3.1	Grid-Connected PV Systems with DC-Coupled Battery Storage		244
	5.3.2	Grid-Connected PV Systems with AC-Coupled Battery Storage		246
	5.3.3	Grid-Connected PV Systems with AC-Coupled Battery Backup		247
5.4	Stand-Alone Photovoltaic Systems			247
	5.4.1	Direct-Coupled Stand-Alone PV Systems		248
	5.4.2	Stand-Alone Photovoltaic DC Systems with Battery Storage		248
	5.4.3	Stand-Alone Photovoltaic AC Systems with Battery Storage		249
	5.4.4	Stand-Alone Hybrid PV Systems		250
5.5	PV System Sizing and Component Selection			252
	5.5.1	PV Array Sizing and Land Requirements		253
		5.5.1.1	PV Array Selection and Sizing for Stand-Alone Systems	253
	5.5.2	Land Area Requirements for PV Power Plants		270
	5.5.3	DC Combiner and Recombiner Box Sizing and Selection		283
	5.5.4	Battery Sizing and Selection		286
	5.5.5	Charge-Controller Sizing and Selection		298
	5.5.6	Inverter Sizing and Selection		304
		5.5.6.1	Inverter Sizing and Selection for Off-Grid Systems	304
		5.5.6.2	Inverter Sizing and Selection for On-Grid Systems	307
		5.5.6.3	Inverter Sizing and Selection for Hybrid Systems	311
	5.5.7	Protection System Requirements, Sizing, and Selection		312
		5.5.7.1	Fuse Sizing and Selection for Photovoltaic Systems	312
		5.5.7.2	Switch Disconnector Sizing and Selection for Photovoltaic Systems	317
		5.5.7.3	Circuit Breaker Sizing and Selection for Photovoltaic Systems	322
		5.5.7.4	Sizing and Selection of Circuit Breakers against Grid Short Circuits	323
		5.5.7.5	Surge Protection Device Sizing and Specifying for Photovoltaic Systems	341
		5.5.7.6	SPD Selection and Installation Requirements for Communication Lines	365
		5.5.7.7	Residual Current Device Sizing and Selection for Photovoltaic Systems	371
		5.5.7.8	Specification and Selection of Instrument Transformers	384
		5.5.7.9	Specification and Selection of DC Auxiliary Supply System	393
		5.5.7.10	Overcurrent Protection of Grid-Connected PV Systems	399
		5.5.7.11	Cable Sizing and Selection for Photovoltaic Systems	420
		5.5.7.12	Busbar Sizing and Selection for Photovoltaic Systems	431
		5.5.7.13	AC Combiner and Recombiner Sizing and Specifying	434

Contents   xi

		5.5.7.14	Bonding and Grounding Methods in Photovoltaic Systems 436
		5.5.7.15	Lightning System Design for Photovoltaic Power Systems 463
	5.5.8	Identifying Load Requirements: Estimation and Calculations 471	

Problems 475
Bibliography 488

Chapter 6 Photovoltaic System Applications 493

6.1 Introduction 493
6.2 Warning Signals 494
6.3 Lighting Systems 497
6.4 Telecommunication Systems 503
6.5 Water Pumping Systems 511
6.6 Remote Monitoring and Cathodic Protection 525
6.7 Problems 533
Bibliography 536

Chapter 7 Grid Codes for Integration of PV Systems and Power Quality Assessment 539

7.1 Introduction 539
7.2 Terms and Definitions 540
    7.2.1 Point of Common Coupling and Grid Connection Point 540
    7.2.2 Relative Voltage Change (ε) 540
7.3 Steady-State Voltage Deviation Limits 543
7.4 Voltage Flicker Limits 548
7.5 Steady-State Frequency Limits 557
7.6 Active Power Control and Frequency Regulation 558
7.7 Reactive Power Control and Voltage Regulation 559
7.8 Voltage Unbalance Limits 560
7.9 Harmonic Limits 562
7.10 Voltage Ride-through Requirements 563
7.11 Supporting the Network with Reactive Current 564
7.12 Frequency Ride-through Requirements 566
7.13 Problems 567
Bibliography 569

Chapter 8 Economics of Solar Photovoltaic Systems 573

8.1 Introduction 573
8.2 Basic Economic Parameters and Calculations 573
    8.2.1 Cash Flow Models 573
    8.2.2 Present Value Method 574
    8.2.3 Net Present Value Method 575
    8.2.4 Capital Recovery Factor 577
    8.2.5 Life-Cycle Cost 580
    8.2.6 Levelized Cost of Energy 581
    8.2.7 Internal Rate of Return 584
    8.2.8 Simple Payback Period 586
    8.2.9 Discounted Payback Period 587

	8.3	Economic Assessment of Grid-Connected PV Systems	588
	8.4	Economic Assessment of Stand-Alone PV Systems	593
	8.5	Problems	594
	Bibliography		597

Index 599

Foreword

To the Readers of This Book

In addition to the technical content, I believe that the story of the emergence of this book is one of the aspects that make the book special. I guess every reader who reads the foreword will also agree with this assessment. For this reason, I would like to write the preface of the book in two parts.

The Story of the Book

Without touching on the details of my academic resume, I would like to briefly introduce myself. As of 2016, I had a dynamic academic background of 21 years, with the last 5 years being a full-time professor and Head of the Electrical Engineering Department at Yildiz Technical University. I have taught several undergraduate and graduate courses at various universities throughout my academic career. I have also worked as a post-doctoral researcher at the University of Liverpool in the United Kingdom and the University of South Alabama in the United States. I have supervised dozens of graduate and doctoral students so far. I have worked as a principal investigator, consultant, or project member on dozens of different projects. The advancements and diversity in my academic studies continued to increase. In short, I had all the opportunities and conditions ideal for an academician. However, this happiness in my academic life would not last long and our life would soon turn into a nightmare. While I was the Head of the Electrical Engineering Department at Yıldız Technical University in Istanbul, I left unemployed all of a sudden. I had to do something, and I decided to write an international engineering book as one of the best things an academic can do. However, it was clear that my job would not be easy due to the challenges of writing an international book. Because I had to meet an academic demand, and I had to be able to do this in the limited conditions I was in.

In fact, I had no difficulty in determining the subject, scope, and title of the book. Because I was an academician who gave lectures, supervised theses, and developed projects in the field of renewable energy systems for many years. For this reason, I knew the existing books in the literature closely and knew very well how I could contribute to the existing books. I was very confident about this and decided to write a book on Photovoltaic Systems Engineering. However, my work was just beginning, and I had already taken the first step of a journey that would take about 6 years.

It should not be overlooked that while I was writing this book, I was an unemployed academic who struggled to live under stress and had limited opportunities, rather than a comfortable academician with all the opportunities. However, all these negative conditions I was in became a more motivational source for me instead of making me despair. In my book studies, I sometimes used my laptop with a cracked screen, and sometimes I used the internet facilities of public libraries. Sometimes it took days to find an article I might need, and sometimes it took weeks to complete the missing data in an article. When I couldn't find enough time during the day, I was working at night or sacrificed my sleep time. Most of the time, a pen, a blank sheet of paper, and a calculator were sufficient for me to solve the examples in the book. Of course, the manual solution was time-consuming, but it was the most efficient way to debug in the literature. Thus, I have identified many problematic issues in the literature and confirmed the theoretical information given with resolved results.

Days, weeks, and months followed each other in this way, and at the end of six years, this book was completed by the effort as if I raised a child. I would like to take this opportunity to express my gratitude to my beloved wife FATMA and kids YUSUF and ZEYNEP for their moral support throughout this difficult period of my life. Well, let's now get to the point. What makes this book different from its counterparts, what makes this book important, and why academicians, students, and engineers should have this book as a reference source? The answers to these questions and similar ones are available in detail in the technical evaluation section below.

Technical Evaluation of the Book

While writing this book, I did not want to leave the slightest doubt in the reader's mind about the ethical rules, the original value, and the accuracy of the information in this book. In addition, I had to keep the scientific value, the richness of content, and the original aspects of this book at the highest level. Because with this book, I wanted to gain a respectable place in world literature and create a fundamental source that students and academicians in all world universities could refer to in the field of photovoltaic systems. For this purpose, I would like to touch upon some of the principles and issues that I followed while writing this book.

The Text and Language of the Book: None of the texts used in the book, even a single sentence, were copied from any source. In other words, this book was written by the author himself, from the very first line to the last line. A detailed literature reading was completed before each part of the text was written. Then, the notes obtained from the literature readings were transferred to the book from the author's own pen in plain and understandable language with the author's interpretation and synthesis in the context of the subject to be explained. The information in this book has been tried to be given in full consistency. In other words, the author has paid attention not to use any additional sentences that may be considered unnecessary in the context of the subject. In other words, omitting any sentence in the book may cause a lack of expression, while each new sentence to be added to this book may mean unnecessary detail or repetition. Hence, the author has created the text of this book based on this intention.

The Figures: The author has given great importance to the visuals of this book. Because technical issues in engineering become more understandable only when supported by good descriptive visuals. Therefore, each figure was created from scratch by the author himself for better understanding. In this framework, additional information was added to some figures, while some were drawn in three dimensions. Thus, a strong and holistic explanation was created with the help of text-formulation-figure triads.

The Tables: In engineering books, there are various benefits of presenting information/data in the form of tables. The main benefits of using tables are as follows: (i) it is possible to give details that are difficult to express linguistically in tables; (ii) it is possible to summarize long texts in tabulated forms; (iii) it is possible to compare different knowledge and data; and (iv) it is possible to give different variations and large-scale data groups easily and understandably. In this book, the tables have been created for all these needs. While some tables were used to summarize tens of pages, some were preferred for ease of explanation. Likewise, some tables organize the specifications from many company catalogs. One of the most important characteristics of this book is that all datasets needed for calculations are given in full. For example, articles often publish a particular part of their dataset in their publications. It is sometimes necessary to search for dozens of different articles to reach the entire dataset. In this book, the incomplete data sets in different sources have been completed and understandably transferred to tables. Another important point I should underline about the tables is the importance of table footnotes. Sub-table footnotes are used for many of the tables in this book. It is of great benefit to read these footnotes together with the table. Because these footnotes contain exceptional cases, supplementary information, and additional explanations. For these reasons, the readers should read the table footnotes.

The Formulations: This book covers the math behind the theory of photovoltaic systems engineering in detail. This is one of the most important features of this book that distinguishes it from its counterparts on the market. Namely, the number of formulations in an average photovoltaic system book on the market is around 50–60 at the most. In this book, a total of 510 mathematical formulas have been given. Moreover, current photovoltaic system books on the market contain worked examples for very few of the relevant formulations. However, the worked examples in this book are given for all 510 formulas applied to real-life problems. Thus, theory and practice have been given together at every stage to maximize the understanding of explanations.

The Chapters: This book consists of eight chapters in total. Each chapter has its own characteristics and strengths. These sections are described in detail below.

Foreword

In Chapter 1, the concepts of energy and power are explained in the most basic sense. In this context, the different forms of energy and the energy conversions between them are given in an understandable way. Besides, solved examples are given for the energy forms that are related to photovoltaic system engineering. At the end of the chapter, the problems to be studied are given.

The main theme of Chapter 2 is solar geometry and solar radiation calculations. Those who complete this section by understanding can calculate solar radiation values at any location, time, and angle in the world (instant, hourly, daily, and monthly), perform shading analysis, and find sunshine durations. With these calculations, it is possible to conduct site-specific solar feasibility studies. Thus, the first necessity of deciding on design and investment becomes possible through the information in this section. In this chapter, besides solar calculations, the basics of solar radiation measurements and measuring devices are introduced. Again, at the end of the chapter, the problems to be solved are included as similar to the worked examples.

In Chapter 3, the fundamentals of photovoltaic systems, such as electrical models, current-voltage characteristics, performance parameters, photovoltaic equations, and solution methods, have been explained with the help of worked examples. In addition, commercially available and new emerging photovoltaic technologies are comparatively explained with brief explanations. At the end of the chapter, the problems to be worked on are given.

In Chapter 4, the effects of environmental conditions on the performance of PV systems are given. The main environmental effects considered are temperature, angle of incidence, air mass, wind speed, dusting, rain, humidity, snow, ice, shading, and tilt angle. Numerical solutions are provided for each parameter described here. The most important point to be underlined in Chapter 4 is that the optimum tilt angle calculation for PV panels is given in a simple way that anyone can apply. This is a very important point. Because the parameter that most affects the performance of a PV system (in terms of energy output) is the panel tilt angle, which has a greater performance impact than the aggregated effect of all environmental parameters. For this reason, in PV system applications, it is necessary to pay attention to the positioning of the solar panels at the optimum tilt angle. Here in this section, calculations of optimum tilt angles on a daily, monthly, seasonal, and annual basis for any location in the world have been explained in a simple way that anyone could apply.

Chapter 5 is the most detailed and the most crucial part of this book in terms of practical applications. In this chapter, the design procedure of photovoltaic systems of any scale from scratch is given in detail at each phase, with the help of worked real-life examples. In this context, the topics covered are as follows: (i) PV array configurations, specifications, formation styles, sizing, and selections; (ii) the charges, converters, and inverters used in the PV system, their technical specifications, sizing, and selection criteria; (iii) AC and DC combiner boxes, technical specifications, sizing, and selections; (iv) protection elements (fuses, disconnectors, circuit breakers, switches, contactors, residual current devices, surge protection devices, etc.), technical specifications, sizing, and selections; (v) sizing and selection of other system components (such as energy storage, monitoring, net metering, cable systems, etc.); (vi) short-circuit analysis and its use in component selection; (vii) land-area requirements and site sizing; (viii) design of grid integration and protection cabinet (protection relay, measurement transformers, etc.), determination of protection functions/set values, and element selection; and (ix) designing the grounding and lightning protection systems. All these topics, the details of which are included in Chapter 5, are explained by applying them to real-field examples.

In Chapter 6, special PV system applications are included. Namely, this chapter describes the application of the previously described design criteria to various areas. In this context, signaling, lighting, telecommunication, monitoring, cathodic protection, and water pumping applications are explained with solved examples.

In Chapter 7, the grid codes required for grid integration of the PV system have been explained and different country codes have been evaluated comparatively. In this context, voltage deviations, frequency deviations, active/reactive power regulation limits, voltage unbalance, voltage flickers, and system harmonic limits are defined in detail and numerical examples are given.

In the last chapter of this book, photovoltaic system investments have been evaluated in terms of economic criteria. In this context, net present value, life-cycle cost, levelized cost of energy, internal rate of return, and payback period methods are explained. Finally, numerical examples were given separately for grid-connected and stand-alone systems.

Worked Examples: A total of 107 worked examples are included in this book. All these examples, specifically created for this book, are based on real systems and are not found in any other source. These examples also demonstrated the application of the 509 mathematical equations given in this book. Most of these 107 examples were solved step by step, and the solution templates were created for them.

The Problems: To consolidate the theory and worked examples in this book, the problems to be solved are given at the end of each chapter (81 problems in total). Thus, a resource opportunity was created for academics and university students for their homework and project assignments.

The Outcomes: The learning outcomes of this book are already described in the above parts. On the other hand, the potential uses of this book and its possible readerships are evaluated in this section. Namely, (i) this book can be used as a textbook in undergraduate, graduate, and doctoral programs of universities. In this context, it is a reference work for academicians and students; (ii) it is a theoretical and practical guide for all engineers and technical staff working in the field of solar systems and photovoltaics; (iii) it is a reference work for solar system engineers working interdisciplinary in fields, such as physics, computers, meteorology, mechanical, electrical, and energy engineering; and (iv) many engineers in the market only do simulation and software-based PV system designs. Mathematical solutions for every design phase of solar PV systems are presented in this book. Therefore, these engineers can be able to verify their designs with the help of this book; (v) this book explains in detail how to implement electrical engineering systems and equipment in the field, specific to the photovoltaic system. For this reason, this book also contributes greatly to the discipline of electrical engineering.

In the broadest sense, the readership of this book will be academics, university students, field engineers, and investors.

I hope that this book will make a noteworthy contribution to the literature and be useful to everyone who reads it.

Mugdesem TANRIOVEN
March 2023/ISTANBUL
E-mail: mtanrioven@gmail.com

Author

The author, **Mugdesem Tanrioven**, was born on September 11, 1970 in Kayseri, Turkey. He received undergraduate, graduate, and doctorate degrees from Yıldız Technical University, Department of Electrical Engineering. He studied his first post-doctoral research at the University of Liverpool, England. Afterward, he returned to Turkey and started to work as an Assistant Professor in the Department of Electrical Engineering at Yildiz Technical University. He worked for the second time as a post-doctoral researcher at the University of South Alabama between 2003 and 2005. In addition to the research studies at the University of South Alabama, he taught at the University of South Alabama as a part-time Instructor in 2005. Then he returned to Yildiz Technical University at the end of 2005 and was promoted to associate professor in 2016. Between 2011 and 2016 years, he also worked as a full-time professor at Yıldız Technical University. During the same period he worked as a professor, he also served as the Head of the Electrical Engineering Department. He is the founder of the Alternative Energy Systems (AES) Division at Yıldız Technical University. At the same time, he is the founder of the Energy Efficiency, Energy Systems, and Smart Grids laboratories in the AES Division.

He has conducted dozens of research and industry projects, as well as dozens of postgraduate theses. He has received many awards from IEEE, university, and industry organizations. As of 2023 August, his articles published in Science Citation–indexed journals have received more than 860 citations. He is the Chair of the "International Conference on Electric Power and Energy Conversion Systems (EPECS)" held at Yildiz Technical University for the third time in 2013. He also took part in the technical, organizational, and scientific committees of many international conferences. He served as a referee/reviewer for dozens of scientific journals/articles.

His main research interests and areas of expertise are energy systems, reliability analysis, lighting system designs, wind and solar power systems, high-voltage systems, energy efficiency, smart grids, distributed generation, fuel cells, and load management. He is married and has two children.

1 Energy and Energy Conversion

1.1 WORK, ENERGY, AND POWER

Due to an applied force, displacement or movement of an object in the direction of the force is called work. Energy is defined as the capacity or capability to perform this work. Considering these definitions, power can be defined as the rate of using energy to perform the work or the work done in a unit of time. If the amount of work done (or used energy) in the time interval Δt is ΔW, then the average power in the interested period can be given as:

$$P_{ave} = \frac{\Delta W}{\Delta t} \quad (1.1)$$

Hence instantaneous power: $P = \lim_{\Delta t \to 0} (P_{ave})$

$$P = \lim_{\Delta t \to 0} (P_{ave}) = \lim_{\Delta t \to 0} \frac{\Delta W}{\Delta t} = \frac{dW}{dt} \quad (1.2)$$

According to the International System of Units (SI), the unit of energy is Joule / second [J/s], also called Watt [W]. Electrical installations often have large power ratings. Therefore, the kW or MW scales are often used in practice. The British Standard "Horsepower" (HP) is also frequently used to make it easier for people to perceive. Note that one HP equals 746 W.

Since the work done (or energy consumed/generated) per unit of time is called power, the amount of energy can be calculated by integrating the power.

$$W = \int P \cdot dt \quad (1.3)$$

If the power is constant over the time t:

$$W = P \cdot t \quad (1.4)$$

It would be more appropriate to use E instead of W when dealing with the "Power-Energy" relationship instead of "Power-Work." Since this book focuses on the "Power-Energy" relationship, the symbol E will be used for energy.

1.2 ENERGY CONVERSION AND DIFFERENT FORMS OF ENERGY

Energy is generally converted from one form to another during its use. This process is called energy conversion. In this context, various technologies have developed for energy conversions. Table 1.1 gives different forms of energy, while energy conversion technologies and their efficiencies are outlined in Table 1.2.

1.2.1 Electrical Energy

Electrical energy is a kind of clean energy that can be converted from other energy forms such as mechanical, chemical, and thermal. It is often used to describe the energy amount consumed by an electrical circuit. The kilowatt-hour (kWh) is the most commonly used unit for electrical energy.

TABLE 1.1
Different Forms of Energy

Electrical energy	Electromagnetic energy
Thermal energy	Mechanical energy (rotational kinetic energy)
Photon (solar) energy	Nuclear energy
Chemical energy	Potential energy
Kinetic energy (linear motion energy)	Sound energy

TABLE 1.2
Energy Conversion Forms and Efficiency of Energy Conversion Technologies

Energy Conversion Forms	Conversion Technologies	Efficiency
Mechanical to electrical energy	Electric generator	95%–98%
Chemical to electrical energy	Battery	90%
	Fuel cell	60%
Light (solar) to electric energy	Photovoltaic solar cell	15%–20%
Thermal to electrical energy	Thermocouple	5%–7%
Mechanical to mechanical energy	Gearbox	75%–85%
Kinetic to mechanical energy	Hydraulic turbine	95%
	Wind turbine	30%–45%
Electrical to mechanical energy	Electric motor	65%–95%
Chemical to Thermal Energy	Gas oven/stove	85%
	Steam boiler	87%
	Diesel engine	38%
	Gas turbine	35%–38%
	Internal combustion Engine	25%
Electrical to thermal energy	Resistance heaters	100%
	Arc furnace	35%–45%
	Induction furnace	50%–75%
Electrical to light energy	Incandescent lamp	Efficacy factor: 8–16 Lm/W
	Gas discharge lamp	Efficacy factor: 45–130 Lm/W
	LED	Efficacy factor: 20–140 Lm/W
Solar to thermal energy	Solar collector	42%–45%

1.2.2 THERMAL ENERGY

Thermal energy is expressed as the energy transferred from a system with a high temperature to a system with a low temperature due to the heat difference and is denoted by Q.

$$Q = m \cdot c \cdot \Delta T = m \cdot c \cdot (T_2 - T_1) \qquad (1.5)$$

where

Q: The thermal energy required to change the heat of an object $[\text{cal}]$

m: The mass of the object that is heated or cooled $[\text{gr}]$

c: Specific heat of the object (the heat required to change the temperature of the unit mass by 1°C) $[\text{cal}/(\text{gr} \cdot °C)]$

ΔT: The change in the temperature of the object $[°C]$

Energy and Energy Conversion

In general, all energy units can also be used as thermal energy units. However, the most common units in thermal energy are calorie (cal) and BTU (British thermal unit). One calorie is the heat required to raise the temperature of 1 g of water by 1°C, while 1 BTU is expressed as the amount of heat needed to raise the temperature of 453 g (1 pound) of water by 1°F. In additon, 1 calorie equals approximately 4.184 Joules.

Example 1.1: Thermal Energy Balance

A new mixture is obtained by mixing oil with mass m_1, specific heat c_1, temperature T_1, and material with mass m_2, specific heat c_2, and temperature T_2 ($T_2 > T_1$). Calculate the final temperature (T_m) of the mixture. Note that thermal losses are neglected.

Solution

Since ($T_2 > T_1$), the mass m_2 loses heat, while the mass m_1 gains heat. If the losses are neglected, the thermal energy lost and the thermal energy gained will be equal to each other.

$$Q_1 = Q_2$$

$$m_1 \cdot c_1 \cdot (T_m - T_1) = m_2 \cdot c_2 \cdot (T_2 - T_m)$$

$$T_m = \frac{m_1 c_1 T_1 + m_2 c_2 T_2}{m_1 c_1 + m_2 c_2}$$

1.2.3 Light (Solar) Energy

Light is a form of energy that comes from the Sun to the Earth as light and heat. Solar irradiance, also called solar radiation or radiation intensity, is expressed in W/m². Solar radiation is approximately constant outside the atmosphere and is about 1370 W/m². However, the amount of radiation reaching the Earth varies between 0 and 1100 W/m².

Light generally refers to the visible portion of solar radiation. For the human eyes, the visible light spectrum is between 380 and 780 nm wavelengths. The most fundamental particle of all kinds of electromagnetic radiation as well as solar radiation is known as the photon. Photons are characterized by their wavelength, frequency, and energy.

$$c = \lambda \cdot f \tag{1.6}$$

The photon energy is:

$$E = h \cdot f = h \cdot \frac{c}{\lambda} \tag{1.7}$$

where
- c: Light speed [3×10^8 m/s]
- f: Frequency [Hz]
- λ: Wavelength [m]
- h: Planck constant [6.626×10^{-34} J·s]
- E: Photon energy [J]

As stated earlier, solar radiation is a source of light and heat. For this reason, various technologies have been developed to take advantage of solar energy's heat and light. These technologies include solar thermal collectors, photovoltaic solar cells, concentrated solar cells, and artificial photosynthesis systems.

1.2.4 Chemical Energy

A certain amount of energy may be released due to chemical reactions. This released energy can then be converted into other forms of energy with the help of different technologies. For example, internal combustion engines, rockets, fuel cells, batteries, steam boilers, gas turbines, and gas furnaces are the main technologies that convert chemical energy into other forms.

For example, batteries are used as energy storage or energy source in electrical systems. Batteries store electrical energy in the form of chemical energy and convert this chemical energy back into electrical energy when needed. The main performance parameters of a battery storage system are energy density, charging efficiency, depth of discharge, state of charge, and cycle life. Lead-acid and lithium-ion batteries are the most common types among commercially available technologies.

1.2.5 Linear Kinetic Energy

Kinetic energy is a form of energy that an object has due to its motion. The motion of an object is generally linear or circular. In this section, the kinetic energy of an object with linear motion is considered.

Suppose that we apply a constant force F to an object. If the object is displaced by $d\ell$ in the direction of the applied force, then the work done per unit distance will be:

$$W = F \cdot d\ell \tag{1.8}$$

where
F: Force [Newton, N]
$d\ell$: (Differential) distance [m]
W: Work [N·m = Joule, J]

If an object with a mass of m is moving in a straight path with a linear velocity v, then the instantaneous velocity of the object is:

$$v = \frac{d\ell}{dt} \tag{1.9}$$

If $d\ell$ in equation (1.9) is substituted in equation (1.8):

$$W = F \cdot v \cdot dt \tag{1.10}$$

If an object with a mass of m moves with a velocity of v, the generated force is:

$$F = m \cdot a \tag{1.11}$$

where the acceleration is denoted by a in [m/s^2]. Since it is expressed as the change of velocity with time:

$$a = \frac{dv}{dt} = \frac{d}{dt}\left(\frac{d\ell}{dt}\right) = \frac{d^2\ell}{dt^2} \tag{1.12}$$

By combining equations (1.11) and (1.12), we get:

$$F = m\frac{dv}{dt} \tag{1.13}$$

Energy and Energy Conversion

By inserting equation (1.13) into (1.10):

$$W = mvdv \tag{1.14}$$

Assuming the object's velocity at $t = 0$ is 0 m/s and at t is v, the kinetic energy E_k is then calculated as follows:

$$E_k = W_k = \int_0^v mvdv = \frac{1}{2}mv^2 \tag{1.15}$$

1.2.6 Electromagnetic Energy

The energy stored in the magnetic field is called magnetic energy, and the energy stored in the electric field is called electric field energy. In general terms, electric field energy and/or magnetic field energy is called electromagnetic energy. For example, the amount of magnetic energy stored in the magnetic field of a coil (E_{coil} in Joule), carrying a current I[A] and having an inductance L[H], is:

$$E_{coil} = \frac{1}{2}LI^2 \tag{1.16}$$

A capacitor can be given as an example of the stored energy in the electric field. The amount of stored energy in the electric field of a capacitor (E_c in Joule), having a DC volt V[volt] and a capacitance C[F], is:

$$E_c = \frac{1}{2}CV^2 \tag{1.17}$$

Example 1.2: Stored Energy by a Coil and Capacitor

Assume that a 2 µF capacitor is connected in parallel to a 12 V battery. The same amount of energy stored by the capacitor is wanted to be taken from a copper coil carrying a 2 A current. Calculate the required inductance value.

Solution

Since the stored energy amounts will be the same ($E_c = E_{coil}$):

$$\frac{1}{2}CV^2 = \frac{1}{2}L \cdot I^2$$

$$\frac{1}{2}(2 \times 10^{-6})12^2 = \frac{1}{2}L \times 2^2 \Rightarrow L = 72\,\mu H$$

1.2.7 Mechanical Energy (Rotational Kinetic Energy)

Many energy conversion systems require mechanical rotational energy. To understand the basic principles of mechanical rotation, let's assume that an object with a mass of m [kg] rotates around a circular path with a radius of r [m] and angular velocity of ω [rad/s], as depicted in Figure 1.1. In this case, the linear velocity (tangential velocity) v [m/s] of the object with a mass of m is calculated as in equation (1.18). In this case, a tangential force (F in Newton) acts on the rotating object, and the torque (T in N·m) given by equation (1.19) is determined.

$$v = \omega r \tag{1.18}$$

$$T = Fr = mar \tag{1.19}$$

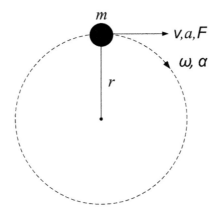

FIGURE 1.1 The circular motion of an object of mass m.

Since the acceleration is the change of velocity with time:

$$a = \frac{dv}{dt} = \frac{d(\omega r)}{dt} = r\frac{d\omega}{dt} \qquad (1.20)$$

The moment (T, Tork) expression can be obtained as below by combining equations (1.19) and (1.20).

$$T = mr^2\frac{d\omega}{dt} = mr^2\alpha = J\alpha = J\frac{d\omega}{dt} \qquad (1.21)$$

The α [rad/s²] in equation (1.21) depends on the change of the angular velocity with time. Therefore, it is called angular acceleration. Also, the expression mr^2 in (1.21) expresses the polar moment of inertia (J in kgm²). By combining equations (1.15) and (1.18), the rotational kinetic energy of the object (E_k in J) is obtained as in (1.22).

$$E_k = \frac{1}{2}mv^2 = \frac{1}{2}m(\omega r)^2 = \frac{1}{2}mr^2\omega^2 = \frac{1}{2}J\omega^2 \qquad (1.22)$$

Example 1.3: Rotational Kinetic Energy of an Electrical Machine

The polar moment of inertia of an electric machine rotor is 24 kgm², and it rotates at a constant speed of 1500 rpm. Calculate the rotational kinetic energy of the electric machine.

Solution

To calculate the rotational kinetic energy of the machine, its angular velocity is required to be known. The relationship between angular velocity ω and n (rpm) is given by $\omega = 2\pi n / 60$.

$$\omega = \frac{2\pi n}{60} = \frac{2\pi \cdot 1500}{60} = 5\pi \text{ rad/s}$$

So, the rotational kinetic energy from equation (1.22):

$$E_k = \frac{1}{2}J\omega^2 = \frac{1}{2}24(5\pi)^2 = 2960.7 \text{ Joule}$$

1.2.8 Nuclear Energy

Nuclear energy is a form of hidden energy within the nucleus structure of the atom. This energy in the atomic structure is released by nuclear reactions such as fusion (combination of two lighter nuclei), fission (splitting of a heavy nucleus into two lighter nuclei), and nuclear decay. This released energy can be explained by the formula E [J] $= m$ [kg] $\cdot c^2$ [Light Speed, m/s], which expresses the conversion of a mass into energy. Steam is produced with the help of heat released as a result of the nuclear chain reactions occurring in nuclear reactors. This produced steam is then converted into electrical energy through steam turbines in nuclear power plants.

1.2.9 Potential Energy

In general terms, potential energy is expressed as the energy possessed by an object due to its physical position relative to other objects. For example, potential energy is stored in situations such as lifting an object to a higher place, collecting water in dams, compressing a spring, or transferring electrical charges onto an object. Therefore, it is possible to divide the potential energy into different types. For example, gravitational potential energy, electrical potential energy, elastic potential energy, and chemical potential energy are the main types of potential energy.

Gravitational potential energy is the energy held by an object in a gravitational field. Gravitational potential energy E_p [Joule] depends on the object's mass m [kg], height h [m], and gravitational acceleration g [9.81 m/s^2] and is expressed by equation (1.23).

$$E_p = mgh \tag{1.23}$$

The electrostatic potential energy E_p [Joule] between a charge Q [Coulomb] and a point charge q, placed at a distance r [m] in the electric field of the respective Q, is given by equation (1.24).

$$E_p = k \frac{Q \cdot q}{r} \tag{1.24}$$

where $k = \dfrac{1}{4\pi\varepsilon_0} = 9 \times 10^9$ Nm^2/C^2 is the Coulomb constant and $\varepsilon_0 = 8.854 \times 10^{-12}$ [F/m] is the dielectric constant of air.

Let us consider the example of the spring to explain the elastic potential energy. The energy accumulated in the spring is equal to the work done to compress or extend the spring. Therefore, the potential energy of this spring compressed or extended by x[m] is given as (1.25).

$$E_p = -\int_0^x \vec{F} \cdot \overrightarrow{dx} = -\int_0^x -kx\,dx \Rightarrow E_p = \frac{1}{2}kx^2 \tag{1.25}$$

where k [N/m] is the spring constant and E_p [Joule] is the potential energy in the spring.

Chemical potential energy refers to the amount of energy absorbed or released during a chemical reaction. This energy may differ according to the type of chemical reaction.

Example 1.4: Potential Energy of Water Falling at Height, h

Obtain the potential power expression of water at a head height h in a dam's reservoir due to the gravitational acceleration. Obtain the energy balance expression for the case of water at height h falling down with a velocity of v through the penstocks. Assuming that the power loss in the penstock pipes is 10%, calculate the highest velocity that the water can reach during the free fall. Note that the mass of water flowing per unit time is $\dot{m} = Q \cdot \rho$ [kg / s], where Q = water flaw rate [m^3 / s] and ρ = water density [1000 kg/m^3].

Solution

Considering Figure 1.2, the potential energy of water of mass m at height h is $E_p = mgh$ [Ws].

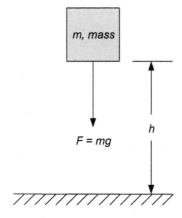

FIGURE 1.2 A mass at height h above the ground.

If the water flow is constant:

$$\text{Potential Power of Water in the Reserve } (P) = \frac{E_p}{\text{time}} = \frac{mgh}{\text{time}} = \frac{m}{\text{time}} gh = \dot{m}gh$$

Since $\dot{m}$ [kg/s] is the mass flow per unit of time and $\dot{m} = Q\rho$, the power expression:

$$P = Q\rho gh \, [\text{W}]$$

The potential energy is converted into kinetic energy as a result of the water at a height h falling down through penstocks. Considering these losses in penstock pipes, the energy balance equation:

$$E_p = E_k + \Delta E_{\text{LOSS}}$$

$$mgh = \frac{1}{2}mv^2 + \Delta E_{\text{LOSS}}$$

Since ΔE_{LOSS} was given as equal to $0.1 \times mgh$,

$$0.9 \times mgh = \frac{1}{2}mv^2$$

Hence, the maximum velocity that water can reach is:

$$v_{\max} = \sqrt{1.8 \times gh}$$

1.2.10 Sound Energy

Sound energy is a type of energy that occurs due to the vibration of a substance. The sound energy E_s [Joule] in a volume of interest V [m³] can be calculated by integrating the summations of kinetic and potential energy densities over that volume.

$$E_S = E_p + E_k = \frac{1}{2}\left(\int \frac{p^2}{\rho_0 c^2} dV + \int \rho v^2 \, dV\right) \quad (1.26)$$

where
- p: Sound pressure [Pascal],
- v: Particle velocity [m/s],
- c: Speed of the sound [m/s],
- ρ: Density of the medium [kg/m³],
- ρ_0: Density of the medium before the sound [kg/m³].

1.3 ENERGY UNITS AND CONVERSION FACTORS

The most commonly used unit system in science, technology, and engineering is the International System of Units (SI). This section summarizes the energy systems units of the SI system and the conversion factors between units. The fundamental units are given with their symbols and abbreviations in Table 1.3. Other units derived by mathematical operations depending on the basic units are listed in Table 1.4.

If the calculated quantities are very large or very small, it is much more difficult to make sense of decimal numbers or numbers with lots of zeros. For this reason, the use of multipliers in quantities will make it easier to interpret those numbers. The multipliers and prefixes in SI units are given in Table 1.5. The conversion factors used between power and energy units are given in Table 1.6.

TABLE 1.3
Fundamental Units (SI)

Quantity	Unit	Symbol
Length	Meter	m
Mass	Kilogram	kg
Time	Second	s
Electrical current	Ampere	A
Temperature	Kelvin	K

TABLE 1.4
Derived Units (SI)

Quantity	Unit	Symbol
Speed (v)	Meter/second	m/s
Area (A)	Meter square	m²
Force (F)	Newton	N or kgm/s²
Energy, Work (E,W)	Joule (Newton-meter)	J or N·m
Power (P)	Watt	W or J/s

TABLE 1.5
Upper/Lower Multipliers and Prefixes in SI Units

Prefix	Symbol	Multiplier	Prefix	Symbol	Multiplier	Prefix	Symbol	Multiplier
Exa	E	10^{18}	Hecto	h	10^2	Micro	μ	10^{-6}
Peta	P	10^{15}	Deka	da	10^1	Nano	n	10^{-9}
Tera	T	10^{12}	Desi	d	10^{-1}	Pico	p	10^{-12}
Giga	G	10^9	Centi	c	10^{-2}	Femto	f	10^{-15}
Mega	M	10^6	Mili	m	10^{-3}	Atto	a	10^{-18}
Kilo	k	10^3						

TABLE 1.6
Power and Energy Conversion Factors

Quantity	Unit	Its Equivalent
Power	1 Watt (W)	1 Joule/second (J/s) = 0.001341 HP
	1 Kilowatt (kW)	1000 W = 1.34 HP
	1 Horsepower (HP)	745.7 W = 550 ft·lb/s
Energy	1 Joule	1 Watt-second (W·s)
	1 Electron Volt (eV)	1.602×10^{-19} J
	1 Kilowatt-Hour (kWh)	3.6×10^6 J = 3.412×10^3 BTU
Thermal Energy	1 Calorie (cal)	4.1868 J
	1 BTU	1055.06 J = 778.169 ft·lb
	1 Therm	10^5 BTU = 29.3 kWh = 1.05506×10^8 J

1.4 PROBLEMS

P1.1. An object of mass m falls freely at height h. Obtain the kinetic and potential energies of the object at heights h, $h/2$, and $h = 0$.

P1.2. An object of mass 24 kg is rotating around a circle of radius 1 m at a constant speed of 1800 rpm. Calculate the rotational kinetic energy of the object.

P1.3. Assume that a 4×10^{-6} H coil is carrying 3A. The same amount of energy stored by the coil is wanted to be taken from a capacitor connected in parallel to a 12V battery. Calculate the required capacitance value in Farad.

P1.4. Hot oil in a closed tank is cooled by mixing with an impeller attached to the shaft of an electric motor. The initial internal energy of the oil in the tank is 1800 kJ. During the cooling process, the heat of 1350 kJ is emitted to the surrounding environment. Besides, a 250 kJ of work is done to mix the oil. Find the final internal energy of the oil. Note that the work required to mix the oil takes place in the tank and the relevant amount of energy is transferred to the oil.

P1.5. A British gallon (4.54609 liters) in the form of a closed container is filled with pure water. All of the water was heated homogeneously, and its temperature was increased by 30°C. Calculate the internal energy of the water in the container.

P1.6. The planned capacity of the Mersin Akkuyu Nuclear Power Plant is 5000 MW. Obtain the equivalents of this power in Horsepower (HP), Kilowatts, and Joule/s.

BIBLIOGRAPHY

1. W. Shepherd, D. W. Shepherd, *"Energy Studies"*, Imperial College Press, London, Second Edition, 2003.
2. U.S. Department of Energy, *"Advanced Melting Technologies: Energy Saving Concepts and Opportunities for the Metal Casting Industry"*, BSC Incorporated, Columbia, 2005.
3. A. Ünal, *"Aydınlatma Tasarımı ve Proje Uygulamaları"*, Birsen Yayıncılık, İkinci Baskı, Turkey, 2014.
4. Y. Chu, P. Meisen, *"Review and Comparison of Different Solar Energy Technologies"*, GENI, Global Energy Network Institute, San Diego, CA, 2011.
5. G. M. Masters, *"Renewable and Efficient Electric Power Systems"*, John Wiley&Sons, New Jersey, 2004.
6. I. S. Grant, W. R. Phillips *"Electromagnetism"*, Manchester Physics Series, Manchester, 2008.
7. R. A. Serway, J. W. Jewett, *"Physics for Scientists and Engineers"*, Brooks/Cole Cengage Learning, Boston, 2010.
8. F. Alton Everest, K. C. Pohlmann, *"Master Handbook of Acoustics"*, The McGraw Hill Companies, New York, 2009.
9. ODTÜ Mühendislik Fakültesi, *"SI Uluslararası Birimler Sistemi"*, Ankara, Turkey, 1981.

2 The Sun and Solar Radiation

2.1 THE SUN AS AN ENERGY SOURCE

One of the most important energy sources among the clean and renewable energies is the Sun itself. Seventy-three percent of the Sun's mass is made up of hydrogen and 25% of helium. The remaining 2% of mass consists of other elements. During the nuclear reactions in the Sun, the hydrogen turns into helium, and a certain amount of energy is released during these reactions. This basic chain of reactions taking place in several steps is summarized in Table 2.1.

As can be seen from Table 2.1, the mass at the output of the reaction is less than the mass of the input atoms. This decrease in mass turns into energy according to the formula $E = mc^2$. The total mass of the Sun is about 2×10^{30} kg. In addition, the conversion speed of the solar mass to energy is around 4×10^9 kg/s. Therefore, the Sun is expected to continue to produce energy likewise for about 5×10^9 years.

The instantaneous electromagnetic power dispersed into space as a result of nuclear reactions in the Sun is around 3.8×10^{20} MW. The portion of these electromagnetic radiations outside of the Earth's atmosphere is between 1325 and 1420 W/m². However, the amount of radiation reaching the Earth's surface varies between 0 and 1100 W/m².

2.2 SOLAR SPECTRUM

Each object, depending on its structure, emits electromagnetic radiation of different wavelengths at temperatures above absolute zero $(T > 0°K)$. The most basic way to determine the amount of radiation emitted by an object is to compare it with the radiation emitted by a reference object. This reference object represents an ideal structure called a black body. Because this ideal black object is an excellent emitter and absorber. As a perfect emitter, the energy emitted by the unit area of the black body is always greater than the energy emitted by the unit area of a real object at the same temperature. Likewise, as a perfect absorber, the black body absorbs all the radiation incident on it, and no part of the radiation is reflected or transmitted to the opposite side of the surface. The wavelength of the radiation emitted by an ideal black body is given as a function of temperature by *Planck's Law* (equation 2.1).

$$E(\lambda, T) = \frac{2\pi h c^2}{\lambda^5 \cdot \left[e^{\left(\frac{hc}{\lambda k T}\right)} - 1 \right]} = \frac{3.7468 \times 10^{10}}{\lambda^5 \cdot \left[e^{\left(\frac{14404}{\lambda T}\right)} - 1 \right]} \qquad (2.1)$$

where

$E(\lambda, T)$: Radiation $\left[W/(m^2 \cdot \mu m) \right]$
λ: Wavelength $[m \to \mu m]$
T: Absolute temperature of the object $[°K]$
h: Planck's Constant $= 6.626 \times 10^{-34}$ $[J \cdot s]$
k: Boltzmann Constant $= 1.380 \times 10^{-23}$ $[J/K]$
c: Light speed $(3 \times 10^8$ m/s$)$

TABLE 2.1
The Basic Chain of Reactions Occurs in the Sun

Reaction Chain	The Reactions
# 1	Hydrogen + Hydrogen → Deuterium + Neutron
# 2	Deuterium + Hydrogen → Helium − 3
# 3	Helium − 3 + Helium − 3 → Helium − 4 + Hydrogen + Hydrogen
# 4	4 Hydrogen → Helium − 4 + 2 Neutrons

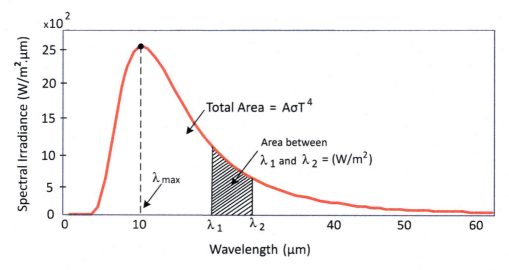

FIGURE 2.1 Spectral distribution of power emitted from one square meter of a black body ($A = 1 \text{ m}^2$).

If the spectral distribution of the blackbody is plotted for 15°C (288.15°K) using equation (2.1), the graph in Figure 2.1 is obtained. The area between any two wavelengths on the graph gives the power value emitted from a unit square meter area between those wavelengths. Likewise, the total area under the graph will give the total amount of power radiated from the unit square meter of the black body (equation 2.2).

$$E = A\sigma T^4 \tag{2.2}$$

where $\sigma = 5.67 \times 10^{-8} \left[\text{W/m}^2 \cdot \text{K}^4 \right]$ is the Stefan-Boltzmann constant, and A [m²] is the surface area of the black body. From the graph, the wavelength corresponding to the maximum spectral power (λ_{max} [μm]) can be determined by equation (2.3).

$$\lambda_{max} = \frac{2898}{T \, [°K]} \tag{2.3}$$

Since the radiation reaching the Earth's surface is exposed to reflection and absorption as it passes through the atmosphere, it is of lower value than extraterrestrial radiation. The amount of radiation absorbed or reflected by the Earth's atmosphere varies in proportion to the path that the radiation travels through the atmosphere. The air mass ratio (AM) is a factor used to calculate the amount of solar radiation reduction in the atmosphere (equation 2.4 and Figure 2.2).

$$AM = \frac{L}{L_0} = \frac{1}{\cos(\beta)} = \frac{1}{\sin(\alpha)} \tag{2.4}$$

The Sun and Solar Radiation

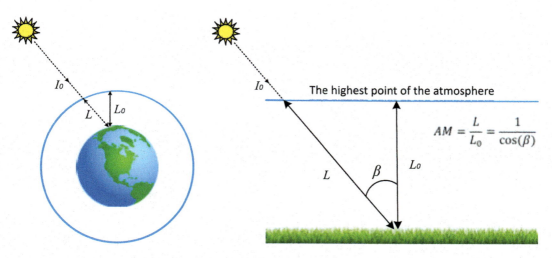

FIGURE 2.2 The radiation path in the atmosphere and the air mass (AM) factor.

where

L_0: Zenith distance relative to sea level (surface normal)
L: The radiation path in the atmosphere
β: Zenith angle (the angle between incident radiation and surface normal) [°C].
α: Elevation angle (the angle between incident radiation and horizontal plane) [°C]

The symbol $AM\emptyset$ denotes the case where the Air Mass Factor is 0 (zero), and the symbol $AM1$ means the Air Mass Factor is 1. In other words, $AM\emptyset$ corresponds to the extraterrestrial state, while $AM1$ corresponds to the case of solar radiation being perpendicular to the Earth's atmosphere. On the other hand, the average solar radiation at the Earth's surface is represented by $AM1.5$. In this context, the effects of different AM values on solar radiation are shown in Figure 2.3.

In general, the higher the zenith angle, the larger the AM factor. The AM factor in equation (2.4) is accurate for the values of the zenith angle up to 70°C ($\beta \leq 70°$). However, for $\beta > 70°$ cases, the AM factor is expressed by equation (2.5).

$$AM = \frac{1}{\cos(\beta) + 0.50572 \times (96.07995 - \beta)^{-1.6364}} \quad (2.5)$$

Solar radiation is the measure of radiation perpendicular to any part of the Earth's surface ($AM = 1$) in [W/m²]. The amount of radiation in a region varies depending on different factors such as latitude, altitude, humidity, air pollution, amount of fog, and local elevations of the region. In addition, equation (2.5a), expressed as a function of AM, provides a good approximation for calculating the amount of solar radiation at sea level (I [W/m²]).

$$I = 1.1 \times I_0 \times 0.7^{(AM)^{0.678}} \quad (2.5a)$$

where $I_0 = 1367$ [W/m²] is the radiation at the top of the atmosphere.

Example 2.1: Solar Radiation and Wavelengths Emitted by Earth and Sun Surfaces

Assume that the surfaces of the Earth and the Sun reflect a black body feature and the average surface temperatures of the Earth and the Sun are 16°C and 5778°K, respectively.

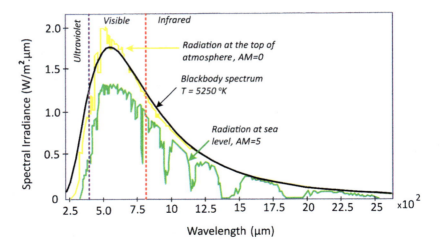

FIGURE 2.3 Solar spectrum distributions for extraterrestrial, sea level, and a blackbody.

a. Find the amount of power that is emitted from the Earth and the Sun's surface areas of 1 m².
b. Find the maximum wavelengths that can occur in these emitted amounts of power.

Solution

a. Using equation (2.2)

$$E = A\sigma T^4 \Rightarrow \frac{E}{A} = \sigma T^4 \; [\text{W/m}^2]$$

Accordingly, the Earth radiates:

$$\frac{E}{A} = \left(5.67 \times 10^{-8} \; \text{W/m}^2 \cdot \text{K}^4\right) \left[(273.15 + 16) \; \text{K}\right]^4 = 1.64 \times 10^{-5} \; \text{W/m}^2$$

Similarly, the Sun radiates:

$$\frac{E}{A} = \left(5.67 \times 10^{-8} \; \text{W/m}^2 \cdot \text{K}^4\right) \left[(5778) \text{K}\right]^4 = 3.27 \times 10^{-4} \; \text{W/m}^2$$

b. Using equation (2.3), the maximum wavelength of the Earth's surface is:

$$\lambda_{\max}(\text{Earth}) = \frac{2898}{T[°K]} = \frac{2898}{289.15} = 10.02 \; \mu\text{m}$$

Similarly, the surface of the Sun's wavelength:

$$\lambda_{\max}(\text{Sun}) = \frac{2898}{T[°K]} = \frac{2898}{5778} = 0.502 \; \mu\text{m}$$

2.3 SOLAR GEOMETRY

Our planet, Earth receives the 94% of its energy use from the Sun. The radiation intensity on the solar surface is approximate 6.33×10^7 W/m². Since the radiation is gradually decreasing with the square of the distance, the radiation falls on 1 m² of the atmospheric surface area is reduced to 1367 W/m², as shown in Figure 2.4. The radiation intensity sent by the Sun is relatively constant.

The Sun and Solar Radiation

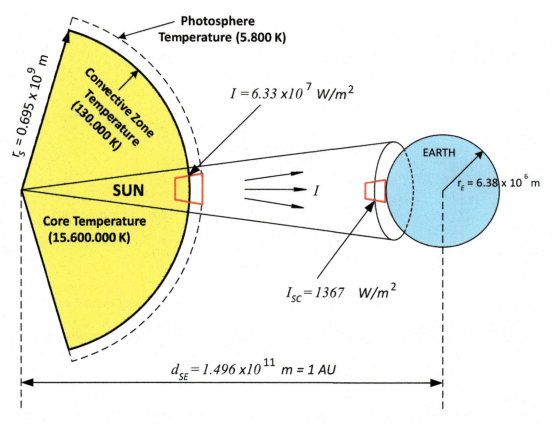

FIGURE 2.4 The departure of energy from the Sun to the Earth.

Accordingly, the solar radiation intensity at a distance of 1 Astronomic Unit (AU) is called the solar constant (I_{SC}) and accepted as 1367 W/m². Since the distance between the Earth and the Sun continuously changes with the motion of the Earth around the Sun, the value of the solar constant, I_{SC} also changes depending on the day number of the year (N). Hence, solar radiation on the surface of the atmosphere at an incidence angle $\theta = 0°$ is expressed as a function of the solar constant and day number of the year as equation (2.6).

$$I_0 = I_{SC}\left[1 + 0.034 \cdot \cos\left(\frac{360N}{365}\right)\right] \quad (2.6)$$

where I_0 is called extraterrestrial insolation in [W/m²], I_{SC} is called the solar constant, and N is the day number of the year. Figure 2.5 gives the solar constant variation during an entire year.

2.3.1 THE ORBIT OF EARTH

The Earth rotates around the Sun in an elliptical orbit at an average distance of 149.5 million km and completes its entire rotation in one year (365.256 days). Figure 2.6 shows this ecliptic plane, which causes the seasons of the World. The elliptical distance variation between Earth and Sun is expressed by equation (2.7).

$$d = 1.5 \times 10^8 \left[1 + 0.017 \cdot \sin\left(\frac{360 \cdot (N-93)}{365}\right)\right] \quad (2.7)$$

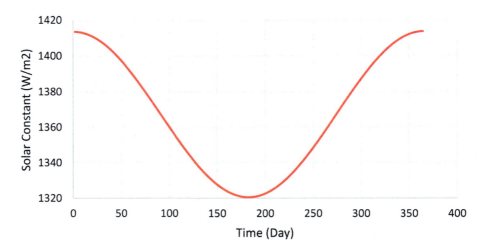

FIGURE 2.5 Solar constant variation during a year.

where N is the day number with reference to January 1 and d is the distance of Earth from the Sun in [km].

Due to the oval path of the Earth around the Sun, Earth comes sometimes nearer to the Sun and sometimes moves away from the Sun. Earth's nearest approach to the Sun, called perihelion, occurs on January 2 and is about 147 Million km. On the other side, the farthest point is about 152 million km from the Sun, which is called aphelion. The aphelion occurs on July 3 (see Figure 2.6).

The axis of Earth's rotation around itself is tilted at an angle of 23.45° with respect to the plane of the orbit around the Sun. The axis is always orientated toward the Pole Star. The angle between the equatorial plane of the Earth and a line linking the centers of the Sun and the Earth is called the *solar declination angle* (δ). For the reason that the axis of the Earth's rotation is always pointing to the Pole Star, the declination angle constantly changes between −23.45° and +23.45°.

Example 2.2: Solar Radiation

Calculate the following during January.

   a. The solar radiation change per square meter just at the outside of Earth's atmosphere.
   b. The change of the elliptical distance between the Earth and the Sun.
   c. Calculate the proportional change between the elliptical distance and the solar radiation.

Solution

   a. Firstly, let us do the calculation of extraterrestrial radiation for the first day of January. Then repeat the same calculation for other days of January. Note that $N = 1, 2, 3, ..., 31$ for January.
      Using equation (2.6),

$$I_0 = 1367\left[1 + 0.034 \cdot \cos\left(\frac{360 \cdot 1}{365}\right)\right] = 1413.471 \text{ W/m}^2$$

   All the results of extraterrestrial radiation for January are given in Table 2.2.
   b. Similarly, as in (a), let us do the calculation of elliptical distance for the first day of January and then repeat the calculation for other days of January.

The Sun and Solar Radiation

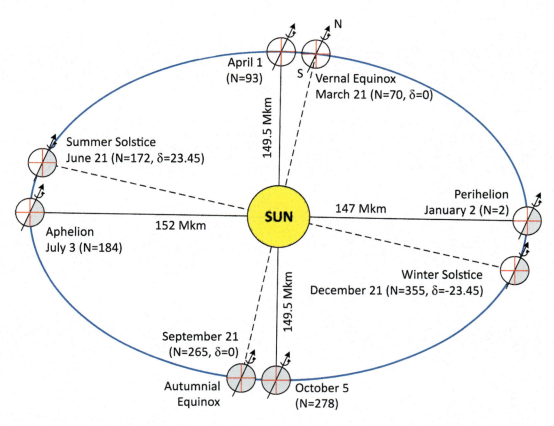

FIGURE 2.6 The orbit of the Earth around the Sun.

TABLE 2.2
Extraterrestrial Radiation for January

Day	I_0 [W/m²]	Day	I_0 [W/m²]	Day	I_0 [W/m²]	Day	I_0 [W/m²]
1	1413.47	9	1412.92	17	1411.5	25	1409.24
2	1413.45	10	1412.79	18	1411.26	26	1408.90
3	1413.42	11	1412.65	19	1411.01	27	1408.55
4	1413.37	12	1412.49	20	1410.75	28	1408.18
5	1413.31	13	1412.32	21	1410.47	29	1407.81
6	1413.23	14	1412.13	22	1410.18	30	1407.42
7	1413.14	15	1411.94	23	1409.88	31	1407.02
8	1413.04	16	1411.73	24	1409.57		

Using equation (2.7),

$$d = 1.5 \times 10^8 \left[1 + 0.017 \cdot \sin\left(\frac{360 \cdot (1 - 93)}{365} \right) \right]$$

$$= 147.45 \times 10^6 \text{ km} = 147.45 \text{ M km}$$

All the results of elliptical distance for January are given in Table 2.3.
  c. The ratio of extraterrestrial irradiances found in (a) to elliptical distances found in (b) is given in the graphic below (Figure 2.7).

TABLE 2.3
Elliptical Distances for January

Day	d [M km]	Day	d [M km]	Day	d [M km]	Day	d [M km]
1	147.45	9	147.47	17	147.54	25	147.65
2	147.45	10	147.48	18	147.55	26	147.67
3	147.45	11	147.48	19	147.56	27	147.69
4	147.45	12	147.49	20	147.57	28	147.71
5	147.45	13	147.50	21	147.59	29	147.73
6	147.46	14	147.51	22	147.60	30	147.75
7	147.46	15	147.52	23	147.62	31	147.77
8	147.46	16	147.53	24	147.63		

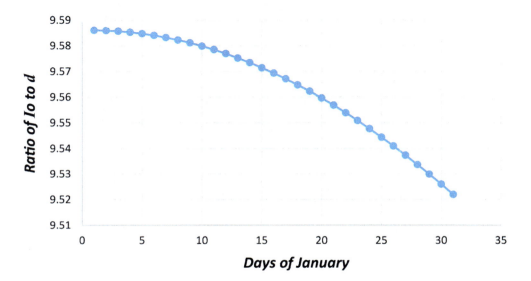

FIGURE 2.7 The ratio of extraterrestrial irradiances to elliptical distances during January.

2.3.2 THE EARTH'S LATITUDE AND LONGITUDE

Any point on the Earth's surface can be identified by the pairs of latitude and longitude, which are called coordinates. The former one, i.e., latitude is defined as a measurement of distance north or south of the Equator. It is generally measured in degrees, minutes, and seconds. There is 90° of latitude from the Equator to each of the poles, north, and south. Hence, there are 180 imaginary lines parallel to the Equator that form circles around the Earth from East to West. These lines are also known as parallels. One degree of latitude covers approximately 111 kilometers (69 miles). Because of the Earth's curving shape, latitude circles close to the Equator are of large radius, while those of farther distant ones are of small radius. Figure 2.8 shows the representation of Earth's latitudes.

To identify the positions more precisely, latitudinal degrees are divided into 60 minutes and those minutes are divided into 60 seconds. Hence one minute of latitude covers approximately 1.8 kilometers and one second of latitude covers about 32 meters.

The latter, i.e., the longitude specifies the east-west position of a point on the Earth's surface with 360 imaginary vertical lines. These lines, known as meridians, run around the Earth and meet at the North and South Poles (see Figure 2.9).

The Sun and Solar Radiation   19

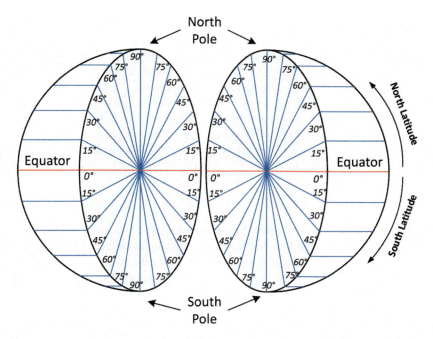

FIGURE 2.8 The lines of latitudes around the Earth.

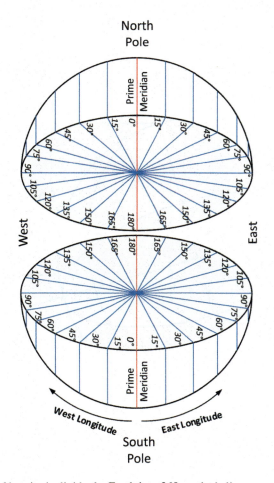

FIGURE 2.9 The lines of longitude divide the Earth into 360 vertical slices.

Longitudinal measurement takes the prime meridian as a reference. The prime meridian is accepted as the line of 0° longitude that runs through Greenwich, England. There are 180° of longitude from the prime meridian to each half of the Earth, east and west. The Eastern Hemisphere is measured as 180° east of the prime meridian and the Western Hemisphere is 180° west of the prime meridian. Similar to latitude, longitude is also measured usually in degrees, minutes, and seconds. Degrees of longitude is divided into 60 minutes and minutes are divided into 60 seconds. One degree of longitude covers approximately 111 kilometers (69 miles) at its widest. While the distance between the longitudes is the maximum at the equatorial plane, the distance between the longitudes toward the poles gradually decreases. Notice that all longitude lines meet at the North and South Poles.

2.3.3 SOLAR ANGLES

The relative position of the Sun with respect to any point on the surface of Earth changes continuously. For instance, it is very well known that the Sun in the summer is higher in the sky than in winter. However, the relative motions of the Sun and Earth are not coincidental, but they are systematic. Thus, the relative positions of the Sun and Earth with respect to each other can be calculated mathematically. For this purpose, it is vital to recognize solar angles in detail.

The first solar angle shown in Figure 2.10 to be explained is the declination angle, which is denoted by δ. The solar declination is the angle between the Earth's Equator plane and a line that connects the centers of the Earth and the Sun. The declination angle changes on a seasonal base between −23.45° and +23.45° due to the Earth's tilt angle of 23.45° and the Earth's rotation around the Sun (see Figure 2.6). The declination angle is equivalent to 0° only at the spring and fall equinoxes.

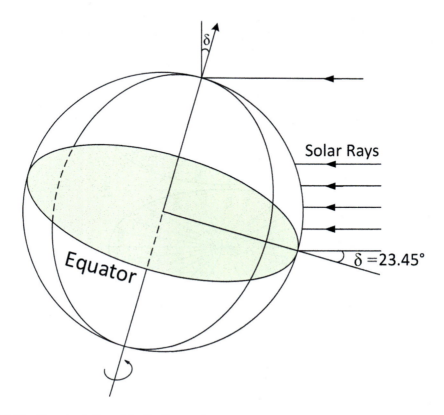

FIGURE 2.10 Representation of declination angle.

The declination angle (δ) in degrees can approximately be calculated by equation (2.8) for any day of the year (N).

$$\delta = -23.45 \cdot \cos\left[\frac{360}{365}(N+10)\right] \tag{2.8}$$

where N is the day of the year with January 1 as $N = 1$. In the literature, there are alternative formulations to calculate solar declination angle variations during an entire year. The equations are given in (2.9) and (2.10), respectively.

$$\delta = 23.45 \cdot \sin\left[\frac{360}{365}(N+284)\right] \tag{2.9}$$

$$\delta = 23.45 \cdot \sin\left[\frac{360}{365}(N-81)\right] \tag{2.10}$$

From any point on the Earth's surface, the Sun's position in the sky can be defined using various angles. Altitude (elevation) angle (α), zenith angle (β), and hour angle (ω) are easily defined in terms of a viewpoint facing the Earth (Figure 2.11).

The altitude angle denoted by α in Figure 2.11 is the angle between a horizontal plane and the Sun in the sky. The altitude angle varies during the day. The altitude angle is 0° at sunrise and 90° at the Equator at noon of spring and fall equinoxes. The mathematical expression for the solar altitude angle is:

$$\sin(\alpha) = \cos(\beta) = \sin(\delta)\sin(L) + \cos(\delta)\cos(L)\cos(\omega) \tag{2.11}$$

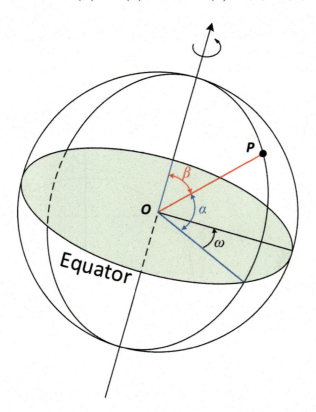

FIGURE 2.11 Representation of altitude (α), zenith (β), and hour angles (ω).

where L is the local latitude. It is clear from Figure 2.11 that the zenith angle β is the angle between the Sun and the vertical plane. Hence, the altitude and zenith angles complement each other by 90°.

$$\alpha + \beta = 90° \quad (2.12)$$

This maximum altitude angle occurs at solar noon and depends on the latitude and declination angle in Figure 2.12 and by equation (2.13) below.

$$\alpha = 90° - L + \delta \quad (2.13)$$

As shown in Figure 2.11, the hour angle for any point P on the Earth's surface is the angle between the meridian having point P and the meridian parallel to the Sun's rays. The hour angle is negative in the morning and becomes zero at solar noon (at 12.00) and afterward, the hour angle goes positive in the afternoon. One should recall that the hour angle at any specific time is the same for all points on that specific meridian (longitude). As the Erath completes its rotation around itself in 24 hours, the hour angle changes by (360/24) 15° every hour (equation 2.13).

$$\omega = 15°(t_{LS} - 12) \quad (2.13)$$

where t_{LS} is the local solar time and ω is the hour angle in degrees.

The last solar angle we need to define is the azimuth angle (z). There are different assumptions to identify the azimuth angle. In some references, the true south is accepted as a reference, while some other references use true north for the azimuth angle reference plane. By convention, the true south plane is used for the azimuth reference plane in this textbook. Moving from the true south plane in the counter-clockwise direction on a 360-degree circle, the south has an azimuth of 0°, east 90°, north 180°, and west 270°. It is alternatively assumed that solar azimuth angles to the east of due south are positive with due east having an azimuth of −90° and solar azimuth angles to the west of due south are negative with true west having an azimuth of +90°. Figure 2.13 shows the azimuth angle representation with reference to due south (true south).

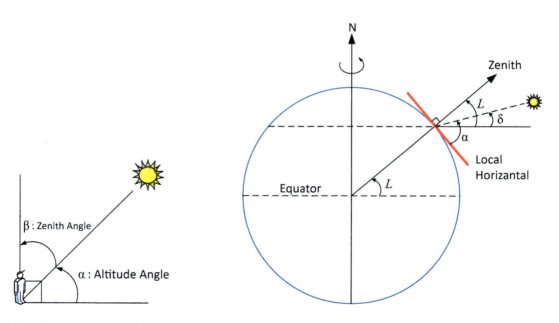

FIGURE 2.12 The altitude angle of the Sun at solar noon (α).

The Sun and Solar Radiation

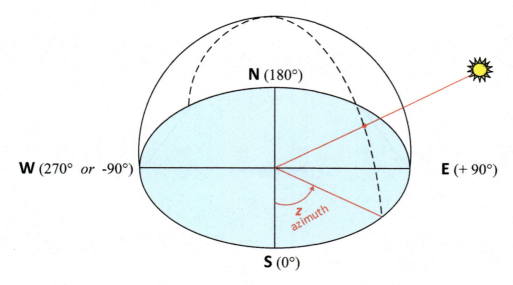

FIGURE 2.13 The solar azimuth angle (z) with reference to true south.

In general, the azimuth angle varies with the location and time of year. The mathematical expression for the solar azimuth angle is given in equation (2.14).

$$\sin(z) = \frac{\cos(\delta)\sin(\omega)}{\cos(\alpha)} \quad (2.14)$$

Notice that for sine functions $\sin(x) = \sin(180-x)$. Hence, there are two possibilities in the inverse of the sine function. That's why a test should be applied to equation (2.14) to decide which of these two options is correct. Such a test is given as equation (2.15).

$$\text{If } \cos(\omega) \geq \frac{\tan(\delta)}{\tan(L)}, \quad \text{then } |z| \leq 90°; \quad \text{otherwise } |z| > 90° \quad (2.15)$$

Example 2.3: Declination Angle Variation

Using the three formulas (equations 2.8, 2.9, and 2.10) given above, calculate the change in declination angle over a year and compare the results.

Solution

Firstly, let us calculate the declination angles for the first day of January. Afterward, we shall repeat the same calculation for other days of the year in sequences.

Using equations (2.8), (2.9), and (2.10) and for ($N = 1$),

$$\delta_1 = -23.45 \cdot \cos\left[\frac{360}{365}(1+10)\right] = -23.0308°$$

$$\delta_2 = 23.45 \cdot \sin\left[\frac{360}{365}(1+284)\right] = -23.0116°$$

$$\delta_3 = 23.45 \cdot \sin\left[\frac{360}{365}(1-81)\right] = -23.0116°$$

Let us draw graphs of the results obtained for 365 days, which are given below (Figure 2.14).

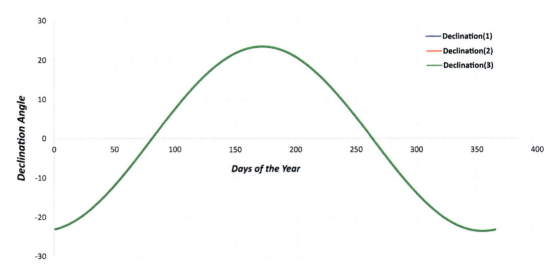

FIGURE 2.14 Declination angle variation through a year.

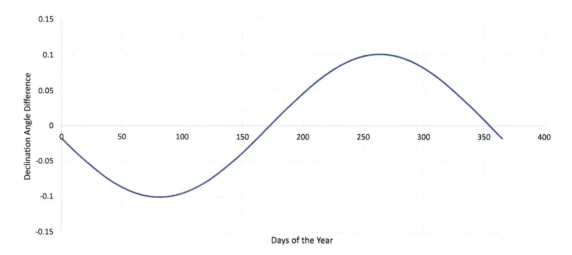

FIGURE 2.15 The variation of difference between declination angles through a year.

The results of δ_2 and declination δ_3 are equal to each other. The maximum angle difference is about 0.1 degrees between δ_1 and the others. The difference between these two angle groups also corresponds to a sinusoidal change. The change in the one-year difference between declination angles is shown in the following graph (Figure 2.15).

Example 2.4: Solar Hour Calculation

Calculate the solar hour angle at 10.30 and 15.00 solar time.

Solution

Using equation (2.13), hour angle at 10.30 local solar time:

$$\omega = 15°(t_{LS} - 12) = 15(10.5 - 12) = -22.5°$$

For 15.00 local solar time:

$$\omega = 15°(t_{LS} - 12) = 15(15 - 12) = 45°$$

Example 2.5: Solar Position Angles: Azimuth and Altitude Angles

a. Calculate the solar azimuth and altitude angles in London $(51.5074°\,N, 0.1278°\,W)$ and Los Angeles $(34.0522°\,N, 118.2437°\,W)$ on April 15 at 15.00 local solar time.
b. Obtain the solar positions on April 15 for the cities of London and Los Angeles between (8 AM and 7 PM) and plot the Sun path diagram for the given period.

Solution

a. On April 15, $N = 105$ and from equation (2.9):

$$\delta = 23.45 \cdot \sin\left[\frac{360}{365}(105 + 284)\right] = 9.415°$$

The hour angle at 15.00 local solar time:

$$\omega = 15°(t_{LS} - 12) = 15(15 - 12) = 45°$$

From equation (2.12), the solar altitude angle (α) for London is:

$$\alpha = \sin^{-1}\left[\sin(9.415)\sin(51.507) + \cos(9.415)\cos(51.507)\cos(45)\right] = 34.21°$$

Similarly, the solar altitude angle (α) for Los Angeles is:

$$\alpha = \sin^{-1}\left[\sin(9.415)\sin(34.052) + \cos(9.415)\cos(34.052)\cos(45)\right] = 42.03°$$

From equation (2.31), the solar azimuth angle for London City is:

$$z = \sin^{-1}\left[\frac{\cos(9.415)\sin(45)}{\cos(34.21)}\right] = 57.51°$$

Similarly, the solar azimuth angle (z) for Los Angeles is:

$$z = \sin^{-1}\left[\frac{\cos(9.415)\sin(45)}{\cos(42.03)}\right] = 69.91°$$

b. By repeating the calculations performed in (a) for the hours given in the period, the following plot (Figure 2.16) of solar positions is obtained for the city of London and Los Angeles.

Example 2.6: Solar Zenith and Altitude Angles

Calculate the solar zenith and altitude angles of London $(51.5074°\,N, 0.1278°\,W)$ and Los Angeles $(34.0522°\,N, 118.2437°\,W)$ at solar noon on July 20.

Solution

Day number, $N = 201$ on July 20, and from equation (2.9):

$$\delta = 23.45 \cdot \sin\left[\frac{360}{365}(201 + 284)\right] = 20.636°$$

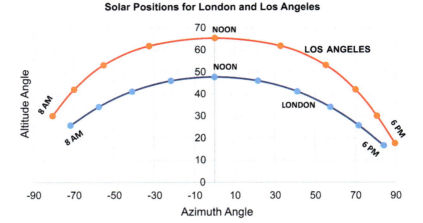

FIGURE 2.16 Solar positions for London and Los Angeles on April 15.

Maximum altitude angle occurs at solar noon, and from equation (2.13):

$$\alpha_{max}(\text{London}) = 90° - 51.507° + 20.636° = 59.129°$$

$$\alpha_{max}(\text{Los Angeles}) = 90° - 34.052° + 20.636° = 76.584°$$

Hence the solar zenith angle of the Sun at noon is given by equation (2.12):

$$\beta_{min}(\text{London}) = 90° - 59.129° = 30.871°$$

$$\beta_{min}(\text{Los Angeles}) = 90° - 76.584° = 13.416°$$

2.4 THE SOLAR INCIDENCE ANGLE

For PV applications, it is necessary to calculate radiation on a tilted surface. For this purpose, the solar incidence angle needs to be calculated. As an example, let us consider a PV panel positioned at a tilt angle b in a horizontal plane (Figure 2.17).

The solar incidence angle, θ, for any surface on the Earth is defined as the angle between the Sun's ray and the normal plane. For horizontal surfaces, the incidence angle becomes equal to the zenith angle, β. For PV applications, azimuth and tilt angles are typically known for fixed surfaces. Hence the third main angle, i.e., incidence angle needs to be calculated. The general expression for the solar incidence angle is specified by the following equation (2.16).

$$\begin{aligned}\cos(\theta) = &\sin(L)\sin(\delta)\cos(b) - \cos(L)\sin(\delta)\sin(b)\cos(z) + \cos(L)\cos(\delta)\cos(b)\cos(\omega) \\ &+ \sin(L)\cos(\delta)\sin(b)\cos(z)\cos(\omega) + \cos(\delta)\sin(b)\sin(z)\sin(\omega)\end{aligned} \quad (2.16)$$

where

L: Latitude angle $-90° \leq L \leq 90°$ (L is positive in Northern Hemisphere)
δ: Declination angle
b: Tilt angle (angle between horizon and surface ($0° \leq b \leq 180°$))
z: Azimuth angle
ω: Hour angle (positive before noon and negative in the afternoon)

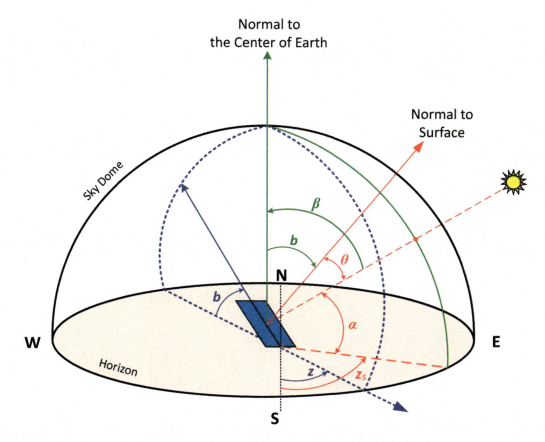

FIGURE 2.17 Solar angles with reference to the tilted surface (b: tilt angle, z: azimuth angle of PV panel, z_s: solar azimuth angle, θ: solar incedence angle, α: altitude angle, β: zenith angle).

Equation (2.16) has relatively complex and can be simplified for different special cases. For example, since tilt angle is zero for horizontal planes ($b = 0$, $\cos b = 1$, $\sin b = 0$), equation (2.16) is simplified to equation (2.17).

$$\cos(\theta) = \sin(L)\sin(\delta) + \cos(L)\cos(\delta)\cos(\omega) \tag{2.17}$$

For a south-facing surface in the northern hemisphere ($z = 0$, $\cos z = 1$, $\sin z = 0$),

$$\cos(\theta) = \sin(\delta)\sin(L-b) + \cos(\delta)\cos(\omega)\cos(L-b) \tag{2.18}$$

Similarly, for a south-facing surface in the southern hemisphere ($z = 180$, $\cos z = -1$, $\sin z = 0$),

$$\cos(\theta) = \sin(\delta)\sin(L+b) + \cos(\delta)\cos(\omega)\cos(L+b) \tag{2.19}$$

In the last case, the incidence angle for vertical surfaces ($b = 90$, $\sin b = 1$, $\cos b = 0$),

$$\cos(\theta) = \sin(L)\cos(\delta)\cos(z)\cos(\omega) + \cos(\delta)\sin(z)\sin(\omega) - \cos(L)\sin(\delta)\cos(z) \tag{2.20}$$

The sunrise and sunset hour angles (ω_{st}) for a tilted surface can be obtained by putting $\theta = 90°$ in equation (2.18) and solving for ω:

$$\omega_{st} = \cos^{-1}\left[-\tan\delta\tan(L-b)\right] \tag{2.21}$$

Notice that after converting ω_{st} into hours, it is added to 12 for sunset time and it is subtracted from 12 for sunrise time. However, since ω_{st} for tilted surfaces cannot have values greater than $\omega_s = \cos^{-1}(-\tan\delta \tan L)$ for horizontal surfaces, the general equation for ω_{st} is given by:

$$\omega_{st} = \min\{\omega_s, \cos^{-1}[-\tan\delta \tan(L-b)]\} \tag{2.22}$$

Finally, the sunrise and sunset angles (ω_{sr_t}, ω_{ss_t}) for a tilted surface not facing true south may not be symmetrical with respect to solar noon and can be obtained from equation (2.16) by setting the incidence angle ($\theta = 90°$). This solution results in two values for ω depending on the surface orientation.

For $z < 0$,

$$\omega_{sr\_t} = \min\left\{\omega_s, \cos^{-1}\left[\frac{A \cdot B - \sqrt{(A^2 - B^2 + 1)}}{A^2 + 1}\right]\right\} \tag{2.23}$$

$$\omega_{ss\_t} = \min\left\{\omega_s, \cos^{-1}\left[\frac{A \cdot B + \sqrt{(A^2 - B^2 + 1)}}{A^2 + 1}\right]\right\} \tag{2.24}$$

For $z > 0$,

$$\omega_{sr\_t} = \min\left\{\omega_s, \cos^{-1}\left[\frac{A \cdot B + \sqrt{(A^2 - B^2 + 1)}}{A^2 + 1}\right]\right\} \tag{2.25}$$

$$\omega_{ss\_t} = \min\left\{\omega_s, \cos^{-1}\left[\frac{A \cdot B - \sqrt{(A^2 - B^2 + 1)}}{A^2 + 1}\right]\right\} \tag{2.26}$$

where

$$A = \frac{\cos L}{\sin z \tan b} + \frac{\sin L}{\tan z} \tag{2.27}$$

$$B = \frac{\cos L \tan \delta}{\tan z} - \frac{\sin L \tan \delta}{\sin z \tan b} \tag{2.28}$$

An important note here is that in the above equations the azimuth angle of the inclined surface is measured positive from true south to westward and measured negative from true south to eastward.

An alternative approximation to solar sunrise and sunset times is given below based on trigonometry and analytic geometry. The formula given by (2.29) provides simple and a fairly good set of approximations for solar sunrise and sunset times. The t value in (2.29) gives the number of minutes after noon that sunset occurs. Similarly, $-t$ value represents the number of minutes before noon that sunrise occurs.

$$t = \frac{1440}{2\pi} \cos^{-1}\left\{\frac{R - r \cdot \sin\left[\frac{2\pi(N-80)}{365.25}\right] \sin\delta \sin L}{\sqrt{r^2 - \left\{r \cdot \sin\left[\frac{2\pi(N-80)}{365.25}\right] \sin\delta\right\}^2} \cos L}\right\} \pm 5 \tag{2.29}$$

The Sun and Solar Radiation

where $R = 6378$ km (Radius of the Earth) and $r = 149598000$ km (the Earth's mean distance from the Sun). The ± 5 is an approximate adjustment factor which is negative for sunrise and positive for sunset.

Example 2.7: Incidence Angle Calculation

A PV panel with 30° of inclination is placed at an angle of 10° east of south in London $(51.5074° \text{N}, 0.1278° \text{W})$ and Los Angeles $(34.0522° \text{N}, 118.2437° \text{W})$ on June 15 at 11.00 local solar time. Calculate the angle of incidence for the PV panel.

Solution

Declination and hour angle for June 15 ($N = 166$) at 11.00 local solar time:

$$\delta = 23.45 \cdot \sin\left[\frac{360}{365}(166 + 284)\right] = 23.314°$$

$$\omega = 15°(t_{LS} - 12) = 15(11 - 12) = -15°$$

From equation (2.16), the inclination angle for London and Los Angeles:

$$\cos(\theta) = \sin(51.507)\sin(23.314)\cos(30) -$$
$$\cos(51.507)\sin(23.314)\sin(30)\cos(10) +$$
$$\cos(51.507)\cos(23.314)\cos(30)\cos(-15) +$$
$$\sin(51.507)\cos(23.314)\sin(30)\cos(10)\cos(-15) +$$
$$\cos(23.314)\sin(30)\sin(10)\sin(-15) = 0.946° \Rightarrow \theta(\text{London}) = 18.853°$$

$$\cos(\theta) = \sin(34.052)\sin(23.314)\cos(30) -$$
$$\cos(34.052)\sin(23.314)\sin(30)\cos(10) +$$
$$\cos(34.052)\cos(23.314)\cos(30)\cos(-15) +$$
$$\sin(34.052)\cos(23.314)\sin(30)\cos(10)\cos(-15) +$$
$$\cos(23.314)\sin(30)\sin(10)\sin(-15) = 0.866° \Rightarrow \theta(\text{Los Angeles}) = 27.015°$$

2.5 SUN'S POSITION, SOLAR TIME, AND CLOCK TIME

Both the azimuth angle and the elevation angle throughout the day determine the solar position in the sky. The variation of both angles during the daytime can be calculated using "solar time." For this purpose, local solar time is found based on the positional variation of the Sun in the sky first. Afterward, the elevation and azimuth angles are calculated. These angles at solar noon are also the two key angles to orient photovoltaic modules. From equation (2.11), the elevation (altitude) angle can be calculated as:

$$\alpha = \sin^{-1}\left[\sin(\delta)\sin(L) + \cos(\delta)\cos(L)\cos(\omega)\right] \qquad (2.30)$$

Similarly, the azimuth angle is obtained from equation (2.14)

$$z = \sin^{-1}\left[\frac{\cos(\delta)\sin(\omega)}{\cos(\alpha)}\right] \qquad (2.31)$$

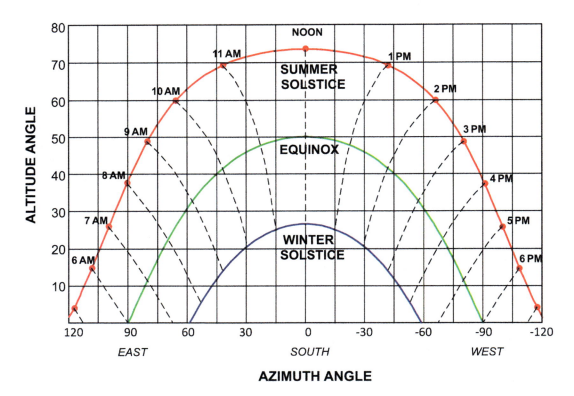

FIGURE 2.18 The plot of the solar path for different days of the year at a latitude of 40°N.

The plot of elevation versus azimuth angle for a specific daytime and latitude visualizes the solar position in the sky, which is called the Sun path diagram. Figure 2.18 shows the solar path diagram for the latitude of 40°N. The Sun path diagrams show how high the Sun will be in the sky on a certain day of a year based on the azimuth angle determined by the time of the day.

In the above equations, the declination angle for any day of the year can be calculated by any of equations (2.8), (2.9), and (2.10). As for the hour angle ω, the expression $\omega = 15°(t_{LS} - 12)$ can be used. Here t_{LS} is the local solar time and can be found by using the parameters, time correction factor (t_C), and the local time (t_L). Note that Daylight Saving Time is used in spring and summer (between March/April and September). Standard time is used in late fall and winter when daytime is shorter. Greenwich Mean Time (GMT) is the standard by which all country's clocks are set. An important note here is that GMT never switches to daylight time. This fact should be considered when converting solar time to local time.

$$t_{LS} = t_L + \frac{t_C}{60} \tag{2.32}$$

The time correction factor in (2.13) has a unit of minute and is a function of longitude (LNG), local standard time meridian (t_{LSM}) and equation of time (EoT).

$$t_C = \text{EoT} + 4|\text{LNG} - t_{LSM}| \tag{2.33}$$

The factor of 4 minutes in equation (2.33) comes from the fact that the Earth rotates 1° every 4 minutes. This factor is used for offsetting the difference between the observer's longitude and that of the Central Meridian. For example, the local zone meridian for Austin, Texas is 90° West and Austin's longitude is 97.8° West. Since the Sun's relative motion is from East to West, the meridian offset setting for Austin is (97.8° − 90 = 7.8° West). Hence, the minutes (7.8 · 4 = 31.2 minutes) should be subtracted from the solar time when converting solar time to clock time (civil time). The local time meridians for the Pacific Time zone (North American standard time zones) are listed in Table 2.4 and the meridian offset setting for a sample list of cities is given in Table 2.5.

The function (EoT) is an empirical equation (in minutes) that corrects for the eccentricity (non-circularity) of the Earth's orbit and the Earth's axial tilt.

$$\text{EoT} = 9.87\sin\left[\frac{720(N-81)}{365}\right] - 7.53\cos\left[\frac{360(N-81)}{365}\right] - 1.5\sin\left[\frac{360(N-81)}{365}\right] \quad (2.34)$$

where N is the day number of the year since January 1. Figure 2.19 gives the variation of EoT over one year.

The factor of t_{LSM} in equation (2.33) is calculated according to equation (2.35),

$$t_{\text{LSM}} = 15°\Delta T \quad (2.35)$$

where ΔT is the difference of the Local Time from Greenwich Mean Time in hours. Notice that prime meridian (longitude = 0°) is used for Greenwich Mean Time. Similar to the prime meridian, the local standard time meridian (t_{LSM}) given in degrees is also used as a reference meridian (ideally in the middle of the zone) for a specific time zone (local time zone or standard time zone). Standard time zones are defined by subdividing the surface of the Earth into 24 zones. Each time zone is bordered by meridians each 15° of longitude in width. For instance, the standard meridian for France is 15° and for Turkey is 45°. Note that time zone boundaries on territories are usually changed to match political boundaries.

TABLE 2.4
North American/Pacific Time Zones

Time Zone	Zone Meridian (West)	Standard Time Offset from UTC (Hours)	Daylight Time Offset from UTC (Hours)
Atlantic	60°	−4	−3
Eastern	75°	−5	−4
Central	90°	−6	−5
Mountain	105°	−7	−6
Pacific	120°	−8	−7
Alaska	135°	−9	−8
Hawaii-Aleutian	150°	−10	−9

TABLE 2.5
Sample List of Cities, Their Longitude, Time Zones, and the Required Meridian Offset Settings

City	Longitude (West)	Time Zone	Meridian Offset Setting
New York	74.0°	Eastern/75°	1.0° E
Chicago	87.7°	Central/90°	2.3° E
Salt Lake City	111.9°	Mountain/105°	6.9° W
Las Vegas	115.1°	Pacific/120°	4.9° E
Honolulu	157.9°	Hawaii-Aleutian/150°	7.9° W

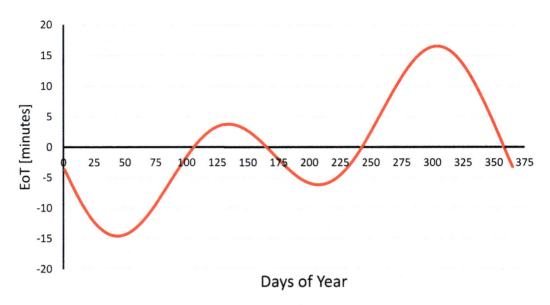

FIGURE 2.19 The plot of EoT over one year.

By using the above equations, sunrise and sunset times can also be calculated, which is important to find the daylight duration for a particular day. The elevation angle in equation (2.11) is set to zero to find the hour angle (ω), which is used in equation (2.13) together with equation (2.32). Afterward, the following equations can be derived from rearranging the above equations to give sunrise and sunset times in hours.

$$t_L(\text{Sunrise}) = 12 - \frac{1}{15°}\cos^{-1}(-\tan\delta\tan L) - \frac{t_C}{60} \qquad (2.36)$$

$$t_L(\text{Sunset}) = 12 + \frac{1}{15°}\cos^{-1}(-\tan\delta\tan L) - \frac{t_C}{60} \qquad (2.37)$$

Daylight Duration (DD) in hours can be obtained from the difference between sunrise and sunset times and it is given by equation (2.38).

$$\text{DD} = \frac{1}{7.5}\cos^{-1}(-\tan\delta\tan L) \qquad (2.38)$$

The above-given equations regarding sunrise and sunset times are geometric relationships with reference to the center of the Sun and these formulas are absolutely sufficient for any kind of solar work. However, they are not exactly the same as the sunrise and sunset times on the calendars. The difference between sunrise/sunset times by meteorologists and geometric sunset/sunrise times is the result of two main factors. The first reason for this difference is based on the principle that the sunbeam is refracted in the atmosphere and as a result of this, the Sun's ray is bent, making the Sun's first appearance approximately 2.5 minutes sooner than geometric calculations and the Sun sinks 2.5 minutes later according to geometric results. The second reason for the difference is based on the computational assumptions between geometric methods and weather service definitions of sunrise and sunset. The geometric method takes into consideration the central point of the Sun both at sunset and at sunrise. However, meteorologists calculate the sunrise and sunset times based on the upper top of the Sun. Accordingly, a difference of 4–8 minutes occurs due to this second reason. Consequently, a total of about ±10 minutes difference between geometric methods and meteorological approaches

The Sun and Solar Radiation

can occur. Hence a correction factor (Q) developed by the U.S. Department of Energy can be used to overcome these complications. The unit of Q in (2.39) is the minute.

$$Q = \frac{3.467}{\sin \omega \cos L \cos \delta} \tag{2.39}$$

Based on the above calculations regarding correction factor (Q), geometric sunrise time can be calculated as $(12 - \omega/15° - Q)$, while sunset is $(12 + \omega/15° + Q)$.

Example 2.8: Sunrise and Sunset Times

A south-facing PV panel is placed horizontally in London $(51.5074°\,N, 0.1278°\,W)$ and Los Angeles $(34.0522°\,N, 118.2437°\,W)$ on June 15. Calculate the solar sunrise and sunset times for this PV panel.

Solution

From example 2.7, we have $\delta = 23.314°$. The sunrise and sunset hour angles (ω_s) *for horizontal surfaces* can be found by setting the solar altitude angle (α) in equation (2.11) equal to zero. Hence, sunset hours for London and Los Angeles:

$$\omega_{s(\text{London})} = \cos^{-1}(-\tan\delta \tan L) = \cos^{-1}[-\tan(23.314)\tan(51.507)] = 122.79°$$

$$\omega_{s(\text{Los Angeles})} = \cos^{-1}[-\tan(23.314)\tan(34.052)] = 106.887°$$

The hour angles found above can be converted to hours by dividing 15°, and thus:

$$t(\text{Sunrise, London}) = 12 - \frac{122.79}{15} = 3.814 \text{ hours} = 3:48 \text{ AM}$$

$$t(\text{Sunrise, Los Angeles}) = 12 - \frac{106.887}{15} = 4.875 \text{ hours} = 4:53 \text{ AM}$$

$$t(\text{Sunset, London}) = 12 + \frac{122.79}{15} = 20.186 \text{ hours} = 20:11 \text{ PM}$$

$$t(\text{Sunset, Los Angeles}) = 12 + \frac{106.887}{15} = 19.125 \text{ hours} = 19:08 \text{ PM}$$

Note that current sunset and sunrise hours can be found by adding Daylight Saving Time (1 hour) to the calculated hours for June 15.

Example 2.9: Incidence Angle Calculation

Assume that the PV panel in Example 2.8 is placed with an inclination of 30°. Calculate the solar sunrise and sunset times for this PV panel on June 15.

Solution

From example 2.8, we have $\delta = 23.314°$, $\omega_{s(\text{London})} = 122.79°$ and $\omega_{s(\text{Los Angeles})} = 106.887°$.
And from equation (2.21) for tilted surfaces:

$$\omega_{st(\text{London})} = \cos^{-1}[-\tan(23.314)\tan(51.507 - 30)] = 99.77°$$

$$\omega_{st(\text{Los Angeles})} = \cos^{-1}[-\tan(23.314)\tan(34.052 - 30)] = 91.75°$$

According to equation (2.22), the solar sunset angle on the collector is the smaller of the two calculated values. Hence, the solar sunrise and sunset angles on the PV panel for London and Los Angeles are 99.77° and 91.75°, respectively.

The hour angles found above for the inclined PV panel can be converted to hours by dividing 15°, which yields to:

$$t(\text{sr}, \text{London}) = 12 - \frac{99.77}{15} = 5.348 \text{ hours} = 5:21 \text{ AM}$$

$$t(\text{sr}, \text{Los Angeles}) = 12 - \frac{91.75}{15} = 5.883 \text{ hours} = 5:53 \text{ AM}$$

$$t(\text{ss}, \text{London}) = 12 + \frac{99.77}{15} = 18.651 \text{ hours} = 18:39 \text{ PM}$$

$$t(\text{ss}, \text{Los Angeles}) = 12 + \frac{91.75}{15} = 18.117 \text{ hours} = 18:07 \text{ PM}$$

Notice that current sunset and sunrise hours can be found by adding Daylight Saving Time (1 hour) to the calculated hours for June 15.

Example 2.10: Daylight Duration

A PV panel is located in London $(51.5074°\text{ N}, 0.1278°\text{ W})$ and Los Angeles $(34.0522°\text{ N}, 118.2437°\text{ W})$ 118.2437° W) with a tilt angle of 40° and facing 10° east of south. Calculate the daylight duration that the Sun shines on the PV Panels on March 21.

Solution

The spring equinox is on March 21 and we have $\delta = 0°$ for March 21 ($N=81$). Since the PV panel is tilted and does not face true south, from equations (2.27) and (2.28), we have the following for London and Los Angeles:

$$A_{(\text{London})} = \frac{\cos 51.507}{\sin(-10)\tan 40} + \frac{\sin 51.507}{\tan(-10)} = -8.7105$$

$$B_{(\text{London})} = \frac{\cos 51.507 \tan 0}{\tan(-10)} - \frac{\sin 51.507 \tan 0}{\sin(-10)\tan 40} = 0$$

$$A_{(\text{Los Angeles})} = \frac{\cos 34.052}{\sin(-10)\tan 40} + \frac{\sin 34.052}{\tan(-10)} = -8.8618$$

$$B_{(\text{Los Angeles})} = \frac{\cos 34.052 \tan 0}{\tan(-10)} - \frac{\sin 34.052 \tan 0}{\sin(-10)\tan 40} = 0$$

Since $\tan \delta = 0$ in equation $\omega_s = \cos^{-1}(-\tan \delta \tan L)$, we have:

$$\omega_{s(\text{London})} = \omega_{s(\text{Los Angeles})} = 90°$$

Since $z < 0$, the sunrise and sunset hour angles for the tilted PV panel in London are given by using equations (2.23) and (2.24), respectively.

The Sun and Solar Radiation

$$\omega_{sr\_t} = \min\left\{\omega_s, \cos^{-1}\left[\frac{A\cdot B - \sqrt{(A^2 - B^2 + 1)}}{A^2 + 1}\right]\right\} = \min\{90, 96.55\} = 90°$$

$$\omega_{ss\_t} = \min\left\{\omega_s, \cos^{-1}\left[\frac{A\cdot B + \sqrt{(A^2 - B^2 + 1)}}{A^2 + 1}\right]\right\} = \min\{90, 83.45\} = 83.45°$$

Similarly, sunrise and sunset hour angles for Los Angeles:

$$\omega_{sr\_t} = \min\left\{\omega_s, \cos^{-1}\left[\frac{A\cdot B - \sqrt{(A^2 - B^2 + 1)}}{A^2 + 1}\right]\right\} = \min\{90, 96.44\} = 90°$$

$$\omega_{ss\_t} = \min\left\{\omega_s, \cos^{-1}\left[\frac{A\cdot B + \sqrt{(A^2 - B^2 + 1)}}{A^2 + 1}\right]\right\} = \min\{90, 83.56\} = 83.56°$$

Therefore, the sunrise and sunset hour angles on the tilted PV panel are 90° and 83.56°, respectively. Finally, the day length for the inclined PV panel in London is $(90+83.45)/15 = 11.5563$ hours. And for horizontal surfaces, the day length in London is $(90+90)/15 = 12$ hours. Similarly, the day length for the inclined PV panel in Los Angeles is $(90+83.56)/15 = 11.57$ hours. And for horizontal surfaces, the day length in Los Angeles is $(90+90)/15 = 12$ hours.

Example 2.11: Daylight, Sunrise, and Sunset Calculations

a. Calculate the local time of sunset, sunrise, and day length in Las Vegas $(36.169°\,\text{N}, 115.14°\,\text{W})$ on June 15.
b. Calculate the local time of sunset, sunrise, and day length in Los Angeles $(34.0522°\,\text{N}, 118.2437°\,\text{W})$ on June 15.

Solution

a. Day number $N = 166$ on June 15, and from equation (2.9):

$$\delta = 23.45 \cdot \sin\left[\frac{360}{365}(166 + 284)\right] = 23.314°$$

$$\omega_{s(\text{Las Vegas})} = \cos^{-1}(-\tan\delta\tan L) = \cos^{-1}[-\tan(23.314)\tan(36.169)] = 108.36°$$

The solar sunrise and sunset hours for Las Vegas:

$$t(\text{Sunrise, Las Vegas}) = 12 - \frac{108.36}{15} = 4.776 \text{ hours} = 4:47\text{ AM}$$

$$t(\text{Sunset, Las Vegas}) = 12 + \frac{108.36}{15} = 19.224 \text{ hours} = 19:13\text{ PM}$$

In order to find the local time of sunset and sunrise, from equation (2.33), we have:

$$t_C = 4|\text{LNG} - t_{\text{LSM}}| + \text{EoT}$$

Now let us write all the values in the equation for Las Vegas, and from equation (2.33) on June 15, we have EoT = −0.192 minutes. In addition, the local time zone for Las Vegas is 120° W. Thus, we have a time correction:

$$t_C = 4|120 - 115.14| - (-0.192) = 19.632 \text{ minutes}$$

Hence local standard sunrise and sunset times for Las Vegas:

$$t_L(\text{Sunrise, Las Vegas}) = 4.776 - \frac{19.632}{60} = 4.448 = 4:27 \text{ AM}$$

$$t_L(\text{Sunset, Las Vegas}) = 19.224 - \frac{19.632}{60} = 18.896 \text{ hours} = 18:54 \text{ PM}$$

Finally, from (2.39) correction factor to adjust the irregularities:

$$Q = \frac{3.467}{\sin 108.36 \cos 36.169 \cos 23.314} = 4.92 \text{ minutes}$$

Hence final clock time of sunrise and sunset for Las Vegas:

$$t_L(\text{Sunrise, Las Vegas}) = 4:27 \text{ AM} - 4.92 \text{ minutes} = 4:22 \text{ AM}$$

$$t_L(\text{Sunset, Las Vegas}) = 18:54 \text{ PM} + 4.92 \text{ minutes} = 18.59 \text{ PM}$$

Let's add 1 hour to adjust for Daylight Saving Time; thus, sunrise will be 5:22 AM, and sunset at 19.59 PM. The sunrise/sunset times are the same as civil times on June 15 for Las Vegas.

For daylight duration (DD) in hours:

$$\text{DD} = \frac{108.36}{7.5} = 14.45 \text{ hours}$$

b. Day number $N = 166$ on June 15, and from equation (2.9):

$$\delta = 23.45 \cdot \sin\left[\frac{360}{365}(166 + 284)\right] = 23.314°$$

$$\omega_{s(\text{Los Angeles})} = \cos^{-1}(-\tan\delta \tan L) = \cos^{-1}[-\tan(23.314)\tan(34.052)] = 106.93°$$

The solar sunrise and sunset hours for Los Angeles:

$$t(\text{Sunrise, Los Angeles}) = 12 - \frac{106.93}{15} = 4.871 \text{ hours} = 4:52 \text{ AM}$$

$$t(\text{Sunset, Los Angeles}) = 12 + \frac{106.93}{15} = 19.128 \text{ hours} = 19:13 \text{ PM}$$

To find the local time of sunset and sunrise, from equation (2.33), we have:

$$t_C = 4|\text{LNG} - t_{\text{LSM}}| + \text{EoT}$$

The Sun and Solar Radiation

Now let us write all the values in the equation for Las Vegas, and from equation (2.33) on June 15, we have EoT = −0.192 minutes. In addition, the local time zone for Las Vegas is 120° W. Thus, we have a time correction:

$$t_C = 4|120 - 118.24| - (-0.192) = 7.232 \text{ minutes}$$

Hence local standard sunrise and sunset times for Los Angeles:

$$t_L(\text{Sunrise, Los Angeles}) = 4.871 - \frac{7.232}{60} = 4.750 \text{ hours} = 4:45 \text{ AM}$$

$$t_L(\text{Sunset, Las Vegas}) = 19.224 - \frac{7.232}{60} = 19.103 \text{ hours} = 19:06 \text{ PM}$$

Finally, from (2.39) correction factor to adjust the irregularities:

$$Q = \frac{3.467}{\sin 106.93 \cos 34.052 \cos 23.314} = 4.76 \text{ minutes}$$

Hence final clock time of sunrise and sunset for Las Vegas:

$$t_L(\text{Sunrise, Los Vegas}) = 4:45 \text{ AM} - 4.76 \text{ minutes} = 4:40 \text{ AM}$$

$$t_L(\text{Sunset, Las Vegas}) = 19:06 \text{ PM} + 4.76 \text{ minutes} = 19.11 \text{ PM}$$

Let's add 1 hour to adjust for Daylight Saving Time; thus, sunrise will be 5:40 AM, and sunset 20.11 PM. The sunrise/sunset times are almost the same as civil times on June 15 for Los Angeles.

For daylight duration (DD) in hours:

$$DD = \frac{106.93}{7.5} = 14.26 \text{ hours}$$

2.6 SOLAR RADIATION CALCULATIONS

The amount of solar insolation on a surface at a specific tilt angle changes at all times based on the Sun's position in the atmosphere all through the year. Solar insolation on a tilted surface depends not only on direct beam radiation but also on reflected and diffused radiations. Figure 2.20 shows solar radiation components on a collector surface.

The solar radiation on a tilted surface can be calculated as a function of latitude and day number of the year. These calculations are essential in determining the power incident on a PV module. When the PV module surface and the sunbeams are perpendicular to each other, the power density on the PV surface will always be at its maximum value. Direct beam radiation on a Sun-facing surface is the biggest portion of all incoming radiations.

2.6.1 CALCULATION OF CLEAR-SKY RADIATION

As the Sun's beam passes through the atmosphere, a significant portion of the rays is absorbed by the atmosphere or dispersed by the air molecules in the atmosphere. Hence, the rays with decreasing amounts reach the ground surface in three components. These are direct radiation, reflected radiation, and diffused radiation. The reduction of incoming direct solar radiation (I_D) in the atmosphere is generally expressed with an exponentially decreasing function, which is given as equation (2.40).

$$I_D = Ae^{-K \cdot AM} \tag{2.40}$$

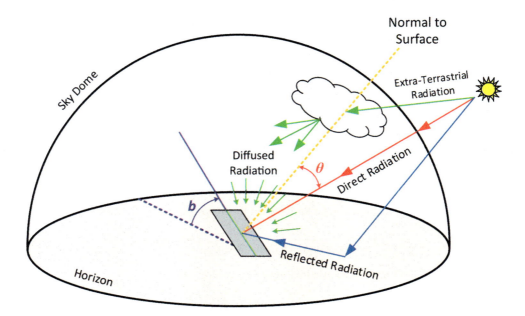

FIGURE 2.20 Solar radiation components on a tilted surface.

where I_D is the direct component of the incoming radiation on the Earth's surface, A is called apparent extraterrestrial radiation, and K is the optical depth parameter. The variable AM is the air mass ratio, which is already expressed as equation (2.4). The parameters of A and K are given below:

$$A = 1160 + 75 \sin\left[\frac{360(N-275)}{365}\right] \quad (2.41)$$

$$K = 0.174 + 0.035 \sin\left[\frac{360(N-100)}{365}\right] \quad (2.42)$$

where N is the day number as defined earlier.

In order to perform a reasonably accurate solar workout, the total insolation amount on the module surface should be known. For this purpose, all components of incoming solar radiation, namely direct, diffused, and reflected beams are needed to be calculated for tilted and horizontal surfaces. The amount of direct component of solar radiation on a tilted surface, which is parallel to the normal of the module's surface can be obtained from the incoming direct beam of the Sun. Figure 2.21 shows the relations between the tilted surface and horizontal surface to calculate direct beam radiation on a collector surface.

The amount of direct component of incoming solar radiation on a tilted module surface (I_{DM}) is a function of incidence angle (θ) and it is given by equation (2.43). We suppose that the radiation distribution at a given place and time is isotropic.

$$I_{DM} = I_D \cdot \cos\theta = I_D \cdot \sin(\alpha + b) \quad (2.43)$$

Similarly, solar beam insolation on the horizontal surface (I_{DH}) is given by

$$I_{DH} = I_D \cdot \cos(90 - \alpha) = I_D \cdot \sin\alpha \quad (2.44)$$

The Sun and Solar Radiation

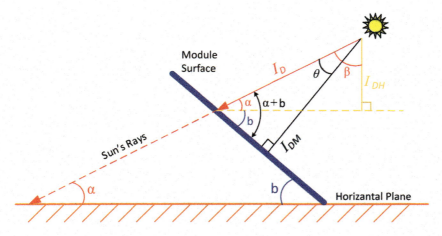

FIGURE 2.21 Solar radiation relations on a tilted surface.

The incidence angle (θ) can also be expressed at any particular time as a function of module orientation (b, z_M), solar azimuth angle (z_S) and altitude angle of the Sun (α).

$$\cos\theta = \cos(\alpha) \cdot \cos(z_S - z_M) \cdot \sin(b) + \sin(\alpha) \cdot \cos(b) \qquad (2.45)$$

where module azimuth angle (z_M) with reference to due south is positive for the southeast direction and negative for the southwest direction.

For the diffused component of incident radiation on the tilted surface of the module, a sky diffuse factor developed by Threlked and Jordon (1958) is used to calculate the diffused portion of the incoming beam. Hence, the equations relating to diffused radiation on horizontal surfaces (I_{DFH}) and on tilted module surface (I_{DFM}) are given by the following equations:

$$I_{DFH} = DF \cdot I_D \qquad (2.46)$$

$$I_{DFM} = I_{DFH} \cdot \left(\frac{1+\cos b}{2}\right) = DF \cdot I_D \cdot \left(\frac{1+\cos b}{2}\right) \qquad (2.47)$$

where DF is the sky diffuse factor and it is expressed based on the day number of the year (N) as equation (2.48).

$$DF = 0.095 + 0.04 \cdot \sin\left[\frac{360(N-100)}{365}\right] \qquad (2.48)$$

Finally, the reflected beam on a tilted module is expressed as a function of ground reflectance (ρ_G), tilt angle (b), direct and diffused beam on the horizontal plane (I_{DH} and I_{DFH}). The ground reflectance is also called albedo and it is the fraction of reflected solar radiation incident on the ground. A typical ground reflectance value for grassy land is about 20% and for snowy areas, it may have a reflectance as high as 80%.

The equation relating reflected radiation on tilted module surface (I_{RM}) is given by the following equations:

$$I_{RM} = \rho_G (I_{DH} + I_{DFH}) \left(\frac{1-\cos b}{2}\right) \qquad (2.49)$$

It is clear from equation (2.49) that I_{RM} becomes zero for the horizontal plane ($b = 0 \Rightarrow \cos b = 1$), which means that no reflected radiation on horizontal module surfaces. For vertical surfaces ($b = 90 \Rightarrow \cos b = 0$), the module sees half of the reflected radiation.

Substituting (2.44) and (2.46) into (2.49) gives the following equation for the reflected radiation on the tilted module.

$$I_{RM} = \rho_G I_D (DF + \sin \alpha) \left(\frac{1 - \cos b}{2} \right) \quad (2.50)$$

By combining the three components of insolations (2.44), (2.47), and (2.50), the total insolation on the tilted module (I_M) on an isotropic clear-sky day can be calculated by equation (2.51).

$$I_M = I_{DM} + I_{DFM} + I_{RM} \quad (2.51)$$

Example 2.12: Extraterrestrial Solar Insolation

a. Calculate the daily total extraterrestrial solar insolation on a horizontal surface on March 15 for (36.169° N Longitude).
b. Calculate the hourly extraterrestrial solar radiation value on March 15 for the hour 13:00–14:00 solar time.

Solution

a. Day number $N = 75$ on March 15, and from equation (2.9):

$$\delta = 23.45 \cdot \sin \left[\frac{360}{365} (75 + 284) \right] = -2.417°$$

Using the sunset hour angle equation, ω_s is:

$$\omega_s = \cos^{-1}(-\tan \delta \tan L) = \cos^{-1}[-\tan(-2.417)\tan(36.169)] = 88.23°$$

Daily extraterrestrial irradiation on the horizontal surface can then be calculated from (2.72) as:

$$I_{d0H} = \frac{24}{\pi} 1367 \left[1 + 0.034 \cdot \cos\left(\frac{360 \cdot 75}{365} \right) \right] \cdot \left[\begin{array}{l} \frac{88.23\pi}{180} \sin(-2.417)\sin(36.169) \\ +\cos(-2.417)\cos(36.169)\sin(88.23) \end{array} \right]$$

$$= 8094 \text{ Wh}/(\text{m}^2 \cdot \text{day})$$

b. Solar hour angles for 13:00 and 14:00 are $\omega_1 = 15°$ and $\omega_2 = 30°$, respectively. From (2.71), the hourly extraterrestrial radiation on a horizontal surface is:

$$I_{h0H} = \frac{12}{\pi} \cdot 1367 \cdot \left[1 + 0.034 \cdot \cos\left(\frac{360 \cdot 75}{365} \right) \right] \cdot \left[\begin{array}{l} \frac{\pi(30-15)}{180} \sin(-2.417) \cdot \sin(36.169) \\ +\cos(-2.417) \cdot \cos(36.169) \cdot (\sin 30 - \sin 15) \end{array} \right]$$

$$= 990.92 \text{ W/m}^2$$

2.6.2 Radiation Calculation Based on Atmospheric Transmission

The radiation calculations given in the above section are given according to isotropic clear-sky conditions. In fact, the sky is not generally 100% clear due to clouds, humidity, dust, and air pollution.

The Sun and Solar Radiation

To take this effect into solar radiation calculation, factors such as *Clearness Index* (K) and *Diffuse Index* (K_D) can be used. Let us modify equations (2.43) to (2.49) to demonstrate the use of the clearness index. From (2.43) and (2.44), we have a tilt conversion factor of the direct beam (C_D):

$$C_D = \frac{I_{DM}}{I_{DH}} = \frac{\sin(\alpha+b)}{\sin \alpha} = \frac{\cos \theta}{\cos \beta} = \frac{\cos \theta}{\sin(\delta)\sin(L)+\cos(\delta)\cos(L)\cos(\omega)} \quad (2.52)$$

From (2.47), we have the tilt conversion factor of the diffused beam (C_{DF}):

$$C_{DF} = \frac{I_{DFM}}{I_{DFH}} = \frac{1+\cos b}{2} \quad (2.53)$$

Similarly, from (2.49), we have the tilt conversion factor of the reflected beam (C_R):

$$C_R = \frac{I_{RM}}{(I_{DH}+I_{DFH})\rho_G} = \frac{I_{RM}}{I \cdot \rho_G} = \frac{1-\cos b}{2} \quad (2.54)$$

Thus, the total solar irradiance on the tilted module surface I_M from the sum of three terms:

$$I_M = C_D I_{DH} + C_{DF} I_{DFH} + C_R I \rho_G \quad (2.55)$$

Considering equations (2.52) to (2.54), equation (2.55) can be rewritten as:

$$I_M = C_D I_{DH} + I_{DFH}\left(\frac{1+\cos b}{2}\right) + I\rho_G\left(\frac{1-\cos b}{2}\right) \quad (2.56)$$

Dividing both sides of equation (2.56) by I and considering $I = I_{DH} + I_{DFH}$, following equation can be written:

$$\frac{I_M}{I} = \left(1 - \frac{I_{DFH}}{I}\right)C_D + \frac{I_{DFH}}{I}\left(\frac{1+\cos b}{2}\right) + \rho_G\left(\frac{1-\cos b}{2}\right) \quad (2.57)$$

where the ratio (I_{DFH}/I) is called diffuse index K_{DF}, defined as the ratio of diffused radiation on a horizontal surface (I_{DFH}) to the global solar radiation (I) on that surface. The diffuse index K_{DF} is generally expressed empirically based on the sky clearness index (K), which is defined as the ratio of radiation on a horizontal surface (I) to the extraterrestrial radiation (I_0) on that surface. The equations relating to the diffuse index K_{DF} and clearness index K are given below.

$$K_{DF} = \frac{I_{DFH}}{I} \quad (2.58)$$

$$K = \frac{I}{I_0} = \frac{I_{DH}+I_{DFH}}{I_0} \quad (2.59)$$

If *clearness index* K is defined as the ratio of mean hourly radiation on a horizontal surface (I_h) to the mean hourly extraterrestrial radiation (I_{h0}) on that surface, it is denoted by K_h and equal to (I_h/I_{h0}). Similarly, If the *clearness index* K is defined as the ratio of mean daily radiation on a horizontal surface (I_d) to the mean daily extraterrestrial radiation (I_{d0}) on that surface, it is denoted by K_d and equal to (I_d/I_{d0}). Finally, If the *clearness index* K is defined as the ratio of mean monthly radiation on a horizontal surface (I_m) to the mean monthly extraterrestrial radiation (I_{m0}) on that surface, it is denoted by K_m and equal to (I_m/I_{m0}).

2.6.3 Calculation of Hourly, Daily, and Monthly Radiation

Hourly, daily, and monthly radiations on a tilted module surface can be calculated depending on the clearness index value. This index is considered to be the radiation reduction factor of the atmosphere. Normally it is a stochastic parameter and it varies as a function of year time, season, climatic condition, and geographic location. In general, the clearness index is empirically formulized depending on the measurement data.

The developed equation in (2.57) can be used for the calculation of the total hourly, daily and monthly irradiation on a tilted surface. Since the diffuse and ground reflected radiations in the isotropic model are not dependent on the incidence angle, tilt conversion factors of (2.53) and (2.54) are the same for all hourly, daily, and monthly solar radiation calculations. Hence considering (2.56), total hourly, daily, and monthly solar irradiations on a tilted surface can be written as below, respectively:

$$\frac{I_{M(h)}}{I_{(h)}} = \left(1 - \frac{I_{\text{DFH}(h)}}{I_{(h)}}\right) C_D + \frac{I_{\text{DFH}(h)}}{I_{(h)}} \left(\frac{1+\cos b}{2}\right) + \rho_G \left(\frac{1-\cos b}{2}\right) \quad (2.60)$$

$$\frac{I_{M(d)}}{I_{(d)}} = \left(1 - \frac{I_{\text{DFH}(d)}}{I_{(d)}}\right) C_D + \frac{I_{\text{DFH}(d)}}{I_{(d)}} \left(\frac{1+\cos b}{2}\right) + \rho_G \left(\frac{1-\cos b}{2}\right) \quad (2.61)$$

$$\frac{I_{M(m)}}{I_{(m)}} = \left(1 - \frac{I_{\text{DFH}(m)}}{I_{(m)}}\right) C_{D(m)} + \frac{I_{\text{DFH}(m)}}{I_{(m)}} \left(\frac{1+\cos b}{2}\right) + \rho_G \left(\frac{1-\cos b}{2}\right) \quad (2.62)$$

where the sub-indices (h, d, and m) are, respectively, used for hourly, daily, and monthly values. The tilt factor of C_D is the same for (2.60) and (2.61), while $C_{D(m)}$ in (2.62) is given by (2.63) for monthly average solar irradiance. Note that equations (2.63) and (2.64) will be usable for the surfaces in the northern hemisphere (azimuth angle, $z = 0°$) if the ($\mp$) sign is replaced by the minus ($-$) sign only. Otherwise, the calculations will be valid for the surfaces in the southern hemisphere ($\mp$ sign is replaced by "+" and $z = 180°$).

$$C_{D(m)} = \frac{\cos(L \mp b) \cdot \cos\delta \cdot \sin\omega_{ss} + \frac{\pi}{180}\omega_{ss} \cdot \sin(L \mp b) \cdot \sin\delta}{\cos L \cdot \cos\delta \cdot \sin\omega_s + \frac{\pi}{180}\omega_s \cdot \sin L \cdot \sin\delta} \quad (2.63)$$

where ω_s is the sunset hour angle and ω_{ss} is the sunset hour angle for the tilted surface, which is given by the following equation:

$$\omega_{ss} = \min\left\{\omega_s, \cos^{-1}\left[-\tan(L \mp b) \cdot \tan\delta\right]\right\} \quad (2.64)$$

The hourly diffuse index $K_{\text{DF}(h)} = \dfrac{I_{\text{DFH}(h)}}{I_{(h)}}$ in (2.60) is given based on the *hourly clearness index* (K_h) values and it is given by equation (2.65).

$$K_{\text{DF}(h)} = \begin{cases} 1 - 0.09 K_h & K_h \leq 0.22 \\ 0.9511 - 0.1604 K_h + 4.388 K_h^2 - 16.638 K_h^3 + 12.336 K_h^4 & 0.22 < K_h < 0.8 \\ 0.165 & K_h \geq 0.8 \end{cases} \quad (2.65)$$

For the daily irradiance calculation of (2.62), the daily diffuse index $K_{\text{DF}(d)} = \dfrac{I_{\text{DFH}(d)}}{I_{(d)}}$ is obtained based on the clearness index, K.

$$K_{DF(d)} = \begin{cases} 1 - 0.2227K + 2.4495K^2 - 11.9514K^3 + 9.3879K^4 & K < 0.715 \quad \omega_s < 81.4° \\ 0.143 & K \geq 0.715 \quad \omega_s < 814° \\ 1 - 0.2227K + 2.4495K^2 - 11.9514K^3 & K < 0.722 \quad \omega_s \geq 81.4 \\ 0.175 & K \geq 0.722 \quad \omega_s \geq 81.4 \end{cases} \quad (2.66)$$

Finally, for the monthly irradiance calculation of (2.61), the daily diffuse index $K_{DF(m)} = \dfrac{I_{DFH(m)}}{I_{(m)}}$ is obtained based on the clearness index K_m, is defined as the ratio of the monthly average daily of total irradiation on a horizontal surface to the monthly average daily of total extraterrestrial radiation on that surface. According to Liu and Jordan, the ratio $K_{DF(m)}$ can be calculated using (2.67).

$$K_{DF(m)} = \frac{I_{DFH(m)}}{I_{(m)}} = 1.390 - 4.027 K_m + 5.531 K_m^2 - 3.108 K_m^3 \quad (2.67)$$

2.6.4 Extraterrestrial Radiation on Horizontal Surfaces

Extraterrestrial solar radiation on the horizontal surface is a function of day number of the year, declination, solar hour angle, and location latitude. The relation of extraterrestrial radiation I_0 [W/m²] with extraterrestrial radiation on horizontal surface (I_{0H}) is represented as in Figure 2.22.

The hourly extraterrestrial radiation on a horizontal surface can be calculated approximately by assuming the zenith angle (β) corresponds to the midpoint of the considered period.

$$I_{0H} = I_0 \cdot \cos\beta = I_{SC}\left[1 + 0.034 \cdot \cos\left(\frac{360N}{365}\right)\right](\sin\delta\sin L + \cos\delta\cos L\cos\omega) \quad (2.68)$$

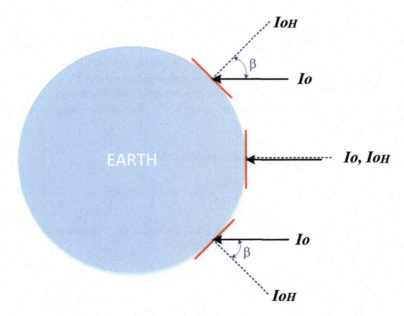

FIGURE 2.22 Extraterrestrial radiation is the radiation incident on the surface tangent to the outer surface of the atmosphere.

Notice that from (2.11), we have substituted the $(\cos\beta = \sin\delta\sin L + \cos\delta\cos L\cos\omega)$ into (2.68). The extraterrestrial radiation on a horizontal surface (I_{0H}) for a specific period can be obtained by integrating (2.68).

$$I_{0H} = I_{SC}\left[1+0.034\cdot\cos\left(\frac{360N}{365}\right)\right]\int(\sin\delta\sin L + \cos\delta\cos L\cos\omega)dt \qquad (2.69)$$

In (2.69), the time (t) is in hours and it is given by $t = (180/\pi)(\omega/15)$. Thus, the differential time element $dt = [180/(15\pi)]d\omega$. By substituting dt in (2.69), the hourly extraterrestrial radiation on a horizontal surface (I_{h0H}) can be obtained by integrating (2.68) for a period between hour angles ω_1 and ω_2.

$$I_{h0H} = \frac{12}{\pi}I_{SC}\left[1+0.034\cdot\cos\left(\frac{360N}{365}\right)\right]\int_{\omega_1}^{\omega_2}(\sin\delta\sin L + \cos\delta\cos L\cos\omega)d\omega \qquad (2.70)$$

After integration, the hourly extraterrestrial radiation on a horizontal surface is

$$I_{h0H} = \frac{12}{\pi}I_{SC}\left[1+0.034\cdot\cos\left(\frac{360N}{365}\right)\right]\cdot\left[\frac{\pi(\omega_2-\omega_1)}{180}\sin\delta\sin L + \cos\delta\cos L(\sin\omega_2 - \sin\omega_1)\right] \qquad (2.71)$$

where ω_1 and ω_2 are given in degrees, I_{SC} in W/m^2 and I_{h0H} in W/m^2. Notice that I_{h0H} can be converted into Joul/m^2 by multiplying 3600.

Similarly, the daily extraterrestrial radiation on a horizontal surface (I_{d0H}) can be obtained by integrating (2.68) over the period from sunrise (ω_{sr}) to sunset (ω_{ss}).

$$I_{d0H} = \frac{24}{\pi}I_{SC}\left[1+0.034\cdot\cos\left(\frac{360N}{365}\right)\right]\cdot\left[\frac{\pi\omega_s}{180}\sin\delta\sin L + \cos\delta\cos L\sin\omega_s\right] \qquad (2.72)$$

where ω_s in degrees is calculated by $\omega_s = \cos^{-1}(-\tan\delta\tan L)$.

Example 2.13: Hourly Total Solar Radiation

A PV panel is planned to be installed in Las Vegas $(36.169°\text{N}, 115.14°\text{W})$ at a slope of 40° in the south direction. Calculate the hourly total solar radiation value for this PV panel on June 15 for the hour 13:00–14:00 solar time. Note that the ground reflectance factor (ρ_G) in the area is 0.3 and the measured hourly total solar radiation on the horizontal surface is 650 Wh/m^2 in the area.

Solution

Day number $N = 166$ on June 15, and from equation (2.9):

$$\delta = 23.45\cdot\sin\left[\frac{360}{365}(166+284)\right] = 23.314°$$

Solar hour angles for 13:00 and 14:00 are $\omega_1 = 15°$ and $\omega_2 = 30°$, respectively. From (2.71), the hourly extraterrestrial radiation on a horizontal surface is:

$$I_{h0H} = \frac{12}{\pi}\cdot 1367\cdot\left[1+0.034\cdot\cos\left(\frac{360\cdot 166}{365}\right)\right]\cdot\left[\begin{array}{l}\frac{\pi(30-15)}{180}\sin 23.314\cdot\sin 36.169 \\ +\cos 23.314\cdot\cos 36.169\cdot(\sin 30 - \sin 15)\end{array}\right]$$

$$= 1198 \text{ W/m}^2$$

The Sun and Solar Radiation

As an approximate alternative method, I_{h0H} can also be calculated from the relation (2.68) for one hour from ω_1 (13:00) to ω_2 (14:00). In this case, the solar hour angle (ω) should be taken as the midpoint of the considered period, assuming that the irradiance is constant for the required period of one hour. Thus, $\omega = 22.5°$. By writing the all-known values of δ, ω, and L in (2.68):

$$I_{h0H} = 1367 \cdot \left[1 + 0.034 \cdot \cos\left(\frac{360 \cdot 166}{365}\right)\right](\sin 23.314 \sin 36.169 + \cos 23.314 \cos 36.169 \cos 22.5) = 1200 \text{ W/m}^2$$

The hourly clearness index using (2.59),

$$K_h = \frac{650}{1198} = 0.54$$

The ratio, $\frac{I_{DFH(h)}}{I_{(h)}}$ is calculated below by replacing the K_h in (2.65),

$$K_{DF(h)} = \frac{I_{DFH(h)}}{I_{(h)}} = 0.9511 - 0.1604 \times 0.54 + 4.388 \times 0.54^2 - 16.638 \times 0.54^3 + 12.336 \times 0.54^4 = 0.5731$$

Hence,

$$I_{DFH(h)} = 0.5731 \times 650 = 372.5 \text{ Wh/m}^2$$

$$I_{(h)} = (1 - 0.5731) \times 650 = 0.4269 \times 650 = 277.5 \text{ Wh/m}^2$$

The tilt conversion factor in (2.60) can be calculated using (2.52) and (2.18)

$$C_D = \frac{I_{DM}}{I_{DH}} = \frac{\cos\theta}{\cos\beta} = \frac{\sin(\delta)\sin(L-b) + \cos(\delta)\cos(\omega)\cos(L-b)}{\sin(\delta)\sin(L) + \cos(\delta)\cos(L)\cos(\omega)}$$

$$C_D = \frac{\sin(23.314)\sin(36.169 - 40) + \cos(23.314)\cos(22.5)\cos(36.169 - 40)}{\sin(23.314)\sin(36.169) + \cos(23.314)\cos(36.169)\cos(22.5)} = \frac{0.82}{0.92} = 0.891$$

Using (2.56), hourly total irradiance on tilted module surface is:

$$I_{M(h)} = 0.891 \times 277.5 + 372.5\left(\frac{1 + \cos 40}{2}\right) + 650 \times 0.3 \times \left(\frac{1 - \cos 40}{2}\right) = 598.98 \text{ Wh/m}^2$$

Example 2.14: Monthly Total Solar Radiation

A PV panel is planned to be installed in Las Vegas $(36.169° \text{N}, 115.14° \text{W})$ at a slope of 40° in the south direction. Using the isotropic diffuse model, calculate the average monthly total solar irradiation on inclined PV panels for January. Use the ground reflectance factor (ρ_G) in the area as 0.3. Assume that monthly average daily of the total irradiation on a horizontal surface for January is 62.7 kWh/m².

Solution

For the mean January day, the solar declination can be calculated according to the 16th day of the year. Thus for $N = 16$, equation (2.9) gives:

$$\delta = 23.45 \cdot \sin\left[\frac{360}{365}(16 + 284)\right] = -21.105°$$

Using the sunset hour angle equation, ω_s is:

$$\omega_s = \cos^{-1}(-\tan\delta \tan L) = \cos^{-1}\left[-\tan(-21.105)\tan(36.169)\right] = 73.61°$$

For January, the monthly average daily extraterrestrial irradiation can then be calculated from (2.72) as:

$$I_{d0H} = \frac{24}{\pi} 1367 \left[1 + 0.034 \cdot \cos\left(\frac{360 \cdot 16}{365}\right) \right] \cdot \left[\begin{array}{l} \frac{73.61\pi}{180} \sin(-21.105)\sin(36.169) \\ + \cos(-21.105)\cos(36.169)\sin(73.61) \end{array} \right]$$

$$= 4847.86 \text{ Wh/(m}^2 \cdot \text{day)}$$

There are 31 days in January; hence, monthly average daily extraterrestrial irradiation will be $4847.86 \times 31 = 151.31$ kWh/m². The clearness index K_m is obtained below:

$$K_m = \frac{62.7}{151.31} = 0.414$$

The value of the clearness index K_m is used to calculate $K_{DF(m)} = \frac{I_{DFH(m)}}{I_{(m)}}$ from equation (2.67)

$$\frac{I_{DFH(m)}}{I_{(m)}} = 1.390 - 4.027 \times 0.414 + 5.531 \times 0.414^2 - 3.108 \times 0.414^3 = 0.45$$

Sunset hour angle for the tilted surface ω_{ss} is calculated using (2.64).

$$\omega_{ss} = \min\left\{73.61, \cos^{-1}\left[-\tan(36.169-40) \cdot \tan(-21.105)\right]\right\} = \min(73.61, 91.3) = 73.61$$

$C_{D(m)}$ is calculated by (2.63)

$$C_{D(m)} = \frac{\cos(36.169-40) \cdot \cos(-21.105) \cdot \sin(73.61) + \frac{\pi \times 73.61}{180} \cdot \sin(36.169-40) \cdot \sin(-21.105)}{\cos(36.169) \cdot \cos(-21.105) \cdot \sin(73.61) + \frac{\pi \times 73.61}{180} \cdot \sin(36.169) \cdot \sin(-21.105)} = 2.045$$

The mean monthly total solar irradiation for the tilted surface is now then calculated for January by using (2.62). The values in (2.62) are as follows:

$$I_{(m)} = 62.7 \text{ kWh/m}^2$$

$$I_{DFH(m)} = 0.45 \times 62.7 = 28.215 \text{ kWh/m}^2$$

$$C_{D(m)} = 2.045$$

$$\frac{I_{DFH(m)}}{I_{(m)}} = 0.45$$

$$\rho_G = 0.3$$

$$b = 40°$$

Now let us substitute these values into (2.62) to calculate $I_{M(m)}$.

$$I_{M(m)} = \left[I_{(m)} - I_{DFH(m)}\right] C_{D(m)} + I_{DFH(m)} \left(\frac{1+\cos b}{2}\right) + I_{(m)} \rho_G \left(\frac{1-\cos b}{2}\right)$$

$$I_{M(m)} = [62.7 - 28.215] \times 2.045 + 28.215 \times \left(\frac{1+\cos 40}{2}\right) + 62.7 \times 0.3 \left(\frac{1-\cos 40}{2}\right)$$

$$= 70.52 + 24.914 + 2.20 = 97.634 \text{ kWh/m}^2$$

2.7 INSOLATION ON SOLAR TRACKING SURFACES

In addition to constant angle solar systems, Sun-tracking systems are also quite common in practice. Sun-tracking systems have two types, single-axis and two-axis. Single-axis trackers have one degree of freedom that tracks only one angle. In general, single-axis trackers for PVs have a manually adjustable tilt angle in the south-north direction and a tracking mechanism to rotate the system from East to West as shown in Figure 2.23.

A special case is called a polar-mount solar tracker if the tilt angle of the mount is set equal to the local latitude (Figure 2.24).

It is not only easy to calculate incoming insolation on the PV module surface for the polar-mount system, but also an optimum angle is obtained on a yearly basis. The centerline of the PV panel will always face directly into the Sun if the polar-mount rotates 15°/h around its axis, which is the same rate as the Earth's rotation. In this case, the incidence angle θ will be equal to the solar declination angle δ. Hence, the direct beam radiation on the module surface is given by (2.73).

$$I_{DM} = I_D \cdot \cos\delta \tag{2.73}$$

The tilt angle should be known for the polar-mount tilted surface. Thus, the equations of diffuse and reflected radiation relating to the single-axis polar-mount tilted surface are given below.

$$I_{DFM} = DF \times I_D \cdot \left[\frac{1+\cos(90-\alpha+\delta)}{2}\right] \tag{2.74}$$

$$I_{RM} = \rho_G (I_{DH} + I_{DFH}) \cdot \left[\frac{1-\cos(90-\alpha+\delta)}{2}\right] \tag{2.75}$$

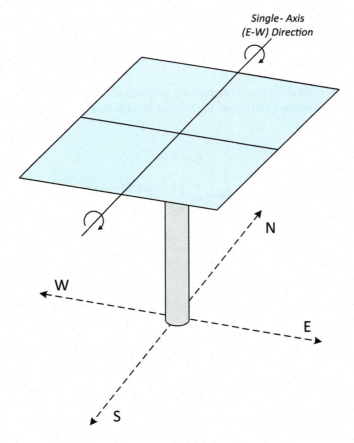

FIGURE 2.23 Single-axis system with east-west tracking.

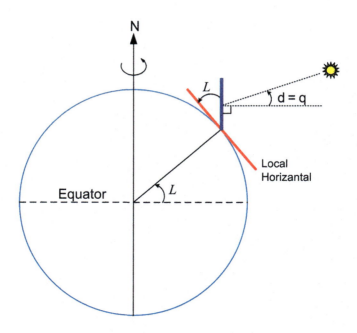

FIGURE 2.24 Single-axis polar-mount tracker with a 15°/h angular rotation.

where the angle $(90 - \alpha + \delta)$ is called effective tilt angle and denoted by b_e, which is defined as the angle between the horizontal plane and normal to the tilted surface.

Two-axis trackers have two degrees of freedom that tracks the azimuth and altitude angles of the Sun to point directly at the Sun all the time. In general, two-axis trackers for PVs have two automatically adjustable angles in the south-north and east-west directions as shown in Figure 2.25.

Since the two-axis tracker surface is constantly perpendicular to the Sun's direct beam, the calculation of the radiation amount falling on the surface is quite straightforward and it is given by the following equations.

$$I_{DM} = I_D \tag{2.76}$$

$$I_{DFM} = DF \times I_D \cdot \left[\frac{1 + \cos(90 - \alpha)}{2} \right] \tag{2.77}$$

$$I_{RM} = \rho_G (I_{DH} + I_{DFH}) \cdot \left[\frac{1 - \cos(90 - \alpha)}{2} \right] \tag{2.78}$$

where the tilt angle b is equal to $(90 - \alpha)$, which is defined as the complement of solar altitude angle.

Example 2.15: Solar Insolation on Single-Axis and Two-Axis Tracking Panels

a. Calculate the clear-sky insolation on a polar-mount single-axis PV module at solar noon on June 21 for Las Vegas (36.169° N Longitude). Ignore the ground reflectance.
b. Calculate the clear-sky insolation on a two-axis PV module at solar noon on June 21 for Las Vegas (36.169° N Longitude). Ignore the ground reflectance.

The Sun and Solar Radiation

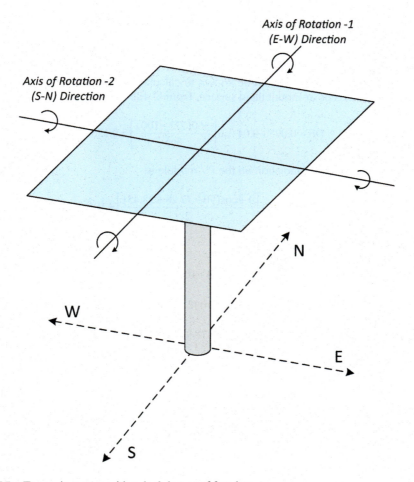

FIGURE 2.25 Two-axis system with a dual degree of freedom.

Solution

a. The parameters of A, K, and AM in (2.40) are given for June 21 ($N = 172$) as below:

$$A = 1160 + 75\sin\left[\frac{360(172-275)}{365}\right] = 1086.5$$

$$K = 0.174 + 0.035\sin\left[\frac{360(172-100)}{365}\right] = 0.207$$

In order to calculate AM, we need to find the altitude angle from (2.13) as below:

$$\alpha = 90 - L + \delta = 90 - 36.169 + 23.45 \Rightarrow \alpha = 77.28°$$

$$AM = \frac{1}{\sin(77.28)} = 1.025$$

Hence, clear-sky direct beam insolation from (2.40) is obtained as below:

$$I_D = 1086.5 \times e^{-0.207 \times 1.025} = 878.94 \text{ W/m}^2$$

Consequently, direct beam insolation on the module surface from (2.73) is

$$I_{DM} = I_D \cdot \cos\delta = 878.94 \times \cos(23.45) = 806.34 \text{ W/m}^2$$

The sky diffuse factor DF should be known to calculate diffuse and reflected radiation on the single-axis polar-mount tilted surface. From (2.48), we have:

$$DF = 0.095 + 0.04 \cdot \sin\left[\frac{360(172-100)}{365}\right] = 0.133$$

Using (2.74), the diffuse radiation on the PV module is:

$$I_{DFM} = 0.133 \times 878.94 \times \left[\frac{1+\cos(90-77.28+23.45)}{2}\right] = 105.65 \text{ W/m}^2$$

The total insolation on the PV module is

$$I_M = I_{DM} + I_{DFM} = 806.34 + 105.65 = 911.99 \text{ W/m}^2$$

b. For the two-axis solar tracking system, from (2.76) and (2.77), we have:

$$I_{DM} = I_D = 878.94 \text{ W/m}^2$$

$$I_{DFM} = 0.133 \times 878.94 \times \left[\frac{1+\cos(90-77.28)}{2}\right] = 115.46 \text{ W/m}^2$$

The total insolation on the two-axis PV module is

$$I_M = I_{DM} + I_{DFM} = 878.94 + 115.46 = 994.4 \text{ W/m}^2$$

The insolation on the two-axis PV module is 9% higher than on the single-axis mount.

2.8 SHADING AND SHADOW ANALYSIS

The performance of solar PV systems on the roof of buildings as well as onsite PV plants may be affected considerably due to shadow variations of surroundings. That's why a shadow analysis in the planning phase is essential to minimize or avoid power losses due to the shading effect for stand-alone and grid-connected PV systems. Shadow types should be known in order to evaluate shading effects accurately.

2.8.1 Shadow Types

Shadow casts on a PV system should be eliminated as much as possible. For this purpose, evaluations for each of the shadow types should be made according to shadow types. The most common shadow types are temporary shading due to soiling, snow, bird droppings, etc., surrounding shading like buildings and trees, and finally self-shading due to the system itself. Temporary shadings are out of scope for this section and their effect on PV performance will be discussed in detail in the later sections of the book.

Any shadow can be analyzed by dividing it into core shadow and partial shadow. This separation is very important to evaluate the performance losses of PV systems. Because the core shadow cast approximately reduces the energy incidence by 60%–80% on the PV cell, a partial shade leads to a reduction of up to 50%. This irradiation reduction on the PV cell depends on the distance between the shadow-casting object and the PV cell. Therefore, the smaller the distance between the

The Sun and Solar Radiation

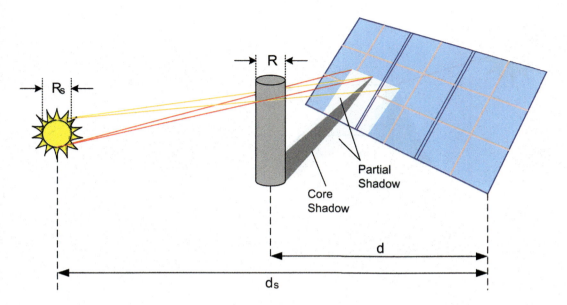

FIGURE 2.26 Shadow cast of an object on a PV Module and formation of core/partial shadow.

shadow-casting object and the PV cell, the darker the shadow on the PV will be. Because as this distance increases, the diffuse radiation on the PV will be more. Figure 2.26 shows the formation of core shadow and partial shadow on a photovoltaic panel.

In order to avoid core shadow on the PV panel, the optimum distance from the PV module can be calculated as below using the similarity relations of triangles.

$$d_{opt} = \frac{R \times d_s}{R_s} \qquad (2.79)$$

where d_s is the distance between Earth and the Sun (150×10^9 m), R_s is the diameter of the Sun (1.39×10^9 m), R is the thickness of the shadow-casting object, and d_{opt} is the optimum distance between the PV panel and the shadow-casting object. For example, a wooden pole with a diameter of 30 cm should be at least 32.37 m ($0.3 \times 150 \times 10^9 / 1.39 \times 10^9$) away to eliminate the core shading on the PV panel.

2.8.2 Shadow Geometry

Shadow geometry should be defined here to evaluate shadow types in a PV site. The dimensions of shadow-casting objects and the distance from the PV panel are needed to be calculated to make an energy-efficient site planning. Hence, the azimuth angle (z) of the PV panel and the altitude angle (α) are shown in Figure 2.27 for roof-mounted PV modules with surrounding objects such as trees.

The altitude angle α can be obtained based on Figure 2.27:

$$\tan \alpha = \frac{H_2 - H_1}{d} = \frac{\Delta H}{d} \Rightarrow \alpha = \tan^{-1}\left(\frac{\Delta H}{d}\right) \qquad (2.80)$$

where H_1 is the height of the PV system from the ground, H_2 is the height of the shading object and d is the distance between the PV system and the shading object. Note that the altitude angle should be worked out for all shadow-casting objects in the surrounding area of the PV system.

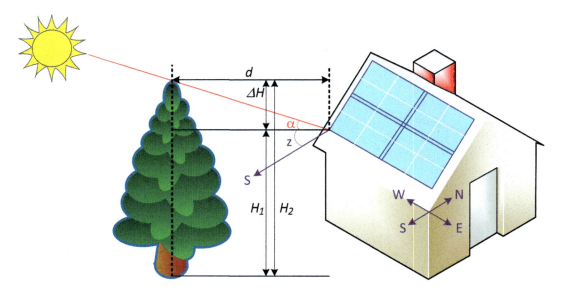

FIGURE 2.27 Calculation of a PV module's azimuth angle and altitude angle for site planning.

Another shading geometry should be taken into consideration for self-shaded PV arrays, which is the case generally for free-standing tilted PV arrays mounted on flat areas. As the tilt of the PV arrays decreases, the shading rates of each other will also decrease and the rate of area exploitation will increase. However, as the slope approaches the horizontal, the operating efficiency of PV systems will be lower. For this reason, PV arrays are generally mounted at a slope of 30–40 degrees in the area. Figure 2.28 shows the shading of PV arrays in a solar field.

The distance between the rows of PV arrays (d) depends upon the module widths (H), tilt angle (b), and shading angles (altitude angle, α) and it is obtained from the sinus theorem by the following relation:

$$\frac{d}{\sin(180-b-\alpha)} = \frac{H}{\sin(\alpha)} \Rightarrow d = H \times \frac{\sin(180-b-\alpha)}{\sin(\alpha)} \quad (2.81)$$

It is suggested for greater energy yield that the modules should remain shade-free at least at noon on December 21, which is the shortest day of the year. Hence, if the altitude angle at noon on the winter solstice is taken as the shading angle, an optimal "no shading distance" can be obtained from (2.81) for flat solar areas as below.

$$d = H \times \frac{\sin(180-b-\alpha_{\text{Dec. 21, 12:00 PM}})}{\sin(\alpha_{\text{Dec. 21, 12:00 PM}})} = H \times \frac{\sin(90-b+\beta_{\text{Dec. 21, 12:00 PM}})}{\cos(\beta_{\text{Dec. 21, 12:00 PM}})} \quad (2.82)$$

Notice that shading angle is not only the factor to decide the distance between PV rows in the planning and installation phase. Factors such as field costs, field utilization factors (u_f), wind and snow loads, soiling, self-cleaning effect, and material costs play important roles in the design of the PV arrangement. Considering all these factors, equation (2.82) can be used to calculate the level of field utilization factor, which is defined as the ratio of modules' (H) to the distance between the rows (d).

$$u_f = \frac{H}{d} \quad (2.83)$$

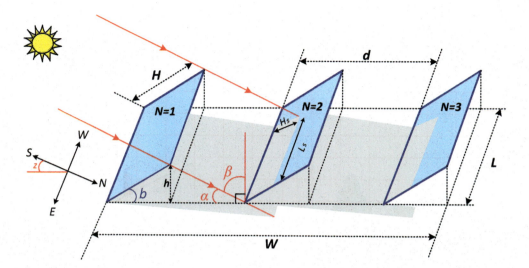

FIGURE 2.28 Shading of PV arrays in a solar field.

It is important to know the relatively shaded area to evaluate how the PV modules are affected by the shaded area. The relatively shaded area (a_s) given in (2.83) is defined as the ratio of the shaded area on the module surface to the total module surface.

$$a_s = \frac{H_s \times L_s}{H \times L} \quad (2.84)$$

2.8.3 Shadow Analysis

Shading analysis should be performed for solar PV systems in order to evaluate the shading resulting from PV location. For this purpose, the objects surrounding the PV system are determined according to a specific reference point on the PV system. This reference point is usually taken as the center point of the PV array for small-scale PV system applications. For large-scale applications, the reference points may be several points for greater accuracy in shading analysis.

Fundamental shading analysis for the surrounding area of the PV system can simply be performed by using either a site plan or using Sun path diagram.

If a site-plan approach is used for shading analysis, the distance and dimensions of the shadow-casting object to the PV array with azimuth and altitude angles should be determined. Shadow calculations should then be made using formulas (2.80)–(2.84).

One of the methods used to evaluate shading is to use Sun path diagrams. For this purpose, the altitude and azimuth angles are plotted as the axes of the angular grid, which includes the obstacles surrounding the PV system as shown in Figure 2.29.

The observer records the position of the obstacles by looking at the point of reference and accordingly the altitude and azimuth angles are determined. Afterward, the corresponding obstacles are included in the Sun path diagram to estimate the periods of shading at the PV site. Figure 2.30 shows a typical Sun path diagram with a shadow outline.

Now that it is possible to read the shading periods for a particular time of the year. For example, the location, in general, is approximately 50% shaded on the winter solstice and no shading occurs on the summer solstice and in equinox times. In addition to graphical approaches, simulation programs such as Solar Pathfinder, Solarius, PVSol, PVSyst, and PVcad are commercially available for shading analysis.

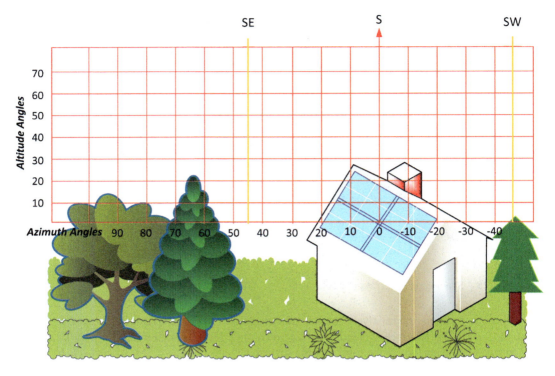

FIGURE 2.29 Surrounding objects with angular grid for PV shading analysis.

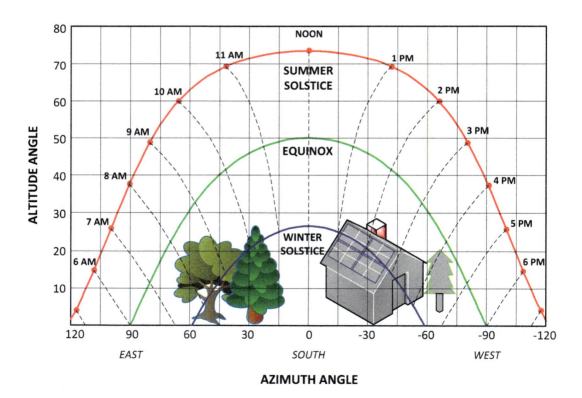

FIGURE 2.30 A typical Sun path diagram with obstructions inclusion.

Example 2.16: Distance between Two PV Rows

A PV power plant is planned to be installed in Las Vegas $(36.169°\,N, 115.14°\,W)$ at a slope of 36° in the south direction. The PV array configuration is given in Figure 2.31.

In order to avoid inter-row shading, calculate the required distance between rows of PV arrays.

Solution

The distance between two rows of modules is calculated based on the noon time on December 21, which is depicted below (Figure 2.32).

Using equation (2.11), the altitude angle at noon on December 21 for latitude 36.169°N:

$$\sin(\alpha) = \sin(-23.45)\sin(36.169) + \cos(-23.45)\cos(36.169)\cos(0)$$

$$\sin(\alpha) = 0.419 \Rightarrow \alpha = 24.48°$$

From equation (2.82), optimal "no shading distance" for the given PV arrays,

$$d = H \times \frac{\sin(180 - b - \alpha_{\text{Dec. 21, 12:00}})}{\sin(\alpha_{\text{Dec. 21, 12:00}})} = 3 \times \frac{\sin(180 - 36 - 24.48)}{\sin(24.48)} = 6.29 \text{ m}$$

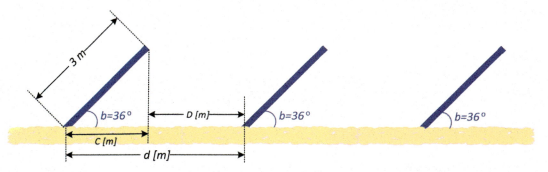

FIGURE 2.31 PV array configuration.

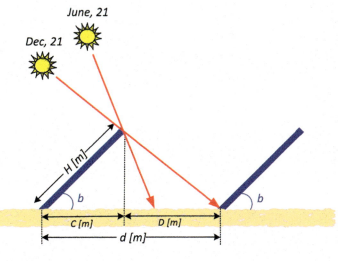

FIGURE 2.32 Calculation of distance between two rows.

Considering the above figure, the distance D in meters:

$$D = d - H \times \cos(b) = 6.29 - 3 \times \cos(36) = 3.86 \text{ m}$$

Example 2.17: Apparent Motion of the Sun

Draw the apparent motion of the Sun on the equinoxes and the solstices with reference to a fixed observer at latitude 36° in the northern hemisphere. Show the position of the Sun at solar noon on each solar path of these days.

Solution

The plane of the Sun's apparent orbit lies at an angle equal to the latitude from the observer's vertical on any day of the year. At spring and fall equinoxes (March 21 and September 23), the Sun rises true east and sets true west. Therefore, the solar altitude at solar noon on the equinoxes equals 90° minus the latitude. At the summer and winter solstices for the northern hemisphere, the solar altitude angle at solar noon increases by the Earth's declination, which is $\delta = 23.45°$. According to these explanations, the apparent motion of the Sun relative to a fixed observer can be plotted for latitude 36° as below (Figure 2.33).

Example 2.18: Sun Positions Throughout the Year

Plot the position of the Sun at noon each day throughout the year against the equation of time is called the analemma graph. Hence, plot this graph to find the solar declination of the vertical solar noon for each day of the year.

Solution

Using equations (2.8) and (2.34), the analemma graph is plotted throughout the year as below (Figure 2.34). Using this graph, one can find the declination of the Sun at solar noon on any given day.

2.9 SOLAR RADIATION MEASUREMENTS

Measurement of solar radiation is essential for photovoltaic solar design, sizing, performance evaluation, applications, and research. Basic solar measurements include global solar radiation, direct radiation, diffuse radiation, albedo, and sunlight duration. The instruments for measuring the abovementioned solar parameters are described in the following subsections.

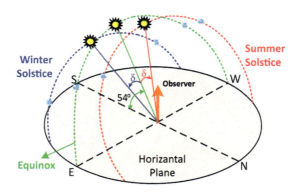

FIGURE 2.33 Apparent motion of the Sun relative to a fixed observer at latitude 36°.

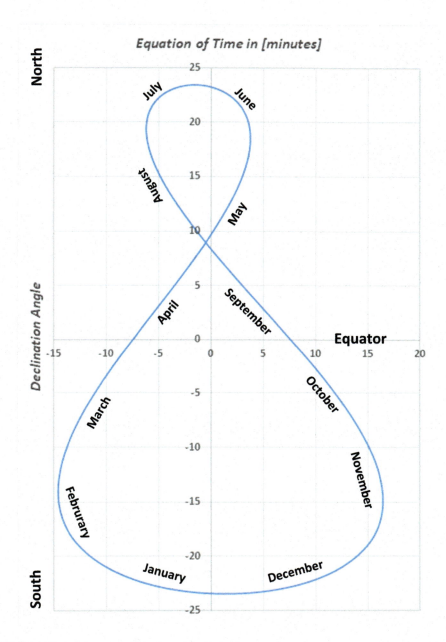

FIGURE 2.34 The analemma graph plotted for the entire year.

2.9.1 Pyranometer

As one of the solar radiation sensors, the measures global radiation, which is the sum of direct and diffuse components of solar irradiance at the point of measurement. Most pyranometers nowadays work based on the principle of either thermoelectric or the photoelectric effect.

The pyranometer based on the thermoelectric effect, which is called thermopile pyranometer, includes a thermal detector and converts incoming solar radiation from a hemisphere to a voltage signal through series or series-parallel connected thermocouples. Firstly, the radiant energy is absorbed by a black surface disk, which results in a temperature increase on the disk surface. The generated heat flows through a thermal resistance to the pyranometer body (heat sink). Afterward, the temperature difference between the two ends of the thermal detector is converted to a voltage signal, which is proportional to the absorbed temperature. In order to prevent temperature absorption from outside effects such as wind and rain, the detector is shielded by two glass domes. These glass domes allow equal transmittance of direct solar beam for every Sun position through any given daytime. The basic working principles of a thermopile pyranometer are shown in Figure 2.35, while a schematic diagram of pyranometer sections is given in Figure 2.36.

The pyranometer based on the photoelectric effect includes a sensing element, e.g., a silicon photodiode, which converts incoming solar radiation from a solid angle 2π to electrical current with a wide spectral response. The output voltage is taken from the sensor's ends and the console displays the solar irradiance and also calculates the total incident energy by integrating irradiance values over the period.

If a pyranometer is equipped with a special shadow archway to block the glass dome of the pyranometer from the direct beam, it can also be used for the measurement of diffuse irradiance.

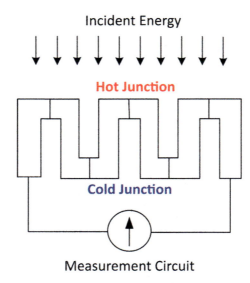

FIGURE 2.35 Basic working principles of a thermopile pyranometer.

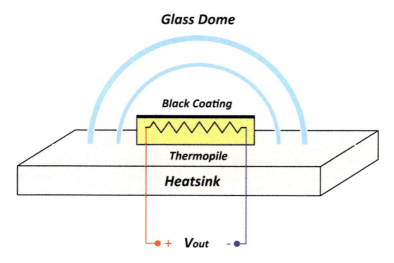

FIGURE 2.36 Schematic diagram of pyranometer sections.

The Sun and Solar Radiation

The shadow archway should be mounted with its axis titled at a latitude angle of the site so that the plane of the archway is to be parallel to the plane defined by the daily solar path. Figure 2.37 shows a pyranometer with a shadowing archway to measure the diffuse solar radiation.

2.9.2 Pyrheliometer

It is an instrument designed to measure the direct component of solar irradiance. The sensing element of the pyrheliometer such as a thermocouple is placed at the bottom of a cylindrical tube, which allows only the direct sunbeam through the cylindrical tube onto a thermopile at the bottom of the tube. Due to direct incidence on the thermopile, which converts the created heat to an electrical signal. The pyrheliometer is used with a two-axis solar tracking system to keep the instrument aimed directly at the Sun. The general structure of a pyrheliometer is given in Figure 2.38. Measuring the direct beam is often very important for concentrated solar applications.

2.9.3 Albedometer

It generally consists of two standard pyranometers with thermopile sensors and measures global and reflected solar radiation as well as the solar albedo. Solar albedo (also called solar reflectance) is defined as the ratio of the reflected radiation to the global radiation. The general structure of an albedometer is given in Figure 2.39. This application requires horizontal leveling for measurement accuracy.

2.9.4 Sunlight Recorder

The sunshine recorder is an instrument to measure sunshine duration. The duration of sunshine is defined as the length of time that the ground surface is irradiated by a direct solar beam. Two types of sunshine recorders are widely used. The first one works based on the principle of the photoelectric effect, while the second one works based on the photoelectric effect. The focusing type of sunshine recorder consists of a glass sphere that is mounted in a section of a spherical bowl. The sunshine recorder concentrates the sunlight through the glass sphere onto a recording paper band placed at its focal point. The length of the burnt trace on the recording band represents the sunshine duration for the day. The structure of the sunshine recorder is shown in Figure 2.40.

The photoelectric effect type sunshine recorder consists of two identical photovoltaic cells. The first of two PV cells is exposed to direct solar radiation while the second one with a shading ring is positioned to expose only diffuse radiation. Normally, the direct beam radiation produces higher

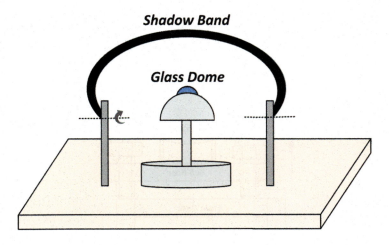

FIGURE 2.37 Pyranometer with shadow band to measure the diffuse solar radiation.

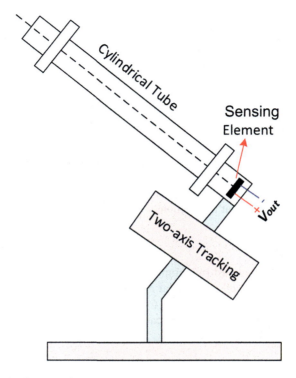

FIGURE 2.38 Schematic diagram of pyrheliometer sections.

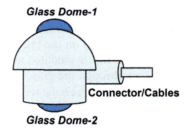

FIGURE 2.39 The general structure of an albedometer.

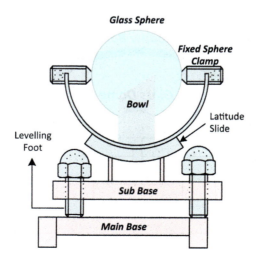

FIGURE 2.40 The general arrangement of the sunshine recorder.

The Sun and Solar Radiation

output than the shaded one. In the case of only diffuse radiation exist, the two cells will have the same output. Hence, the radiation difference between the two identical PV cells gives the measurement of sunshine duration.

2.10 PROBLEMS

P2.1. Within a year, find the dates on which the distance of the Earth from the Sun is the lowest and highest.

P2.2. A south-facing PV panel is placed at a tilt angle of 30° in Los Angeles (34.0522° N, 118.2437° W). Find the angle between the normal of the PV to the Sun's ray at solar noon on March 21, January 1, and September 1.

P2.3. Calculate the solar azimuth and altitude angles at latitudes of 30°, 40°, and 50° on December 21 at 10.00 AM local solar time.

P2.4. Considering the following solar site (Figure 2.41), calculate the height of the tree and azimuth angle with respect to the PV site. How many days are there in a year that the shadow of a tree falls on the PV site?

P2.5. A PV panel with 40° of inclination is placed at an angle of 10° west of south at a latitude of 40° N. Calculate the angle of incidence at 11.00 local solar time on March 21 for the PV panel.

P2.6. A south-facing PV panel is placed horizontally at latitude (40° N). Calculate the solar sunrise and sunset times for this PV panel on June 21.

P2.7. Assume that the PV panel in Problem 2.6 is placed with an inclination of 30°. Calculate the solar sunrise and sunset times for this PV panel on June 21.

P2.8. Considering the newspaper given below (Figure 2.42), calculate the following:
   a. What time (clock time) is the solar noon.
   b. By ignoring the Q correction given by (2.39), estimate the latitude.
   c. By taking the Q correction into account, find the clock time at which geometric sunrise occurs.
   d. Determine the observer's latitude.

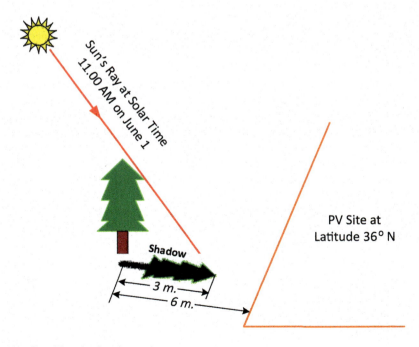

FIGURE 2.41 Considered solar site.

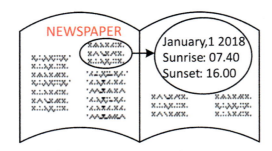

FIGURE 2.42 A newspaper on January 1, 2018.

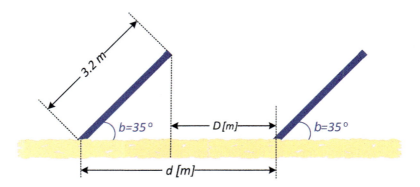

FIGURE 2.43 Considered PV array configuration.

P2.9. A south-facing PV panel is placed with an inclination of 30° latitude (35° N). Calculate the direct insolation, diffuse radiation, and reflected radiation on the PV surface at solar noon on March 21. Assume that the ground reflectance factor is 0.3 at the PV site and the measured hourly total solar radiation on the horizontal surface is 700 Wh/m² in the area.

P2.10. Two options are being considered to meet the annual energy requirement of 5000 kWh for a latitude located in Las Vegas (36° N, 115° W). The first one of these is a polar-mount single-axis tracker system. The second option is to use a double-axis tracker system. Assuming 12% efficiency when converting sunlight into electricity, how large a PV array is needed from each option (Assume that 70% of the radiation on the single-axis PV module at noon on January 1 is equal to the average daily radiation on an annual basis). Also assume that the annual sunshine duration for the region is 2800 hours/year and the ground reflectance is ignored.

P2.11. If the Earth's atmosphere is neglected, calculate the amount of daily solar radiation falling on the horizontal surface on January 15 at latitude 36° N. Note that the solar constant is 1367 W/m².

P2.12. Calculate the amount of extraterrestrial radiation falling on a horizontal surface between the solar hours 9.00 and 10.00 AM on January 15 at a latitude of 36° N. Note that the solar constant is 1367 W/m².

P2.13. A PV power plant consisting of (3.2 m × 10 m) PV arrays is planned to be installed at latitude $(40°\,N)$ at a slope of 35° in the south direction. The PV array configuration is given in Figure 2.43.

In order to avoid inter-row shading, calculate the required distance between rows of PV arrays.

BIBLIOGRAPHY

1. G. M. Masters, *"Renewable and Efficient Electric Power Systems"*, John Wiley & Sons, New Jersy, 2004.
2. V. Quaschning, *"Understanding Renewable Energy Systems"*, Earthscan, London, 2005.
3. F. Kasten and A. T. Young, "Revised optical air mass tables and approximation formula", *Appl. Opt.*, vol. 28, pp. 4735–4738, 1989.
4. C. Fröhlich, and R. W. Brusa, "Solar radiation and its variation in time", *Solar Phys.*, vol. 74, pp. 209–215, 1981.
5. M. Iqbal, *"An Introduction to Solar Radiation"*, Academic Press, Toronto, 1983.
6. J. E. Braun and J. C. Mitchell. "Solar geometry for fixed and tracking surfaces", *Solar Energy*, vol. 31, pp. 439–444, 1983.
7. C. A. Gueymard, "The sun's total and spectral irradiance for solar energy applications and solar radiation models", *Solar Energy*, vol. 76, no. 4, pp. 423–453, 2004.
8. S. A. Klein, "Calculation of monthly average insolation on tilted surfaces", *Solar Energy*, vol. 19, no. 4, pp. 325–329, 1977.
9. A. E. Roy, *"Orbital Motion"*, 3rd Edition, Taylor & Francis, UK, 1988.
10. U.S. Department of Energy, *"On the Nature and Distribution of Solar Radiation"*, U.S. Department of Energy, Washington, DC, 1978.
11. J. L. Threlked and R. C. Jordon, "Direct radiation available on clear days", *ASHRAE Transaction*, vol. 64, p. 45, 1958.
12. B. Y. H. Liu and R. C. Jordan, "The interrelationship and characteristic distribution of direct, diffuse, and total solar radiation", *Solar Energy*, vol. 4, no. 3, pp. 1–19, 1960.
13. J. A. Duffie and W. A. Beckman, *"Solar Engineering of Thermal Processes"*, John Wiley & Sons, New Jersey, 2006.
14. F. Antony and C. Durschner, *"Photovoltaic for Professionals"*, ISBN-13: 978-3-93459543-9, Earth Scan, 2007.
15. J. K. Page, M. Albuisson, and L. Wald, "The European solar radiation atlas: A valuable digital tool", *Solar Energy*, vol. 71, no. 1, pp. 81–83, 2001.
16. F. Kuik, "Global Radiation Measurements in the Operational KNMI Meteorological Network," Technical report 197, KNMI, Royal Dutch Meteorological Institute, 1997.
17. International Energy Agency, *"An Introduction to Meteorological Measurements and Data Handling for Solar Energy Applications"*, IEA Solar R&D, Paris, 1980.
18. K. Van den Bos and E. Hoeksema, "An Introduction to Atmospheric Radiation Measurement", Kipp & Zonen / Sci-Tek Instruments Technical Paper 970529, Delft, 1997.

3 Introduction to Photovoltaic Systems and Photovoltaic Technologies

3.1 INTRODUCTION

The photovoltaic effect is described as the potential difference generation at the junction of two different materials when it is illuminated by solar radiation. Due to the photovoltaic effect, solar radiation is directly converted to electricity on the solar cells, which is often described as photovoltaic (PV) energy conversion. Solar cell technologies, in general, are classified as wafer-based crystalline silicon solar cell technology, thin-film solar cell technology, and other novel emerging technologies. Figure 3.1 gives the classification of solar cells based on PV technologies.

Among all other solar PV technologies, the most commonly available PV module technologies on the existing market are wafer-based crystalline silicon PV cells made from either single or polycrystalline materials and thin-film photovoltaic cells coated on a glass or a metal substrate. Although there are so many different photovoltaic technologies, PV market leadership belongs to wafer-based crystalline silicon technology. Si-wafer-based PV technology still has a share of more than 90% of total production. The share of polycrystalline technology in total production is now about 70%. After Si-wafer technology, the second largest share belongs to thin-film technology at about 6%. However, it is expected that the share of thin-film technology on the market will increase progressively in the following years, while new emerging technologies will start to show up in the market.

3.2 SEMICONDUCTORS AND *P-N* JUNCTION

Sunlight is converted to electricity in photovoltaics through semiconductors. This technology is strictly related to conventional solid-state technology used in the fabrication of semiconductor elements such as diodes and transistors. Pure silicon and germanium are the two most common elements used as a starting point in the fabrication of semiconductors and photovoltaics. Each has four valence electrons (the electrons in the outermost shell of an atom), although silicon has 14

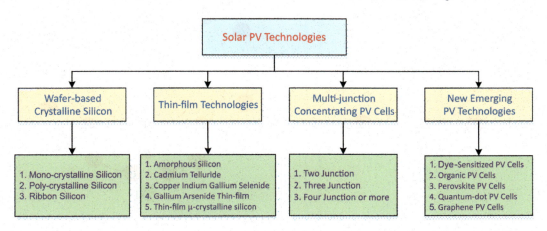

FIGURE 3.1 Classification of solar PV technologies.

DOI: 10.1201/9781003415572-3

orbital electrons and germanium has 32 electrons. The atomic structure of silicon (Si) and germanium (Ge) is shown in Figure 3.2.

Other chemical elements such as boron (B), phosphorus (P), arsenic (As), cadmium (Cd), and tellurium (Te) and indium (In) are used in the fabrication process of most PVs. For example, the PV technology in which cadmium (Cd) and tellurium (Te) are used is called CdTe PV cells, whereas the PVs in which gallium (Ga) and arsenic (As) are used are called GaAs PV cells.

3.2.1 Energy Bands and Silicon Lattice

In a pure crystallized silicon lattice, each silicon atom forms covalent bonds with four adjacent atoms in a diamond lattice. Figure 3.3 illustrates the silicon covalent bonds with a simplified sketch and diamond lattice form.

To simplify the understanding, the silicon lattice can be drawn on a two-dimensional plane as a basic diagram as shown in Figure 3.4. Normally, these simplified diagrams do not reflect the true nature of silicon. The point here is that a silicon atom has four valance electrons, which are shared in covalent bonds with neighboring atoms.

Materials can be classified as conductors, insulators, and semiconductors according to the number of electrons in the outermost orbit of the material. If the last orbit has 1, 2, or 3 electrons, the material is called a conductor. If there are 5, 6, 7, or 8 electrons in the last shell of the atom, the material is called an insulator. Finally, if the last shell of the atom has 4 electrons, it is called a semiconductor. A useful way to describe the difference between conductors, insulators, and semiconductors is to plot the energy band diagrams such as those given in Figure 3.5.

According to quantum theory, there is a specific band energy level corresponding to each electron group in the atomic shells. The top energy band in the atomic structure is called the conduction band of which electrons contribute to current flow. It is clear from Figure 3.5 that the conduction band for conductors (metals) is partially filled by the valance electrons, but for the insulators and semiconductors at absolute zero temperature, the conduction band is empty (not filled by electrons). However, at room temperature, there may be a single electron in the conduction band of silicon out of ten billion electrons.

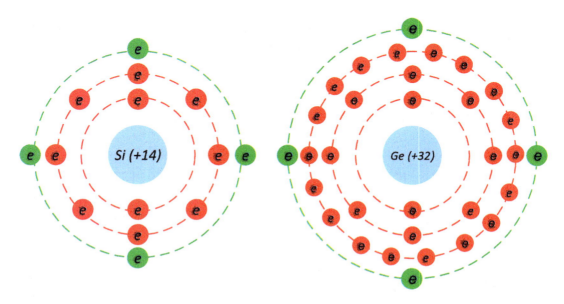

FIGURE 3.2 Atomic structure of silicon and germanium.

Photovoltaic Systems and Photovoltaic Technologies   67

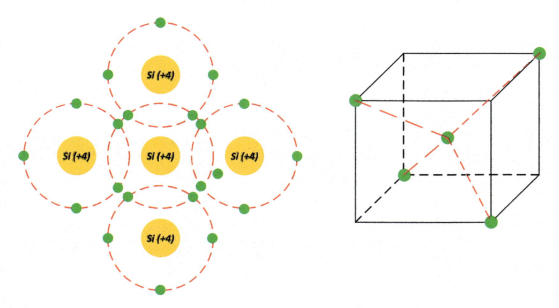

FIGURE 3.3 The illustration of shared electrons of a covalent bond in pure crystalline silicon and simplified three-dimensional tetrahedral pattern (diamond lattice).

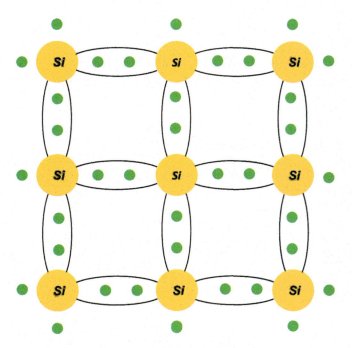

FIGURE 3.4 The simplified diagrams of silicon lattice in a two-dimensional flat plane.

The gaps between allowed energy bands are called band gaps or forbidden bands. The energy difference between the valence band and conduction band is called bandgap energy, which is denoted as E_{gap}. In other words, if an electron acquires bandgap energy, it can jump across the forbidden gap to the conduction band. The units for bandgap energy are usually electron-volts (eV) and 1 eV equal to 1.6×10^{-19} J.

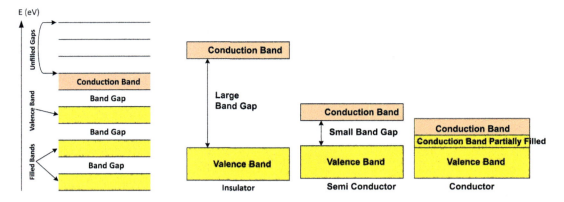

FIGURE 3.5 (a) The energy band structure of electrons for a material. The band above the valence band is known as the conduction band. (b) The difference between conductors, insulators, and semiconductors. The energy difference between the valence and conduction bands shows a distinguished pattern to describe the material types.

The required minimum bandgap energy (E_{gap}) for silicon is 1.12 eV at room temperature to free an electron to jump into the conduction band. The question needed to be asked here is where the energy will come from. For photovoltaics, the source of this energy is the electromagnetic energy of photons coming from the sun. If a photon with more than 1.12 eV of electromagnetic energy is penetrated into a photovoltaic cell, a single electron can pass to the conduction band. If an electron is separated from the valence band by the photon radiation, the silicon nucleus still has a +4 charge, but there are three electrons connected to it. This means that there is a net positive charge associated with this silicon nucleus, called a hole, as shown in Figure 3.6. By the separation of the photon mixture from the silicon structure, the holes and electrons recombine. If there is no other effect to keep the electron in the conduction band, the holes and electrons will be finally recombined (Figure 3.6). The energy of the electron in the conduction band is released as a photon in the course of hole-electron recombination. This principle is also the same as the working principle of light-emitting diodes (LEDs).

It is important to note here that not only does the negatively charged electron in the conduction band moves freely but also the positively charged hole formed by the leaving of the electron starts to move as well. Here, the holes formed by the separation of electrons are filled by the valence electrons of the nearby atoms. Thus, the holes will also move relatively. Afterward, each new hole formed is filled by another electron. And this process, given in Figure 3.7, goes on in the same way as long as photon radiation continues.

Note that electric current in a semiconductor is carried not only by negatively charged electrons but also by positively charged holes. The photon energy in joules (E) creating hole-electron pairs can be expressed by their wavelengths in meters (λ) and their frequency in hertz (f) as below:

$$E = h \cdot f = h \cdot \frac{c}{\lambda} \tag{3.1}$$

where h is Planck's constant (6.626×10^{-34} J·s) and c is the light speed (3×10^8 m/s). Notice that c is expressed as below:

$$c = \lambda \cdot f \tag{3.2}$$

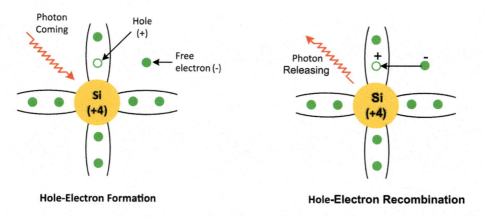

FIGURE 3.6 (a) Formation of hole-electron pair due to photon radiation. (b) Recombination of an electron with a hole and releasing a photon of energy.

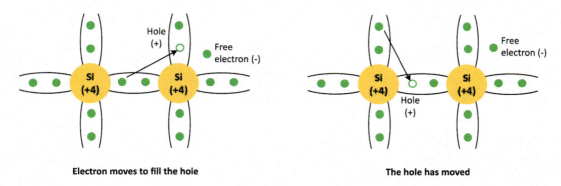

FIGURE 3.7 Filling of the hole by an electron and relative movement of the hole.

Example 3.1: Photon Energy versus Wavelength

It is known that the required minimum band gap energy (E_{gap}) for silicon at absolute zero (0°K) is 1.17 eV and 1.11 eV at 300°K. What is the maximum wavelength to create hole-electron pairs in the silicon for each temperature level? Calculate the required minimum photon frequency in these cases. Draw the obtained results on the graph of photon energy versus wavelength. Lastly, interpret the plotted graph. Note that 1 eV equal to 1.6×10^{-19} J.

Solution

From equation (3.1) for absolute zero (0°K), we have:

$$\lambda \leq h \cdot \frac{c}{E} = 6.626 \times 10^{-34} \cdot \frac{3 \times 10^8}{1.17 \times 1.6 \times 10^{-19}} = 1.062 \times 10^{-6} \text{ m} = 1.062 \text{ µm}$$

and for (300°K), we have:

$$\lambda \leq h \cdot \frac{c}{E} = 6.626 \times 10^{-34} \cdot \frac{3 \times 10^8}{1.11 \times 1.6 \times 10^{-19}} = 1.113 \times 10^{-6} \text{ m} = 1.113 \text{ µm}$$

Using equation (3.2) for absolute zero (0°K), the frequency should be more than

$$f \geq \frac{c}{\lambda} = \frac{3 \times 10^8}{1.062 \times 10^{-6}} = 2.82 \times 10^{14} \text{ Hz}$$

Similarly, for (300°K), the frequency should be more than

$$f \geq \frac{c}{\lambda} = \frac{3 \times 10^8}{1.113 \times 10^{-6}} = 2.69 \times 10^{14} \text{ Hz}$$

If we draw the graph of equation (3.1) and show the results on it, the diagram in Figure 3.8 will be obtained. The graph shows the energy lost, the usable energy range, and the corresponding wavelengths.

As can be seen from the figure, the photon energy is inversely proportional to the wavelength. Notice that the essential comments are given on the graph.

3.2.2 The P-N Junction

Semiconductors can be divided into two groups, intrinsic and extrinsic. Fully pure semiconductors without any impurities inside the crystal lattice are called intrinsic semiconductors, which have equal free holes and free electrons at any temperature. In applications, semiconductors are doped to get a larger number of holes or electrons to modify the electrical characteristics of the semiconductors. Hence, the doped semiconductors (impurities added) in a well-controlled manner are called extrinsic semiconductors.

Assume that the silicon crystals are doped with the chemical elements such as phosphorus or arsenic from the fifth column of the periodic table of which the valence electron number is five (also called pentavalent). So, these negatively doped semiconductors with phosphorus or arsenic are called *n*-type materials. Since these atoms have extra electrons, they donate electrons and so are called donors.

Again let's consider that the silicon crystals are doped with the chemical element from the third column of the periodic table such as boron (also called trivalent). Accordingly, this positively doped semiconductor with boron is called *p*-type material. Since boron atoms have holes, they accept electrons and so they are called acceptors.

Figure 3.9 shows the donor and acceptor atoms in Si crystal. The four silicon atoms in Figure 3.9 form a covalent bond with four of five electrons of pentavalent donor atoms. Hence four electrons out of five are tightly bound, but the fifth electron is left free to roam around in the crystal lattice. Similarly, the trivalent boron atoms in Figure 3.9 form a covalent bond with three silicon atoms, leaving the fourth one not filled, which results in positively charged holes in the crystal lattice.

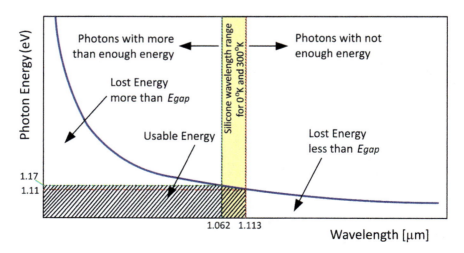

FIGURE 3.8 Photon energy versus wavelength of a silicon to show the energy lost, the usable energy range, and the corresponding wavelengths.

Photovoltaic Systems and Photovoltaic Technologies

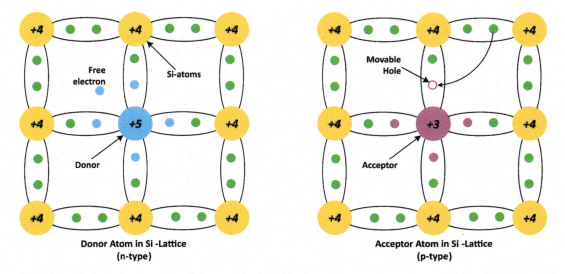

FIGURE 3.9 The donor and acceptor atoms form *n*-type and *p*-type materials.

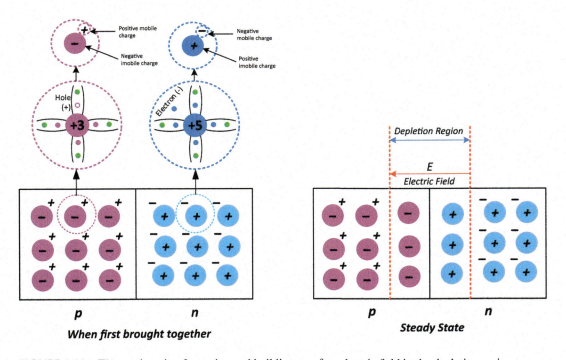

FIGURE 3.10 The *p-n* junction formation and building up of an electric field in the depletion region.

Now suppose we are forming a *p-n* junction by bringing *n*-type material and *p*-type material together. Mobile (free and movable) electrons in the *n*-type material carry away by diffusion across the junction. Meanwhile, the mobile holes in the *p*-type material dragged away across the junction in the opposite direction as shown in Figure 3.10.

When a mobile electron on the *n*-region passes through the *p*-region to fill a hole there, the electron leaves an immobile positive charge in the *n*-type silicon while creating an immobile negative charge in the *p*-region. These immobile charged atoms in the *p* and *n* regions create an electric field, which works against the movement of electrons and holes across the junction. As the diffusion process continues, the electric field in the junction region against the hole-electron motion continuously increases until all further movement of charge carriers across the junction stops. The immobile charges creating electric fields in the junction region form the depletion region, which is shown in Figure 3.10. Mobile charges are kept away from the depletion region by pushing a positive charge to the *p*-region while repelling the electrons back into the *n*-region.

3.2.3 THE *P-N* JUNCTION DIODE

Because of common electrical characteristics, it is useful to discuss the *p-n* diode structure before examining the photovoltaic *p-n* junction exposed to sunlight. The *p-n* junction of a diode has 3 band structures, namely, equilibrium (zero voltage bias), forward bias, and reverse bias. A *p-n* junction diode without applying any external voltage is under equilibrium between diffusion and drift currents, which is described earlier.

For a *p-n* junction of a diode at equilibrium, the Fermi levels are equal on both sides of the junctions. Negatively charged electrons and positively charged holes reach an equilibrium at the *p-n* junction and form a depletion zone as shown in Figure 3.11.

Consider that an external voltage is now applied across to the *p-n* junctions as shown in Figure 3.12, which is called forward biasing. In forward biasing, the holes are moved from the *p*-side to the *n*-side as the electrons are transferred from *n*-side to the *p*-side. At the junction, the electrons and holes are combined to maintain continuous current.

The application of a reverse voltage as shown in Figure 3.13 to the *p-n* junction causes a transient current to flow until the potential formation at the widened depletion layer. When the potential

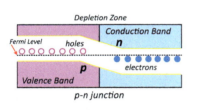 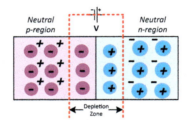

FIGURE 3.11 The *p-n* junction at equilibrium.

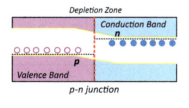 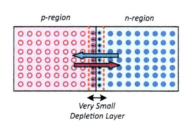 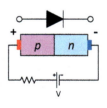

FIGURE 3.12 Forward-biased *p-n* junction.

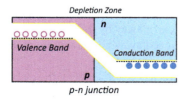 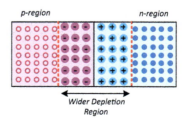 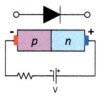

FIGURE 3.13 Reverse-biased *p-n* junction.

Photovoltaic Systems and Photovoltaic Technologies

across the depletion layer equals the applied voltage, the transient current due to the pulling away of electrons and holes from the junction will cease except for a small current due to the thermal effect.

It is useful to describe the ideal diode before going to examine the real diode characteristic. An ideal diode is a diode that works like a perfect conductor when the voltage V_d is applied forward biased and it behaves like a perfect insulator when the voltage is applied reverse biased. Thus, an instant current flows from the anode to the cathode once the positive voltage V_d is applied across the diode terminals. The voltage–current characteristic curve for the ideal p-n junction diode is described in Figure 3.14.

Let us now consider a conventional Shockley diode, of which circuit symbol, I–V characteristic curve, and the diode equation, are given by Figure 3.15 and equation (3.3).

$$I_D = I_o \left(e^{\frac{qV_D}{kT}} - 1 \right) \quad (3.3)$$

where I_D is the diode current (A) under forward biasing, V_D is the voltage (volts) across the diode terminals (from p-side to n-side), I_o is the reverse saturation current (A), q is the electron charge

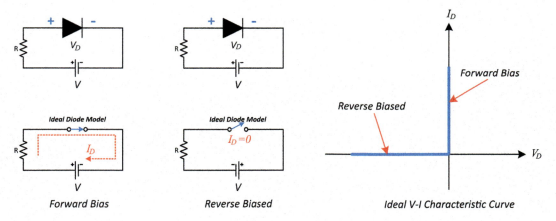

FIGURE 3.14 The ideal diode model, equivalent circuits, and V–I characteristic curve.

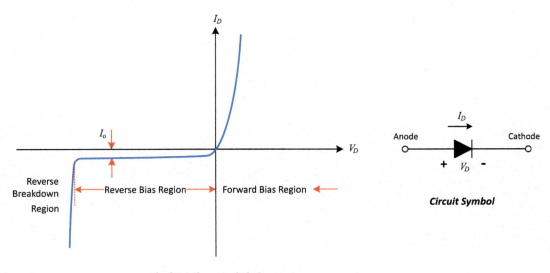

FIGURE 3.15 The V–I characteristic curve of a conventional diode and its circuit symbol.

$(1.602 \times 10^{-19}$ C), k is Boltzmann's constant $(1.381 \times 10^{-23}$ J/K), T is the junction temperature (K) and n is the ideality factor (typically $1 \leq n \leq 2$). The ideality factor is close to one in most cases, for example, $n = 1.026$ for monocrystalline silicon, $n = 1.025$ for polycrystalline silicon, $n = 1.357$ for thin-film silicon PV modules. Sometimes, the ideality factor is greater than 2 for hetero-junction solar cells, such as $n = 3.086$ for triple-junction a-Si, $n = 3.3$ for a-Si-H-Tandem, and $n = 5$ for a-Si-H-triple). Note that normally ideality factor is related to the material of the PV cell, temperature, and solar radiation. However, the ideality factor for most of the studies is assumed to be related only to the PV cell material, the value is taken at SRC or STC (Standard Reference Condition: 25°C and 1000 W/m² or Standard Test Condition).

Example 3.2: Voltage Drop across a *p-n* Junction Diode

Consider the *p-n* junction diode with the following parameters (Table 3.1).
Calculate the voltage drop across the diode at 25°C and 100°C junction temperatures for the following operating conditions:

a. When it is carrying 1 A.
b. When it is carrying 3 A.
c. When it is carrying 10 A.

Solution

From equation (3.3), we have the constant (qV_D / nkT) in the exponent. Let's calculate the constant for 25°C and 100°C $(n = 1)$.

$$\frac{qV_D}{kT} @ 25°C = \frac{1.602 \times 10^{-19} V_D}{1.381 \times 10^{-23} \times (273 + 25)} = 38.93 V_D$$

$$\frac{qV_D}{kT} @ 100°C = \frac{1.602 \times 10^{-19} V_D}{1.381 \times 10^{-23} \times (273 + 100)} = 31.11 V_D$$

a. The diode voltage V_D can now be calculated by rearranging (3.3) for 1 A:

$$V_D @ 25°C = \frac{1}{38.93} \times \ln\left(\frac{I_D}{I_o} + 1\right) = \frac{1}{38.93} \times \ln\left(\frac{1}{0.5 \times 10^{-3}} + 1\right) = 0.195 \text{ V}$$

$$V_D @ 100°C = \frac{1}{31.11} \times \ln\left(\frac{I_D}{I_o} + 1\right) = \frac{1}{31.11} \times \ln\left(\frac{1}{10 \times 10^{-3}} + 1\right) = 0.148 \text{ V}$$

b. Similarly, the diode voltage V_D for 3 A:

$$V_D @ 25°C = \frac{1}{38.93} \times \ln\left(\frac{I_D}{I_o} + 1\right) = \frac{1}{38.93} \times \ln\left(\frac{3}{0.5 \times 10^{-3}} + 1\right) = 0.223 \text{ V}$$

TABLE 3.1

The Parameters of the *p-n* Junction Diode

Parameter	Junction Temperature (25°C)	Junction Temperature (100°C)
Reverse Saturation Current	0.5×10^{-3} A	10×10^{-3} A

$$V_D @100°C = \frac{1}{31.11} \times \ln\left(\frac{I_D}{I_o}+1\right) = \frac{1}{31.11} \times \ln\left(\frac{3}{10\times10^{-3}}+1\right) = 0.183\ V$$

c. Finally, the diode voltage V_D for 10 A:

$$V_D @25°C = \frac{1}{38.93} \times \ln\left(\frac{I_D}{I_o}+1\right) = \frac{1}{38.93} \times \ln\left(\frac{10}{0.5\times10^{-3}}+1\right) = 0.254\ V$$

$$V_D @100°C = \frac{1}{31.11} \times \ln\left(\frac{I_D}{I_o}+1\right) = \frac{1}{31.11} \times \ln\left(\frac{10}{10\times10^{-3}}+1\right) = 0.222\ V$$

Notice that the diode voltage $V_D = 0$ for open-circuit conditions (at no current). Let us now draw the *I–V* curve for the forward biasing (Figure 3.16):

Note that despite the large variations in diode current, the voltage drop across the diode remains at very low levels.

3.3 THE PV CELL

The photovoltaic (PV) cells convert sunlight directly into electricity without causing any sort of pollution, which is illustrated in Figure 3.17. The PV cells are made of two semiconductor layers, one of which is positively charged and the other is negatively charged. Once the sunlight reaches the PV cell,

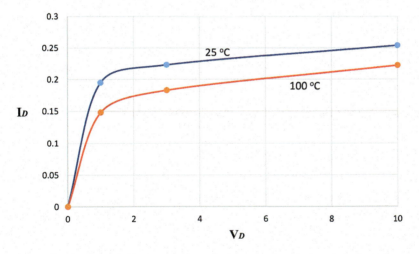

FIGURE 3.16 *I–V* curve of the diode for the forward biasing.

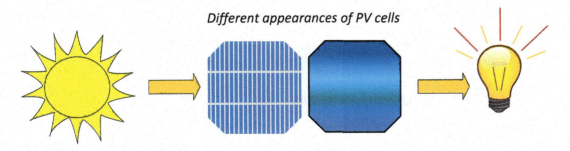

FIGURE 3.17 Illustration of direct conversion of solar radiation to electricity via photovoltaic cells.

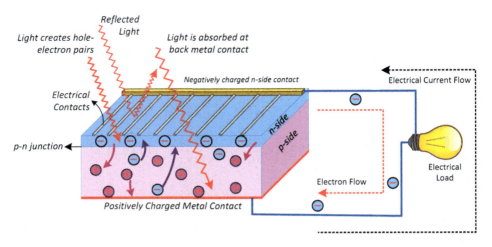

FIGURE 3.18 Creation of hole-electron pairs and electrical current flow through an external circuit.

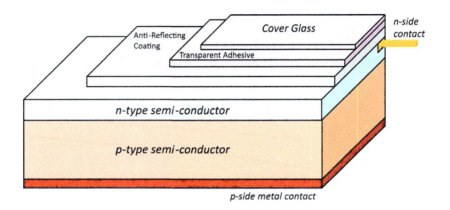

FIGURE 3.19 The lamination layers of a PV solar cell.

some of the photons are absorbed by the semiconductor materials and free some of the electrons in the negative layer as shown in Figure 3.18. Then the free electrons flow through an external circuit to the positive layer. This flow of electrons produces an electric current as represented in Figure 3.18.

Photovoltaic cells are laminated with various layers to protect them from environmental influences. Figure 3.19 shows the lamination layers of a PV solar cell. As can be seen from the figure, there are an anti-reflecting coating layer and a transparent adhesive layer between *p-n*-type material and cover glass. After this lamination process, the PV structure is framed with aluminum.

Typically, a photovoltaic cell has about 100 cm^2 area and produces an open-circuit voltage of about 0.5–0.6 volts at room temperature. Serial and parallel multiplication of PV cells is performed to obtain greater voltage and power levels, which is explained in the below section.

3.4 THE PV MODULE AND PV ARRAY

A crucial phase in designing the PV system is to decide the required number of modules in series and arrays in parallel to deliver the requested energy demand. The photovoltaic cells are connected in series to form photovoltaic modules. Then the modules are assembled in series to create PV panels or PV strings. The PV strings are wired in parallel to create photovoltaic arrays. Multiple PV arrays are brought together to form a PV farm on utility scale. The illustration from PV cells to modules and then to arrays is given in Figure 3.20.

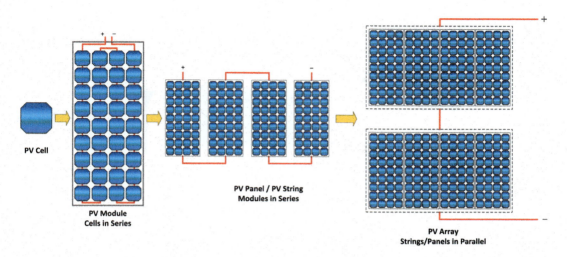

FIGURE 3.20 The PV cell, module, panel, and array for field applications.

A typical 18-V PV module is achieved by serial connection of 36 cells. Nowadays, 36-V modules are widely used and obtained by a series connection of 72 cells. The performance of a PV module is typically rated according to its maximum DC power [watts] under STC, which are discussed comprehensively in the following sections.

3.5 THE PV MODELING AND ELECTRICAL CHARACTERISTICS OF PV SYSTEMS

There are several electrical and environmental factors affecting the output of a photovoltaic system. Considering all PV system parameters in one model increases the complexity of the problem and makes the solution challenging. For this reason, there are various models in the literature from an ideal model to the simplified models and even more sophisticated models. Hence, the commonly used PV models are covered in the following sections.

3.5.1 Ideal PV Model

An ideal circuit model for a PV cell consists of an ideal current source in parallel with a real diode as shown in Figure 3.21. The current source in the model provides current in proportion to incident solar radiation on the cell surface.

Open-circuit voltage (V_{OC}) and short-circuit current (I_{SC}) as illustrated in Figure 3.22 are the two important parameters in determining the performance of a PV and its characteristic equation.

If the terminals of the PV cell are left open, no current flows through the PV terminals. In this case, the voltage across the PV terminals is equal to the open-circuit voltage ($V = V_D = V_{OC}$). Similarly, if the terminals of the PV cell are shorted together, then the short-circuit current will flow through the shorted leads. In this case, the PV current equals to short-circuit current ($I = I_{SC}$).

Now from Kirchhoff's law, we can write the characteristic equation for the ideal equivalent circuit of the PV cell shown in Figure 3.21 as below:

$$I = I_{SC} - I_D \tag{3.4}$$

And then insert equation (3.3) into (3.4), we get output current as a function of output voltage (V).

$$I = I_{SC} - I_o \left(e^{\frac{qV_D}{nkT_c}} - 1 \right) \tag{3.5}$$

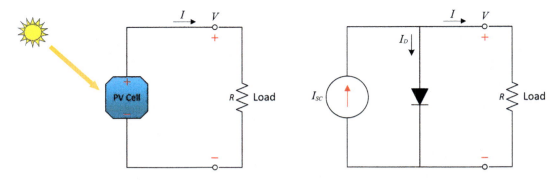

FIGURE 3.21 Ideal circuit model for a PV cell.

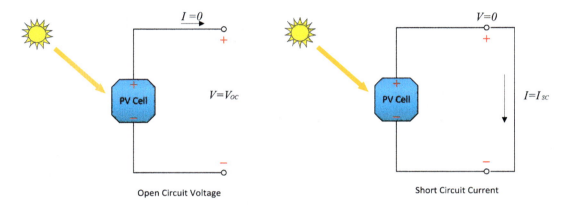

FIGURE 3.22 Open-circuit voltage and short-circuit current for photovoltaics.

Notice that equation (3.5) is a nonlinear $(I-V)$ equation for the PV cell and n is the ideality factor in the equation. It is clear from equation (3.5) that when $V = V_D = 0$ (open circuit), the solved output current I is equal to I_{SC}. The open-circuit voltage V_{OC} can be solved when $I = 0$ as below:

$$V_{OC} = \frac{nkT_c}{q} \times \ln\left(\frac{I_{SC}}{I_o} + 1\right) \tag{3.6}$$

It is noteworthy that $I-V$ characteristic equation in (3.5) has two terms, one of which is short-circuit current and the second term is the negative diode current. That means the $I-V$ characteristic of an ideal PV cell is the short-circuit current minus diode currents under sunlight. The short-circuit current I_{SC} will be zero when there is no light. Hence, the $I-V$ curve, when it is dark, will equal the negative diode current as represented in Figure 3.23.

Example 3.3: Open Circuit of a PV Cell under Solar Radiation

Consider the ideal photovoltaic model for a PV cell with reverse saturation current $I_o = 1 \times 10^{-11}$ A. Under 1000 W/m² solar radiation, the PV cell produces 6 A at room temperature (25°C). Find the open-circuit voltage (V_{OC}) at 1000 W/m² and 250 W/m² solar radiation. Show the results on the $I-V$ plot.

Solution

From equation (3.6), the open-circuit voltage at 25°C and 1000 kW/m²:

Photovoltaic Systems and Photovoltaic Technologies

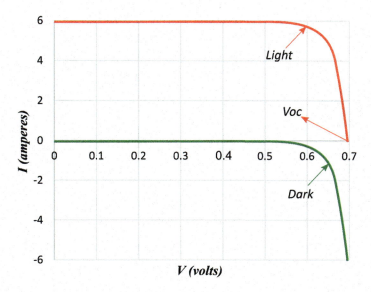

FIGURE 3.23 The current–voltage curve of a photovoltaic cell for dark and sunlight conditions.

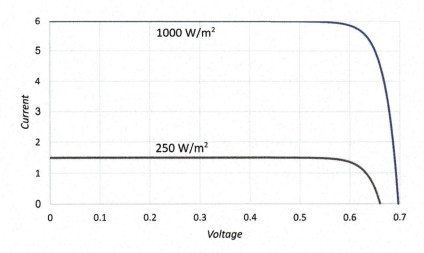

FIGURE 3.24 *I–V* curve of the PV cell for different radiations.

$$V_{OC} = \frac{1.381 \times 10^{-23} \times (273 + 25)}{1.602 \times 10^{-19}} \times \ln\left(\frac{6}{1 \times 10^{-11}} + 1\right) = 0.697\,\text{V}$$

Similarly, the open-circuit voltage at 25°C and 250 W/m² ($I_{SC} = 6/4 = 1.5\,\text{A}$):

$$V_{OC} = \frac{1.381 \times 10^{-23} \times (273 + 25)}{1.602 \times 10^{-19}} \times \ln\left(\frac{1.5}{1 \times 10^{-11}} + 1\right) = 0.661\,\text{V}$$

Following ($I - V$) curve at 25 °C is obtained for 1000 W/m² and 250 W/m² radiation (Figure 3.24).

3.5.2 The PV Cell with Shunt Resistance

A more accurate model can be obtained by adding a parallel leakage resistance R_P to the model as shown in Figure 3.25. In this case, the short-circuit current (I_{SC}) is equal to the sum of three nodal currents, namely, diode current (I_D), parallel branch current (I_P), and load current (I).

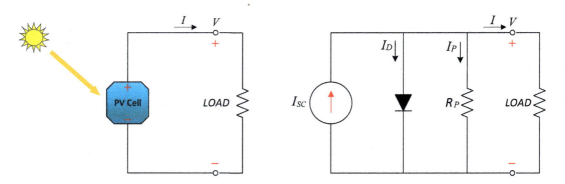

FIGURE 3.25 Equivalent model of single-diode PV cell with leakage shunt resistance.

By applying Kirchhoff's Current Law, the $I-V$ characteristics of the PV cell with single-diode and shunt resistance can be obtained as below:

$$I = I_{SC} - I_D - I_P \tag{3.7}$$

$$I = I_{SC} - I_o \left(e^{\frac{qV}{nkT_c}} - 1 \right) - \frac{V}{R_P} \tag{3.8}$$

Notice that $V = V_D$ and $I_P = V / R_P$ in equation (3.8). Figure 3.26 gives the plot of (3.8) and shows the effect of R_P on the $I-V$ curve. It is clear from the figure that the load current in the ideal model is decreased by V / R_P in the model with an added parallel resistance.

Normally, commercial PV cells have less than 1% power losses caused by their parallel leakage resistance. To achieve this power loss goal, R_P value for a PV cell must be greater than $100 \times V_{OC} / I_{SC}$. For example, the leakage resistance for a PV cell with a short-circuit current of 6 A and open-circuit voltage of 0.6 V must be greater than 10 Ω to have less than 1% power losses ($R_P > 100 \times 0.6 / 6 \Rightarrow R_P > 10\ \Omega$).

3.5.3 THE PV CELL WITH SERIES RESISTANCE

The ideal PV cell model can be modified by adding a series resistance (R_S) caused by contact and semiconductor resistances. The electric circuit model for the PV cell with series resistance is shown in Figure 3.27. In this case, the short-circuit current (I_{SC}) is equal to the sum of diode current (I_D) and load current (I).

By applying Kirchhoff's Current and Voltage Laws, the $I-V$ characteristics of the PV cell with single diode with series resistance can be obtained as below:

$$I = I_{SC} - I_D \tag{3.9}$$

$$I = I_{SC} - I_o \left(e^{\frac{qV_D}{nkT_c}} - 1 \right) \tag{3.10}$$

$$I = I_{SC} - I_o \left(e^{\frac{q(V + I \cdot R_S)}{nkT_c}} - 1 \right) \tag{3.11}$$

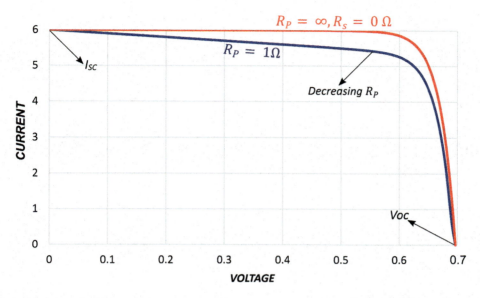

FIGURE 3.26 The $I-V$ characteristics of the PV cell with single diode and shunt resistance, and the effects of the R_P on the curve.

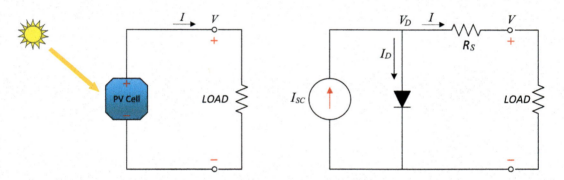

FIGURE 3.27 Equivalent model of single-diode PV cell with series resistance.

Notice that $V_D = V + I \cdot R_S$ in equation (3.11). Figure 3.28 gives the plot of (3.11) and shows the effect of R_S on the $I-V$ curve. It is clear from the figure that the voltage at any given current is shifted to the left by $\Delta V = IR_S$ in the model with an added series resistance. For a PV cell to keep power losses less than 1%, R_S value for a PV cell must be less than about $0.01 \times V_{OC}/I_{SC}$.

3.5.4 The PV Cell with Series and Shunt Resistance

The most common circuit model to describe the PV cell characteristics is given in Figure 3.29, which includes both series and parallel resistances. In this case, the short-circuit current (I_{SC}) is equal to the sum of three nodal currents, namely, diode current (I_D), parallel branch current (I_P), and load current (I).

Now, we can write the following $I-V$ equation for the model:

$$I = I_{SC} - I_o \left(e^{\frac{q(V+I \cdot R_S)}{nkT_c}} - 1 \right) - \frac{V + I \cdot R_S}{R_P} \tag{3.12}$$

Notice that $V_D = V + I \cdot R_S$ in equation (3.12). Figure 3.30 gives the plot of (3.12) and shows the effect of R_S and R_P on the $I-V$ curve.

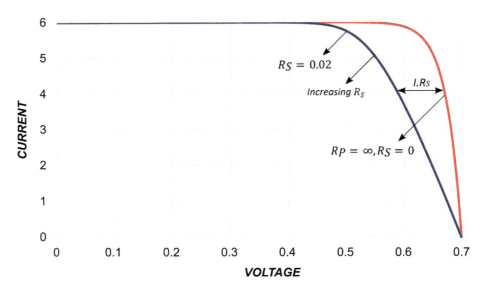

FIGURE 3.28 The $I-V$ characteristics of the PV cell with single diode and series resistance, and the effects of the R_s on the curve.

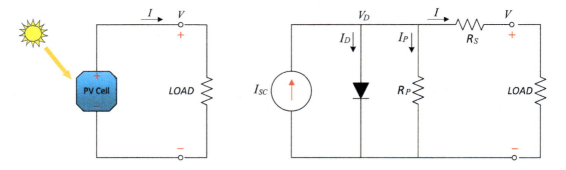

FIGURE 3.29 The most commonly used equivalent PV cell model with series and parallel resistances added.

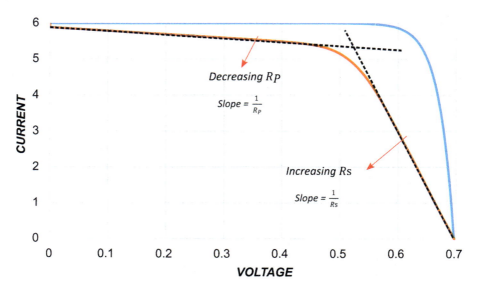

FIGURE 3.30 The effect of series and parallel resistances ($R_P = 1\ \Omega$, $R_S = 0.02\ \Omega$) on the $I-V$ characteristics of the PV cell.

It is clear from the figure that it is possible to estimate the series and parallel resistances from the inverse of the slopes of the $I-V$ curve at V_{OC} and I_{SC} as below:

$$R_P \cong \frac{1}{\text{Slope}} @ I_{SC} \qquad (3.13)$$

$$R_S \cong \frac{1}{\text{Slope}} @ V_{OC} \qquad (3.14)$$

It is also clear that high R_P and low R_S values are needed to improve the PV cell performance.

3.5.5 Two-Diode Model

The accuracy of the dual-diode model taking into account the effect of recombination is higher than the single-diode model. However, it is more difficult to solve the equations of the two-diode model. Figure 3.31 shows the equivalent circuit of the two-diode models.

The $I-V$ equation of the two-diode model for a PV cell can be obtained from Kirchhoff's Current Law as follows:

$$I = I_{SC} - I_{D1} - I_{D2} - I_P \qquad (3.15)$$

where the diode current I_{D1} and I_{D2} are:

$$I_{D1} = I_{o1} \left(e^{\frac{q(V+I \cdot R_S)}{n_1 k T_c}} - 1 \right) \qquad (3.16)$$

$$I_{D2} = I_{o2} \left(e^{\frac{q(V+I \cdot R_S)}{n_2 k T_c}} - 1 \right) \qquad (3.17)$$

After combining of above expressions, equation (3.15) becomes:

$$I = I_{SC} - I_{o1} \left(e^{\frac{q(V+I \cdot R_S)}{n_1 k T_c}} - 1 \right) - I_{o2} \left(e^{\frac{q(V+I \cdot R_S)}{n_2 k T_c}} - 1 \right) - \frac{(V + I \cdot R_S)}{R_P} \qquad (3.18)$$

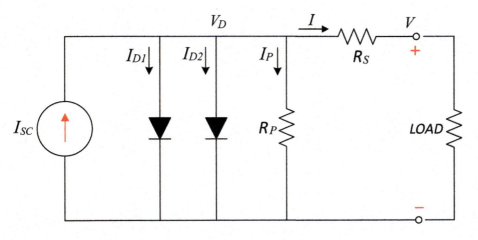

FIGURE 3.31 Two-diode circuit model for a PV cell.

where I_{o1} is the reverse saturation current due to diffusion and I_{o2} is the reverse saturation current due to recombination in the space charge layer. The ideality factors in (3.18) are $n_1 = 1$ and $n_2 \geq 1.2$ for diode-1 and diode-2, respectively.

3.6 CHARACTERISTICS CURVES OF PV CELLS

A generic $I-V$ characteristic of a PV cell is obtained in the above sections. Now let's obtain the $P-V$ curve of a solar module by multiplying the $I-V$ pairs on the same axis system as shown in Figure 3.32. Characteristic points for the PV cell are shown on this curve. The key parameters are given in Figure 3.32 such as open-circuit voltage (V_{OC}), short-circuit current (I_{SC}), current (I_{mp}), and voltage (V_{mp}) at maximum power point (MPP) are necessary points for most of the PV applications. The power at MPP is designated as the maximum power ($P_{mp} = P_{max} = P_m$) or rated power (P_R), which is the product of ($V_{mp} = V_m$) and ($I_{mp} = I_m$). Similarly, current and voltage pairs at MPP are sometimes designated as rated current (I_R) and rated voltage (V_R), of which product gives maximum power.

As indicated in the above sections, the PV modules are amplified by serial and parallel connections to meet the high-power demands. The obtained $I-V$ curves as a result of serial and parallel connections are shown in Figures 3.33–3.35. The voltages are added while the current remains the same in a series of connected modules as shown in Figure 3.33. As for the PV modules in parallel, the currents are summed up while the voltage remains the same (Figure 3.34). For series-parallel connected modules, both the currents of the parallel arms and the voltages of the series-connected modules are summed up as shown in Figure 3.35.

The $I-V$ curves of PV cells change continuously according to the temperature and solar radiation values. For this reason, standard reference conditions (SRC) for PV modules are defined, which enables us to compare different PV technologies. SRC includes solar radiation of 1000 W/m² for solar spectrum at an air mass ratio of 1.5 (*AM* 1.5) and temperature of 25°C. The maximum PV module capacity under STC (1000 W/m², spectrum *AM*1.5 and 25°C) is defined as the peak power of the PV module and it is expressed in Watts Peak (W_p).

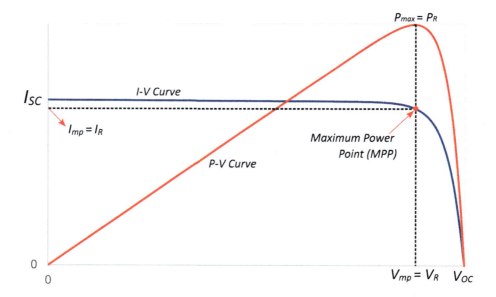

FIGURE 3.32 The *I–V* and *P–V* curves for a PV module and its characteristic points.

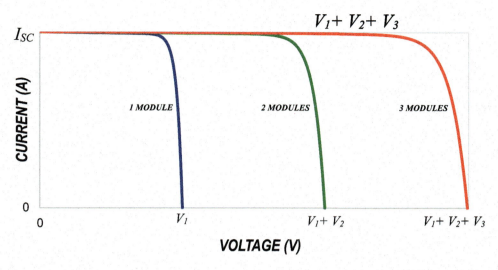

FIGURE 3.33 The *I–V* curve for PV modules in series. The voltages are added at any given current.

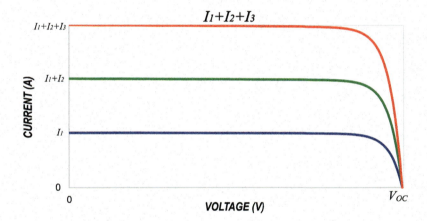

FIGURE 3.34 The *I–V* curve for PV modules in parallel. The currents are added at any given voltage.

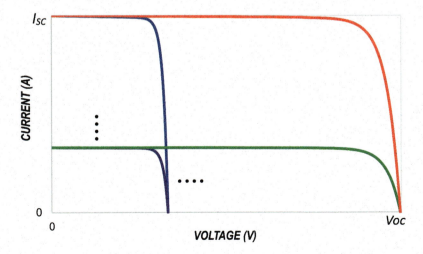

FIGURE 3.35 The *I–V* curve for PV modules in series and parallel connections. Both the voltages and currents are added.

3.7 SPECIAL ISSUES ON *I–V* CURVES

Mismatch Effect (Series Connection): Photovoltaic cells or panels are not generally 100% identical due to differences in production stages and operating conditions. For this reason, the rated current and voltage levels of the photovoltaic panels show slight differences under actual atmospheric conditions. In series-connected photovoltaics, the system current is limited by the component with the lowest current rate. For example, if a 4 A panel is connected in series with a 3 A panel, then the overall current will be dragged down to 3 A. Similarly, in parallel-connected photovoltaics, the system voltage is limited by the component with the lowest output voltage. For example, if a 24 V panel is connected in parallel with a 18 V panel, then the overall system voltage will be dragged down to 18 V. This is known as the mismatch effect in the literature. Hence any differences in $I-V$ curves of PV cells may lead to mismatch losses at some operating point. Comparative *I–V* curves of ideal and non-ideal cells and their operating regimes are represented in Figure 3.36.

Let us consider the serial connection of non-identical PV panels to show the mismatch effect on the *I–V* curve. Short-circuit currents and open-circuit voltages are affected in different ways in the series connection of non-identical PV panels. Figure 3.37 shows the open-circuit voltage mismatch for series-connected PV cells. It is clear from the figure that the overall system current remains the same, while the total system voltage equals the sum of the voltages of the two cells. In this case, the total power of the configuration will be equal to the sum of both PV cells' output.

Now let us consider two PV cells with different short-circuit currents, but the same output voltages. Figure 3.38 shows short-circuit current mismatch for series-connected PV cells. Since series-connected cells carry the same current, each PV cell voltage is adjusted accordingly to deliver the required current. Note that the maximum current available from the series-connected system is affected by the PV component with the lowest current. As a simple approach to finding the short-circuit current of the series-connected mismatched PV cells, let us consider the reflected $I-V$ curve of a PV cell with respect to the current axis as shown in Figure 3.38. As is shown in Figure 3.38 that

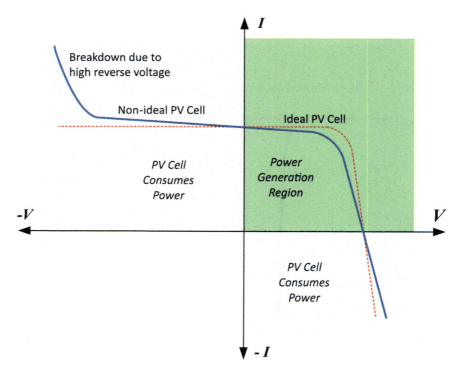

FIGURE 3.36 The comparison of an ideal and non-ideal *I–V* curve.

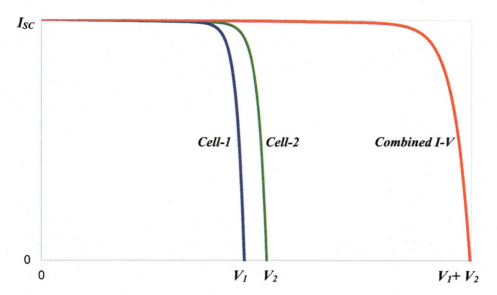

FIGURE 3.37 Series connection of two PV cells with different output voltages.

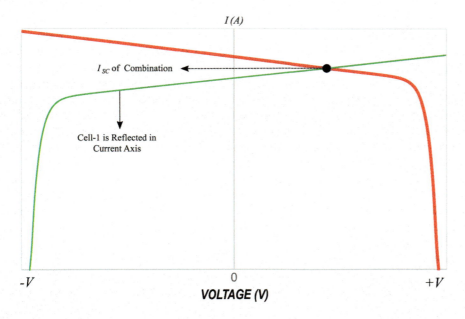

FIGURE 3.38 Determination of I_{SC} point in the series connection of two PV cells with different short-circuit currents but same output voltages.

the short-circuit current of the combination is determined by the current at the point of intersection. Notice that the voltage of the reflected cell is equal to the voltage of the second cell ($|-V|=|+V|$).

The mismatch losses are caused by the series-parallel connection of non-identical PV cells. In this case, the excessive current of the good cell is dissipated in the poor cell, which can cause severe power reduction and result in hot points in the poor cell. Eventually, the module can be damaged as a result of hot spots.

The serial connection of two PV cells with different current and voltage values is shown in Figure 3.39. In this case, the mismatch effect for the (I_{SC}) of combination is determined as the intersection point of two $I-V$ curves, one of which is reflected. For the open voltage of the series-connected system, it is equal to the sum of the voltages of the two PV cells.

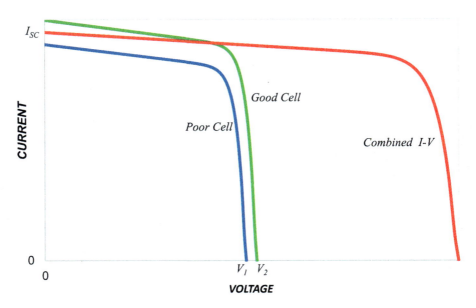

FIGURE 3.39 Series connection of two PV cells with different short-circuit current and output voltages.

Mismatch Effect (Parallel Connection): Photovoltaic cells are connected in series in PV modules. Parallel-connected photovoltaics generally exist in PV panels and array structures. For this reason, the mismatch effect in parallel structures occurs at least at the module level.

Let us consider the parallel connection of non-identical PV panels to show the mismatch effect on the I–V curve. The voltage across the parallel PV module combination is equal at all times and the total current from the parallel combination is the sum of the currents of individual PV modules. Figure 3.40 shows the calculation of open-circuit voltage of mismatched modules in parallel. It is shown in the figure that the $I - V$ curve of one module in the parallel combination is reflected in the voltage axis to determine the (V_{OC}) of the combination.

The parallel connection of two PV cells with different current and voltage values is shown in Figure 3.41. In this case, the mismatch effect for the (V_{OC}) of combination is determined as the intersection point of two $I - V$ curves, one of which is reflected as shown in Figure 3.40. For the short current of a parallel-connected system, it is equal to the sum of the currents of the two PV cells.

Temperature Effect: I–V characteristics of photovoltaic cells are affected by the change in cell temperature. As the cell temperature rises, the open-circuit voltage of the PV decreases noticeably, while the short-circuit current of the PV increases slightly. Hence, as the cell temperature increases, the I–V curve of the PV shifts to the left as shown in Figure 3.42. For example, open-circuit voltage, V_{OC} drops by approximately 0.37% for each increment of PV cell temperature in degree Celsius, while short-circuit current, I_{SC} increases by approximately 0.05% per °C. As a result, available maximum PV output power decreases by about 0.5% for each temperature increase in degree Celsius.

The effect of temperature can also be obtained mathematically. From the ideal modeling of PV cells, the simplest relationship between V_{OC}, I_{SC}, and temperature (T) is given in the above sections by equations (3.5) and (3.6). Let us consider the ideal cell model with series resistance for a more comprehensive temperature effect on the I–V curve of the PV. The reverse saturation component of the diode current is given below by equation (3.19). According to (3.19), the variation of the cell temperature affects the reverse saturation current as of the cube of temperature.

$$I_o = I_{or} \left(\frac{T}{T_r}\right)^3 \cdot e^{\left[\frac{qE_g}{nk}\left(\frac{1}{T_r} - \frac{1}{T_c}\right)\right]} \quad (3.19)$$

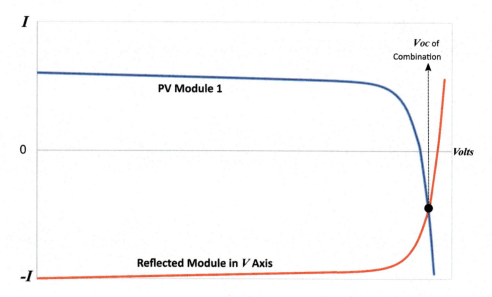

FIGURE 3.40 Determination of V_{OC} point in parallel connection of two PV cells with different short-circuit current and output voltages.

FIGURE 3.41 Parallel connection of two PV cells with different short-circuit current and output voltages.

where E_g is the bandgap energy of the semiconductor ($E_g = 1.12$ eV for Si at 25°C), I_{or} is the reference (or rated) saturation current and T_r is the reference (or rated) cell temperature (25°C). The saturated reverse current I_{or} at reference temperature T_r is given by (3.20).

$$I_{or} = \frac{I_{SCr}}{e^{\left(\frac{V_{OCr}}{nV_t}\right)}} \tag{3.20}$$

where V_t (0.02586 V at 300°K) is the thermal voltage of the PV cell and given by the following equation.

$$V_t = \frac{kT_c}{q} \tag{3.21}$$

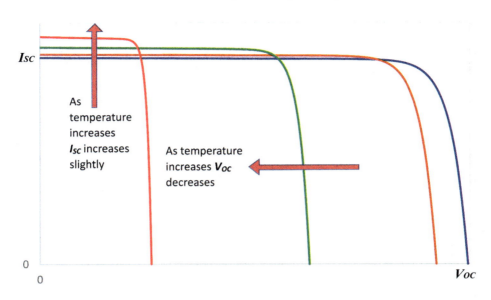

FIGURE 3.42 The influence of the cell temperature on the *I–V* curve.

PV cell temperature depends on various atmospheric parameters such as ambient temperature, solar radiation, and wind speed. There are numerous approaches to estimating the cell temperature in the literature. However, the cell temperature formula only according to the IEC 61215 standard is given herein. The effects of environmental impacts on the cell temperature and system performance will be examined in detail in Section 3.4. The cell temperature according to IEC 61215 is calculated as in (3.22).

$$T_c = T_{amb} + \frac{(T_{NOCT} - 20)}{800} \times G \tag{3.22}$$

where T_{cell} is PV module temperature (°C), T_{amb} is the ambient temperature (°C), T_{NOCT} is nominal operating cell temperature (°C) and G is the solar radiation (W/m²). The short-circuit current and open-circuit voltage of the cell depend on and is also influenced by the temperature according to the following equations.

$$I_{SC} \,@\, T_c = I_{SCr} + \alpha_I \times (T_c - T_r) \tag{3.23}$$

$$V_{OC} \,@\, T_c = V_{OCr} + \beta_V \times (T_c - T_r) \tag{3.24}$$

where $I_{SC}(T_c)$ and $V_{OC}(T_c)$ are short-circuit current and open-circuit voltage values at cell temperature T_c(°C), I_{SCr} and V_{OCr} are reference (or rated) short-circuit current and open-circuit voltage values at a reference temperature T_r(25°C) and α_I and β_V are the manufacturer-supplied temperature coefficients of short-circuit current and open-circuit voltage, respectively. Manufacturers sometimes give temperature coefficients in (%/°C), sometimes in (V/°C), or in (A/°C). Formulas (3.23), (3.24), and (3.25) provided herein are arranged in the case of temperature coefficients given in (V/°C) or in (A/°C). If temperature coefficients are given in terms of (%/°C), formulas (3.24) and (3.25) can be, respectively, written as $I_{SCr} \times [1 + \alpha_I \times (T_c - T_r)]$ and $V_{OCr} \times [1 + \beta_V \times (T_c - T_r)]$. Note that the same logic applies to (3.25).

Solar Radiation Effect: The variation of solar radiation intensity also affects the short-circuit current. The short-circuit current is affected linearly by solar irradiance. Hence equation (3.23) can be modified as follows:

$$I_{SC} \,@\, T_c \text{ and } G = [I_{SCr} + \alpha_I \times (T_c - T_r)] \times \frac{G}{G_r} \tag{3.25}$$

where G and G_r are solar irradiance and reference solar irradiance (1000 W/m²), respectively. From the above equations, it is possible to calculate V_{OC} and I_{SC} values for any given cell temperature/ambient temperature and solar radiation. Hence, the effect of solar radiation variation on $I-V$ curve can be represented in Figure 3.43. Notice that the effect of decreasing I_{SC} on V_{OC} is ignored in Figure 3.43. Normally, V_{OC} decreases slightly as the I_{SC} decreases. This effect is shown in Figure 3.44.

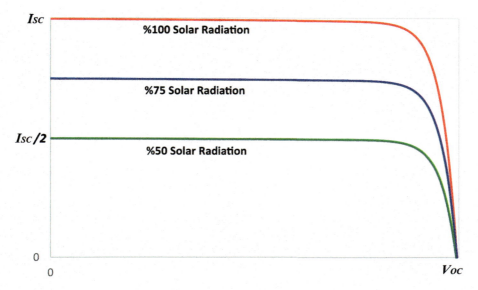

FIGURE 3.43 The influence of solar radiation on the $I-V$ curve (the ambient temperature is considered to be constant and the effect of I_{SC} variation on V_{OC} is ignored).

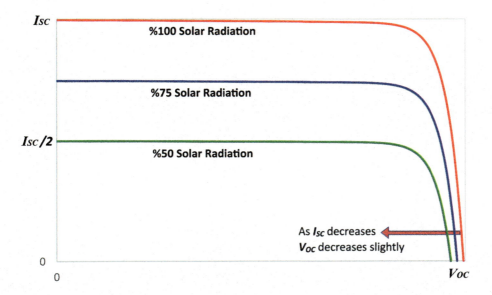

FIGURE 3.44 The influence of solar radiation on the $I-V$ curve with consideration I_{SC} effect on V_{OC}.

3.8 ADDITIONAL PERFORMANCE PARAMETERS OF PV CELLS

Generally, the performance parameters in a photovoltaic cell are related to parameters V_{OC}, I_{SC} (or J_{SC} mA/cm^2), parasitic resistances (series resistance R_S, parallel resistance R_P), efficiency and fill factor (FF). Among these parameters V_{OC}, I_{SC}, R_S, R_P are covered in the above sections. Characteristic resistance, PV cell efficiency, and FF parameters are examined below.

The Characteristic Resistance (R_0): The characteristic resistance of a PV cell is defined as the output resistance of the cell at its MPP. In other words, if the characteristic resistance of the PV is equal to the load resistance, then the maximum power is transmitted to the load. The characteristic resistance is a useful parameter in the analysis and comparison of PV systems. For example, parasitic resistances in PV cells reduce the efficiency of the solar cell by dissipating power in the resistances. Hence, it can be a useful parameter in understanding the impact of the parasitic loss mechanism (resistive effect) on the PV performance. Figure 3.45 shows the characteristic resistance on a $I-V$ curve.

It is clear from Figure 3.45 that the inverse slope of the line gives the characteristic impedance of the PV cell and can be given as (3.26). The characteristic resistance can also be approximated by dividing V_{OC} by I_{SC} for most of the PV cell types.

$$R_0 = \frac{V_{mp}}{I_{mp}} \approx \frac{V_{OC}}{I_{SC}} \qquad (3.26)$$

Fill Factor (FF): The FF can be defined as the ratio of the maximum power from a PV cell to the product of V_{OC} and I_{SC}. Graphically, it is a factor to measure the squareness level of $I-V$ the curve of the PV cell. Namely, it is the ratio of the largest rectangle area under the IV curve to the largest rectangle ($V_{OC} \times I_{SC}$) which covers the $I-V$ curve. Basically, the greater the FF, the performance of the PV cell is better. The FF is graphically illustrated in Figure 3.46.

It is clear from Figure 3.46 that the FF can be expressed mathematically as (3.27).

$$FF = \frac{P_{max}}{V_{OC} \cdot I_{SC}} = \frac{\eta \cdot A_c \cdot G}{V_{OC} \cdot I_{SC}} = \frac{V_{mp} \cdot I_{mp}}{V_{OC} \cdot I_{SC}} \qquad (3.27)$$

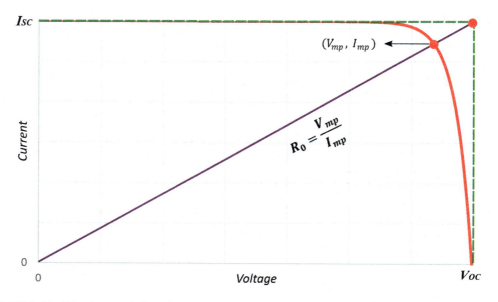

FIGURE 3.45 The characteristic resistance of a PV cell is shown on the I–V curve.

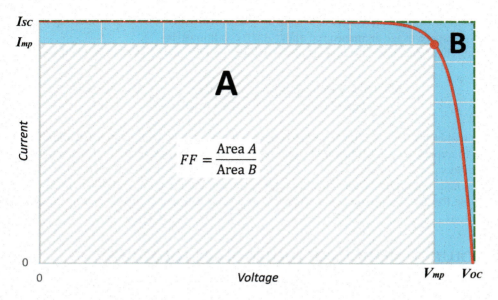

FIGURE 3.46 The illustration of FF on I–V curve.

where η is the cell efficiency, A_c cell area (m²) and G is the solar radiation in [W/m²]. The FF for a silicon PV cell is around 80%, and typically, it can change in the range of 50%–90% for different PV technologies. The voltage corresponding to the maximum theoretical FF value can be found by differentiating the power from a PV cell with respect to its voltage and solving the equation for voltage as shown below.

$$\frac{d(P)}{dV} = \frac{d(IV)}{dV} = 0 \tag{3.28}$$

By solving the (3.28), we get V_{mp} as in (3.29):

$$V_{mp} = V_{OC} - \frac{nkT}{q} \ln\left[\frac{V_{mp}}{\left(\frac{nkT}{q}\right)} + 1\right] = 0 \tag{3.29}$$

The above equation requires an iterative method for the solution of V_{mp} and additional equations are needed to find I_{mp} and FF. Hence, the following commonly used empirical expression is developed to calculate the FF in a simple way.

$$FF = FF_0\left(1 - \frac{R_s}{V_{OC}/I_{SC}}\right) = \left[\frac{\frac{q}{nkT}V_{OC} - \ln\left(\frac{q}{nkT}V_{OC} + 0.72\right)}{\frac{q}{nkT}V_{OC} + 1}\right]\left(1 - \frac{R_s}{V_{OC}/I_{SC}}\right) \tag{3.30}$$

where FF_0 is the fill factor of the ideal PV module without resistive effect.

PV Cell Efficiency (η): The PV cell efficiency is the most commonly used performance parameter to compare the PV cell technologies to each other and it can be defined as the ratio of maximum output power to the input power due to incoming solar radiation on a PV cell surface. As explained above, cell efficiency is not only dependent upon solar radiation but also depends on the solar spectrum, the intensity of incident sunlight, and PV cell temperature. Hence, it is vital to control the conditions under which the efficiency is measured to compare the PV cell systems' performance one to another. For instance, terrestrial PV cells are measured under $AM1.5$ conditions at 25°C, while PV cells for space usage are measured under $AM0$ conditions.

In the light of the above discussions, the efficiency of a PV device (η) can be expressed as (3.31).

$$\eta = \frac{\text{Maximum Electric Power}}{\text{Input Power}} = \frac{P_{max}}{\text{Area} \times \text{Irradiance}} = \frac{V_{OC} \cdot I_{SC} \cdot FF}{A_c \cdot G} \quad (3.31)$$

where η represents the cell efficiency, V_{OC} is the open-circuit voltage, I_{SC} represents the short-circuit current, FF represents the fill factor, A_c represents cell surface area, and G represents the incoming solar radiation.

Some other important factors such as solar radiation, temperature, and wind speed, which are responsible for the performance and efficiency of a solar cell, are comprehensively covered in the next section.

3.9 SOLUTION METHODS OF PHOTOVOLTAIC EQUATIONS

In photovoltaic studies, it is sometimes aimed to find the maximum operating point of the system and it is aimed to find PV model parameters on occasions. For this reason, different solution techniques have been developed in the literature to achieve the desired goal. The most commonly used solution techniques are described in the following sections.

3.9.1 *I–V* Equations for Series-Parallel Connections

The size and variety of photovoltaic applications are increasing progressively and the need for the analysis of different models for PV systems increases as well. As the most common PV applications, the analysis of series, parallel, and series-parallel connections is explained below.

Series-Connected PV Cells: The voltage of the PV modules is increased by connecting the appropriate number of PV cells in series. For this purpose, let us consider the single-diode model of the PV cell. As explained in the previous sections, $I-V$ the characteristics of a single-diode PV cell model are given below as (3.32).

$$I = I_{SC} - I_o \left(e^{\frac{q(V + I \cdot R_S)}{nkT_C}} - 1 \right) - \frac{V + I \cdot R_S}{R_P} \quad (3.32)$$

By substituting the thermal voltage (V_t) equation (3.21) into (3.32), the following equation can be modified as (3.33).

$$I = I_{SC} - I_o \left(e^{\frac{(V + I \cdot R_S)}{nV_t}} - 1 \right) - \frac{V + I \cdot R_S}{R_P} \quad (3.33)$$

Since all parameters in (3.32) and (3.33) are explained in the previous sections, they will not be re-described here. Figure (3.47) shows the equivalent model of N_S identical series-connected PV cells operating under the same conditions.

From Figure 3.46, the output current (I_M) and voltage (V_M) of the series-connected PV module can be written, respectively, as below.

$$I_M = I_{SC} - I_o \left[e^{\frac{V_M + I_M \cdot N_S \cdot R_S}{N_S \cdot nV_t}} - 1 \right] - \frac{V_M + I_M \cdot N_S \cdot R_S}{N_S \cdot R_P} \quad (3.34)$$

$$V_M = N_S \cdot V \quad (3.35)$$

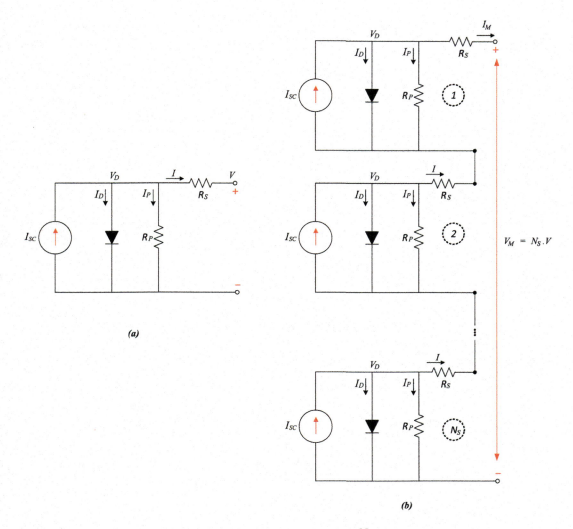

FIGURE 3.47 The equivalent circuit of (a) a single PV cell and (b) N_S series-connected PV cells.

Parallel-Connected PV Cells: When N_P identical PV cells are connected in parallel, the output voltage will equal a single-cell voltage, while the output current I_P will equal to $I \times N_P$. In order to better understand the current–voltage relationship in the parallel connection of PV cells, let's look at the equivalent circuits given in Figure (3.48).

From Figure (3.48), the output current (I_P) and voltage (V_P) of the series-connected PV module can be written, respectively, as below.

$$I_P = N_P \times I_{SC} - N_P \times I_o \left[e^{\frac{V+(I_P/N_P) \cdot R_S}{nV_t}} - 1 \right] - \frac{V+(I_P/N_P) \cdot R_S}{(R_P/N_P)} \quad (3.36)$$

$$V_P = V_{\text{cell}} = V \quad (3.37)$$

Series and Parallel-Connected PV Cells: To build up a PV array, PV modules are connected in series and parallel. Series connection of PV modules in a PV array forms a PV string, while parallel connection of PV strings develops the PV array (see Figure 3.49).

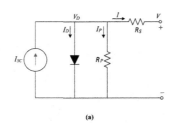

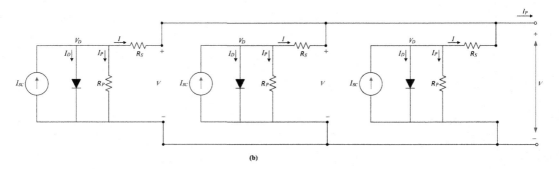

FIGURE 3.48 The equivalent circuit of (a) a single PV cell and (b) N_P parallel-connected PV cells.

Concerning Figure 3.48, the expression for output current I_A and output voltage V_A of a PV array consisting of N_S cells in series and N_P strings in parallel can be written as the following equation.

$$I_A = N_P \times I_{SC} - N_P \times I_o \left[e^{\frac{V_A + I_A \cdot (N_S/N_P) \cdot R_S}{N_S \cdot n V_t}} - 1 \right] - \frac{V_A + I_A \cdot (N_S/N_P) \cdot R_S}{(N_S/N_P) \cdot R_P} \tag{3.38}$$

$$I_A = I_M \cdot N_P \tag{3.39}$$

$$V_M = N_S \cdot V \tag{3.40}$$

3.9.2 Simple Approximation Method

Power system engineers are usually concerned with the current, voltage, and power outputs of photovoltaic systems. Of course, photovoltaic systems are assumed to be operated at maximum power points via MPPTs (Maximum Power Point Trackers). As explained above, photovoltaic $I-V$ equations are nonlinear and need iterative solution methods, which are very complex and there is no precise solution for either voltage V and current I. Hence, a simple but fast and effective solution approach is given here for the solution of $I-V$ equations. This approach is based on finding $I-V$ pairs at the MPP, namely I_m, V_m values. For this purpose, a spreadsheet can easily be prepared for the solution of a single-diode equation (3.12) of PV cells. By substituting diode voltage $V_D = V + I \cdot R_S$ in (3.12) and thermal voltage $V_t = kT_C/q$ following (3.41) equation can be obtained.

$$I = I_{SC} - I_o \left(e^{\frac{V_D}{V_t n}} - 1 \right) - \frac{V_D}{R_P} \tag{3.41}$$

As can be seen from equation (3.41), the current I can be calculated directly if the V_D value is known. The voltage value V across the PV cell can then be calculated from (3.42).

$$V = V_D - I \cdot R_S \tag{3.42}$$

The point of the solution approach is to find the V_D value correctly. We know that a PV cell's junction voltage is around 0.5 volt. Hence let's assume that the V_D value is in the range of 0.30 to 0.60

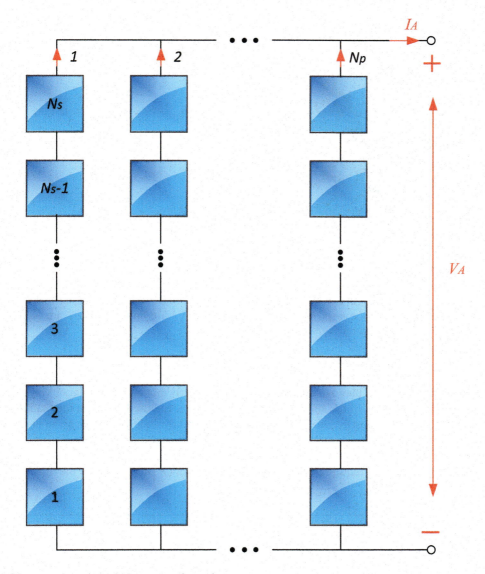

FIGURE 3.49 Forming a PV array configuration.

$(0.30 \leq V_D \leq 0.60)$. Now let us write the V_D on the calculation chart from 0.30 to 0.60 by 1% increments and then calculate I and V from (3.41) and (3.42) for each of the V_D value. After that calculate the output powers from $P = I \times V$ equation. We will notice that the output power reaches the highest value for MPP (I_m, V_m). The solution is the point that makes the power maximum and accordingly the solution set is $\{I_m, V_m, P_m\}$. Note that one can easily solve any PV cell, PV module, and PV array equations using a simple approximation method.

Example 3.4: Voltage, Current, and Power at MPP of a PV Cell

A PV module with an ideality factor of 2.5 has a circuit current of 5 A at STC (25°C and 1 kW/m²). Assume that its reverse saturation current is 1.6×10^{-10} A, parallel resistance is 10 Ω, and series resistance is 0.01 Ω. Using the simple approximation method, find the voltage, current, and power at MPP.

Solution

Assume that V_D values in the following $I-V$ equation (from 3.41) are known and greater than 0.40 ($V_D \geq 0.40$). In the given expression, $\frac{1}{V_t}$ is inserted as 38.9 under STC.

$$I = I_{SC} - I_o \left(e^{\frac{38.9 \times V_D}{n}} - 1 \right) - \frac{V_D}{R_P}$$

For example, let us calculate the current for $V_D = 0.40$.

$$I = 5 - 1.6 \times 10^{-10} \left(e^{\frac{38.9 \times 0.4}{2.5}} - 1 \right) - \frac{0.4}{25} = 4.984 \text{ A}$$

Under these conditions, the output voltage and output power are:

$$V = V_D - I \cdot R_S = 0.4 - 4.984 \times 0.01 = 0.350 \text{ V}$$

$$P = V \times I = 0.350 \times 4.984 = 1.745 \text{ W}$$

Hence, a calculation table in Excel can easily be prepared to find the MPP as below (Table 3.2):
As can be seen from the result that $P_{mp} = 6.618$ with $V_{mp} = 1.313$ and $I_{mp} = 4.698$ are obtained easily from the simple approximation method.

3.9.3 Parameter Estimation of Single-Diode Model

It is needed to find essential photovoltaic parameters to correctly model and simulate PV systems. In this section, the parameters of the single-diode model will be computed based on the manufacturer specification sheets. The most general $I-V$ expression for photovoltaic systems is given in (3.38). In (3.38), there are five unknown parameters at SRC. These unknown parameters are I_{PV} (photovoltaic current), I_0, R_S, R_P and n (ideality factor). Solving for these five parameters using the mathematical model explained below requires a datasheet parameter, which is provided by the manufacturer.

TABLE 3.2
Simple Approximation Method to Find MPP

V_D	$I = I_{SC} - I_o \left(e^{\frac{38.9 \times V_D}{n}} - 1 \right) - \frac{V_D}{R_P}$	$V = V_D - I \cdot R_S$	$P = V \times I$
0.40	4.984	0.350	1.745
0.41	4.984	0.360	1.795
...	...	...	...
1.30	4.850	1.251	6.070
1.31	4.834	1.262	6.098
1.32	4.814	1.272	6.123
1.33	4.791	1.282	6.143
1.34	4.765	1.292	6.158
1.35	4.734	1.303	6.166
1.36	**4.698**	**1.313**	**6.168**
1.37	4.655	1.323	6.161
1.38	4.606	1.334	6.144
1.39	4.549	1.345	6.116
1.40	4.482	1.355	6.074

Photovoltaic Systems and Photovoltaic Technologies

The PV manufacturers provide $I-V$ characteristics as well as the values of selected parameters at SRC except for the NOCT (nominal operating cell temperature), which is given under nominal operating conditions (G is 800 W/m² and AM1.5).

Important Note: If a PV cell is exposed to photon irradiation, a light-generated current occurs in that cell. This photovoltaic current (I_{PV}) is equal to the short-circuit current (I_{SC}) only in the ideal cell. For this reason, the photovoltaic current (I_{PV}) is assumed to be equal to the I_{SC} current ($I_{PV} \approx I_{SC}$) in the photovoltaic models given in the above sections. However, since the parameter estimation requires more precise calculations, the previously accepted assumption ($I_{PV} \approx I_{SC}$) is not considered here. In this context, a single-diode and double-diode photovoltaic model to be used in parameter estimation are considered as given in Figure 3.50. As a result, the single and double-diode $I-V$ equations are modified as (3.43) and (3.44), respectively.

$$I = I_{PV} - I_o \left[e^{\frac{q(V+I \cdot R_S)}{nkT_c}} - 1 \right] - \frac{V + I \cdot R_S}{R_P} \quad (3.43)$$

$$I = I_{PV} - I_{o1} \left[e^{\frac{q(V+I \cdot R_S)}{n_1 kT_c}} - 1 \right] - I_{o2} \left[e^{\frac{q(V+I \cdot R_S)}{n_2 kT_c}} - 1 \right] - \frac{(V + I \cdot R_S)}{R_P} \quad (3.44)$$

Firstly, let us consider the single-diode model equation (3.43). The mathematical model should include a set of five equations with five unknown parameters in order to have a solvable system. The first equation can be obtained from open-circuit conditions at SRC where $I = 0$, and $V = V_{OC}$. Hence equation (3.43) becomes:

$$0 = I_{PV} - I_o \left[e^{\frac{qV_{OC}}{nkT_c}} - 1 \right] - \frac{V_{OC}}{R_P} \quad (3.45)$$

Therefore, I_{PV} equals to:

$$I_{PV} = I_o \left[e^{\frac{qV_{OC}}{nkT_c}} - 1 \right] + \frac{V_{OC}}{R_P} \quad (3.46)$$

The second equation is derived from the short-circuit conditions at SRC where $V = 0$ and $I = I_{SC}$. Thus, equation (3.43) becomes:

$$I_{SC} = I_{PV} - I_o \left[e^{\frac{q \cdot I_{SC} \cdot R_S}{nkT_c}} - 1 \right] - \frac{I_{SC} \cdot R_S}{R_P} \quad (3.47)$$

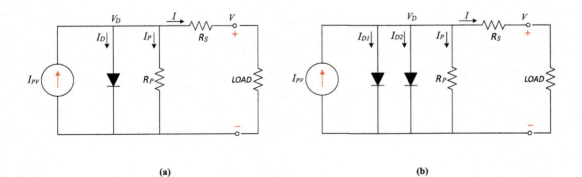

FIGURE 3.50 The equivalent circuit of (a) single-diode PV cell and (b) double-diode PV cell.

Therefore, I_{PV} equals to:

$$I_{PV} = I_{SC} + I_o \left[e^{\frac{q \cdot I_{SC} \cdot R_S}{nkT_c}} - 1 \right] + \frac{I_{SC} \cdot R_S}{R_P} \tag{3.48}$$

Now, I_0 can be found from (3.46) and (3.48) as below:

$$I_0 = \frac{I_{SC} + \left(\frac{I_{SC} \cdot R_S}{R_P}\right) - \left(\frac{V_{OC}}{R_P}\right)}{\left(e^{\frac{qV_{OC}}{nkT_c}}\right) - \left(e^{\frac{q \cdot I_{SC} \cdot R_S}{nkT_c}}\right)} \tag{3.49}$$

I_{PV} can then be found by substituting (3.49) into (3.46):

$$I_{PV} = \left[\frac{I_{SC} + \left(\frac{I_{SC} \cdot R_S}{R_P}\right) - \left(\frac{V_{OC}}{R_P}\right)}{\left(e^{\frac{qV_{OC}}{nkT_c}}\right) - \left(e^{\frac{q \cdot I_{SC} \cdot R_S}{nkT_c}}\right)} \right] \left[e^{\frac{qV_{OC}}{nkT_c}} - 1 \right] + \frac{V_{OC}}{R_P} \tag{3.50}$$

The third equation can be written from MPP under SRC conditions where $V = V_{mp}$ and $I = I_{mp}$. Accordingly, equation (3.43) becomes:

$$I_{mp} = I_{PV} - I_o \left[e^{\frac{q(V_{mp} + I_{mp} \cdot R_S)}{nkT_c}} - 1 \right] - \frac{V_{mp} + I_{mp} \cdot R_S}{R_P} \tag{3.51}$$

More equations can be found by obtaining the derivative of (3.43) with respect to voltage V.

$$\frac{dI}{dV} = -I_o \left[\frac{q}{nkT_c} \cdot \left(1 + \frac{dI}{dV} R_S\right) e^{\frac{q(V + I \cdot R_S)}{nkT_c}} \right] - \frac{1}{R_P} \left(1 + \frac{dI}{dV} R_S\right) \tag{3.52}$$

Again at the open-circuit point on the $I-V$ curve, we have $I = 0$ and $V = V_{OC}$. Therefore,

$$\frac{dI}{dV} = \left.\frac{dI}{dV}\right|_{I=0} \tag{3.53}$$

Applying (3.53) in (3.52),

$$\left.\frac{dI}{dV}\right|_{I=0} = -I_o \left[\frac{q}{nkT_c} \cdot \left(1 + \left.\frac{dI}{dV}\right|_{I=0} R_S\right) e^{\frac{qV}{nkT_c}} \right] - \frac{1}{R_P} \left(1 + \left.\frac{dI}{dV}\right|_{I=0} R_S\right) \tag{3.54}$$

Yet again at the short current point on the $I-V$ curve, we have $V = 0$ and $I = I_{SC}$. Hence,

$$\frac{dI}{dV} = \left.\frac{dI}{dV}\right|_{V=0} \tag{3.55}$$

Applying (3.55) in (3.52),

$$\left.\frac{dI}{dV}\right|_{V=0} = -I_o \left[\frac{q}{nkT_c} \cdot \left(1 + \left.\frac{dI}{dV}\right|_{V=0} R_S\right) e^{\frac{qI_{SC}R_S}{nkT_c}} \right] - \frac{1}{R_P} \left(1 + \left.\frac{dI}{dV}\right|_{V=0} R_S\right) \tag{3.56}$$

Photovoltaic Systems and Photovoltaic Technologies

As known, the power at any point on the $I-V$ curve is given by $P = I \cdot V$ and the derivative of the power expression is:

$$\frac{dP}{dV} = \left(\frac{dI}{dV}\right)V + I \tag{3.57}$$

We know that $(dP/dV = 0)$ at maximum power, therefore (3.57) becomes:

$$\frac{dI}{dV} = -\frac{I_{mp}}{V_{mp}} \tag{3.58}$$

Substituting (3.58) into (3.52),

$$-\frac{I_{mp}}{V_{mp}} = -I_o \left[\frac{q}{nkT_c}\left(1 - \frac{I_{mp}}{V_{mp}}R_S\right)e^{\frac{q(V_{mp}+I_{mp})R_S}{nkT_c}}\right] - \frac{1}{R_P}\left(1 - \frac{I_{mp}}{V_{mp}}R_S\right) \tag{3.59}$$

Equations (3.45), (3.47), (3.51), (3.54), (3.56), and (3.59) obtained in the above analyzes are independent of each other and sufficient to solve for the five unknown parameters I_{PV}, I_0, R_S, R_P and n at reference conditions. These equations can now be solved easily using the nonlinear equation solver "*fsolve*" included in the toolboxes of MATLAB®. The flowchart diagram of the solution algorithm can be given in Figure 3.51. After all PV cell parameters have been found at reference conditions, the performance of the considered PV system can be estimated for different operating conditions by using PV module and PV array models.

For the resulting nonlinear equations, the solution set can also be obtained by using different iterative methods, such as the commonly used Newton–Raphson method. The application of the Newton–Raphson method to the PV equation system is described in the following section.

3.9.4 Newton–Raphson Method

The Newton–Raphson method can be used to solve any nonlinear equation system. The method seeks the roots of nonlinear functions by calculating the Jacobian linearization of the function around an initial guess point. Therefore, it is important to obtain a successful initial value for fast-moving to the nearest zero. A graphical representation of this process is given in Figure 3.52.

Let's give a nonlinear equation $f(x) = 0$. The aim is to find the root x of the equation. As illustrated in Figure 3.52, the Newton–Raphson method is based on the principle that assigning an initial guess x_i as a root of $f(x)$ and then drawing them to the curve at $f(x_i)$, the point x_{i+1} where the tangent crosses the x-axis is an improved estimate of the root. Hence, the slope of a function at $x = x_i$

$$f'_{(x_i)} = \tan(\theta) = \frac{f(x_i) - 0}{x_{i+1} - x_i} \tag{3.60}$$

From (3.60), we have:

$$x_{i+1} = x_i - \frac{f(x_i)}{f'_{(x_i)}} \tag{3.61}$$

Equation (3.61) is called the Newton–Raphson formula for solving nonlinear equations of the form $f(x) = 0$. Accordingly beginning with an initial guess x_i, the next guess x_{i+1} can be found by using (3.61). So, this process is repeated until the root is reached within an applicable tolerance. In short, the Newton–Raphson algorithm can be summarized in four steps as below:

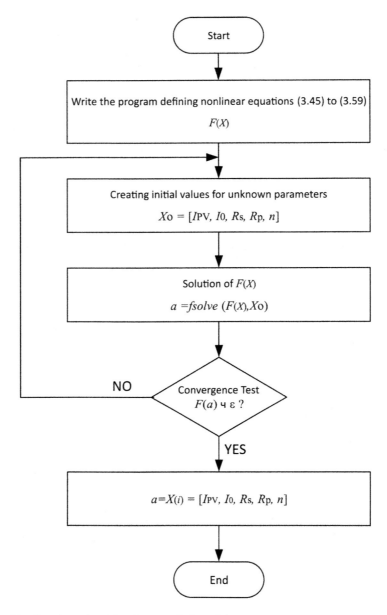

FIGURE 3.51 The flowchart algorithm for the solution of unknown parameters.

- **Step 1.** Take the derivative of $f(x)$ symbolically $f'_{(x)}$.
- **Step 2.** Guess an initial x_i, and use this value to estimate new value of the root x_{i+1} using equation (3.61).
- **Step 3.** Calculate the absolute value of relative error, $|\varepsilon|$ as below:

$$|\varepsilon| = \left| \frac{x_{i+1} - x_i}{x_{i+1}} \right| \times 100 \tag{3.62}$$

- **Step 4.** Compare the absolute relative error found in Step 3 with the reference relative error, ε_r. If $|\varepsilon| > \varepsilon_r$, then go Step 2, else stop the algorithm.

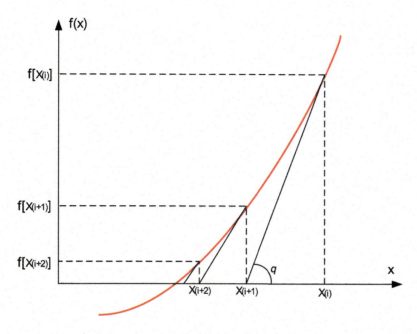

FIGURE 3.52 Graphical illustration of the Newton–Raphson method.

Since the photovoltaic equations analyzed above are more than five, the Newton–Raphson algorithm should be applied to a set of equations in the matrix form. Hence, a set of nonlinear equations in matrix form is given by:

$$F(x) = \begin{bmatrix} f_1(x) \\ f_2(x) \\ \vdots \\ f_N(x) \end{bmatrix} = 0 \tag{3.63}$$

Following the foregoing description, the Newton–Raphson formula can be written as follows for a set of equations.

$$x(i+1) = x(i) - \mathbf{J}^{-1}(i) \cdot F[x(i)] \tag{3.64}$$

where $\mathbf{J}$ is the Jacobian matrix and it is given for each iteration as:

$$\mathbf{J}(i) = \frac{dF}{dx}\bigg|_{x=x(i)} = \begin{bmatrix} \frac{\partial f_1}{\partial x_1} & \frac{\partial f_1}{\partial x_2} & \cdots & \frac{\partial f_1}{\partial x_N} \\ \frac{\partial f_2}{\partial x_1} & \frac{\partial f_2}{\partial x_2} & \cdots & \frac{\partial f_2}{\partial x_N} \\ \vdots & \vdots & \ddots & \vdots \\ \frac{\partial f_N}{\partial x_1} & \frac{\partial f_N}{\partial x_2} & \cdots & \frac{\partial f_N}{\partial x_N} \end{bmatrix} \tag{3.65}$$

Now, we can assemble the Newton–Raphson method to solve five unknown parameters of the PV module as below:

$$\begin{bmatrix} x_1(i+1) \\ x_2(i+1) \\ x_3(i+1) \\ x_4(i+1) \\ x_5(i+1) \end{bmatrix} = \begin{bmatrix} x_1(i) \\ x_2(i) \\ x_3(i) \\ x_4(i) \\ x_5(i) \end{bmatrix} - \begin{bmatrix} \frac{\partial f_1}{\partial x_1} & \frac{\partial f_1}{\partial x_2} & \cdots & \frac{\partial f_1}{\partial x_5} \\ \frac{\partial f_2}{\partial x_1} & \frac{\partial f_2}{\partial x_2} & \cdots & \frac{\partial f_2}{\partial x_5} \\ \vdots & \vdots & \ddots & \vdots \\ \frac{\partial f_5}{\partial x_1} & \frac{\partial f_5}{\partial x_2} & \cdots & \frac{\partial f_5}{\partial x_5} \end{bmatrix}^{-1} \cdot \begin{bmatrix} f_1(x(i)) \\ f_2(x(i)) \\ f_3(x(i)) \\ f_4(x(i)) \\ f_5(x(i)) \end{bmatrix} \qquad (3.66)$$

For example, let us obtain the Jacobian matrix for the five unknown parameters I_{PV}, I_0, R_S, R_P, and n from the above equations (3.45), (3.47), (3.54), (3.56), and (3.59). Note that the following definitions need to be given here for notational convenience. Assume that the slopes of the $I-V$ curve at open circuit and short circuit are represented as R_{SO} and R_{SHO}, respectively. Hence, as R_{SO} and R_{SHO} are defined as below:

$$R_{SO} = \left.\frac{dV}{dI}\right|_{V=V_{OC}} \qquad (3.67)$$

$$R_{SHO} = \left.\frac{dV}{dI}\right|_{I=I_{SC}} \qquad (3.68)$$

The component of the Jacobian matrix is found as follows. Considering (3.45) as $f_1(x)$, we get:

$$\mathbf{J}(1,1) = \frac{\partial f_1}{\partial I_{PV}} = 1$$

$$\mathbf{J}(1,2) = \frac{\partial f_1}{\partial I_0} = -e^{\frac{qV_{OC}}{nkT}} + 1$$

$$\mathbf{J}(1,3) = \frac{\partial f_1}{\partial R_S} = 0$$

$$\mathbf{J}(1,4) = \frac{\partial f_1}{\partial R_P} = -\frac{V_{OC}}{R_P^2}$$

$$\mathbf{J}(1,5) = \frac{\partial f_1}{\partial n} = \frac{qI_0 V_{OC}}{n^2 kT} e^{\frac{qV_{OC}}{nkT}} \qquad (3.69)$$

Considering (3.47) as $f_2(x)$, we get:

$$\mathbf{J}(2,1) = \frac{\partial f_2}{\partial I_{\text{PV}}} = 1$$

$$\mathbf{J}(2,2) = \frac{\partial f_2}{\partial I_0} = -e^{\frac{qI_{\text{SC}}R_S}{nkT}} + 1$$

$$\mathbf{J}(2,3) = \frac{\partial f_2}{\partial R_S} = -\frac{qI_0 I_{\text{SC}} R_s}{nkT} e^{\frac{qI_{\text{SC}}R_S}{nkT}} - \frac{I_{\text{SC}}}{R_P}$$

$$\mathbf{J}(2,4) = \frac{\partial f_2}{\partial R_P} = \frac{I_{\text{SC}} R_S}{R_P^2}$$

$$\mathbf{J}(2,5) = \frac{\partial f_2}{\partial n} = \frac{qI_0 I_{\text{SC}} R_S}{n^2 kT} e^{\frac{qI_{\text{SC}}R_S}{nkT}} \tag{3.70}$$

Considering (3.54) as $f_3(x)$, we get:

$$\mathbf{J}(3,1) = \frac{\partial f_3}{\partial I_{\text{PV}}} = 0$$

$$\mathbf{J}(3,2) = \frac{\partial f_3}{\partial I_0} = -\frac{q\left(1 - \dfrac{R_S}{R_{\text{SO}}}\right)}{nkT} e^{\left(\frac{qV_{\text{OC}}}{nkT}\right)}$$

$$\mathbf{J}(3,3) = \frac{\partial f_3}{\partial R_s} = \frac{qI_0(R_S - R_{\text{SO}})}{nkTR_{\text{SO}}} e^{\left(\frac{qV_{\text{OC}}}{nkT}\right)} + \frac{1}{R_p R_{\text{SO}}}$$

$$\mathbf{J}(3,4) = \frac{\partial f_3}{\partial R_P} = -\frac{1 - \dfrac{R_S}{R_{\text{SO}}}}{R_P^2}$$

$$\mathbf{J}(3,5) = \frac{\partial f_3}{\partial n} = \frac{qI_0(R_{\text{SO}} - R_S)}{n^3 k^2 T^2 R_{\text{SO}}} \left[e^{\left(\frac{qV_{\text{OC}}}{nkT}\right)} \right] (nkT + qV_{\text{OC}}) \tag{3.71}$$

Considering (3.56) as $f_4(x)$, we get:

$$\mathbf{J}(4,1) = \frac{\partial f_4}{\partial I_{\text{PV}}} = 0$$

$$\mathbf{J}(4,2) = \frac{\partial f_4}{\partial I_0} = -\frac{q\left(1 - \dfrac{R_S}{R_{\text{SO}}}\right)}{nkT} e^{\left(\frac{qI_{\text{SC}}R_S}{nkT}\right)}$$

$$\mathbf{J}(4,3) = \frac{\partial f_4}{\partial R_s} = \frac{qI_0(R_{\text{SHO}} - R_S)}{nkTR_{\text{SHO}}} e^{\left(\frac{qI_{\text{SC}}R_S}{nkT}\right)} - \frac{q^2 I_0 \left(1 - \dfrac{R_S}{R_{\text{SHO}}}\right) I_{\text{SC}}}{n^2 k^2 T^2} e^{\left(\frac{qI_{\text{SC}}R_S}{nkT}\right)} + \frac{1}{R_P R_{\text{SHO}}}$$

$$\mathbf{J}(4,4) = \frac{\partial f_4}{\partial R_P} = \frac{R_{SHO} - R_S}{R_P^2 R_{SHO}}$$

$$\mathbf{J}(4,5) = \frac{\partial f_4}{\partial n} = \frac{qI_0(R_{SHO} - R_S)}{n^3 k^2 T^2 R_{SHO}} \left[e^{\left(\frac{qI_{sc}R_S}{nkT}\right)} \right] (nkT + qI_{SC}R_S) \tag{3.72}$$

Considering (3.59) as $f_5(x)$ we get:

$$\mathbf{J}(5,1) = \frac{\partial f_5}{\partial I_{PV}} = 0$$

$$\mathbf{J}(5,2) = \frac{\partial f_5}{\partial I_0} = -\frac{q\left(1 - \frac{I_{mp}R_S}{V_{mp}}\right)}{nkT} e^{\left[\frac{q(V_{mp}+I_{mp}R_S)}{nkT}\right]}$$

$$\mathbf{J}(5,3) = \frac{\partial f_5}{\partial R_S} = \frac{qI_0 I_{mp}}{nkT V_{mp}} e^{\left[\frac{q(V_{mp}+I_{mp}R_S)}{nkT}\right]} - \frac{q^2 I_0 \left(1 - \frac{I_{mp}R_S}{V_{mp}}\right) I_{mp}}{n^2 k^2 T^2} e^{\left[\frac{q(V_{mp}+I_{mp}R_S)}{nkT}\right]} + \frac{I_{mp}}{R_P V_{mp}}$$

$$\mathbf{J}(5,4) = \frac{\partial f_5}{\partial R_P} = \frac{1 - \frac{I_{mp}R_S}{V_{mp}}}{R_P^2}$$

$$\mathbf{J}(5,5) = \frac{\partial f_5}{\partial n} = \frac{qI_0(V_{mp} - I_{mp}R_S)[nkT + q(V_{mp} + I_{mp}R_S)]}{n^3 k^2 T^2 V_{mp}} e^{\left[\frac{q(V_{mp}+I_{mp}R_S)}{nkT}\right]} \tag{3.73}$$

Important Note: (1) One can easily obtain the Jacobian matrix in MATLAB® by using the symbolic toolbox and a numerical solution can be achieved using the optimization toolbox. (2) The correct estimation of initial values in the application of the Newton–Raphson method is very important. Because if the initial values are not selected correctly, the Newton–Raphson method will not converge or the matrix may give a singularity error.

3.9.5 Parameter Estimation of Double-Diode Model

The analysis and parameter estimation of the single-diode model explained above is also applicable to double-diode analysis. The double diode of the PV model is already given in Figure 3.50 and equation (3.44). At the open-circuit point ($I = 0$ and $V = V_{OC}$), therefore from (3.44), we get:

$$0 = I_{PV} - I_{o1}\left[e^{\frac{qV_{OC}}{n_1 kT_c}} - 1\right] - I_{o2}\left[e^{\frac{qV_{OC}}{n_2 kT_c}} - 1\right] - \frac{V_{OC}}{R_P} \tag{3.72}$$

From (3.72), the expression I_{PV} can be written as:

$$I_{PV} = I_{o1}\left[e^{\frac{qV_{OC}}{n_1 kT_c}} - 1\right] + I_{o2}\left[e^{\frac{qV_{OC}}{n_2 kT_c}} - 1\right] + \frac{V_{OC}}{R_P} \tag{3.73}$$

A short-circuit point ($I = I_{SC}, V = 0$), therefore from (3.44), we get:

$$I_{SC} = I_{PV} - I_{o1}\left[e^{\frac{q \cdot I_{SC} \cdot R_S}{n_1 k T_c}} - 1\right] - I_{o2}\left[e^{\frac{q \cdot I_{SC} \cdot R_S}{n_2 k T_c}} - 1\right] - \frac{I_{SC} \cdot R_S}{R_P} \qquad (3.74)$$

Again I_{PV} can be written from (3.74) as:

$$I_{PV} = I_{SC} + I_{o1}\left[e^{\frac{q \cdot I_{SC} \cdot R_S}{n_1 k T_c}} - 1\right] + I_{o2}\left[e^{\frac{q \cdot I_{SC} \cdot R_S}{n_2 k T_c}} - 1\right] + \frac{I_{SC} \cdot R_S}{R_P} \qquad (3.75)$$

From equations (3.73) and (3.75), I_{o2} is written as:

$$I_{o2} = \frac{I_{SC} + + \frac{I_{SC} \cdot R_S}{R_P} - \frac{V_{OC}}{R_P} + I_{o1}\left[\left(e^{\frac{q \cdot I_{SC} \cdot R_S}{n_1 k T_c}} - 1\right) - \left(e^{\frac{qV_{OC}}{n_1 k T_c}} - 1\right)\right]}{\left[\left(e^{\frac{qV_{OC}}{n_2 k T_c}} - 1\right) - \left(e^{\frac{q \cdot I_{SC} \cdot R_S}{n_2 k T_c}} - 1\right)\right]} \qquad (3.76)$$

Let us substitute (3.76) into (3.73)

$$I_{PV} = I_{o1}\left[e^{\frac{qV_{OC}}{n_1 k T_c}} - 1\right] + \frac{I_{SC} + \frac{I_{SC} \cdot R_S}{R_P} - \frac{V_{OC}}{R_P} + I_{o1}\left[\left(e^{\frac{q \cdot I_{SC} \cdot R_S}{n_1 k T_c}} - 1\right) - \left(e^{\frac{qV_{OC}}{n_1 k T_c}} - 1\right)\right]}{\left[\left(e^{\frac{qV_{OC}}{n_2 k T_c}} - 1\right) - \left(e^{\frac{q \cdot I_{SC} \cdot R_S}{n_2 k T_c}} - 1\right)\right]\left[e^{\frac{qV_{OC}}{n_2 k T_c}} - 1\right]^{-1}} + \frac{V_{OC}}{R_P} \qquad (3.77)$$

Another equation can be found from the MPP conditions by writing $V = V_{mp}$ and $= I_{mp}$:

$$I_{mp} = I_{PV} - I_{o1}\left[e^{\frac{q(V_{mp} + I_{mp} \cdot R_S)}{n_1 k T_c}} - 1\right] - I_{o2}\left[e^{\frac{q(V_{mp} + I_{mp} \cdot R_S)}{n_2 k T_c}} - 1\right] - \frac{(V_{mp} + I_{mp} \cdot R_S)}{R_P} \qquad (3.72)$$

As explained in the single-diode model, more equations can be obtained depending on the number of parameters to be solved. Since these equations are explained in detail in Sections 3.9.3 and 3.9.4, the information given in this section is considered to be sufficient for the double-diode model.

3.10 PHOTOVOLTAIC TECHNOLOGIES

As one of the emerging technologies, different PV systems are increasingly used to meet global energy consumption. Dozens of different PV technology are commercially available in the market. The general classification of PV technologies is given in Figure 3.3 in the introduction part of this chapter. This section aims to introduce the most basic features of PV technologies and to give information about performance and techno-economical parameters to allow one to make a comparison between PV technologies.

3.10.1 CRYSTALLINE SILICON TECHNOLOGIES

Among the photovoltaic technologies, crystalline silicon photovoltaics are the most widely used technology with a global market share of about 90%. There are two types of PV cells used in crystalline silicon photovoltaics. These are monocrystalline silicon and polycrystalline silicon cells.

Monocrystalline PV cells are produced by slicing high-purity cylindrically-shaped ingots into wafers. Polycrystalline PV cells are produced by cutting a cast block of silicon into bars and then into wafers. Here, pieces of different silicon crystals are melted together to form polycrystalline ingots and wafers. In general, the efficiency of monocrystalline cells is better than polycrystalline PV cells. The maximum energy conversion efficiency reported so far is about 25% for monocrystalline PV cells. The energy conversion efficiency of polycrystalline cells is less than 22% so far. However, the difference between efficiency rates is decreasing gradually.

3.10.1.1 Monocrystalline Silicon

Monocrystalline PV cells, also called single-crystalline cells, can easily be distinguished by their color, uniform look, and rounded edges. Because these cells are produced from a pure single crystal and its color is black. As explained above, monocrystalline PV cells are not only the most efficient among other crystalline-based PV cells, but they are also the most space-efficient in terms of power per unit area. Another advantage of monocrystalline cells is that their lifetime is longer than other cell types. Most of the manufacturers offer 25 years warranty for these types of PV systems. One more important advantage of monocrystalline cells is that their performance is less affected by temperature change. Therefore, the difference in the performances of monocrystalline PV cells is less due to seasonal changes. The most important disadvantage of monocrystalline PV cells is their prices. In fact, they are the most expensive of all solar cells. For this reason, polycrystalline and thin-film cells are often more preferred choices than monocrystalline cells.

3.10.1.2 Polycrystalline Silicon

Polycrystalline PV cells, also called multi-crystalline cells, can easily be distinguished by their multiple color and rectangular shape. Because the silicon is melted and poured into a square mold and the shape ingots are cut into square shapes after cooling. In this way, the silicon is not wasted as in the monocrystalline manufacturing process. The efficiency of the polycrystalline cell is lower than that of the monocrystalline cell, but the cost of the polycrystalline cell is lower due to the simpler process of producing polycrystalline silicon. One of the reasons for preferring polycrystalline cells is that they have an aesthetic appearance like thin-film technology. Although their highest energy conversion efficiency is about 22%, polycrystalline solar PV systems are operated at an average efficiency of 13%–16% in the field. Hence, it has a lower space efficiency compared to monocrystalline cells. Another drawback of the polycrystalline cell is that they have higher temperature coefficients compared to monocrystalline cells, which means that the output powers of the polycrystalline cells decrease more at high temperatures. High temperatures adversely affect the performance of the polycrystalline cell, resulting in a lower lifetime when it is compared to monocrystalline cells. However, their lifespan is not under 20 years.

3.10.1.3 String Ribbon Silicon

String Ribbon Si (silicon) PV panels are made from polycrystalline Si. In their manufacturing process, temperature-resistant wires are pulled through molten (melted) silicon to obtain very thin silicon ribbons. The PV panels obtained in this way are very similar in appearance to the conventional polycrystalline panels. The main advantage of String Ribbon Si panels is the use of less silicon in production. Almost half of the silicones used in the manufacturing of monocrystalline cells, which contributes to lower cost. Due to their low efficiency at about 13%–17%, string ribbon silicon panels have the lowest space efficiency among the main types of crystalline-based solar panels.

3.10.2 THIN-FILM TECHNOLOGIES

The thin-film solar cell manufacturing process is based on the principle of coating one or more photovoltaic materials on a substrate. Amorphous silicon (a-Si), Cadmium telluride (CdTe), and Copper indium gallium selenide (CIS/CIGS) are the main types of thin-film technologies. Thin-film panels have an efficiency of 7%–13% depending on the technology they are made of. Commercially available thin-film panels with a global market share of about 6%–7% usually have an efficiency in the range of 8%–10%. Future thin-film technologies are expected to reach about 16% efficiency. Currently, thin-film solar cells have poor space efficiency due to their low energy conversion efficiency. Another drawback is that thin-film PV panels have a shorter lifespan than their crystalline counterparts.

Thin-film photovoltaic modules are ideal for building applications due to their excellent aesthetics. The main advantage of thin-film photovoltaic modules is that they can generate consistent power at higher temperatures, on cloudy days, and at low sun angles. Their output power also drops by a small amount under partial shadowing. These properties give thin-film PV modules superior design flexibility for building applications. Mass production of thin-film modules is much easier, which results in relatively cheaper modules when compared to crystalline-based solar panels.

3.10.2.1 Amorphous Silicon

The rate of silicon used in the production of amorphous silicon (a-Si) cells is quite low, such that about 1%–2% of silicon used in the crystalline silicon cell is sufficient for amorphous silicon solar cells. Due to their low output power, a-Si cells are not preferred in large-scale applications. In contrast, amorphous silicon cells are used for smaller-scale applications such as calculators, travel lights, camping equipment, etc. The efficiency of a-Si cells can be increased by the process called stacking, which contains several layers of amorphous silicon cells. The efficiency of a-Si cells is increased up to 8% by the stacking process. However, they are still relatively costly due to the expensive stacking process. It is expected that a-Si cells will be beneficial as well for some large-scale applications because of ongoing innovations.

3.10.2.2 Cadmium Telluride

Cadmium telluride (CdTe) is a PV technology based on the use of a thin film of CdTe. CdTe solar panels are operated at an average efficiency of 9%–11% in the field. Although its highest energy conversion efficiency is about 16%, CdTe is the only thin-film solar panel technology that is cost-competitive with crystalline silicon models. CdTe is growing rapidly and after silicon-based technology, CdTe currently represents the second most utilized solar cell material around the World. CdTe panels have several advantages over traditional silicon technology. The major advantage of this technology is that the panels can be manufactured at lower costs than silicon-based solar panels due to its ease of manufacturing and the abundance of cadmium, which is produced as a by-product of other important industrial metals such as zinc.

3.10.2.3 Copper Indium Gallium Selenide (CIGS-CIS)

Copper Indium Gallium Selenide materials with different bandgaps are sandwiched between semi-conductors to form CIGS PV cells having hetero-junction structures. The top layer of CIGS PV cells is made from cadmium sulfide (CdS) material and sometimes zinc is added to the layer to improve the transparency. Although its highest energy conversion efficiency is about 20% under laboratory conditions, CIGS solar panels are operated at an average efficiency of 10%–12% in the field, to some extent comparable to crystalline technologies. This technology is still an emerging technology and compared to the other thin-film technologies above, CIGS PV cells have shown the highest energy efficiency potential. Hence, it is expected to be used in larger or commercial applications in the near future. The most important disadvantage of CIGS technologies is their higher costs since the material is more difficult to work with. The most important advantages in terms of performance are that CIGS PV panels show good performance under indirect radiation and high-temperature conditions. For this reason, they are attractive in the applications facing the north, applications in the building, and secure places.

3.10.2.4 Gallium Arsenide Thin Film

Thin-film Gallium Arsenide (GaAs) PV cell is made from a mixture of two elements, namely gallium and arsenide. GaAs PV cells are still emerging technologies and have a significant potential among the PV systems. Because The GaAs band gap is 1.43 eV and is suitable for use in multi-junction and high-efficiency solar cells. GaAs PV cells have important advantages over silicon-based cells, such that GaAs is the highest efficiency solar material with a low-temperature coefficient and shows good performance under low light. Since they are expensive, flexible, lightweight, and resistant to moisture/ultra-violet light, they are suitable for use in space technologies and military applications.

3.10.2.5 Thin-Film Microcrystalline (Polycrystalline) Silicon Cell

As one of the newly emerging solar technologies, thin-film microcrystalline silicon (μc-Si) PV cells are composed of tiny grains of crystalline silicon, which is used as a power-generating layer. The maximum conversion efficiency is about 10%, but it is expected to be further increased by innovations. Studies on μc-Si are still fairly new and laboratory-scale studies are ongoing.

3.10.2.6 Multi-Junction Concentrated PV Cells

Differing from flat-plate PV technologies, concentrated PV (CPV) systems use lenses and curved mirrors to focus solar radiation onto small but highly efficient PV cells. There are mainly two types of CPV, namely high concentrated PV (HCPV) and low concentrated PV (LCPV). HCPV systems concentrate the sunlight by a factor of between 300 and 1000 times onto a small cell area. The cells used in HCPV are comparatively expensive multi-junction solar cells based on III − V semiconductors obtained by combining group III elements with group V elements from the periodic table. LCPV systems with a concentration ratio up to 100 times use crystalline silicon PV cells (or other cells) and generally single-axis tracking systems (dual-axis tracking can also be used).

CPV systems have several advantages over other PV technologies, such that: They have low-temperature coefficients and high efficiency for direct irradiance. The efficiency of HCPV systems has exceeded 40% in recent studies. CPV systems have waste heat and this heat can be used for different purposes, such as active cooling purposes. They have better and more stable energy production for the daytime due to their tracking systems. CPV systems have greater potential for efficiency increase in the future compared to single-junction flat-plate PV systems.

CPV systems also have several weaknesses, such that: HCPV systems cannot utilize diffuse radiation, although LCPV can only use a fraction of diffuse radiation. They are highly expensive due to the additional costs of optical and tracking systems.

3.10.3 NEW EMERGING PV TECHNOLOGIES

Newly emerging PV technologies aim to open up new potentials for photovoltaics with reduced cost and increased efficiency. The major newly emerging PV technologies are introduced below and their main features are given.

3.10.3.1 Dye-Sensitized PV Cells

A dye-sensitized solar cell (DSC) is an advanced thin-film solar cell. DSCs are based on a semiconductor formed between a photo-sensitized anode (transparent anode) and an electrolyte plus a conductive plate. These solar cells have numerous advantages over their counterparts. Here are some of the advantages of DSCs. The overall efficiency of DSCs is about 11%, so they are suitable mostly for low-density applications. The DSCs are made of low-cost materials and are cheaper than many of the best thin-film cells. Hence the price-to-performance ratio obtained through DSCs is superior to others. The dye used in DCSs can absorb diffused sunlight and fluorescent light. Therefore, DSCs also work under cloudy and low-light conditions with little effect on efficiency. Another advantage of DSCs is their longer life when compared to their counterparts. DSCs do not degrade so much in sunlight, so they do not require frequent replacements.

3.10.3.2 Organic PV Cells

Organic PV (OPV) cells are made from carbon-based molecules, and most of them are polymer solar cells. Although organic cells are cheap and easy to produce, their efficiency and lifespan are considerably lower than other PV technologies. Other disadvantages of OPV cells are their low stability and low strength compared to their inorganic counterparts. However, the interest of researchers in OPVs has increased with an exceeding 10% efficiency in polymer solar cells.

3.10.3.3 Perovskite PV Cells

A perovskite solar cell is a type of solar cell which includes a perovskite-structured compound. Perovskite material offers reduced cost, excellent light absorption, and a direct optical band gap of around 1.5 eV, which results in high device efficiencies. The efficiency level achieved so far is around 20%. As a result, Perovskite PVs indeed hold a promising future for the PV market entry.

3.10.3.4 Quantum-Dot PV Cells

Quantum dots (QDs) are made from electro-optically active materials, of which photon energy can be controlled and their absorption wavelength can be tuned by changing the QD size to deliver multiple electron-hole pairs. With this process, it is possible to operate the QDs cell at around 10% efficiency. QDs-based cells offer two important benefits, i.e., versatility and flexibility in their use. This great advantage makes them excellent candidates for future developments.

3.10.3.5 Graphene PV Cells

Graphene, as a material, is made of a single layer of carbon atoms that are bonded together in hexagons. It is a highly strong, transparent, and flexible material with amazingly good conductive characteristics, but it is not very good at collecting the electrical current produced inside the solar cell. Therefore, scientists are in search of finding applicable ways to adapt graphene for this purpose. Graphene-based PV cells are not presently commercially available. As an abundant and relatively low-priced material, graphene has boundless potential for future applications.

3.10.3.6 Hybrid PV Cells

Recently, one of the most important trends in the PV industry is to produce hybrid silicon cells and several research studies have been focusing on exploring ways of combining different materials to make solar cells with better efficiency, longer life, and reduced costs. For example, an amorphous silicon layer is recently set down on top of single crystal wafers to form a hybrid solar cell. The result has shown an improved performance with a lower temperature coefficient as well as an improved efficiency under indirect light.

3.10.3.7 Bifacial PV Cells

Bifacial PV is a promising technology that increases the electricity generation per unit area of PV modules by using the reflected light absorption from the albedo. In other words, bifacial PV cells can accept the solar radiation from both surfaces to generate electrical energy. Bifacial PV panels can be produced using different technologies. Most of them use monocrystalline cells, but there are also polycrystalline designs. *P*-type polycrystalline silicon, *n*-type monocrystalline silicon, heterojunction monocrystalline, and HIT (hetero-junction with intrinsic thin layer) PV technologies are examples of bifacial PV devices. Bifacial modules come in various designs. For example, some are framed, while others are frameless. Some are dual-glass, and others use back sheets. In bifacial PV cells, a transparent rear can generate additional energy, ranging from 5% to 30% of the energy generated by only the front side. The efficiency of bifacial PV cells increased above 22% depending on the technology type.

The main difference between bifacial PV cells from mono-facial designs is related to modeling. Bifacial solar cells can be modeled electrically as two mono-facial cells in parallel, based on the single-diode or double-diode equivalent circuit. In this context, the equivalent circuit of a typical bifacial PV cell can be obtained as in Figure 3.53. After this stage, all techniques regarding the solution of bifacial PV cells can be performed by following the same procedures as in mono-facial cells.

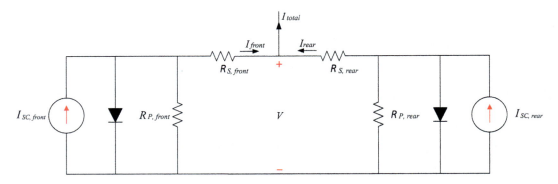

FIGURE 3.53 Typical equivalent circuit for a bifacial PV cell.

3.11 TECHNICAL SPECIFICATIONS OF DIFFERENT PV TECHNOLOGIES

Analyzing the characteristics parameters of different PV technologies and commercial PV modules is important for market characterization and technologies comparison. Therefore, datasheet information of commercially available PV products is an important reference database for end-users and investors. The considered technical specifications can directly be obtained from the PV module datasheet, which includes:

- Technology type (mono/polycrystalline, thin-film, etc.)
- Efficiency (%)
- Peak power and tolerance (W)
- Temperature coefficients (%/°C)
- Nominal operating cell temperature (°C)
- Bypass diodes
- Package type
- Dimensions/weight

Efficiency: Sample efficiency values, varying from 5% to 19%, from commercial PV modules are given in Table 3.3.

Peak Power and Tolerance: The maximum power of a module is defined as the Peak Power or Rated Power (Watts Peak, W_p) under STC. The power tolerance provides a percentage change in its Peak Power. For example, if the peak power of a module is 100 W and the power tolerance is ±5%, this means that the peak power output could be either 105 W or 95 W under rated conditions. Clearly, a high positive power tolerance is advantageous for the end-user.

Temperature Coefficients: The loss in power production due to high operating temperatures is an important loss factor, which is described by the temperature coefficients such as current temperature coefficient (+ % / °C), voltage temperature coefficient (− % / °C), and power temperature

TABLE 3.3
Sample Efficiency Values from Different Commercial PV Technologies

PV Technology	Module Efficiency	Surface Area for 3 kWp System
Monocrystalline Si	14%–19%	20–25 m²
Polycrystalline Si	12%–17%	22–26 m²
Copper-Indium-Selenide	13%–15%	23–30 m²
Amorphous silicon	5%–10%	46–57 m²

Photovoltaic Systems and Photovoltaic Technologies

coefficient (– % / °C). The negative power temperature coefficient provides a measure of the decrease in produced power due to per temperature increase. For example, the output of a PV module decreases more at higher temperatures (over 25°C) when the temperature coefficient is –0.45% / °C, compared to another module with a power temperature coefficient of –0.40% / °C.

Nominal Operating Cell Temperature: Nominal Operating Cell Temperature (NOCT) of a PV module provides an indication associated with the thermal operation, which is defined under 800 W / m², 1 m / s wind speed and 20°C ambient temperature when the module is mounted at an open rack at an inclination of 45° from the horizontal. When selecting PV modules, it is important to choose the PV panel with a low NOCT value. Because the module with a NOCT value of 47°C is expected to heat up more in the warmer conditions compared to a module with a NOCT value of 45°C.

Bypass Diodes: PV modules include bypass diodes to protect the cells from hot spots if the module is partly shaded, either from soiling or a physical obstacle. Hence, bypass diodes in a module allow the non-shaded cells to continue power generation while protecting the shaded cells from being overheated and damaged.

Package Type: Typically, PV modules are composed of a series of interconnected cells that are encapsulated and electrically isolated using an encapsulant material and enclosed between a front glass surface and back plate material. Most modules also have frames that provide a rigid stricture to the module. Hence, packaging information regarding encapsulant, glass, back sheet, and the frame is supplied in the datasheet of the module.

Dimensions/Weight: PV modules are produced in a variety of sizes depending on the technology. For instance, typical c-Si modules with 250 W_p have dimensions: 1665 × 991 × 30 mm (Length × Width × Height). The weight of such modules ranges from 18 to 25 kg.

Table 3.4 shows performance data of different PV technologies based on their datasheet information. The chief parameter here is rated power. This power is DC and we will calculate the actual AC power delivered by the combination of inverter and PV panel.

Example 3.5: Performance Calculation of Mono-Crystal and Poly-Crystal PV Modules

Considering Table 3.4, estimate open-circuit voltage, short-circuit current, and maximum power output of mono-crystal and poly-crystal PV modules under 750 W/m² and ambient temperature of 35°C.

Solution

(a) Mono-crystal PV Module: Using (3.22) with 750 W/m², cell temperature is estimated to be

$$T_c = T_{amb} + \frac{(T_{NOCT} - 20)}{800} \times G = 35 + \frac{(47 - 20)}{800} \times 750 = 60.3125°C$$

From Table 3.4 for the mono-crystal module, the temperature coefficient of I_{SC} and V_{OC} are +0.056%/°C and –0.35%/°C, respectively. I_{SC} and V_{OC} values are 9.08 A and 38.2 V at 25°C. Hence the new I_{SC} and V_{OC} values from (3.25) and (3.24) will be about:

$$I_{SC} = 9.08 \times [1 + (0.00056) \times (60.3125 - 25)] \times \frac{750}{1000} = 6.945 \, A$$

$$V_{OC} = 38.2 \times [1 - 0.0035 \times (60.3125 - 25)] = 33.479 \, V$$

From Table 3.4, maximum power is expected to drop by about –0.45% / °C. Therefore, a 265 W_p PV module at its maximum power will deliver:

$$P_{max} = 265 \times [1 - (0.0045) \times (60.3125 - 25)] \times \frac{750}{1000} = 167.17 \, W$$

TABLE 3.4
Sample Performance Parameter of Commercial PV Panels under STC

Manufacturer	Mitsubishi	SunModule	First-Solar	Solar Frontier
PV Model Material	Mono-crystal	Poly-crystal	CdTe-Thin Film	CIS-Thin Film
Rated Power (W_p)	265	140	122.5	145
Module Efficiency (%)	16	13.7	17	11.8
Voltage at Max. Power (V)	31.7	18	71.5	81
Current at Max. Power (A)	8.38	7.85	1.71	1.80
Open-Circuit Voltage (V)	38.2	22.1	88.7	107
Short-Circuit Current (A)	9.08	8.35	1.85	2.20
NOCT (°C)	47	46	45	47
Power Tolerance	±3%	±5%	−0/+5W	−0/+5W
Temperature Coefficient of I_{SC}	+0.056%/°C	+0.034%/K	+0.04%/°C	+0.01%/K
Temperature Coefficient of V_{OC}	−0.35%/°C	−0.35%/K	−0.28%/°C	−0.30%/K
Temperature Coefficient of P_{max}	−0.45%/°C	−0.45%/K	−0.28%/°C	−0.31%/K
Length (mm)	1625	1508	1200	1257
Height (mm)	1019	680	600	977
Width (mm)	46	34	6.8	31
Weight (kg)	20	11.8	12	20
Product Warranty	10 years	10 years	10 years	5 years
Output Warranty[a]	25 years	25 years	25 years	25 years

[a] 80% of Rated Power.

(b) Poly-crystal PV Module: Using (3.22) with 750 W/m², the cell temperature is estimated to be

$$T_c = T_{amb} + \frac{(T_{NOCT} - 20)}{800} \times G = 35 + \frac{(46 - 20)}{800} \times 750 = 59.375 °C$$

From Table 3.4 for the poly-crystal module, the temperature coefficient of I_{SC} and V_{OC} are +0.034%/K and −0.35%/K, respectively. I_{SC} and V_{OC} values are 8.35 A and 22.1 V at 25°C. Hence the new I_{SC} and V_{OC} values from (3.25) and (3.24) will be about:

$$I_{SC} = 8.35 \times [1 + (0.00034) \times (59.375 - 25)] \times \frac{750}{1000} = 6.335 \, A$$

$$V_{OC} = 22.1 \times [1 - 0.0035 \times (59.375 - 25)] = 19.44 \, V$$

From Table 3.4, maximum power is expected to drop by about −0.45% / K. Therefore, 140 W_p PV module at its maximum power will deliver:

$$P_{max} = 140 \times [1 - (0.0045) \times (59.375 - 25)] \times \frac{750}{1000} = 88.76 \, W$$

Note that the temperature differences in degrees, or Kelvin will be the same. Therefore, there is no complication in doing the calculations in degrees.

Photovoltaic Systems and Photovoltaic Technologies

Example 3.6: Fill Factor Calculation

Considering Table 3.4, calculate the FFs of given PV modules under STC.

Solution

Reading the V_{OC}, I_{SC}, and P_{max} values from Table 3.4, the FF values for the given PV devices are:

$$\text{FF}(\text{Mono}) = \frac{P_{max}}{V_{OC} \cdot I_{SC}} = \frac{265}{38.2 \times 9.08} = 76.40\%$$

$$\text{FF}(\text{Poly}) = \frac{P_{max}}{V_{OC} \cdot I_{SC}} = \frac{140}{22.1 \times 8.35} = 75.86\%$$

$$\text{FF}(\text{CdTe}) = \frac{P_{max}}{V_{OC} \cdot I_{SC}} = \frac{122.5}{88.7 \times 1.85} = 74.65\%$$

$$\text{FF}(\text{CIS}) = \frac{P_{max}}{V_{OC} \cdot I_{SC}} = \frac{145}{107 \times 2.20} = 61.59\%$$

Example 3.7: Calculation of Performance Parameters (Max Power, FF, and Efficiency)

A PV module with an area of $1.98 \, m^2$ has $V_{OC} = 42.7 \, V$, $I_{SC} = 8.64 \, A$, $V_{mp} = 36.9 \, V$, $I_{mp} = 8.12 \, A$ under STC. Calculate the maximum power output, FF, and efficiency of the given PV module.

Solution

All necessary parameters are given; therefore, we have:

$$P_{max} = V_{mp} \times I_{mp} = 36.9 \times 8.12 = 299.62 \, W$$

$$\text{FF} = \frac{P_{max}}{V_{OC} \cdot I_{SC}} = \frac{299.62}{42.7 \times 8.64} = 81.21\%$$

$$\eta = \frac{P_{max}}{\text{Area} \times \text{Irradiance}} = \frac{299.62}{1.98 \cdot 1000} = 15.13\%$$

Example 3.8: Output Power Calculation of Series-Connected PV Modules

Consider a PV array consisting of series-connected eight identical PV modules, each having an output power of 260 W ($V_{mp} = 31.6 \, V$, $I_{mp} = 8.24 \, A$) under STC. Calculate the total power output of the array when I_{mp} of the Module-3 drops by 10% and I_{mp} of Module-7 drops by 25%.

Solution

Normally, the total output power of the PV array having eight identical modules at STC is equal to $260 \times 8 = 2080 \, W$. However, there are drops in the currents of Module-3 and Module-7.

In series-connected photovoltaics, the system current is limited by the component with the lowest current rate. The current on Module-7 drops the most, so the system current drops by 25%. Therefore total output power of the array under derated conditions will be $260 \times 8 \times 0.75 = 1560 \, W$. In this condition, the array $I_{mp} = 8.24 \times 0.75 = 6.18 \, A$ and array voltage $V_{mp} = 8 \times 31.6 = 252.8 \, V$.

Example 3.9: Series, Parallel, and Complex-Connected PV Modules

Consider a PV panel consisting of four identical PV modules with three possible configurations, which are (i) all PV modules are connected in series, (ii) all PV modules are connected in parallel, and (iii) four PV modules are connected as $2S \times 2P$ (2 − series × 2 − parallel). Assume that each PV panel configuration will deliver a 6Ω load as specified in the following Figure 3.54.

  a. Calculate the powers delivered to 6Ω load at STC for each PV panel combination. Explain which combination would be the best.
  b. Calculate the characteristic resistances for each option and find the delivered powers under characteristic resistance loading.

Solution

  a. Firstly, let us draw the *I–V* curves for all three combinations to be able to calculate the delivered powers to the load. Figure 3.55 gives the *I–V* curve of the first combination.
  As can be seen from Figure 3.55, the operating point for the 6 Ω load is (8 A, 48 V). Hence, the delivered power for the option-1 is $8 \times 48 = 384\,W$. Now, Figure 3.56 gives the *I–V* curve of second combination.

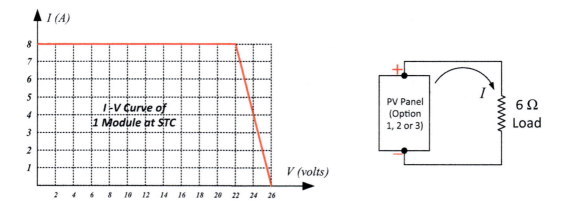

FIGURE 3.54 *I–V* curve of PV module and load supply.

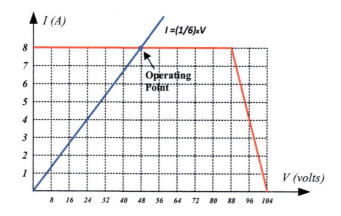

FIGURE 3.55 *I–V* curve of the first combination.

Photovoltaic Systems and Photovoltaic Technologies

As can be seen from Figure 3.56, the operating point (intersection point) can be obtained from the equality of two equations:

$$-8V + 208 = (1/6)V$$

$$\Rightarrow V = 25.47 \text{ volts}$$

and $I = 4.245$ A. Hence, the delivered power for the option-2 is $4.245 \times 25.47 = 108.12$ W.

As can be seen from Figure 3.57, the operating point (intersection point) can be obtained from the equality of two equations:

$$-2V + 104 = (1/6)V$$

$\Rightarrow V = 48$ V and $I = 8$ A. Hence, the delivered power for the option-3 is $8 \times 48 = 384$ W.

It is clear from the result that option-1 and option-3 yield the same result, while option-2 gives the lowest output power. However, the best result was never achieved in any combination.

b. To calculate the characteristic resistance for all three configurations, let us calculate the voltage and current pairs at MPP, which are summarized in the following table (Table 3.5).

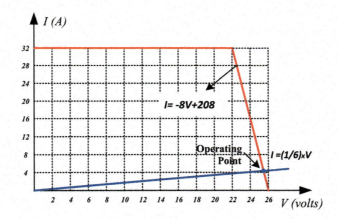

FIGURE 3.56 *I–V* curve of the second combination.

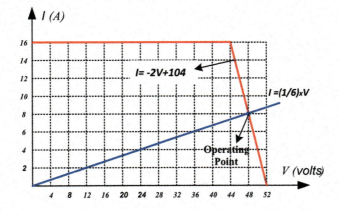

FIGURE 3.57 *I–V* curve of the third combination.

TABLE 3.5
Voltage and Current Pairs at Each MPP

PV Panel Combination	I_{mp}	V_{mp}	$R_0 = \dfrac{V_{mp}}{I_{mp}}$	$P_{mp} = V_{mp} \times I_{mp}$
Four-series (4S)	8 A	88 V	11 Ω	704 W
Four-parallel (4P)	32 A	22 V	0.6875 Ω	704 W
$2S \times 2P$	16 A	44 V	2.75 Ω	704 W

Example 3.10: Solution of the PV Equation with Newton–Raphson Technique

Consider the photovoltaic model, of which the equation is given below:

$$3.16 = 3.4 - 6 \times 10^{-10}\left(e^{38.9 \times V} - 1\right) - \frac{V}{6}$$

Calculate the voltage V at the MPP point by using the Newton–Raphson method in MATLAB®.

Solution

To apply the Newton–Raphson method, we need to write the equation in the format of $f(V) = 0$. Therefore, we get:

$$0.24 - 6 \times 10^{-10}\left(e^{38.9 \times V} - 1\right) - \frac{V}{6} = 0$$

To apply the Newton–Raphson iterations, we need $f'(V)$. Hence, from the above equation, we get:

$$f'(V) = -233.4 \times 10^{-10} \times e^{38.9 \times V} - \frac{1}{6}$$

We can adapt equation (3.61) for the unknown variable V as below:

$$V_{i+1} = V_i - \frac{f(V_i)}{f'(V_i)}$$

Now, we can solve this equation easily by writing a simple MATLAB® code. Note that the initial value for V is taken as 0.6 for the Newton–Raphson iteration.

```
clear all;
clc;
V=0.6;
for m=1:10
    fV=0.24-6*10^(-10)*(exp(38.9*V)-1)-V/6;
    dfV=-233.4*10^(-10)*exp(38.9*V)-1/6;
    V=V-fV/dfV;
    disp(V)
end
```

Once we run the program, the following iteration result will be found. As can be seen from the result that the equation converged to the solution point after the eighth iteration. So, the solution for V is 0.4983.

$$V(1) = 0.5747; V(2) = 0.5503; V(3) = 0.5279; V(4) = 0.5102$$

$$V(5) = 0.5006; V(6) = 0.4984; V(7) = 0.4983; V(8) = 0.4983$$

$$V(9) = 0.4983; V(10) = 0.4983$$

Alternatively, one can easily solve this nonlinear equation using *"fsolve"* command in MATLAB® as below.

```
% Function definition

function F=solar(V)

F=[0.24-6*10^(-10)*(exp(38.9*V)-1)-V/6];

end
  Once the function is determined,
  Write the following in the command window
V0 =[1]; % Initial value for V
fsolve(@solar, V0)
Then the solution will be:       0.4983
```

Example 3.11: Solution of Unknown PV Parameters with MATLAB®

The $I-V$ characteristic of a photovoltaic panel based on a single-diode model is given in Figure 3.58.

Considering the $I-V$ curve, estimate the five unknown parameters I_{PV}, I_0, R_S, R_P and n by using the nonlinear equation solver *"fsolve"* and the Newton–Raphson method. Write a MATLAB® code for these methods and compare the results.

Solution

Firstly, let us write the general equation for the single-diode model to identify the unknown parameters.

$$I = I_{PV} - I_o \left(e^{\frac{q(V+I \cdot R_S)}{nkT_c}} - 1 \right) - \frac{V + I \cdot R_S}{R_P}$$

Since we have five unknown parameters, we need five equations to find the unknowns. The five points can be determined as shown on the $I-V$ curve given below (Figure 3.59).

The first point corresponds to the initial point of the curve near I_{SC} where $I = I_1$ and $V = V_1$. Similarly, the fifth point corresponds to the final point of the curve near V_{OC} where $I = I_5$ and $V = V_5$. The third corresponds to the MPP P_{mp} with $V = V_3$ and $I = I_3$. The second and fourth points can then be selected arbitrarily as indicated in Figure 3.59. The values of the current–voltage pairs at the selected points are as follows.

$I_1 = 5.997; V_1 = 0.075;$
$I_2 = 5.973; V_2 = 0.675;$
$I_3 = 5.733; V_3 = 1.525;$
$I_4 = 5.275; V_4 = 1.60;$
$I_5 = 4.495; V_5 = 1.65;$

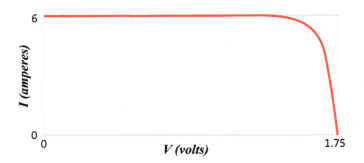

FIGURE 3.58 $I-V$ characteristic of a considered photovoltaic panel.

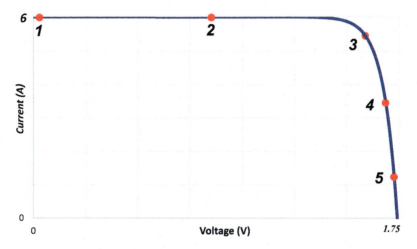

FIGURE 3.59 The five points to find the unknown parameters.

Firstly, let's assign x variables to unknown parameters to work symbolically easier. Therefore,

$$\begin{bmatrix} x_1 \\ x_2 \\ x_3 \\ x_4 \\ x_5 \end{bmatrix} = \begin{bmatrix} I_{PV} \\ I_0 \\ R_S \\ n \\ R_p \end{bmatrix}$$

Then let's rewrite the given equation as a function of x variables:

$$I = x_1 - x_2 \left(e^{\frac{V_T(V+I \cdot x_3)}{x_4}} - 1 \right) - \frac{V + I \cdot x_3}{x_5}$$

where $V_T = \dfrac{q}{kT_c}$ (reverse of thermal voltage, V_t) and equal to 38.9 at 25°C. Now let's write the equations in the format of $F(x) = 0$. Therefore, we get:

$$f_1(x) = x_1 - x_2 \left[e^{\frac{V_T(V_1+I_1 \cdot x_3)}{x_4}} - 1 \right] - \frac{V_1 + I_1 \cdot x_3}{x_5} - I_1 = 0$$

$$f_2(x) = x_1 - x_2 \left[e^{\frac{V_T(V_2+I_2 \cdot x_3)}{x_4}} - 1 \right] - \frac{V_2 + I_2 \cdot x_3}{x_5} - I_2 = 0$$

$$f_3(x) = x_1 - x_2\left[e^{\frac{V_T(V_3+I_3 \cdot x_3)}{x_4}} - 1\right] - \frac{V_3 + I_3 \cdot x_3}{x_5} - I_3 = 0$$

$$f_4(x) = x_1 - x_2\left[e^{\frac{V_T(V_4+I_4 \cdot x_3)}{x_4}} - 1\right] - \frac{V_4 + I_4 \cdot x_3}{x_5} - I_4 = 0$$

$$f_5(x) = x_1 - x_2\left[e^{\frac{V_T(V_5+I_5 \cdot x_3)}{x_4}} - 1\right] - \frac{V_5 + I_5 \cdot x_3}{x_5} - I_5 = 0$$

Firstly, let us solve the above equations using "*fsolve*" function in MATLAB®. For this purpose, the following simple code can be generated to solve the equations for unknowns.

```
% Function definition
function F=solar5p(x)

F=[x(1)-x(2)*(exp(38.9*(0.075+5.997*x(3))/x(4))-1)-
(0.075+5.997*x(3))/x(5)-5.997;

    x(1)-x(2)*(exp(38.9*(0.675+5.973*x(3))/x(4))-1)-
(0.675+5.973*x(3))/x(5)-5.973;

    x(1)-x(2)*(exp(38.9*(1.525+5.733*x(3))/x(4))-1)-
(1.525+5.733*x(3))/x(5)-5.733;

    x(1)-x(2)*(exp(38.9*(1.6+5.275*x(3))/x(4))-1)-
(1.6+5.275*x(3))/x(5)-5.275;

    x(1)-x(2)*(exp(38.9*(1.65+4.495*x(3))/x(4))-1)-
(1.65+4.495*x(3))/x(5)-4.495];

end
```

Once the function is determined,

Write the following in the command window

```
X0 =[5,    0.5,    0.1,    2.5,    5]; % Initial value for unknown parameters
fsolve(@solar5p, X0)
```

Then the solution will be:

$$\begin{bmatrix} x_1 \\ x_2 \\ x_3 \\ x_4 \\ x_5 \end{bmatrix} = \begin{bmatrix} I_{PV} \\ I_0 \\ R_S \\ n \\ R_p \end{bmatrix} \approx \begin{bmatrix} 6 \\ 1.06 \times 10^{-11} \\ 0 \\ 2.5039 \\ 25.0005 \end{bmatrix}$$

Important Note: The correct estimation of initial values in the applications of *"fsolve"* and the Newton–Raphson method is very important. Because it is normally possible to obtain an infinite number of *I–V* curves by changing five parameters. It is, therefore, important to assign initial values close to the solution points of the curve of interest. If the initial values are not selected correctly, the *"fsolve"* function and the Newton–Raphson method may jump onto other $I-V$ curves or will not converge to the desired solution points.

Given the equations, it is possible to write the equation system to solve the problem based on the (3.66). The elements of the Jacobian matrix are found as follows. Considering $f_1(x)$, $f_2(x)$, $f_3(x)$, $f_4(x)$, and $f_5(x)$, we get:

$$\mathbf{J}(1,1) = \frac{\partial f_1}{\partial I_{PV}} = \frac{\partial f_1}{\partial x_1} = 1$$

$$\mathbf{J}(1,2) = \frac{\partial f_1}{\partial I_0} = \frac{\partial f_1}{\partial x_2} = e^{\frac{V_T(V_1 + I_1 \cdot x_3)}{x_4}} - 1$$

$$\mathbf{J}(1,3) = \frac{\partial f_1}{\partial R_S} = \frac{\partial f_1}{\partial x_3} = -\left[\frac{V_T \cdot I_1 \cdot x_2}{x_4}\right] e^{\frac{V_T(V_1 + I_1 \cdot x_3)}{x_4}} - \frac{I_1}{x_5}$$

$$\mathbf{J}(1,4) = \frac{\partial f_1}{\partial n} = \frac{\partial f_1}{\partial x_4} = -\left[\frac{V_T \cdot x_2 \cdot (V_1 + I_1 \cdot x_3)}{x_4^2}\right] e^{\frac{V_T(V_1 + I_1 \cdot x_3)}{x_4}}$$

$$\mathbf{J}(1,5) = \frac{\partial f_1}{\partial R_P} = \frac{\partial f_1}{\partial x_5} = \frac{V_1 + I_1 \cdot x_3}{x_5^2}$$

Similarly, the Jacobian matrix can be written as follows.

$$\mathbf{J} = \begin{bmatrix} \left(e^{\frac{V_T(V_1+I_1x_3)}{x_4}} - 1\right) & \left(-\frac{V_T I_1 x_2 e^{\frac{V_T(V_1+I_1x_3)}{x_4}}}{x_4} - \frac{I_1}{x_5}\right) & -\frac{V_T x_2 (V_1+I_1 x_3)}{x_4^2} e^{\frac{V_T(V_1+I_1x_3)}{x_4}} & \frac{V_1+I_1 x_3}{x_5^2} \\ \left(e^{\frac{V_T(V_1+I_1x_3)}{x_4}} - 1\right) & \left(-\frac{V_T I_1 x_2 e^{\frac{V_T(V_1+I_1x_3)}{x_4}}}{x_4} - \frac{I_1}{x_5}\right) & -\frac{V_T x_2 (V_1+I_1 x_3)}{x_4^2} e^{\frac{V_T(V_1+I_1x_3)}{x_4}} & \frac{V_1+I_1 x_3}{x_5^2} \\ \left(e^{\frac{V_T(V_1+I_1x_3)}{x_4}} - 1\right) & \left(-\frac{V_T I_1 x_2 e^{\frac{V_T(V_1+I_1x_3)}{x_4}}}{x_4} - \frac{I_1}{x_5}\right) & -\frac{V_T x_2 (V_1+I_1 x_3)}{x_4^2} e^{\frac{V_T(V_1+I_1x_3)}{x_4}} & \frac{V_1+I_1 x_3}{x_5^2} \\ \left(e^{\frac{V_T(V_1+I_1x_3)}{x_4}} - 1\right) & \left(-\frac{V_T I_1 x_2 e^{\frac{V_T(V_1+I_1x_3)}{x_4}}}{x_4} - \frac{I_1}{x_5}\right) & -\frac{V_T x_2 (V_1+I_1 x_3)}{x_4^2} e^{\frac{V_T(V_1+I_1x_3)}{x_4}} & \frac{V_1+I_1 x_3}{x_5^2} \end{bmatrix}$$

Now, we can solve the Newton–Raphson method for the five unknown parameters using (3.66). To solve the system, the following MATLAB® code is generated. Note that the initial values for the Newton–Raphson iteration.

```
clear all;
clc;
disp('Newton-Raphson Method for 5 parameter estimation of PV
  Module:')

%Defining the known parameter
I1= 5.997; V1= 0.075;
I2= 5.973; V2= 0.675;
I3= 5.733; V3= 1.525;
I4= 5.275; V4= 1.60;
I5= 4.495; V5= 1.65;

VT=38.9;

% Initializing the unknowns
x1=5; x2=1*10^(-12); x3=0.1; x4=2.5; x5=25;

X=[x1; x2; x3; x4; x5];
for m=1:4
    % Calculation of Function Values

    f1=x1-x2*(exp(VT*(V1+I1*x3)/x4)-1)-(V1+I1*x3)/x5-I1;
    f2=x1-x2*(exp(VT*(V2+I2*x3)/x4)-1)-(V2+I2*x3)/x5-I2;
    f3=x1-x2*(exp(VT*(V3+I3*x3)/x4)-1)-(V3+I3*x3)/x5-I3;
    f4=x1-x2*(exp(VT*(V4+I4*x3)/x4)-1)-(V4+I4*x3)/x5-I4;
    f5=x1-x2*(exp(VT*(V5+I5*x3)/x4)-1)-(V5+I5*x3)/x5-I5;
    % Calculation of elements of Jacobian matrix

    J11= 1;
    J12= exp(VT*(V1+I1*x3)/x4)-1;
```

```
J13= -(VT*x2*I1/x4)*exp(VT*(V1+I1*x3)/x4)-I1/x5;
J14= -(VT*x2*(V1+I1*x3)/(x4)^2)*exp(VT*(V1+I1*x3)/x4);
J15= (V1+I1*x3)/(x5)^2;


J21= 1;
J22= exp(VT*(V2+I2*x3)/x4)-1;
J23= -(VT*x2*I2/x4)*exp(VT*(V2+I2*x3)/x4)-I2/x5;
J14= -(VT*x2*(V1+I1*x3)/(x4)^2)*exp(VT*(V1+I1*x3)/x4);
J15= (V1+I1*x3)/(x5)^2;


J21= 1;
J22= exp(VT*(V2+I2*x3)/x4)-1;
J23= -(VT*x2*I2/x4)*exp(VT*(V2+I2*x3)/x4)-I2/x5;
J24= -(VT*x2*(V2+I2*x3)/(x4)^2)*exp(VT*(V2+I2*x3)/x4);
J25= (V2+I2*x3)/(x5)^2;


J31= 1;
J32= exp(VT*(V3+I3*x3)/x4)-1;
J33= -(VT*x2*I3/x4)*exp(VT*(V3+I3*x3)/x4)-I3/x5;
J34= -(VT*x2*(V3+I3*x3)/(x4)^2)*exp(VT*(V3+I3*x3)/x4);
J35= (V3+I3*x3)/(x5)^2;


J41= 1;
J42= exp(VT*(V4+I4*x3)/x4)-1;
J43= -(VT*x2*I4/x4)*exp(VT*(V4+I4*x3)/x4)-I4/x5;
J44= -(VT*x2*(V4+I4*x3)/(x4)^2)*exp(VT*(V4+I4*x3)/x4);
J45= (V4+I4*x3)/(x5)^2;


J51= 1;
J52= exp(VT*(V5+I5*x3)/x4)-1;
J53= -(VT*x2*I5/x4)*exp(VT*(V5+I5*x3)/x4)-I5/x5;
```

```
J54= -(VT*x2*(V5+I5*x3)/(x4)^2)*exp(VT*(V5+I5*x3)/x4);

J55= (V5+I5*x3)/(x5)^2;


J=[J11 J12 J13 J14 J15; J21 J22 J23 J24 J25; J31 J32 J33 J34
J35;

    J41  J42 J43 J44 J45; J51 J52 J53 J54 J55];


%Updation of unknown variables


X=X-inv(J)*[f1; f2; f3; f4; f5];

x1=X(1);    x2=X(2);    x3=X(3);    x4=X(4);    x5=X(5);

disp(X)


end
```

Once we run the MATLAB® code, the following solution will be obtained from the Newton–Raphson method:

$$\begin{bmatrix} x_1 \\ x_2 \\ x_3 \\ x_4 \\ x_5 \end{bmatrix} = \begin{bmatrix} I_{PV} \\ I_0 \\ R_S \\ n \\ R_p \end{bmatrix} \approx \begin{bmatrix} 6.0239 \\ 1.38 \times 10^{-11} \\ 0.0992 \\ 2.483 \\ 24.9171 \end{bmatrix}$$

As can be seen, similar results are obtained from both methods.

3.12 PROBLEMS

P3.1. Bandgap energies of some photovoltaic materials at 300 K temperatures are given below (Table 3.6).
Calculate the maximum wavelength of solar energy to create hole-electron pairs in the materials given. Note that 1 eV equal to 1.6×10^{-19} J.

P3.2. Consider the *p-n* junction diode with the following parameters (Table 3.7).
Calculate the reverse saturation currents at 25°C and 100°C when the diode is carrying 5 A.

P3.3. Consider an ideal photovoltaic model for a 144 cm² PV cell with a reverse saturation current of $I_o = 1 \times 10^{-12}$ A/cm². Under 1000 W/m² solar radiation, the PV cell has an open-circuit voltage of 0.7 V at room temperature (25°C).
(a) Find the short-circuit current (I_{SC}) at 1000 W/m² solar radiation.
(b) Calculate the load current, delivered power, when the cell output voltage is 0.6 V.

TABLE 3.6
Bandgap Energy of Different PV Materials

Material	Symbol	Band Gap Energy
Gallium Arsenide	GaAs	1.43 eV
Cadmium Telluride	CdTe	1.5 eV
Gallium Phosphide	GaP	2.26 eV

TABLE 3.7
The Parameters of the p-n Junction Diode

Parameter	Junction Temperature (25°C)	Junction Temperature (100°C)
Diode Voltage	25×10^{-2} V	10×10^{-2} V

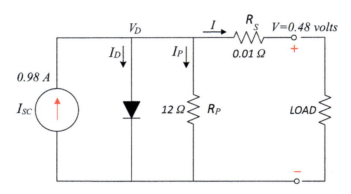

FIGURE 3.60 Equivalent circuit of the PV cell.

P3.4. A PV module with an ideality factor of 1.0245 has a circuit current of 1.75 A at STC (25°C and 1 kW/m²). Assume that its reverse saturation current is 4.41×10^{-8} A, parallel resistance is 60 Ω, and series resistance is 0.2 Ω. Using the simple approximation method, find the voltage, current, and power at MPP.

P3.5. Consider the equivalent circuit of a 0.0045 m² PV cell with a reverse saturation current of $I_o = 1 \times 10^{-9}$ given below (Figure 3.60). Calculate the cell efficiency at STC (25°C and 1 kW/m²).

P3.6.

(a) Considering Table 3.4, estimate the maximum power output of CdTe and CIS PV modules under 1000 W/m² at ambient temperatures of 0°C and 35°C, respectively.

(b) Assuming that the NOCT values are 5°C greater than the Table 3.4 values, calculate the output power of the respective PV modules under the conditions given in (a).

P3.7. A PV module consists of 48 series-connected PV cells, each having an area of 0.045 m². Each PV cell has $V_{OC} = 0.86$ V, $I_{SC} = 8.5$ A, $V_{mp} = 0.75$ V, $I_{mp} = 8$ A under STC. Calculate the maximum power output, FF, and efficiency of the given PV module.

P3.8. Consider a PV panel consisting of two identical PV modules with two possible configurations, which are (i) all PV modules are connected in series and (ii) all PV modules are connected in parallel. Assume that each PV panel configuration will deliver a 10 Ω load as specified in Figure 3.61.

Photovoltaic Systems and Photovoltaic Technologies

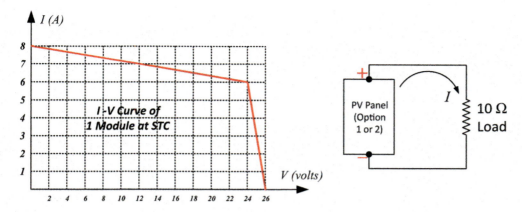

FIGURE 3.61 *I–V* curve of PV module and load supply.

(a) Calculate the powers delivered to 10 Ω load at STC for each PV panel combination.
(b) Calculate the characteristic resistances for each option and find the delivered powers under characteristic resistance loading.

P3.9. Consider the photovoltaic model, of which the equation is given below:

$$f_1(x) = 6 - 1.2 \times 10^{-11} \left[e^{\frac{38.9 \times (1.525 + 5.733 \cdot x)}{2.5}} - 1 \right] - \frac{1.525 + 5.733 \cdot x}{25} - 5.733 = 0$$

Calculate the unknown parameter *x* by using the nonlinear equation solver "*fsolve*" and the Newton–Raphson method. Create a MATLAB® script for these methods and compare the results.

BIBLIOGRAPHY

1. A. Luque and S. Hegedus, *"Handbook of Photovoltaic Science and Engineering"*, John Wiley & Sons, 2002.
2. Fraunhofer Institute for Solar Energy Systems, "Photovoltaics Report", www.ise.fraunhofer.de, February 2018.
3. J. J. Liou and J. S. Yuan, *"Semiconductor Device Physics and Simulation"*, Plenum Press, New York, 1998.
4. H. L. Tsai, C. S. Tu, and Y. J. Su, "Development of generalized photovoltaic model using MATLAB/SIMULINK," in *Proceedings of the World Congress on Engineering and Computer Science*, San Francisco, USA, 2008, vol. 2008, pp. 1–6.
5. N. Sheikh, T. Daim, D. F. Kocaoglu, "Use of multiple perspectives and decision modeling for PV technology assessment", Green Energy and Technology, January 2011.
6. M. A. Green, K. Emery, Y. Hishikawa, W. Warta, and E. D. Dunlop, "Solar cell efficiency tables (Version 49)", Progress in Photovoltaics: Research and Applications, 2017.
7. J. Perko, M. Nidarec, and T. Danijel, "Comparative analysis of electricity production from different technologies of PV modules", in *10th International Conference on Sustainable Energy and Environmental Protection*, 2017, Bled, Slovenia.
8. T. C. Wang and S. Y. Tsai, "Solar panel supplier selection for the photovoltaic system design by using fuzzy multi-criteria decision making (MCDM) approaches", *Energies*, vol. 11, no. 8, p. 1989, 2018.
9. S. Chakraborty, R. Kumar, A. K. Haldkar, and S. Ranjan, "Mathematical method to find best-suited PV technology for different climatic zones of India", *Int. J. Energy Environ. Eng.*, vol. 8, pp. 153–166, 2017.
10. L. R. D. Reis, J. R. Camacho, D. F. Novacki, "The Newton Raphson method in the extraction of parameters of PV modules", in *International Conference on Renewable Energies and Power Quality*, Malaga (Spain), April 2017.

11. J. A. del Cueto, "Comparison of energy production and performance from flat-plate photovoltaic module technologies deployed at fixed tilt comparison of energy production and performance from flat-plate photovoltaic module technologies deployed at fixed tilt", in *29th IEEE PV Specialists Conference*, New Orleans, Louisiana, May 2002.
12. A. M. Bagher, M. M. A. Vahid, and M. Mohsen, "Types of solar cells and application", *Am. J. Opt. Photonics*, vol. 3, no. 5, pp. 94–113, 2015.
13. R. Djamila and M. Ernest, *"Optimization of Photovoltaic Power Systems: Modelization, Simulation and Control"*, Springer, Belgium, 2012.
14. A. M. Humada, M. Hojabri, S. Mekhilef, and H. M. Hamada, "Solar cell parameter extraction based on single and double-diode models: A review", *Renew. Sustain. Energy Rev.*, vol. 56, pp. 494–509, 2016.
15. P. P. Biswas, P. N. Suganthan, G. Wu, and G. A. J. Amaratunga, "Parameter estimation of solar cells using datasheet information with the application of an adaptive differential evolution algorithm", *Renew. Energy*, vol. 132, pp. 425–438, 2018.
16. V. Jha and U. S. Triar "A novel approach for evaluation of parameters of photovoltaic modules", *Int. J. Appl. Eng. Res.*, vol. 12, no. 21, pp. 11167–11178, 2017.
17. G. Dzimano, "Modeling of Photovoltaic Systems", Master Thesis, The Ohio State University, 2008.
18. H. Tian, F. Mancilla-David, K. Ellis, E. Muljadi, and P. Jenkins, "A Detailed Performance Model for Photovoltaic Systems", NREL, July 2012.
19. A. Hussein, "A simple approach to extract the unknown parameters of PV modules", *Turk. J. Electr. Eng. Comput. Sci.*, 25, pp. 4431–4444, 2017.
20. A. M. Dizqaha, A. Maheria, and K. Busawon, "An accurate method for the PV model identification based on a genetic algorithm and the interior-point method", *Renew. Energy*, vol. 72, pp. 212–222, 2014.
21. R. Chenni, M. Makhlouf, T. Kerbache, and A. Bouzid, "A detailed modeling method for photovoltaic cells", *Energy*, vol. 32, no. 9, pp. 1724–1730, 2007.
22. M. G. Villalva, J. R. Gazoli, and E. R. Filho, "Comprehensive approach to modeling and simulation of photovoltaic arrays", *IEEE Trans. Power Electron.*, vol. 24, no. 5, pp. 1198–1208, 2009.
23. B. A. Ikyo, A. Johnson, and F. Gbaorun, "Analytical solution to nonlinear photovoltaic diode equation", *Mind Sourcing*, vol. 2017, 6 pages.
24. D. Bonkoungou, Z. Koalaga, and D. Njomo, "Modelling and simulation of photovoltaic module considering single-diode equivalent circuit model in MATLAB", *Int. J. Emerg. Technol. Adv. Eng.*, vol. 3, no. 3, pp. 493–502, 2013.

4 Effect of Environmental Conditions on the Performance of Photovoltaic Systems

4.1 INTRODUCTION

The photovoltaic output power depends on several different aspects such as type of PV material, module orientation, incident solar radiation, and its spectral range, geographical location, cell operating temperature, parasitic resistances, and environmental conditions including cloudiness, sky clearness, rainfall, snowfall, icing, humid, shading, and soiling. Besides, electrical system losses as well as efficiency of electrical system components such as inverter efficiency and power transmission losses, etc., also affect the output power of PV modules. This chapter only aims to cover the effect of environmental conditions on the performance of solar PV systems. Some of the remaining factors affecting PV system performance have been examined in the above chapters and others will be examined in the following chapters.

4.2 FACTORS AFFECTING PV MODULE/PV CELL OPERATING TEMPERATURE

To estimate the energy generation of PV modules, it is crucial to predict the cell temperature in a PV module. Ambient temperature, wind speed, wind direction, total irradiance, and relative humidity are the key parameters affecting the PV cell operating temperature. This section presents the most common mathematical models to predict the module temperature as a function of ambient temperature, wind speed, wind direction, and relative humidity.

The most general approach in the literature to estimate the PV module operating temperature is based on a mathematical model including five parameters of ambient temperature, wind speed, wind direction, global irradiance, and relative humidity. This equation to predict the PV module temperature (T_m) is given in (4.1).

$$T_m = c_1 \cdot T_{amb} + c_2 \cdot G + c_3 \cdot v + c_4 \cdot W_d + c_5 \cdot \text{RH} + C \tag{4.1}$$

where C is a constant and c_1, c_2, c_3, c_4 and c_5 are the respective coefficients regarding the ambient temperature (T_{amb}), global irradiance (G), wind speed (v), wind direction (W_d), and relative humidity (RH). All these coefficients vary depending on the PV technology type, and Table 4.1 shows the five coefficients for the most common PV technologies. These coefficients are nothing but coefficients and may change according to manufacturers and product versions.

In many studies in the literature, the PV module temperature equation with three parameters is used instead of (4.1), which is a function of ambient temperature, irradiance, and wind speed. The equation to predict the PV module temperature is given as follows:

$$T_m = c_1 \cdot T_{amb} + c_2 \cdot G + c_3 \cdot v + C \tag{4.2}$$

DOI: 10.1201/9781003415572-4

TABLE 4.1
Five Input Parameters for the Operating Temperature of PV Modules

PV Technology	T_{amb} c_1	Irradiance $c_2 \left[\frac{°C \cdot m^2}{W} \right]$	Wind Speed $c_3 \left[\frac{°C \cdot s}{m} \right]$	Wind Direction $c_4 \left[\frac{°C}{deg} \right]$	Humidity $c_5 \left[\frac{°C}{RH\%} \right]$	Constant $C [°C]$
Mono c-Si	0.961±0.008	0.029±0.003	−1.457±0.1	0.001±0.003	0.109±0.06	2.3±2.2
Poly c-Si	0.954±0.009	0.030±0.002	−1.629±0.15	−0.005±0.003	0.088±0.05	4±1.5
a-Si	0.964±0.02	0.026±0.003	−1.406±0.25	−0.002±0.003	0.01±0.080	2±4
CIS	0.969±0.02	0.029±0.001	−1.451±0.18	−0.001±0.005	0.055±0.065	3.7±2.4
CdTe	0.964±0.04	0.030±0.003	−1.488±0.24	−0.005±0.001	0.01±0.026	5±2.7

Note: The coefficient values are within the range of up to plus or minus tolerance values.

TABLE 4.2
Three Input Parameters for the Operating Temperature of PV Modules

PV Technology	T_{amb} c_1	Irradiance $c_2 \left[\frac{°C \cdot m^2}{W} \right]$	Wind Speed $c_3 \left[\frac{°C \cdot s}{m} \right]$	Constant $C [°C]$
Mono c-Si	0.942±0.020	0.028±0.003	−1.509±0.09	4±0.5
Poly c-Si	0.930±0.018	0.030±0.002	−1.666±0.1	5.1±0.5
a-Si	0.943±0.015	0.027±0.004	−1.450±0.17	4±1.5
CIS	0.960±0.02	0.029±0.002	−1.517±0.15	4.1±0.7
CdTe	0.953±0.022	0.030±0.003	−1.658±0.12	5±1.1

Note: The coefficient values are within the range of up to plus or minus tolerance values.

where C is a constant and $c_1, c_2,$ and c_3 are the respective coefficients regarding the ambient temperature (T_{amb}), global irradiance (G), and wind speed (v). All these coefficients vary depending on the PV technology type and Table 4.2 shows the three coefficients for the most common PV technologies.

Based on the results tested in the literature, it can be concluded that the derived coefficients in Table 4.2 for ambient temperature, wind speed, and irradiance are fairly independent of site location and there are slight differences between PV technology types. For example, it is possible to write the following equation for the mono c-Si PV module.

$$T_m = 0.942 \cdot T_{amb} + 0.028 \cdot G - 1.509 \cdot v + 4 \qquad (4.3)$$

Referring to Table 4.2 values, PV module temperature equations can similarly be written for other PV technologies as well.

In literature, there are many different approaches to predicting PV cell temperature, ranging from simpler regression models to more complex models. The simplest approach is the model considering the ambient temperature and irradiance. Thus, the PV cell temperature:

$$T_c = T_{amb} + 0.035 \cdot G \qquad (4.4)$$

TABLE 4.3
Empirical Coefficients Used in Sandia Thermal Model

Module Type	Mounting Configuration	a	b	ΔT
Glass/cell/glass	Open rack	−3.47	−0.0594	3
Glass/cell/glass	Close roof mount	−2.98	−0.0471	1
Glass/cell/polymer sheet	Open rack	−3.56	−0.0750	3
Glass/cell/polymer sheet	Insulated back	−2.81	−0.0455	0
Polymer/thin-film/steel	Open rack	−3.58	−0.113	3

As stated in Chapter 3, one of the most common formulas for estimating PV cell temperature is the approach specified in IEC 61215 standard. The cell temperature according to IEC 61215 takes into account NOCT and global irradiance G and it is given by (3.22).

Another important model developed by Sandia National Laboratories considers the empirically obtained exponential function, which is based on solar irradiance, ambient air temperature, and wind speed measured at a standard 10 m height. Hence, the Sandia thermal model is given as shown in (4.5) and (4.6).

$$T_{mb} = G \cdot \left(e^{a+b \cdot v}\right) + T_{amb} \quad (4.5)$$

$$T_c = T_{mb} + \frac{G}{1000} \cdot \Delta T \quad (4.6)$$

where G is the incident solar irradiance measured on the module surface (W/m²), v is the wind speed measured at standard 10 m height (m/s), T_c is the cell temperature (°C), ΔT is the temperature difference measured between the cell and the module back surface at an irradiance level of 1000 W/m² (typically 2°C–3°C for flat-plate modules on an open rack). In the above equations, a is an empirically determined coefficient establishing the upper limit for module temperature at low wind speeds and high solar irradiance, and the coefficient b is an empirically determined coefficient establishing the rate at which module temperature drops as wind speed increases. Table 4.3 represents these empirical coefficients used in Sandia thermal model (4.5) and (4.6).

Another significant PV module temperature model based on a simple heat transfer concept is developed by David Faiman (2008). The temperature model considers the solar irradiance, ambient temperature (T_{amb}, °C) and wind speed (v, m/s) to predict the module temperature (T_m, °C), which is given by (4.7).

$$T_m = T_{amb} + \frac{G}{u_0 + u_1 \cdot v} \quad (4.7)$$

where G is the global irradiance on the module plane in (W/m²) and u_0 is the constant heat transfer coefficient in (W/m² °C), which describes the effect of the radiation on the module temperature. The coefficient u_1 is the convective heat transfer component in (Ws/m³ °C), which describes the cooling effect of the wind. The two coefficients u_0 and u_1 for polycrystalline silicon modules are given in Table 4.4.

The two coefficients u_0 and u_1 can also be given with the dimensions of $\left[\text{W/m}^2\text{K}\right]$ and $\left[\text{Ws/m}^3\text{K}\right]$. In this case, Table 4.5 gives the coefficient values for thin-film and polycrystalline technologies.

According to tested results in the literature, Faiman parameters in (4.7) are site-specific and more sensitive to climatic changes and module types.

TABLE 4.4
Values of Coefficients u_0 and u_1 Used in Equation (4.7)

Module Type	u_0 [w/m²C]	u_1 [w/m²C]	Site
Poly c-Si	35.7	8.22	Cologne, DE
Poly c-Si	41.9	3.95	Ancona, IT
Poly c-Si	32.1	6.08	Tempe, US
Poly c-Si	39.7	3.06	Thuwal, SA
Poly c-Si	30.1	4.75	Chennai, IN

TABLE 4.5
Values of Coefficients u_0 and u_1 Considering the Temperature in K

Module Type	u_0 [w/m²K]	u_1 [w/m²K]	Site
CdTe	21.68	3.27	Not given
CIS	19.63	3.68	Not given
a-Si/μc-Si	24.61	5.30	Not given
HIT	22.91	4.65	Not given
Poly c-Si	23.58	4.52	Not given
Mono c-Si	30.02	6.28	Not given

A popular model based on the Faiman approach for module temperature is the one used in the PVsyst software. Hence the cell temperature (T_c) according to the PVsyst model is given below.

$$T_c = T_{amb} + \frac{G}{u_0 + u_1 \cdot v} \alpha_m (1 - \eta_m) \qquad (4.8)$$

where α_m is the absorption coefficient of the module (typical value is 0.9) and η_m is the module efficiency (typically 0.08 – 0.2). Other parameters in (4.8) are the same as that of (4.7).

There are tens of different correlations used in the literature to predict cell temperature. However, almost all of them are based on the approaches outlined above. Therefore, the approaches to cell temperature models given in this section are considered to be sufficient.

Example 4.1: Temperature Effect on PV Performance (Sandia Model)

a. It is known that the coefficients u_0 and u_1 used in PVsyst thermal model is site-specific and technology-dependent parameters. For this reason, the coefficients u_0 and u_1 in the cell temperature prediction of PVsyst model need to be measured or calculated for the considered site in order to be more precise. Now assume that one has the Sandia parameters and wants to test the cell temperature using PVsyst model. Calculate the u_0 and u_1 coefficients as a function of Sandia model parameters by matching the Sandia thermal model to the PVsyst model.

b. Consider a 140 W poly-crystal PV module with a 16% efficiency rate and (−0.45% / °C) temperature coefficient of P_{max}. Estimate the maximum power output of the module under 750 W/m² and ambient temperature of 35°C based on the Sandia and PVsyst

Effect of Environmental Conditions on the Performance of Photovoltaic Systems

thermal models. Use the Table 4.3 values of glass-cell-backsheet module on an open rack mounting for the Sandia model. In addition, use an absorption coefficient, $\alpha_m = 0.9$ and wind speed $v = 2$ m/s.

Solution

a. Firstly, let us try to arrange the Sandia and PVsyst models in the same form to be able to mutually equalize the parameters. From PVsyst model given by equation (4.8), we have:

$$T_c = T_{amb} + G \frac{\alpha_m (1-\eta_m)}{u_0 \left(1 + \frac{u_1}{u_0} \cdot v\right)}$$

$$T_c = T_{amb} + G \frac{\alpha_m (1-\eta_m)}{u_0} \cdot \frac{1}{\left(1 + \frac{u_1}{u_0} \cdot v\right)}$$

To equate this equation with the Sandia model, let's try to write the equation term $\dfrac{1}{\left(1 + \dfrac{u_1}{u_0} \cdot v\right)}$ in the form of $(a + b \cdot v)$. To process the math easy, let's assign the variable k to $\left(\dfrac{u_1}{u_0}\right)$. Hence,

$$k = \frac{u_1}{u_0} \Rightarrow \frac{1}{1+kv} = a + bv$$

in the above equation, the parameters a and b are unknown. So, we need to find a and b. By re-writing the above equation, we have:

$$bkv^2 + bv + a = -akv + 1$$

We want to write the above equation in linear form. Hence, we have the following equivalences.

$$a = 1$$

$$b = -k = -\frac{u_1}{u_0}$$

$$b \cdot k = -\left(\frac{u_1}{u_0}\right)^2 \approx 0, \text{ for } u_0 \gg u_1$$

Therefore, the equation term $\dfrac{1}{\left(1 + \dfrac{u_1}{u_0} \cdot v\right)}$ can be approximated as:

$$\frac{1}{\left(1 + \frac{u_1}{u_0} \cdot v\right)} \approx 1 - \frac{u_1}{u_0} \cdot v, \text{ for } u_0 \gg u_1$$

As a result, PVsyst model can be approximated as below:

$$T_c \approx T_{amb} + G \frac{\alpha_m(1-\eta_m)}{u_0} \cdot \left(1 - \frac{u_1}{u_0} \cdot v\right)$$

$$T_c \approx T_{amb} + G \cdot \left[\frac{\alpha_m(1-\eta_m)}{u_0} - \frac{\alpha_m(1-\eta_m)u_1}{u_0^2} \cdot v\right]$$

Similarly, the Sandia cell temperature model (equations 4.5 and 4.6) can also be approximated as below:

$$T_c = T_{amb} + \frac{G}{1000} \cdot \Delta T + G \cdot \left(e^{a+bv}\right)$$

$$T_c \approx T_{amb} + G \cdot \left[\frac{\Delta T}{1000} + e^a(1+bv)\right]$$

$$T_c \approx T_{amb} + G \cdot \left[\frac{\Delta T}{1000} + e^a + e^a bv\right]$$

Equating mutual terms gives the following equations, which relate the PVsyst model parameters ($\alpha_m, \eta_m, u_0, u_1$) to the Sandia model parameters ($a, b, \Delta T$):

$$T_{amb} + G \cdot \left[\frac{\alpha_m(1-\eta_m)}{u_0} - \frac{\alpha_m(1-\eta_m)u_1}{u_0^2} \cdot v\right] = T_{amb} + G \cdot \left[\frac{\Delta T}{1000} + e^a + e^a bv\right]$$

$$\frac{\alpha_m(1-\eta_m)}{u_0} = \frac{\Delta T}{1000} + e^a \Rightarrow u_0 = \frac{\alpha_m(1-\eta_m)}{\frac{\Delta T}{1000} + e^a}$$

$$-\frac{\alpha_m(1-\eta_m)u_1}{u_0^2} = e^a b \Rightarrow u_1 = \frac{e^a b \alpha_m(1-\eta_m)}{\frac{\Delta T}{1000} + e^a}$$

b. All empirical coefficients used in Sandia thermal model are given ($a = -3.56$, $b = -0.0750$, $\Delta T = 3$). In addition, the parameters ($\alpha_m = 0.9, \eta_m = 0.16$) are given for PVsyst model. The parameters (u_0, u_1) can then be calculated from the expression obtained in (a).

$$u_0 = \frac{\alpha_m(1-\eta_m)}{\frac{\Delta T}{1000} + e^a} = \frac{0.9(1-0.16)}{\frac{3}{1000} + e^{(-3.56)}} = 24.05$$

$$u_1 = \frac{e^a b \alpha_m(1-\eta_m)}{\frac{\Delta T}{1000} + e^a} = \frac{e^{(-3.56)}(-0.0750)(0.9)(1-0.16)}{\frac{3}{1000} + e^{(-3.56)}} = -0.0512$$

Now we can calculate the cell temperatures based on the Sandia and PVsyst models. From the Sandia thermal model,

$$T_c = T_{amb} + \frac{G}{1000} \cdot \Delta T + G \cdot \left(e^{a+bv}\right) = 35 + \frac{750}{1000} \cdot 3 + 750 \cdot \left(e^{-3.56-0.0750 \times 2}\right) = 55.6025 \text{ °C}$$

From the given data, the maximum power is expected to drop by about $-0.45\%/°C$. Therefore 140 W_p PV module at its maximum power will deliver:

$$P_{max} = 140 \times [1-(0.0045) \times (55.6025-25)] \times \frac{750}{1000} = 90.54 \text{ W}$$

From the PVsyst thermal model, we have:

$$T_c = T_{amb} + G \frac{\alpha_m(1-\eta_m)}{u_0\left(1+\frac{u_1}{u_0}\cdot v\right)} = 35 + 750 \frac{0.9 \times (1-0.16)}{24.05\left(1-\frac{0.0512}{24.05}\cdot 2\right)} = 58.67°C$$

Consequently, 140 W_p module based on the PVsyst model will deliver at its maximum power:

$$P_{max} = 140 \times [1-(0.0045) \times (58.67-25)] \times \frac{750}{1000} = 89.09 \text{ W}$$

Example 4.2: Temperature Effect Calculations on CdTe and Poly c-Si PV Modules

Consider the *Poly c-Si* and *CdTe* PV modules with the following parameters (Tables 4.6 and 4.7).

Calculate the cell temperatures and output power of the PV modules according to the six different methods described above (simple two-parameter method, three-parameter method, NOCT method, Faiman method, Sandia model, PVsyst model).

Solution

All necessary parameters to predict cell temperature and output power under specified conditions (ambient temperature, solar irradiance, wind speed) are given. Let's first calculate the cell temperatures for each model.

First Model (Simple approach with two parameters): From equation (4.4), we have,

TABLE 4.6
Parameters of a Poly c-Si PV Module

	P_{max} [W]	η_m [%]	P_{max} Coefficient	NOCT [°C]	T_{amb} [°C]	G [W/m²]	v [m/s]	Sandia Parameters $a/b/\Delta T$	α_m
Poly c-Si	120	13.7	−0.45 %/°C	47	30	600	2	−3.56 / −0.075 / 3	0.9
CdTe	120	17	−0.28 %/°C	45	30	600	2	−3.56 / −0.075 / 3	0.9

TABLE 4.7
Parameters of a CdTe PV Module

| | Three Parameters Model | | | |
PV Type	c_1	c_2	c_3	C
Poly c-Si	0.930	0.030	−1.666	5.1
CdTe	0.953	0.030	−1.658	5

$$T_c = T_{amb} + 0.035 \cdot G = 30 + 0.035 \cdot 600 = 51°C$$

The result obtained by the simple approach is the same for both PV technologies.
Second Model (Three parameters model): From equation (4.2), we have,

$$T_m \text{ (Poly c - Si)} = 0.93 \cdot 30 + 0.030 \cdot 600 - 1.666 \cdot 2 + 5.1 = 47.668°C$$

$$T_m \text{ (CdTe)} = 0.953 \cdot 30 + 0.030 \cdot 600 - 1.658 \cdot 2 + 5 = 48.274°C$$

Third Model (NOCT method): From equation (3.22), we have,

$$T_c \text{ (Poly c - Si)} = 30 + \frac{(47-20)}{800} \times 600 = 50.25$$

$$T_c \text{ (CdTe)} = 30 + \frac{(45-20)}{800} \times 600 = 48.75$$

Forth Model (Sandia method): From equations (4.5 and 4.6), we have,

$$T_c = T_{amb} + \frac{G}{1000} \cdot \Delta T + G \cdot \left(e^{a+bv}\right) = 30 + \frac{600}{1000} \cdot 3 + 600 \cdot \left(e^{-3.56 - 0.075 \times 2}\right) = 46.486°C$$

Since the parameters are the same, the result obtained from the Sandia model is equal to each other.
Fifth Model (PVsyst method): The u_0 and u_1 coefficients in PVsyst model are not known. Hence, firstly we need to find these parameters. From the result of Example 4.1, we have the following equations.

$$u_0 \text{ (Poly c - Si)} = \frac{\alpha_m (1 - \eta_m)}{\frac{\Delta T}{1000} + e^a} = \frac{0.9(1 - 0.137)}{\frac{3}{1000} + e^{(-3.56)}} = 24.71$$

$$u_1 \text{ (Poly c - Si)} = \frac{e^a b \alpha_m (1 - \eta_m)}{\frac{\Delta T}{1000} + e^a} = \frac{e^{(-3.56)} (-0.075)(0.9)(1 - 0.137)}{\frac{3}{1000} + e^{(-3.56)}} = -0.052$$

$$u_0 \text{ (CdTe)} = \frac{\alpha_m (1 - \eta_m)}{\frac{\Delta T}{1000} + e^a} = \frac{0.9(1 - 0.17)}{\frac{3}{1000} + e^{(-3.56)}} = 23.76$$

$$u_1 \text{ (CdTe)} = \frac{e^a b \alpha_m (1 - \eta_m)}{\frac{\Delta T}{1000} + e^a} = \frac{e^{(-3.56)} (-0.075)(0.9)(1 - 0.17)}{\frac{3}{1000} + e^{(-3.56)}} = -0.0506$$

From the PVsyst thermal model, we have:

$$T_c \text{ (Poly c - Si)} = T_{amb} + G \frac{\alpha_m (1 - \eta_m)}{u_0 \left(1 + \frac{u_1}{u_0} \cdot v\right)} = 30 + 600 \frac{0.9 \times (1 - 0.137)}{24.71 \left(1 - \frac{0.052}{24.71} \cdot 2\right)} = 48.93°C$$

$$T_c \text{ (CdTe)} = T_{amb} + G \frac{\alpha_m (1 - \eta_m)}{u_0 \left(1 + \frac{u_1}{u_0} \cdot v\right)} = 30 + 600 \frac{0.9 \times (1 - 0.17)}{23.76 \left(1 - \frac{0.0506}{23.76} \cdot 2\right)} = 48.94°C$$

Sixth Model (Faiman method): The calculated coefficients u_0 and u_1 can also be used for Faiman thermal model. From (4.7), we have:

$$T_m(\text{Poly c - Si}) = T_{\text{amb}} + \frac{G}{u_0 + u_1 \cdot v} = 30 + \frac{600}{24.71 - 0.052 \cdot 2} = 54.39°C$$

$$T_m(\text{CdTe}) = T_{\text{amb}} + \frac{G}{u_0 + u_1 \cdot v} = 30 + \frac{600}{23.76 - 0.0506 \cdot 2} = 55.36°C$$

Now, we can calculate the output power of the PV modules according to the six methods given above. From the given data, the maximum power is expected to drop about –0.45% / °C for the poly c-Si module and –0.28% / °C for CdTe module. Therefore 120 W_p modules at their maximum power will deliver:

$$P_{\max}^1(\text{Poly c - Si}) = 120 \times [1-(0.0045) \times (51-25)] \times \frac{600}{1000} = 63.576 \text{ W}$$

$$P_{\max}^2(\text{Poly c - Si}) = 120 \times [1-(0.0045) \times (47.668-25)] \times \frac{600}{1000} = 64.65 \text{ W}$$

$$P_{\max}^3(\text{Poly c - Si}) = 120 \times [1-(0.0045) \times (50.25-25)] \times \frac{600}{1000} = 63.819 \text{ W}$$

$$P_{\max}^4(\text{Poly c - Si}) = 120 \times [1-(0.0045) \times (46.486-25)] \times \frac{600}{1000} = 65.038 \text{ W}$$

$$P_{\max}^5(\text{Poly c - Si}) = 120 \times [1-(0.0045) \times (48.93-25)] \times \frac{600}{1000} = 64.246 \text{ W}$$

$$P_{\max}^6(\text{Poly c - Si}) = 120 \times [1-(0.0045) \times (54.39-25)] \times \frac{600}{1000} = 62.477 \text{ W}$$

And for CdTe module:

$$P_{\max}^1(\text{CdTe}) = 120 \times [1-(0.0028) \times (51-25)] \times \frac{600}{1000} = 66.758 \text{ W}$$

$$P_{\max}^2(\text{CdTe}) = 120 \times [1-(0.0028) \times (48.274-25)] \times \frac{600}{1000} = 67.307 \text{ W}$$

$$P_{\max}^3(\text{CdTe}) = 120 \times [1-(0.0028) \times (48.75-25)] \times \frac{600}{1000} = 67.212 \text{ W}$$

$$P_{\max}^4(\text{CdTe}) = 120 \times [1-(0.0028) \times (46.486-25)] \times \frac{600}{1000} = 67.668 \text{ W}$$

$$P_{\max}^5(\text{CdTe}) = 120 \times [1-(0.0028) \times (48.94-25)] \times \frac{600}{1000} = 67.173 \text{ W}$$

$$P_{\max}^6(\text{CdTe}) = 120 \times [1-(0.0028) \times (55.36-25)] \times \frac{600}{1000} = 65.879 \text{ W}$$

It is clear from the calculation that the results are close to each other and the methods do not have a noticeable advantage over one another.

4.3 THE ANGLE OF INCIDENCE EFFECT ON PV PERFORMANCE

The incidence angle, θ defined as the angle between solar irradiance and the normal of a PV cell, has a negative effect on the performance when it is compared to smaller incident angles. The simple relationship between the incidence angle and the output current can be expressed by cosine law as below:

$$I_\theta = I_0 \cdot \cos\theta \tag{4.9}$$

where I_0 is the PV output current in the case of solar radiation is perpendicular to the surface of the module, which means that the incidence angle between solar beam and surface normal is zero. I_θ is the PV output current when the incidence angle is θ. It is assumed that equation (4.9) is valid if the total global radiation values in their directions are the same (solar radiation of G_θ on the tilted surface of θ is equal to G_0 on the horizontal surface). Figure 4.1 shows the graphical illustration of incidence angle and solar radiation.

The cosine law expressed here gives reliable results when the incidence angle is between 0° and 30° degrees. At values above 30°, the actual electrical output deviates notably from the output power according to the cosine law. The PV cell produces almost no power at $\theta \geq 85°$. This relationship between the angle of incidence and relative output current is known as the Kelly cosine law, which considers the optical losses and cosine law together. Kelly cosine law for silicon-based photovoltaic cells, which is shown in Figure 4.2 and Table 4.8.

One can easily develop a mathematical model based on Kelly cosine values, which calculates the relative output current of a PV module as a function of incidence angle (θ). This relative output current expressions (I_θ) are given below by equation (4.10).

$$I_\theta = \begin{cases} I_0 \cdot \cos\theta, & \text{for } 0° \leq \theta \leq 30° \\ -9 \cdot 10^{-5} \cdot \theta^2 - 0.0049 \cdot \theta + 1.103, & \text{for } 30° < \theta \leq 85° \\ 0, & \text{for } 85° < \theta \leq 90° \end{cases} \tag{4.10}$$

An empirical fifth-order regression formula as a function of solar incidence angle (θ), which takes into account only the effect of PV optical losses on relative I_{SC}, is given below.

$$\text{Relative } I_{SC}(\theta) = b_0 + b_1(\theta) + b_2(\theta)^2 + b_3(\theta)^3 + b_4(\theta)^4 + b_5(\theta)^5 \tag{4.11}$$

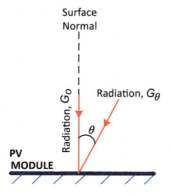

FIGURE 4.1 Graphical illustration of incidence angle (θ).

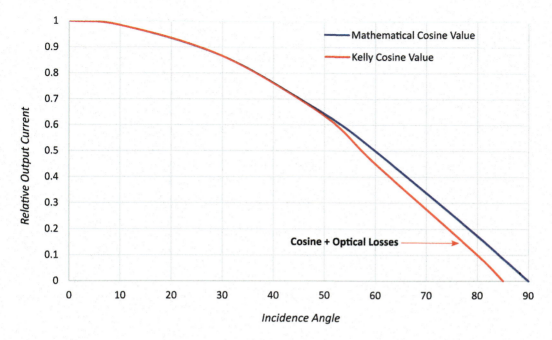

FIGURE 4.2 Kelly cosine curve for the PV cell at solar incidence angle from 0° to 90°.

TABLE 4.8
Kelly Cosine Values for the Relative Output Current of Silicon PV Cells

Solar Incidence Angle (Degrees)	Mathematical Cosine Value	Kelly Cosine Value
10°	0.984	0.984
30°	0.866	0.866
50°	0.643	0.635
60°	0.500	0.450
80°	0.174	0.100
85°	0.087	0.000

The predicted coefficients in (4.11) for various PV modules are given in Table 4.9 and Figure 4.3 illustrates the effect of optical losses on the relative short-circuit current as a function of solar incidence angle θ for two different CdTe-type photovoltaic devices.

It is clear from Figure 4.3 that relative I_{SC} response to θ is very similar for *CdTe thin-film* modules (with and without anti-reflective coating).

Example 4.3

Consider the horizontally mounted *Poly c-Si* PV module given below (Figure 4.4). The figure shows the radiation values on the PV module at different local solar times in the given directions.

Besides, the short-circuit current of the PV module is measured 4 A at 10.00 o'clock (local solar time). Estimate the PV module short-circuit currents at 12.00 and 15.00 local times based on Kelly cosine values.

TABLE 4.9
Polynomial Coefficients of Solar Incidence Angle θ for Various PV Cells

Module Type	b_0	b_1	b_2	b_3	b_4	b_5
Thin-Film c-Si (CIS/CdTe) (with ARC[a])	0.999	−0.00610	0.000812	−0.0000338	5.65×10^{-7}	-3.37×10^{-9}
Thin-Film c-Si (CIS/CdTe) (without ARC)	1	−0.00244	0.000310	−0.0000125	2.11×10^{-7}	-1.36×10^{-9}
CdTe (with ARC)	1	−0.00193	0.000258	−0.0000112	1.99×10^{-7}	-1.28×10^{-9}
CdTe (without ARC)	1	−0.00167	0.000218	−0.0000934	1.64×10^{-7}	-1.07×10^{-9}
Mono c-Si-1 (with ARC)	1	−0.00244	0.000310	−0.0000125	2.11×10^{-7}	-1.36×10^{-9}
Mono c-Si-2 (with ARC)	1	−0.00556	0.000653	−0.0000273	4.64×10^{-7}	-2.82×10^{-9}
Triple junction a-Si-1 without ARC)	1	−0.00244	0.000310	−0.0000125	2.11×10^{-7}	-1.36×10^{-9}
Triple junction a-Si-2 (without ARC)	1	−0.00565	0.000725	−0.0000293	4.97×10^{-7}	-2.74×10^{-9}
Triple junction a-Si (with ARC)	1	−0.00502	0.000584	−0.0000230	3.83×10^{-7}	-2.31×10^{-9}
Poly c-Si	0.998	−0.01212	0.001439	−0.0000557	8.77×10^{-7}	-4.91×10^{-9}

[a] ARC: Anti-Reflective Coating

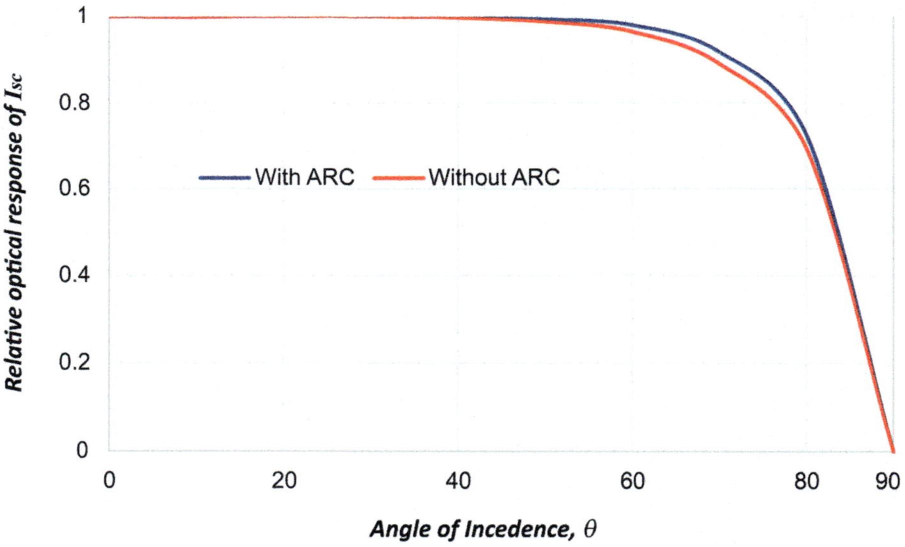

FIGURE 4.3 Optical relative response of relative short-circuit current with respect to solar incidence angle θ for two different CdTe-type photovoltaic modules.

Solution

Since the incidence angle at 10:00 is 22.5°, from equation (4.10), we have:

$$I_{SC}(22.5°) = I_{SC}(0°) \cdot \cos(22.5) \Rightarrow I_{SC}(0°) = \frac{I_{SC}(22.5°)}{\cos(22.5)} = \frac{4}{0.923} = 4.333 \, A$$

$$I_{SC}(22.5°)@(900 \, W/m^2) = \frac{900}{600} \cdot 4.333 = 6.5 \, A$$

Effect of Environmental Conditions on the Performance of Photovoltaic Systems

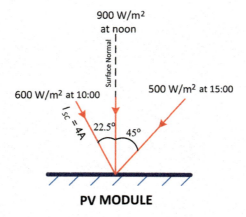

FIGURE 4.4 Radiation values on a horizontal PV module at different solar times.

Note that the short-circuit currents are directly proportional to the solar radiation value.

Similarly, the relative short-circuit current for the incidence angle of 45° can be calculated with reference to the short-circuit current at noon (incidence angle is 0°). So, from equation (4.10), we have:

$$I_{SC}(45°) = I_{SC}(0°) \times \left(-9 \cdot 10^{-5} \cdot 45^2 - 0.0049 \cdot 45 + 1.103\right)$$

$$\Rightarrow I_{SC}(45°) = 6.5 \times (0.70025) = 4.551 \, A$$

$$I_{SC}(45°)@(500 \, W/m^2) = \frac{500}{900} \cdot 4.551 = 2.528 \, A$$

4.4 AIR MASS EFFECT ON PV PERFORMANCE

As previously discussed, the electrical current generated by a PV cell is influenced by the solar spectral distribution. As known, the solar spectrum is influenced by a number of variables such as absolute air mass (AM_a), turbidity and cloudiness of atmosphere, dustiness, fogginess, ground albedo, etc. Among these factors, air mass (AM) is a term used to include the effect of solar spectral distribution on the conversion efficiency of PV cells. Air mass is a function of the zenith angle (sun's location) and it can accurately be calculated for the given site on any day of the year, and time of the day. Besides, the term AM_a is used to distinguish between sites at altitudes other than sea level. The AM_a can be obtained from multiplying AM by the ratio of (P/P_o), which represents the ratio of the site's atmospheric pressure (P) to the pressure at sea level (P_o).

The AM function was expressed in Chapter 2 by equations (2.4), and (2.5) for $\beta > 70°$. Another simple approach to obtaining the AM function is based on the shadow length (s) of a vertical pole. Consider Figure 4.5 to calculate the AM from the Pythagorean (also known as Pythagoras) theorem (Equation 4.12).

$$AM = \frac{d}{h} = \sqrt{\frac{h^2 + s^2}{h^2}} = \sqrt{1 + \left(\frac{s}{h}\right)^2} \qquad (4.12)$$

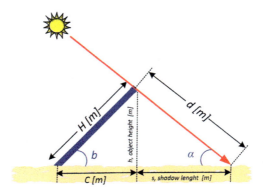

FIGURE 4.5 A simple model for AM calculation.

A reasonable approximation to *AM* can also be obtained from modeling the atmosphere as a simple spherical shell as:

$$AM = \sqrt{(708.6 \times \cos\beta)^2 + 1418.2} - 708.6 \times \cos\beta \quad (4.13)$$

where β is the zenith angle, the constant 708.6 is the ratio of Earth radius (6378 km) to effective atmosphere height (approximately 9 km), and the constant 1418.2 is obtained from $(2 \times 708.6 + 1)$. The relationship between *AM* and AM_a is given as below:

$$AM_a = \left(\frac{P}{P_o}\right) AM = \left(e^{-11.84 \times 10^{-5} \times H}\right) \cdot AM \quad (4.14)$$

where $(P/P_o) = \left(e^{-11.84 \times 10^{-5} \times H}\right)$, H is the altitude of the site in *meters* and P_o is the sea level pressure, which is 760 mmHg.

As explained above, the energy conversion efficiency of a solar cell is influenced by the air mass factor, which affects the solar spectrum distribution. As a generally accepted approach, a fourth-order regression is used to determine the effect of the solar spectrum on the relative I_{SC} value as a function of AM_a.

$$\text{Relative } I_{SC}(AM_a) = a_0 + a_1(AM_a) + a_2(AM_a)^2 + a_3(AM_a)^3 + a_4(AM_a)^4 \quad (4.15)$$

The predicted coefficients in (4.15) for various PV modules are given in Table 4.10 and Figure 4.6 illustrates the relative short-circuit current as a function of AM_a for different photovoltaic devices. It is clear from Figure 4.6 that relative I_{SC} response to AM_a is very similar for *thin-film c-Si* and *poly c-Si* modules, while AM_a has a greater impact on the triple-junction PV modules.

As can be seen from the above explanations, the short-circuit current value is dependent upon the operating conditions including irradiance (G), cell temperature (T_c), absolute air mass (AM_a), and solar angle of incidence (θ). Therefore, in the most general form, the I_{SC} current can be written as a function of (G, T_c, AM_a and θ) as follows.

$$I_{SC}(G,T_c,AM_a,\theta) = \left(\frac{G}{G_r}\right) \cdot [I_{SC}(AM_a)] \cdot [I_{SC}(\theta)] \cdot [I_{SCr} + \alpha_{I_{SC}} \times (T_c - T_r)] \quad (4.15)$$

where $I_{SC}(G,T_c,AM_a,\theta)$ is operating conditions dependent on short-circuit current, $I_{SC}(AM_a)$ is relative I_{SC} as a function of AM_a, $I_{SC}(\theta)$ is the relative optical response of I_{SC}, I_{SCr} is reference

TABLE 4.10
AM_a Polynomial Coefficients to Consider the Spectral Effect for Various PV Cells

Module Type	a_0	a_1	a_2	a_3	a_4
Thin-Film c-Si (CIS/CdTe) (with ARC[a])	0.938	0.0622	−0.01500	0.001220	−0.0000340
Thin-Film c-Si (CIS/CdTe) (without ARC)	0.939	0.0552	−0.01090	0.000813	−0.0000235
Poly c-Si-1	0.941	0.0527	−0.00958	0.000676	−0.0000181
Poly c-Si-2	0.918	0.0862	−0.02445	0.002816	−0.0001260
Mono c-Si-1 (with ARC)	0.931	0.0674	−0.01690	0.001530	−0.0000552
Mono c-Si-2 (with ARC)	0.936	0.0543	−0.00868	0.000527	−0.0000110
Triple junction a-Si-1 (without ARC)	0.925	0.0689	−0.01390	0.001150	−0.0000383
Triple junction a-Si-2 (without ARC)	0.982	0.0588	−0.03730	0.004120	−0.0001470
Triple junction a-Si (with ARC)	1.100	−0.0614	−0.00443	0.000632	−0.0000192
Triple junction (a-Si/a-Si/a-si:Ge)	1.047	0.00082	−0.02590	0.003173	0.00011026

[a] ARC: Anti-Reflective Coating

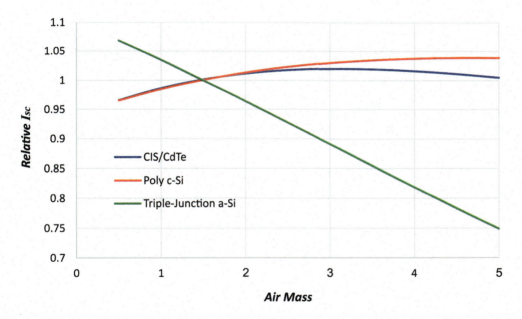

FIGURE 4.6 Relative short-circuit current versus AM_a for a different photovoltaic modules.

(or rated) short-circuit current at the reference temperature T_r (25°C), G_r is reference radiation (1000 W/m²) and $\alpha_{I_{SC}}$ is temperature coefficient in (A/°C) for short-circuit current.

As it is well known, there are four important points on the $I-V$ curve of a PV module, which are $I_{SC}, V_{OC}, I_{mp}, V_{mp}$. Equation (4.15) only defines the I_{SC} formula. The remaining three points (V_{OC}, I_{mp}, V_{mp}) can also be determined as a function of irradiance, cell temperature, absolute air mass, and solar angle of incidence. However, these points can be expressed more easily by making a new definition called effective irradiance (G_E). Accordingly, the functions of $G_E, V_{OC}, I_{mp}, V_{mp}$ are given as:

$$G_E = \frac{I_{SC}(G, T_c = T_r, AM_a, \theta)}{I_{SCr}} \tag{4.16}$$

TABLE 4.11
Empirically Determined Coefficients in Equations (4.17)–(4.19) for Different Types of PV Modules

Module Type	C_0	C_1	C_2	C_3
Mono c-Si	1.000	0.003	−0.538	−21.438
Poly c-Si	1.014	−0.005	−0.321	−30.201
Thin-Film c-Si (CIS/CdTe)	0.0961	0.037	0.232	−9.429
Triple junction a-Si	1.072	−0.098	−1.846	−5.176

$$I_{mp}(G_E, T_c) = C_0 + G_E \cdot \left[I_{mpr} + \alpha_{I_{mp}} \times (T_c - T_r)\right] \tag{4.17}$$

$$V_{OC}(G_E, T_c) = V_{OCr} + C_1 \cdot \ln(G_E) + \beta_{V_{OC}} \times (T_c - T_r) \tag{4.18}$$

$$V_{mp}(G_E, T_c) = V_{mpr} + C_2 \cdot \ln(G_E) + C_3 \cdot \ln^2(G_E) + \beta_{V_{mp}} \times (T_c - T_r) \tag{4.19}$$

where G is the global irradiance on the array plane, G_E is the effective irradiance (dimensionless, or in "Suns"), G_r is the reference irradiance in plane-of-array (one sun, or 1000 W/m²), $\alpha_{I_{mp}}$ is I_{mp} temperature coefficient in (A/°C), $\beta_{V_{OC}}$ is V_{OC} temperature coefficient in (V/°C) and $\beta_{V_{mp}}$ is V_{mp} temperature coefficient in (V/°C). C_0, C_1, C_2, C_3 are empirically determined coefficients relating $I_{mp}, V_{OC},$ and V_{mp}. Typical C_0, C_1, C_2, C_3 coefficients for different types of PV modules are listed in Table 4.11.

Note that the parameters at reference operating conditions are $G_r = 1000$ W/m², $T_c = T_r$ (typically 25°C or 50°C), $AM_a = 1.5, \theta = 0°$. Hence, the parameters $I_{SCr}, I_{mpr}, V_{OCr},$ and V_{mpr} in the above equations can be obtained as below:

$$I_{SCr} = I_{SC}\left(G = 1000 \text{ W/m}^2, T_c = T_r \text{ °C}, AM_a = 1.5, \theta = 0°\right) \tag{4.20}$$

$$I_{mpr} = I_{mp}\left(G_E = 1, T_c = T_r \text{ °C}\right) \tag{4.21}$$

$$V_{OCr} = V_{OC}\left(G_E = 1, T_c = T_r \text{ °C}\right) \tag{4.22}$$

$$V_{mpr} = V_{mp}\left(G_E = 1, T_c = T_r \text{ °C}\right) \tag{4.23}$$

Example 4.4: Air Mass Calculations

Considering the following polycrystalline PV module (Figure 4.7), calculate the air mass and absolute air mass values at sea level and 1000 m altitude.

Solution

Air mass value can be calculated based on the different approaches given above. Let's use the (4.12) to calculate AM. But, firstly, we need to calculate object height, h. From the left-hand side triangle,

$$h = 1 \text{ m} \times \sin 26 = 0.438 \text{ m}$$

Effect of Environmental Conditions on the Performance of Photovoltaic Systems

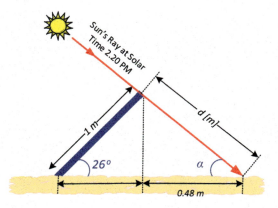

FIGURE 4.7 Considered PV module.

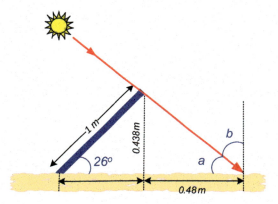

FIGURE 4.8 Zenit angle calculation.

and shadow length is given as $s = 0.48$ m. Therefore, from (4.12) we have:

$$AM = \sqrt{1+\left(\frac{0.48}{0.438}\right)^2} = 1.483$$

Air mass values can also be calculated based on (2.4) and (4.13). The zenith angle needs to be calculated to find the air mass value. The zenith angle from the given geometry can easily be calculated as below (see Figure 4.8).

Altitude angle, α

$$\alpha = tg^{-1}\left(\frac{0.438}{0.48}\right) = 42.38°$$

Zenith angle, β

$$\beta = 90 - 42.38 = 47.62°$$

From (4.13),

$$AM = \sqrt{(708.6 \times \cos 47.62)^2 + 1418.2} - 708.6 \times \cos 47.62 = 1.482$$

From (2.4),

$$AM = \frac{1}{\cos(\beta)} = \frac{1}{\cos(47.62)} = 1.483$$

As can be seen from the results, air mass formulas yield the same result. So, the absolute air mass value can be found from (4.14):
As known, $AM = AM_a$ at sea level. At 1000 m altitude,

$$AM_a = \left(e^{-11.84 \times 10^{-5} \times 1000}\right) \cdot 1.483 = 1.317$$

Example 4.5: Altitude Effect on PV Performance

PV panels with two different technologies (*Poly c-Si and CIS*) have an inclination of 30° are placed at an angle of 10° east of south in Los Angeles $(34.0522°\,N, 118.2437°\,W)$ on June 15. Calculate the PV performance parameters I_{SC}, I_{mp}, V_{OC}, V_{mp} at sea level and 1000 m altitude based on the parameters given below (Tables 4.12 and 4.13). Use Tables 4.9–4.11 for polynomial coefficients and equation (4.2) for the prediction of cell temperature.

Solution

In order to calculate panels' performance parameters as a function of irradiance, cell temperature, air mass, and angle of incidence at local solar times 11:00, 12:00, and 14:00. Firstly, we need to find the unknown parameters of cell temperature (T_c), absolute air mass (AM_a), and angle of incidence (θ) for the given local solar times.

TABLE 4.12
The Parameters of Two Different PV Panels

Performance Parameters under STC	Poly-crystal	CIS-Thin Film
Rated Power (W_p)	140	145
Voltage at Max. Power (V)	18	81
Current at Max. Power (A)	7.85	1.80
Open-Circuit Voltage (V)	22.1	107
Short-Circuit Current (A)	8.35	2.20
Temperature Coefficient of I_{SC}	+0.034%/°C	+0.01%/°C
Temperature Coefficient of V_{OC}	−0.35%/°C	−0.30%/°C
Temperature Coefficient of I_{mp}	+0.0028%/°C	+0.00056%/°C
Temperature Coefficient of V_{mp}	−0.37%/°C	−0.31%/°C

TABLE 4.13
Operating Conditions

Operating Conditions	Local Solar Time		
	11:00	12:00	14:00
Solar Radiation	800 W/m²	900 W/m²	750 W/m²
Ambient Temperature at Sea Level (wind speed = 1 m/s)	15°C	25°C	20°C
Ambient Temperature at 1000 m Altitude (wind speed = 3 m/s)	10°C	18°C	14°C

Effect of Environmental Conditions on the Performance of Photovoltaic Systems

TABLE 4.14
The Constants to Calculate the Cell Temperature

PV Technology	T_{amb} c_1	Irradiance c_2	Wind Speed c_3	Constant c
Poly c-Si	0.930	0.030	−1.666	5.1
CIS	0.960	0.029	−1.517	4.1

For the cell temperature calculation, equation (4.2) can be used. The constants used in (4.2) are taken from Table 4.2 as below (Table 4.14):

Thus, cell temperatures are predicted for Poly C-Si as below:

$$T_c(\text{at } 11:00 \text{ and sea level}) = 0.930 \cdot T_{amb} + 0.030 \cdot G - 1.666 \cdot v + 5.1 \Rightarrow$$

$$T_c(800 \text{ W/m}^2, 1 \text{ m/s}) = 0.930 \cdot 15 + 0.030 \cdot 800 - 1.666 \cdot 1 + 5.1 = 41.384°C$$

$$T_c(\text{at } 12:00 \text{ and sea level}) = 0.930 \cdot T_{amb} + 0.030 \cdot G - 1.666 \cdot v + 5.1 \Rightarrow$$

$$T_c(900 \text{ W/m}^2, 1 \text{ m/s}) = 0.930 \cdot 25 + 0.030 \cdot 900 - 1.666 \cdot 1 + 5.1 = 53.684°C$$

$$T_c(\text{at } 14:00 \text{ and sea level}) = 0930 \cdot T_{amb} + 0.030 \cdot G - 1.666 \cdot v + 5.1 \Rightarrow$$

$$T_c(750 \text{ W/m}^2, 1 \text{ m/s}) = 0.930 \cdot 20 + 0.030 \cdot 750 - 1.666 \cdot 1 + 5.1 = 44.534°C$$

Let's calculate the cell temperatures at 1000 m altitude for Poly C-Si.

$$T_c(\text{at } 11:00 \text{ and } 1000 \text{ m altitude}) = 0.930 \cdot T_{amb} + 0.030 \cdot G - 1.666 \cdot v + 5.1 \Rightarrow$$

$$T_c(800 \text{ W/m}^2, 3 \text{ m/s}) = 0.930 \cdot 10 + 0.030 \cdot 800 - 1.666 \cdot 3 + 5.1 = 33.402°C$$

$$T_c(\text{at } 12:00 \text{ and } 1000 \text{ m altitude}) = 0.930 \cdot T_{amb} + 0.030 \cdot G - 1.666 \cdot v + 5.1 \Rightarrow$$

$$T_c(900 \text{ W/m}^2, 3 \text{ m/s}) = 0.930 \cdot 18 + 0.030 \cdot 900 - 1.666 \cdot 3 + 5.1 = 43.842°C$$

$$T_c(\text{at } 14:00 \text{ and } 1000 \text{ m altitude}) = 0930 \cdot T_{amb} + 0.030 \cdot G - 1.666 \cdot v + 5.1 \Rightarrow$$

$$T_c(750 \text{ W/m}^2, 3 \text{ m/s}) = 0.930 \cdot 14 + 0.030 \cdot 750 - 1.666 \cdot 3 + 5.1 = 35.622°C$$

By performing the same calculations for the CIS module, the following cell temperatures are obtained for the given local times and altitudes.

$$T_c\left(\text{at } 11:00 \text{ and sea level}\right) = 0.960 \cdot T_{\text{amb}} + 0.029 \cdot G - 1.517 \cdot v + 4.1 \Rightarrow$$

$$T_c\left(800 \text{ W/m}^2, 1 \text{ m/s}\right) = 0.960 \cdot 15 + 0.029 \cdot 800 - 1.517 \cdot 1 + 4.1 = 40.183°C$$

$$T_c\left(\text{at } 12:00 \text{ and sea level}\right) = 0.960 \cdot T_{\text{amb}} + 0.029 \cdot G - 1.517 \cdot v + 4.1 \Rightarrow$$

$$T_c\left(900 \text{ W/m}^2, 1 \text{ m/s}\right) = 0.960 \cdot 25 + 0.029 \cdot 900 - 1.517 \cdot 1 + 4.1 = 52.683°C$$

$$T_c\left(\text{at } 14:00 \text{ and sea level}\right) = 0.960 \cdot T_{\text{amb}} + 0.029 \cdot G - 1.517 \cdot v + 4.1 \Rightarrow$$

$$T_c\left(750 \text{ W/m}^2, 1 \text{ m/s}\right) = 0.960 \cdot 20 + 0.029 \cdot 750 - 1.517 \cdot 1 + 4.1 = 43.533°C$$

$$T_c\left(\text{at } 11:00 \text{ and } 1000 \text{ m altitude}\right) = 0.960 \cdot T_{\text{amb}} + 0.029 \cdot G - 1.517 \cdot v + 4.1 \Rightarrow$$

$$T_c\left(800 \text{ W/m}^2, 3 \text{ m/s}\right) = 0.960 \cdot 10 + 0.029 \cdot 800 - 1.517 \cdot 3 + 4.1 = 32.349°C$$

$$T_c\left(\text{at } 12:00 \text{ and } 1000 \text{ m altitude}\right) = 0.960 \cdot T_{\text{amb}} + 0.029 \cdot G - 1.517 \cdot v + 4.1 \Rightarrow$$

$$T_c\left(900 \text{ W/m}^2, 3 \text{ m/s}\right) = 0.960 \cdot 18 + 0.029 \cdot 900 - 1.517 \cdot 3 + 4.1 = 42.929°C$$

$$T_c\left(\text{at } 14:00 \text{ and } 1000 \text{ m altitude}\right) = 0.960 \cdot T_{\text{amb}} + 0.029 \cdot G - 1.517 \cdot v + 4.1 \Rightarrow$$

$$T_c\left(750 \text{ W/m}^2, 3 \text{ m/s}\right) = 0.960 \cdot 14 + 0.029 \cdot 750 - 1.517 \cdot 3 + 4.1 = 34.739°C$$

Now, the calculated cell temperatures can then be summarized in Table 4.15.

TABLE 4.15
The Results of PV Cell Temperatures

Local Solar Times	11:00 ($G=800$W/m^2)		12:00 ($G=900$W/m^2)		14:00 ($G=750$W/m^2)	
Altitude	Sea Level	1000 m	Sea Level	1000 m	Sea Level	1000 m
	($v=1$ m/s)	($v=3$ m/s)	($v=1$ m/s)	($v=3$ m/s)	($v=1$ m/s)	($v=3$ m/s)
Ambient Temperature	15°C	10°C	25°C	18°C	20°C	14°C
Poly c-Si Module	41.384°C	33.402°C	53.684°C	43.842°C	44.534°C	35.622°C
CIS Module	40.183°C	32.349°C	52.683°C	42.929°C	43.533°C	34.739°C

Effect of Environmental Conditions on the Performance of Photovoltaic Systems

Secondly, we need to calculate the angle of incidence θ and $I_{SC}(\theta)$ on June-15 from (2.16) and (4.11) for the PV module with an inclination of 30°, which is placed at an angle of 10° east of south in Los Angeles $(34.0522° \text{N}, 118.2437° \text{W})$.

Declination and hour angle for June 15 ($N = 166$) at 11.00 local solar time:

$$\delta = 23.45 \cdot \sin\left[\frac{360}{365}(166 + 284)\right] = 23.314°$$

$$\omega = 15°(t_{LS} - 12) = 15(11 - 12) = -15°$$

From equation (2.16), the inclination angle for Los Angeles:

$$\cos(\theta) = \sin(34.052)\sin(23.314)\cos(30) - \cos(34.052)\sin(23.314)\sin(30)\cos(10)$$
$$+ \cos(34.052)\cos(23.314)\cos(30)\cos(-15)$$
$$+ \sin(34.052)\cos(23.314)\sin(30)\cos(10)\cos(-15)$$
$$+ \cos(23.314)\sin(30)\sin(10)\sin(-15) = 0.866°$$
$$\Rightarrow \theta(\text{Los Angeles at } 11:00) = 27.015°$$

By performing the same calculations, the following incidence angles are obtained for the given local times.

$$\theta(\text{Los Angeles at } 12:00) = 19.506°$$

$$\theta(\text{Los Angeles at } 14:00) = 30.653°$$

$I_{SC}(\theta)$ values for Poly c-Si and CIS modules can be calculated from (4.11). The coefficients used in (4.11) are taken from Table 4.9 as below (Table 4.16):

Accordingly, $I_{SC}(\theta)$ values for Poly c-Si and CIS modules are obtained for the given solar times as below:

$$I_{SC}(\theta = 27.015, \text{Poly}) = b_0 + b_1(\theta) + b_2(\theta)^2 + b_3(\theta)^3 + b_4(\theta)^4 + b_5(\theta)^5$$

$$I_{SC}(\theta = 27.015, \text{Poly}) = 0.998 - 0.01212 \cdot (27.015) + 0.001439 \cdot (27.015)^2$$
$$- 0.0000557 \cdot (27.015)^3 + 8.77 \cdot 10^{-7} \cdot (27.015)^4 - 4.91 \times 10^{-9} \cdot (27.015)^5 = 1.0190$$

TABLE 4.16
Incidence Angle Coefficients for CIS and Poly c-Si Modules

Module Type	b_0	b_1	b_2	b_3	b_4	b_5
Thin-Film c-Si (CIS)	0.999	−0.00610	0.000812	−0.0000338	5.65×10^{-7}	-3.37×10^{-9}
Poly c-Si	0.998	−0.01212	0.001439	−0.0000557	8.77×10^{-7}	-4.91×10^{-9}

Similarly, for the remaining solar times:

$$I_{SC}(\theta = 19.506, \text{Poly}) = 1.0088$$

$$I_{SC}(\theta = 30.653, \text{Poly}) = 1.0157$$

By performing the same calculations, the following $I_{SC}(\theta)$ values for CIS modules are obtained for the given local times.

$$I_{SC}(\theta = 27.015, \text{CIS}) = 1.0071$$

$$I_{SC}(\theta = 19.506, \text{CIS}) = 1.0092$$

$$I_{SC}(\theta = 30.653, \text{CIS}) = 0.9982$$

Thirdly, we need to calculate AM_a values at sea level and 1000 m altitude, and then $I_{SC}(AM_a)$ values should be calculated from (4.15) for Poly C-Si and CIS modules.

From (2.4),

$$AM(\text{Sea Level at } 11:00) = \frac{1}{\cos(\beta)} = \frac{1}{\cos(27.015)} = 1.1224$$

$$AM(\text{Sea Level at } 12:00) = \frac{1}{\cos(\beta)} = \frac{1}{\cos(19.506)} = 1.0608$$

$$AM(\text{Sea Level at } 14:00) = \frac{1}{\cos(\beta)} = \frac{1}{\cos(30.653)} = 1.1624$$

Absolute air mass values are equal to air masses at sea level, $AM = AM_a$. Absolute air masses at 1000 m altitude can then be found from (4.14):

$$AM_a(1000 \text{ m altitude at } 11:00) = \left(e^{-11.84 \times 10^{-5} \times 1000}\right) \cdot 1.1224 = 0.9970$$

$$AM_a(1000 \text{ m altitude at } 12:00) = \left(e^{-11.84 \times 10^{-5} \times 1000}\right) \cdot 1.0608 = 0.9423$$

$$AM_a(1000 \text{ m altitude at } 14:00) = \left(e^{-11.84 \times 10^{-5} \times 1000}\right) \cdot 1.1624 = 1.0325$$

Consequently, $I_{SC}(AM_a)$ values for Poly c-Si and CIS modules are obtained from (4.15) for the given solar times and altitudes. The coefficients used in (4.15) are taken from Table 4.10 as below (Table 4.17):

TABLE 4.17
Air Mass Coefficients for CIS and Poly c-Si Modules

Module Type	a_0	a_1	a_2	a_3	a_4
Thin-Film c-Si (CIS)	0.938	0.0622	−0.01500	0.001220	−0.0000340
Poly c-Si	0.941	0.0527	−0.00958	0.000676	−0.0000181

Effect of Environmental Conditions on the Performance of Photovoltaic Systems

From (4.15), we have for Poly c-Si:

$$I_{SC}(AM_a \text{ at } 11:00, \text{Sea level}) = a_0 + a_1(AM_a) + a_2(AM_a)^2 + a_3(AM_a)^3 + a_4(AM_a)^4 \Rightarrow$$

$$I_{SC}(AM_a \text{ at } 11:00, \text{Sea level}) = 0.941 + 0.0527 \cdot (1.1224) - 0.00958 \cdot (1.1224)^2$$
$$+ 0.000676 \cdot (1.1224)^3 - 0.0000181 \cdot (1.1224)^4 = 0.9890$$

Similarly, for the remaining solar times:

$$I_{SC}(AM_a \text{ at } 12:00, \text{Sea level}) = 0.9869$$

$$I_{SC}(AM_a \text{ at } 14:00, \text{Sea level}) = 0.9903$$

By performing the same calculations, the remaining $I_{SC}(AM_a)$ values for Poly c-Si and CIS modules are obtained. So, all $I_{SC}(AM_a)$ values can be listed in Table 4.18.

Now that all parameters are known to calculate I_{SC} under given operating conditions. The required parameters to calculate I_{SC} are summarized in Table 4.19:

Other parameters needed to calculate I_{SC} are specified in the problem, which is deducted as in Table 4.20.

TABLE 4.18
$I_{SC}(AM_a)$ Values for Poly c-Si and CIS Modules

Local Solar Times	11:00		12:00		14:00	
$I_{SC}(AM_a)$ Function	$I_{SC}(AM_a)$ Sea Level	$I_{SC}(AM_a)$ 1000 m	$I_{SC}(AM_a)$ Sea Level	$I_{SC}(AM_a)$ 1000 m	$I_{SC}(AM_a)$ Sea Level	$I_{SC}(AM_a)$ 1000 m
Poly c-Si Module	0.9890	0.9847	0.9869	0.9827	0.9903	0.9859
CIS Module	0.9906	0.9863	0.9885	0.9843	0.9919	0.9875

TABLE 4.19
$I_{SC}(AM_a)$ Values for Poly c-Si and CIS modules

Module Type	Sea Level at 11:00	Sea Level at 12:00	Sea Level at 14:00	1000 m Altitude at 11:00	1000 m Altitude at 12:00	1000 m Altitude at 14:00
Poly c-Si Module	$G=800\,W/m^2$ $T_C=41.384\,°C$ $I_{SC}(\theta)=1.0190$ $I_{SC}(AM_a)=0.9890$	$G=900\,W/m^2$ $T_C=53.684\,°C$ $I_{SC}(\theta)=1.0088$ $I_{SC}(AM_a)=0.9869$	$G=750\,W/m^2$ $T_C=44.534\,°C$ $I_{SC}(\theta)=1.0157$ $I_{SC}(AM_a)=0.9903$	$G=800\,W/m^2$ $T_C=33.402\,°C$ $I_{SC}(\theta)=1.0190$ $I_{SC}(AM_a)=0.9847$	$G=900\,W/m^2$ $T_C=43.842\,°C$ $I_{SC}(\theta)=1.0088$ $I_{SC}(AM_a)=0.9827$	$G=750\,W/m^2$ $T_C=35.622\,°C$ $I_{SC}(\theta)=1.0157$ $I_{SC}(AM_a)=0.9859$
CIS Module	$G=800\,W/m^2$ $T_C=40.183\,°C$ $I_{SC}(\theta)=1.0071$ $I_{SC}(AM_a)=0.9906$	$G=900\,W/m^2$ $T_C=52.683\,°C$ $I_{SC}(\theta)=1.0092$ $I_{SC}(AM_a)=0.9885$	$G=750\,W/m^2$ $T_C=43.533\,°C$ $I_{SC}(\theta)=0.9998$ $I_{SC}(AM_a)=0.9919$	$G=800\,W/m^2$ $T_C=32.349\,°C$ $I_{SC}(\theta)=1.0071$ $I_{SC}(AM_a)=0.9863$	$G=900\,W/m^2$ $T_C=42.929\,°C$ $I_{SC}(\theta)=1.0092$ $I_{SC}(AM_a)=0.9843$	$G=750\,W/m^2$ $T_C=34.739\,°C$ $I_{SC}(\theta)=0.9998$ $I_{SC}(AM_a)=0.9875$

TABLE 4.20
Performance Parameters for Poly-crystal and CIS-Thin Film

Performance Parameters under STC	Poly-crystal	CIS-Thin Film
Short-Circuit Current (A)	8.35	2.20
Temperature Coefficient of I_{SC}	+0.034%/°C	+0.01%/°C
Reference Cell Temperature	25°C	25°C

We can now apply the (4.15) equation to predict the I_{SC}. However, we need to make a slight modification in equation (4.15). Because the formula provided herein is arranged in the case of temperature coefficients given in $(A/°C)$. But, the unit of temperature coefficient given in this problem is in terms of $(\%/°C)$. In this case, I_{SCr} will be proportional to $[1+\alpha_{Isc} \times (T_c - T_r)]$. For this reason, equation (4.15) is modified as follows.

$$I_{SC}(G, T_c, AM_a, \theta) = \left(\frac{G}{G_r}\right) \cdot [I_{SC}(AM_a)] \cdot [I_{SC}(\theta)] \cdot I_{SCr} \cdot [1 + \alpha_{Isc} \times (T_c - T_r)]$$

Hence, I_{SC} values for Poly c-Si Module at sea level are:

$$I_{SC}(800 \text{ W/m}^2, T_c = 41.384, AM_a = 1.1224, \theta = 27.015)$$

$$= \left(\frac{800}{1000}\right) \cdot (0.989) \cdot (1.019) \cdot (8.35) \cdot [1 + 0.00034 \times (41.384 - 25)] = 6.769 \text{ A}$$

$$I_{SC}(900 \text{ W/m}^2, T_c = 53.684, AM_a = 1.1608, \theta = 19.506)$$

$$= \left(\frac{900}{1000}\right) \cdot (0.9869) \cdot (1.0088) \cdot (8.35) \cdot [1 + 0.00034 \times (53.684 - 25)] = 7.554 \text{ A}$$

$$I_{SC}(750 \text{ W/m}^2, T_c = 44.534, AM_a = 1.1624, \theta = 30.653)$$

$$= \left(\frac{750}{1000}\right) \cdot (0.9903) \cdot (1.0157) \cdot (8.35) \cdot [1 + 0.00034 \times (44.534 - 25)] = 6.3409 \text{ A}$$

As for CIS Module:

$$I_{SC}(800 \text{ W/m}^2, T_c = 40.183, AM_a = 1.1224, \theta = 27.015)$$

$$= \left(\frac{800}{1000}\right) \cdot (0.9906) \cdot (1.0071) \cdot (2.20) \cdot [1 + 0.0001 \times (40.183 - 25)] = 1.758 \text{ A}$$

$$I_{SC}(900 \text{ W/m}^2, T_c = 52.683, AM_a = 1.1608, \theta = 19.506)$$

$$= \left(\frac{900}{1000}\right) \cdot (0.9885) \cdot (1.0092) \cdot (2.20) \cdot [1 + 0.0001 \times (52.683 - 25)] = 1.980 \text{ A}$$

$$I_{SC}(750 \text{ W/m}^2, T_c = 43.533, AM_a = 1.1624, \theta = 30.653)$$

$$= \left(\frac{750}{1000}\right) \cdot (0.9919) \cdot (0.9998) \cdot (2.20) \cdot [1 + 0.0001 \times (43.533 - 25)] = 1.639 \text{ A}$$

TABLE 4.21
I_{sc} Results for Poly c-Si and CIS Modules

Module Type	Sea Level at 11:00 (800 W/m²)	Sea Level at 12:00 (900 W/m²)	Sea Level at 14:00 (750 W/m²)	1000 m Altitude at 11:00 (800 W/m²)	1000 m Altitude at 12:00 (900 W/m²)	1000 m Altitude at 14:00 (750 W/m²)
I_{SC} Poly c-Si	6.769 A	7.554 A	6.3409 A	6.882 A	7.497 A	6.293 A
I_{SC} CIS	1.758 A	1.980 A	1.639 A	1.749 A	1.970 A	1.627 A

The remaining results at 1000 m altitude for CIS and poly c-Si modules are given in Table 4.21.

Before calculating the other PV performance parameter (I_{mp}, V_{mp}, V_{OC}) under given operating conditions, we need to modify equations (4.17)–(4.19) in order to make them compatible with temperature coefficients given in (%/°C). Accordingly, the modified equation can be written as below:

$$I_{mp}(G_E, T_c) = C_0 + G_E \cdot I_{mpr} \cdot \left[1 + \alpha_{I_{mp}} \times (T_c - T_r)\right]$$

$$V_{OC}(G_E, T_c) = C_1 \cdot \ln(G_E) + V_{OCr} \cdot \left[1 + \beta_{V_{OC}} \times (T_c - T_r)\right]$$

$$V_{mp}(G_E, T_c) = C_2 \cdot \ln(G_E) + C_3 \cdot \ln^2(G_E) + V_{mpr} \cdot \left[1 + \beta_{V_{mp}} \times (T_c - T_r)\right]$$

where G_E is given below:

$$G_E = \frac{I_{SC}(G, T_c = T_r, AM_a, \theta)}{I_{SCr}}$$

Here, a sample calculation will be shown for (I_{mp}, V_{mp}, and V_{OC}) and the results of the remaining calculations for the other operating conditions will be summarized in tabular form. Let's begin with the sea level calculation at 11:00 for the Poly c-Si module. Firstly, the effective irradiance (G_E) is needed to be calculated, which is given below:

$$G_E = \frac{I_{SC}(G, T_c = T_r, AM_a, \theta)}{I_{SCr}} = \left(\frac{G}{G_r}\right) \cdot \left[I_{SC}(AM_a)\right] \cdot \left[I_{SC}(\theta)\right]$$

So, effective irradiance for poly c-Si can be calculated at 11:00 as below:

$$G_E = \left(\frac{800}{1000}\right) \cdot (0.989) \cdot (1.019) = 0.8062$$

Other coefficients in $I_{mp}(G_E, T_c)$ formula are $C_0 = 1.014$, $T_c = 41.384$, $I_{mpr} = 7.85$ A, and $\alpha_{I_{mp}} = 0.0028\%/°C$. After all this data, $I_{mp}(G_E, T_c)$ value for the poly c-Si module is found as below:

$$I_{mp}(G_E, T_c) = C_0 + G_E \cdot I_{mpr} \cdot \left[1 + \alpha_{I_{mp}} \times (T_c - T_r)\right]$$

$$= 1.014 + 0.8062 \cdot 7.85 \cdot \left[1 + 0.000028 \times (41.384 - 25)\right] = 7.3458$$

Similarly, $V_{OC}(G_E, T_c)$ and $V_{mp}(G_E, T_c)$ values for the poly c-Si module can then be calculated based on the previously defined coefficients.

TABLE 4.22
I_{mp}, V_{OC}, and V_{mp} Results for Poly c-Si and CIS Modules

	Altitude	θ (deg)	G (W/m²)	C_E	C_0	C_1	C_2	C_3	$I_{mp}(G_E, T_c)$	$V_{OC}(G_E, T_c)$	$V_{mp}(G_E, T_c)$
Poly c-Si Module	Sea Level	27.01	800	0.806	1.014	−0.005	−0.321	−30.201	7.345 A	20.833 V	14.312 V
	Sea Level	19.50	900	0.896	1.014	−0.005	−0.321	−30.201	8.053 A	19.881 V	14.447 V
	Sea Level	30.65	750	0.754	1.014	−0.005	−0.321	−30.201	6.939 A	20.590 V	13.081 V
	1000 m	27.01	800	0.802	1.014	−0.005	−0.321	−30.201	7.316 A	21.451 V	14.855 V
	1000 m	19.50	900	0.892	1.014	−0.005	−0.321	−30.201	8.021 A	20.643 V	15.138 V
	1000 m	30.65	750	0.751	1.014	−0.005	−0.321	−30.201	6.911 A	21.279 V	13.670 V
CIS Module	Sea Level	27.01	800	0.798	0.0961	0.037	0.232	−9.429	1.532 A	102.11 V	77.564 V
	Sea Level	19.50	900	0.897	0.0961	0.037	0.232	−9.429	1.712 A	98.109 V	74.859 V
	Sea Level	30.65	750	0.742	0.0961	0.037	0.232	−9.429	1.432 A	101.039 V	76.386 V
	1000 m	27.01	800	0.794	0.0961	0.037	0.232	−9.429	1.526 A	104.632 V	79.463 V
	1000 m	19.50	900	0.894	0.0961	0.037	0.232	−9.429	1.705 A	101.240 V	77.251 V
	1000 m	30.6	750	0.739	0.0961	0.037	0.232	−9.429	1.426 A	103.862 V	78.517 V

$$V_{OC}(G_E, T_c) = C_1 \cdot \ln(G_E) + V_{OCr} \cdot \left[1 + \beta_{V_{OC}} \times (T_c - T_r)\right]$$
$$= -0.005 \cdot \ln(0.8062) + 22.1 \cdot \left[1 - 0.0035 \times (41.384 - 25)\right] = 20.8338$$

$$V_{mp}(G_E, T_c) = C_2 \cdot \ln(G_E) + C_3 \cdot \ln^2(G_E) + V_{mpr} \cdot \left[1 + \beta_{V_{mp}} \times (T_c - T_r)\right]$$
$$= -0.321 \cdot \ln(0.8062) - 30.201 \cdot \ln^2(0.8062) + 18 \cdot \left[1 - 0.0037 \times (41.384 - 25)\right] = 14.3128$$

The remaining results regarding (I_{mp}, V_{OC} and V_{mp}) for CIS and poly c-Si modules are given in Table 4.22.

4.5 WIND EFFECT ON PV SYSTEMS

The PV panels can be placed either on building roofs or in the fields, always exposed to wind. The wind sometimes has a positive effect on the PV array performance and sometimes has a negative effect. Due to the ventilating effect of the wind, it cools down the cell temperature of the PV panels and thus increases the PV performance. The effect of wind on the PV cell temperature is examined comprehensively in the above sections. Besides, air movement (wind) reduces the relative humidity by a certain limit, which also causes to increase in PV performance a little bit. However, the wind spreads out the sand and dust in the air, which causes to accumulate the dust and sands on the surface of the PV panel.

The wind flow has also negative effects on the mechanical system of the PV panels. Due to the random behavior of the wind, the pressure distribution is varying all the time and it is different at each point of the panel surface. Particularly the wind flows over 80 km/h can cause damage to PV panels. Due to the fluctuating wind loads, a complete breakdown can happen if a certain level of vibration frequency is reached. As discussed in the literature, many wind patterns are circular (vortexes), and cause lift force under the PV panels. Hence, the most vulnerable parts of a PV array are the corners, and they often depend on the position and orientation of the panels in the environment. That's why all PV panels must be tested to withstand certain wind loads. Since the wind loads, ice/snow loads, ground survey, and other mechanical calculations will be examined in Chapter 13, the discussion regarding the wind effect is considered to be sufficient here.

4.6 SOILING/DUST EFFECT ON PV SYSTEMS

Sand and dirt particles are collected on the surface of PV modules which affect the photovoltaic performance negatively. The soiling effect becomes more influential for the PV modules with lower tilt angles. Dust deposition on PV modules has a relationship with the location, climate, the modules' tilt angle, and the frequency of rain at the site.

In the literature, numerous experiments and observations have been made to estimate the power losses caused by soiling. For example, annual system losses due to soiling were found to be approximately 5% in a study conducted in Los Angeles. According to another study conducted in Mexico City during the 60 days without rain, dust deposition rate on horizontal surface varied between ($24 \text{ g/m}^2 \text{day} - 102 \text{ g/m}^2 \text{day}$). Accordingly, it was found that the performance ratio of the modules has reduced by about 15% at the end of 60 days without rain. According to this study by Bernd Weber et al., the decrease in the daily performance ratio (PR_{Day}) of PV modules is expressed by equation (4.24) and Figure 4.9, depending on the number of days (N) without rain.

$$PR_{Day} = -0.0024N + 0.8378 \qquad (4.24)$$

Notice that the performance ratio in Figure 4.9 is 0.8378 for ($N = 0$), which means that the factor 0.8378 in (4.24) represents the decrease in the performance ratio due to the temperature effect.

To estimate the annual loss of photovoltaic energy generation, frequency of rainfall occurrence (f_0) at the considered site needs to be measured throughout the year. It is more appropriate to take 10-year average values from the Institute of Meteorology regarding the frequency of rainfall occurrence. For example, a typical f_0 values versus rainless periods (in days) are plotted in Figure 4.10.

Daily dust performance (DDP) is another factor needed to calculate the annual loss of photovoltaic performance due to soiling, which is given as (4.25).

$$DDP = D_0 \left[\frac{\%m^2}{g}\right] \cdot \cos b \cdot m_{dust} \left[\frac{g}{m^2 \cdot day}\right] \qquad (4.25)$$

where D_0 is a site-specific constant regarding the dust deposition rate (e.g., $0.0039 \frac{\%m^2}{g}$), b is the tilt angle of the PV panel and m_{dust} is the mass of dust deposited on the surface per square meter per day (typical value can be in the range of $20-120 \frac{g}{m^2 \cdot day}$).

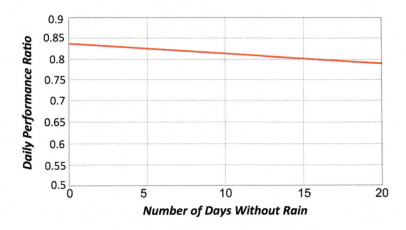

FIGURE 4.9 Daily performance ratio due to dust deposition for Mexico City.

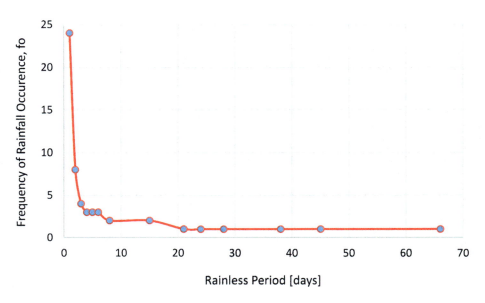

FIGURE 4.10 Typical frequency of rainless day occurrence.

Hence, the annual loss of energy generation can now be estimated as a function of DDP and f_0 as below:

$$\text{Energy Loss Ratio} = \frac{E_{\text{real}}}{E_{\text{clean}}} = 1 - \frac{\text{DDP}}{365} \sum_{i}^{N} f_{0,i} \cdot \frac{i}{2}(i-1) \quad (4.26)$$

where E_{real} is the annual real electrical energy generation out of the inverter, E_{clean} is the annual theoretical energy generation by the inverter under clean conditions, and the index i is the rainless day number corresponding to the frequency of rainfall occurrence.

Example 4.6: Dust Deposition Effect on PV Performance (Sandia Model)

A PV panel is planned to be installed in Las Vegas at a slope of 40° in the south direction. Estimate the annual energy loss ratio for the site with an average dust deposition of $65\,\text{gm}^{-2}\text{d}^{-1}$. Assume that the coefficient D_0 in (4.25) is $0.0039\%\,\text{m}^2\text{g}^{-1}$ and frequency of rainfall occurrence is given as Table 4.23.

Solution

To estimate the annual loss of energy generation ratio, we need to calculate DDP value first. From (4.25), we have:

$$\text{DDP} = D_0 \cdot \cos b \cdot m_{\text{dust}} = \frac{0.0039}{100} \times \cos 40 \times 65 = 0.001941$$

TABLE 4.23
Frequency of Rainfalls

Number of Rainless Days, i	1	2	3	4	5	6	8	15	21	24	28	38	45	66
Frequency of Occurrence, f_0	24	8	4	3	3	3	2	2	1	1	1	1	1	1

The energy loss ratio can be estimated from (4.26) as below:

$$\text{Energy Loss Ratio} = 1 - \frac{\text{DDP}}{365} \sum_{i}^{N} f_{0,i} \cdot \frac{i}{2}(i-1)$$

$$= 1 - \frac{0.001941}{365} \left[24 \cdot \frac{1}{2}(1-1) + 8 \cdot \frac{2}{2}(2-1) + \cdots + 1 \cdot \frac{66}{2}(66-1) \right]$$

$$= 1 - \frac{0.001941}{365}(5081) = 0.972$$

As a result, the expected annual performance loss from the PV system is estimated to be 2.7% under the specified conditions.

4.7 EFFECT OF RAIN AND HUMIDITY ON PV SYSTEMS

Rainfall has a significant impact on the performance of PV systems due to its dust removal and cooling effects on PV panels. The cleaning effect due to the rainfall will be higher where the panel tilt angle is greater than 10°. Therefore, it is strongly important to clean the panels manually to reduce soiling losses during long dry periods. According to the studies in the literature, it is recommended to clean the PV panels once a week or every two weeks, depending on the rate of soiling amount.

One negative effect of rainfall on the PV panel is related to relative humidity. In general, rainfall locally increases the relative humidity as a result of evaporation. The main effect of humidity on PV solar cells is corrosion. Naturally, higher temperatures and higher humidity levels accelerate the corrosion process. The corrosive effect mostly happens on the titanium-silver contact of PV cells in long-term humid environments.

Increased relative humidity also causes the surface to become sticky, and accordingly collects dirt and dust on the surface of the PV panel. So, a sticky surface due to increased humidity levels results in drops in the conversion efficiency of modules. It has been proven that the relative humidity in the vicinity of PV sites causes degradation in PV cell current, voltage, power, and conversion efficiency. Although many studies are showing this effect in the literature, there is no generalized formulation for the humidity effect on different PV technologies. However, according to the studies in the literature, the humidity effect on PV performance can be significant. For instance, a 50% decrease in cell efficiency can happen if relative humidity reaches from 60% to 95%.

4.8 SNOW AND ICING EFFECT ON PV SYSTEMS

The existence of snow covering the PV modules can reduce the energy output of the photovoltaics based on the snow density at the site. If the snow is light and melts easily, negligible energy losses can occur in the PV systems. However, more significant losses can occur when the snow covering is thick and does not easily melt. The power losses due to snow covering are based on the factors such as the type of snow, the weather conditions, and the tilt angle of the panels.

The negative effects that may occur in the PV panels due to snowfall are itemized briefly as below.

- In some cases, snow shadowing of a single cell can greatly drop the efficiency of an entire PV module.
- The most significant factor affecting snowing energy loss is the tilt angle of the PV panel. As the tilt angle increases, the panel self-cleaning time decreases. For example, in an experimental study, the energy losses of a 40° tilted panel were found to be approximately 50% less than that of a 30° tilted panel for the snow covering less than one inch.

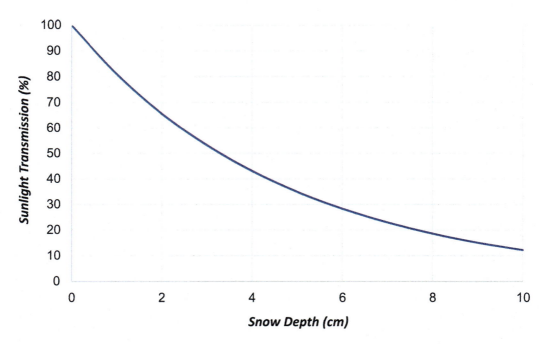

FIGURE 4.11 Typical sunlight transmission through the snow.

- Snow density could be a factor to compare the sites to each other. As known, the denser the snow, the more sunlight will get through the panel surface. Studies have shown that transmitted sunlight falls off exponentially through snow and ice. Typical sunlight transmission versus snow depth is plotted in Figure 4.11.
- As the snow melts, its water content increases and causes a non-uniform snow density. The water-snow interface causes a greater degree of sunlight to scatter on the photovoltaic cells, generating heat as they absorb the solar radiation. Accordingly, this heat increases the amount of melted snow on the module surface.
- Another important factor contributing to energy loss at the site is the cloudy days following heavy snow. Because, even if the ambient temperature is at freezing temperature, sunlight reaching the surface increases the cell temperature, which causes the surface snow to be melted. However, the panel does not receive enough sunlight to generate heat for melting the snow in cloudy situations.

Consequently, it would be useful to measure the amount of energy loss due to snow cover at the site with heavy snowfall. Hence, site-specific solutions are needed to be found to minimize these losses.

4.9 SHADING AND BYPASS DIODES EFFECT ON PV SYSTEM

Shading is one of the most significant factors, which can affect the PV system's performance and safety. Many factors affect the shading phenomenon, such as heavy clouding, hazing, soiling, bird droppings, environmental plant covering, buildings, and roof-top structures. Generally, PV output power, voltage, and current are affected differently by shading, which results in complications in addressing true derated levels of PV characteristic parameters. Different levels of shading (partial or full shading) will lead to different performance variations in the entire PV array. This is mainly because of the activation of the bypass diodes enclosed inside the module junction box.

Effect of Environmental Conditions on the Performance of Photovoltaic Systems

4.9.1 Partial Shading and Hotspot Phenomenon

The PV arrays in an outdoor environment are exposed to non-uniform insolation due to partial shading conditions as in Figure 4.12. During partial shading, PV cells in non-shaded parts of the PV modules continue to operate at normal efficiency under uniform irradiance.

Since the current flowing through every cell in a series configuration is naturally unchanged, the shaded cells are operated with a reverse bias voltage to deliver the same current as of the illuminated cells. However, the polarity of the resulting reverse power causes power consumption in the shaded cells, reducing the maximum output power of the PV module. Hence exposing the shaded cells to reverse bias over-voltage, hotspots (in one cell or a group of the cell) happen on the partially shaded module. The hotspot phenomenon due to physical damage may create an open circuit in the entire PV module. The hotspot problem is often resolved with the inclusion of a bypass diode to a specific number of cells in the series circuit as in Figure 4.13. These diodes are connected in parallel to a specific group of series-connected cells to create an alternative current path in case of shadowed or broken cell conditions.

Figure 4.14 shows the I–V curves of PV cells operating under forward and reverse bias to illustrate the effect of hotspot heating as a function of parallel leakage resistance R_P (shunt resistance of PV cells).

It is clear from Figure 4.14 that the decrease in shunt resistance (R_P) affects the $I - V$ slope of a PV cell in the reverse direction. Figure 4.15a shows a bypass diode covering n PV cells; one of them is working under shaded conditions, while the rest are free of shadow. Figure 4.15b represents the hotspot visualization in a series configuration and I–V characteristics of a combined series configuration with one cell shadowed. Note that the effect of bypass diode operation is not included in the I–V curves in Figure 4.15b.

4.9.2 Estimation of Bypass Diodes Included in a Junction Box

As a result of dissipated power in the reverse bias mode, hotspots can cause a breakdown in the shaded cells. Each PV cell in the reverse breakdown region can dissipate power losses up to a specific level without hotspot damage. In reverse bias operation, the current flowing through the PV cell does not increase drastically until the breakdown voltage is reached. If we determine a safety factor of 80% for breakdown voltage (V_b), the dissipated power by the shaded PV cell will stay under

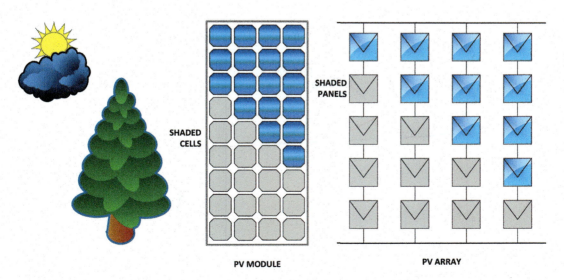

FIGURE 4.12 PV system under partially shaded conditions.

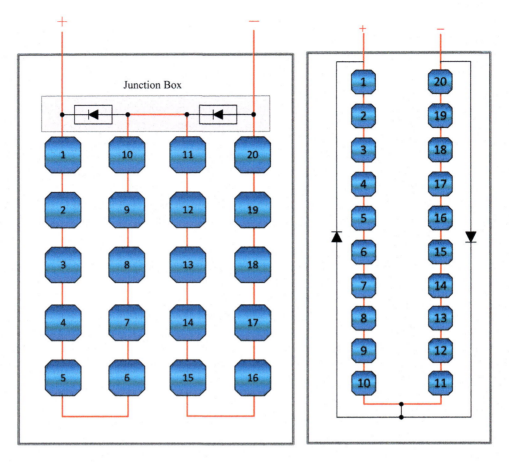

FIGURE 4.13 A typical application of bypass diodes in a PV module.

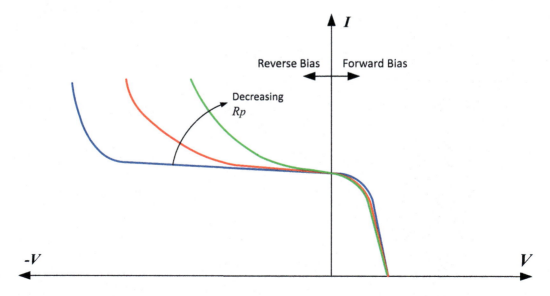

FIGURE 4.14 Illustration of I–V curves of PV cells operating under reverse bias mode and the effect of shunt resistances.

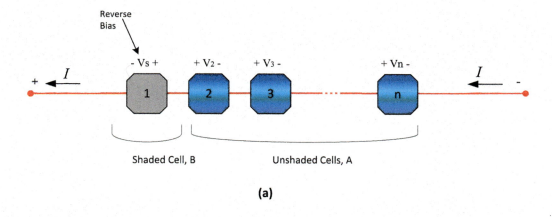

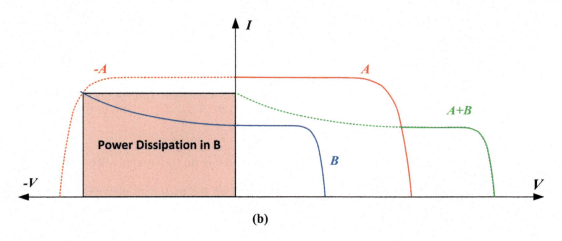

FIGURE 4.15 (a) The electrical connection of n identical PV cells with a bypass diode (b) I–V curves of shaded, non-shaded, and combined series configuration (with no bypass diode).

the maximum allowed power. Considering Figure 4.15a, the reverse voltage across the shaded cell (V_{shaded}) must provide the following condition:

$$V_{shaded} \geq V_D + \sum_{i=1}^{n-1} V_i \tag{4.27}$$

Since the maximum voltage level in forwarding bias mode is open-circuit voltage (V_{OC}), equation (4.27) can be rewritten as follows:

$$V_{shaded} \geq V_D + \sum_{i=1}^{n-1} V_{OC} \tag{4.28}$$

Considering all PV cells are identical, equation (4.28) is modified as below:

$$V_{shaded} \geq V_D + (n-1)V_{OC} \tag{4.29}$$

By using the safety factor of 80% for breakdown voltage ($V_{shaded} = 0.8 \times V_b$) as the maximum allowed voltage at the shaded PV cell (reverse bias operation), the maximum number of PV cells (n_{max}) to be protected by a bypass diode can be calculated as (4.30).

$$n_{max} \leq \frac{0.8 \times V_b - V_D}{V_{OC}} + 1 \qquad (4.30)$$

Note that bypass diode reverse voltage (V_{Dr}) should be bigger than the maximum open circuit of the PV panel divided by the number of the bypass diode (n_D) included in the junction box.

$$V_{Dr} > \frac{V_{OC,max}}{n_D} \qquad (4.31)$$

where $V_{OC,max}$ is the open-circuit voltage at the minimum ambient temperature with maximum irradiance, which can be expressed as (4.32).

$$V_{OC,max} @ T_{c,min} = V_{ocr} + \beta_V \times (T_{c,min} - T_r) \qquad (4.32)$$

where V_{OCr} are reference (or rated) open-circuit voltage at the reference temperature T_r (25°C) and β_V are the manufacturer-supplied temperature coefficients of the open-circuit voltage in (V/°C). In the worst case, $T_{c,min}$ can be calculated from (3.22) based on the ambient temperature of (−40°C).

Example 4.7: Bypass Diode Calculation

A 36-volt PV module consisting of 60 cells, each having 100 cm² (10 cm × 10 cm), will be designed based on the following parameters (Table 4.24).

Determine the possible PV module dimension based on the maximum number of bypass diodes to protect the series-connected cells in the module.

Solution

Based on the given data, the maximum number of PV cells to bridge with bypass diodes can be calculated according to (4.30).

$$n_{max} \leq \frac{0.8 \times V_b - V_D}{V_{OC}} + 1 = \frac{0.8 \times 18 - 1}{0.7} + 1 = 20.1 \Rightarrow n_{max} \leq 20.1$$

In this way, the maximum number of cells per one bypass diode is found to be 20. Hence, the number of bypass diodes (n_D) in the module:

$$n_D = \frac{60}{20} = 3$$

As a result, the dimensions of the PV module suitable for lamination can be designed as Figure 4.16.

TABLE 4.24
The PV Cell's Parameters

V_{mp} of a Cell	V_{OC} of a Cell	Breakdown Voltage of a Cell (V_b)	Diode Voltage (V_D)
0.6 volt	0.7 volt	18 volt	1 volt

Effect of Environmental Conditions on the Performance of Photovoltaic Systems

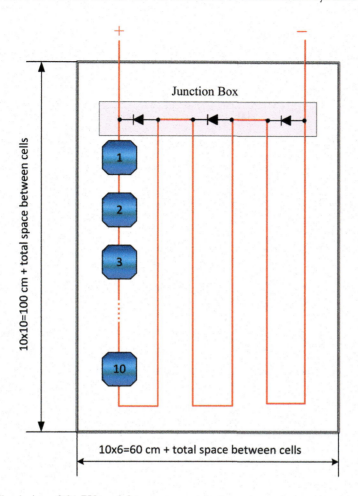

FIGURE 4.16 The design of the PV module.

Example 4.8: Bypass Diode Sizing

A PV module consisting of 68 cells will be designed for the site with a minimum ambient temperature of (−30°C) and maximum irradiance of 1000 W/m². The cell parameters under standard reference conditions are given in Table 4.25.

Calculate the number of bypass diodes to protect the series-connected cells in the module. Determine the minimum required reverse breakdown voltage of a bypass diode used in the junction box.

TABLE 4.25
The PV Cell Parameters under STC

V_{mp}	V_{OC}	Breakdown Voltage V_m	Diode Voltage V_D	NOCT	Temperature Coefficient $\beta_{V_{OC}}$
0.6 volt	0.7 volt	18 volt	1 volt	47°C	−0.0025 V/°C

Solution

Initially, the cell temperature under given atmospheric conditions should be calculated. Based on the (3.22), we have:

$$T_c = T_{amb} + \frac{(T_{NOCT} - 20)}{800} \times G = -30 + \frac{(47-20)}{800} \times 1000 = 3.75°C$$

Maximum open-circuit voltage can then be estimated from (4.32) as below:

$$V_{OC,max} = V_{OCr} + \beta_V \times (T_{c,min} - T_r) = 0.7 - 0.0025 \times (3.75 - 25) = 0.753 \, V$$

Now, the maximum number of PV cells to be bridged with bypass diodes can be calculated from (4.30) as below.

$$n_{max} \leq \frac{0.8 \times V_b - V_D}{V_{OC}} + 1 = \frac{0.8 \times 18 - 1}{0.753} + 1 = 18.79 \cdot n_{max} \leq 18$$

So, the maximum number of cells per one bypass diode is found to be 18. Hence, the number of bypass diodes (n_D) in the module:

$$n_D = \frac{72}{18} = 4$$

As a result, reverse breakdown voltage (V_{Dr}) of a bypass diode can be selected based on (4.31) as below:

$$V_{Dr} > \frac{V_{OC,max}}{n_D} > \frac{72 \times 0.753}{4} > 13.554 \, V$$

If we select a safety factor of 1.2, V_{Dr} value for a bypass diode can be selected as $V_{Dr} > 16.2 \, V$.

4.9.3 Mathematical Modeling of Partially Shaded PV System

PV system models are needed to include all essential parameters to analyze the normal operating as well as partial shading conditions. The widely known photovoltaic cell electrical equivalent circuit for a PV cell is the classical one-diode model, which is sufficient to analyze the shaded PV cells. Figure 4.17 demonstrates the circuit model of n series-connected modules with their bypass diodes in a PV array.

Considering the circuit model in Figure 4.17, it is possible to analyze the shaded and unshaded modes of operation. It is vital to note that the characteristic curves of a PV module with bypass diodes are different than that of a PV module without bypass diodes. Because bypass diodes create an alternate path that bridges the shaded cells. For that reason, partially shaded cells no longer carry the same current. Therefore, multiple peak points (maxima) may occur in the power-voltage ($P-V$) curve of a PV module with shaded cells and bypass diodes. An example $P-V$ curve is shown in Figure 4.18.

If the generated current (I_{PV}) of i th module during the shaded condition drops to less than array current (I_A), the bypass diode limits the reverse voltage to be smaller than the breakdown voltage of the PV cells. In other words, the i th bypass diode shown in Figure 4.17 will be activated when ($I_{PV}(i) < I_A$). In order to better explain the purpose of bypass diode usage in PV modules, the operating phases are illustrated in Figure 4.19. The bypass diode is blocked when all cells are illuminated. In this case, the bypass diode voltage, $V_D(i)$ is equal to ($-n \cdot V_{cell}$). If one or several cells are shadowed, the bypass diode will conduct. In this case, the bypass diode voltage, $V_D(i)$ is equal to ($V_{shaded} - (n-1) \cdot V_{cell}$) where n is the number of cells per bypass diode.

Effect of Environmental Conditions on the Performance of Photovoltaic Systems

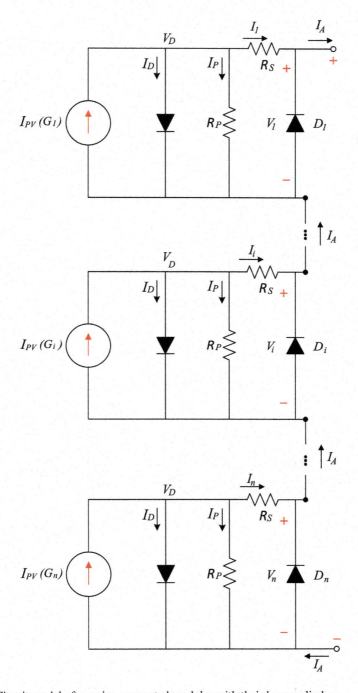

FIGURE 4.17 Circuit model of n series-connected modules with their bypass diodes.

It is possible to model a partially shaded module/array under two groups as shaded cells/modules and unshaded cells/modules, each group exposed to a different level of irradiance. Let's first assume that there is no bypass diode used in the PV module. In this case, the circuit model for the partially shaded module can be illustrated in Figure 4.20.

The module is composed of n_s cells connected in series. Among these PV cells, r shaded cells receive irradiance G_1 and $(n_s - r)$ unshaded cells receive irradiance G_2. Hence, the output current

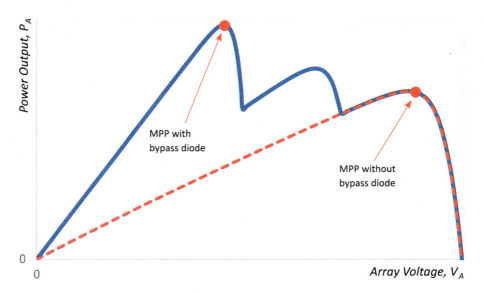

FIGURE 4.18 The power-voltage curve of a PV array under partial shading conditions.

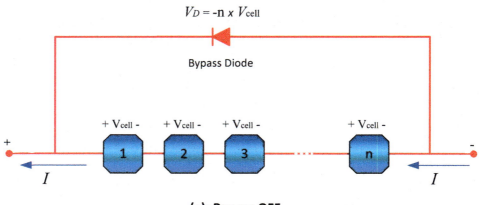

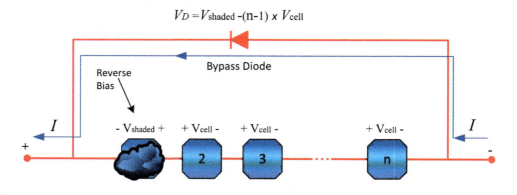

FIGURE 4.19 Bypass diode operating phases.

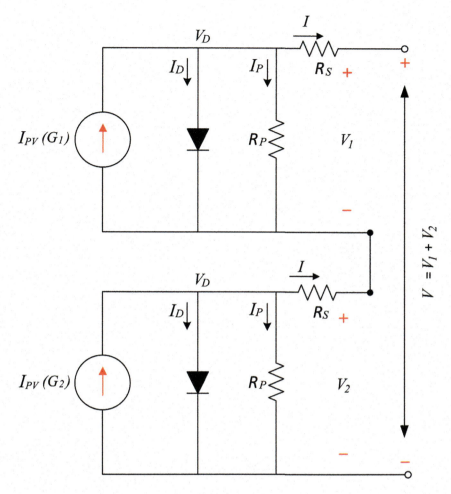

FIGURE 4.20 Circuit model of a partially shaded module without bypass diodes.

(I_{shaded}) and voltage (V_{shaded}) of partially shaded PV module/array without bypass diode application are equal to the following mathematical statements.

$$I_{shaded} = \min(I_1, I_2) \tag{4.33}$$

$$V_{shaded} = V_1 + V_2 \tag{4.34}$$

where subscripts 1 and 2 refer to the cells under irradiance of G_1 and G_2, respectively.

Secondly, let's consider the PV module/array with bypass diodes. In this case, the $I-V$ curve of the series-connected PV system consists of two zones (unshaded and shaded zones) associated with the unshaded and shaded parts. The output current (I_{system}) and voltage (V_{system}) of a partially shaded series-connected PV system consisting of bypass diodes can be written as the following expression.

$$I_{system} = \begin{cases} I_1 & \text{for Zone-1,} \quad I > I_{PV}(G_1) \\ I_2 & \text{for Zone-2,} \quad I < I_{PV}(G_2) \end{cases} \tag{4.35}$$

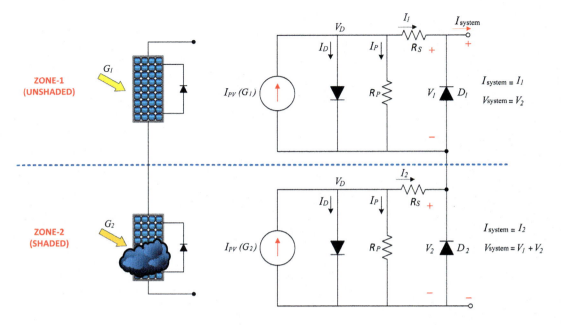

FIGURE 4.21 Block diagram and circuit model of series PV assembly with bypass diodes and partially shaded mode.

$$V_{system} = \begin{cases} V_1 & \text{for Zone-1,} \quad I > I_{PV}(G_1) \\ V_1 + V_2 & \text{for Zone-2,} \quad I < I_{PV}(G_2) \end{cases} \quad (4.36)$$

where the PV output currents I_1 and I_2 refer to the modules currents under irradiance of G_1 and G_2, respectively. The output current equations can be modified based on the (3.34) as below:

$$I_1 = I_{PV}(G_1) - I_{o1}\left[e^{\frac{V_1 + I \cdot N_{S1} \cdot R_S}{N_{S1} \cdot nV_t}} - 1\right] - \frac{V_1 + I \cdot N_{S1} \cdot R_S}{N_{S1} \cdot R_P} \quad (4.37)$$

$$I_2 = I_{PV}(G_2) - I_{o2}\left[e^{\frac{V_2 + I \cdot N_{S2} \cdot R_S}{N_{S2} \cdot nV_t}} - 1\right] - \frac{V_2 + I \cdot N_{S2} \cdot R_S}{N_{S2} \cdot R_P} \quad (4.38)$$

The circuit model and block diagram of series-connected configuration operating under partially shaded and unshaded conditions are given as illustrated in Figure 4.21. The Zone-1 (unshaded) and Zone-2 (shaded) formations defined in equations (4.35) and (4.36) are shown on the $(I-V)$ and $(P-V)$ curves as illustrated in Figure 4.22.

By following the similar methodology explained above, the output characteristics of the partially shaded PV system can easily be obtained for the occurrence of three and more zones.

Example 4.9: Shading Effect on the Performance of PV Modules

The $I-V$ characteristic of a PV module consisting of a 72-series connected cell is given below under 1000 W/m² radiation (Figure 4.23). Each cell in the module has a 6 Ω shunt resistance and they produce 0.5 volt (open-circuit voltage) at full sun (1000 W/m²) conditions.

Effect of Environmental Conditions on the Performance of Photovoltaic Systems 169

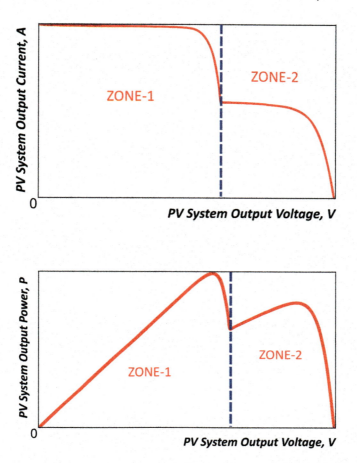

FIGURE 4.22 Output characteristics (*I–V* and *P–V* characteristics) of partially shaded series PV system with bypass diodes.

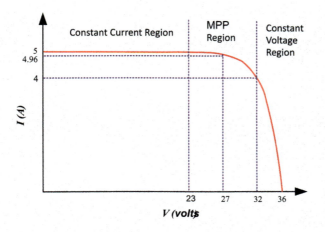

FIGURE 4.23 The PV cell's parameters.

a. When one cell and two cells are completely shaded, plot the module $I-V$ curves respectively. Assume that there is no bypass diode application in the module.
b. Assume that the PV module is charging a 24 V battery bank. Calculate the expected charging currents under the given conditions (full sun radiation, one cell-shaded, two cell-shaded).
c. Find the MPP points when one cell is fully shaded.
d. Suppose that every 18 cells in a given PV module possess a bypass diode. Draw the $I-V$ curve of the module when only one cell in the module is totally shaded.

Solution

a. Let us first calculate the voltage drop at the module caused by the shaded cell. Now consider the nth cell in the module is completely shaded (Figure 4.24).

The generated current in series-connected configuration flows through $(R_P + R_S)$ of the shaded cell. Hence, the shaded cell becomes reverse biased and its terminal shows a negative voltage equal to $-I(R_P + R_S)$. So, the output voltage of the module with one cell-shaded (V_{Shaded}) is

$$V_{Shaded} = (n-1)V_{cell} - I(R_P + R_S)$$

where $V_{cell} = V/n$ and V is module output under unshaded operation. As a result, the voltage drop due to the shaded cell:

$$\Delta V = V - V_{Shaded} = V - (n-1)\frac{V}{n} + I(R_P + R_S)$$

$$\Delta V = V - V_{Shaded} = \frac{V}{n} + I(R_P + R_S)$$

Since R_S is neglected in the given example,

$$\Delta V \cong \frac{V}{n} + IR_P$$

So, the voltage drop effect caused by a shaded cell can be drawn on the $I-V$ curve given as in Figure 4.25.

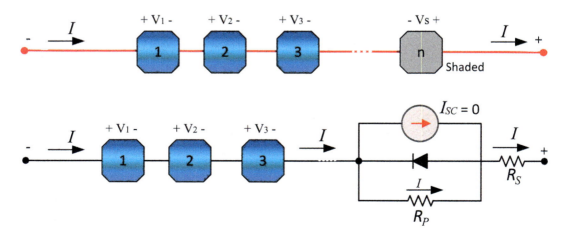

FIGURE 4.24 A shaded cell in a PV module.

Effect of Environmental Conditions on the Performance of Photovoltaic Systems

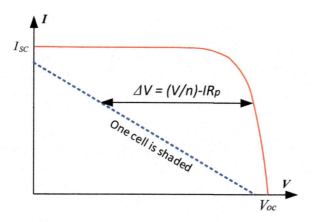

FIGURE 4.25 The effect of a shaded cell on the I–V curve.

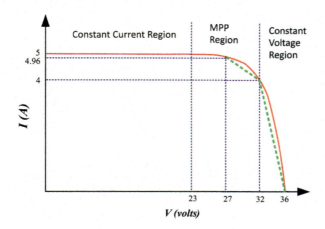

FIGURE 4.26 Linearization of I–V curve.

TABLE 4.26
Simplified Calculation Table for the Voltage Drops due to Shaded Cells

	1-Cell Shaded $R_p = 6\,\Omega$						2-Cell Shaded $R_p = 12\,\Omega$		
$I(A)$	0	4	4.96	5	5	5	0	4	4.96
V (volts)	36	32	27	18	5	0	36	32	27
$\Delta V = \dfrac{V}{n} + IR_p$	0.5	24.88	30.51	30.5	30.13	30	1	48.88	60.27
$V - \Delta V$ (volts)	35.5	7.12	−3.51	−12.5	−25.13	−30	35	−16.88	−33.27

To perform the hand calculation simpler, let's make a slight simplification (linearization) on the $I-V$ curve as shown in Figure 4.26.

In order to obtain the I–V curves of the module in the case of one-cell and two-cell shadings, voltage drops due to shaded cells should be calculated. For this purpose, a simple calculation table (Table 4.26) is created as below. The calculation points in the table are chosen based on the simplified $I-V$ curve given in (Figure 4.26).

If we specify the data calculated above on the graph, the $I-V$ curves for shaded cell conditions are obtained as in Figure 4.27.

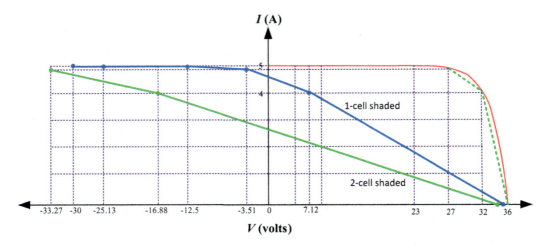

FIGURE 4.27 I–V curves of shaded modules.

b. Battery charging current can easily be obtained from the $I-V$ equations including 24 V operating point. For this purpose, $I-V$ equations for shaded conditions can be written as follows:

$$I = \begin{cases} -0.1409 \cdot V + 5, & \text{for one-cell shaded} \\ -0.0771 \cdot V + 2.6985, & \text{for two-cell shaded} \end{cases}$$

Hence, the corresponding battery charging currents are:

$$I = \begin{cases} 1.618\,\text{A}, & \text{for one-cell shaded} \\ 0.848\,\text{A}, & \text{for two-cell shaded} \end{cases}$$

c. MPP points for a single cell-shaded condition can be obtained from the expression $\frac{dP}{dV} = 0$. For one-cell shaded condition, the $I-V$ equation from (c) is:

$$I = -0.1409 \cdot V + 5 \Rightarrow P = V \cdot I = -0.1409 \cdot V^2 + 5V$$

$$\frac{dp}{dV} = -0.2818V + 5 = 0 \Rightarrow V = 17.74 \text{ volts}$$

$$\Rightarrow I = 2.5\,\text{A}$$

So, $P_{max} = 17.74 \times 2.5 = 44.35$ W

d. Voltage drop due to single shaded cell is calculated above. Hence, the $I-V$ curve of the module affected by a shaded cell can be plotted as Figure 4.28 (with bypass diodes).

4.9.4 BLOCKING DIODES AND PARTIAL SHADING

To minimize the shading/mismatching losses, additional diodes, called "blocking diodes" are utilized in the photovoltaic strings in addition to the use of bypass diodes. Another purpose of using blocking diodes is to prevent the PV system from possible current flows from the battery

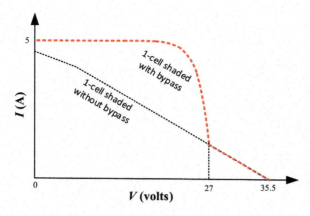

FIGURE 4.28 *I–V* curves of shaded modules with bypass diodes.

banks at night times. The application of blocking diodes along with bypass diode usage is shown in Figure 4.29. As can be seen in the figure, each string connected in parallel should have a blocking diode, which prevents the current from flowing from the ones with higher voltages.

As in the series-connected mismatching case, it is also possible to divide the partially shaded PV array into two groups as shaded and unshaded strings. Assume that each group is exposed to the different levels of irradiance as G_1 and G_2. In this case, the circuit model for the partially shaded array is illustrated in Figure 4.30.

Considering the array configuration in Figure 4.30, it is possible to analyze the effect of blocking diode on the characteristic curves of partially shaded operation. In this case, the $I-V$ curve of the parallel-connected PV array is formed based on two zones (unshaded and shaded zones) associated with the unshaded and shaded strings. The output current (I_{Array}) and voltage (V_{Array}) of partially shaded parallel-connected PV system consisting of blocking diodes can be written as the following expression.

$$I_{\text{Array}} = \begin{cases} I_{\text{string}(1)} + I_{\text{string}(2)} & \text{for Zone-1,} \quad 0 \leq V \leq V_{\text{string}(2)} \\ I_{\text{string}(1)} + I_{\text{string}(2)} & \text{for Zone-2,} \quad V_{\text{string}(2)} \leq V \leq V_{\text{string}(1)} \end{cases} \quad (4.39)$$

$$V_{\text{Array}} = \begin{cases} V_{\text{string}(2)} & \text{for Zone-1,} \quad 0 \leq V \leq V_{\text{string}(2)} \\ V_{\text{string}(1)} & \text{for Zone-2,} \quad V_{\text{string}(2)} \leq V \leq V_{\text{string}(1)} \end{cases} \quad (4.40)$$

where the PV output currents $I_{\text{string}(1)}$ and $I_{\text{string}(2)}$ refer to the PV string currents under unshaded and shaded conditions, respectively. Considering Figure 4.30, the Zone-1 (unshaded) and Zone-2 (shaded) formations defined in equations (4.39) and (4.40) are shown on the $(I-V)$ and $(P-V)$ curves as in Figure 4.31.

It is very important to draw attention to an important point here. Normally, the partial shading/mismatching effect is a very complicated phenomenon, and output characteristic curves are affected differently based on many factors such as PV cell technology type, non-uniform shading/soiling, PV assembly configuration (series, parallel, series-parallel, total cross-tied, and bridge-linked), non-equivalency in PV modules, non-uniform illumination, number of bypass diode in the modules, etc. Therefore, the above-mentioned formulations and characteristic curves are given for the specified conditions. However, using the principles identified in this section, it is possible to perform accurate shading analyzes and performance evaluations of the PV systems.

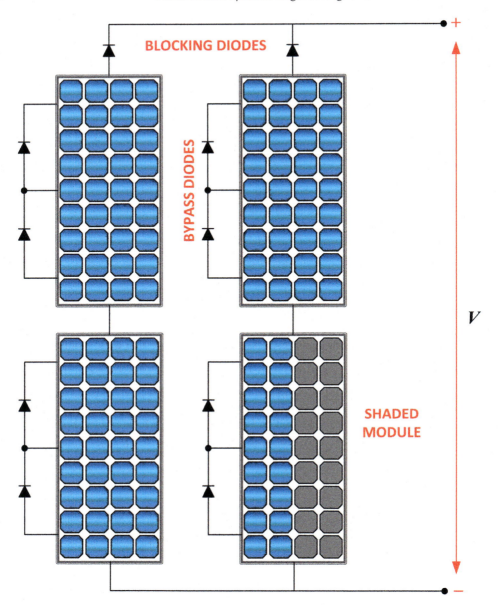

FIGURE 4.29 PV array configuration with bypass and blocking diodes. The blocking diode on the shaded string prevents the current flow from the unshaded parallel string into the shaded string.

Example 4.10: Shading Effect Considering Bypass and Blocking Diodes

The $I-V$ characteristics of partially shaded strings in a PV array with bypass/blocking diodes are given below (Figure 4.32). Draw the output $I-V$ characteristic for the entire PV array.

Solution

It is understood from $I-V$ characteristic curves of each string that three different zones will be formed for the entire PV array. Considering the maximum voltage range across the parallel-connected strings, the $I-V$ characteristic curve for the entire PV system can be plotted below (Figure 4.33).

Effect of Environmental Conditions on the Performance of Photovoltaic Systems

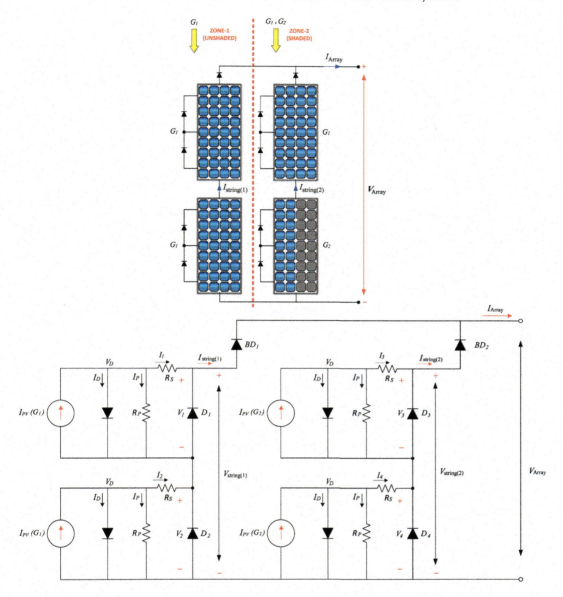

FIGURE 4.30 PV array configuration and circuit model under irradiances of G_1 and G_2 (bypass and blocking diodes included).

4.10 TILT ANGLE DETERMINATION FOR STAND-ALONE AND GRID-CONNECTED SYSTEMS

As known, the performance of PV systems is significantly affected by the orientation and tilt angle of the panels. The PV panel azimuth should be faced to true south (south in the northern hemisphere and north in the southern hemisphere) for fixed-tilt angle assemblies. On the other hand, adjusting the tilt angle for photovoltaics is much more complex and depends on the structure of the PV system, the place of its use, and the purpose of its operation. Putting aside the physical constraints in adjusting the tilt angle, we have two different systems that need to be addressed and resolved. These are grid-connected and stand-alone PV systems. The reason that makes the difference in adjusting the tilt angle is hidden behind their operating purposes. For example, large-scale grid-connected

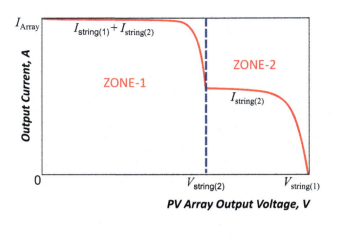
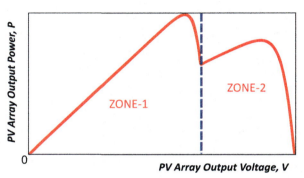

FIGURE 4.31 Output characteristics (*I–V* and *P–V* characteristics) of partially shaded parallel PV strings with blocking diodes (based on Figure 4.30).

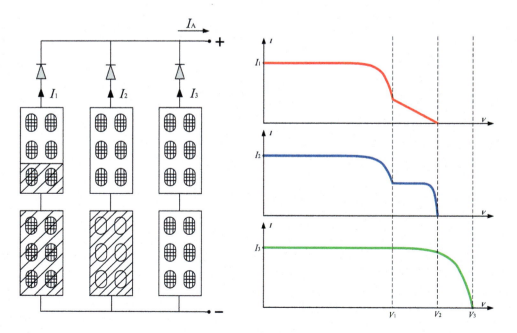

FIGURE 4.32 *I–V* characteristics of partially shaded strings in a PV array having bypass and blocking diodes.

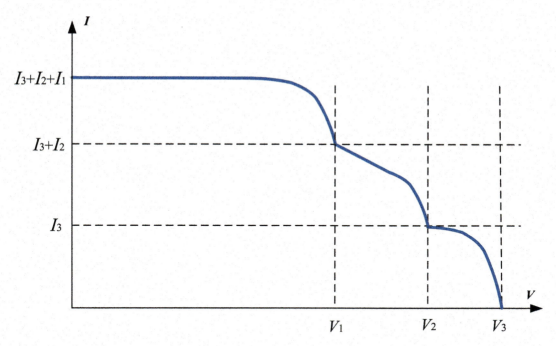

FIGURE 4.33 *I–V* characteristics of the entire PV array.

PV systems should be positioned with the optimum tilt angle to maximize the monthly, seasonal, or annual energy outputs. However, the combination of load and solar radiation should be considered together for stand-alone PV systems. This is because stand-alone systems should have the ability to feed the load even under the worst-case conditions. These issues are explained in the following sections.

4.10.1 Tilt Angle for Grid-Connected Systems

Large-scale grid-connected PV systems deliver all the generated electrical energy to the grid. For this reason, the tilt angle of the PV panels should be determined to maximize the annual energy output. In other words, the optimal tilt angle for any solar site should be calculated by maximizing the solar radiation incident on a sloped surface. Therefore, total irradiation falling on a tilted surface can be found by varying the tilt angle "b" from 0° to 90°. In fixed-tilt systems, the optimum tilt angle can be calculated on an hourly, daily, monthly, seasonal, or annual basis, depending on the design purpose of the PV system. To determine the optimal tilt angle, total solar insolation on a tilted surface must be known on an hourly, daily, and monthly basis.

4.10.1.1 Hourly Tilt Angle

As known, the total incident radiation on a tilted surface is the sum of the direct (or beam), diffuse, and ground-reflected radiations. However, diffuse and ground-reflected components are site-specific and negligible. Therefore, direct beam radiation on a tilted surface given by (4.48) can be used to find the optimum tilt angle on an hourly basis.

$$I_{db} = C_D I_b \qquad (4.41)$$

where C_D is called tilt conversion factor, and can be expressed as the ratio of hourly extraterrestrial incident radiation on a tilted surface (at angle b) to that of a horizontal surface.

$$C_D = \frac{I_{0b}}{I_0} \approx \frac{\cos\theta}{\cos\beta} \tag{4.42}$$

where θ is the solar incidence angle and β is the zenith angle. From equations (2.11) and (2.16), the tilt conversion factor C_D can be arranged as (4.43).

$$C_D = \frac{(A+C)\cos(b)+(E+D-B)\sin(b)}{F+G} \tag{4.43}$$

where

$A = \sin(L)\sin(\delta)$

$B = \cos(L)\sin(\delta)\cos(z)$

$C = \cos(L)\cos(\delta)\cos(\omega)$

$D = \sin(L)\cos(\delta)\cos(z)\cos(\omega)$

$E = \cos(\delta)\sin(z)\sin(\omega)$

$F = \sin(\delta)\sin(L)$

$G = \cos(\delta)\cos(L)\cos(\omega)$

To find the optimum tilt angle, equation (4.43) should be differentiated with respect to the tilt angle b and set to zero. So, let us first find the optimum tilt angle of equation (4.43) for a fixed PV orientation. In short, the following equation (4.48) must be solved for the optimum tilt angle.

$$\frac{d}{db}[C_D] = 0 \tag{4.48}$$

From the solution of equation (4.48), b_{opt} value for any instantaneous time on Earth is found as follows.

$$b_{\text{opt}} = \tan^{-1}\left[\frac{\cos(\delta)\sin(z)\sin(\omega)+\sin(L)\cos(\delta)\cos(z)\cos(\omega)-\cos(L)\sin(\delta)\cos(z)}{\sin(L)\sin(\delta)+\cos(L)\cos(\delta)\cos(\omega)}\right] \tag{4.49}$$

The parameters in the above expression are previously defined in Chapter 2. Therefore, it is not repeated herein. For $z = 0°$ from equation (4.49), the optimum tilt angle b_{opt} for south-facing surfaces in the northern hemisphere can be found as follows ($\sin 0 = 0$, and $\cos 0 = 1$).

$$b_{\text{opt}} = \tan^{-1}\left[\frac{\sin(L)\cos(\delta)\cos(\omega)-\cos(L)\sin(\delta)}{\sin(L)\sin(\delta)+\cos(L)\cos(\delta)\cos(\omega)}\right] \tag{4.50}$$

Effect of Environmental Conditions on the Performance of Photovoltaic Systems

Similarly, for $z = -180°$ from equation (4.49), the optimum tilt angle b_{opt} for north-facing surfaces in the southern hemisphere can be found as follows $[\sin(-180) = 0$, and $\cos(-180) = -1]$.

$$b_{opt} = \tan^{-1}\left[\frac{\cos(L)\sin(\delta) - \sin(L)\cos(\delta)\cos(\omega)}{\sin(L)\sin(\delta) + \cos(L)\cos(\delta)\cos(\omega)}\right] \quad (4.51)$$

From equations (4.50) and (4.51), hourly $b_{opt(h)}$ value can be obtained by considering the midpoint of a particular hourly period. For example, for the period from 13.00 to 14.00, the hour angles are 15° and 30°, respectively. Therefore, the hourly optimum tilt angle can be calculated by taking the ω hour angle $(15+30)/2 = 22.5°$ for this particular period. Therefore, a general expression for the hourly optimum tilt angle can be written for a particular hourly period, covering from ω_1 to ω_2 as below.

$$b_{opt\,(h)} = b_{opt}\big|_{\omega = (\omega_1 + \omega_2)/2} \quad (4.52)$$

Example 4.11: Optimum Tilt Angle at Solar Noon

Calculate the optimum tilt angles at solar noon for the 15th day of each month for latitudes from 20 to 60.

Solution

Let us consider latitudes 20, 30, 40, 50, and 60 for the calculations. To calculate the optimum tilt angle at solar noon on January 15th, the parameters in equation 4.50 should be determined as a first step.

Julian day number, $N = 15$ on January 15, and from equation (2.9):

$$\delta = 23.45 \cdot \sin\left[\frac{360}{365}(15+284)\right] = -21.269°$$

The hour angle at solar noon is 0°, and from (4.50) for $L = 20$, we get:

$$b_{opt} = \tan^{-1}\left[\frac{\sin(20)\cos(-21.269)\cos(0) - \cos(20)\sin(-21.269)}{\sin(20)\sin(-21.269) + \cos(20)\cos(-21.269)\cos(0)}\right]$$

$$b_{opt} = \tan^{-1}\left[\frac{0.342 \times 0.931 - 0.939 \times (-0.362)}{0.342 \times (-0.362) + 0.939 \times 0.931}\right] = 41.26°$$

Note that b_{opt} value at solar noon is given by the following equation for the northern hemisphere.

$$b_{opt} = \tan^{-1}\left[\frac{\sin(L)\cos(\delta) - \cos(L)\sin(\delta)}{\sin(L)\sin(\delta) + \cos(L)\cos(\delta)}\right] = \tan^{-1}\left[\frac{\sin(L-\delta)}{\cos(L-\delta)}\right] = \tan^{-1}\left[\tan(L-\delta)\right] = L - \delta$$

Similarly, from (4.51), b_{opt} value at solar noon is given by the following equation for the southern hemisphere.

$$b_{opt} = \tan^{-1}\left[\frac{\cos(L)\sin(\delta) - \sin(L)\cos(\delta)}{\sin(L)\sin(\delta) + \cos(L)\cos(\delta)}\right] = \tan^{-1}\left[\frac{-\sin(L-\delta)}{\cos(L-\delta)}\right]\tan^{-1}\left[-\tan(L-\delta)\right] = \delta - L$$

From the simplified equations above, the b_{opt} value for the northern hemisphere will yield the same result.

$$b_{opt} = L - \delta = 20 - (-21.26) = 41.26°$$

Similarly, the b_{opt} value for the southern hemisphere will yield:

$$b_{opt} = \delta - L = -21.26 - 20 = -41.26°$$

All other calculations to find optimum tilt angles can similarly be done at solar noon. Therefore, only the calculated results are given in Table 4.27. In addition, Figure 4.34 shows the optimum tilt angle variation versus the 15th day of each month and latitudes.

TABLE 4.27
Optimum Tilt Angles (in Degrees) at Solar Noon of Different Latitudes

Considered Day	Julian Day Number, N	$L = 20$ b_{opt}	$L = 30$ b_{opt}	$L = 40$ b_{opt}	$L = 50$ b_{opt}	$L = 60$ b_{opt}
January 15	15	41.27	51.27	61.27	71.27	81.27
February 15	46	33.62	43.62	53.62	63.62	73.29
March 15	74	22.42	32.42	42.42	52.42	62.82
April 15	105	10.59	20.59	30.59	40.59	50.59
May 15	135	1.21	11.21	21.21	31.21	41.21
June 15	166	−3.27	6.73	16.73	26.73	36.69
July 15	196	−1.67	8.33	18.33	28.33	38.48
August 15	227	5.57	15.57	25.57	35.57	46.22
September 15	258	16.58	26.58	36.58	46.58	57.78
October 15	288	28.48	38.48	48.48	58.48	69.60
November 15	319	38.17	48.17	58.17	68.17	79.15
December 15	349	43.12	53.12	63.12	73.12	83.34

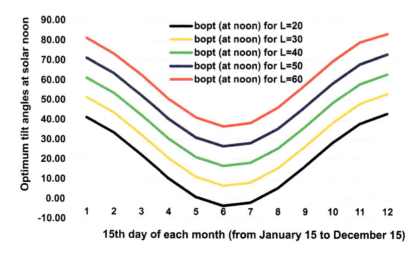

FIGURE 4.34 Optimum tilt angle variation versus the 15th day of each month for different latitudes.

Effect of Environmental Conditions on the Performance of Photovoltaic Systems

An important point needs to be emphasized here. It is clear from the results that the optimum tilt angle may take negative values in particular periods of the year. The tilt angle mechanisms of PV panels are usually made to operate within a certain range (typically in the range of 0°–90°). Therefore, it would be a practical approach to set the tilt angle as zero degrees for the days and months when the optimum tilt angle is negative.

Example 4.12: Optimum Tilt Angle at Different Local Times

Calculate the optimum tilt angles at local solar time 14.30 for the 15th day of each month for latitudes from 20 to 60. Assume that the PV surface is oriented to the true south ($z = 0°$).

Solution

Examples 4.11 and 4.12 are the same, except for the difference between the hour angles. The hour angle at the solar time is $(14.5-12) \times 15° = 37.5°$, and from (4.49) for $L = 20$ and $z = 0°$, we get:

$$b_{opt} = \tan^{-1}\left[\frac{\cos(-21.269)\sin(0)\sin(37.5) + \sin(20)\cos(-21.269)\cos(0)\cos(37.5) - \cos(20)\sin(-21.269)\cos(0)}{\sin(20)\sin(-21.269) + \cos(20)\cos(-21.269)\cos(37.5)}\right]$$

$$= 46.14$$

Similar to the above, the calculated results are given in Table 4.28 for the local solar time of 14.30.

Example 4.13: Optimum Tilt Angle for a Given Azimuth Angle

Calculate the optimum tilt angles questioned in Examples 4.11 and 4.12 for the azimuth angles $z = 30°$ and $z = -30°$.

Solution

The solution steps are already given in Examples 4.11 and 4.12. Therefore, only the solved results will be given here. Optimum tilts obtained by substituting the given parameters into (4.49) are given in Tables 4.29 and 4.30, respectively.

TABLE 4.28

Optimum Tilt Angles (in Degrees) at Local Solar Time 14.30 for Different Latitudes

Considered Day	Julian Day Number, N	$L = 20$ b_{opt}	$L = 30$ b_{opt}	$L = 40$ b_{opt}	$L = 50$ b_{opt}	$L = 60$ b_{opt}
January 15	15	46.14	56.14	66.14	76.14	86.14
February 15	46	36.58	46.58	56.58	66.58	76.58
March 15	74	23.55	33.55	43.55	53.55	63.55
April 15	105	8.19	18.19	28.19	38.19	48.19
May 15	135	−3.21	6.79	16.79	26.79	36.79
June 15	166	−8.51	1.49	11.49	21.49	31.49
July 15	196	−6.43	3.57	13.57	23.57	33.57
August 15	227	2.82	12.82	22.82	32.82	42.82
September 15	258	17.21	27.21	37.21	47.21	57.21
October 15	288	32.03	42.03	52.03	62.03	72.03
November 15	319	43.64	53.64	63.64	73.64	83.64
December 15	349	48.54	58.54	68.54	78.54	88.54

TABLE 4.29
Optimum Tilt Angles at Solar Noon of Different Latitudes for $z = 30°$ and $z = -30°$

Considered Day	Julian Day Number, N	$z = 30°$					$z = -30°$				
		$L=20$ b_{opt}	$L=30$ b_{opt}	$L=40$ b_{opt}	$L=50$ b_{opt}	$L=60$ b_{opt}	$L=20$ b_{opt}	$L=30$ b_{opt}	$L=40$ b_{opt}	$L=50$ b_{opt}	$L=60$ b_{opt}
January 15	15	37.24	47.20	57.67	68.62	79.94	37.24	47.20	57.67	68.62	79.94
February 15	46	29.62	39.21	49.27	59.84	70.88	29.62	39.21	49.27	59.84	70.88
March 15	74	20.02	29.18	38.75	48.79	59.33	20.02	29.18	38.75	48.79	59.33
April 15	105	9.19	18.02	27.11	36.57	46.50	9.19	18.02	27.11	36.57	46.50
May 15	135	1.05	9.74	18.57	27.68	37.18	1.05	9.74	18.57	27.68	37.18
June 15	166	−2.87	5.80	14.55	23.52	32.83	−2.87	5.80	14.55	23.52	32.83
July 15	196	−1.31	7.36	16.14	25.17	34.55	−1.31	7.36	16.14	25.17	34.55
August 15	227	5.39	14.14	23.10	32.38	42.10	5.39	14.14	23.10	32.38	42.10
September 15	258	15.52	24.53	33.88	43.67	53.96	15.52	24.53	33.88	43.67	53.96
October 15	288	26.20	35.62	45.50	55.88	66.76	26.20	35.62	45.50	55.88	66.76
November 15	319	35.18	45.04	55.40	66.26	77.52	35.18	45.04	55.40	66.26	77.52
December 15	349	39.25	49.32	59.89	70.93	82.32	39.25	49.32	59.89	70.93	82.32

TABLE 4.30
Optimum Tilt Angles at Solar Time 14.30 for $z = 30°$ and $z = -30°$

Considered Day	Julian Day Number, N	$z = 30°$					$z = -30°$				
		$L=20$ b_{opt}	$L=30$ b_{opt}	$L=40$ b_{opt}	$L=50$ b_{opt}	$L=60$ b_{opt}	$L=20$ b_{opt}	$L=30$ b_{opt}	$L=40$ b_{opt}	$L=50$ b_{opt}	$L=60$ b_{opt}
January 15	15	54.43	62.35	70.40	78.57	86.81	22.00	33.92	47.89	64.23	82.61
February 15	46	47.74	55.41	63.20	71.12	79.14	10.47	20.81	32.81	47.05	63.94
March 15	74	38.49	45.95	53.50	61.17	68.97	−2.30	6.55	16.43	27.83	41.38
April 15	105	26.75	34.21	41.68	49.22	56.86	−14.29	−6.32	2.18	11.50	22.05
May 15	135	16.94	24.61	32.19	39.77	47.41	−21.89	−14.15	−6.11	2.42	11.71
June 15	166	11.93	19.79	27.48	35.12	42.79	−25.20	−17.47	−9.54	−1.23	7.69
July 15	196	13.94	21.72	29.36	36.97	44.63	−23.91	−16.18	−8.21	0.17	9.23
August 15	227	22.27	29.81	37.31	44.85	52.48	−17.97	−10.14	−1.91	6.97	16.84
September 15	258	33.80	41.22	48.71	56.30	64.02	−7.57	0.82	10.00	20.37	32.48
October 15	288	44.55	52.13	59.82	67.64	75.57	5.67	15.40	26.56	39.72	55.49
November 15	319	52.67	60.52	68.50	76.60	84.78	18.78	30.26	43.71	59.56	77.72
December 15	349	56.12	64.12	72.25	80.48	88.78	25.22	37.58	52.03	68.77	87.23

As can be seen from the results in Table 4.29, optimum tilt angles at solar noon give the same results for $z = 30°$ and $z = -30°$. Because both angles are symmetric to each other with respect to the solar noon. However, it is seen that the results according to solar hour 14.30 are quite different from each other because the azimuth angles are not symmetrical with respect to the 14.30 solar hour.

4.10.1.2 Daily Tilt Angle

Total extraterrestrial solar insolation on a tilted surface (tilt angle is b relative to a horizontal plane) I_b [W/m^2] is given by equation (4.53) in the literature.

$$I_b(N,L,b,z) = \left(\frac{12}{\pi} I_{SC}\right) \cdot C(N) \cdot \{A_1(\omega_{ss} - \omega_{sr}) + A_2[\sin(\omega_{ss}) - \sin(\omega_{sr})] - A_3[\cos(\omega_{ss}) - \cos(\omega_{sr})]\} \quad (4.53)$$

where

$$I_{SC} = 1367 \text{ W/m}^2$$

$$C(N) = \left[1 + 0.034 \cdot \cos\left(\frac{360N}{365}\right)\right]$$

$$A_1 = \sin(\delta)\sin(L)\cos(b) - \sin(\delta)\sin(b)\cos(L)\cos(z)$$

$$A_2 = \cos(\delta)\cos(L)\cos(b) - \cos(\delta)\sin(b)\sin(L)\cos(z)$$

$$A_3 = \cos(\delta)\sin(b)\sin(z)$$

$$\omega_{ss} = \min\left\{\cos^{-1}(-\tan\delta\tan L); \cos^{-1}\left(-\frac{A_1}{A_4}\right) + \sin^{-1}\left(\frac{A_3}{A_4}\right)\right\}$$

$$\omega_{sr} = \max\left\{\cos^{-1}(-\tan\delta\tan L); \cos^{-1}\left(-\frac{A_1}{A_4}\right) + \sin^{-1}\left(\frac{A_3}{A_4}\right)\right\}$$

$$A_4 = \sqrt{A_2^2 + A_3^2}$$

Remember that the parameters in (4.53) were already explained in Chapter 2. Therefore, they are not repeated herein. From (4.53), the average extraterrestrial solar insolation over a particular period can be obtained from (4.54).

$$\overline{I_b} = \frac{\sum_{N=N_1}^{N=N_2} I_b(N,L,b,z)}{(N_2 - N_1 + 1)} \quad (4.54)$$

where N_1 and N_2 are the first and last Julian days of the considered period, respectively. Note that the average solar insolation on a tilted surface can be calculated for a single day by taking $N_1 = N_2$. From (4.54), the optimum tilt angle for any given period can be obtained from the solution of equation (4.55).

$$\sum_N \frac{\partial}{\partial b}[I_b(N,L,b,z)] = 0 \quad (4.55)$$

When equation 4.55 is differentiated with respect to tilt angle b and set to zero, the expression 4.56 can be written. Hence, the optimum tilt angle b_{opt} is obtained from the solution of equation 4.56.

$$\sum C(N) \cdot \left\{ \frac{\partial A_1}{\partial b}(\omega_{ss} - \omega_{sr}) + A_1 \left(\frac{\partial \omega_{ss}}{\partial b} - \frac{\partial \omega_{sr}}{\partial b} \right) + \frac{\partial A_2}{\partial b} \left[\sin(\omega_{ss}) - \sin(\omega_{sr}) \right] \right.$$
$$+ A_2 \left[\cos(\omega_{ss}) \frac{\partial \omega_{ss}}{\partial b} - \cos(\omega_{sr}) \frac{\partial \omega_{sr}}{\partial b} \right] - \frac{\partial A_3}{\partial b} \left[\cos(\omega_{ss}) - \cos(\omega_{sr}) \right]$$
$$\left. - \frac{\partial A_3}{\partial b} \left[\cos(\omega_{ss}) - \cos(\omega_{sr}) \right] + A_3 \left[\sin(\omega_{ss}) \frac{\partial \omega_{ss}}{\partial b} - \sin(\omega_{sr}) \frac{\partial \omega_{sr}}{\partial b} \right] \right\} \quad (4.56)$$

If $z = 0$ for south-facing surfaces in the northern hemisphere, equation 4.56 is simplified to 4.57. Note that $\omega_{ss} = -\omega_{sr}$ for south-facing PV panels.

$$\omega_{ss} \frac{\partial A_1}{\partial b} + A_1 \frac{\partial \omega_{ss}}{\partial b} + \frac{\partial A_2}{\partial b} \sin(\omega_{ss}) + A_2 \cos(\omega_{ss}) \frac{\partial \omega_{ss}}{\partial b} = 0 \quad (4.57)$$

For the period from September 22 to March 21, ω_{ss} becomes independent of the tilt angle b. In general terms, equation 4.57 is a nonlinear algebraic function. Therefore, the solution to this function can be obtained if ω_{ss} is independent of b. When ω_{ss} is independent of b, equation 4.57 turns into 4.58.

$$\omega_{ss} \frac{\partial A_1}{\partial b} + \frac{\partial A_2}{\partial b} \sin(\omega_{ss}) = 0 \quad (4.58)$$

If the expressions A_1 and A_2 are substituted into 4.58 for $z = 0$, the daily optimum tilt angle $b_{opt(d)}$ for the northern hemisphere is obtained as equation 4.59.

$$b_{opt(d)} = \tan^{-1} \left[\frac{\cos(\delta)\sin(L)\sin(\omega_{ss}) - \omega_{ss}\cos(L)\sin(\delta)}{\omega_{ss}\sin(L)\sin(\delta) + \cos(L)\cos(\delta)\sin(\omega_{ss})} \right] \quad (4.59)$$

When solving equation 4.59, attention should be paid to the following points.

- It is sufficient to take $\omega_{ss} = \cos^{-1}(-\tan\delta \tan L)$ for the period from September 22 to March 21 when ω_{ss} is independent of b.
- For the period from March 22 to September 21, the ω_{ss} is dependent of b. So for this interval, $\omega_{ss} = \cos^{-1}[-\tan\delta \tan(L-b)]$. In this case, the daily optimum tilt angle $b_{opt(d)}$ can be solved iteratively. For example, the Newton iteration method offers a simple and effective solution.
- If ω_{ss} is calculated in degrees, the ω_{ss} in equation 4.59 must be converted to radians by multiplying by $\pi/180$. In other words, if ω_{ss} is considered in degrees, the equation 4.59 will be in the following format.

$$b_{opt(d)} = \tan^{-1} \left[\frac{\cos(\delta)\sin(L)\sin(\omega_{ss}) - \frac{\pi}{180}\omega_{ss}\cos(L)\sin(\delta)}{\frac{\pi}{180}\omega_{ss}\sin(L)\sin(\delta) + \cos(L)\cos(\delta)\sin(\omega_{ss})} \right] \quad (4.60)$$

- By following similar steps given above, the daily optimum tilt angle $b_{opt(d)}$ for the southern hemisphere can also be found as equation 4.61. Note that ω_{ss} is considered again in degrees ($z = -180°$).

$$b_{opt(d)} = \tan^{-1}\left[\frac{\frac{\pi}{180}\omega_{ss}\cos(L)\sin(\delta) - \cos(\delta)\sin(L)\sin(\omega_{ss})}{\frac{\pi}{180}\omega_{ss}\sin(L)\sin(\delta) + \cos(L)\cos(\delta)\sin(\omega_{ss})}\right] \quad (4.61)$$

As can be seen from the above equations (4.59–4.61), there is a clear relation between optimum tilt angle and latitude. Therefore, many researchers have made simple correlations based on the latitude angle to determine the tilt angle. Accordingly, numerous studies have suggested the tilt angle (b) based on site latitude (L). For example, the yearly optimum tilt angle is taken equal to local latitude ($b = L$) for simplicity by several researchers. Some of the researchers have recommended the tilt-angle formula as $(L+15) \mp 15°$. Besides, it is concluded that optimal tilt angles on a seasonal or monthly basis are usually within the range of $(L \mp 15°)$. Note that the negative sign is used for summer and the positive sign is used for winter months in these formulas. There are important differences between latitude-based tilt angle expressions. This is because most researchers have given the tilt angles considering local conditions, while others have given them for a specific season. Although these formulas provide a simple estimate for tilt angle determination, they cannot be generalized. Therefore, the optimum tilt angle for a given site should be determined based on local solar data and b_{opt} formula as specified above.

Example 4.14: Optimum Daily Tilt Angle on Different Days

Calculate the optimum daily tilt angles for the 15th days of January, April, July, and October months for latitudes 30 and 50.

Solution

January and October are within the period of September 22 to March 21, while April and July are in the March 22 to September 21 period. Therefore, for January 15 and October 15, ω_{ss} is independent of b and given by $\omega_{ss} = \cos^{-1}(-\tan\delta \tan L)$. For April 15 and July 15, ω_{ss} is a function of b and given by $\omega_{ss} = \cos^{-1}[-\tan\delta \tan(L-b)]$.

Julian day number, $N = 15$ on January 15, and from equation (2.9):

$$\delta = 23.45 \cdot \sin\left[\frac{360}{365}(15+284)\right] = -21.269°$$

So, the ω_{ss} value for $L = 30$ is calculated as follows.

$$\omega_{ss} = \cos^{-1}(-\tan\delta \tan L) = \cos^{-1}[-\tan(-21.269°)\tan(30°)] = 54.73°$$

Now all the necessary parameters for the $b_{opt(d)}$ are known. Thus, from 4.60 we get:

$$b_{opt(d)} = \tan^{-1}\left[\frac{\cos(-21.269)\sin(30)\sin(54.73) - \frac{\pi}{180}(54.73)\cos(30)\sin(-21.269)}{\frac{\pi}{180}(54.73)\sin(30)\sin(-21.269) + \cos(30)\cos(-21.269)\sin(54.73)}\right] \cong 55.6°$$

Likewise, the daily tilt for $L = 50$ on January 15 is calculated as follows.

$$b_{opt(d)} = \tan^{-1}\left[\frac{\cos(-21.269)\sin(50)\sin(54.73) - \frac{\pi}{180}(54.73)\cos(50)\sin(-21.269)}{\frac{\pi}{180}(54.73)\sin(50)\sin(-21.269) + \cos(50)\cos(-21.269)\sin(54.73)}\right] \cong 73.2°$$

Similar to the calculations above, the optimum tilt angles for latitudes 30 and 50 on October 15 will yield the following results.

$$b_{opt(d)} = 43.6°, \text{ for } L = 30 \text{ on October 15}$$

$$b_{opt(d)} = 62.8°, \text{ for } L = 50 \text{ on October 15}$$

Let us substitute the expression $\omega_{ss} = \cos^{-1}[-\tan\delta\tan(L-b)]$ into 4.60 to find the daily optimum tilt solution for April 15 and July 15.

$$b_{opt(d)} = \tan^{-1}\left\{\frac{\cos(\delta)\sin(L)\sin\left(\cos^{-1}[-\tan\delta\tan(L-b)]\right) - \frac{\pi}{180}\cos^{-1}[-\tan\delta\tan(L-b)]\cdot\cos(L)\sin(\delta)}{\frac{\pi}{180}\cos^{-1}[-\tan\delta\tan(L-b)]\cdot\sin(L)\sin(\delta) + \cos(L)\cos(\delta)\sin\left(\cos^{-1}[-\tan\delta\tan(L-b)]\right)}\right\}$$

Julian day number, $N = 105$ on April 15, and from equation (2.9):

$$\delta = 23.45\cdot\sin\left[\frac{360}{365}(105+284)\right] = 9.415°$$

We can use the Newton iteration method to find the optimum tilt solution. Let's start the iteration by giving zero degrees as the initial value of $b_{opt(0)}$. Accordingly, the result of the first iteration can be calculated as follows ($L = 30$).

$b_{opt(1)}$

$$= \tan^{-1}\left\{\frac{\cos(9.4)\sin(30)\sin\left(\cos^{-1}[-\tan(9.4)\tan(30-0)]\right) - \frac{\pi}{180}\cos^{-1}[-\tan(9.4)\tan(30-0)]\cdot\cos(30)\sin(9.4)}{\frac{\pi}{180}\cos^{-1}[-\tan(9.4)\tan(30-0)]\cdot\sin(30)\sin(9.4) + \cos(30)\cos(9.4)\sin\left(\cos^{-1}[-\tan(9.4)\tan(30-0)]\right)}\right\}$$

$= 14.483$

If the iterations are continued similarly, the results in Table 4.31 are obtained. As can be seen from the results in Table 4.31, equality is achieved in the third iteration. So, the $b_{opt(d)} = 14.991$.

Likewise, if iterations are carried out for July 15, the results are obtained as in Table 4.32. It is clear from the iteration results that equality is achieved in the fourth iteration. So, the daily $b_{opt(d)} = -0.234$.

4.10.1.3 Monthly, Seasonally, and Yearly Tilt Angles

In the above section, the methodology for calculating the daily optimum tilt angle for any latitude of the Earth, including both the northern and southern hemispheres, is given in detail. Accurate calculation of the daily optimum tilt angle is the most critical step in determining the monthly, seasonal, or yearly tilt angles of fixed PV panel applications. Because the optimum tilt angle for any

Effect of Environmental Conditions on the Performance of Photovoltaic Systems

TABLE 4.31
Newton Iteration Results for the Daily Optimum Tilt on April 15

	L = 30		L = 50	
Iteration	$b_{opt(i)}$	$f \to b_{opt}(L, \delta, b)$	$b_{opt(i)}$	$f \to b_{opt}(L, \delta, b)$
0	0	14.483	0	33.334
1	14.483	14.976	33.334	34.942
2	14.976	14.991	34.942	34.990
3	**14.991**	**14.991**	34.990	34.991
4			**34.991**	**34.991**

TABLE 4.32
Newton Iteration Results for the Daily Optimum Tilt on July 15

	L = 30		L = 50	
Iteration	$b_{opt(i)}$	$f \to b_{opt}(L, \delta, b)$	$b_{opt(i)}$	$f \to b_{opt}(L, \delta, b)$
0	0	−0.185	0	7.385
1	−0.185	−0.224	7.385	10.593
2	−0.224	−0.232	10.593	11.618
3	−0.232	−0.234	11.618	11.913
4	**−0.234**	**−0.234**	11.913	11.996
5			11.996	12.019
6			12.019	12.025
7			12.025	12.027
8			**12.027**	**12.027**

period can easily be expressed as the arithmetic means of the individual optimum angles within that period.

$$b_{opt(\text{for any period})} = \frac{1}{n} \cdot \sum_{i=1}^{n} b_{opt(i)} \quad (4.62)$$

For example, when determining the optimum tilt angle for January, the monthly optimum tilt angle can be calculated by averaging the daily optimum tilt angles from January 1 to January 31.

$$b_{opt(\text{January})} = \frac{1}{31} \cdot \sum_{i=1}^{31} b_{opt(i,\ \text{january})} \quad (4.63)$$

A simpler approach could also be used in determining the monthly optimum tilt angle. According to this simple approach, the monthly optimum tilt angle can be estimated based on the principle of determining the optimum tilt angle for the day at the midpoint of the relevant period. In other words, the daily optimum tilt angle at the midpoint will exactly be equal to the monthly optimum tilt angle for the period from September 22 to March 21, when ω_{ss} is independent of b. However, the daily optimum tilt angle at the midpoint will be equal to the monthly optimum tilt angle with a negligible error for the period from March 22 to September 21. After these explanations, it is clear that the 15th and 16th days should be taken into account for the months with 31 days, and only the 15th day for the months with 30 days. Of course, for February, this date should be February 14. In this context, the monthly optimum tilt angles ($b_{opt(m)}$) for the months with 30 and 31 days can be given respectively by the equations of 4.64 and 4.65.

$$b_{opt(m)} \cong b_{opt(15,d)} \text{ for the months with 30 days} \tag{4.64}$$

$$b_{opt(m)} \cong \frac{b_{opt(15,d)} + b_{opt(16,d)}}{2} \text{ for the months with 31 days} \tag{4.65}$$

The difference in the daily optimum tilt angle between two consecutive days is small enough to be neglected. Therefore, it is also acceptable to calculate only for day 15 or day 16 in equation 4.65.

Applying similar logic, the seasonal and yearly optimum tilt angles ($b_{opt(y)}$, $b_{opt(s)}$) can also be expressed by the following equations, respectively.

$$b_{opt(s)} = \frac{1}{3} \cdot \sum_{i=1}^{3} b_{opt(m_i)} \tag{4.66}$$

$$b_{opt(y)} = \frac{1}{12} \cdot \sum_{i=1}^{12} b_{opt(m_i)} \tag{4.67}$$

where the parameter m_i represents the ith month in the relevant period. For example, for the winter season, m_i represents the months of December, January, and February. It represents the months of March, April, and May for the spring season.

Example 4.15: Optimum Tilt Angle on Monthly, Seasonally, and Yearly Basis

Calculate the optimum tilt angles on a monthly, seasonally, and yearly basis for latitudes 30 and 50.

Solution

As explained above, the monthly optimum tilt angle can be considered equal to the daily tilt angle of the 15th day of the respective month. Seasonal and yearly tilt angles can also be calculated by taking the average of the monthly tilt angles in the relevant periods. Since the methodology to calculate the optimum tilt angle on a daily basis has already been detailed in Example 4.14, only the calculated results are given here in Table 4.33.

An important point regarding the results in Table 4.33 needs to be highlighted herein. As can be seen, the optimum tilt angle for $L = 30$ gives negative values in May, June, July, and the summer season. It would be an impractical approach to create a tilt-changing mechanism of PV frames to cover negative angles. Therefore, the tilt-changing taps are constructed to be adjustable between zero degrees and the maximum tilt angle. Accordingly, tilt angles can be modified as shown in Table 4.34.

4.10.2 TILT ANGLE FOR STAND-ALONE SYSTEMS

The design month to size the PV array of an off-grid system should be the month with the worst-case scenario. For this purpose, the month having the lowest insolation and the highest load condition should be taken into consideration when sizing the PV array. However, for the optimum tilt angle determination, the entire period of interest should be considered. In other words, to keep

TABLE 4.33
Monthly, Seasonally, and Yearly Optimum Tilt Angles for Latitudes 30 and 50

Months	L = 30 $b_{opt(m)}$	L = 50 $b_{opt(m)}$	L = 30 $b_{opt(s)}$	L = 50 $b_{opt(s)}$	L = 30 $b_{opt(y)}$	L = 50 $b_{opt(y)}$
December (31 days)	57.23	74.60	53.58	71.50	26.46	45.70
January (31 days)	55.60	73.23				
February (28 days)	47.92	66.67				
March (31 days)	34.31	54.23	15.82	35.79		
April (30 days)	14.991	34.991				
May (31 days)	−1.838	18.161				
June (30 days)	−13.074	6.925	−4.86	15.16		
July (31 days)	−7.971	12.027				
August (31 days)	6.453	26.542				
September (30 days)	26.515	46.515	41.30	60.35		
October (31 days)	43.60	62.84				
November (30 days)	53.77	71.70				

TABLE 4.34
Modified Results of Optimum Tilt Angles for Latitudes 30 and 50

Months	L = 30 $b_{opt(m)}$	L = 50 $b_{opt(m)}$	L = 30 $b_{opt(s)}$	L = 50 $b_{opt(s)}$	L = 30 $b_{opt(y)}$	L = 50 $b_{opt(y)}$
December (31 days)	57.23	74.60	53.58	71.50	27.10	45.70
January (31 days)	55.60	73.23				
February (28 days)	47.92	66.67				
March (31 days)	34.31	54.23	15.82	35.79		
April (30 days)	14.991	34.991				
May (31 days)	0	18.161				
June (30 days)	0	6.925	0	15.16		
July (31 days)	0	12.027				
August (31 days)	6.453	26.542				
September (30 days)	26.515	46.515	41.30	60.35		
October (31 days)	43.60	62.84				
November (30 days)	53.77	71.70				

the availability higher, the entire period for which the PV array is designed should be taken into account when determining the optimum tilt angle, while solar insolation of the design month at the optimum tilt should be considered when sizing the PV array. For example, if the load is not varying, that means constant throughout the year. The design month for the PV array sizing will be the month with the lowest irradiance. In other cases, if load demand is variable throughout the year, the month having the highest ratio of "*load to solar insolation*" is considered as the design month for sizing the PV array. It should be noted that the solar insolation of the design month at the optimum tilt angle should be considered when sizing the PV array. However, the solar insolation data at the respective tilt-angle is often not available. For example, average PSH (peak solar hours) values are generally available for horizontal surfaces. Therefore, when sizing the PV array, the horizontal insolation data of the design month should be converted to the insolation data for a surface tilted at the optimum angle.

To find the insolation conversion factor from a horizontal surface to a tilted surface, it is sufficient to know the daily extraterrestrial insolation values on the relevant surfaces. The daily extraterrestrial radiation on a horizontal surface (I_{d0H}) is already given by (2.72). The daily extraterrestrial insolation on a tilted south-facing surface (I_{d0b}) can also be expressed for the northern hemisphere as (4.68).

$$I_{d0b} = \frac{24}{\pi} I_{SC} \left[1 + 0.034 \cdot \cos\left(\frac{360N}{365}\right) \right] \cdot \left[\frac{\pi \omega_{ss}}{180} \sin\delta \sin(L-b) + \cos\delta \cos(L-b) \sin\omega_{ss} \right] \quad (4.68)$$

By taking the ratio between (4.68) and (2.72), the insolation conversion ratio (C_b) can be found as below for south-facing surfaces, tilted at an angle b.

$$C_{b(\text{sf})} = \frac{I_{d0b}}{I_{d0H}} = \frac{\dfrac{\pi \omega_{ss}}{180} \sin\delta \sin(L-b) + \cos\delta \cos(L-b) \sin\omega_{ss}}{\dfrac{\pi \omega_s}{180} \sin\delta \sin L + \cos\delta \cos L \sin\omega_s} \quad (4.69)$$

where $C_{b(\text{sf})}$ is the insolation conversion factor for south-facing surfaces, tilted at an angle b. Since the other parameters in (4.68) are already explained in the previous sections, they are not explained here again. Similarly, the insolation conversion factor for north-facing surfaces ($C_{b(\text{nf})}$) can also be found as (4.70).

$$C_{b(\text{nf})} = \frac{I_{d0b}}{I_{d0H}} = \frac{\dfrac{\pi \omega_{ss}}{180} \sin\delta \sin(L+b) + \cos\delta \cos(L+b) \sin\omega_{ss}}{\dfrac{\pi \omega_s}{180} \sin\delta \sin L + \cos\delta \cos L \sin\omega_s} \quad (4.70)$$

Using equations (4.69) and (4.70), the solar insolation data given daily, monthly, seasonally, or yearly on a horizontal surface can easily be converted to the solar data for any tilt angle. In fact, the above-given conversion factors are obtained based on the daily insolation values. However, it is easy to obtain monthly, seasonally, or yearly results using daily-based conversion factors. For example, extraterrestrial daily insolation data at the midpoint of each month exhibit characteristic similarities for the month it is in. It is also possible to obtain seasonal and yearly solar data by taking the average of the monthly values within the period of interest. Note that the solar data here can be either daily insolation (kWh/m²day) or peak solar hours (hours/day).

In conclusion, a stand-alone PV system sized according to the criteria described above (criteria for optimal tilt angle & load to insolation ratio) will have the ability to meet the load demand while keeping the battery fully charged during the worst-case scenario of the year. It should be noted that the tilt angle, estimated with incorrect or incomplete solar data, will be misleading and may cause design errors. Therefore, it is crucial to obtain long-term solar data for the relevant site to get a good performance from the PV system. If possible, locally measured solar data on the tilted surface should be preferred. If this is not possible, the optimum tilt angle for the design period can be obtained from the methods explained above.

Example 4.16: Monthly and Yearly Insolation on a Tilted PV

The average PSH (peak sun hours) values on the horizontal surface for a solar site located at latitude 20 are given on a monthly basis as below (Table 4.35).
Estimate the monthly and yearly insolation values in $(\text{kWh/m}^2\text{year})$ for a PV surface tilted at a 15° angle in the site.

TABLE 4.35
PSH Values on a Monthly Basis

Months	Jan.	Feb.	Mar.	Apr.	May	Jun.	Jul.	Aug.	Sep.	Oct.	Nov.	Dec.
PSH (hours/day)	4.39	5.25	6.59	7.50	7.66	7.73	7.76	7.84	6.73	6.62	5.10	4.46

TABLE 4.36
Monthly and Yearly Insolation on the Tilted Surface

Months	Day Number (DN)	PSH$_h$ (h/day) on Horizontal Surface	C$_{b(sf)}$	PSH$_b$ on Tilted Surface at b PSH$_b$ = C$_{b(sf)}$ × PSH$_h$	Monthly Tilted Insolation = PSH$_b$ · DN · 1kW/m² (kWh/m² month)	Yearly Tilted Insolation (kWh/m² year)
January	31	4.39	1.27	5.563	172.442	2475.761
February	28	5.25	1.18	6.187	173.233	
March	31	6.59	1.08	7.138	221.284	
April	30	7.5	0.99	7.409	222.263	
May	31	7.66	0.92	7.043	218.348	
June	30	7.73	0.89	6.854	205.622	
July	31	7.76	0.90	6.982	216.448	
August	31	7.84	0.96	7.493	232.298	
September	30	6.73	1.04	7.016	210.488	
October	31	6.62	1.14	7.565	234.521	
November	30	5.1	1.24	6.333	189.977	
December	31	4.46	1.29	5.769	178.837	

Solution

As explained above, the monthly tilt factor (or insolation conversion factor) can be considered equal to the daily tilt factor on the 15th day of the respective month. Some of the required parameters are already calculated in Example 4.15. By substituting all the necessary parameters into (4.69) $C_{b(sf)}$, can be found for January 15 as below.

$$\omega_s = \omega_{ss} = \cos^{-1}\left[-\tan(-21.27)\tan(20°)\right] = 81.85$$

$$C_{b(sf)} = \frac{\frac{\pi \times 81.85}{180}\sin(-21.27)\sin(20-15)+\cos(\delta)\cos(20-15)\sin(81.85)}{\frac{\pi \times 81.85}{180}\sin(-21.27)\sin 20 + \cos(-21.27)\cos 20 \sin(81.85)} \cong 1.27$$

Similarly, if $C_{b(sf)}$ values are calculated for all remaining months, then the estimated insolation results on the tilted surface can be given as in Table 4.36.

4.11 PROBLEMS

P4.1. Estimate the cell temperature, short-circuit current, open-circuit voltage, and delivered power values for a 90 W_p ($V_{OC} = 21.2\,V$, $I_{SC} = 5.2\,A$) PV module with the following conditions. Assume that the module has 0.5%/°C power loss, −8 mV/°C open voltage coefficient, and 0.74 mA/°C open voltage coefficient.
  a. 1000 W/m² Solar radiation, 0°C ambient temperature, and NOCT = 50°C.
  b. 500 W/m² Solar radiation, 35°C ambient temperature, and NOCT = 50°C.
  c. 1000 W/m² Solar radiation, 0°C ambient temperature, and NOCT = 45°C.
  d. 500 W/m² Solar radiation, 35°C ambient temperature, and NOCT = 45°C.

P4.2. Consider the *Poly c-Si* and *CdTe* PV modules with the following parameters (Table 4.37).
  Calculate the cell temperatures and output power of the PV modules for the following operating conditions. Use all possible methods for the solutions and compare the results.
  a. 1000 W/m² Solar radiation, 0°C ambient temperature, and 2 m/s wind speed.
  b. 500 W/m² Solar radiation, 35°C ambient temperature, and 2 m/s wind speed.

P4.3. Consider the horizontally mounted *Poly c-Si* PV module given below. Figure 4.35 shows the radiation values on the PV module at different local solar times in the given directions.
  Estimate the PV module short-circuit currents at 12.00 and 15.00 local times based on Kelly cosine values.

P4.4. Calculate the air mass and absolute air mass values for a PV site in Los Angeles $(34.0522°\,N, 118.2437°\,W)$ on June 15 for the following conditions.
  a. At solar times 10.00, 12.00, and 16.00 at sea level.
  b. At solar times 10.00, 12.00, and 16.00 at 1000m altitude.

TABLE 4.37
The Parameters of the Given PV Modules

					Temperature Coefficients for Different Models					
	P_{max}	NOCT	η_m	P_{max}	Sandia Parameters		Three Parameters Model			
PV Types	[W]	[°C]	[%]	Coefficient	a/b/ΔT	α_m	c_1	c_2	c_3	C
Poly c-Si	100	50	15	−0.48 %/°C	−3.56/−0.075/3	0.9	0.930	0.030	−1.666	5.1
CdTe	100	45	16	−0.30 %/°C	−3.56/−0.075/3	0.9	0.953	0.030	−1.658	5

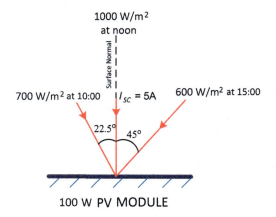

FIGURE 4.35 Horizontally mounted PV module under different solar times.

Effect of Environmental Conditions on the Performance of Photovoltaic Systems

P4.5. A south-facing poly c-Si PV panel with an inclination of 20° is placed in Istanbul $(41.01°\text{N}, 28.94°\text{W})$ on July 20. Calculate the PV performance parameters I_{SC}, I_{mp}, V_{OC}, and V_{mp} at solar noon at 500 m altitudes under 900 W/m² radiation and 18°C. Use Tables 4.9–4.11 for polynomial coefficients and use equation (4.2) for the cell temperature prediction. Assume that wind speed is 3 m/s at 500 m altitude (Table 4.38).

P4.6. A south-facing PV panel is planned to be installed in Athens at a slope of 35° in the south direction. Estimate the annual energy loss ratio for the site with an average dust deposition of 50 gm^{-2}d^{-1}. Assume that the coefficient D_0 in (4.25) is 0.0032% m² g^{-1} and frequency of rainfall occurrence is given as Table 4.39.

P4.7. A 48-volt PV module consisting of 96 cells, each having 64 cm² (8 cm × 8 cm), will be designed based on the following parameters (Table 4.40).
Determine the possible PV module dimension based on the maximum number of bypass diodes to protect the series-connected cells in the module.

P4.8. A PV module consisting of 72 cells will be designed for the site with a minimum ambient temperature of (−40°C) and maximum irradiance of 1000 W/m². The cell parameters under standard reference conditions are given in Table 4.41.

Calculate the number of bypass diodes to protect the series-connected cells in the module. Determine the minimum required reverse breakdown voltage of a bypass diode used in the junction box.

TABLE 4.38
PV Panel's Parameters

Performance Parameters under STC	Poly-crystal	Temperature Coefficients
Voltage at Max. Power (V)	18	−0.37%/°C
Current at Max. Power (A)	7.85	+0.0028%/°C
Open-Circuit Voltage (V)	22.1	−0.35%/°C
Short-Circuit Current (A)	8.35	+0.034%/°C

TABLE 4.39
Frequency of Rainfalls

Number of Rainless Days, i	1	2	3	4	5	6	8	15	21	24	28	38	45	66
Frequency of Occurrence, f_0	24	8	4	3	3	3	2	2	1	1	1	1	1	1

TABLE 4.40
PV Cell's Parameters

V_{mp} of a Cell	V_{OC} of a Cell	Breakdown Voltage of a Cell (V_b)	Diode Voltage (V_D)
0.5 volt	0.6 volt	16 volt	1 volt

TABLE 4.41
PV Cell's Parameters

V_{mp}	V_{OC}	Breakdown Voltage V_b	Diode Voltage V_D	NOCT	Temperature Coefficient β_{Voc}
0.5 volt	0.6 volt	15 volt	1.2 volt	45°C	−0.002 V/°C

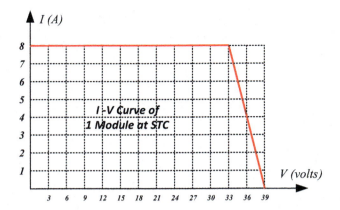

FIGURE 4.36 *I–V* curve of the PV module.

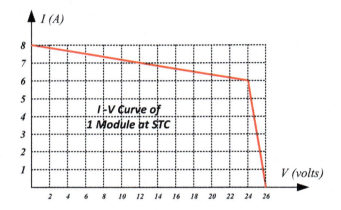

FIGURE 4.37 *I–V* curve of the PV module.

P4.9. The $I-V$ characteristic of a PV module consisting of a 66-series connected cell is given below under 1000 W/m² radiation (Figure 4.36). Each cell in the module has a 5 Ω shunt resistance and they produce 0.59 volt (open-circuit voltage) at full sun (1000 W/m²) conditions.
  a. When one cell and two cells are completely shaded, plot the module $I-V$ curves respectively. Assume that there is no bypass diode application in the module.
  b. Find the MPP points when one cell is fully shaded.
  c. Suppose that every 22 cells in a given PV module possess a bypass diode. Draw the $I-V$ curve of the module when only one cell in the module is totally shaded.

P4.10. Consider a PV module with the following $I-V$ curve at STC. The module consists of 3 bypass diodes, each of them protecting 16 cells out of 48 cells. Assume that the bypass diodes are ideal and there are no voltage drops across them during the conduction phase (Figure 4.37).

Draw the new $I-V$ curves of the module for the following shaded conditions. Assume that each cell in the module has a 3Ω shunt resistance.
  a. Only one bypass diode is active due to the full shading of one cell in the module.
  b. Two bypass diodes are operating due to the full shading of a single cell from each of the active diode zones.

    c. Suppose that only one bypass diode is active due to 50% shading of one cell in the module.
P4.11. Calculate the optimum tilt angles at solar noon for the first day of each month for latitudes from −20 to +20.
P4.12. Calculate the optimum tilt angles at local solar time 14.30 for the 15th day of each month for latitudes from −60 to +20. Assume that the PV surface is oriented to true south ($z = 0°$) for the northern hemisphere, and oriented to true north ($z = -180°$) for the southern hemisphere.
P4.13. Calculate the optimum tilt angles questioned in P.4.11 and P.4.12 for the azimuth angles $z = 30°$ and $z = -30°$.
P4.14. Calculate the optimum tilt angles on a monthly, seasonally, and yearly basis for latitudes from −30 and +30.
P4.15. The yearly average PSH (peak sun hours) value on the horizontal surface is 7.45 h/day for a solar site located at latitude 10. Estimate the yearly insolation value in $\left(\text{kWh/m}^2\text{year}\right)$ for a PV surface tilted at 0-5-10-15 and 20 degrees. Assume that yearly $C_{b(\text{sf})}$ value is equivalent to the $C_{b(\text{sf})}$ value on September 15.

BIBLIOGRAPHY

1. D. Faiman, "Assessing the outdoor operating temperature of photovoltaic modules", *Prog. Photovolt.*, vol. 16, no. 4, pp. 307–315, 2008.
2. M. Koehl, M. Heck, S. Wiesmeier, and J. Wirth, "Modeling of the nominal operating cell temperature based on outdoor weathering", *Solar Energy Mater., Solar Cells*, vol. 95, pp. 1638–1646, 2011.
3. M. R. Patel, *"Wind and Solar Power System"*, CRC Press, Boca Raton, FL, 1999.
4. B. Weber, A. Quiñones, R. Almanza, and M. Dolores Duran, "Performance reduction of PV systems by dust deposition", *Energy Procedia*, vol. 57, pp. 99–108, 2014.
5. M. W. Davis, A. Hunter Fanney, and B. P. Dougherty. "Measured versus predicted performance of building integrated photovoltaics", *J. Solar Energy Eng.*, vol. 125, pp. 21–27, no. 1, 2003.
6. M. Hernandez Velasco, "Performance Evaluation of Different PV-Array Configurations under Weak Light Conditions and Partial Shadings", Master Thesis, European Solar Engineering School, Dalarna University, 2012.
7. B. Marion, "Preliminary Investigation of Methods for Correcting for Variations in Solar Spectrum under Clear Skies", National Renewable Energy Laboratory, Technical Report NREL/TP-520-47277, March 2010.
8. A. K. Damral, "Increase the efficiency of PV module by using low cost tracking system", *Int. Res. J. Eng. Technol.*, vol. 02, pp. 1150–1160, no. 07, 2015.
9. D. L. King, "Photovoltaic Module and Array Performance Characterization Methods for All System Operating Conditions", in *Proceeding of NREL/SNL Photovoltaics Program Review Meeting*, November 18–22, 1996, Lakewood, CO, AIP Press, New York, 1997.
10. P. Gilman, "SAM Photovoltaic Model Technical Reference", National Renewable Energy Laboratory, Technical Report NREL/TP-6A20-64102 May 2015.
11. K. Soga and H. Akasaka, "Influences of solar incident angle on power generation efficiency of PV modules under field conditions", *J. Asian Archit. Build. Eng.*, vol. 2, no. 2, pp. 43–48, 2018.
12. T. Huld and A. M. Gracia Amillo, "Estimating PV module performance over large geographical regions: The role of irradiance, air temperature, wind speed and solar spectrum", *Energies*, vol. 8, pp. 5159–5181, 2015.
13. D. L. King, W. E. Boyson, and J. A. Kratochvil, "Photovoltaic Array Performance Model", Sandia National Laboratories, SAND2004-3535, August 2004.
14. B. Knisely, S. V. Janakeeraman, J. Kuitche, and G. T. Mani, "Angle of Incidence Effect on Photovoltaic Modules", Arizona State University, Photovoltaic Reliability Laboratory, March 2013.
15. A. Hunter Fanney, M. W. Davis, B. P. Dougherty, D. L. King, W. E. Boyson, and J. A. Kratochvil, "Comparison of photovoltaic module performance measurement", *Trans. ASME*, vol. 128, pp. 152–159, 2006.
16. S. V. Janakeeraman, "Angle of Incidence and Power Degradation Analysis of Photovoltaic Modules", Master Thesis, Arizona State University, May 2013.

17. D. S. Ryberg and J. Freeman, "Integration, Validation, and Application of a PV Snow Coverage Model in SAM", National Renewable Energy Laboratory, Technical Report NREL/TP-6A20-68705, August 2017.
18. M. C. Failla, "Snow and Ice on Photovoltaic Devices: Analysis of a Challenge and Proposals for Solutions", Master Thesis, Norwegian University of Science and Technology, September 2016.
19. A. Haque and N. Sheth, "Energy loss in solar photovoltaic systems under snowy conditions", *J. Electr. Electron. Eng.*, vol. 5, no. 6, pp. 209–214, 2017.
20. S. Dezso and B. Yahia, "On the impact of partial shading on PV output power", in *Aalborg University, Proceedings of RES'08*, WSEAS Press, Denmark, 2008.
21. C. Barreiro, P. M. Jansson, A. Thompson, and J. L. Schmalzel, "PV By-Pass Diode Performance in Landscape and Portrait Modalities", *IEEE*, 978-1-4244-9965-6/11, 2011.
22. A. M. Muzathik, "Photovoltaic modules operating temperature estimation using a simple correlation", *Int. J. Energy Eng.*, vol. 4, no. 4, pp. 151–158, 2014.
23. G. TamizhMani, L. Ji, Y. Tang, and L. Petacci, "Photovoltaic Module Thermal/Wind Performance: Long-Term Monitoring and Model Development for Energy Rating", in *NCPV and Solar Program Review Meeting*, Denver, Colorado, 2003.
24. V. E. Sunday, O. Simeon, and U. M. Anthony, "Multiple linear regression photovoltaic cell temperature model for PVSyst simulation software", *Int. J. Theor. Appl. Math.*, vol. 2, no. 2, pp. 140–143, 2017.
25. D. Riley, C. Hansen, and M. Farr, "A Performance Model for Photovoltaic Modules with Integrated Micro-inverters", Sandia National Laboratories, Sandia Report: 2015-0179, January 2015.
26. R. E. Cabanillas and H. Munguía, "Dust accumulation effect on the efficiency of Si photovoltaic module", *J. Renew. Sustain. Energy*, vol. 3, p. 043114, 2011.
27. M. Seyedmahmoudian, S. Mekhilef, R. Rahmani, R. Yusof, and E. T. Renani, "Analytical modeling of partially shaded photovoltaic systems", *Energies*, vol. 6, pp. 128–144, 2013.
28. G.-A. Migan, "Study of the operating temperature of a PV module", Project Report 2013/MVK160 Heat and Mass Transfer May 16, 2013, Sweden.
29. A. Pavgi, "Temperature Coefficients and Thermal Uniformity Mapping of PV Modules and Plants", Master Thesis, Arizona State University, August 2016.
30. J. S. Stein, "PV Performance Modeling Methods and Practices Results from the 4th PV Performance Modeling Collaborative Workshop", Sandia National Laboratories, IEA PVPS Task 13, Subtask-2 Report IEA-PVPS T13-06:2017 March 2017.
31. K. V. Vidyanandan, "An overview of factors affecting the performance of solar PV systems", *Energy Scan*, vol. 27, pp. 2–8, 2017.
32. H. A. Kazem and M. T. Chaichan, "Effect of humidity on photovoltaic performance based on experimental study", *Int. J. Appl. Eng. Res.*, vol. 10, no. 23 pp. 43572–43577, 2015.
33. Z. Ahmed Darwish, H. A. Kazem, K. Sopian, M. A. Alghoul, and M. T Chaichan, "Impact of some environmental variables with dust on solar photovoltaic (PV) performance: Review and research status", *Int. J. Energy Environ.*, vol. 7, pp. 152–159, no. 4, 2013.
34. M. K. Panjwani and G. B. Narejo, "Effect of humidity on the efficiency of solar cell (photovoltaic)", *Int. J. Eng. Res. General Sci.*, vol. 2, pp. 499–503, no. 4, 2014.
35. H. K. Elminir, V. Benda, and J. Tousek, "Effects of solar irradiation conditions and other factors on the outdoor performance of photovoltaic modules", *J. Electr. Eng.*, vol. 52, no. 5–6, pp. 125–133, 2001.
36. B. Weber, A. Quinones, R. Almanza, and M. Dolores Duran, "Performance reduction of PV systems by dust deposition", *Energy Procedia*, vol. 57, pp. 99–108, 2014.
37. A. Rao, R. Pillai, M. Mani, and P. Ramamurthy, "Influence of dust deposition on photovoltaic panel performance", *Energy Procedia*, vol. 54, pp. 690–700, 2014.
38. L. Micheli, D. Ruth, and M. Muller, "Seasonal Trends of Soiling on Photovoltaic Systems", National Renewable Energy Laboratory. NREL/CP-5J00-68673, 2017.
39. B. L. Brench, "Snow-Covering Effects on the Power Output of Solar Photovoltaic Arrays", Department of Energy, December 1979.
40. S. N. Nnamchi, C. O. C. Oko, F. L, Kamen, and O. D. Sanya, "Mathematical analysis of interconnected photovoltaic arrays under different shading conditions", *Cogent Eng.*, vol. 5, p. 1507442, 2018.
41. S. Silvestre, A. Boronat, and A. Chouder, "Study of bypass diode configuration on PV modules", *Appl. Energy*, 86, pp. 1632–1640, 2009.
42. S. Sathish Kumar and C. Nagarajan, "Global maximum power point tracking (MPPT) technique in DC–DC boost converter solar photovoltaic array under partially shaded condition", *WSEAS Trans. Power Syst.*, vol. 12, pp. 49–62, 2017.

43. B. A. Alsayid, S. Y. Alsadi, J. S. Jallad, and M. H. Dradi, "Partial shading of PV system simulation with experimental results", *Smart Grid Renew. Energy*, vol. 4, pp. 429–435. 2013.
44. S. E. Boukebbous and D. Kerdoun, "Study, modeling and simulation of photovoltaic panels under uniform and non-uniform illumination conditions", *Revue des Energies Renouvelables*, vol. 18, no. 2, pp. 257–268, 2015.
45. R. E. Hanitsch, D. Schulz, and U. Siegfried, "Shading effects on output power of grid connected photovoltaic generator systems", *Rev. Energ. Ren. : Power Eng.*, vol. 1, pp. 93–99, 2001.
46. M. Benghanem, "Optimization of tilt angle for solar panel: Case study for Madinah, Saudi Arabia", *Appl. Energy*, vol. 88, no. 4, pp. 1427–1433, 2011.
47. Y. P. Chang, "Optimal the tilt angles for photovoltaic modules in Taiwan", *Int. J. Electr. Power Energy Syst.*, vol. 32, no. 9, pp. 956–964, 2010.
48. H. Darhmaoui and D. Lahjouji, "Latitude based model for tilt angle optimization for solar collectors in the Mediterranean region", *Energy Procedia*, vol. 42, pp. 426–435, 2013.
49. M. M. El-Kassaby, "Monthly and daily optimum tilt angle for south-facing solar collectors; theoretical model, experimental and empirical correlations", *Sol. Wind Technol.*, vol. 5, no. 6, pp. 589–596, 1988.
50. M. M. Elsayed, "Optimum orientation of absorber plates", *Sol. Energy*, vol. 42, no. 2, pp. 89–102, 1989.
51. E. A. Handoyo and D. Ichsani, "The optimal tilt angle of a solar collector", *Energy Procedia*, vol. 32, pp. 166–175, 2013.
52. J. Kern and I. Harris, "On the optimum tilt of a solar collector", *Sol. Energy*, vol. 17, no. 2, pp. 97–102, 1975.
53. S. A. Klein, "Calculation of monthly average insolation on tilted surfaces", *Solar Energy*, vol. 19, pp. 325–329, 1976.
54. S. A. M. Maleki, H. Hizam, and C. Gomes, "Estimation of hourly, daily and monthly global solar radiation on inclined surfaces: Models re-visited", *Energies*, vol. 10, no. 1, 2017.
55. M. A. A. Mamun, R. Md Sarkar, M. Parvez, J. Nahar, and M. Sohel Rana, "Determining the optimum tilt angle and orientation for photovoltaic (PV) systems in Bangladesh", in *2nd International Conference on Electrical and Electronics Engineering (ICEEE 2017)*, Rajshahi, Bangladesh, no. December, pp. 1–4, 2018.
56. M. A. A. Mamun, M. M. Islam, M. Hasanuzzaman, and J. Selvaraj, "Effect of tilt angle on the performance and electrical parameters of a PV module: Comparative indoor and outdoor experimental investigation", *Energy Built Environ.*, vol. 3, pp. 278–290, 2022.
57. J. Modarresi and H. Hosseinnia, "Worldwide daily optimum tilt angle model to obtain maximum solar energy", *IETE J. Res.*, vol. 0, no. 0, pp. 1–9, 2020.
58. A. M. Mujahid, "Optimum tilt angle for solar collection systems", *Int. J. Sol. Energy*, vol. 14, no. 4, pp. 191–202, 1994.
59. R. Sirous, R. Lopes, and S. Sirous, "Application of soft systems methodology to trigger solar energy in Iranian buildings architect", *J. Solar Energy*, vol. 1, no. 1, pp. 35–43, 2016.
60. K. Skeiker, "Optimum tilt angle and orientation for solar collectors in Syria", *Energy Convers. Manag.*, vol. 50, no. 9, pp. 2439–2448, 2009.
61. S. S. Soulayman, "On the optimum tilt of solar absorber plates", *Renew. Energy*, vol. 1, no. 3–4, pp. 551–554, 1991.
62. S. Soulayman and W. Sabbagh, "Comment on 'Optimum tilt angle and orientation for solar collectors in Syria' by Skeiker, K.", *Energy Convers. Manag.*, vol. 89, pp. 1001–1002, 2015.
63. G. N. Tiwari and M. J. Ahmad, "Optimization of tilt angle for solar collector to receive maximum radiation", *Open Renew. Energy J.*, vol. 2, no. 1, pp. 19–24, 2009.
64. T. Som, A. Sharma, and D. Thakur, "Effect of solar tilt angles on photovoltaic module performance: A behavioral optimization approach", *Artif. Intell. Evol.*, vol. 1, pp. 90–102, 2020.
65. A. K. Yadav and S. S. Chandel, "Tilt angle optimization to maximize incident solar radiation: A review", *Renew. Sustain. Energy Rev.*, vol. 23, pp. 503–513, 2013.

5 Photovoltaic Power Systems
Designing and Sizing

5.1 INTRODUCTION

Photovoltaic power systems can generally be classified as grid-connected systems (on-grid) and stand-alone systems (off-grid). Functional/operational requirements, component configurations, equipment connection interactions with other power sources, and supplied electrical load types are the main factors in the classification of PV systems. The following block diagram given in Figure 5.1 summarizes the general classification of PV power systems.

It is clear from Figure 5.1 that PV power systems can be designed to provide DC and/or AC power services either in on-grid or off-grid modes. Hence, photovoltaic power systems consist of several components such as PV modules/arrays, converter (DC-DC), inverter (DC-AC), control system (maximum power point tracker: MPPT), distribution panel, energy storage systems, combiner boxes, protection devices and loads, etc. As can be seen, PV power systems are very diverse and each system has different design requirements. Therefore, the design and sizing steps of each PV system are explained in the following sections.

5.2 GRID-CONNECTED PHOTOVOLTAIC POWER SYSTEMS

Photovoltaic power systems' topologies are very diverse for numerous reasons. For example, the size of the PV system, PV systems' single-phase or three-phase structure, the hybrid configuration with other energy sources, PV array configuration, PCC (point of common coupling) voltage level, PV system operating plan, local grid constraints, and inverter topologies can affect the design steps and configuration of grid-connected PV power systems. For this reason, firstly, a common structure of a grid-connected PV system will be considered in the designing and sizing steps, then differences in other configurations will be taken into consideration individually.

Typical single-line diagrams for grid-connected solar PV systems, which are directly grid-connected systems, grid-connected systems via local loads, and grid-connected systems with diesel generator backup, are illustrated in Figures 5.2–5.4. Among these grid-connected systems, the first

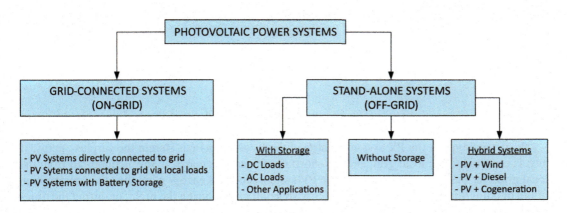

FIGURE 5.1 Classification of photovoltaic power systems.

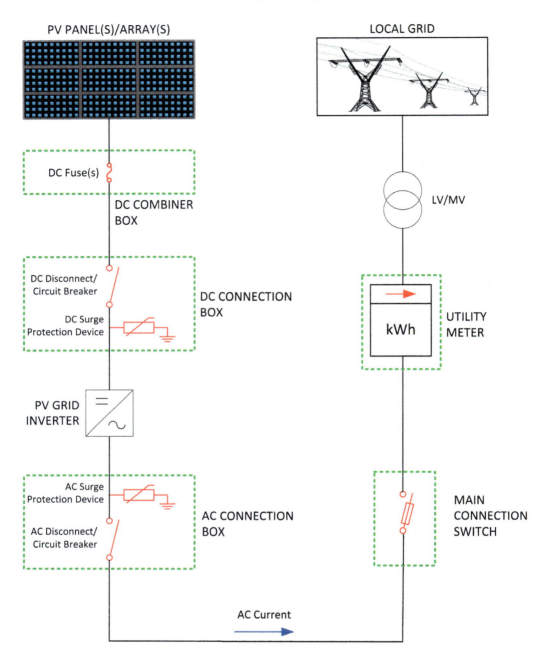

FIGURE 5.2 Typical wiring diagram of a grid-connected photovoltaic power system.

type is the PV system that feeds solar energy directly to the grid. Since the grid-connected PV systems have no battery storage, they will not work during a grid outage.

The second type is the PV systems that feed the local loads first. Herein the surplus energy, if any, is exported to the local grid, and the shortfall of energy demand, if any, is imported from the local grid. Surplus energy occurs when the PV system produces energy more than the consumption of the building/local loads. In contrast, energy shortfall exists during the night, or during the times when the energy demand of local loads/buildings exceeds PV generation.

Photovoltaic Power Systems

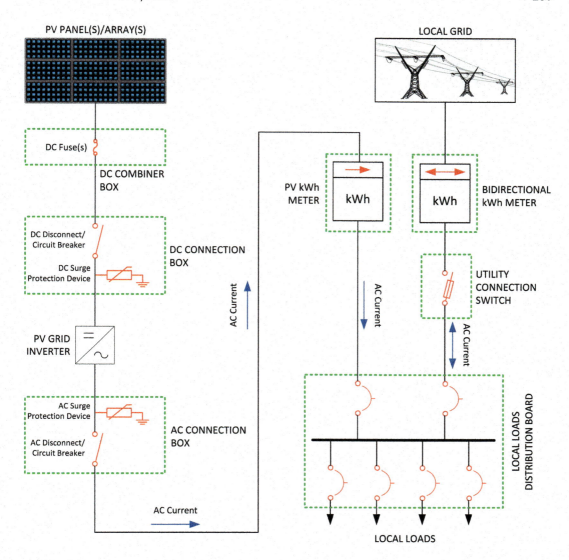

FIGURE 5.3 Typical wiring diagram of a grid-connected photovoltaic power system via local loads.

In the third group of PV systems, a diesel generator shall be used as a backup for local loads and buildings. Note that the PV grid inverter cannot run in parallel with the diesel generator. Hence PV grid inverter needs to be connected to a consumer distribution board on the grid side through an automatic or manual change-over switch as shown in Figure 5.4.

Each component in the grid-connected PV systems may have different topologies according to the arrangements of PV power systems. Hence, system components such as solar PV arrays, combiner boxes, distribution boards, solar grid inverters, energy generation meters, protection devices, cables, and transformers are needed to be explained systematically for specifying the requirements of high-standard PV projects.

5.2.1 Photovoltaic Array Configurations

As shown in Figures 5.2–5.4, the PV array structures are the preliminary components of all photovoltaic power systems. Each solar array in a photovoltaic field is composed of PV modules that

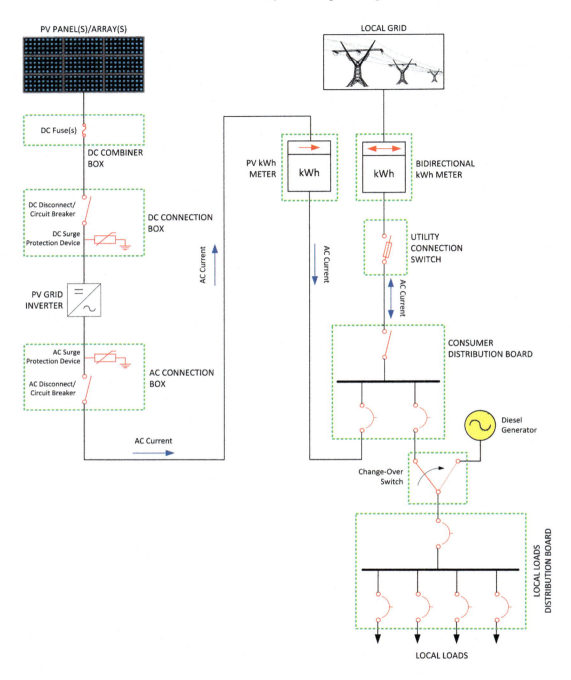

FIGURE 5.4 Typical wiring diagram of a grid-connected photovoltaic power system with a diesel generator backup.

are interconnected in series and parallel. In addition, solar PV arrays in a PV site are assembled in different configurations to meet the project requirements. The most commonly used PV array architectures are explained here in this section.

Photovoltaic arrays are usually configured to match the inverter technology used. It is possible to categorize the photovoltaic inverter/converter technologies into six types, which are module inverters (microinverters), string inverters, multi-string inverters, inverters with a master-slave concept, team concept inverters (a master-slave concept with central inverter unit), and central inverters. Naturally, PV array architectures in a solar site will be created based on the inverter technology used. In this section, we will only focus on possible PV array topologies. Design characteristics, advantages, and limitations of different inverter technologies will be explained below under the sub-section of *photovoltaic inverters*.

5.2.1.1 Photovoltaic Array for Module-Inverter Concept

The use of a module (micro) inverter adds significant flexibility to the photovoltaic array design and improves the operational efficiency of the PV module. Because every PV module with a micro-inverter could ideally be operated at its MPP, especially when the PV modules come with their inverter. Hence, these combined module-inverter units can also be called AC PV modules. Figure 5.5 represents the general layout for module-inverter-based PV generator formation. As can be seen from the figure that the micro-inverter-based PV units are suitable for the gradual system expansion over time and they are generally preferred for building-integrated structures, either for rooftop applications or building facade applications.

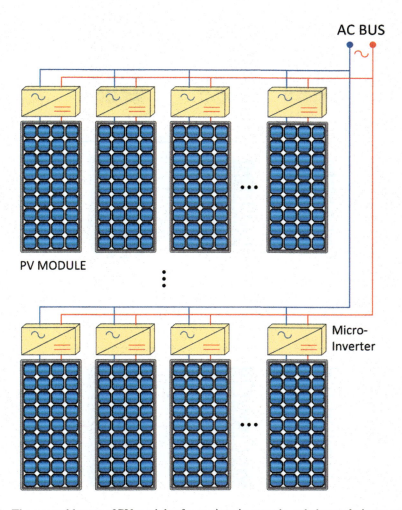

FIGURE 5.5 The general layout of PV modules for a micro-inverter-based photovoltaic power systems.

5.2.1.2 Photovoltaic Array for String Inverter Concept

In the areas where the PV site is asymmetrical or exposed to non-homogenous irradiation, the string inverter or array inverter concept provides more flexibility in designing the architecture of the PV array. Figure 5.6 shows the typical drawings of the string inverter concept. In some cases, it is not necessary to arrange PV modules in series configuration for a string inverter concept. Instead, a combined series and parallel configurations (PV strings in parallel) can be used to form a PV array. In this case, each PV array can be connected to the system via a single-string inverter. Figure 5.7 shows the typical connection diagrams of the PV array concept with string inverters.

5.2.1.3 Photovoltaic Array for Multi-String Inverter Concept

The design architecture of PV arrays with a multi-string inverter concept is almost the same as the string inverter concept except for the power electronic system designs. A multi-string inverters system has an independent DC-DC converter with an MPP tracker, which feed the solar energy into a

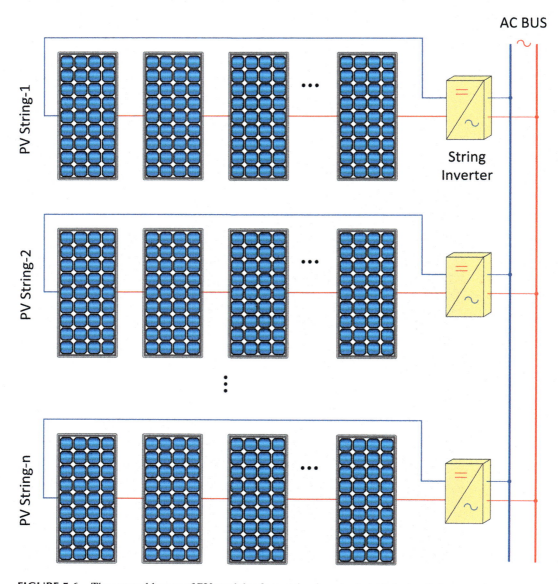

FIGURE 5.6 The general layout of PV modules for a string-inverter-based photovoltaic power systems.

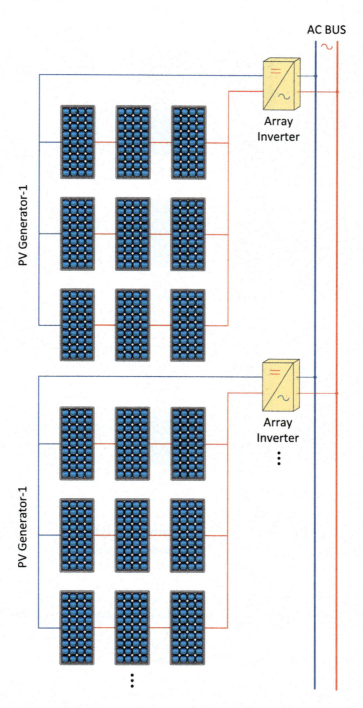

FIGURE 5.7 Typical connection diagrams of PV array concept with string inverters.

common DC-AC inverter. In the multi-string inverter case, it is possible to connect PV strings with different specifications, sizes, and technology to one common inverter, since each string is capable of working at its maximum power point. Hence, the multi-string inverter concept can be illustrated in Figure 5.8.

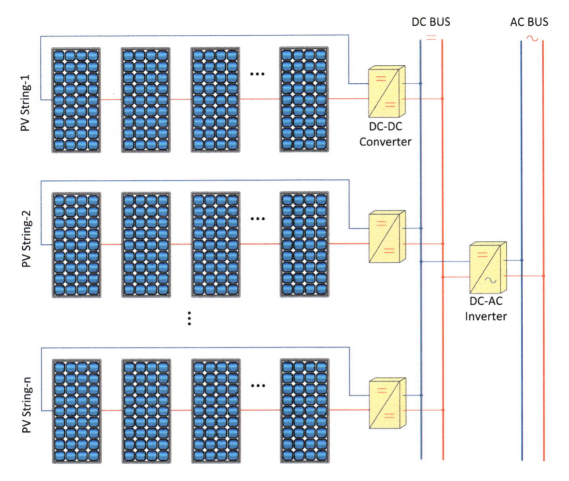

FIGURE 5.8 Typical connection diagrams of PV arrays with multi-string inverter concept.

5.2.1.4 Photovoltaic Array for Master-Slave Concept Inverters

All types of inverters must be designed to operate at maximum efficiency under their rated conditions. To benefit from this feature in low radiation conditions, PV panels are connected to a group of PV inverters, one of them is a master device, while the others are slave inverters. Under a low irradiance level, the master inverter becomes active, while the slave inverters will start to operate respectively with the increasing irradiance. Hence, the master-slave inverter concept can be demonstrated as shown in Figure 5.9.

5.2.1.5 Photovoltaic Array for Team Concept Inverters

Team concept topology is normally used for utility-scale PV power systems and it includes both the advantages of the master-slave concept and the string inverter concept. During the conditions of low solar radiation, all the PV arrays are connected to master-slave topology and for higher irradiance conditions, all the PV strings are independently connected with the individual inverters to track their maximum power point more precisely. Figure 5.10 shows the team concept and its connection to the PV array structure.

5.2.1.6 Photovoltaic Array for Central Inverter Concept

In the central inverter topology, only one inverter is used for the entire PV system as shown in Figure 5.11. The most important problem of this topology is its lower reliability. Because the inverter

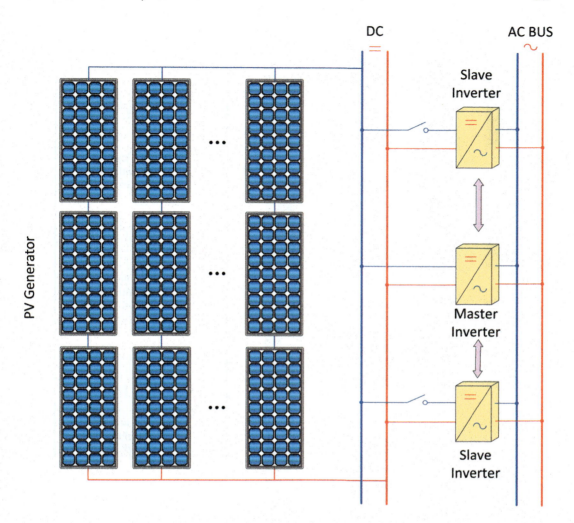

FIGURE 5.9 Typical connection diagrams of PV arrays with master-slave inverter concept.

fails, the entire PV system will go down. Besides, mismatch power loss for the PV array is an important problem in this topology and the accuracy of the MPPT system is reduced due to a single MPPT with a high power rating. It is possible to use the central inverter concept for low-voltage (short string) and high-voltage (longer string) designs. One advantage of using a short string concept over a longer string is that the shadowing effect is less when compared to a longer string concept.

5.2.1.7 Alternative Photovoltaic Array Configurations

The traditional series-parallel connection configuration of photovoltaic arrays is sensitive to non-uniform irradiance levels or partial shading on PV modules of the solar PV system. It is possible to use various alternative topologies to reduce unavoidable partial shade effects. Figure 5.12 shows modified module interconnections inside a PV array to increase the system performance by reducing the mismatch losses due to partial shading. The total cross-tied (TCT) and bridge-linked (BL) structures, shown in Figure 5.12, are newly introduced connection topologies. In the TCT topology, each module is in series and parallel with another one. However, in the BL topology, half of the interconnections in TCT are canceled. For this reason, cable losses and wiring time are less in BL topology. Experimental results in the literature show that TCT topology seems to be the most efficient topology for decreasing the mismatch losses among the alternative array configurations.

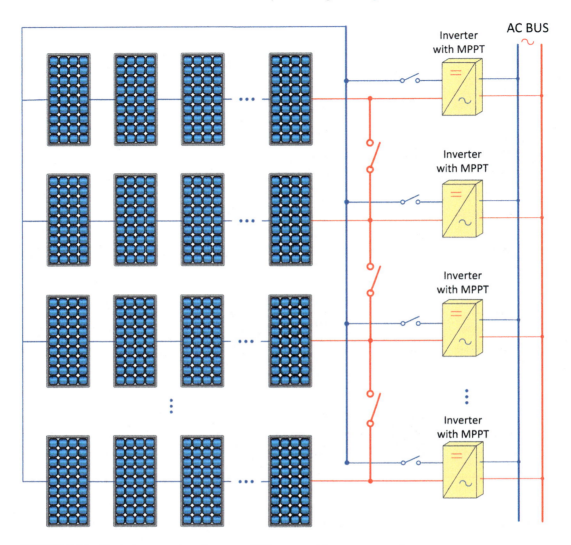

FIGURE 5.10 Typical connection diagrams of PV arrays with team concept inverters.

An important point should be noted that TCT topology has no adverse effect on overall system efficiency during non-shaded conditions.

5.2.2 DC Combiner Box and DC Distribution Box

Photovoltaic DC combiner box and DC distribution box are the boards used between PV Inverter and a photovoltaic array. The role of these boxes is to bring the output of several solar strings together while providing system safety and protection. Normally, the functions of combiner and distribution boxes are different. However, by combining their features, they can physically be designed as a single box, which can also be called a photovoltaic DC switchboard. The main features that can be integrated into the combiner/distribution boxes are as follows:

- String conductors on the negative and positive sides are protected by DC fuses.
- DC monitoring units to measure the string current, voltage, PV array temperature and the state of the lightning protection device can be installed.
- Lightning and Surge protection devices are included.

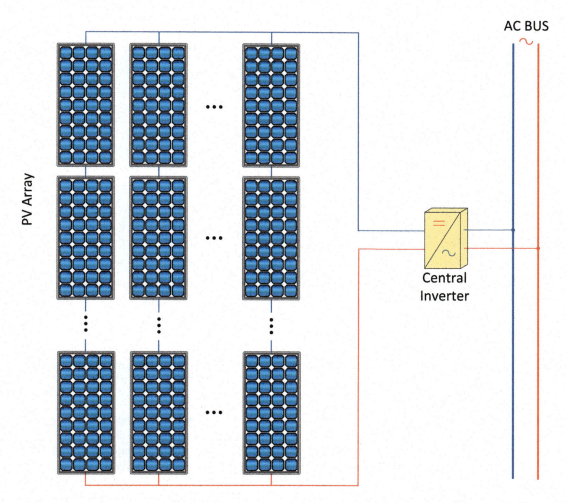

FIGURE 5.11 The general architecture of PV arrays with team concept inverters.

- A circuit breaker is added to clear the conditions of over-temperature and/or high PV string current.
- A disconnector (disconnect switch or isolator switch) is used with the box.
- Remote rapid shutdown devices can also be included.
- Earthing bus is provided.

Considering the above-mentioned functions, the electrical diagram of the combiner/distribution box can then be drawn as in Figure 5.13.

As can be understood from Figure 5.13, the series connection of the modules is made on the modules themselves, while the parallel connection of the strings is made inside the combiner boxes. The combiner boxes are manufactured according to standardized voltage and current values for a specific number of PV strings. Accordingly, the principle layout diagram of a DC combiner box can be drawn as in Figure 5.14.

The use of a single combiner box (also called string combiner box or array combiner box) is insufficient for large-scale PV systems. In this case, several combiner boxes and a sub-combiner box (also called a sub-array combiner box or recombiner box) are used together to connect all PV strings as shown in Figure 5.15.

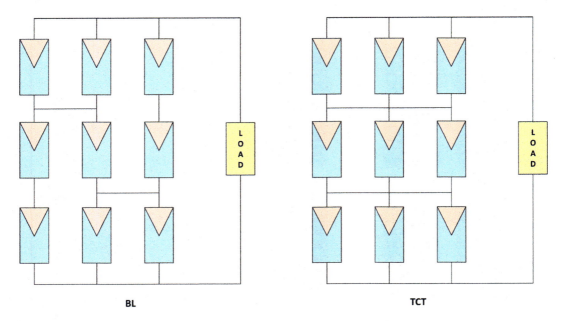

FIGURE 5.12 Alternative PV array configurations, BL: Bridged linked, TCT: Total cross-tied.

5.2.3 Photovoltaic Inverters

It is possible to divide the photovoltaic inverters into three main groups according to the operational modes, which are grid-tied (grid-interactive or grid-connected) inverters, stand-alone inverters, and bimodal (battery-based interactive) inverters. Stand-alone inverters simply operate via batteries and supply power to the load independent of the utility grid. However, grid-tied inverters operate from PV arrays and supply the AC power in parallel with the grid. As for the bimodal inverters, they can operate in both grid-connected and stand-alone modes. During the energized grid conditions, bimodal inverters generate AC power in proportion to PV array production while charging the battery bank. The inverters automatically passed to the stand-alone mode during grid voltage loss and supply powers to the critical loads. The following sections examine these PV inverter types in detail.

5.2.3.1 Photovoltaic Grid Inverters

It is possible to classify the grid-tied photovoltaic inverters in different topological aspects such as structural topologies, transformerless inverter topologies, and multilevel inverter topologies. We will examine the PV inverters in this section only by structural topologies. As discussed in the above sections, structural topologies of photovoltaic inverters are divided into six different types such as module inverters (microinverters), string inverters, and multi-string inverters, inverters with a master-slave concept, team concept inverters, and central inverters. Figure 5.16 shows the grid connection configurations for each inverter topology. The design/operating characteristics, advantages, and limitations of these inverter technologies are also summarized in Table 5.1 for further comparison. These grid-tied inverters use PV arrays as a DC input and supply the grid with synchronized AC output power. Site AC loads may also be served by the inverter output. All listed utility-interactive inverters have an anti-islanding feature to de-energize the inverter output in case of a grid outage. Inverter specifications can be different according to different applications. Hence, inverter specification sheets are crucial in selecting and specifying the suitable inverters for a given application.

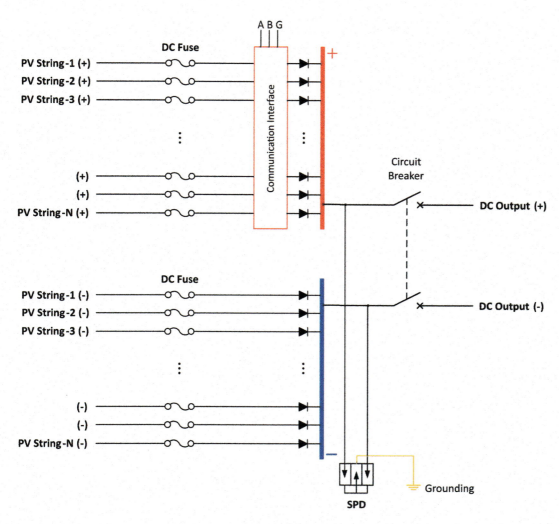

FIGURE 5.13 Electrical diagram of a PV array combiner/distribution box at the DC side.

All types of inverters are rated for their maximum continuous AC power and current output over a specified temperature range. Hence, inverters' power ratings will be restricted by switching elements if temperature limits are exceeded. DC input voltage, AC output voltage, AC power ratings, and power conversion efficiency are the standard specifications for all types of inverters. Besides, an MPPT (maximum power point tracking) function is added to all grid-interactive inverters to extract the maximum output from PV arrays. Other inverter specifications and sample technical data regarding the grid-interactive inverters are listed in Table 5.2.

As can be seen from the tables and explanations above, there are various important considerations for the inverters. The points related to design and installation issues are explained below briefly. The inverter characteristics curves, performance parameters, and the criteria for selecting the best PV inverter will be described in the next coming sections. The most basic output arrangements of grid-tied PV inverters are structured according to common electrical system configurations such as TN-S, TN-C, TT, and IT grid connections. Figure 5.17 shows the common grid types and 3-phase inverter connections.

The input arrangements of grid-tied PV inverters can be designed based on PV array configurations and structural types of solar inverters. We will give herein the block diagrams of the most commonly used string and central inverter circuits as an example. String inverters can be designed

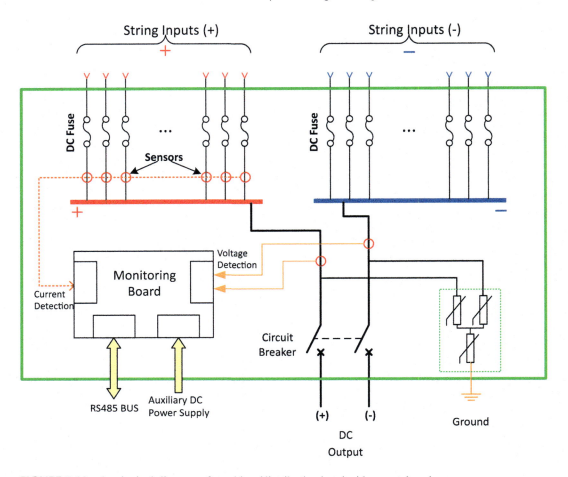

FIGURE 5.14 A principal diagram of combiner/distribution box inside a metal enclosure.

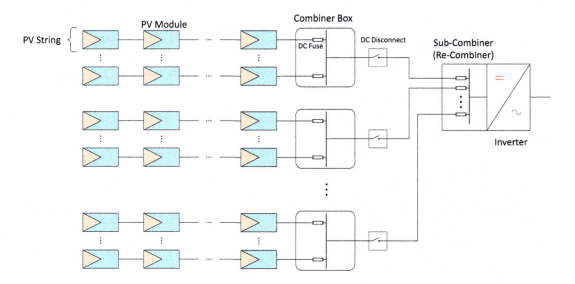

FIGURE 5.15 Combiner and Sub-combiner boxes in a large-scale PV system.

Photovoltaic Power Systems

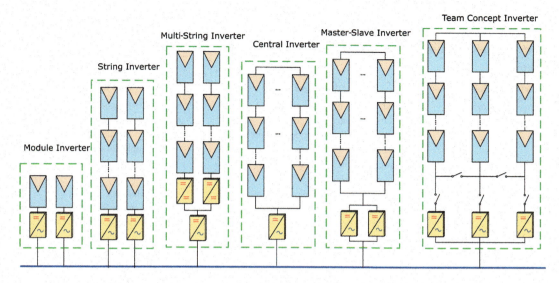

FIGURE 5.16 Grid-connected PV array configurations according to inverter topologies.

TABLE 5.1
General Specifications, Advantages, and Limitations of Grid-Connected PV Inverter Topologies

Inverter Topology	Advantages	Disadvantages and Applications
Module inverter	• No array mismatch loss • Flexible application due to modular structure • Reduced maintenance cost • High accuracy in MPP tracking	• Higher installation cost • Reduced overall efficiency due to higher voltage amplification • Suitable for residential applications
String inverter	• Reduced cost due to mass production • High reliability in terms of PV panel failure • High accuracy in MPP tracking • Maintenance and panel replacement easier • Higher overall efficiency • No losses in string diodes • Flexible design	• There is no obvious drawback. • Suitable for large-scale and residential applications
Multi-string inverter	• High accuracy in MPP tracking • Flexible design • Reduced cost • Higher overall efficiency • The advantages of string and module integrated inverter is combined	• Low inverter reliability due to a single centralized inverter • Suitable for large-scale applications
Master-slave inverter	• Higher reliability • Reduced mismatch power loss • Higher overall efficiency • Design flexibility • High accuracy in MPP tracking	• More costly than a central inverter • Suitable for large-scale applications

(Continued)

TABLE 5.1 (*Continued*)
General Specifications, Advantages, and Limitations of Grid-Connected PV Inverter Topologies

Inverter Topology	Advantages	Disadvantages and Applications
Team concept inverter	• The advantages of the string concept and master-slave concept are combined • High overall efficiency under low and high irradiance • High accuracy in MPP tracking • High inverter reliability	• Relatively higher cost • Control system is more complex • Suitable for large-scale/utility-scale applications
Central inverter	• Easy installation for small-scale applications • Relatively reduced cost when compared to team/master-slave concepts	• Low reliability • Low accuracy in MPP tracking • Non-flexible design • Poor power quality with current harmonic contents • Requires high voltage DC cable • High mismatch loss • Suitable up to several Megawatts

TABLE 5.2
Specifications of Grid-Interactive Inverter and Sample Technical Data for Different Inverters

Data Types	Technical Specifications	8 kW String Inverter (ABB)	100 kW Central Inverter (ABB)
DC input	• Nominal PV Power (P_{PV})	8100 Watt	—
	• (Recommended) Maximum Array Power ($P_{PV,max}$)	8900 Watt	120 kWp
	• Maximum DC Voltage ($V_{DC,max}$: Open Circuit, Cold)	900 V	900 V
	• Start Voltage and Operating Range (V_{DC})	250–800 V	250 to 750 V
	• MPPT Voltage Range (V_{DC})	335–800 V	450 to 750 V
	• Nominal Voltage (V_N)	480 V	480 V
	• Maximum DC Current ($I_{DC,max}$)	25.4 A	245 A
	• Number of DC Inputs (parallel)	4 with quick connectors	4 ±1 (80 A each)
AC output	• Nominal AC Output Power (P_{AC})	8000 Watt	100 kW
	• Nominal AC Output Current ($I_{AC,nom}$)	34.8 A	195 A
	• Nominal AC Output Voltage ($V_{AC,nom}$)	230 V	300 V
	• Operating Range, Grid Voltage	180–276 V	±10%
	• Operating Range, Grid Frequency	47–63 Hz	48–63 Hz
	• Harmonic Distortion of Grid Current	< 3%	< 3%
	• Power Factor	1	Adjustable Cos(φ)
	• Grid Connection	Single-phase: L, N, PE	Three Phase: TN/IT
Performance	• Maximum Efficiency	97.1%	98.0%
	• Weighted Efficiency (Euro-eta efficiency)	96.6%	97.5%
	• Standby losses	< 12 W	< 55 W
	• Nighttime Consumption	< 1 W	-
	• Own Consumption in Operation	< 30 W	< 350 W
	• Ambient Temperature Range	-25C°–60C°	-15C°–50C°
	• Nominal Power up to	50C°	40C°

(*Continued*)

TABLE 5.2 (Continued)
Specifications of Grid-Interactive Inverter and Sample Technical Data for Different Inverters

Data Types	Technical Specifications	8 kW String Inverter (ABB)	100 kW Central Inverter (ABB)
Protection	• Monitoring and Communication Interface	EIA-485, Modbus, Ethernet, Inverter to Inverter	Modbus, PROFIBUS, Ethernet, Control Panel, Digital, and Analog Inputs/Outputs
	• Ground Fault and Arc Fault Detection	Yes	Yes
	• DC and AC Disconnect	Yes	Yes
	• Fuse Ratings	12 A	—
	• Anti-islanding protection	Yes	Yes
	• Overload and Over Temperature	Yes	Yes
Other features	• Size (Width/Height/Depth, mm)	W:392/H:581/D:242	1030×2130×644
	• Weights (kg)	29 kg	550 kg
	• Enclosure Types	Outdoor IP55 enclosure	IP22 or IP42 (optional)
	• Maximum Operating Altitude	2000 m	2000 m
	• Standard and Extended Warranties	5 years, up to 20 years	5 years, up to 20 years

with a single MPPT, or they can be designed with multiple MPPT structures. In the case of multiple MPPTs, there are two options to operate MPPTs, either in parallel (single MPPT concept) or independent (dual MPPT concept). String inverters may also include a bulk capacitor group after DC/DC stage to have a constant DC bus before the main inverter unit. Figure 5.18 shows the independent and parallel connection of two MPPT units in a solar string inverter.

As can be understood from Figure 5.18, the dual MPPT concept can be used for PV arrays with different orientations, different string lengths, and with dissimilar PV modules. For example, it would be appropriate to prefer the dual MPPT concept for photovoltaic roof applications with different zones. However, the single MPPT concept is suitable for identical PV string applications in terms of shading, orientation, and preferred PV technology types. It should also be noted that using the dual MPPT function would be a better choice even for identical PV string applications. Because one of the strings can be damaged and/or can be exposed to higher soiling/shading rates, which would affect the output of the entire solar array and result in lowered overall efficiency for the string inverters with a single MPPT function. In addition to the MPPT components, the string inverters include components such as bulk capacitors, line filters, control/monitoring units, surge protection devices, and different switches. Figure 5.19 shows a common block diagram of the string inverter design with auxiliary components.

Another commonly used photovoltaic inverter type is the central inverter. Central inverters are generally installed to integrate large-scale PV plants into the AC grid. These PV power plants are combined using the series-parallel connection of PV strings. Each PV string of PV modules connected to a centralized inverter may also include independent MPP trackers. Central inverters are installed without an internal transformer and connected to the grid via three-winding or two-winding transformers. Central inverters also have a DC-DC converter as an intermediate stage. Considering the above explanations, the conceptual block diagram of the photovoltaic central inverter and its grid connection via a three-winding transformer is given in Figure 5.20.

As another example, the central inverters can also be connected to the grid by paralleling two-central inverters through two-winding transformers as illustrated in Figure 5.21.

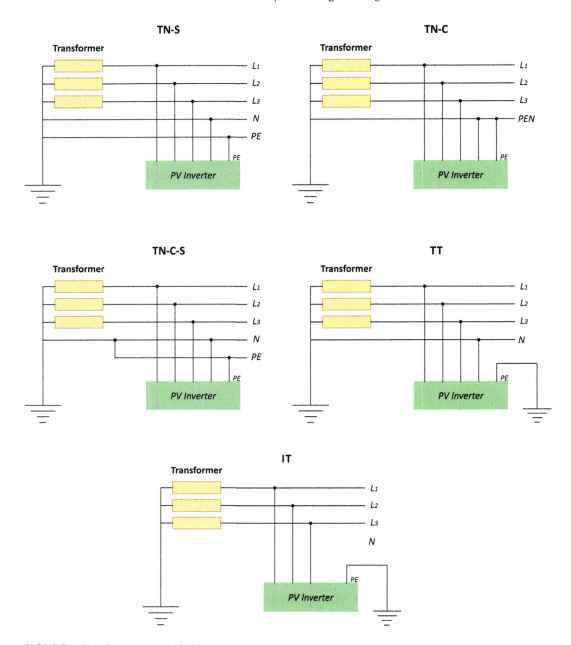

FIGURE 5.17 Grid-types and PV inverter connections.

5.2.3.2 Stand-Alone Inverters

The general structure of a stand-alone inverter (off-grid inverter) for PV system application is shown in Figure 5.21. As shown, the main components of a stand-alone PV system are the photovoltaic array, power conditioning units (control equipment, charger, and MPPT), DC disconnect, AC disconnect, energy storage, and the loads. The working principle of stand-alone inverters can be briefly explained as follows. During the daytime, the system will convert the energy from the battery and PV array to meet the load demand while charging the battery at the same time. During the nighttime, the system will convert the energy of the battery to give the power to load through the inverter. Stand-alone inverters can also be placed in different topologies with their surrounding components.

Photovoltaic Power Systems

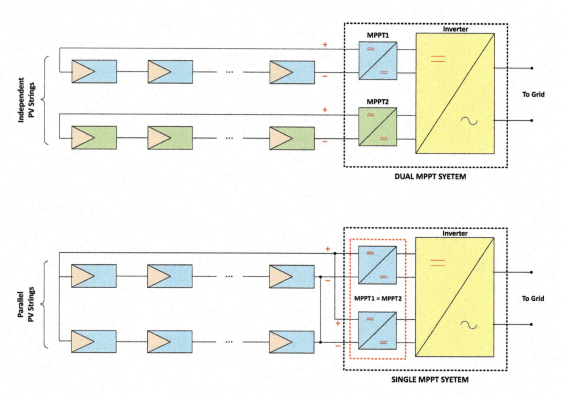

FIGURE 5.18 Dual and single MPPT concepts for string inverters.

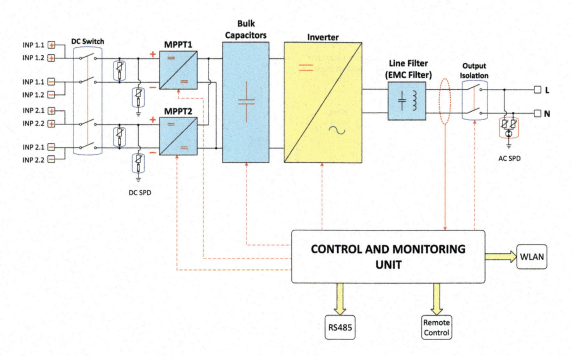

FIGURE 5.19 A general block diagram of a string inverter with auxiliary components.

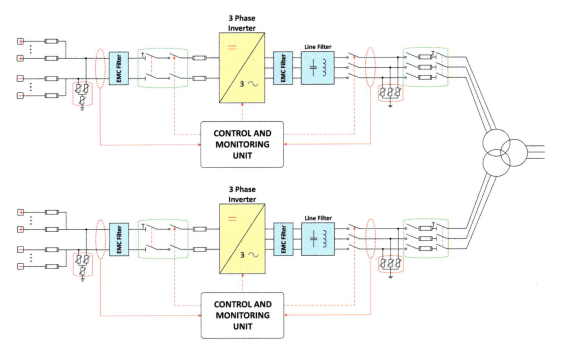

FIGURE 5.20 Block diagram of PV central inverter and its grid connection via a three-winding transformer.

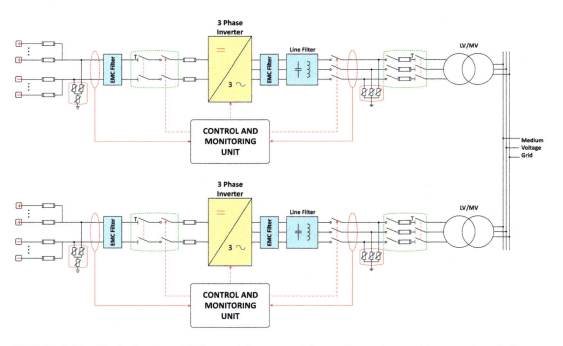

FIGURE 5.21 Block diagram of PV central inverter and its medium voltage grid connection via 2 two-winding transformers.

For example, it is possible to connect a backup generator to a stand-alone inverter as an alternative option. Besides, it is possible to connect these inverters to AC distribution boards via an isolation transformer. These topologies and other configurations will be covered in detail under the section "Stand-alone Photovoltaic Power Systems."

Photovoltaic Power Systems

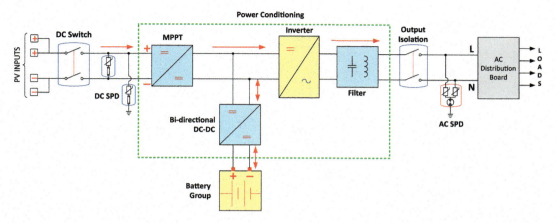

FIGURE 5.22 Block diagram of a stand-alone inverter with auxiliary components.

5.2.3.3 Bimodal (Hybrid) Inverters

Hybrid (bi-direction) solar Inverters are designed to operate in both stand-alone mode and grid-connected mode. The general structure of a bimodal inverter for PV system application is illustrated in Figure 5.23. During the daytime, excess energy will be transferred to the grid after the batteries are fully charged and the load demand has been met. Batteries continue to deliver power to AC load after sunset. If the battery is insufficient for the load demand, electrical energy will be supplied from the grid. When the grid fails, the system will automatically switch to islanding mode and use the battery-stored energy and PV energy as long as they are available.

5.2.4 PHOTOVOLTAIC SYSTEMS FOR THREE-PHASE CONNECTION

Depending on the application area, photovoltaic power systems can be designed in single-phase or three-phase architectures. As known, the key component connecting a PV system to the utility is the inverter. The PV Inverters are commercially available as single-phase and three-phase structured. In this perspective, the PV systems can be connected to the three-phase network over three-phase inverters or via single-phase inverters in the numbers of multiple 3. The simple idea behind the

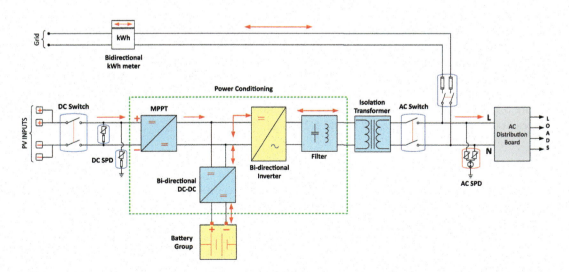

FIGURE 5.23 Block diagram of a bi-directional (hybrid or bimodal) inverter with auxiliary components.

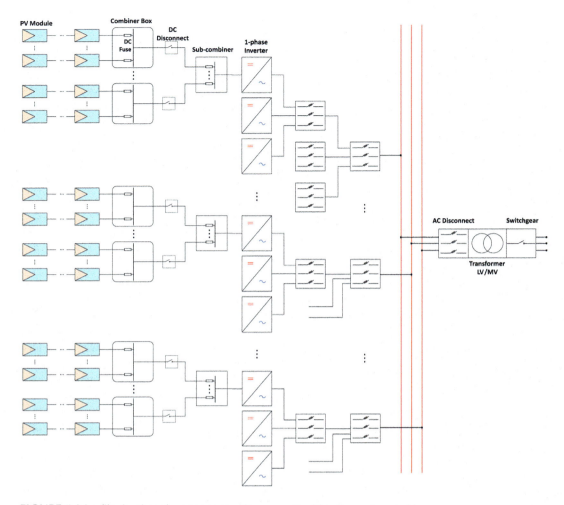

FIGURE 5.24 Single-phase inverter connection concept with a three-phase grid.

single-phase inverter connections with a three-phase grid is to group the single-phase inverters in a three arrangement in aiming to assemble the three-phase system. Accordingly, the PV array structure is also needed to be configured depending on the number and type of inverters. Figure 5.24 shows the basic idea of the connection of a single-phase inverter system to a three-phase grid. In this perspective, a single-phase string inverter connection with a three-phase system is illustrated in Figure 5.25. As for the three-phase inverter connection with the three-phase system, a centralized inverter concept is demonstrated in Figure 5.26 as an example.

5.2.5 AC Distribution Box and Utility Connection

The photovoltaic AC distribution box receives the AC power from the PV inverter and transfers it to the local grid for grid-connected applications, or transfers it to the AC loads through the distribution board for off-grid applications. The photovoltaic AC distribution box should include necessary components for safe operation, which are surge protection devices (SPD), circuit breaker (CB), fuse connectors, overcurrent protection, and residual current devices. SPDs are used to protect the PV inverters from high voltage, while circuit breakers are used in AC distribution boxes to protect the system from excessive currents or short-circuit currents.

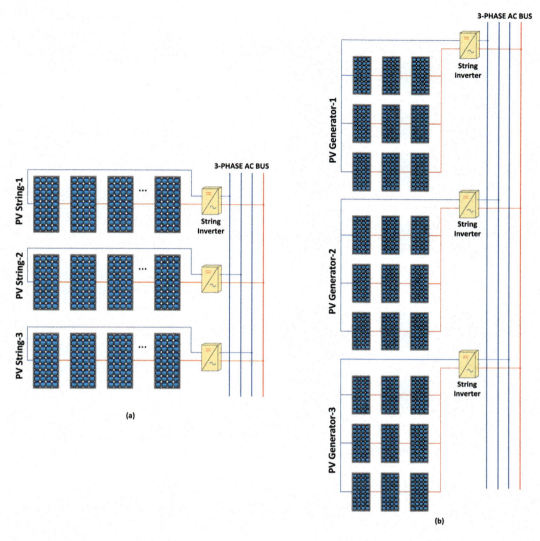

FIGURE 5.25 Single-phase string inverter connection with three-phase system: (a) PV string concept and (b) PV array concept with string inverters.

Solar PV Systems can be connected to distribution grids, either at low-voltage (LV) or medium-voltage (MV) connection points. Depending on the voltage level of the connection points, there may be differences in the protection requirements and utility connection schemes. A typical AC side utility connection scheme for grid-tied PV systems is given in Figure 5.27. As can be seen from Figure 5.27, the fundamental components of photovoltaic AC side structure are circuit breakers, residual current devices, kWh meters, switch disconnectors, fuse disconnectors, and surge protective devices. In addition to basic protection schemes, different arrangements such as contactors, insulation, string monitoring devices, AC distribution boards, transformers/substations as well as switchgear can be added to photovoltaic AC side assemblies based on the system requirements. These photovoltaic connection topologies will be studied in detail with sample examples given below in this section. In addition, the characteristics and selection of the PV power system components will be covered later in the following sections.

Note that the symbols used in the PV system drawings may vary depending on the type of device being used. For example, a circuit breaker can be represented by different symbols depending on its

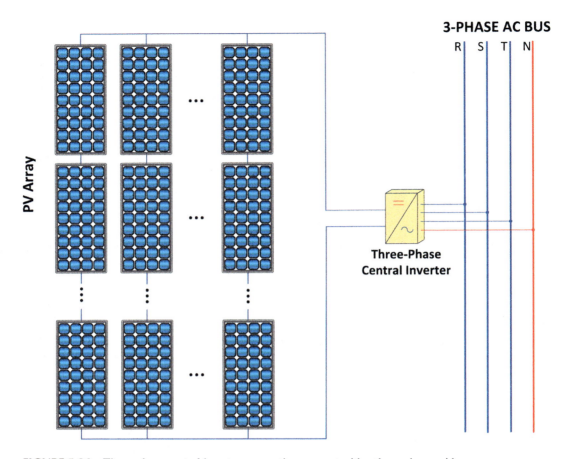

FIGURE 5.26 Three-phase central inverter connection concept with a three-phase grid.

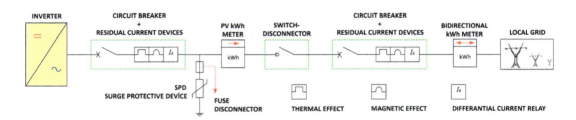

FIGURE 5.27 AC side grid connection example for small-scale grid-tied PV systems.

type such as vacuum/air/oil circuit breakers, miniature circuit breakers (MCB), molded case circuit breakers (MCCB), etc. Besides, additional signs are added to CB breaker symbols to identify the trip properties of CBs such as thermal, magnetic, and thermomagnetic effects. Therefore, as shown in Figure 5.27, the explanations regarding the symbols used in the system drawings are defined on the figure if necessary.

5.2.6 Substation Layout for Utility-Connected PV Systems

A photovoltaic substation is used for utility-connected PV systems to transfer the photovoltaic power reliably to the grid. The main components of a substation are transformer(s), circuit switches/breakers, protection equipment, control system, and communication network. The substation components may vary depending on the voltage level of the connection points. For example, the structure and operating principle of circuit breakers can differ according to their voltage levels such as low voltage, medium voltage, high voltage, and extra-high voltage. In this section, we will only consider the substation at low and medium voltage levels where PV systems connect to the grid. The first step in planning/designing a PV substation is the preparation of a single-line diagram that simply shows the switching/protection schemes, as well as the transformer(s) with incoming photovoltaic supply and outgoing feeders. Hence, the single-line diagram should include lines, switches, circuit breakers, and transformers. The transformers are used in PV power plants to connect different voltage levels between PV source and utility. The incoming lines in a substation should have a disconnect switch and a circuit breaker. As known, the circuit breaker is used as protective equipment to interrupt fault currents or overload currents automatically, while a disconnect switch is used to provide isolation at no-load conditions. These breakers and switches have a control circuitry to clear the faults arising from the failure of the components. Figure 5.28 shows a protection system and basic components in a grid-connected PV substation. The battery group in Figure 5.28 supplies the control circuitry, signal, and alarm circuits. The other ancillary devices in the protection system include secondary relays, signal lights and audible alarm devices (horns), etc. A potential transformer can also be added to the protection system for under-voltage trips.

In the past, electromechanical relays were used in substation protection systems. However, multifunction digital relays are preferred in today's modern applications. These digital relays have input channels to read both current and voltage data out of measurement transformers to produce necessary signals to operate the desired protection functions. In addition, these relays have communication channels to be controlled remotely and monitored via the SCADA (Supervisory Control and Data Acquisition) systems.

5.2.7 Protection Devices in PV Systems

Photovoltaic systems are composed of PV panels, power inverters, cables, batteries and switches, and protection devices such as lightning, surge protection, and earth leakage devices (residual current device: RSD). The purpose of this section is to describe the protection components on the DC and AC sides to determine a proper protection system for photovoltaic applications. It is important to note that, unlike the conventional AC system applications, the available short-circuit current inside a PV system is limited. Therefore, it is possible to use a fuse-based or breaker-based solution for the protection of PV applications. For this purpose, let us consider the photovoltaic system drawing in Figure 5.29, which shows the protection elements and their locations on a PV-powered distribution network.

It is clear from Figure 5.29 that the PV strings can be protected by DC fuses if the strings are capable of generating enough fault current to damage the conductors, equipment, or modules. It should be remembered that a DC fuse link may not be used in the case of the smaller size PV applications (up to two strings in parallel), provided that the conductors are sized properly. Depending on the designed PV system capacity, there may be numerous PV strings connected in parallel to reach higher current and PV power output. As the power rating increase, it is more proper to use molded case switch protection instead of disconnect switch solutions. As known, the molded case circuit breakers shown in Figure 5.29b can be controlled remotely to shut down the PV systems immediately in case of critical contingency conditions.

5.2.7.1 Surge Protection Devices

Another important protection requirement in photovoltaic applications is overvoltage surges. As known overvoltage surges in a grid-tied PV system can happen due to grid overvoltages, lightning strikes, and ground faults. Hence, surge protection devices (SPDs) should be installed properly in the PV systems to avoid high-voltage damage. To install the SPDs correctly, all wiring entering and exiting the system should be coupled to the ground through surge protection devices, and also all conductive surfaces are needed to be grounded strictly.

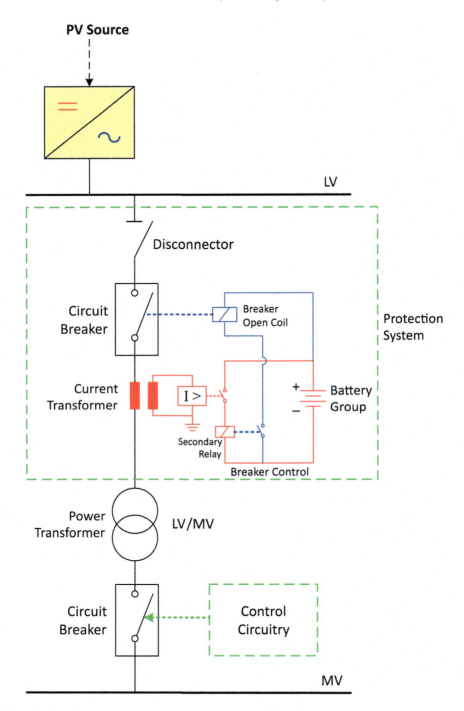

FIGURE 5.28 Protection system and basic components in a grid-connected PV substation.

It is also important to note that overvoltage surge protection requirements are versatile and dependent upon the system configuration, geographic location, and grid integration issues. The overvoltage surge protection should be performed on both the DC and AC sides. In addition, communication lines in the system must also be protected with external SPDs. Because they are conductive cables and they may provide a path for voltage surges to the inverter. Figure 5.30 shows a typical

Photovoltaic Power Systems 225

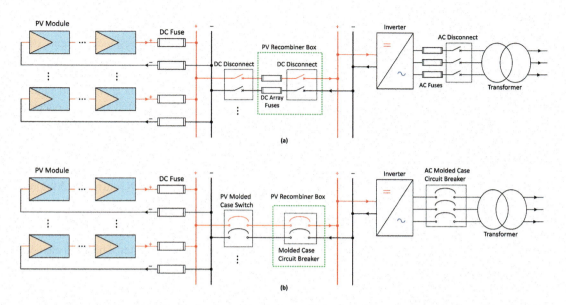

FIGURE 5.29 Protection devices and basic components in a grid-tied PV system: (a) Fuse solution and (b) Breaker solution.

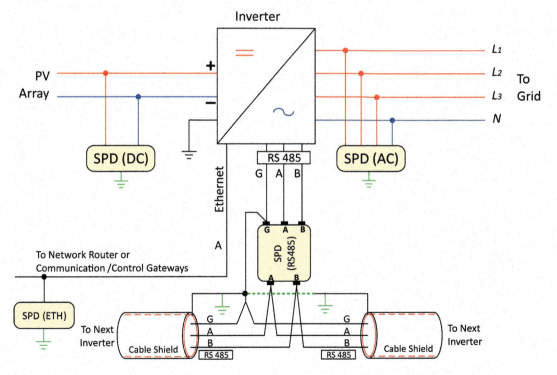

FIGURE 5.30 A sample SPD wiring diagram for a typical PV application [G: Ground, A: Data A (+), B: Data B (−)].

application of SPDs on the DC and AC sides as well as communication lines. The communication cables used for RS485 can be three-wire or four-wire shielded and twisted cables such as CAT5 or CAT6 cables.

5.2.7.2 Residual Current Devices

Another important protection area in PV systems is the protection against residual currents (the difference between incoming current and outgoing current). There may be several reasons causing residual currents in PV systems such as parasitic capacitances, earth leakage currents, and lightning discharges. The schematic of leakage current, often as a result of ground faults is shown in Figure 5.31a. The leakage currents (residual currents) above a threshold value are detected and disconnected from the PV system by residual current devices (RCDs). The residual current may be AC, DC, or mixed (combination of AC and DC). Therefore, the types of RCDs to be used should be sensitive to the possible residual current types (AC, DC, or mixed) in the system. The IEC 60755 standard specifies three different types of RCDs, which are Type-AC (sensitive to AC currents), Type-A (sensitive to AC and pulsed DC currents), and Type-B (sensitive to AC, pulsed DC, and smooth DC currents). As noted above, PV systems are needed to be provided with RCDs, which are often integrated into PV inverters as shown in Figure 5.31b. As illustrated in Figure 5.31, a ground fault in the DC side generates DC residual current that can flow from the ground into the AC circuit through AC neutral-to-ground linking, and again back to the DC circuit through the non-isolated photovoltaic inverter. The RCD element

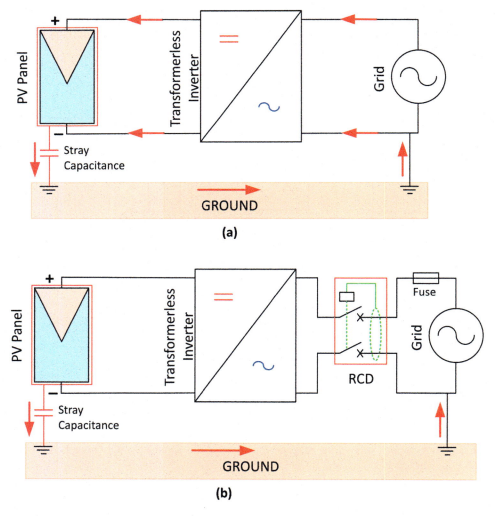

FIGURE 5.31 (a) Schematic of leakage current (residual current) due to PV(+) to Earth Fault and (b) an RCD protection scheme for a typical PV application.

Photovoltaic Power Systems

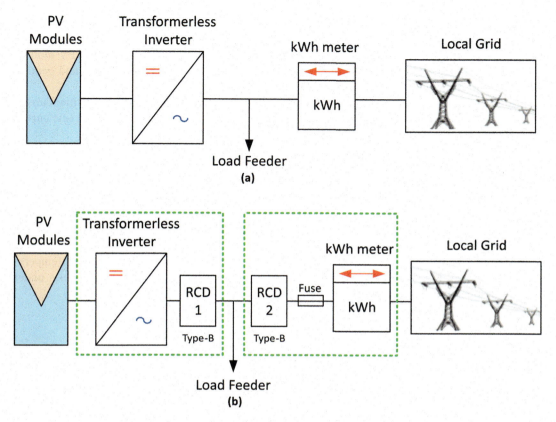

FIGURE 5.32 (a) A small-size grid-connected PV system with a local load feeder and (b) RCD application in the given PV system.

in the PV system detects this leakage current and opens the circuit on the AC side. An important point should be underlined herein. It is clear from Figure 5.31 that parasitic capacitances may create a natural leakage current loop between the DC side and AC side if both the PV array and transformerless inverter (on the AC side) are grounded. Hence, the PV array and frame cannot be connected directly to the earth in case of inverter usages with no-galvanic isolation. Note that under normal operating conditions, parasitic capacitances of PV generators may still leak currents to the earth due to contamination, dew formation, rain, and snow.

The type and number of RCDs to be used in photovoltaics can be different depending on the PV system configuration. For example, let us consider a grid-connected PV system with a local load branch as shown in Figure 5.32a. If a transformerless inverter is selected for the given system, then it is necessary to use two RCDs as shown in Figure 5.32b, the first RCD before the load point and the second one after the load point. As a simple approach, the Type-B RCD can generally be preferred because it covers the properties of other RCD types. Note that a single RCD after load point would be sufficient if an isolated inverter (with transformer) was used instead of a transformerless one.

It is vital to understand the ratings, operating characteristics, standards of protection, and other auxiliary components in photovoltaics to design and install the PV systems properly. Therefore, the structures, operating principles, characteristic values, and application areas of these components are examined in detail in the following sections.

5.2.7.3 Photovoltaic DC Fuses

DC fuses are commonly used as overcurrent protection devices in residential, commercial/industrial, and utility-scale photovoltaic applications. DC fuses are manufactured in different body types with different operating characteristics such as amperage rating (A), interrupting rating (kA), DC voltage level, etc. Common application areas of DC fuses in photovoltaic systems are PV string/array level protection, combiner/recombiner applications, in-line PV module protection, inverter DC input protection, and battery charge controllers. The types and general specifications of the DC fuses are summarized in Table 5.3.

The selectivity and sizing of photovoltaic DC fuses are carried out based on the time-current characteristics of the fuses. Figure 5.33 gives an example of the time-current characteristic curves of the DC fuses at different rated currents.

As it is well known, the ambient temperature affects the current-carrying capacity of the conductors. Hence, the temperature derating coefficients should also be taken into account in the photovoltaic fuse selection procedure. Figure 5.34 shows sample temperature derating characteristics of the fuses at different rated currents. Note that derating values can be found from the manufacturer's published curves. For example, Table 5.4 shows the sample ambient temperature derating coefficients (K_t) for NH-type fuses. The experimental studies show that derating coefficients may vary slightly according to fuse types.

It is understood from Table 5.4 that the derating factor will rise above "1" at ambient temperatures below zero. In other words, the fuses can carry the current above their rated values under negative temperature conditions.

TABLE 5.3
Specifications of Photovoltaic DC Fuses and Sample Technical Data for Different Fuse Types

Body	Fuse Type	Fuse View (Sample)	Rated Current (A)	Rated Voltage (V_{DC})	Interrupting Rating (kA)
Cylindrical	Ferrule		Up to 32 A	Up to 1500 V_{DC}	Up to 50 kA
	Bolted		Up to 32 A	Up to 1500 V_{DC}	Up to 50 kA
	In-line		Up to 20 A	Up to 1000 V_{DC}	Up to 50 kA
UL RK5[a]	Ferrule		Up to 400 A	Up to 600 V_{DC}	Up to 50 kA
	Bladed				
Square Body (NH)	Bladed		Up to 600+ A	Up to 1500 V_{DC}	Up to 50 kA
	Bolted		Up to 600+ A	Up to 1500 V_{DC}	Up to 50 kA

[a] Time Delay Fuse

Photovoltaic Power Systems

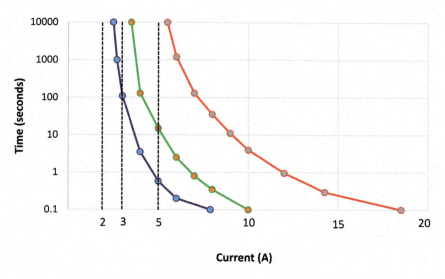

FIGURE 5.33 Time-current characteristics of different DC fuses (2 A, 3 A, and 5 A).

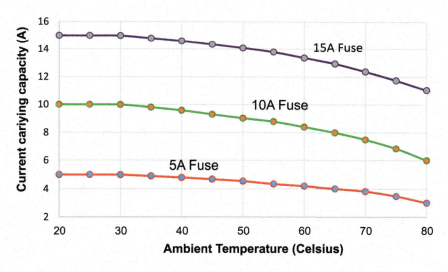

FIGURE 5.34 Temperature derating curves of different DC fuses (5 A, 10 A, and 15 A).

TABLE 5.4
Sample Ambient Temperature Derating Coefficients (K_t) for an NH Type Fuse

Ambient temperature (°C)	20	40	45	50	55	60	65	70	75	80
Derating coefficient (K_t)	1	0.92	0.90	0.87	0.85	0.82	0.79	0.76	0.72	0.69

5.2.7.4 Disconnectors

The disconnectors, also known as isolating switches, are manually or automatically operating mechanical devices to fulfill designated isolation functions under no-load conditions. They are mainly composed of a handle, a contact knife, a static socket, and an insulating bottom plate. Both AC and DC isolators are commercially available and applicable to PV system installations.

While the DC isolator switch is used between the PV array and inverter, the AC isolator is used between the inverter and AC distribution board. Photovoltaic disconnectors are mainly characterized by their rated voltage, continuous operating current, and service conditions like ambient temperature range and humidity level. Isolator switches can be either single poles or multiple poles (two poles to eight poles). Note that the term "poles" used herein refers to isolator terminals. The isolator poles can be connected either in series or in parallel. Figure 5.35 shows an internal connection of a typical isolator switch with two poles in series. Isolator switch configurations also vary according to PV system applications. Switch configurations and their connection diagrams are often provided by the manufacturers. The most common switch configurations in isolators are those in which the poles are connected in series or parallel. Besides, asymmetrical configurations are also possible such as three poles on the positive leg and one pole on the negative leg.

The required isolator current and voltage ratings should be determined based on the type of inverter used and array configuration. Since the details of PV system sizing are covered in the "Designing, Sizing and Selection" chapter below, the requirements for DC isolator sizing are not studied herein.

A new switch arrangement can be obtained by adding serial fuses to the disconnect switch, which is called a fused disconnect switch (or fused disconnector). Fused disconnect switches in PV systems are designed to open PV system circuitry in the event of a short-circuit or overloading conditions. As a consequence, the switch itself provides an opportunity to manually shut off the power on the circuit. And then, the serial fuse in the equipment disconnects the circuit automatically if the current in the circuit exceeds the fuse's rating.

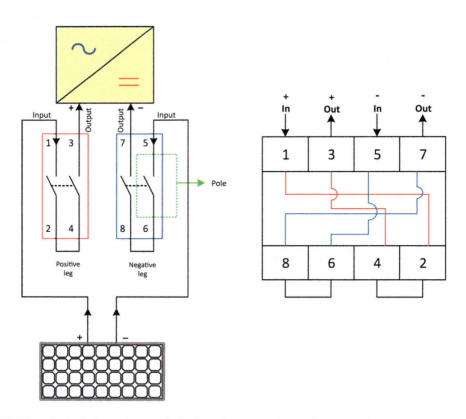

FIGURE 5.35 An isolating switch configuration of two poles in a series connection.

Photovoltaic Power Systems

5.2.7.5 Load Switches

The load switch (load break switch or only "switch") is a mechanical device capable of making, carrying, and breaking under normal load conditions. Load switches are also designed to carry specified overloading conditions such as short-circuit currents for specific periods (e.g., 10 kA short-circuit currents for 50 ms). They are often sized at different current and voltage levels such as from 32 A to 3200 A up to 1500 VDC. As in the other type of switches, the load switches are manufactured with several poles in different wiring configurations to perform various PV circuit connections. The load break switches can also be used for maintenance and emergency switching conditions like fire and electric shock. The range and connection diagram details can be found in the manufacturers' catalogs.

5.2.7.6 Switch Disconnectors

Switch disconnectors combine the properties of (load) switches and disconnectors. Hence, they can be used in PV systems to isolate the PV generator and/or PV inverter for safety/maintenance reasons, but they can also be used to make and break the PV system under loading/overloading conditions. Normal switch disconnectors and fused switch disconnectors are the two main switch disconnectors types used in PV systems. Three basic parameters, that is, rated insulation voltage, rated operating voltage, and rated operating current should be considered to choosing proper switch disconnectors for PV applications. Improper sizing on any of these parameters may lead to risky conditions for the installation itself and the end-user. These ratings and their selection process will be discussed in-depth "Designing, Sizing and Selection" chapter below.

As in the other switches and fuses, operational temperature range beyond normal conditions causes derating in the rated current of the switch disconnectors. In addition, the multipole switch configurations described above are also valid for switch disconnectors. Therefore, the same points will not be repeated herein.

5.2.7.7 Circuit Breakers

Circuit breakers are automatic switching devices that can turn on and turn off load current under normal, overload, and short-circuit conditions. Additional isolation functions such as under-voltage and leakage protection can also be added to circuit breakers with the help of their control systems. Photovoltaic systems often use a low-voltage circuit breaker, which is composed of a contact, an arc extinguishing device, an operating mechanism, and a protection device. As in the other switch types, circuit breakers can also be manufactured with multiple poles. Typically three- or four-pole circuit breakers are often used in PV applications.

Circuit breakers must be sized for interrupting ratings equal to or greater than the highest short-circuit current at the rated voltage at the system point, where the circuit breaker is applied to. Typical interrupting ratings for photovoltaic applications can vary from 3 kA to 10 kA dependent upon the rated voltage such as 600 V_{DC} and 1000 V_{DC}. Besides, the classical amperage ratings of circuit breakers may differ from 50 A to 500 A. The circuit breaker configurations and their connection diagram details as well as other technical specifications can be found in the manufacturers' catalogs.

The time-current curves are the key characteristics for the sizing and selection of the circuit breakers. Figure 5.36 gives examples of tripping (time-current) curves of a circuit breaker for thermal-magnetic and electronic protective schemes.

As described in Figure 5.36, circuit breakers may have two types of tripping units, namely thermomagnetic trip, and electronic tripping units. As its name implies, the thermomagnetic trip unit consists of two parts: the thermal trip unit and the magnetic trip unit. The thermal trip unit includes a bimetal thermal element to actuate the circuit breaker opening with a time delay for the protection against overloads. The magnetic trip unit consists of an electromagnetic device to trigger the circuit breaker opening with a constant trip time (about some tens of milliseconds) for the protection

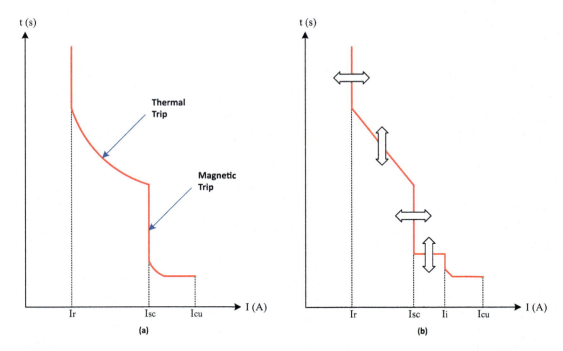

FIGURE 5.36 Time-current characteristics of different circuit breakers, (a) thermal-magnetic protection and (b) electronic protection. (I_r: current setting for overload trip, I_{sc}: current setting for short circuit trip, I_i: short circuit instantaneous current setting, I_{cu}: breaking capacity).

against short-circuit conditions. This electromagnetic device can be either a fixed or adjustable instantaneous trip value.

Electronic tripping units in circuit breakers use a microprocessor to activate the circuit breakers opening by processing the current signal digitally. Electronic tripping units can provide various protection functions such as short-time/long-time overcurrent protection, instantaneous/delayed short-circuit protection, ground-fault trip function, and residual current protection (differential current interrupter).

5.2.7.8 Contactors

Depending on application requirements, the DC- or AC-type contactors (or contactor relays) can be used in PV systems for many reasons. For example, contractors can be used to isolate the local loads between the PV generator and the inverter. Contactors can also be used with the interface system between the inverter and the output line. Changing the PV string configurations as well as disconnection the inverter from the PV strings is the other application of contactors in PV systems. An important point should be noted herein that PV strings are disconnected from the inverter when the output power is too low. Besides, string configurations are changed to optimize the system's overall efficiency, generally at low output conditions. The following diagram in Figure 5.37 shows possible application points of the contactors in PV systems.

The contactors in general are suitable for remote control and frequent open-close cycle under load up to 1000 A. The contactor relays used at point-1 can simultaneously cut off the positive and negative terminals at DC voltages up to 1500 V. Switching needs at points-1 and -2 are generally required in disaster situations such as electrical shock and fire, and in cases of a decrease in total system generation due to reasons of panel defect, low irradiance, and shading conditions. Besides, the PV system can be isolated for maintenance purposes by switching the relays/contactors at point-2. As it is clear from Figure 5.37, the relays at point-3 provide battery charging/discharging

Photovoltaic Power Systems

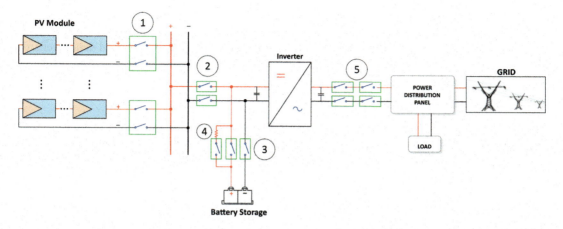

FIGURE 5.37 Possible contactor/relay application points in photovoltaic power generation systems (1: for the cut-off of the strings, 2: for DC safety cut-off, 3: battery charge/discharge switching and for DC safety cut-off, 4: for inrush current prevention, 5: for AC safety cut-off).

switching, while relays at point-4 are used for inrush current prevention. Finally, the power contactor relays at point-5 make available AC safety cut-off when necessary.

It is important to note that solid-state relays are used instead of electromechanical relays in battery charging/discharging applications that require frequent on/off switching. The advantages of solid-state relays used in battery charging/discharging systems are:

- The current flow in both directions on the same line is shown in Figure 5.38.
- Over-charge/discharge control is easier when compared to electromechanical relays. For example, the relay on the charge control side is turned off to prevent overcharging, while the relay on the discharge control side is turned off for over-discharge prevention.

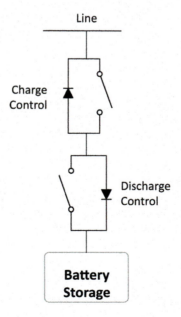

FIGURE 5.38 Typical battery charge control switches for solid-state-based relays.

The contactors can also be used with series fuses to increase their protection effectiveness against fault currents. By adding fuses in series to the contractors, they are called fused contractors (the contactors with series fuses), which are suitable for PV applications. Other technical details regarding the use of contractors can be found in the manufacturer's catalog.

5.2.8 Other Balance of System Components

The balance of system (BOS) refers to all mechanical and electrical equipment, which are necessary to integrate the photovoltaic power with the grid and/or building system. Hence, several components are needed for assembling, controlling, and protecting the photovoltaic system. Examples of BOS components include:

- Cabling and wiring systems for DC and AC parts such as conductors and conduits (raceways)
- Manual switches and automatic protection devices such as fuses, disconnect switches, and circuit breakers
- The mounting structures and hardware such as junction and combiner boxes
- DC/AC inverters and other power converters (if necessary)
- The battery bank and charge controller (if energy storage is required)

Sufficient discussion regarding most of the above components is provided in the above sections. The remaining BOS components are considered below.

5.2.8.1 Inverters

The solar inverters are designed to convert DC photovoltaic power into AC power that can be sent to the local grid and/or loads. All inverters are rated in watts for their maximum continuous AC power and current output over a specified temperature range. Typically, the MPPT units are embedded into PV inverters to allow them to operate at their optimal output voltages to get maximum power generation. Environmental factors such as solar radiation and ambient temperature affect the inverter performance, and these parameters vary all the time throughout the whole year. Hence, the inverters should be compatible with all operating conditions range at all times. Note that operating conditions not only involve the environmental parameters but are also related to system design and installation requirements. That is why the inverter specification sheet is crucial for selecting and specifying the best inverter for a given PV application. The most important performance characteristic is the inverter efficiency curve, which can be found in the manufacturers' catalogs. Figure 5.39 shows the sample test result of the efficiency curve for a typical solar inverter. As in Figure 5.39, the efficiency values of commercially available inverters generally vary in the range from 92% to 98%.

As can be seen from Figure 5.39, the inverter efficiency is not a constant value and it varies depending on the power drawn from the inverter. A solar inverter efficiency can be calculated as the ratio of DC power out of solar panels (or batteries if available) to AC output power. However, two efficiency factors should be considered for solar inverters to calculate the overall efficiency more accurately, which are conversion efficiency (η_{conv}) and MPPT efficiency (η_{MPPT}). The conversion efficiency can also be of two types, namely peak and weighted efficiency. Weighted efficiency can further be defined as Euro (EU) and California Energy Commission (CEC) weighted efficiencies. While Figure 5.40 illustrates the concept of overall inverter efficiency, equations (5.1)–(5.5) express the efficiency calculations regarding the aforementioned efficiency factors.

$$\eta_{conv} = \frac{\int_0^{T_0} P_{AC}(t)dt}{\int_0^{T_0} P_{DC}(t)dt} \tag{5.1}$$

Photovoltaic Power Systems

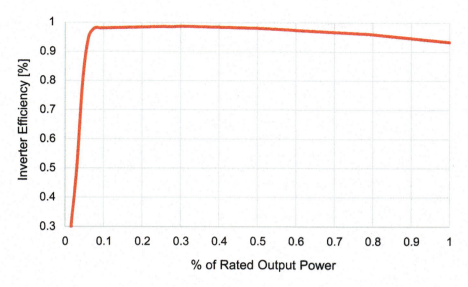

FIGURE 5.39 The test result of the efficiency curve for a typical solar inverter.

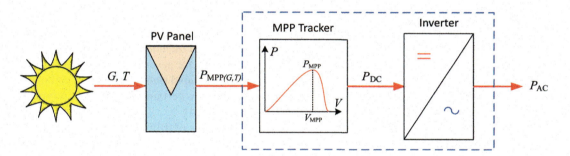

FIGURE 5.40 The illustration for the concept of overall PV inverter efficiency.

$$\eta_{\text{MPPT}} = \frac{\int_0^{T_0} P_{\text{DC}}(t)\,dt}{\int_0^{T_0} P_{\text{MPP}}(t)\,dt} \tag{5.2}$$

$$\eta_{\text{overall}} = \eta_{\text{conv}} \cdot \eta_{\text{MPPT}} = \frac{\int_0^{T_0} P_{\text{AC}}(t)\,dt}{\int_0^{T_0} P_{\text{MPP}}(t)\,dt} \tag{5.3}$$

where T_0 is the test period (measuring period), $P_{\text{AC}}(t)\,dt$ is the instantaneous power at the AC terminal, $P_{\text{DC}}(t)\,dt$ is the instantaneous power at the DC terminal, and $P_{\text{MPP}}(t)\,dt$ is the instantaneous value of MPP power supplied by the PV array.

Due to solar irradiance variation throughout the day, PV inverters do not always perform their optimal performance. Therefore weighted efficiency calculation is needed for the inverter to specify their performance more realistically. As discussed above, the two common weighted conversion efficiencies, the European efficiency formula (η_{EU}) and American efficiency formula (η_{CEC}), are given in the below equations. These weighted conversion efficiency formulas also calculate the inverter output performance through the range of its capacity.

$$\eta_{EU} = 0.03 \times \eta_{5\%} + 0.06 \times \eta_{10\%} + 0.13 \times \eta_{20\%} + 0.10 \times \eta_{30\%} + 0.48 \times \eta_{50\%} + 0.20 \times \eta_{100\%} \quad (5.4)$$

$$\eta_{CEC} = 0.04 \times \eta_{10\%} + 0.05 \times \eta_{20\%} + 0.12 \times \eta_{30\%} + 0.21 \times \eta_{50\%} + 0.53 \times \eta_{75\%} + 0.05 \times \eta_{100\%} \quad (5.5)$$

An important point should be noted that the formula η_{CEC} considers the higher irradiation region in more detail and η_{EU} formula emphasizes the region with relatively lower radiation. However, both formula yields better annual estimation for a solar inverter when compared to the peak efficiency-based performance calculation.

Photovoltaic inverter types and their specifications and wiring diagrams were covered in the above sections. Besides, the details of power electronics related to inverters will be examined in the chapter "Power Electronics Converters in Photovoltaic Systems and Grid Connection Criteria" later in this book. For this reason, the discussion about inverters is considered to be enough for now.

5.2.8.2 PV Rapid Shutdown System

PV rapid shutdown systems are generally preferred for building-integrated PV systems and provide safe protection to remove the voltages under rapid shutdown conditions. The shutdown happens automatically when the inverter or grid power is lost or when the power of the PV control vault is turned off. The stop button on the rapid shutdown boxes also provides manual control to disconnect the PV panels from the inverter. Most rapid shutdown system include a resistor to charge the input capacitor of the inverter, which can maintain high voltages for several minutes. Some rapid shutdown systems can also protect against fire. In case of fire, the rapid shutdown devices automatically shut down the system when the panel temperature rises above a certain critical value. A typical wiring diagram of the rapid shutdown system is given in Figure 5.41.

There are different rapid shutdown systems used in PV systems of different configurations. For example, rapid shutdown systems can combine two or four PV strings and provide single or more outputs depending on the number of MPPTs in the inverter.

5.2.8.3 Monitoring and Communication Systems

A monitoring system is an essential tool for solar PV plant applications to operate them reliably and optimally. As known, several reasons such as soiling, dirt, adverse weather conditions, and failures may lead to a reduction in photovoltaic power generation. Monitoring equipment can detect these abnormal system statuses locally or remotely and then enable the necessary actions to restore the

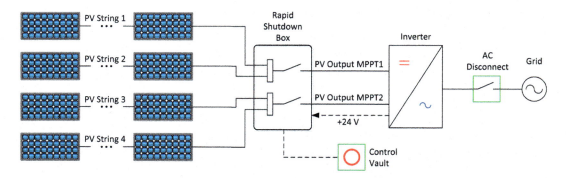

FIGURE 5.41 Typical wiring diagram of the rapid shutdown system for a small-scale (building integrated) PV system.

power generation. The simplest monitoring system reads the current, voltage, and power values at the DC and AC sides of the inverter and displays them on an LCD screen. For more sophisticated monitoring systems, environmental data such as PV module temperature, ambient temperature, solar radiation, and wind speed can also be collected, monitored, stored, and uploads the information over the internet to the PV plant management platforms. The collected and transferred data are then analyzed for the detection of any failure and inefficiencies so that one can fix the problem and enhance the system operation timely.

Data coming from the monitoring systems are usually transferred via communication protocols to a central location, for remote control and data analysis purposes. At this stage, remote control and monitoring can be accomplished by different remote connections such as an analog modem, ISDN (Integrated Services Digital Network), GSM (Global System for Mobile Communications), etc. The most common connections used in local and/or remote control systems are:

- USB or RS232 for local monitoring,
- RS485 and power line for inverter interconnection,
- Bluetooth and Wi-Fi for wireless connection,
- Connection to external SCADA through TCP/IP connection.

As it can be understood from the above explanations, it is possible to divide the monitoring system into two groups, local and remote monitoring. The most common parameters, which are locally monitored, are current, voltage, and power values for PV arrays and grids. These parameters can easily be monitored locally by displaying them from the inverter's memory or the external data loggers by utilizing such as a local PC via RS232 and/or a wireless system. Additional sensors will be required if other environmental parameters such as module/ambient temperature, global radiation, and wind speed are required to be monitored.

Remote control and communication between solar PV inverters can be performed with several connection types such as wireless connection (Bluetooth or Wi-Fi), RS485, and power-line communication systems. Dozens of inverters can be connected in a chain and communicated with each other over RS485 up to distances of 1200 meters and these inverters can be monitored at the same time. The concept of multiple-string inverter monitoring including remote access is given in Figure 5.42.

The most common communication tools used for remote monitoring are ethernet, internet, dial-up access, GSM, etc., which can send status messages and e-mail alerts to the management and control centers of PV power plants. As can be seen in the above explanations, the connection types in the monitoring systems can be wired and/or wireless. Figure 5.43 shows a typical photovoltaic monitoring system architecture for PV applications and its wiring diagram on a grid-connected photovoltaic power plant.

The data acquisition system shown in Figure 5.43 can be used to measure photovoltaic power plants regardless of their size and configuration. The monitoring system software is used to manage, display and store the collected data in a PC system and allow the users to access the data via the internet and mobile phones.

5.2.8.4 Bidirectional Net Metering

It will be sufficient to use conventional one-way energy meters in grid-connected PV systems if the generated energy out of PV arrays is only delivered to the grid. However, the use of bidirectional energy meters in grid-connected PV systems containing the battery and/or local loads is a technical requirement. As known, a net meter (also known as a bidirectional meter) measures the electric power flow in two directions. It measures the imported energy from the grid to meet the load shortage and records the surplus energy exported to the grid. Hence, a net meter measures the net power flow in kWh, which is the difference between imported and exported electrical energy. If exported

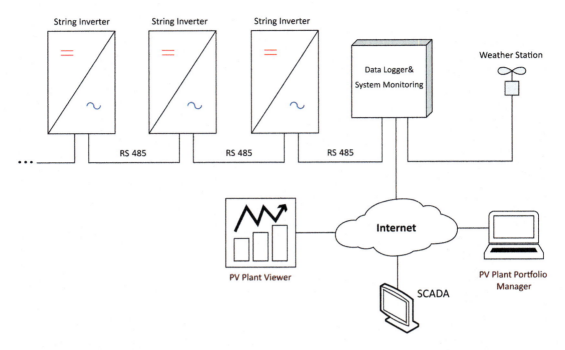

FIGURE 5.42 The wiring concept of multiple string inverter monitoring for PV applications.

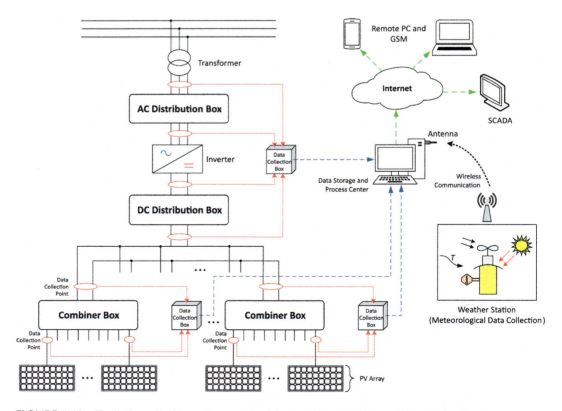

FIGURE 5.43 Typical monitoring system architecture for PV applications and its wiring diagram on a grid-connected PV power plant.

Photovoltaic Power Systems

energy is greater than imported energy, then the net energy shown on the meter will be positive. Otherwise, the net energy amount shown on the meter will be negative.

Single-phase and three-phase bidirectional net meters with different rated currents are available on the market for consumers who are willing to install their solar PV systems in parallel with the local utility. These smart net meters can also record the energy amounts for different billing periods separately. Besides other billing parameters such as cumulative active and apparent powers, average power factor, power-on hours, and maximum power demand in kW, etc. are also stored in the meters as displaying data.

5.2.8.5 Photovoltaic Cables and Cable Systems

The cables used in photovoltaic power plants can be grouped under three main headings, which are power cables (DC and AC cables), communication cables, and grounding wires/cables. The corresponding cable groups are illustrated as shown in Figure 5.44.

Power Cables: PV power plants under actual environmental conditions are exposed to high temperatures, ultraviolet rays, and adverse weather conditions such as snow, rain, ice, wind, sand, etc. Therefore, the insulation of the cables used in PV power plants must be able to withstand both adverse weather conditions and severe mechanical loads throughout the entire year. Besides, they should satisfy some electrical requirements such as voltage drop and power loss, which must be considered when selecting and sizing the cables. Furthermore, the physical flexibility and economic value of the cables are the other parameters that need to be well thought out when choosing DC and AC cables. Importantly note that these requirements apply to both DC and AC power cables.

One of the most important processes in setting up a PV system is to determine appropriate cable sizing. For this purpose, cables are characterized by their cross-sectional areas and ampacity levels (current-carrying capacity) at different rated voltages. The ampacity of the cables varies depending on the material of the cable core (copper and aluminum), the cross section, and the operating temperature. Besides, operating temperature ranges (generally −40°C to +90°C) and maximum permissible voltage level are the other important parameters that need to be considered when choosing the best suitable cable configuration. Both single-core and twin-core cables can be used at the DC side of the PV systems. In the case of grid connection via a single-phase inverter, a three-core cable is used for connection to the grid. If there is a three-phase feed-in, then a five-core AC cable can be used for the grid connection.

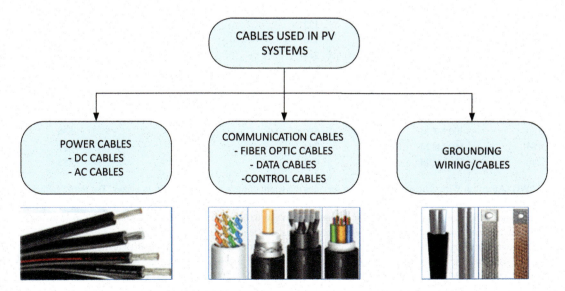

FIGURE 5.44 Classification of cables used in PV Systems.

Conductor sizes are commonly expressed in square millimeters (mm²), AWG (American Wire Gauge), and circular mils (cmil).Table 5.5 shows the universal wire sizes and basic specifications for XLPE insulated copper cables. The technical data given in Table 5.5 is for a typical XLPE insulated copper cable, which differs from manufacturer to manufacturer. Besides, other factors such as using PVC insulation, aluminum wire, and operating temperature also affect the current-carrying capacity of cables. Therefore, it would be more appropriate to use the manufacturers' catalog during the application phase. Table 5.6 shows sample DC and AC photovoltaic power cables and their basic characteristics.

It is important to draw attention to an important point herein. As known, the ampacity of a cable (measured in amperes) is the maximum current-carrying capacity of that cable without exceeding insulation and temperature limitations. Hence, to control the heat flow in a power cable, the ampacity ratings are needed to be adjusted according to the cables' installation environment. Accordingly, power engineers should look for the adjustment factors table, usually available in the manufacturers' catalog, for the percentage ampacity derating values versus ambient temperatures.

Communication Cables: In addition to the aforementioned DC and AC power cables, communication cables also play an important role in PV Power Plants. The main purpose of using communication cables in PV Power Plants is to monitor the generation and reliability performance of the system and to ensure communication between the solar inverters. For this purpose, the analog data obtained from the inverters, solar sensors, and energy meters are firstly converted to digital data,

TABLE 5.5
Universal Wire Size and Basic Technical Data for XLPE Insulated Copper Cables

Cross-sectional Area		Typical Ampacity (A)		
[mm²]	AWG[a] or cmil[b] Equivalent	Single Cable in Air [Cu]	Single Cable on Surface [Cu]	2-Adjacent Cable on Surface [Cu]
1.5	~16 AWG	30	29	24
2.5	~14 AWG	41	39	33
4	~12 AWG	55	52	44
6	~10 AWG	70	67	57
10	~8 AWG	98	93	79
16	~6 AWG	132	125	107
25	~4 AWG	176	167	142
35	~2 AWG	218	207	176
50	~1 AWG	274	260	219
70	~00 (2/0) AWG	406	386	325
95	~000 (3/0) AWG	491	467	393
120	~0000 (4/0) AWG	576	547	461
150	~300 MCM	670	637	536
185	~350 MCM	784	745	627
240	~500 MCM	944	897	755

[a] AWG (American Wire Gauge) metric system specifies the width of the wire diameter (D) in inches. The term "gauge" in the AWG system means diameter, ranging from 0000 (4/0) to 40. As can be understood from the table, the diameter decreases as the wire gauge gets larger. For example, 30 AWG is 0.01 inches in diameter, and (4/0) AWG is 0.46 inches in diameter. Hence, the cross-section area (A) of the wire is calculated as $A = (\pi/4) D^2$.

[b] Cmil (circular mil) is equal to the area of a circle with a diameter of one mil (one-thousandth of an inch). A circular mil (CM) is generally used to denote the cross-sectional area of a wire (or cable) in North America. Note that 1000 Cmil = 1 MCM (kcmil) = 0.5067 mm². For example: 500 MCM = 500 · 0.5067 = 253 mm² ≈ 240 mm².

TABLE 5.6
Different Types of DC and AC Cables and Their Technical Data for PV System Applications

Cable Types and Sample View	Physical/Mechanical Characteristics	Electrical Characteristics
 **PV1-F: DC Power Cable** **Application Area:** It is the DC cable used between the photovoltaic panel and the inverter.	**Conductor:** Electrolytic tinned copper. **Core Insulation:** Temperature resistant and halogen-free Co-Polyolefin, electron beam cross-linked **Outer Sheath:** UV-resistant, flame retardant and halogen-free Co-Polymer, electron beam cross-linked. **Cross-sections:** All standard cross-sections from 2.5 mm² to 240 mm² are available. **Weights:** From 40 kg/km to 2320 kg/km **Outer Diameters:** From 4.5 mm to 26.8 mm	**Rated Voltage:** 0.6 / 1.0 kV **Max. PV System Voltage:** Up to 2000 V DC possible **Max. Permissible Operating Voltage:** 0.7 / 1.2 kV AC or 0.9 / 1.8 kV DC **Ampacity:** From 41 A to 775 A **Temperature Range:** −40°C to 90 °C
 **NAYY-O:** DC or AC Cable (1 ~ and 3 ~ cables). Options for # of Cores: 1, 2, 3, 4, and 5. **Application Area:** It can be used as the main DC cable or low-voltage AC cable.	**Conductor:** Aluminum. **Core Insulation:** PVC (Polyvinyl Chloride) **Outer Sheath:** PVC (Polyvinyl Chloride) **Cross sections:** All standard cross sections from 16 mm² to 800 mm² are available for single-core cables. **Weights:** From 145 kg/km to 3120 kg/km (single-core cables) **Outer Diameters:** From 10.5 mm to 45 mm (single-core cables)	**Rated Voltage:** 0.6 / 1.0 kV **Max. Permissible Operating Voltage:** 0.7 kV AC (1 ~) / 1.2 kV AC (3 ~) or 1.8 kV DC **Ampacity:** From 87 A to 1080 A (in the air) (single-core cables) **Temperature Range:** −5°C to 70°C
 **NYY-J: AC Power Cable** **Application Example:** It can be used as a power cable for single-phase inverters and security system installations.	**Conductor:** (3-core) solid copper. **Core Insulation:** PVC (Polyvinyl Chloride) **Outer Sheath:** Flame Retardant -PVC **Cross-sections:** All standard cross-sections from 1.5 mm² to 95 mm² are available for three core cables. **Weights:** From 166 kg/km to 4488 kg/km **Outer Diameters:** From 10.4 mm to 40 mm	**Rated Voltage:** 0.6 / 1.0 kV **Max. Permissible Operating Voltage:** 0.7 kV AC (1 ~) / 1.2 kV AC (3 ~) or 1.8 kV DC **Ampacity:** From 18 A to 282 A (in the air) **Temperature Range:** −15°C to 70°C
 **YE3SV: AC Power Cable** **Application Area:** Medium voltage power distribution and installations	**Conductor:** Copper. **Core Insulation:** XLPE Insulation **Cross sections:** All standard cross sections from 35 mm² to 630 mm² are available for single-core cables. **Weights:** From 1300 kg/km to 8100 kg/km **Outer Diameters:** From 10.4 mm to 40 mm	**Rated Voltage:** 20.3 / 35 kV **Ampacity:** From 238 A to 1120 A (in the air) **Temperature Range:** −40°C to 90°C

and then the relevant signals are made available for the internet environment via a modem so that one can monitor the PV system status remotely.

There are three main types of communication cables used in photovoltaic applications. These cables are fiber optic cables, twisted pair cables (unshielded and shielded types such as CAT5 to CAT8), and coaxial cables. The twisted cables are generally used for weak current installations of PV power plants such as telephone, cameras, alarm systems, and computer networks. Also, PVC-insulated control cables with a low cross-sectional area are used for measuring and control purposes in PV systems. The cable flexibility and withstanding medium-level mechanical stresses are important criteria when choosing these cables for PV installations. Table 5.7 summarizes the basic characteristics of commonly used communication cables.

Grounding Wiring/Cables: As in conventional electrical installations, non-voltage metal parts of a PV system must be connected to the ground via conductors and earthing electrodes. Properly grounded and bonded PV modules and arrays have always equipotential lines so that the system components and human beings are significantly protected from lightning and other electrical failures.

TABLE 5.7
Different Types of Communication Cables and Their Basic Technical Data

Cable Types and Sample View	Physical/Mechanical Characteristics and Application Area
  **Self-supporting** **Outside fiber optic cables:** Loose tube single (or dual) jacket (w/o armored) fiber optic cables are available for PV applications. A self-supporting feature can also be added to the fiber optic cables for outdoor applications.	Fiber optic cables are generally used for long-distance data transmission and outside-type fiber optic cables are suitable for photovoltaic applications. **Fiber Count Range:** 2–312 **# of Loose Tube Range:** 2–26 **Cable Diameter Range:** 10 mm–27 mm **Weight Ranges:** 70 kg/km–442 kg/km **Temperature Range:** −40°C–70°C
 **Cross-sectional Area:** $4 \times 2 \times 0.56 \text{ mm}^2$ **CAT-7 Cable:** 24 AWG (0.56 mm²) and four-pair solid copper insulated with polyolefin.	**CAT-7 (Category 7)** cables are used for the Ethernet system cabling for monitoring purposes in PV applications. Each twisted pairs in CAT-7 cable is shielded, which is the main difference between CAT-7 and preceding Ethernet cables (CAT-6 and CAT7). **Maximum Operating Frequency:** 600 MHz **Weight:** 67 kg/km; **Cable Diameter:** 7.9 mm **Temperature Range:** −20°C–60°C
 **LICYC Cable:** Data Cable **LICYC Cable:** Multi-wire cable (from 0.14 mm^2 to 1.5 mm^2) Strands are made of bare copper wires Core insulation made of PVC Tinned-copper braiding	**LICYC cables** are designed for data transmission and can also be used in control and signal lines of control equipment and computer systems. **Conductor Number Range:** 2–50 **One Conductor Cross-sectional Area Range:** 0.14 mm²–1.5 mm² (0.14-0.25-0.34-0.5-0.75-1.0-1.5) mm² **Cable Diameter Range:** 4 mm–18 mm **Weight Ranges:** 12 kg/km– 550 kg/km **Temperature:** −40°C to 80°C (for fixed installations)

Photovoltaic Power Systems

The earth connection and bonding are made with earthing rods/electrodes and cables/wires. The most commonly used grounding/bonding conductors are copper strip conductors, aluminum strip conductors and lead-covered conductors, etc. Besides many connectors, conductor terminals and clamps in different types and sizes are available for grounding and bonding purposes. In this section, only general information about PV system grounding/bonding is introduced. Grounding and protection details will be examined under the chapter "PV System Protection and Fault Analysis" later in this book.

Photovoltaic Cable Terminals, Connectors, and Conduits: Cable terminals, connectors, and conduits of many different types and sizes are used in PV system installation for organizing cabling systems and for making secure, flexible, and fast connections. The elementary cable connection components and their usage purposes are listed below.

- **Cable Couplers and Receptacles:** They are typically used to make a quick and reliable connection for photovoltaic cabling and wiring. The PV cable couplers are generally composed of two parts, namely male and female couplers. Cable couplers are designed in different types and sizes depending on their intended use and where they are used. For example, cable couplers and sockets may also be suitable for in-line fusing and pluggable diode applications.
- **Branch Plugs, Sockets, and Splitters:** They are typically used for the series and parallel connection of PV panels or strings quickly and reliably. They are available in different sizes and types such as Y-type, T-type, and E-type splitters and branch plugs.
- **Pre-assembled Leads:** Pre-assembled cable links are manufactured in different lengths with voltage ratings up to 1500 VDC. These cable links are mostly available in two colors (red leads for positive terminals and black cable for negative terminals) and are suitable for free flexible and fixed installations.
- **Conduit Systems:** PV conduit systems made of metal or plastic are used for mechanical, thermal, and ultraviolet protection of photovoltaic cables. More specifically, PV cable conduits are suitable for hazardous environments such as wind, dust, and lightning. The conduits also protect grazing animals, gnawing animals, vermin attacks, and cable theft.

The typical photovoltaic cable connector systems summarized above are illustrated in Figure 5.45, and Figure 5.46a shows a PV installation example using the relevant cabling systems mentioned above.

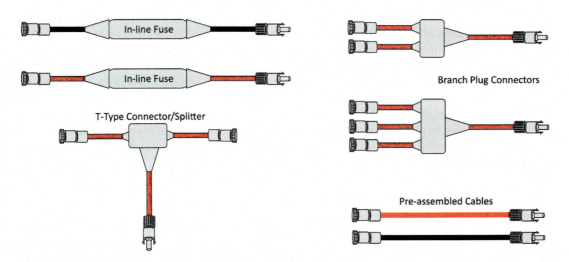

FIGURE 5.45 Typical cable connector systems used in PV installations.

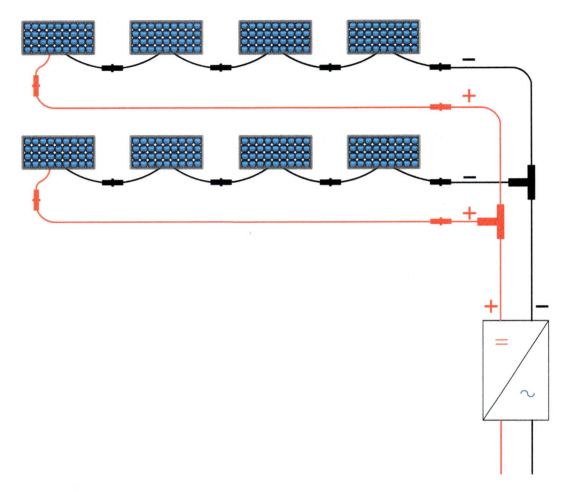

FIGURE 5.46a A PV system installation example based on cabling systems.

5.3 GRID-CONNECTED PV SYSTEMS WITH BATTERY STORAGE

A significant portion of the generated energy is exported to the utility in grid-connected domestic photovoltaic applications. However, instead of transferring some of this energy to the grid, it would be more appropriate to store it for future use when the energy tariff is expensive or when consumption is higher than the generation. Besides, this stored energy can be used as a backup in the event of grid failure. Hence, adding a battery storage unit to the grid-connected PV systems enables us to use the excess generation more cost-effectively. The types of energy storage systems in grid-connected PV applications can be classified into three groups, DC-coupled energy storage, AC-coupled energy storage, and AC-coupled energy storage with battery backup. These energy storage system types are charted in Figure 5.46b.

5.3.1 Grid-Connected PV Systems with DC-Coupled Battery Storage

As explained above, the self-consumption rate from residential PV generation can be increased by adding battery storage units. As one of the typical topologies, DC-coupled battery systems can be used for energy storage in grid-connected PV applications. Figure 5.47 shows the basic wiring diagram of the DC-coupled PV systems for grid-connected domestic applications.

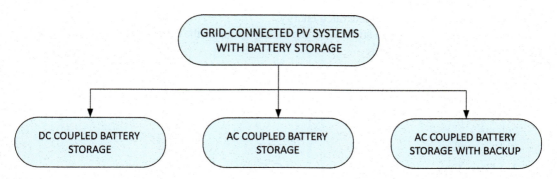

FIGURE 5.46b Types of energy storage systems in grid-connected PV applications.

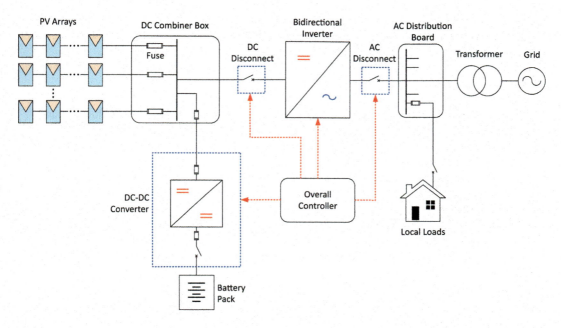

FIGURE 5.47 Typical wiring diagram of a grid-connected PV system with DC-coupled battery units. A DC-coupled system uses only one bi-directional inverter.

There could be different kinds of energy management and operation strategies for the battery storage system given in Figure 5.47. The principal aim is to store the excess energy and use it during demand shortages. For this purpose, a bidirectional DC-DC converter is required for charging and discharging cycles. Depending on the system size and constraints, multiple DC-DC converters and battery packs can also be connected to the DC bus (shown in Figure 5.47) for managing the charging and discharging cycles. Other strategies such as enhancing battery life and maximizing the profits of system owners can also be added to energy management strategies. The main benefits and drawbacks of a DC-coupled battery storage system in grid-connected photovoltaic applications are:

- Lower initial cost than AC-coupled systems. However, it could be more complex and expensive for PV systems above 5 kW due to the requirements of multiple higher voltage chargers and multiple parallel strings.
- Very high battery charging efficiency (up to 99% using MPPT).
- Highly efficient for powering DC loads. However, slightly lower efficiency if powering large AC loads due to the triple stage conversion (DC_{PV} to $DC_{Battery}$ to AC).

- Increases the value of PV-generated energy.
- The generated energy can still flow to the battery group in the event of inverter failure (off-line inverter) or grid failure.
- DC-DC converter-based battery storage systems can store the available PV energy in the early morning and late evening as well as cloud coverage times. Note that PV inverters typically require a minimum wake-up DC voltage (threshold DC voltage) to operate.
- It is possible to apply the DC-coupled battery storage system to utility-scale PV power plants.

5.3.2 Grid-Connected PV Systems with AC-Coupled Battery Storage

AC-coupled PV systems use an inverter/charger unit (multi-mode inverter or hybrid inverter), which is integrated into the AC side of the solar inverter for managing the battery charging and discharging cycles. There are multiple conversion steps between DC and AC while charging and discharging the battery. Figure 5.48 shows the elementary components and typical wiring diagram of the AC-coupled PV systems for grid-connected applications.

The main benefits and drawbacks of an AC-coupled battery storage system in grid-connected photovoltaic applications are:

- Compared with a DC-coupled system, AC-coupled systems may have slightly lower round-trip efficiency due to the extra conversion between DC and AC.
- AC-coupled systems allow the upgrading option for the PV and battery separately. Because these systems are independent of each other.
- It allows more flexibility in selecting the location of batteries and other equipment. Because AC-couples systems have separated PV arrays and battery systems with two or more inverters.
- Possible to charge the battery at a low-priced electricity tariff.
- AC-coupled battery units can be used as backup energy storage with additional switchgear.
- Adding AC-coupled battery storage to an existing PV system is more suitable and practical for most retrofit applications.

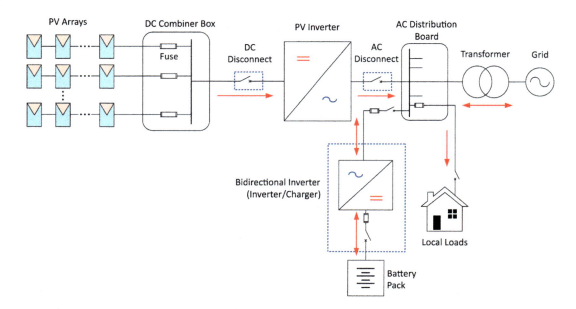

FIGURE 5.48 Typical wiring diagram of a grid-connected PV system with AC-coupled battery units.

Photovoltaic Power Systems

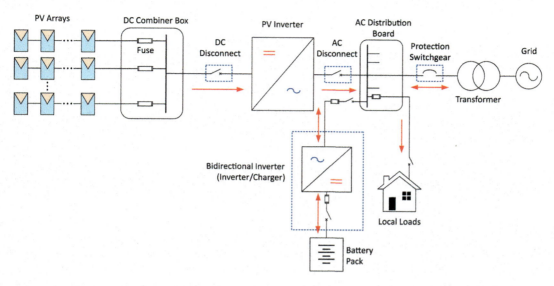

FIGURE 5.49 Typical wiring diagram of a grid-connected PV system with AC-coupled battery backup.

5.3.3 GRID-CONNECTED PV SYSTEMS WITH AC-COUPLED BATTERY BACKUP

The component types of the AC-coupled battery backup system may be the same as those of the AC-coupled energy storage system. The only difference is in the connection topology due to the requirement of additional protection relays in the AC-coupled battery backup systems. Besides, the battery pack can be designed in larger sizes since the battery storage system is planned to be operated as a backup in the event of a power cut. Figure 5.49 shows a typical connection diagram of a grid-connected PV system with AC-coupled battery backup.

The main benefits and drawbacks of an AC-coupled battery backup system in grid-connected photovoltaic applications are:

- They can operate in the event of an electrical power cut.
- Additional protection devices and new neutral-to-ground connections are required to supply AC loads either from the grid or battery units safely.

An imported detail should be noted herein that the storage functions, as well as power trading options described above, can be solved by a single inverter (called **hybrid inverter or multi-mode inverter, or all-in-one inverter**) in the present smart applications. As known, the main task of a PV inverter is to convert photovoltaic DC power to AC power. However, a hybrid inverter is a bidirectional inverter and can convert power from DC to AC and vice-versa. Besides, hybrid inverters can work with batteries to store the excess energy and discharge the stored energy on demand. The hybrid inverters can also be stayed as grid-connected and allow multiple energy sources for charging the batteries and balancing the loads.

5.4 STAND-ALONE PHOTOVOLTAIC SYSTEMS

Stand-alone (or off-grid) photovoltaic systems operate independently of the grid to feed certain DC and/or AC types of electrical loads. For this reason, stand-alone PV systems are designed and sized according to the electrical requirements of given loads. In this context, the vast majority of stand-alone PV applications use battery storage systems to ensure energy demand flows continuously. The types of stand-alone PV systems can be grouped as indicated in Figure 5.50.

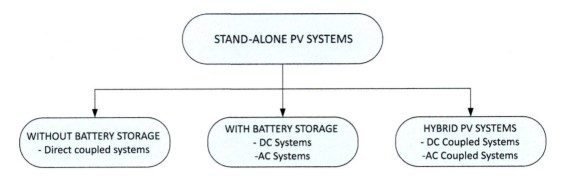

FIGURE 5.50 Types of stand-alone PV systems.

5.4.1 DIRECT-COUPLED STAND-ALONE PV SYSTEMS

Direct-coupled photovoltaic systems are the most basic type of stand-alone PV systems to supply DC loads. Since the system operates only under solar radiation and does not use any battery storage, the impedance of the DC load should be matched to the maximum power output of the PV array for better performance. Figure 5.51 shows the schematic diagram and simplified electrical layout of a direct-coupled PV system.

The directly coupled PV systems are generally common in remote areas of developing countries for the applications of water pumping, ventilation fans, and small-scale circulating pumps of solar thermal heating systems. The most important design criterion in all these applications is the matching of the load with the MPP point of the array.

5.4.2 STAND-ALONE PHOTOVOLTAIC DC SYSTEMS WITH BATTERY STORAGE

The application of stand-alone photovoltaic DC systems with battery storage facilities has been increasing worldwide to supply power to certain DC loads. Typically, a stand-alone photovoltaic DC system consists of PV arrays, battery storage (mostly lead-acid battery), charger, and other direct current appliances. The battery storage herein is used to allow electrical energy for powering the DC loads during the nights or at times of insufficient irradiance levels. Besides, a maximum power point tracking (MPPT) algorithm should be included in the battery charger unit to supply specific load types (such as positive-displacement water pumps) more efficiently. Note that the chargers usually include charge and load control circuits (discharge control) to protect batteries from overcharging/discharging damages. In this context, the schematic diagram and typical energy management layout of a stand-alone photovoltaic DC system can be depicted as in Figure 5.52.

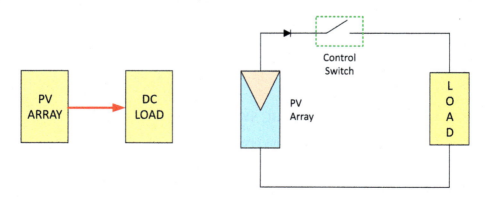

FIGURE 5.51 The schematic diagram and simplified electrical layout of a direct-coupled PV system.

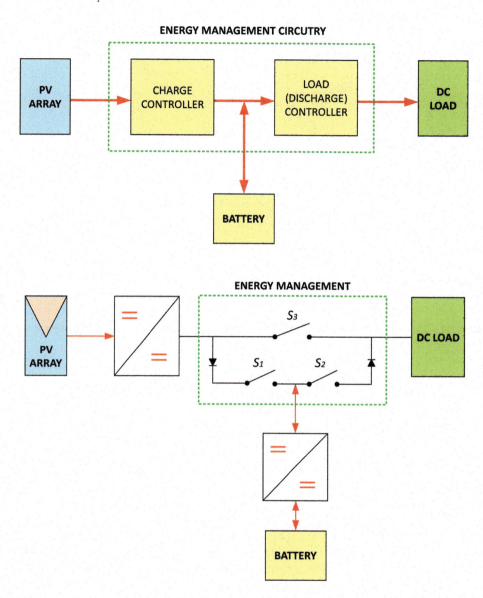

FIGURE 5.52 The schematic diagram and typical energy management layout of a stand-alone photovoltaic DC system including battery storage. Here, the switch S_1 controls the charging rate and the switch S_2 controls the discharging rate. Besides, the switch S_3 is a transfer switch, which allows the power transfer to the DC load directly.

5.4.3 Stand-Alone Photovoltaic AC Systems with Battery Storage

A stand-alone photovoltaic AC system includes four main components, which are PV array, battery storage, off-grid inverter, and AC loads. An off-grid inverter (also called a stand-alone inverter), with the help of a battery storage system can supply electrical power consistently to the local AC loads without a grid connection. The battery charging and energy management system herein are very similar to those of stand-alone photovoltaic DC applications. The only difference here is the use of an off-grid inverter instead of a DC-DC converter. In this context, the schematic diagram and typical electrical wiring diagram of a stand-alone photovoltaic AC system can be represented in Figure 5.53.

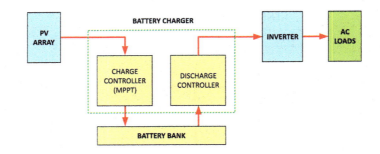

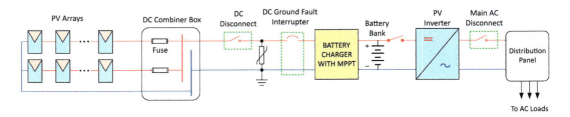

FIGURE 5.53 The schematic diagram and typical wiring diagram of a stand-alone PV system to supply AC loads.

In some cases, stand-alone PV systems can be designed to feed both DC and AC loads. Such PV systems combine the above-described stand-alone AC and DC systems. Figure 5.54 shows a general stand-alone PV system block diagram and its simplified electrical layout.

5.4.4 Stand-Alone Hybrid PV Systems

In some cases, a diesel generator as well as other sources such as wind turbines and fuel cells can also be operated in parallel to the PV system or can be used as backup generators in stand-alone systems. The aim of using such additional sources is to feed the loads continuously in the event of failures and energy generation shortages. Such PV systems, in which a second source is used, are called stand-alone hybrid PV systems. In this context, the main components of a stand-alone hybrid PV system should include PV arrays, battery storage, inverter, generator(s), and AC loads. Generally, diesel generators are often used as a backup source (or parallel operating source) in hybrid PV systems. As known, diesel generators can be operated by burning different fuel types such as propane, petroleum, and gasoline. Therefore, keeping the fuel tank ready at all times is very important to ensure the system supply's reliability.

An important point should be underlined that DC- or AC-coupled hybrid topology may be possible according to the output characteristics of the added generator. For example, a DC-coupled connection type can be preferred for Fuel Cells, DC generators, and wind turbines. However, AC diesel generators with a 50/60 Hz output can be integrated into the stand-alone hybrid systems via the AC-coupled connection. It is clear from these explanations that numerous power electronic topologies are possible to integrate the hybrid system components. Accordingly, a typical schematic diagram for DC- and AC-coupled stand-alone hybrid PV systems can be represented here as in Figure 5.55.

Besides, only two simplified connection drawings for Diesel-PV and Wind-PV hybrid systems are given in Figures 5.56a and 5.56b as examples of AC- and DC-coupled applications.

Another application example could be the use of three sources in a stand-alone hybrid PV system. For example, let us consider a PV array, a wind turbine, and a diesel generator utilization together

Photovoltaic Power Systems

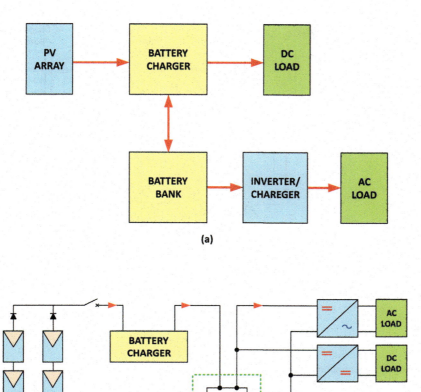

FIGURE 5.54 The schematic block diagram (a) and simplified electrical layout of a general stand-alone PV system (b) supply both DC and AC loads.

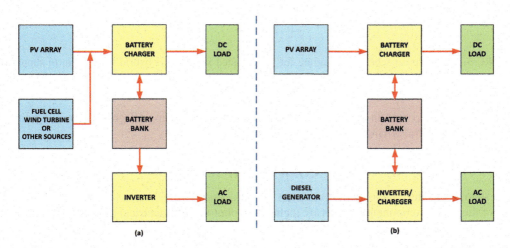

FIGURE 5.55 Typical schematic diagrams for DC and AC coupled stand-alone hybrid PV systems to supply both DC and AC loads.

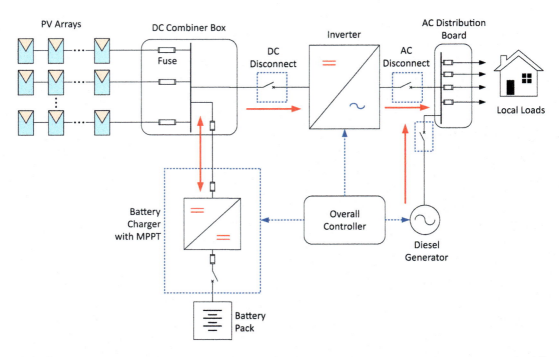

FIGURE 5.56a Typical schematic diagram for AC-coupled stand-alone hybrid PV system to supply AC loads.

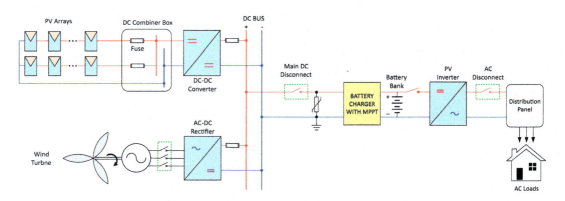

FIGURE 5.56b Typical schematic diagram for DC-coupled stand-alone hybrid PV system to supply AC loads.

in an off-grid power system application. In this regard, the following off-grid system diagram can be drawn as a general wiring example as shown in Figure 5.57.

5.5 PV SYSTEM SIZING AND COMPONENT SELECTION

PV system sizing and component selection is a vital process to satisfy the certain performance objectives of the major components, which are included in photovoltaic electrical design. Component ratings such as voltage and current capacity of each component are determined in the sizing process to meet the load requirements continuously in a safe way. The photovoltaic sizing process may be versatile, depending upon the type of PV systems and their functional requirements. Hence, grid-connected and stand-alone PV systems require different sizing principles. For example, determining

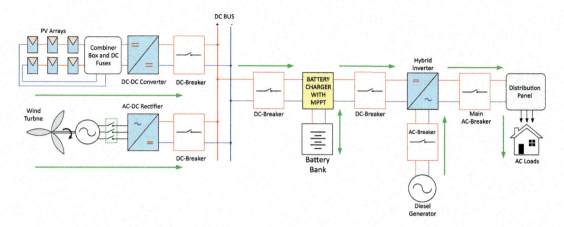

FIGURE 5.57 A general schematic diagram of a stand-alone hybrid system including a PV array, wind turbine, and diesel generator.

the maximum PV array output and determining the inverter size according to this maximum PV power are the two most basic steps in sizing the grid-connected systems having no battery storage. However, the sizing principles for any type of stand-alone PV system are dependent on providing an energy balance between supply and demand. Besides, the stand-alone operated PV array should be sized so that it can meet the load requirements plus system losses under a worst-case scenario. At this stage, the sizing of the battery, which should be sized according to the desired days of storage, is very critical. Note that the above-given principles are general and will vary depending on the system requirements and application constraints. For instance, the available area, the type of PV module, array layout, local weather conditions, and project budget are other constraints, which can affect the system design and sizing procedures. Therefore, the sizing and component selection steps are needed to be handled systematically for the given PV application. In this context, the design, sizing, and component selection procedures for each of the PV system components are described in detail below.

5.5.1 PV Array Sizing and Land Requirements

As known, PV modules with different types and sizes will generate different amounts of photovoltaic power. Besides, there may be different system constraints and design objectives in various PV system applications. Hence, a systematic methodology is needed to find out the correct PV array sizing for both stand-alone and grid-connected photovoltaic system applications. Hereby, sizing methodologies for the corresponding PV system types are described separately.

5.5.1.1 PV Array Selection and Sizing for Stand-Alone Systems

PV array sizing is to find out the required number of PV modules and their connection configuration to meet the load requirements and system losses under the worst-case solar radiation scenario. PV array sizing is also affected by the selection of the most appropriate PV technology, which is also very important in improving the performance, cost-effectiveness, and lifetimes of stand-alone PV systems. To make the correct sizing and match the design objectives, the factors/system constraints affecting the sizing should be considered. Accordingly, the main factors affecting PV array sizing in stand-alone systems are:

- Local solar radiation and shading effect
- The types of stand-alone PV systems
- The type and cost of PV module technology

- Availability of storage in the stand-alone system
- The battery type (such as lead-acid, nickel-cadmium, and others) and battery capacity
- The use of MPPT in battery chargers
- The load requirements and system losses
- The load availability level (loss of load level or minimum critical load level)

Secondly, the main constraints affecting the PV array sizing in stand-alone applications can be written as follows:

- Available area
- Limitations in PV module selection
- Limitations in the PV module orientation (e.g., azimuth and/or tilt angles can be constant due to architectural difficulties)
- Project budget

As stated above, PV array sizing differs according to the types of stand-alone PV systems. Therefore PV array sizing problem is examined below in the following sections separately.

5.5.1.1.1 PV Array Sizing for Direct-Coupled Stand-Alone Systems

As explained above, direct-coupled stand-alone PV systems do not include battery storage and they only operate under solar radiation, which continuously changes during the daytime. Therefore, the impedance of the DC load will only match the maximum output power of the PV array under certain solar radiation conditions. Since there is no MPPT usage in the direct-coupled systems, the maximum output power of the PV array cannot be transferred to the load in the remaining solar radiation periods. Therefore, the array sizing for direct-coupled stand-alone PV systems is very crucial to get better performance all through the considered periods. For this purpose, there are different optimization solutions in the literature. However, such optimization approaches require detailed data analysis, complicated mathematical computations, and good engendering services. In addition, even if the optimal PV array size were obtained based on the considered data, the calculated optimal value would not be the same in real field conditions due to the random behavior of solar radiation. Besides, the estimated PV array output value will be different from the real value for reasons of panel soiling, ambient temperature, and shading effects. As a result, applying a complicated engineering process will not be either cost-effective or practical for small-size direct-coupled PV applications. For this reason, a simple but effective method has been developed in this book, which gives nearly optimum solutions for the direct-coupled stand-alone PV systems. This simplified PV array sizing method given below is developed based on the solar peak hours (or sun peak hours) concept, of which the definition is presented as follows.

Peak Solar Hours (PSH): Solar insolation can be measured in PSH, which is equivalent to 1000 Watts (1 kW) of solar energy falling on an area of 1 m^2 for a certain number of hours. In other words, peak solar hours are defined as the equivalent number of hours per day under the assumption of constant irradiance of 1000 W/m^2. This concept is illustrated in Figure 5.58.

It is clear from Figure 5.58 that one way of expressing solar insolation is to use the average number of PSH available each day during any given period in $\left[kWh/(m^2 \times day) \right]$. The PSH values of a location in a given month can be used to calculate average daily solar radiation, which is often sufficient for basic system analysis. Since the peak solar radiation under STC is 1 kW/m^2, the number of peak sun hours occurring over the daytime is numerically equal to the average daily solar insolation. Because the rated values of PV modules are defined under 1 kW/m^2 radiation, the PSH approach is quite a practical and effective method for calculating the expected amount of energy generation from PV modules. Now that we can write the mathematical equation of daily PSH as in (5.4).

Photovoltaic Power Systems

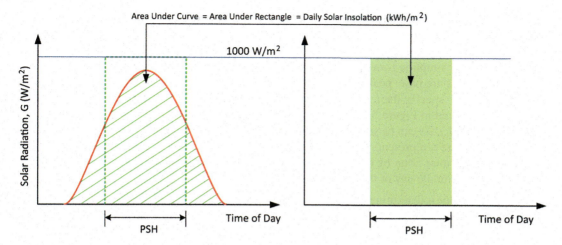

FIGURE 5.58 Peak solar hour (PSH) concept to calculate daily solar radiation

$$\text{PSH} = \frac{\int_{t(\text{sun-rise})}^{t(\text{sun-set})} G(t)\,dt \left[\text{kWh}/(\text{m}^2 \cdot \text{day})\right]}{\text{Peak Sun Insolation at STC}\left[1\ \text{kW/m}^2\right]} \tag{5.4}$$

where $G(t)$ refers to the instantaneous solar radiation as a function of time. It is clear from (5.4) that the PSH value is measured in [sun-hours]. For instance, assume that a location receives 5 kWh/m² solar insolation per day. In this case, the site can be said to have a PSH value of 5(sun-hours) per day under 1 kW/m² solar radiation.

Now that we can move on to the PV array sizing technique in direct-coupled stand-alone systems. The most important PV sizing criterion in this type of PV system is the matching of the load impedance with the MPP of the array. However, solar radiation is not constant during the daytime and changes continuously, which results in variation in the arrays' MPP as well. As a result, it is critical to note that the high radiation periods where PV modules provide the most output should be a dominant factor in array sizing. The following steps are designed for array sizing in direct-coupled PV systems.

Step-1: Select the PV module types and specify the module's parameter and rated $I-V$ curve from the datasheet.

Step-2: Considering the PV panel orientations and site location, calculate the daily peak sun hours of your panel (PSH_1, PSH_2, ..., PSH_n) for the specified period. Because the PSH values can be different for different solar panel orientations and site locations. Remember that the radiation calculation falling on a surface was described in detail in Chapter 2 for any desired period and any panel orientations. Please look into this section for the corresponding calculations. Note that one can also use peak-sun-hours maps for the average PSH values.

Step-3: Calculate the average peak sun hours of your panel based on the PSH results from Step-2 and equation (3.5) given below. Remember that average peak sun hours generally vary between 3 and 6.

$$\text{PSH}_{\text{avg}} = \frac{\sum_{i=1}^{n} \text{PSH}_i}{n} \tag{5.5}$$

where n is the number of days in a considered period. Alternatively, the PSH value (PSH_m) occurring on the middle day of the time of interest can be used instead of calculated PSH_{avg} through

Step-2 and Step-3. It is crucial to note that the peak sun hours on equinox days (March 20/21 and September 22/23) can also be used as a PSH_{avg} value for direct-coupled stand-alone PV systems, which are to be used all year round.

Step-4: In the simplified method described herein, three different solar radiation values are determined. These are base, peak, and reference radiations (G_b, G_p, and G_r). The main purpose of this approach is to specify the reference radiation (G_r) for which the PV array is to be sized. This concept is illustrated in Figure 5.59. Firstly, the PSH_{avg} value from Step-3 is used to determine the base irradiance (G_b). As can be seen from Figure 5.59, the (G_b) can be found in the function $G(t)$ by writing $t = t_1$, where t_1 represents the point corresponding to the distance on the left-hand (or right-hand) side of the solar noon by ($PSH_{avg}/2$ or $PSH_m/2$) value, where PSH_m is the peak sun hours occurring on the middle day of the given period.

Once the G_b value is determined, the next step is to specify the G_r by the equation of $G_r = \dfrac{G_b + G_p}{2}$, where G_p is the peak solar irradiance (solar noon irradiance) on the day with the highest insolation within the considered period.

Step-5: Determine the $I-V$ curve of the selected PV module under the calculated reference irradiance of G_r found in Step-4. Generally, most manufacturers in their datasheets provide the $I-V$ and $P-V$ curves of PV modules under different radiation conditions, ranging from 20% sun to 100% sun irradiance levels. It is clearly shown from these datasheets that V_{mp} (voltage level at P_{max}) point of the module is almost constant between 50% sun and 100% sun radiation. In this model, V_{mp} on the $I-V$ curve is assumed to be constant within the corresponding irradiation range since the variation of V_{mp} between 500 W/m² and 1000 W/m² is negligible. As for the I_{mp} (current at P_{max}) point

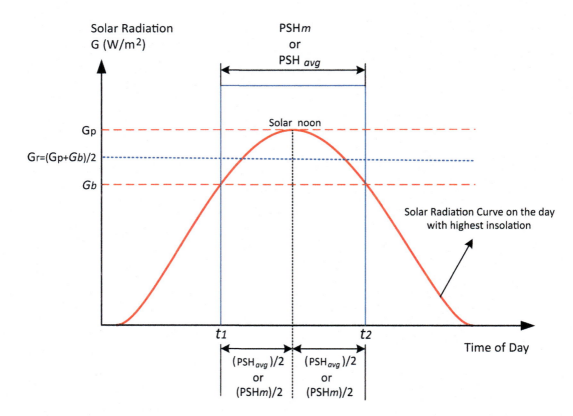

FIGURE 5.59 Determining the reference solar radiation level (G_r) by considering the day with the highest insolation. [PSH_m is the peak sun hours on the middle day of the given period.]

Photovoltaic Power Systems

of the module, it varies in proportional to the solar radiation. Under these assumptions, V_{mp} and I_{mp} values corresponding to P_{max} can be written as (5.6).

$$\begin{cases} V_{mp} @ G_r = V_{mp} @ 100\% \text{ Sun, for } 600 \text{ W/m}^2 \leq G \leq 1000 \text{ W/m}^2 \\ I_{mp} @ G_r = \dfrac{G_r \left[\text{W/m}^2\right]}{1000} \times \left(I_{mp} @ 100\% \text{ Sun}\right) \end{cases} \quad (5.6)$$

Note that the value G_r will not fall below 500 W/m² in the vast majority of the solar sites. Therefore, the approach given by (5.6) will produce results with sufficient accuracy.

Step-6: Through serial and parallel connections of PV modules, the V_{mp} and I_{mp} values of the PV array under the calculated reference irradiance of G_r should be equalized to the rated current and voltage values of the load as close as possible. Note that the resistance of the DC load (R_L) should nearly equal to the R_0 under the calculated reference irradiance of G_r. This equivalence ($R_0 @ G_r \approx R_L$) can be used to test the accuracy of the process. As a result, the PV array sizing approach under the specified Gr irradiation can be detailed below.

- Firstly, the rated voltage and current (V_r, I_r) of the DC load are specified.
- Secondly, the PV module voltage is then equalized to the rated voltage level of the load. If necessary, the PV modules are connected in series until the voltage level of the PV array is matched to the rated load voltage.
- Thirdly, the PV modules are connected in parallel until the current level of the PV array (under G_r radiation) is equalized to the rated load current of the load.
- Afterward, the equivalent resistance of the DC load is calculated as $R_L = (V_r / I_r)$.
- Finally, compare the R_L and $R_0 @ G_r$ to check the accuracy of the applied process. Now that the characteristic resistance of the PV array under solar radiation $G_r (R_0 @ G_r)$ should approximately be equal to the equivalent resistance of the DC load (R_L) as shown in (5.7)

$$R_0 @ G_r = \dfrac{V_{mp} @ G_r}{I_{mp} @ G_r} \approx \dfrac{V_r}{I_r} \quad (5.7)$$

Step-7: Determine the designed PV array configuration and give the final specifications such as the number of series and parallel-connected PV modules in the PV array, PV array dimensions and its area, PV array power at rated conditions and load power, etc. Finally, check the PV array size to see if it matches the load demand.

Example 5.1: PV Array Sizing for Direct-Coupled Stand-Alone Systems

A PV module, of which specifications are given below (Table 5.8), will be used to supply a resistive heater of a water tank with a rated power of 2750 W ($V_r = 220$ V, $I_r = 12.5$ A).

Assume that the water circulation in the tank is fast enough to keep the thermostat in the on position continuously. The PV system to supply the water heater will be a direct-coupled stand-alone system with no battery storage and it is assumed to be operated throughout the entire year. The other design information regarding the solar irradiance and site are summarized below (Table 5.9):

  a. Calculate the PV array size and draw the system design diagram considering the basic protection requirements.

TABLE 5.8
PV Module Specifications

PV Module Specifications: 255 Wp Mono Crystalline Module			Module Efficiency	15.58%
Dimensions (mm)	Voltage at P_{max}	Current at P_{max}	Open Circuit Voltage	Short Circuit Current
$1500 \times 1000 \times 40$	30.8 V	8.28 A	38 V	8.92 A

TABLE 5.9
PV Module Specifications

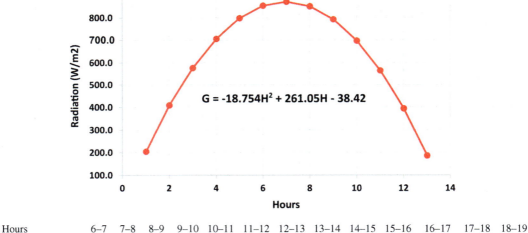

	Specified Design Information	
The tilt of PV panels		34°
The azimuth of PV Panels		South facing (0°)
Daily average PSH value at given tilt angle on equinox dates		6.30 PSH
Site location (Los Angles)		34.0522° N, 118.2437° W
Temperature range		−10°C to 42°C

Hourly solar radiation profile at given tilt on highest insolation day (June 20/21) (W/m2)

$G = -18.754H^2 + 261.05H - 38.42$

Hours	6–7	7–8	8–9	9–10	10–11	11–12	12–13	13–14	14–15	15–16	16–17	17–18	18–19
Radiation (W/m²)	203.9	408.7	575.9	705.7	798	852.7	870	849.7	792	696.7	563.9	393.6	185.8

b. The $I-V$ characteristic of the module used in the PV array is given in Figure 5.60. Considering this characteristic curve, calculate the power that can be transferred from the PV array to the load under 1000 W/m², 800 W/m², 600 W/m², and 400 W/m² radiations.

Solution

a. Since solar radiation is variable throughout the year, the steps explained above can be applied to determine the size of the PV array for a direct-coupled stand-alone system. As described earlier, the hourly radiation profile for the day with peak insolation can be taken as the summer solstice day on June 20/21. Now that the following table showing all calculation steps can be created as below (Table 5.10):

Based on the calculations in Table 5.10, the PV system design can be drawn as below (Figure 5.61). Blocking diodes, surge protector devices, and grounding are added to the system as basic protection equipment.

Photovoltaic Power Systems

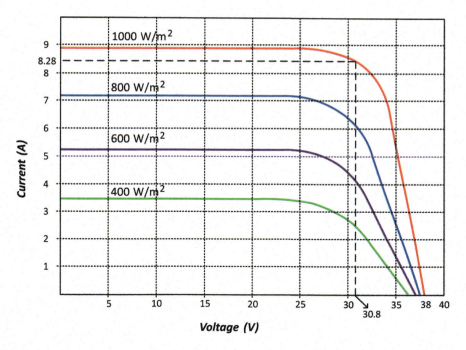

FIGURE 5.60 $I-V$ Characteristic of the module under different radiations.

TABLE 5.10
Calculation Steps for PV Array Configuration

	Explanation	Calculation and Results
Step-1	PV module selection	255 Wp Mono Crystalline Module (V_{mp} = 30.8 V, I_{mp} = 8.28 A)
Step-2	PSH_{avg} at given tilt	Daily Average PSH Value at Given Tilt on Equinox Dates is
Step-3		6.30
Step-4	Calculation of reference radiation (G_r) value	From the given hourly solar radiation profile curve, the solar noon value corresponds to the point $H=7$ (midpoint of the daytime). Since $PSH_{avg} = 6.30$, then G_b is determined for $H = 6.5 - \dfrac{6.30}{2} = 3.35$. Hence using the given Gb equation, $G_b = -18.754 \times 3.35^2 + 261.05 \times 3.35 - 38.42 = 625 \text{ W/m}^2$ Now, G_r can be found as: $G_r = \dfrac{G_b + G_p}{2} = \dfrac{625 + 870}{2} = 747.5 \text{ W/m}^2,$
Step-5	Specifying the V_{mp} and I_{mp} at G_r	From equation (5.6), $V_{mp} \approx 30.8$ V, $I_{mp} = \dfrac{747.5}{1000} \times 8.28 \text{ A} = 6.189 \text{ A}$

(Continued)

TABLE 5.10 (*Continued*)
Calculation Steps for PV Array Configuration

	Explanation	Calculation and Results
Step-6	The V_{mp} and I_{mp} of the PV array at G_r are matched to the V_r and I_r of the load.	# of Series Connected PV Module $= \dfrac{V_r}{V_{mp} @ G_r} = \dfrac{220}{30.8} \approx 7$ # of Parallel Connected Branch $= \dfrac{I_r}{I_{mp} @ G_r} = \dfrac{12.5A}{6.189A} \approx 2$ Final PV Array Configuration: (7 Series × 2 Parallel) Module. Now, let's check the $R_0 @ G_r$ to see if it matches the load resistance. $\dfrac{V_{mp} @ G_r}{I_{mp} @ G_r} \approx \dfrac{V_r}{I_r} \Rightarrow \dfrac{30.8 \times 7}{6.189 \times 2} \approx \dfrac{220}{12.5} \Rightarrow 17.4\,\Omega \approx 17.6$
Step-7	Final PV array configuration	PV Array Configuration: (7s × 2p) PV Array Power at %100 Sun = 3570 Wpeak PV Array Power at %87 Sun = 0.87 × 3570 = 3106 Wpeak PV Array Power at % 62.25 Sun = 0.6225 × 3570 = 2222 Wpeak $(1500 \times 2) \times (1000 \times 7) = 21\,m^2$ PV Array Dimension = or $(1500 \times 7) \times (1000 \times 2) = 21\,m^2$

b. To calculate the transferred power, it is enough to intersect the $I-V$ curve of the PV array and the $I-V$ curve of the load. Based on the given $I-V$ curve of the module, the PV Array's ($7s \times 2p$) characteristic curve can be plotted as follows. The $I-V$ curve of the load can also be shown on the same axis, which is shown in Figure 5.62.

Assume that the I–V curve of the load cuts the irradiation curves of 800 W/m², 600 W/m², and 400 W/m² at points of P_1, P_2, and P_3, respectively. Current and voltage pairs at P_1, P_2, and P_3 are calculated approximately and the results are given in the above Figure. Accordingly, the power values delivered to the load under the relevant solar radiation conditions are:

$$P_1 @ 800\,W/m^2 = 217.5\,V \times 12.1\,A = 2632\,Watts$$

$$P_2 @ 600\,W/m^2 = 182\,V \times 10.36\,A = 1886\,Watts$$

$$P_3 @ 400\,W/m^2 = 123.8\,V \times 7.136\,A = 883\,Watts$$

5.5.1.1.2 PV Array Sizing for Stand-Alone Systems

This section aims to provide a general procedure to size the PV array based on energy balance and PSH (peak-sun-hours) approaches. As known, cost-effective stand-alone PV systems with improved performance and lifetime can be obtained by sizing and selecting the components properly. Therefore, solar radiation and operating conditions, system parameters as well as future load growth plans should accurately be identified for suitable PV array sizing. The purpose of the sizing process herein is to determine the required number of PV modules in a PV array with the connection configuration (number of series and parallel-connected modules in the array).

The PV array sizing process for a stand-alone system should be based on a worst-case scenario regarding solar radiation, load consumption, and system losses. There are various parameters,

Photovoltaic Power Systems

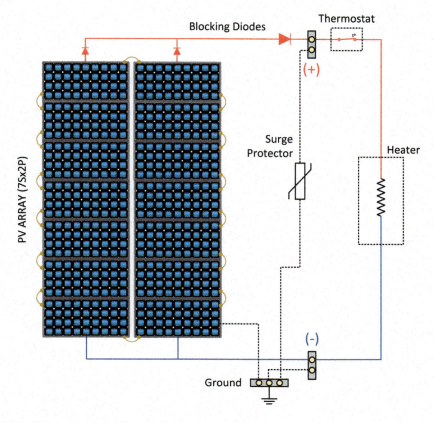

FIGURE 5.61 PV system drawing.

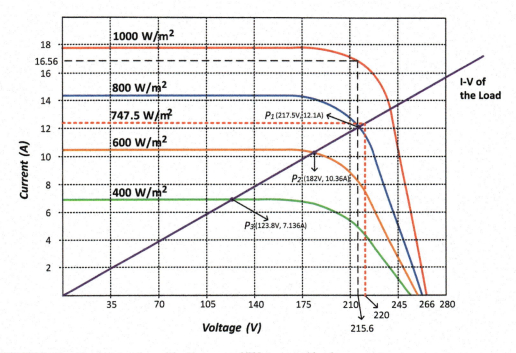

FIGURE 5.62 The intersection of $I-V$ curves of PV array and load.

which have different effects on the PV array sizing methodology. In general, these parameters can be listed as follows:

- Charge-controller types such as standard charge controller (on/off type series or shunt regulator), PWM charge controller, and MPPT-based charge controller.
- PV array losses such as mismatch, dirt, and shading losses.
- PV system losses in the components such as round-trip battery efficiency, inverter efficiency, cabling/wiring losses, etc.
- Temperature effect on the PV array outputs.

Apart from the above considerations, the possible load growth in the future is also a parameter needed to be well thought in the PV array sizing methodology. Importantly note that the PV array sizing for a stand-alone system should be based on average solar radiation during the period of use. In this context, the general PV array sizing steps for the stand-alone systems are given below:

Step-1: Calculate the average daily load consumption in Ah (or kWh). Note that the load calculations in this book are given under a separate sub-section below. Hence, the required load amount in this section is considered to be known in the PV array sizing procedure.

Step-2: Determine the solar radiation data for the given azimuth and tilt angle. Next, calculate the average PSH value of the given site for the designed azimuth and tilt angle. Alternatively, one can also use the global solar maps (atlas) for the desired PSH values.

Step-3: Select the PV module technology and give the technical specifications of the module. To choose the module for a PV system, it is necessary to know the climatic conditions of the site (solar radiation and ambient temperature data), load power, and its operating voltage range. Based on this information, it is crucial to design the array voltage range (minimum and maximum voltage levels for the given site) to ensure that the load can be operated under all climatic conditions of the site. Note that the engineers must also consider all voltage drops in the system to see if the output voltage of the PV array still meets the load requirements.

Step-4: Specify the basic design parameters such as system DC and AC voltages (V_{DC}, V_{AC}). If any and necessary, specify the other operating conditions. For example, the user may say that the PV system will be used during the winter months at ambient temperatures from zero to minus 20 degrees, or will be used during the holiday period from May to October, etc.

Step-5: Specify the stand-alone PV system topology and estimate the system losses. For example, the type of charge controller (standard, PWM, or MPPT-based charge controller), the efficiency value of the bidirectional inverter, and the battery technology/capacity in the PV system should be specified. In the second group, PV array losses (shading, mismatch, temperature derating, soiling, etc.) and DC/AC cable losses should also be specified. Lastly, if there is a certain load increase plan in the future, this value should be estimated and included in the PV array sizing procedure.

Step-6: Once the above-mentioned parameters in (Step-1 to Step-5) are determined, PV array sizing for stand-alone systems can be performed by following the mathematical formulations described below.

The first starting point is to determine the average daily energy needed to meet the load requirement. For this purpose, PV array derating factors (e.g., dirt, mismatch, and temperature derating), power losses in cabling/wiring, and balance of the system efficiencies (round-trip battery efficiency, inverter efficiency, etc.) should be considered to calculate the overall energy need as given by (5.8).

$$E_{\text{overall}} = \frac{E_{\text{Load}}}{(\Delta P_{\text{PV}}) \cdot \eta_{\text{bat}} \cdot \eta_{\text{inv}} \cdot \eta_{\text{ch}} \cdot \eta_{\text{cab}}} \quad (5.8)$$

where E_{Load} is the average daily energy load demand in (Wh/day or Ah/day), ΔP_{PV} are the total deratings for the PV array (which is between zero and one), η_{bat} is the coulombic battery efficiency (due to its internal resistance), η_{inv} is the inverter efficiency, η_{ch} is the charger efficiency, and η_{cab}

is the cabling/wiring efficiency in the system. Herein, the ΔP_{PV} system deratings expressed as in (5.9) occur due to the factors of manufacturing tolerance, temperature, soiling, mismatching, and shading.

$$\Delta P_{PV} = \left(f_{\text{Manufacturing Tolerance}}\,[\%]\right) \times \left(f_{\text{Temperature Losses}}\,[\%]\right) \times \left(f_{\text{Mismatch Losses}}\,[\%]\right) \times \left(f_{\text{Soiling Losses}}\,[\%]\right)$$
$$\times \left(f_{\text{Shading Losses}}\,[\%]\right) \tag{5.9}$$

Since the derating factors considered in equation (5.9) are examined in detail under Chapter 4, the PV array losses will not be discussed here again. If a certain load growth rate is planned in the future, it will be useful to take the load growth factor (L_{gf}) into account. In this case, equation (5.8) can be rewritten as shown below:

$$E_{\text{overall}} = \frac{E_{\text{Load}} \cdot L_{gf}}{(\Delta P_{PV}) \cdot \eta_{\text{bat}} \cdot \eta_{\text{inv}} \cdot \eta_{\text{ch}} \cdot \eta_{\text{cab}}} \tag{5.10}$$

Typically, L_{gf} ranges from 1.1 to 1.5 for residential and small-scale commercial applications. After the daily energy requirement from the solar PV array is determined, the peak power of the PV array ($P_{\text{Peak(PV)}}$) can be determined by equation (5.11).

$$P_{\text{Peak(PV)}} = \frac{E_{\text{overall}}}{\text{PSH}_{\text{avg}}} \tag{5.11}$$

where PSH_{avg} is the average peak sun hours at a given tilt and azimuth of the site. Once the peak power of the PV array is calculated, the total DC current of the array (I_{DC}) is then obtained can be determined by dividing the result of (5.11) by system DC voltage (V_{DC}).

$$I_{DC} = \frac{P_{\text{Peak(PV)}}}{V_{DC}} \tag{5.13}$$

Since the system-rated DC voltage (V_{DC}) and module-rated voltage (V_{mp}) are determined from the steps described above, the number of series-connected PV modules (N_s) in a string can be calculated by equation (5.14).

$$N_s = \frac{V_{DC}}{V_{mp}} \tag{5.14}$$

If the N_s from (5.14) is not a whole number, then the result should be rounded up to the nearest integer number. Alternatively, a different module (e.g., 265 W module instead of 250 W module) can also be selected if its voltage fits better with the system's nominal DC voltage.

Here, an important detail regarding the DC system output voltage formed by the series-connected PV modules should be pointed out. As known, there are two common solar charger types in the market, which are PWM and MPPT-based solar chargers. If a PWM-based solar charger is to be used (although not recommended due to its low operating efficiency), equation (5.14) can be used to find out the number of series-connected modules in the array. However, if an MPPT-based solar charger is to be used, the output voltage of the PV array under varying cell temperature ranges must comply with the MPP tracking voltage range. To achieve this aim, the number of series-connected modules in PV systems in which MPPT-based solar chargers are used can be obtained from equation (5.15).

$$N_s = 0.95 \times \frac{V_{\text{max\_ch}}}{V_{mp}\,@\,T_{\text{cell}}} \tag{5.15}$$

where the coefficient of 0.95 is used to take into account the safety margin of 5%, V_{max_ch} is the maximum input voltage of the MPPT charger and $V_{mp}@T_{cell}$ is the MPP voltage of the one module under effective cell temperature. After rounding the result of (5.15) to the upper whole number, one should check the array voltage to see if $N_s \times V_{mp}$ is less than or equal to V_{max_ch} ($N_s \times V_{mp} \leq V_{max\_ch}$). Wherein the effective cell temperature (T_{cell}) and the voltage V_{mp} corresponding to this T_{cell} can be calculated by any of the methods described in Chapter 4.

After determining the number of modules in a string, the number of parallel-connected PV strings (N_p) can be calculated by equation (5.16).

$$N_p = \frac{I_{DC}}{I_{mp}} = \frac{\frac{P_{Peak(PV)}}{V_{DC}}}{I_{mp}} = \frac{\frac{P_{Peak(PV)}}{N_s \cdot V_{mp}}}{I_{mp}} = \frac{P_{Peak(PV)}}{N_s \cdot P_{module}} \tag{5.16}$$

where I_{mp} is the PV module current (one module-rated current) at maximum power. If the N_p from (5.16) is not a whole number, then the result should be rounded up to the nearest integer number. Note that the designer should search for alternative modules that may result in a more cost-effective solution for the considered site. Now that the total number of modules (N_T) in the PV array is then determined by multiplying the number of series-connected modules by the number of parallel-connected modules as below:

$$N_T = N_s \times N_p \tag{5.17}$$

Example 5.2: Array Sizing for Stand-Alone PV Systems

The electricity requirement of a small house in a rural area is planned to be met by a stand-alone PV system. The selected PV module is a polycrystalline silicon 185 W_p with the specifications of $V_{oc} = 30.4$ V, $V_{mp} = 24.2$ V, $I_{sc} = 8.09$ A, and $I_{mp} = 7.65$ A. The average daily energy demand of this house is estimated to be 20 kWh/day approximately. Assume that the average daily peak sun hours (PSH_{avg}) through the entire year is 5.2 at the given tilt and azimuth of the site.

a. If the designed DC system voltage is 48 V, calculate the PV array size and draw the system wiring diagram considering the PWM-based solar charger.
b. Calculate the PV array size and configuration in case of MPPT solar charger usage. Draw the PV system wiring diagram and show the basic protection schemes on the design. Note that the basic specification of the MMPT solar charger is given below (Table 5.11).

Solution

a. Let us follow the steps described above regarding the PV array sizing for stand-alone PV systems. For this purpose, the following table showing all calculation steps can be created as below (Table 5.12):

TABLE 5.11
The Parameters of Solar Chargers

Basic Specifications of the MPPT Solar Charger	
MPP tracking voltage range	65-150 Volt DC
Maximum charging current	30 A
Nominal battery voltage	48 Volt DC
Maximum open circuit voltage of the PV array	192 Volt DC

TABLE 5.12
Calculation Steps for PV Array Sizing

	Explanation	Calculation and Results
Step-1	The average daily energy demand of the load [kWh]	It is given as 20 kWh/day
Step-2	PSH_{avg} at given tilt [h]	Daily average PSH value at given tilt and azimuth (PSH_{avg}) is 5.2 hours
Step-3	PV module selection	185 Wp Polycrystalline module (V_{oc} = 30.4 V, V_{mp} = 24.2 V, I_{sc} = 8.09 A and I_{mp} = 7.65 A)
Step-4	Basic design parameters of the system	Nominal battery voltage, V_{DC} = 48 V Nominal AC voltage (inverter output), V_{AC} = 220 V
Step-5	System topology: PV array losses/deratings: System losses and efficiencies:	Array→DC Cabling→Combiner→Charger→Inverter→AC Cabling→Load **Note that PV system losses and efficiency values are not given. Hence, typical values are used in this example.** **PV Array Losses:** $(f_{Manufacturing} : 0.98), (f_{Temp} : 0.92),$ $(f_{Mismatch} : 0.97), (f_{Soiling} : 0.95), (f_{Shading} :$ no shading effect) **System Losses & Efficiencies:** $\eta_{bat} : 0.85$, $\eta_{inv} : 0.98$, $\eta_{ch} : 0.97$, $\eta_{cab} : 0.98$
Step-6	PV array sizing calculations	$E_{overall} = \dfrac{E_{Load}}{(\Delta P_{PV}) \cdot \eta_{bat} \cdot \eta_{inv} \cdot \eta_{ch} \cdot \eta_{cab}}$ $= \dfrac{20 \text{ kWh}}{(0.98 \times 0.92 \times 0.97 \times 0.95) \times (0.85 \times 0.98 \times 0.97 \times 0.98)}$ $= 30.4 \text{ kWh}$ $P_{Peak(PV)} = \dfrac{E_{overall}}{PSH_{avg}} = \dfrac{30.4 \text{ kWh}}{5.2 \text{ h}} = 5.85 \text{ kW}_p$ $I_{DC} = \dfrac{P_{Peak(PV)}}{V_{DC}} = \dfrac{5850 \text{ W}}{48 \text{ V}} \approx 122 \text{ A}$ $N_s = \dfrac{V_{DC}}{V_{mp}} = \dfrac{48}{24.2} = 1.98 \approx 2$ $N_p = \dfrac{I_{DC}}{I_{mp}} = \dfrac{122}{7.65} = 15.94 \approx 16$ $N_T = N_s \times N_p = 2s \times 16p$

Considering the array sizing above, the PV system design can be drawn as follows. Note that all metal parts in the PV system should also be grounded (Figure 5.63).

b. As known, the operating voltage range of the MPPT solar charger is different than that of PWM. Besides, the maximum current-carrying capacity of the given MPPT charger is limited to 30 A. Therefore, the use of multiple MPPT chargers connected in parallel should also be considered in the PV array design to meet the total load demand.

In PV array sizing calculations, N_s and N_p values should be recalculated in case of MPPT charger consideration. From (5.15),

$$N_s = 0.95 \times \dfrac{V_{max\_ch}}{V_{mp} @ T_{cell}} = 0.95 \times \dfrac{150}{24.13} = 5.9 \approx 6$$

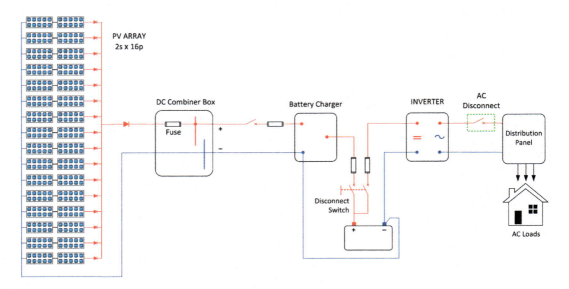

FIGURE 5.63 The PV system design.

Note that the value of $V_{mp}@T_{cell}$ is calculated to be 24.13 for NOCT = 45°C and voltage coefficient of (−0.35% / °C). In addition, it should be checked whether the PV output voltage is within the MPP tracking range. If the requirement $N_s \times V_{mp} \leq V_{max\_ch}$ is checked here:

$$6 \times 24.2 \leq 150 \Rightarrow 145.2 \leq 150$$

Hence, $N_s = 6$ is appropriate for the given MPPT-based charger. Accordingly, the number of parallel-connected PV strings (N_p) from (5.16):

$$N_p = \frac{P_{Peak(PV)}}{N_s \cdot P_{module}} = \frac{5.85}{6 \times 0.185} = 5.27 \approx 6$$

As a result, $6s \times 6p$ module is appropriate for the PV system having MPPT solar charger. However, the current value must also be checked. Since there are six parallel PV strings in the array, the total PV current can be as high as 45.9 A (6 × 7.65 = 45.9 A). Since the maximum charge current of a single MPPT charger is 30 A, it is appropriate to use 2 MPPT chargers in parallel. In this case, the PV system should include two separate arrays ($6s \times 3p$ plus $6s \times 3p$) but one battery group as shown in Figure 5.64.

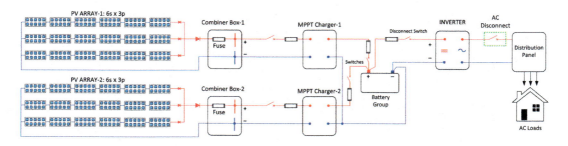

FIGURE 5.64 The PV system design with two MPPT chargers.

5.5.1.1.3 PV Array Sizing and Configuration for Grid-Connected Systems

The types of PV modules and inverter should be taken into account when sizing and configuring the PV array of grid-connected systems. Because matching the PV array output to the inverter operational range is crucial for a better PV system performance. Hence, the following basic steps can be used to size the array for grid-connected PV systems.

Step-1: Decide the PV module type and array capacity in [kW] that will be connected to the grid. Importantly note that there may be different criteria for determining the amount of PV array to be connected to the grid. For example, the budget allocated for the investment, the license for grid connection capacity of the PV power plant, the inverter type available, and the land area owned or hired by the investors could be the factors for determining the large-scale PV array capacity. Similarly, effective roof area, load demand, and solar radiation level can be the other parameters for deciding the building-integrated PV array capacity. Here, we will assume that PV array capacity has been decided for this step. The above-mentioned parameters as well as other effects are discussed separately under the relevant sections of this book.

Step-2: Decide the inverter type and match the PV array to the inverter voltage range. The array voltage for minimum and maximum temperature of the location should not exceed the maximum DC voltage of the selected inverter. It should also be checked that the minimum voltage of the PV array does not fall below the inverter's lower voltage limit. These criteria can then be used to find the minimum number of modules in a string. As known, the voltage of PV modules decreases as the weather gets warmer. Hence, the photovoltaic array has to be designed so that the MPP voltage of the PV array at maximum operational temperature does not go lower than the minimum MPPT voltage level of the inverter. Therefore, the MPP voltage of a PV module for maximum temperature conditions ($V_{mp} @ T_{cell\_max}$) must initially be calculated as below.

$$V_{mp} @ T_{cell\_max} = V_{mp,r} + \beta_{V_{mp}} \times (T_{cell\_max} - T_r) \tag{5.18}$$

where $V_{mp,r}$ is the MPP voltage at reference conditions (STC: standard testing conditions), $\beta_{V_{mp}}$ is the temperature coefficient of V_{mp}, T_{cell_max} is the module temperature at the maximum ambient weather temperature, and T_r is the temperature at reference conditions (25°C). Importantly note that the coefficient $\beta_{V_{mp}}$ in (5.18) takes values less than zero ($\beta_{V_{mp}} < 0$).

It is not correct to obtain the minimum V_{mp} voltage ($V_{mp\,min}$) only based on the highest ambient temperature conditions. Other factors affect the PV module voltage under actual operating conditions. For example, manufacturing tolerances, low lighting conditions, and shading/soiling factors cause reductions in the module voltages. In addition, inverters may not always operate at the ideal MPP point. With all these factors in mind, a safety margin of 5%–10% would be appropriate in determining the minimum MPP voltage of the PV modules. Accordingly, the minimum module voltage that may occur across the module can be obtained by multiplying equation (5.18) by a site-specific safety factor (f_{safety}).

$$V_{mp\,min} = f_{safety} \times (V_{mp} @ T_{cell\_max}) \tag{5.19}$$

Note that typical safety factors vary between 0.90 and 0.95, and it should be estimated based on the site conditions and project parameters. After this stage, the minimum number of modules ($N_{s(min)}$) constituting a PV string can be calculated by equation (5.20).

$$N_{s(min)} = \frac{V_{inv\_lower}}{V_{mp\,min}} \tag{5.20}$$

where V_{inv_lower} is the lower voltage value of the inverter's MPP voltage window. The value calculated by (5.20) should be rounded up to the nearest full number.

In the second phase of inverter's voltage specifications matching, the highest voltage value of the PV module must be considered to find the maximum number of modules ($N_{s(\max)}$) in the array. As known, the highest open-circuit voltage value of the module ($V_{oc} @ T_{cell\_min}$) occurs on the coldest day of the site. Therefore, the maximum open-circuit voltage ($V_{oc\,\max}$) of a PV module at minimum weather temperature conditions has to be calculated as (5.21).

$$V_{oc\,\max} = V_{oc} @ T_{cell\_min} = V_{oc,r} + \beta_{V_{mp}} \times (T_{cell\_min} - T_r) \tag{5.21}$$

where $V_{oc,r}$ is the open-circuit voltage at reference conditions (STC: standard testing conditions), $\beta_{V_{oc}}$ is the temperature coefficient of V_{oc}, T_{cell_min} is the module temperature at the minimum ambient weather temperature, and T_r is the temperature at reference conditions (25°C). Importantly note that the coefficient $\beta_{V_{oc}}$ in (5.21) takes values less than zero ($\beta_{V_{oc}} < 0$). A safety factor of 5% can also be taken into account due to the manufacturing tolerances, calculation approximations, and possible higher voltages. Now that the maximum number of modules ($N_{s(\max)}$) in a PV string can be calculated by equation (5.22).

$$N_{s(\max)} = \frac{0.95 \times V_{inv\_\max}}{V_{oc\,\max}} \tag{5.22}$$

where $V_{inv\_\max}$ is the maximum input voltage value of the inverter (no-exceed DC voltage value). The value calculated by (5.22) should be rounded down to the nearest whole number to avoid producing overvoltages beyond the maximum input voltage of the inverter.

Step-3: Match the PV array to the current rating of the inverter. The goal of this step is to ensure that the maximum array current is lower than the maximum input current of the inverter. To be able to calculate the number of parallel strings in an array of grid-connected PV systems, short-circuit current (I_{sc}) of a module is required.

$$I_{sc} @ T_{cell} = I_{scr} + \alpha_{I_{sc}} \times (T_{cell} - T_r) \tag{5.23}$$

where $I_{sc} @ T_{cell}$ is a short-circuit current at cell temperature T_{cell} (°C) a, I_{scr} is the reference (or rated) short-circuit current at the reference temperature T_r (25°C) (), $\alpha_{I_{sc}}$ is the manufacturer-supplied temperature coefficient of short-circuit current, and T_{cell} is the cell temperature at a specified ambient temperature (generally maximum cell temperature in the field). Note that the result of (5.23) gives the module short-circuit current at reference (STC) irradiance and high cell temperature conditions. It is possible to exceed this short-circuit current in real operating conditions if the irradiation on the module surface becomes higher than 1000 W/m². If over-irradiance is the case for the given site, it should then be taken into account for the maximal short-circuit calculation of the module. Once the maximum short-circuit current of the module is calculated from (5.23), the number of parallel strings in the array (N_p) is calculated by equation (5.24).

$$N_p = \frac{I_{inv\_\max DC\ input}}{I_{sc} @ T_{cell}} \tag{5.24}$$

where $I_{inv\_\max DC\ input}$ is the maximum DC input current of the inverter. The value calculated by (5.24) should be rounded down to the nearest whole number to avoid over-currents in the inverter.

Step-4: Match the PV array to the inverter power rating. The goal of this step is to ensure that the PV array is sized correctly in terms of current, voltage, and power ratings. The above calculations regarding the current and voltage matching give the limits only for the number of modules in a string and string number in an array. Due to roundings made above, the array power may exceed the inverter power or remain lower. For all these reasons, power, current, and voltage calculations should be evaluated together when performing a correct matching of an array to an inverter.

Accordingly, the maximum number of modules in the array for the selected inverter can be calculated by (5.25).

$$\text{Array Size} = \frac{\text{Inverter Maximum Rated Power}}{\text{Module Rated Power}} \qquad (5.25)$$

The result of (5.25) is rounded down to the nearest whole number. Note that the power calculations are crucial to check that the ($N_s \times N_p$) from the above calculations is less than the array size from (5.25). As a result, we can be sure that the PV array is not oversized ($N_s \times N_p \leq$ Array Size from (5.25)).

Example 5.3: PV Array Configuration and Sizing for Grid-Connected Systems

A grid-connected PV power plant to be installed in Los Angeles is expected to have an installed capacity of 450 kW. The characteristic values of the selected PV module in this project are given in Table 5.13.

Two different inverter options for the PV array design are considered. These inverters are string-type and central-type inverters. The basic specifications of these inverters are given in Table 5.14.

Assuming −10°C and +70°C as the minimum and maximum cell temperatures of the modules, calculate the following.

a. Size and configure the PV array in case of string inverter usage. Draw the array configuration.
b. Size and configure the PV array in case of central inverter usage. Draw the array configuration.

TABLE 5.13
The PV Module Parameters

PV Module Specifications: 250 Wp Poly Crystalline Module			Module Efficiency: 15.58%		
Voltage at P_{max}	Current at P_{max}	Open Circuit Voltage	Short Circuit Current		Power Tolerance
30.7 V	8.15 A	37.4 V	8.63 A		−0 / +3%
		Temperature Coefficients			
$\alpha_{I_{sc}}$	$\beta_{V_{mp}}$	$\beta_{V_{oc}}$	P_{max} Coefficient		NOCT
+0.004 A / °C	−0.0757 V / °C	−0.075 V / °C	−0.44 W / °C		45 ± 2°C

TABLE 5.14
Inverter Parameters

Option-1: String Inverter (# of Independent MPPT: 6 / (# of DC Pairs for Each MPPT: 4)				
MPP Voltage Range	Max DC Voltage	Max DC Current	Max Input Power for Each MPPT	Nominal Power
480 V – 850 V	1000 V	36 A	17.5 kW$_p$	100 kW
Option-2: Central Inverter				
MPP Voltage Range	Max DC Voltage	Max DC Input Current for Each MPPT	Max Input Power	Nominal Power
450 V – 825 V	1000 V	1145 A	600 kW$_p$	500 kW

Solution

a. Let us follow the steps described above regarding the PV array sizing for grid-connected PV systems. For this purpose, the following table showing all calculation steps can be created below (Table 5.15):
   Following the calculations in Table 5.15, a string inverter-based PV array design can be drawn as below (Figure 5.65).
b. Let us follow the steps described above in the case of using a central inverter. For this purpose, the following worksheet showing all calculation steps is created similar to that of the string inverter in (a) (Table 5.16).
   From the calculations in Table 5.16, the central inverter-based PV array design can be drawn as below (Figure 5.66).

5.5.2 Land Area Requirements for PV Power Plants

The power generation of a PV system is affected by different factors such as solar irradiance, cell operating temperature, tilt angle, shading, etc. The most important one among these factors is solar radiation. Therefore, any barrier that results in irradiance reduction such as shading on the PV arrays of the grid-connected system should be avoided during the layout planning of the project. Therefore, matching the photovoltaic power plant capacity to the available land area is crucial to getting a successful feasibility study and cost-effective solution for the PV system. Another important concern for the required land area is the selection of PV modules. As known, the higher the efficiency of the PV module, the less the land area required.

The PV Field with Fixed Panel Installation: To estimate the required land area for the PV system installation, let us consider the two basic PV power plant layouts. The first layout shown in Figure 5.67 is a PV field with a fixed-angle panel installation. As can be seen from Figure 5.67, a PV farm consists of sub-generators, and sub-generators consist of array rows. Each array row consists of PV modules. Besides, there is also a circumference zone surrounding all PV sub-generators for maintenance and accessibility.

Let us consider a PV sub-generator layout given in Figure 5.68 to calculate the required land area for an entire PV field. Starting from a PV module size, the total land area occupied by the PV system can easily be calculated by following the below calculations. The total surface area of the panels in a row of a PV sub-generator (A_{Row}) is calculated by (5.26).

$$A_{Row} = (W_m \times L_m) \cdot (n_s \times n_p) \tag{5.26}$$

where W_m is the width of a module, L_m is the length of a module, and $n_s \times n_p$ is the total number of modules in a row (see Figure 5.68).

Considering the distance (D) between the two rows of PV arrays (see Figure 5.68), the total required land area for a PV sub-generator (A_{SubG}) is calculated by equation (5.27).

$$A_{SubG} = N_{P(Row)} \cdot (W_{Row} \times \cos b + D) \cdot (L_{Row} + D) \tag{5.27}$$

where $N_{P(Row)}$ is the number of rows in a PV sub-generator, b is the tilt angle of the array, and L_{Row} is the length of a row. In the given array arrangement (see Figure 5.68), $L_{Row} = W_m \times n_s$ and $W_{Row} = L_m \times n_p$. However, the arrangement of modules on the panel rack may vary depending on the applications. The approach given here can easily be adapted to other possible applications with different module arrangements. Note that the distance (D) between two rows of arrays should be calculated by taking into account the inter-shading effect between PV panels. The relevant calculation regarding the distance between two rows of modules is explained in detail in Chapter 2 (see Section 2.8.2-Shadow Geometry). Therefore, it will not be repeated here in this section.

TABLE 5.15
Calculation Steps for PV Array Sizing

	Explanation	Calculation and Results
Step-1	Decide the PV module type and array capacity in [kW]	Installation capacity is given as 450 kW and the selected PV module is a *250 Wp* polycrystalline module. The specifications of the module are given in the example above.
Step-2	Decide inverter type and determine the minimum and maximum numbers of series-connected PV modules in a string.	The selected inverter type is a string inverter and its specifications are given in the example above. **Calculation of minimum number of modules in the string:** $V_{mp} @ T_{cell\_max} = V_{mp,r} + \beta_{mp} \times (T_{cell\_max} - T_r) = 30.7 - 0.0757 \cdot (70 - 25) = 27.3$ V By assuming $f_{safety} = 0.90$, $V_{mp\ min} = 0.90 \times 27.3 = 24.57$ V and $N_{s(min)} = \dfrac{V_{inv\_lower}}{V_{mp\ min}} = \dfrac{480}{24.57} = 19.53 \approx 20$ modules **Calculation of the maximum number of modules in the string:** $V_{oc\ max} = V_{oc,r} + \beta_{mp} \times (T_{cell\_min} - T_r) = 37.4 - 0.075 \cdot (-10 - 25) = 40.025$ V $N_{s(max)} = \dfrac{0.95 \times V_{inv\_max}}{V_{oc\ max}} = \dfrac{0.95 \times 850}{40.025} = 20.17 \approx 20$ module
Step-3	Match the PV array to the inverter's current rating.	**Calculate the number of parallel strings in the array:** $I_{sc} @ T_{cell\ max} = I_{scr} + \alpha_{isc} \times (T_{cell\ max} - T_r) = 8.63 + 0.004 \times (70 - 25) = 8.81$ A Now that: $N_p = \dfrac{I_{inv\_max\ DC\ input}}{I_{sc} @ T_{cell}} = \dfrac{36}{8.81} = 4.08 \approx 4$ modules
Step-4	Match the PV array to the inverter power rating.	**Calculation of maximum number of modules in the array:** Given string inverter has six independent MPPT units and each has 17.5 kW_p rated power. Besides, each MPPT unit has 4 pairs of DC input. So, we can do the calculations for one MPPT unit and the results can be multiplied by 6 to find the total array size for one string inverter. $\text{Array Size} = \dfrac{\text{Inverter Maximum Rated Power}}{\text{Module Rated Power}} = \dfrac{17.5\ kW}{0.25\ kW} = 70$ modules for one MPPT unit. From the above calculations we have $N_s \times N_p = 20 \times 4 = 80$ modules, which is bigger than 70 modules. Normally, $N_s \times N_p$ should be smaller than the Array Size (80 modules). Therefore, the number 80 must be reduced. Since the number of modules in a string is 20, the array size must be multiple of 20 in the reconfiguration. As a result, the final array configuration for a single MPPT unit of the inverter can be $N_s \times N_p = 20s \times 3p = 60$. Since each MPPT unit has 4 pairs of DC input, 3 strings in the array will be connected to an MPPT module via 3 pairs of DC input. So, 1 DC pair at the input of each MPPT will remain empty. In this case, the PV power connected to a string inverter is $20 \times 3 \times 6 \times 0.25 = 90$ kW. To obtain a total of 450kW (6×90), 6 string inverters are required. Each of the six inverters is connected to a 90 kW_p PV array.

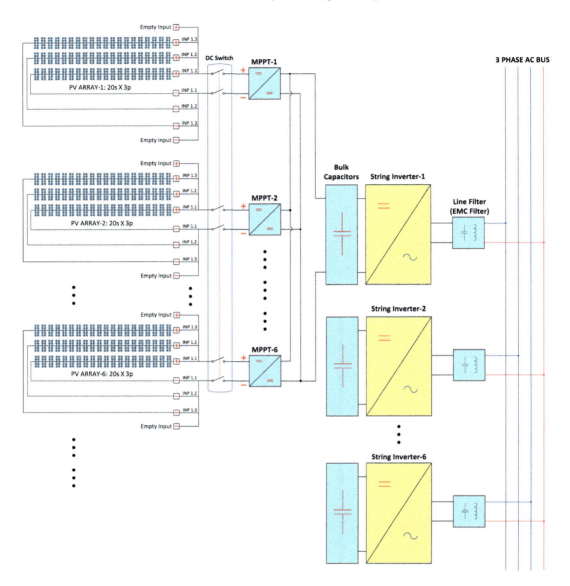

FIGURE 5.65 PV array design.

Considering the distance (D_{SubG}) between PV sub-generators (see Figure 5.69), the total required land area of the overall PV generator ($A_{PV\ Gen}$) is then calculated by equation (5.28).

$$A_{PV\ Gen} = N_{SubG} \cdot (W_{SubG} + D_{SubG}) \cdot (L_{SubG} + D_{SubG}) \tag{5.28}$$

where N_{SubG} is the total number of PV sub-generators in the solar field, W_{SubG} is the width of a PV sub-generator, and L_{SubG} is the length of a PV sub-generator. In the given array arrangement (see Figure 5.69), $L_{SubG} = N_{P(Row)} \times (W_{Row} \times \cos b + D)$ and $W_{SubG} = L_{Row} + D$. However, the arrangement of PV sub-generators herein is formed with the assumption that the land is flat and quadrangular. The approach given above can easily be adapted to other possible applications with non-flat and/or non-quadratic fields. An important note is needed to be underlined here. The distance (D_{SubG}) can be optional depending on the distances and dimensions of the project layout. If the D distance obtained

Photovoltaic Power Systems

TABLE 5.16
Calculation Steps for PV Array Sizing

	Explanation	Calculation and Results
Step-1	Decide the PV module type and array capacity in [kW]	Installation capacity is given as 450 kW and the selected PV module is a *250 W_p* polycrystalline module. The specifications of the module are given in the example above.
Step-2	Decide inverter type and determine the minimum and maximum numbers of series-connected PV modules in a string.	The selected inverter type in this option is a central inverter and its specifications are given in the example above. **Calculation of minimum number of modules in the string:** $V_{mp} @ T_{cell\_max} = V_{mp,r} + \beta_{V_{mp}} \times (T_{cell\_max} - T_r) = 30.7 - 0.0757 \cdot (70 - 25) = 27.3$ V By assuming $f_{safety} = 0.90$, $V_{mp\,min} = 0.90 \times 27.3 = 24.57$ V and $N_{s(min)} = \dfrac{V_{inv\_lower}}{V_{mp\,min}} = \dfrac{450}{24.57} = 18.31 \approx 19$ modules **Calculation of the maximum number of modules in the string:** $V_{oc\,max} = V_{oc,r} + \beta_{V_{mp}} \times (T_{cell\_min} - T_r) = 37.4 - 0.075 \cdot (-10 - 25) = 40.025$ V $N_{s(max)} = \dfrac{0.95 \times V_{inv\_max}}{V_{oc\,max}} = \dfrac{0.95 \times 825}{40.025} = 19.58 \approx 19$ module
Step-3	Match the PV array to the inverter's current rating.	**Calculate the number of parallel strings in the array:** $I_{sc} @ T_{cell\,max} = I_{scr} + \alpha_{I_{sc}} \times (T_{cell\,max} - T_r) = 8.63 + 0.004 \times (70 - 25) = 8.81$ A Now that: $N_p = \dfrac{I_{inv\_max\,DC\,input}}{I_{sc} @ T_{cell}} = \dfrac{1145}{8.81} = 129.96 \approx 129$ modules
Step-4	Match the PV array to the inverter power rating.	**Calculation of the maximum number of modules in the array:** Array Size $= \dfrac{\text{Inverter Maximum Rated Power}}{\text{Module Rated Power}} = \dfrac{600 \text{ kW}}{0.25 \text{ kW}} = 2400$ modules From the above calculations we have $N_s \times N_p = 19 \times 129 = 2451$ modules, which is bigger than 2400 modules. Normally, $N_s \times N_p$ should be smaller than the Array Size (2400 modules). Therefore, the number 2451 must be reduced. Since the number of modules in a string is 19, the array size must be multiples of 19 in the reconfiguration. As a result, the maximum array configuration for the central inverter can be $N_s \times N_p = 19s \times 126p = 2394$. In this case, the maximum power that can be connected to the central inverter is $2394 \times 0.25 = 598.5$ kW. However, the installed capacity of the PV project is assumed to be 450 kW. In this case, the number of PV modules required is 450 kW$/0.25$ kW $= 1800$ modules. Since each string has 19 modules, a total of $1800/19 = 94.7$ parallel strings are needed. Since there are enough tolerances here, let's round up the number of modules. Hence, the array configuration for the supplied central inverter can be $N_s \times N_p = 19s \times 95p = 1805$. Since the project limits are still appropriate, 96 parallel strings can be implemented in the project to get connection symmetry. Assume that the combiner boxes with 8 inputs will be used in the project. Therefore, $8 \times 12 = 96$ strings are required for the connection symmetry. As a result, a total of 456 kW ($= 96 \times 19 \times 0.25$) installed capacity, 12 combiner boxes, and one re-combiner box can be used for the central inverter.

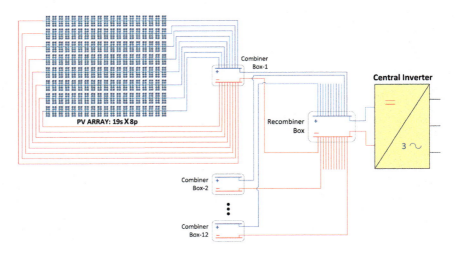

FIGURE 5.66 PV array design.

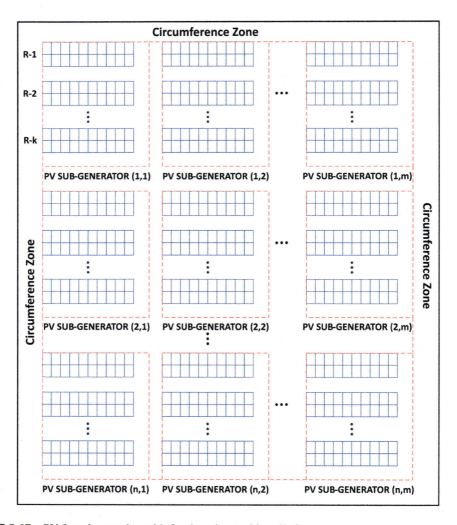

FIGURE 5.67 PV farm layout plan with fixed-angle panel installation.

Photovoltaic Power Systems

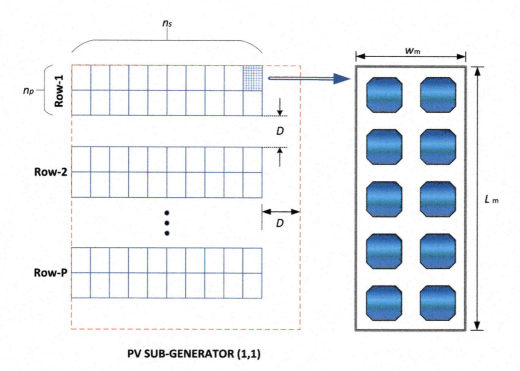

FIGURE 5.68 Layout geometry of a PV sub-generator.

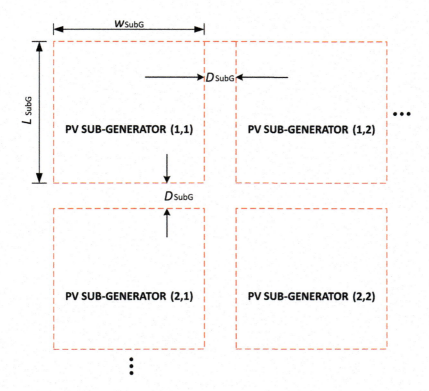

FIGURE 5.69 Layout geometry for the grouping of PV sub-generators.

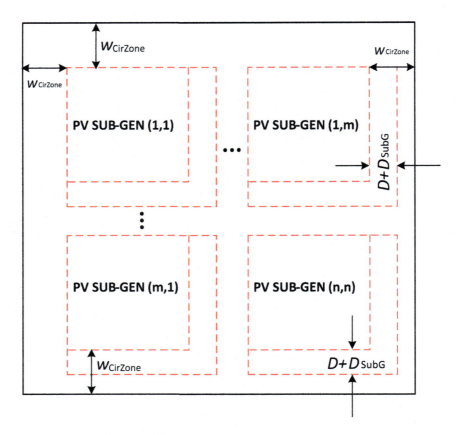

FIGURE 5.70 Geometry of entire PV field based on sub-generators layout.

from the design dimensioning is sufficient for the project requirements such as maintenance, repair service, accessibility, and vehicle entry, an additional D_{SubG} distance may then not be used. However, if the distance D between rows is not sufficient to respond to the maintenance and repair services, a few meters of additional distance can be used between the PV sub-generators. Typically, the distance D_{SubG} ranges from 1 to 3 m.

We can now calculate the total PV field area needed. In this case, let us consider the entire PV field layout given in Figure 5.70.

Considering the circumference zone surrounding all PV sub-generators (see Figure 5.70), the total required land area of the PV field ($A_{PV\ Field}$) can be calculated by (5.29).

$$A_{PV\ Field} = \left[m \times W_{SubG} + W_{CirZone} + \max\left(0, W_{CirZone} - D - D_{SubG}\right) \right]$$
$$\cdot \left[n \times L_{SubG} + W_{CirZone} + \max\left(0, W_{CirZone} - D - D_{SubG}\right) \right] \quad (5.29)$$

where m is the total number of PV sub-generators in one row of the solar field, n is the total number of PV sub-generators in one column of the solar site and $W_{CirZone}$ is the width of the circumference zone around the PV generator system. Typically, the $W_{CirZone}$ is in the range of 2–4 m.

An Important Note: The land area requirements for the PV fields with a single-axis PV tracking system (either in east-west or south-north axis direction) are the same as those of fixed-angle panels. The only difference between land area requirements of the PV fields with fixed-angle and single-axis panels is the shading rates, which affect the distance of D. Therefore, the PV field layout design

Photovoltaic Power Systems

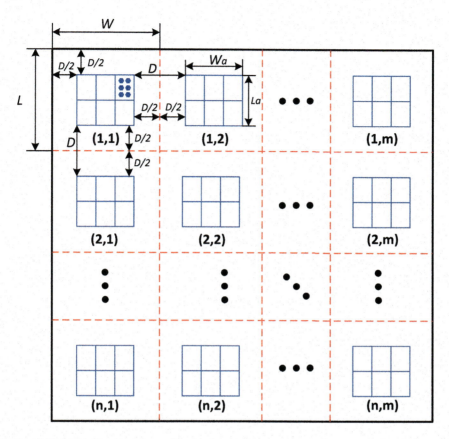

FIGURE 5.71 PV field layout with two-axis sun trackers.

steps given above are also valid for PV fields with single-axis panels. Hence, see Chapter 2 (Section 2.8.2-Shadow Geometry) for the calculation of distance D based on given conditions.

The PV Field with Two-axis (Sun-Tracking) PV System: Shading occurring in every direction on a two-axis sun-tracking PV system is of crucial importance in terms of system output performance. Hence, the shading effect between PV panels/arrays is also a dominant factor in dimensioning the required land area of a PV field with a dual-axis sun-tracking system. To estimate the required land area for a PV field with a dual-axis sun-tracking system, let us consider the layout shown in Figure 5.71.

As can be seen from Figure 5.71, the design consideration herein is based on the assumption that the land is flat and the panel arrangement is in a matrix format. Other assumptions considered in this PV field design are as follows:

- Since the PV tower heights are much higher than that of the fixed-angle system, the distance D between the PV towers will be much higher. Hence the distances between PV towers are considered to be sufficient for maintenance, repair, and service accessibility. Therefore, no additional space between PV arrays is considered.
- The land area occupied by each PV tower is calculated according to the horizontal position of the PV panels.

The layout designs given in Figures 5.67 and 5.71 are the same. The basic difference between the two systems (the PV field with fixed panel and sun-tracking systems) is that each PV array is mounted on an independent tall rack in the sun-tracking-based PV field. Compared to the PV arrays with a fixed angle, the independent PV arrays of the two-axis system have relatively lower output capacity and are smaller in size.

Considering the distance (D) between the PV towers of the two-axis sun-tracking system (see Figure 5.71), the total required land area for the entire PV towers ($A_{\text{PV Field SunTrack}}$) is calculated by equation (5.30).

$$A_{\text{PV Field SunTrack}} = (m \times n) \cdot (W_a + D) \cdot (L_a + D) \tag{5.30}$$

where W_a is the width of a PV array on a two-axis rack, L_a is the length of a PV array mounted on a two-axis rack, m is the total number of PV towers in one row of the solar field, and n is the total number of PV towers in one column of the solar site (see Figure 5.71).

Example 5.4: Required Land Area Calculation for Photovoltaic Power Systems

Consider the string inverter and central inverter-based 450 kW PV applications given in Example 5.3. Assume that the dimension of the *250 W_p* PV module used in Example 5.3 is 1000×1500 mm. Also assume that the site location is Los Angeles with the coordinate of (34.0522° N, 118.2437° W). Calculate the land area requirements for the following three cases.

  a. Calculate the required land area for the string inverter-based application in option (a) of Example 5.3. Suppose that the arrays are south facing and the tilt angle is 34°.
  b. Calculate the required land area for the central inverter-based application in option (b) of Example 5.3. Suppose that the arrays are south facing and the tilt angle is 34°.
  c. Assume that the 450 kW PV system consists of 2-axis PV towers with 2 m heights and each PV tower has a 2 kW$_p$ array. Calculate the required land area for this case.

Solution

  a. According to the design of Example 5.3 (a), there are six sub-generator fields and each has $6 \times (20s \times 3p)$ PV arrays. Let us assume that $20s \times 3p$ modules are installed on the PV rack structure in each row. Therefore six rows from the same rack system are connected to the one string inverter. We can calculate the required land area using equations (5.26)–(5.29). But, let's first create the PV rack system in each row and then calculate the distance (D) between the two rows according to the solar position on December 21, at noon (see Chapter 2, Section 2.8.2 Shadow Geometry).

  According to the panel geometry given in Figure 5.72, the distance between the rows of PV arrays ($d+D$) depends upon the row width (W_{Row}), tilt angle (b), and shading angle (altitude angle, α) and it is obtained from sinus theorem as:

$$d + D = W_{\text{Row}} \times \frac{\sin(180 - b - \alpha)}{\sin(\alpha)}$$

As explained in Chapter 2, shade-free distance is calculated based on December, 21 at noon. Hence, "no-shading distance" can be obtained as below:

$$d + D = W_{\text{Row}} \times \frac{\sin(180 - b - \alpha_{\text{Dec. 21, 12:00 PM}})}{\sin(\alpha_{\text{Dec. 21, 12:00 PM}})}$$

But, we need to know the altitude angle, α on December, 21 at noon for the Los Angeles city (34.0522° N, 118.2437° W). The day number on December, 21 is 356 ($N=356$). So the declination angle from equation (2.9):

$$\delta = 23.45 \cdot \sin\left[\frac{360}{365}(356 + 284)\right] = -23.45°$$

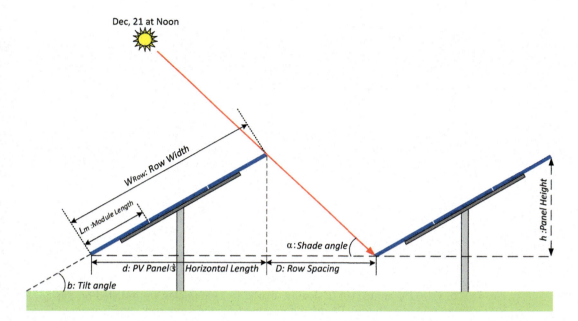

FIGURE 5.72 The PV panel geometry.

The hour angle at noon:

$$\omega = 15°(t_{LS} - 12) = 15(12 - 12) = 0°$$

From equation (2.11), the solar altitude angle (α) for Los Angles is:

$$\alpha = \sin^{-1}\left[\sin(\delta)\sin(L) + \cos(\delta)\cos(L)\cos(\omega)\right]$$
$$= \sin^{-1}\left[\sin(-23.45)\sin(34.05) + \cos(-23.45)\cos(34.05)\cos(0)\right] = 32.47°$$

Now, we can calculate the shade-free distance at noon on December 21 as below:

$$W_{Row} = 1.5 \times 3 = 4.5 \text{ m} \Rightarrow$$

$$d + D = 4.5 \times \frac{\sin(180 - 34 - 32.47)}{\sin(32.47)} = 4.5 \times \frac{0.916}{0.537} = 7.68 \text{ m}$$

So, row spacing between panels:

$$D = 7.68 \text{ m} - d \Rightarrow D = 7.68 \text{ m} - W_{Row} \times \cos b$$

$$D = 7.68 \text{ m} - 4.5 \times \cos 34 = 3.95 \text{ m}$$

From (5.26), the surface area of the PV panels in a row of a sub-generator is:

$$A_{Row} = (W_m \times L_m) \cdot (n_s \times n_p) = (1 \text{ m} \times 1.5 \text{ m}) \cdot (20 \times 3) = 90 \text{ m}^2$$

From (5.27), the total required land area for a PV sub-generator is:

$$A_{SubG} = N_{P(Row)} \cdot (W_{Row} \times \cos b + D) \cdot (L_{Row} + D) = 6 \cdot (4.5 \times \cos 34 + 3.95) \cdot (1 \times 20 + 3.95) = 1103.6 \text{ m}^2$$

By taking D_{SubG} as zero, the total required land area of the overall PV generator from (5.28) is:

$$L_{SubG} = N_{P(Row)} \times (W_{Row} \times \cos b + D) = 6 \cdot (4.5 \times \cos 34 + 3.95) = 46.08 \text{ m}$$

$$W_{SubG} = L_{Row} + D = 1 \times 20 + 3.95 = 23.95 \text{ m}$$

$$A_{PV\,Gen} = N_{SubG} \cdot (W_{SubG} + D_{SubG}) \cdot (L_{SubG} + D_{SubG}) = 6 \cdot (23.95 + 0) \cdot (46.08 + 0) = 6621.7 \text{ m}^2$$

Since the distance between the rows of PV panels is wide enough, $W_{CirZone}$ and D_{SubG} in (5.29) can be taken as zero. In this case, $A_{PV\,Gen} = A_{PV\,Field} = 6621.7 \text{ m}^2$.

b. According to the design of Example 5.3 (b), there are 12 sub-generator fields and each has $4 \times (20s \times 2p)$ PV arrays. Let us assume that $20s \times 2p$ modules are installed on the PV rack structure in each row. Therefore four rows from the same rack system are connected to one combiner box. We can again calculate the required land area using equations (5.26)–(5.29).

Firstly, let us create the PV rack system in each row and then calculate the distance (D) between the two rows according to the solar position on December 21, at noon. For this purpose, consider the rack structure in (a) on which $19s \times 2p$ module is mounted. The new dimensions in the revised rack structure herein are:

$$W_{Row} = 1.5 \times 2 = 3 \text{ m}$$

$$L_{Row} = L_{Row} = W_m \times n_s = 1 \times 19 = 19 \text{ m}$$

The solar altitude angle (shading angle) is the same in both options ($\alpha = 32.47°$). Now, we can calculate the shade-free distance at noon on December 21 as below:

$$d + D = 3 \times \frac{\sin(180 - 34 - 32.47)}{\sin(32.47)} = 3 \times \frac{0.916}{0.537} = 5.11 \text{ m}$$

So, row spacing between panels:

$$D = 5.11 \text{ m} - d \Rightarrow D = 5.11 \text{ m} - W_{Row} \times \cos b$$

$$D = 5.11 \text{ m} - 3 \times \cos 34 = 2.62 \text{ m}$$

From (5.26), the surface area of the PV panels in a row of a sub-generator is:

$$A_{Row} = (W_m \times L_m) \cdot (n_s \times n_p) = (1 \text{ m} \times 1.5 \text{ m}) \cdot (19 \times 2) = 57 \text{ m}^2$$

From (5.27), the total required land area for a PV sub-generator is:

$$A_{SubG} = N_{P(Row)} \cdot (W_{Row} \times \cos b + D) \cdot (L_{Row} + D) = 4 \cdot (3 \times \cos 34 + 2.62) \cdot (1 \times 19 + 2.62) = 441.6 \text{ m}^2$$

By taking D_{SubG} as 0.5 m, the total required land area of the overall PV generator from (5.28) is:

$$L_{SubG} = N_{P(Row)} \times (W_{Row} \times \cos b + D) = 4 \cdot (3 \times \cos 34 + 2.62) = 20.428 \text{ m}$$

$$W_{SubG} = L_{Row} + D = 1 \times 19 + 2.62 = 21.62 \text{ m}$$

$$A_{PV\ Gen} = N_{SubG} \cdot (W_{SubG} + D_{SubG}) \cdot (L_{SubG} + D_{SubG}) = 12 \cdot (21.62 + 0.5) \cdot (20.428 + 0.5) = 5555.12 \text{ m}^2$$

By taking $W_{CirZone}$ as 3 m, the total required land area for the entire PV field from (5.29) is:

$$A_{PV\ Field} = \left[m \times W_{SubG} + W_{CirZone} + \max(0, W_{CirZone} - D - D_{SubG}) \right]$$
$$\cdot \left[n \times L_{SubG} + W_{CirZone} + \max(0, W_{CirZone} - D - D_{SubG}) \right]$$

We know that $m \times n = 12$. So, let's take $m = 12$ and $n = 1$. (Note that the numbers m and n can be taken as any two multiples of 12. The result does not change.)

$$A_{PV\ Field} = \left[12 \times 21.62 + 3 + \max(0, 3 - 2.62 - 0.5) \right] \cdot \left[1 \times 20.428 + 3 + \max(0, 3 - 2.62 - 0.5) \right]$$

$$A_{PV\ Field} = [12 \times 21.62 + 3 + 0] \cdot [1 \times 20.428 + 3 + 0] = 6148.4 \text{ m}^2$$

As can be seen from the result that the second option uses 473 m² less land area than the first option (6621 − 6148 = 473 m²). This result proves that the rack structure system has a significant effect on the required land area.

c. December 21, at noon cannot be taken as a reference plane to determine the "no-shading distance" in dual-axis tracker systems. Because maximum altitude angle for Los Angeles occurs at solar noon (see equation 2.13) on December 21, which is:

$$\alpha = 90 - L + \delta = 90 - 34.05 + (-23.45) = 32.5°$$

Since the PV panel always moves perpendicular to the sunlight, the necessary tilt angle can easily be found from the shadow geometry as illustrated in Figure 5.73.

As can be seen from the figure that the tilt angle of the panel is still so high even at solar noon. Note that the use of a larger tilt angle will greatly increase the required land area in dual-axis sun-tracking systems. As a result, this position can be used as a reference plane to determine the "no-shading distance" between rows of PV panels with trackers. In this case, let's design the PV field to consist of 2 kW$_p$ PV towers as illustrated in Figure 5.74. Since each PV module has a power of 0.25 kW$_p$, 8 modules in total are needed for 2 kW$_p$ power.

According to the panel geometry given above, the distance between the rows of PV towers ($d+D$) can be calculated as below:

$$d + D = 3 \times \frac{\sin(180 - 57.5 - 32.5)}{\sin(32.5)} = 3 \times \frac{1}{0.537} = 5.58 \text{ m}$$

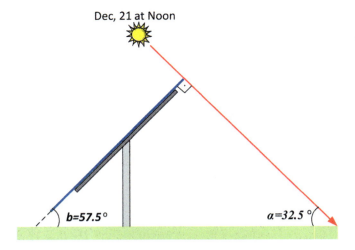

FIGURE 5.73 The PV panel and tilt angle.

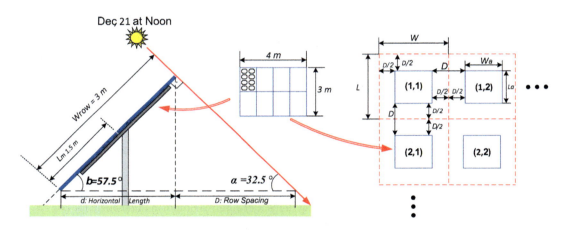

FIGURE 5.74 The PV panel towers in a solar site

So, row spacing between panels:

$$D = 5.58 \text{ m} - d \Rightarrow D = 5.58 \text{ m} - W_{\text{Row}} \times \cos b$$

$$D = 5.58 \text{ m} - 3 \times \cos 57.5 = 3.97 \text{ m}$$

225 PV towers are needed to get 450 kW installed power. From (5.30), the total required land area for the entire PV towers is:

$$A_{\text{PV Field SunTrack}} = (m \times n) \cdot (W_a + D) \cdot (L_a + D) = 225 \cdot (4 + 3.97) \cdot (3 + 3.97) = 12498 \text{ m}^2$$

As can be seen from the above results that the PV field with a two-axis sun-tracking system requires about 2 times more land area than the fixed PV field. Therefore, it can be concluded that the fixed PV field arrangement shall be preferable if the land area impact is of great importance.

5.5.3 DC Combiner and Recombiner Box Sizing and Selection

While the DC combiner box joins the currents of photovoltaic arrays into the next joint string, the combiner boxes are used in larger photovoltaic systems for connecting the first-level combiners into the main output feeder, which then goes to the system inverter. Additionally, the aim of using combiner boxes in the PV systems is to protect against over-currents and lightning while monitoring the system parameters such as current, voltage, temperature, and the state of the lightning protection device.

There is no standardized method for the sizing and selection of the PV combiner boxes. Instead, there are criteria and requirements to choose the right combiner box for the application being considered. So, the following summarizes the most basic requirements for combining several PV strings using a combiner box.

- **Maximum Open-Circuit Voltage** ($V_{oc\ max}$): The switches and/or breakers used in the combiner boxes must be rated for the maximum open-circuit voltage. So, we need to know the maximum V_{oc} of our PV system, whether it is 600 V, 1000 V, or 1500 V DC.
- **Maximum Short-Circuit Current** ($I_{sc\ max}$): Each PV string must have circuit protection. Again, we need to know the maximum short-circuit current I_{sc} of our PV system by finding the I_{sc} of individual PV modules used in the system. The I_{sc} values from all sources (PV strings, batteries, and back-feed from the inverters) should be taken into consideration and the maximum I_{sc} should not exceed the current-carrying capacity (ampacity) of the cables/conductors or the maximum current of the protection device. Hence, the maximum short-circuit current of each parallel string shall be multiplied by a specific safety factor, which could be 1.25, 1.5, or 1.56. The requirements for the multiplier (safety factor) may vary according to PV system size and/or country regulations. For example, suppose that we select 200 W_p module with I_{sc} of 6.08 Amps and a maximum series fuse rating of 10 A. If we need two parallel-connected modules as per PV panel, then an overcurrent protection will be required for each string of PV panel would be 12.16 A exceeding 10 A fuse rating. If the maximum series fuse rating was 20 A, then we could parallel two modules as $6.08 \times 2 \times 1.25 = 15.2$ A. However, in the case of solar system design, the safety factor of 1.25 many times fails to provide a correct rating of fuses or fuses burnt often at sites during the hot periods of seasons. Because solar radiation may rise above 1000 W/m² in summer seasons, which also increases the module temperature further. For this reason, a safety factor of 1.56 (= 1.25×1.25) is used most commonly to size the fuse/circuit breakers correctly. As a result of the above example, $6.08 \times 2 \times 1.56 = 18.97$ A as the output of two modules is still less than the 20 A maximum fuse rating.
- **The Number of Input and Output Pairs:** The number of PV string pairs going into the combiner box and the number of output pairs for the MPPT connections must be known to choose the correct size combiners. Because all array circuits going to each controller/MPPT must have a separate and isolated feeder. Note that one can find the information about the number of MPPTs from the inverter that is used in the system.
- **Fuse Holder Polarity:** It is important to know whether we use a fuse holder only at positive polarity or at both positive and negative polarity to arrange the space requirement for fuse housing in the combiner box. Note that although it is convenient to use fuses in the positive terminal only, the fuse can also be applied to both positive and negative terminals to provide better protection. Generally, fuse only at the positive terminal is used for grounded PV systems, while fuse at positive and negative terminals can be used for ungrounded systems.
- **Monitoring Requirements:** Generally, parameters such as current, voltage, panel temperature, and state of overvoltage/lightning protection device can be monitored. For this purpose, it is appropriate to have a monitoring device in the combiner box as long as the system allows the monitoring of relevant parameters.

- **Other Requirements:** Besides, the combiner boxes for the PV systems should have the following features:
  - Combiner boxes should be suitable for outdoor conditions. For this reason, they must have an IP-65 protection class.
  - They should have a sufficient heat dissipation feature. If necessary, a ventilation system can also be added inside the combiner boxes.
  - They should be resistant to UV radiation.
  - They should be resistant to electrical shocks such as over-currents, overvoltages, spark-overs, and lightning strikes. Hence, they should have good isolation according to these aspects. In a conclusion, the PV array combiner boxes should be total insulated according to the requirements of IEC 61439-1 and tested up to an operating voltage of 1000 V or 1500 V DC.

Example 5.5: DC Combiner and Recombiner Box Sizing

Consider the string and central inverter-based 450 kW PV applications given in Example 5.3.

a. Determine the system combiner box requirements for the string inverter-based application in option (a) of Example 5.3. Select a suitable combiner box from the commercially available products and draw the wiring diagram.
b. Determine the system combiner and recombiner box requirements for the central inverter-based application in option (b) of Example 5.3. Select a suitable combiner and recombiner boxes from the commercially available products and draw the wiring diagram.

Solution

a. According to the design of Example 5.3 (a), we can use one DC combiner box for each string inverter. We know that each string inverter has 6 MPPT and each MPPT has an input of $20s \times 3p$ PV array. Based on the information provided in Example (5.3), we can identify the combiner box requirements for each of the six inverters as below:

Maximum Open-Circuit Voltage (V_{oc}): By assuming $T_{cell\_min} = -10°$, the maximum open-circuit voltage of the module is:

$$V_{oc\ max} = V_{OC,r} + \beta_{V_{mp}} \times (T_{cell\_min} - T_r) = 37.4 - 0.075 \cdot (-10 - 25) = 40.025 \text{ V}$$

Since PV strings consist of 20 series-connected modules, the maximum open-circuit voltage of the string is:

$$V_{oc\ max,\ string} = 20 \times 40.025 \text{ V} = 800.5 \text{ V}$$

Hence, the switches and/or breakers used in the combiner boxes must be rated for a voltage higher than the maximum open-circuit voltage of 800.5 V. In conclusion, the corresponding standard voltage level is 1000 V.

Maximum Short-Circuit Current (I_{sc}): By assuming $T_{cell\_max} = 70°$, maximum short-circuit current of the string at the input of combiner is:

$$I_{sc} @ T_{cell\_max} = I_{scr} + \alpha_{I_{sc}} \times (T_{cell\_max} - T_r) = 8.63 + 0.004 \cdot (70 - 25) = 8.81 \text{ A}$$

By assuming a safety factor of 1.56, string fuses can be sized for $1.56 \times 8.81 \text{ A} = 13.74 \text{ A}$. Hence, the maximum series fuse rating per string can be chosen as 15 A.

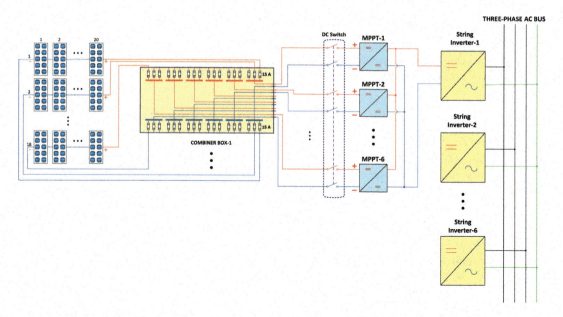

FIGURE 5.75 The wiring diagram of a PV system and combiner box.

The Number of Input and Output Pairs: We know from Example 5.3 (a) that the number of PV string pairs at the input side of the combiner is 18 ($=6\times 3p$) and the number of output pairs is 6 because of 6 MPPT at the input of each string inverter.

Fuse Holder Polarity: Let us assume that we use fuse holders at both positive and negative terminals.

Monitoring Requirements: We have 18 strings to monitor inside each combiner box. The monitoring equipment should also be sized for 15 A and 1000 V ratings as explained above. As a result, the monitoring device for 20 strings with ratings of 1000 V and 15 A is suitable for use in the designed combiner box.

Other aforementioned requirements regarding electrical insulation, heat dissipation, IP protection, and UV radiation must also be met for the relevant combiner box.

As a result of the above analysis and evaluations, the wiring diagram can be drawn below for the string inverter case (Figure 5.75).

b. According to the design of Example 5.3 (b), we can use 12 DC combiner boxes and one recombiner box before the central inverter. Each combiner box has an input of eight DC pairs ($19s \times 8p$ PV array) and recombiner box has an input of 12 DC pairs. Based on the information provided in Example (5.3), we can identify the combiner/recombiner box requirements as below:

Combiner Box Requirements:

Maximum open-circuit voltage (V_{oc}): By assuming $T_{cell\_min} = -10°$, the maximum open-circuit voltage of the module is:

$$V_{oc\,max} = V_{oc,r} + \beta_{V_{mp}} \times (T_{cell\_min} - T_r) = 37.4 - 0.075 \cdot (-10 - 25) = 40.025 \text{ V}$$

Since PV strings consist of 19 series-connected modules, the maximum open-circuit voltage of the string is:

$$V_{oc\,max,\,string} = 19 \times 40.025 \text{ V} = 760.5 \text{ V}$$

Hence, the switches and/or breakers used in the combiner boxes must be rated for a voltage higher than the maximum open-circuit voltage of 760.5 V. In conclusion, the corresponding standard voltage level is 1000 V.

Maximum short-circuit current (I_{sc}): By assuming $T_{cell\_max} = 70°$, maximum short-circuit current of the string at the input of combiner is:

$$I_{sc} @ T_{cell\_max} = I_{scr} + \alpha_{I_{sc}} \times (T_{cell\_max} - T_r) = 8.63 + 0.004 \cdot (70 - 25) = 8.81 \text{ A}$$

By assuming a safety factor of 1.56, string fuses can be sized for 1.56×8.81 A = 13.74 A. Hence, the maximum series fuse rating per string can be chosen as 15 A.

The number of input and output pairs: We know from Example 5.3 (b) that the number of PV string pairs at the input side is 8 (due to $19s \times 8p$ PV array) and the number of output pairs is 1 for each of the 12 combiner boxes.

Fuse holder polarity: Let's this time assume that we use fuse holders at only positive polarity.

Monitoring requirements: We have eight strings to monitor inside each combiner box. The monitoring equipment should also be sized for 15 A and 1000 V ratings as explained above. As a result, the monitoring device for eight strings with ratings of 1000 V and 15 A is suitable for use in the designed combiner box.

Other aforementioned requirements regarding electrical insulation, heat dissipation, IP protection, and UV radiation must also be met for the relevant combiner box.

Recombiner Box Requirements:

Maximum open-circuit voltage (V_{oc}): Maximum V_{oc} of the recombiner box will be the same as the maximum open-circuit voltage of the combiner box. Hence, the recombiner box must also be sized for 1000 V.

Maximum short-circuit current (I_{sc}): Each input pair of the recombiner box collects the eight string currents of 8.81 A. Thus, 8×8.81 A = 70.48 A passes at each input of the recombiner box. If this current is multiplied by a safety coefficient of 1.56, the ampere ratings of the fuse (or breaker) can be chosen as 70.48 A $\times$ 1.56 = 109.95 A. So, we can choose a 110 A standard ampere rating for the fuse in the recombiner box.

The number of input and output pairs: We know from Example 5.3 (b) that the number of PV input pairs at the input side is 12 (due to 12 combiner boxes) and the number of output pairs is 1 for the central inverter input.

Fuse holder polarity: Let's use DC fuses at only positive polarity.

Monitoring requirements: We have 12 PV inputs to monitor inside the recombiner box. The monitoring equipment in the recombiner should also be sized for 110 A and 1000 V ratings as explained above. As a result, the monitoring device for 12 strings with ratings of 1000 V and 110 A is suitable for use in the designed combiner box.

Other aforementioned requirements regarding electrical insulation, heat dissipation, IP protection, and UV radiation must also be met for the relevant recombiner box.

As a result of the above analysis and evaluations, the wiring diagram can be drawn below for the central inverter case (Figure 5.76).

5.5.4 Battery Sizing and Selection

Battery storage systems can be used both in stand-alone and grid-connected systems. The electrical energy generated by stand-alone PV systems is needed to be stored in batteries during high solar radiation times to provide continuous power to the load during periods of low solar radiation. The battery group in grid-connected PV systems can also be used to provide the load during grid failure, which means that the PV system works as a stand-alone during faulty grid conditions. It is necessary to underline an important point herein. Batteries not only provide required energy during the autonomy periods (night times and low solar radiation times) but also have functions to stabilize the voltage and supply the surge current during transient loads.

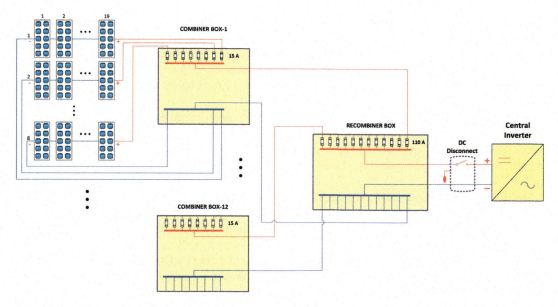

FIGURE 5.76 The wiring diagram of the DC combiner and recombiner boxes.

It is necessary to determine the battery technology type before the sizing process in PV systems. Therefore, the following can be summarized as the crucial points in the selection of PV system batteries.

- **Battery Life Cycle:** It is the number of complete charge/discharge cycles that a battery can work before the nominal capacity decreases to less than 80% of its rated (initial) capacity. After a specific life cycle period, the battery can still work but with reduced capacity. As known, the batteries in PV systems are exposed to repeated charging and discharging process. Since the battery lifetime strongly influences the cost of the system, batteries with a high life cycle are recommended for PV system applications.
- **Battery Cost:** For a cost-effective solution, the initial (purchase) cost plus the annual cost (the maintenance and operation costs) should be included in the battery cost evaluation.
- **Battery Efficiency:** Due to the internal resistances and charge controllers, the battery units have energy losses during charging and discharging cycles. Besides, the batteries have self-discharging losses due to the internal leakage resistances and power losses consumed in external circuits during the unused times. The battery self-discharge rates are directly proportional to the ambient temperature. In other words, the battery self-discharge losses increase as the temperature increases. Hence, battery rooms should be kept at optimum temperatures to reduce self-discharging and improve battery performance. As a result, the average energy efficiency of a battery can be defined as the total charged energy (kWh) divided by the total discharged energy (kWh) over a specified period.
- **Battery Performance under Operating Conditions:** One of the most important features expected from the batteries is showing good reliability under a wide range of operating conditions. For example, the battery features such as low maintenance, wide operating temperature, low self-discharge rate, and robust structure will provide important advantages to the operating performance of the batteries.

From the above explanations, we can easily conclude that an ideal battery for PV systems should have the features of low cost, high energy efficiency, low self-discharge rate, long life cycle, low maintenance, robust construction, high reliability, and wide temperature range.

In addition to the battery selection criteria described above, different battery parameters are of great importance in battery sizing. Hence, the main battery parameters are briefly introduced here as below.

Battery Capacity: The battery capacity is normally specified in Wh or Ah. One can convert watt-hours to ampere-hours by dividing Wh by the battery voltage, or you may convert ampere-hours to watt-hours by multiplying Ah by the battery voltage. To achieve a good operational performance, battery capacities should be sized for a worst-case solar radiation scenario within the period of use.

Battery Voltage/System DC Voltage: The terminal voltage of a battery under operating conditions is known as nominal voltage or working voltage. Depending on the battery design, battery voltages can be 3 V, 6 V, 12 V, or 24 V, etc. For small-scale PV applications, system DC voltage can generally be 12, 24, or 48 V.

Depth of Discharge (DoD)/**State of Charge** (SoC): A battery's DoD gives a measure of discharged capacity as a percentage of its full capacity. However, a battery's SoC indicates the difference between the full charge and the DoD of the battery in percentage. For instance, if the DoD is 30% then the SoC is 70%. As explained below, the maximum (or recommended) DoD value is used in the sizing process of the selected battery type. The recommended DoD value varies depending on the battery temperature. In this respect, the choice of the DoD is very important in applications where the battery will operate at negative temperatures. For example, lead-acid batteries, which are frequently used in PV systems, freeze at about −57°C in 100% discharge state, and at about −8°C in 100% charged state. Accordingly, the maximum DoD value (DoD_{max}) of the battery as a function of battery temperature (T) in [°C] can typically be converged as equation (5.31).

$$DoD_{max}[\%] = 0.0275 \times T^2 + 3.817 \times T + 128.35 \text{ for } -57°C \leq T \leq -8°C \quad (5.31)$$

Note that (5.31) is an approximate equation for deep-cycle lead-acid batteries. More specific data for the selected battery should be obtained from the manufacturers when needed.

Charge/Discharge Rate (C − Rate): A battery's C-rate gives a measure of the rate at which a battery is being charged or discharged. It is represented as C/X, where X is the time in hours for a full charge or discharge. For example, if $X = 5$ h, then C-rating is $C/5$ or $0.2C$. A safe charge or discharge current rate for a battery with a specific C-rating can be found easily by dividing Ah capacity by the X value (total hours of charge or discharge). For instance, for a 50 Ah capacity battery if C-rating is $0.2C$, then the charge or discharge current will be $50/5 = 10$ A. In other words, the discharge current (in this case 10 A) will discharge the entire battery in 5 hours.

Battery Temperature Coefficient: The Ah capacity of a battery depends on both C-rate (discharge rate) and temperature. The effect of these two factors (C-rate and temperature) on Ah capacity must be well known to size the battery unit accurately. Relative battery capacity curves in [% of full capacity] can be determined as a function of battery temperature (T). These curves can be obtained for nominal C-rating as well as for lower and higher C-rating values. For example, suppose that the nominal C-rating value for a battery is $C/20$. In this case, the battery discharged under $C/20$ can use all of its full capacity under 25°C. If the battery is desired to be discharged with a greater current at the same temperature conditions, let's say it is discharged with $C/10$ rating, then the usable capacity of the battery will fall below the rated capacity. Similarly, if the battery is discharged with a lower current, then it is possible to use a higher capacity under the same temperature conditions. So, the Ah capacity of a battery under varying temperatures and C-rating can be represented by the following curve (Figure 5.77).

As can be seen from Figure 5.77, it will be necessary to use a battery temperature correction factor to size the batteries working at temperatures other than nominal conditions. The battery

Photovoltaic Power Systems

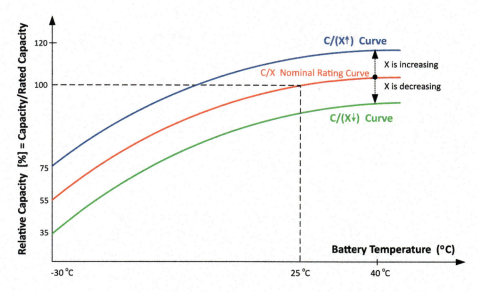

FIGURE 5.77 Relative battery capacity curves as a function of temperature and C-rating.

temperature correction factor ($f_{\text{Temp-b}}$) for any C-rating and temperature (T) can be accepted to be the same as the relative battery capacity as given in equation (5.32).

$$f_{\text{Temp-b}} @ (T, C-\text{rating}) = \frac{\text{Battery Capacity} @ (T, C-\text{rating})}{\text{Battery Rated Capacity}} \quad (5.32)$$

Based on the explanations given above, typical $f_{\text{Temp-b}}$ values for the deep-cycle lead-acid batteries can be written as equation (5.33)–(5.37) for different C-ratings. Equation (5.32) given above is obtained based on the $C/20$ nominal discharge rate. Hence from (5.35), we get $f_{\text{Temp-b}} = 100\%$ for $T = 25°C$.

$$f_{\text{Temp-b}} @ (C/5\,\text{h}) [\%] = -0.0072 \times T^2 + 0.732 \times T + 62.154 \text{ for } -30°C \leq T \leq 40°C \quad (5.33)$$

$$f_{\text{Temp-b}} @ (C/10\,\text{h}) [\%] = -0.0083 \times T^2 + 0.8094 \times T + 84.182 \text{ for } -30°C \leq T \leq 40°C \quad (5.34)$$

$$f_{\text{Temp-b}} @ (C/20\,\text{h}) [\%] = -0.0088 \times T^2 + 0.8168 \times T + 84.793 \text{ for } -30°C \leq T \leq 40°C \quad (5.35)$$

$$f_{\text{Temp-b}} @ (C/48\,\text{h}) [\%] = -0.0099 \times T^2 + 0.7575 \times T + 98.794 \text{ for } -30°C \leq T \leq 40°C \quad (5.36)$$

$$f_{\text{Temp-b}} @ (C/72\,\text{h}) [\%] = -0.0087 \times T^2 + 0.6346 \times T + 103.85 \text{ for } -30°C \leq T \leq 40°C \quad (5.37)$$

Note that the mathematical expressions from (5.31) to (5.37) are approximate equations for typical deep-cycle lead-acid batteries. More specific information about the selected battery should be obtained from the manufacturers' data.

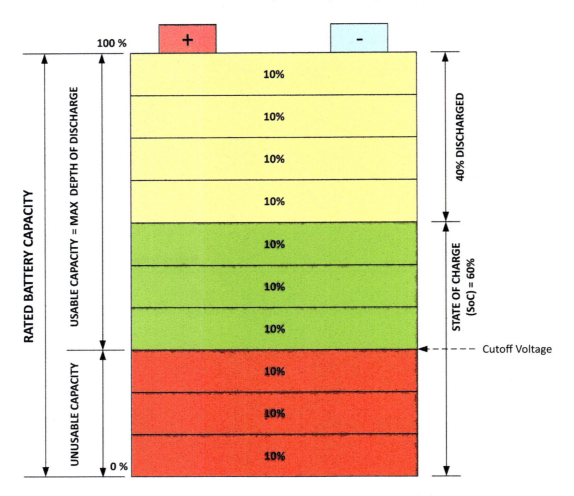

FIGURE 5.78 Illustration of battery characteristic behaviors and its operating limits.

Keeping the above-mentioned battery parameters in mind, let us examine Figure 5.78 for the illustration of battery characteristic behaviors and their operating limits. Suppose that the battery in Figure 5.78 has a DoD_{max} value of 70% and its existing SoC value is 60%.

It is clear from Figure 5.78 that battery sizing is mostly based on the battery operating limits (DoD, f_{Temp-b} etc.), system DC voltage, battery-rated capacity, and average daily energy requirement of the loads. Besides, if the required battery storage is higher than the daily energy consumption, then Days of Autonomy (DoA) should be taken into account as a multiplier in the battery sizing. As known, Days of Autonomy are defined as the expected number of days that the system load will be provided only by the battery group. Hence, the following basic steps can be used to size the battery unit for stand-alone and grid-connected PV systems.

Step-1: Decide the battery technology type, battery-rated voltage (V_b), and the battery capacity (c_b) in [Ah] that will be used in PV systems. Table 5.17 summarizes the basic characteristics and main battery selection criteria for batteries used commonly in solar PV systems. Importantly note that the below table gives a rough comparison of battery characteristics and hence it can be used only as initial guidance for engineers to compare the battery specifications. More specific data should be obtained from the manufacturers' catalogs.

It is possible to reach the following outcomes from the information given in Table 5.17. Lead-acid batteries with deep discharge are commonly used for remote PV applications due to their

TABLE 5.17
Different Battery Types and Their Basic Characteristics

Specifications	Lead Acid Batteries			Nickel (Ni) Batteries		Lithium-Ion Batteries	
	Flooded Batteries	VRLA Gel Batteries	VRLA AGM Batteries	Ni-Cd Batteries	Ni-MH Batteries	LFP Batteries	Li-NMC Batteries
Volumetric energy density	80 Wh/L	100 Wh/L	100 Wh/L	130 Wh/L	250 Wh/L	300 Wh/L	350 Wh/L
Gravimetric energy density	30–40 Wh/kg	40–50 Wh/kg	40–50 Wh/kg	45–80 Wh/kg	60–120 Wh/kg	120 Wh/kg	150
Initial cost[a]	60 $/kWh	110 $/kWh	130 $/kWh	600 $/kWh	540 $/kWh	450 $/kWh	580 $/kWh
Cycle life (@ DoD%)	500@100% 1200@50%	500@80% 1000@50%	400@80% 900@50%	500@85% 200@97%	500@85% 200@97%	2500@80% 5000@50%	2500@80% 5000@50%
Calendar life[b]	2–8 years	3–12 years	2–10 years	3–6 years	2–5 years	6 years	8 years
Self-discharge rate[c]	5–15%/Month	2%/Month	2%/Month	2%/Month	2%/Month	5%/Month	3%/Month
Max DoD	50%	50%	50%	>80%	>80%	>80%	>80%
Temperature range	−30°C to 50°C	−30°C to 50°C	−30°C to 50°C	−20°C to 50°C	−20°C to 50°C	−20°C to 50°C	−20°C to 50°C
Round trip efficiency	70%–90%	75%–92%	75%–92%	70%–90%	80%	>95%	>95%
Relative capacity	100%@C/20	100%@C/20	100%@C/20	100%@C/20	100%@C/20	100%@C/20	100%@C/20
	80%@C/4	80%@C/4	85%@C/4	95%@C/1	95%@C/1	99%@C/4	99%@C/4
	60%@C/1	60%@C/1	65%@C/1	90%@C/0.5	90%@C/0.5	92%@C/1	92%@C/1
Cell voltage	2 v/cell	2 v/cell	2 v/cell	1.2 v/cell	1.2 v/cell	3.3 v/cell	3.7 v/cell
Regular maintenance	YES	NO	NO	YES	YES	NO	NO
On-grid + Off-grid applications	Yes, easy for both	Yes, easy for both	Yes, easy for both	Have potential but not common	Have potential but not common	Mostly On-Grid	Mostly On-Grid

VRLA: Valve Regulated Lead Acid, AGM: Absorbed Glass Mat, MH: Metal Hydride, LFP: Lithium Iron Phosphate, NMC: Nickel, Manganese, and Cobalt

[a] Prices are based on market average and estimate.
[b] Service life varies significantly based on the average battery temperature.
[c] Self-discharge rate varies significantly based on the storage temperature conditions.

maintenance-free operation. Lithium-ion batteries, due to their high energy density and very high efficiency, have also been considered recently as alternative energy storage devices for use in PV system applications.

Step-2: Calculate the average daily load consumption in Ah (or kWh). Note that the load calculations in this book are given under a separate sub-section below. Hence, the required load amount in this section is considered to be known in the battery sizing procedure.

Step-3: For stand-alone systems, calculate the total energy requirement that the battery must supply during the autonomy days. As shown in equation (5.38), the amount of total energy needed (E_{Total}) is equal to the multiplication of average daily load consumption ($E_{Daily\ Avg}$) and days of autonomy (DoA).

$$E_{Total}[\text{Wh}] = E_{Daily\ Avg}[\text{Wh/day}] \times \text{DoA}[\text{day}] \tag{5.38}$$

Importantly note that the total energy needs that the battery must supply during the grid failure can be taken into battery sizing calculations for grid-connected systems. Other steps used in battery sizing will be the same for both systems (On-grid and Off-grid).

Step-4: For battery safety, E_{Total} value calculated in Step-3 should be divided by the battery's maximum depth of discharge (DoD_{max}). In other words, the amount of total energy needed (E_{Total}) should be provided only by the recommended DoD capacity of the battery. Accordingly, the total energy storage capacity of the battery bank ($E_{Battery}$) can be calculated in [Wh] as equation (5.39).

$$E_{Battery}[\text{Wh}] = \frac{E_{Total}[\text{Wh}]}{\text{DoD}_{max}} \tag{5.39}$$

As shown in (5.40), the required storage capacity of the battery bank in [Ah], $C_{Battery}$ can then be found by dividing the $E_{Battery}$ [Wh] by the DC voltage of a selected battery (V_b).

$$C_{Battery}[\text{Ah}] = \frac{E_{Battery}[\text{Wh}]}{V_b} \tag{5.40}$$

Step-5: Net capacity requirement of the battery ($C_{Battery-net}$) should then be calculated by considering the battery charge-discharge efficiency (η_b: round-trip efficiency), and the battery temperature coefficient (f_{Temp-b}) as equation (5.41).

$$C_{Battery-net}[\text{Ah}] = \frac{C_{Battery}[\text{Ah}]}{\eta_b \times f_{Temp-b}} \tag{5.41}$$

By inserting equations (5.38)–(5.40) into (5.41), the net battery capacity requirement can also be written as (5.42).

$$C_{Battery-net}[\text{Ah}] = \frac{E_{Daily\ Avg}[\text{Wh/day}] \times \text{DoA}[\text{day}]}{\text{DoD}_{max} \times V_b \times \eta_b \times f_{Temp-b}} \tag{5.42}$$

Step-6: Once the parameter of $C_{Battery-net}$ [Ah] is determined, the total number of batteries ($N_{Battery}$) included in the battery bank has to be obtained. Assume that Ah capacity of a single-battery unit is denoted by c_b. Hence, the total number of batteries in the battery pack can be found by dividing the $C_{Battery-net}$ [Ah] by the capacity of one battery (c_b [Ah]) as equation (5.43).

$$N_{Battery} = \frac{C_{Battery-net}[\text{Ah}]}{c_b[\text{Ah}]} \tag{5.43}$$

Photovoltaic Power Systems

If the $N_{Battery}$ from (5.43) is not a whole number, then the result should be rounded up to the nearest integer number.

Step-7: Now that the connection configuration of the battery bank can then be calculated easily. Since the system-rated DC voltage (V_{DC}) and battery-rated voltage (V_b) are determined from the steps described above, the number of series-connected battery units (N_{sb}) in the pack can be found as equation (5.44).

$$N_{sb} = \frac{V_{DC}}{V_b} \tag{5.44}$$

After determining the number of batteries in a string of the bank, the number of parallel-connected battery strings (N_{pb}) can be calculated by equation (5.45).

$$N_{pb} = \frac{N_{Battery}}{N_{sb}} \tag{5.45}$$

If the N_{pb} from (5.45) is not a whole number, then the result should be rounded up to the nearest integer number. As a result, the total number of batteries (N_{TB}) in the battery bank is then determined by multiplying the number of series-connected batteries by the number of parallel-connected batteries as shown in (5.46):

$$N_{TB} = N_{sb} \times N_{pb} \tag{5.46}$$

Example 5.6: Battery Sizing for a Small House Powered By Off-Grid PV Systems

The electricity requirement of a small house in a rural area is planned to be met by a stand-alone PV system. The average daily energy demand of this house is estimated to be 20 kWh/day approximately. Suppose that deep-cycle gel batteries, of which specifications are given below, will be used as an energy storage system. Due to the cold winter months, battery storage conditions may drop down to the temperature of −20°C (Table 5.18).

If the designed system DC voltage is 48 V and the number of autonomy days (DoA) is 3, calculate the battery storage size and draw the PV system diagram with battery bank wiring.

Solution

Let us follow the steps described above regarding the battery storage sizing for stand-alone PV systems. For this purpose, the following table showing all the calculation steps can be created below (Table 5.19):

Considering the battery storage sizing in Table 5.19, the PV system wiring diagram with a detailed battery bank can be drawn as follows. As previously determined, the battery bank consists of 12-volt battery units (Figure 5.79).

TABLE 5.18
Battery Parameters

Basic Specifications of the VRLA Gel Battery	
Rated Ah capacity	100 Ah
Nominal battery voltage	12 Volt DC
Temperature range	−30°C to +50°C
Round trip efficiency	85%

TABLE 5.19
Calculation Steps for Battery Sizing

	Explanation	Calculation and Results
Step-1	Battery type and specifications	100 Ah, 12 V gel-type deep-cycle lead-acid battery.
Step-2	The average daily energy demand of the load [kWh]	It is given as 20 kWh/day
Step-3	Total energy requirement [kWh]	From (5.38), the amount of total energy needed (E_{Total}) is: $$E_{Total}[Wh] = E_{Daily\,Avg} \times DoA = 20 \times 3 = 60 \text{ kWh}$$
Step-4	Battery bank energy storage requirement in [Ah]	To calculate battery bank energy storage requirement, the value DoD_{max} must be found firstly for $-20°C$ storage conditions. So from (5.31) we have: $$DoD_{max}[\%] = 0.0275 \times (-20)^2 + 3.817 \times (-20) + 128.35 = 63\%$$ Now, from (5.39) and (5.40) we have: $$E_{Battery}[Wh] = \frac{E_{Total}[Wh]}{DoD_{max}} = \frac{60 \text{ kWh}}{0.63} = 95.24 \text{ kWh}$$ $$C_{Battery}[Ah] = \frac{E_{Battery}[Wh]}{V_b} = \frac{95240 \text{ Wh}}{12 \text{ V}} = 7937 \text{ Ah}$$
Step-5	Net capacity requirement of the battery	Firstly, we need to calculate the battery temperature coefficient for three days of autonomy (discharge period is: 3 days = 72 hours). So from (5.37): $$f_{Temp-b}@(C/72\text{ h}) = -0.0087 \times (-20)^2 + 0.6346 \times (-20) + 103.85 = 87.7\%$$ Round-trip efficiency is also given as $\eta_b = 0.85$, So from (5.41), the net storage requirement in [Ah] is: $$C_{Battery-net}[Ah] = \frac{C_{Battery}[Ah]}{\eta_b \times f_{Temp-b}} = \frac{7937}{0.85 \times 0.877} = 10648 \text{ Ah}$$
Step-6 Step-7	Final Battery sizing calculations	From (5.43), the total number of batteries: $$N_{Battery} = \frac{C_{Battery-net}[Ah]}{c_b[Ah]} = \frac{10648}{100} = 106.48 \approx 107$$ In a nominal system, DC voltage is given as $V_{DC} = 48\,V$, the number of the series-connected battery from (5.44) is: $$N_{sb} = \frac{V_{DC}}{V_b} = \frac{48}{12} = 4$$ The number of parallel-connected battery strings is calculated from (5.45) as: $$N_{pb} = \frac{N_{Battery}}{N_{sb}} = \frac{107}{4} = 26.75 \approx 27$$ As a result, the final battery bank configuration is: $$N_{TB} = N_{sb} \times N_{pb} = 4s \times 27p$$

Photovoltaic Power Systems

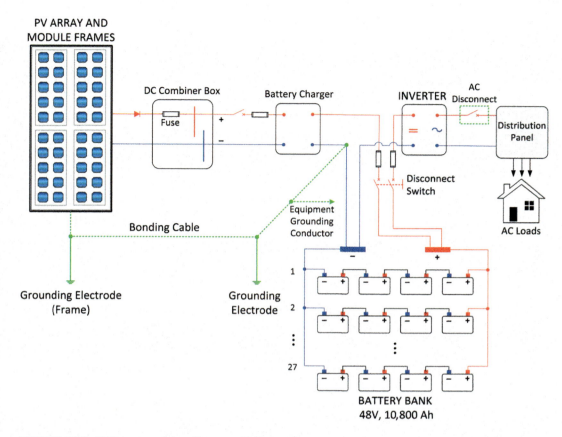

FIGURE 5.79 PV system wiring diagram with battery bank.

Example 5.7: Off-Grid Battery Sizing for a LED-Based Lighting Systems

For security purposes, an off-grid PV system will be designed to feed an LED-based garden lighting system. The energy needed by the lighting system is planned to be supplied by the battery system during the night. The lighting system is planned to consist of 3×20 W light source. The required battery storage sizing will be made for the worst-case scenario, that is, winter months when the temperature falls to −15°C and the nighttime is 14 hours. The other design parameters are given as follows:

- 12 V, 100 Ah deep-cycle lead-acid batteries will be used for energy storage.
- The round-trip efficiency is 85% for the selected battery.
- The predicted DoA value is 3.

If the designed system DC voltage is 24 V, calculate the battery storage size and draw the PV system diagram with battery bank wiring.

Solution

Let us follow the steps described above regarding the battery storage sizing for stand-alone PV systems. For this purpose, the following table showing all the calculation steps can be created below (Table 5.20):

Considering the battery storage sizing in Table 5.20, the PV system wiring diagram with a detailed battery bank can be drawn as follows. Note that a light-sensing photodiode is used to control the lamps (Figure 5.80).

TABLE 5.20
Calculation Steps for Battery Sizing

	Explanation	Calculation and Results
Step-1	Battery type and specifications	100 Ah, 12 V gel-type deep-cycle lead-acid battery.
Step-2	The average daily energy demand of the load [kWh]	It can be calculated as: $E_{Daily}[Wh/day] = 60\ W \times 14\ h/day = 840\ Wh/day$
Step-3	Total energy requirement [kWh]	From (5.38), the amount of total energy needed (E_{Total}) is: $E_{Total}[Wh] = E_{Daily\ Avg} \times DoA = 840 \times 3 = 2520\ Wh$
Step-4	Battery bank energy storage requirement in [Ah]	To calculate battery bank energy storage requirement, the value DoD_{max} must be found firstly for −10°C storage conditions. So from (5.31) we have: $DoD_{max}[\%] = 0.0275 \times (-15)^2 + 3.817 \times (-15) + 128.35 \approx 77.3\%$ Now, from (5.39) and (5.40) we have: $E_{Battery}[Wh] = \dfrac{E_{Total}[Wh]}{DoD_{max}} = \dfrac{2520\ Wh}{0.773} = 3260\ Wh$ $C_{Battery}[Ah] = \dfrac{E_{Battery}[Wh]}{V_b} = \dfrac{3260\ Wh}{12\ V} = 271.66\ Ah$
Step-5	Net capacity requirement of the battery	Firstly, we need to calculate the battery temperature coefficient for three days of autonomy (discharge period is: 3 days = 72 hours). So from (5.37): $f_{Temp-b}@(C/72\ h) = -0.0087 \times (-15)^2 + 0.6346 \times (-15) + 103.85 = 92.3\%$ Round-trip efficiency is also given as $\eta_b = 0.85$, So from (5.41), the net storage requirement in [Ah] is: $C_{Battery-net}[Ah] = \dfrac{C_{Battery}[Ah]}{\eta_b \times f_{Temp-b}} = \dfrac{271.66}{0.85 \times 0.923} = 346.2\ Ah$
Step-6 Step-7	Final Battery sizing calculations	From (5.43), the total number of batteries: $N_{Battery} = \dfrac{C_{Battery-net}[Ah]}{c_b[Ah]} = \dfrac{346.2}{100} = 3.46 \approx 4$ Since nominal DC voltage system is given as $V_{DC} = 24\ V$, the number of the series-connected battery from (5.44) is: $N_{sb} = \dfrac{V_{DC}}{V_b} = \dfrac{24}{12} = 2$ The number of parallel-connected battery strings is calculated from (5.45) as: $N_{pb} = \dfrac{N_{Battery}}{N_{sb}} = \dfrac{4}{2} = 2$ As a result, the final battery bank configuration is: $N_{TB} = N_{sb} \times N_{pb} = 2s \times 2p$

Example 5.8: Battery Sizing for an Off-Grid PV Systems to Feed Flashing Warning Sphere

An off-grid PV system will be designed to feed a flashing aircraft warning sphere, which is used in overhead transmission lines. The flashing system, which is made up of LED-based lighting, is planned to consist of 8 × 5 W bulbs and will operate only during the night times. It is also planned that the flash cycle per minute is 60 and the LEDs run for 20 ms (milliseconds) during each blinking. Also, assume that the lighting system in a warning sphere will work for an average of approximately 12 hours a night. As known, the energy needed by the lighting system will be supplied

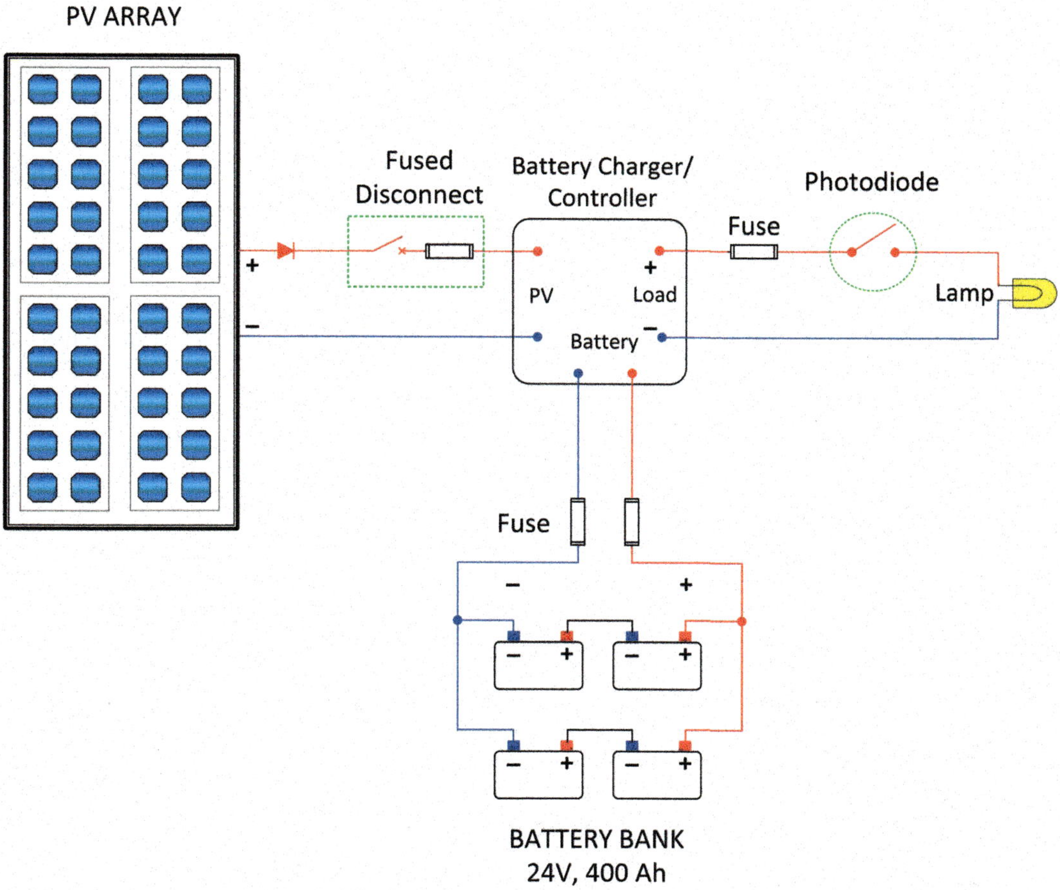

Photovoltaic Power Systems

PV ARRAY

FIGURE 5.80 PV system wiring diagram.

from the lithium iron phosphate (LFP) based battery during the night times. As for the daytime periods, an *80 W$_p$* flexible mono-crystalline PV module mounted on the surface of the sphere will charge the battery. The required battery storage sizing will be made based on the following design parameters:

- The recommended days of autonomy (DoA) is 7.
- The maximum depth of discharge (DoD) is 80%.
- The round-trip efficiency for the LFP battery is 98%.
- Temperature coefficient ($f_{\text{Temp-b}}$) for the LFP battery is 1.

If the storage capacity of the selected LFP battery is 8 Ah and its terminal voltage is 6 V, calculate the battery storage size and its connection configuration based on the 12 V$_{DC}$ system voltage.

Solution

Firstly, let's calculate the daily power consumption of LED lamps. As given in the design criteria, the LED lamps will flash 40 times per minute. In this case, the duty cycle value for the LED lamps is found as $(60 \times 0.2 \text{ s})/60 \text{ s} = 0.2$. As known, the lamps will only work at night times. In the example, the average nighttime is given as 12 hours. Accordingly, the average daily consumption of LED lamps will be:

$$E_{\text{Daily Avg}} \left[\text{Wh/day}\right] = 40 \text{ W} \times 0.2 \times 12 = 96 \text{ Wh/day}$$

Net battery storage requirement can then be obtained from (5.42) as below:

$$C_{\text{Battery-net}}[\text{Ah}] = \frac{E_{\text{Daily Avg}}[\text{Wh/day}] \times \text{DoA}[\text{day}]}{\text{DoD}_{\max} \times V_b \times \eta_b \times f_{\text{Temp-b}}} = \frac{96 \times 7}{0.80 \times 12 \times 0.98 \times 1} = 72 \text{ Ah}$$

From (5.43), the total number of batteries:

$$N_{\text{Battery}} = \frac{C_{\text{Battery-net}}[\text{Ah}]}{c_b[\text{Ah}]} = \frac{72}{8} = 9$$

Since the nominal DC voltage of the system is given as $V_{\text{DC}} = 12$ V, the number of the series-connected battery from (5.44) is:

$$N_{\text{sb}} = \frac{V_{\text{DC}}}{V_b} = \frac{12}{6} = 2$$

The number of parallel-connected battery strings is calculated from (5.45) as:

$$N_{\text{pb}} = \frac{N_{\text{Battery}}}{N_{\text{sb}}} = \frac{9}{2} = 4.5 \approx 5$$

As a result, the final battery bank configuration is:

$$N_{\text{TB}} = N_{\text{sb}} \times N_{\text{pb}} = 2s \times 5p$$

5.5.5 Charge-Controller Sizing and Selection

The function of a battery charger is to control the current flow within the specified limits of operating voltage. The photovoltaic charge controllers are typically rated and sized against PV array current and system voltage. Therefore, choosing and sizing the most suitable controller can be done in two steps.

Step-1 (Voltage Selection): The solar charge controller should be selected to match the voltage of the PV array and batteries. Besides, all battery chargers as specified in their datasheet have an upper voltage limit, which refers to the highest possible voltage value of the solar PV array. Hence, one needs to make sure about the upper voltage limit of the charge controller so that it doesn't exceed the maximum PV system voltage. Therefore, it is necessary to verify that the maximum open-circuit voltage ($V_{\text{oc max PV}}$) at the output of the PV array should be less than or equal to the maximum input voltage withstood by the charge controller ($V_{\text{max charger}}$).

$$V_{\text{oc max PV}} \leq V_{\text{max charger}} \tag{5.47}$$

Remember that equation (5.21) can be used to calculate the value of $V_{\text{oc max PV}}$.

Step-2 (Current Capacity): The solar charge controller is selected to withstand the maximum array current and maximum load current. The maximum array current is equal to the multiplication of a string (or a module) short-circuit current (I_{SC}) and number of parallel strings (N_p). As shown in (5.48), the total current value (I_T) to be referenced in the sizing of the chargers can be found by multiplying the total short-circuit currents of PV strings (or modules) connected in parallel by a safety factor (f_{safe}).

$$I_T = N_p \times I_{\text{SC}} \times f_{\text{safe}} \tag{5.48}$$

Photovoltaic Power Systems

The safety factor in (5.48) is usually taken as $f_{safe} = 1.25$. In some cases, a higher safety factor can also be used due to the addition of new equipment. Depending on the amount of current coming from the PV system, it may be necessary to use more than one charge controller connected in parallel. In this case, the total number of charge controllers ($N_{Charger}$) to be used in the system is found by dividing the maximum current of the total PV array by the amperage rating of the selected charger, $I_{Charger-selected}$ (equation 5.49).

$$N_{Charger} = \frac{I_T}{I_{Charger-selected}} \qquad (5.49)$$

If the result from (5.49) is not a whole number, then the value $N_{Charger}$ should be rounded up to the nearest integer number.

The selection of battery charge controllers in PV systems is of critical importance. Because a proper charge controller may increase the battery's performance and give them as long a life as possible. At the same time, a good charge controller provides more efficient use of PV system energy. In general, there are two types of commercially available controllers:

- Pulse width modulation (PWM)-based charge controller
- Maximum power point tracking (MPPT)-based charge controller

The main differences between these two types of controllers can be given as follows:

- MPPT-based controls are more efficient than PWM-based ones. For example, MPPT-based charge controllers, which are common in current applications, can provide up to 30% more power than PWM controllers.
- The MPPT controllers may allow the PV panels to be connected in series for higher voltages. Thus, the conductor current can be reduced while keeping the wire size smaller. This will be a more economical solution, especially for PV systems that require longer cables.

Table 5.21 summarizes the basic comparisons of PWM and MPPT chargers for use in solar PV systems. In addition, Tables 5.22 and 5.23 specify the main technical parameter of different PWM and MPPT charge controllers for use in solar PV systems. These sample tables (5.22 and 5.23) have been produced following the catalogs of existing commercial products. As can be seen that the current and voltage limits of chargers are summarized so that one can use them as a quick guide for sizing and selection purposes.

TABLE 5.21
Comparison of PWM Chargers and MPPT Chargers (PWM versus MPPT)

PWM-Based Charge Controller	MPPT-Based Charge Controller
PV array and battery voltages must comply with each other.	Array voltage must be higher than the battery voltage. PV array voltage can be even much higher than the battery voltage.
They are generally rated for 12, 24, and 48 volts.	They have a wider input voltage range: 48–600 V_{DC}
They have a lower ampere capacity (up to 60 A).	They have a higher ampere capacity (up to 100 A).
They are inexpensive and compact.	They are more expensive and physically larger.
They are typically recommended for use in a smaller system where boost benefits are minimal.	They can be recommended for 200 W or higher powers to take the advantage of boost benefits more.
They have relatively lower efficiency.	They are more efficient up to 30%.

TABLE 5.22
Technical Specifications of Six Different PWM Chargers for Use in Solar PV Systems

	PWM-Based Charge Controllers—Technical Specifications					
Parameters	Model-1	Model-2	Model-3	Model-4	Model-5	Model-6
Nominal system voltage (battery voltage)	12 V/24 V auto-adapt	12 V/24 V auto-adapt	12 V/24 V auto-adapt	12 V/24 V auto-adapt	12 V/24 V auto-adapt	48 V
Rated charge current (maximum charge current)	5 A	10 A	20 A	30 A	40 A	40 A
Rated load current (maximum output current)	5 A	10 A	20 A	30 A	40 A	40 A
Maximum PV input voltage	<41 V	<41 V	42 V	42 V	47 V	100 V
Maximum PV input power	60 W(12 V) 120 W(24 V)	120 W(12 V) 240 W(24 V)	240 W(12 V) 480 W(24 V)	360 W(12 V) 720 W(24 V)	480 W(12 V) 960 W(24 V)	1920 W

TABLE 5.23
Technical Specifications of Different MPPT Chargers for Use in Solar PV Systems

	MPPT-Based Charge Controllers—Technical Specifications					
Parameters	Model-1	Model-2	Model-3	Model-4	Model-5	Model-6
Nominal system voltage (battery voltage)	12 V/24 V auto-adapt	12/24 V auto-adapt	12/24/36/48 V auto-adapt	12 V/24 V auto-adapt	12/24/48 V auto-adapt	12/24/48 V auto-adapt
Battery input voltage range (MPPT voltage range)	8~32 V	9~15 V @ 12 V 18~30 V @ 24 V	9~15 V @ 12 V 18~30 V @ 24 V 32~40 V @ 36 V 42~60 V @ 48 V	9~15 V @ 12 V 18~30 V @ 24 V	9~15 V @ 12 V 18~30 V @ 24 V 40~67 V @ 48 V	
Maximum PV input voltage	100 V	150 V	150 V	100 V @ 12 V 145 V @ 24 V	150 V	150 V
PV array MPPT voltage range	16~100 V	15~100 V @ 12 V 33~130 V @ 24 V	15~100 V @ 12 V 33~130 V @ 24 V	16~100 V @ 12 V 32~130 V @ 24 V 46~130 V @ 36 V 60~130 V @ 48 V	15~100 V @ 12 V 34~100 V @ 24 V 60~100 V @ 48 V	
Maximum PV input power	130 W(12 V) 260 W(24 V)	260 W(12 V) 520 W(24 V) 1040 W(48 V)	390 W(12 V) 780 W(24 V) 1170 W(36 V) 1560 W(48 V)	600 W(12 V) 1200 W(24 V)	800 W(12 V) 1700 W(24 V) 3400 W(48 V)	1200 W(12 V) 2300 W(24 V) 4600 W(48 V)
Rated charge current (maximum charge current)	10 A	20 A	30 A	40 A	60 A	80 A
Rated load current (maximum output current)	10 A	20 A	30 A	20 A	60 A	60 A

Example 5.9: Charge-Controller Sizing for Different Off-Grid PV Systems

Off-grid PV systems in different configurations will be designed by using the 6 V, 300 Ah batteries and PV modules of which parameters are given in Table 5.24.

a. Design a 12 V off-grid system with four PV modules. Assume that a total of 4 batteries will be used for storage needs. If the maximum load current to be drawn at the DC side is 10 A, specify the required charge-controller size and select the possible charger types based on the technical data given in Tables 5.22 and 5.23.

Photovoltaic Power Systems

b. Design a 24 V off-grid system with four PV modules. Again assume that a total of 4 batteries will be used for storage needs. If the maximum load current to be drawn at the DC side is 13 A, specify the required charge-controller size and select the possible charger types based on the technical data given in Tables 5.22 and 5.24.

Solution

a. Firstly, let us design a 12V off-grid system with four PV modules. In this design, the PV modules will be connected as $1s \times 4p$ and the battery bank configuration will be $2s \times 2p$ for a 12 V_{DC} system. Consequently, the basic system design can be drawn as below (Figure 5.81).

As previously explained, the charge controllers are sized for the voltage and current requirements of the system. As can be seen from the system drawing above, the maximum total PV array current can be as high as:

$$I_T = N_p \times I_{SC} \times f_{safe} = 4 \times 3.7 \times 1.25 = 18.5 \text{ A}$$

Secondly, the maximum PV array input voltage is 14.6 V (the temperature effect is ignored). Hence, from Table 5.22, the PWM-based charge controller given by Model-3 with 12 V and 20 A ratings can be selected for this design.

TABLE 5.24
PV Module Parameters

PV Module Specifications: 40 Wp Poly Crystalline Module			
Voltage at P_{max}	Current at P_{max}	Open Circuit Voltage	Short-Circuit Current
12 V	3.4 A	14.6 V	3.7 A

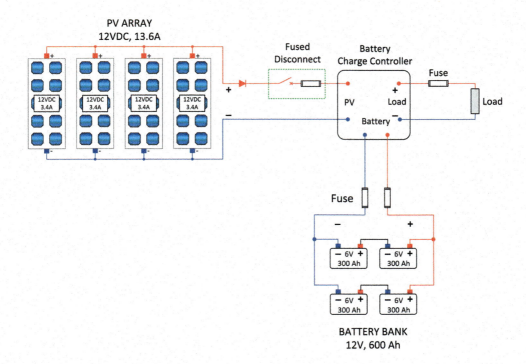

FIGURE 5.81 PV system design.

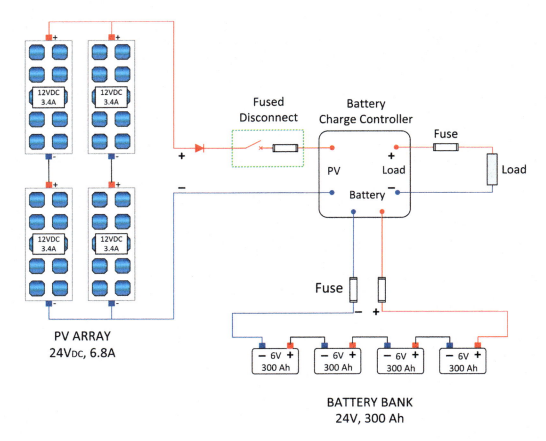

FIGURE 5.82 PV system design.

b. Firstly, let us design a 24 V off-grid system with four PV modules. In this design, the PV modules will be connected as $2s \times 2p$ and the battery bank configuration will be $4s \times 1p$ for a 24 V_{DC} system. Accordingly, the basic system design can be drawn as below (Figure 5.82).

As previously explained, the charge controllers are sized for the voltage and current requirements of the system. As can be seen from the system drawing above, the maximum total PV array current can be as high as:

$$I_T = N_p \times I_{SC} \times f_{safe} = 2 \times 3.7 \times 1.25 = 9.25 \text{ A}$$

Secondly, the maximum PV array input voltage is 29.2 V (the temperature effect is ignored). Hence, from Table 5.22, the PWM-based charge controller given by Model-2 with 12/24 V and 20 A ratings can be selected for this design. Besides, the MPPT-based charge controller given by Model-1 with 12/24 V and 10 A ratings can be selected as well from Table 5.23.

Example 5.10: Charge-Controller Sizing for an Off-Grid PV System

A 48 V off-grid PV system will be designed by using the 12 V, 360 Ah batteries and eight PV modules of which parameters are given in Table 5.25.

Assume that a total of 8 batteries will be used for storage needs. If the maximum load current to be drawn at the DC side is 40 A, specify the required charge-controller size and select the possible charger types based on the technical data given in Tables 5.22 and 5.23.

Solution

Firstly, let us design a 48 V off-grid system with eight PV modules. In this design, the PV modules will be connected as $2s \times 4p$ and the battery bank configuration will be $4s \times 2p$ for a 48 VDC system. Consequently, the basic system design can be drawn as below (Figure 5.83).

As previously explained, the charge controllers are sized for the voltage and current requirements of the system. As can be seen from the system drawing above, the maximum total PV array current can be as high as:

$$I_T = N_p \times I_{SC} \times f_{safe} = 4 \times 10.5 \times 1.25 = 52.5 \text{ A}$$

Secondly, the maximum PV array input voltage is 56.4 V (the temperature effect is ignored). Hence, from Table 5.23, the MPPT-based charge controller given by Model-5 with 48 V and 60 A ratings can be selected for this design.

Alternatively, the MPPT charger given by Model-3 with 48 V and 30 A can be selected from Table 5.23. However, the rated current of the Model-3 charger is smaller than the total PV array current. So, from equation (5.48), we have:

$$N_{Charger} = \frac{I_T}{I_{Charger\text{-}selected}} = \frac{52.5 \text{ A}}{30 \text{ A}} = 1.75 \approx 2$$

Consequently, two parallel-connected PV chargers from Model-3 can also be used for this design.

TABLE 5.25
PV Module Parameters

PV Module Specifications: 40 Wp Poly Crystalline Module			
Voltage at P_{max}	**Current at P_{max}**	**Open Circuit Voltage**	**Short-Circuit Current**
24 V	10.5 A	28.2 V	11.8 A

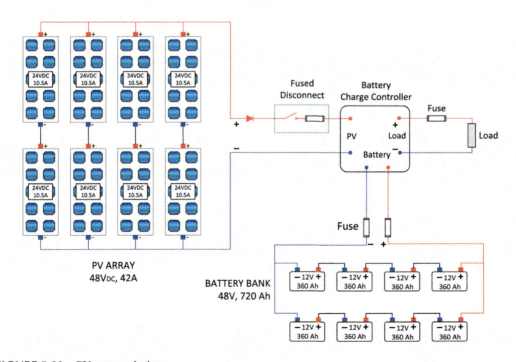

FIGURE 5.83 PV system design.

5.5.6 Inverter Sizing and Selection

The inverter is the key junction between PV array and load in off-grid systems, and between PV array and grid in on-grid systems. Therefore, there are common and different aspects in the inverter sizing and specifying criteria for both on-grid and off-grid systems. For that reason, it would be more appropriate to consider the inverter sizing and selection requirements for on-grid and off-grid systems separately. Except for on-grid and off-grid systems, the inverter selection and sizing procedure for hybrid systems should also be addressed as a third option separately.

5.5.6.1 Inverter Sizing and Selection for Off-Grid Systems

The specification and selection of an inverter in an off-grid PV system are dependent on the current, voltage, and power limitations of the inverter at the input and output. Besides, other factors such as phase type (three-phase or single-phase) and topology type (single inverter, string inverter, multi-string inverter, central inverter, etc.) are the secondary criteria that should be taken into account when choosing an inverter. In this perspective, the general off-grid inverter sizing and specifying steps are given below:

Step 1 – Identifying Load Requirements and Matching Inverter Power Ratings: Firstly, identify all the AC loads that may run simultaneously and sum up their powers to find total AC power ($P_{\text{Total AC Load}}$) in [Watts]. This total AC power determines the minimum continuous power rating limit of the inverter. As a consequence, the following condition (equation 5.50) for the continuous power rating ($P_{\text{inv continuous}}$) of the inverter must be met.

$$P_{\text{inv continuous}} \geq P_{\text{Total AC Load}} \tag{5.50}$$

Off-grid inverters should not be operated continuously at their limits. Besides, possible load increases in the future should also be taken into account. Hence, the nominal power of the inverter can typically be selected as 20%–25% larger than the total continuous AC load.

$$P_{\text{inv rated}} = 1.25 \times \left(P_{\text{Total AC Load}} \right) \tag{5.51}$$

Secondly, if the system includes loads with high start-up (surge) currents such as pumps, compressors, large motors, and other appliances with large inductances, these loads in apparent powers ($S_{\text{Surge Load}}$ [VA]) should also be taken into consideration when selecting the inverter. Among these loads, the largest surge load determines the minimum inverter surge rating, which is given by (5.52).

$$S_{\text{inv surge}} [\text{VA}] \geq \max \left(S_{\text{Surge Load-1}}, S_{\text{Surge Load-2}}, \ldots \right) \tag{5.52}$$

Finally, if your distribution system is based on a 120/240 V secondary system (North American Layout System), then you need to identify any loads that require 240V AC. In this case, it is necessary to use either a 120/240 V transformer (inverter plus autotransformer) or a dual inverter to get both voltage levels (see Figure 5.84).

Step 2 – Identifying Voltage Requirements: PV module open-circuit voltage (V_{oc}) increases as the ambient temperature decreases. For this reason, the inverter input voltage should be rated for the possible lowest temperature conditions to ensure a safe and productive operation. Therefore, it is necessary to verify that the maximum open-circuit voltage ($V_{oc\,\text{max PV}}$) at the output of the PV array should be less than or equal to the maximum input voltage withstood by the inverter ($V_{\text{max inv}}$).

$$V_{oc\,\text{max PV}} \leq V_{\text{max inv}} \tag{5.53}$$

The minimum temperature value at which the maximum open-circuit voltage obtained is normally site-specific. However, the minimum prospective cell temperature is typically taken as −10°C in most applications. Remember that equation (5.21) is used to calculate the value of $V_{oc\,\text{max PV}}$.

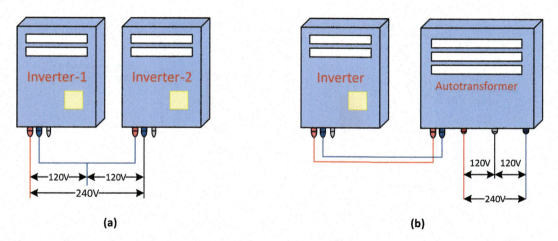

FIGURE 5.84 Generating a 120 / 240 V AC split-phase supply from an inverter (a) inverters operate synchronously and produce 120/240 volts. (b) The system uses a 240 V inverter and an autotransformer to produce 120/240 volts.

Another requirement to be met regarding the voltage requirements is the matching of the inverter's MPPT voltage window to the minimum and maximum voltage ranges generated at the PV output. As it is known, solar inverters are characterized by specific voltage ranges at the input due to their MPP tracking features. Hence, it is necessary to verify that the inverters operate within the specified MPP voltage range under the cell temperatures of (−10°C to +70°C). As a result, (5.54) and (5.55) conditions must be met together. It should be noted that minimum V_{mp} of the PV module ($V_{mp\,min\,PV}$) occurs at the highest temperature conditions, while maximum V_{mp} ($V_{mp\,max\,PV}$) occurs at the lowest temperature conditions. In addition, remember that equation (5.18) can be used to calculate both $V_{mp\,min\,PV}$ and $V_{mp\,max\,PV}$ values.

$$V_{mp\,min\,PV} \geq V_{MPPT\,min\,inv} \tag{5.54}$$

$$V_{mp\,max\,PV} \leq V_{MPPT\,max\,inv} \tag{5.55}$$

where $V_{MPPT\,min\,inv}$ is the minimum acceptable input voltage of the inverter (lower limit of the inverter's MPPT) and $V_{MPPT\,max\,inv}$ is the maximum acceptable input voltage of the inverter (upper limit of the inverter's MPPT).

Considering the inequalities of (5.53)–(5.55), the possible operating voltage limits of the PV array and the designed input voltage limits at the input of the inverter can be represented in Figure 5.85.

Step 3 – Identifying Current Requirements: it is necessary to ensure that the maximum array current under maximum temperature conditions in the field ($I_{sc}\,@\,T_{cell\,max}$) must be lower than the maximum input current of the inverter ($I_{inv\_max\,DC\,input}$). Importantly note that equation (5.23) can be used to calculate the $I_{sc}\,@\,T_{cell\,max}$. Hence, the condition of (5.56) is needed to be met for the safe operation of an inverter.

$$I_{sc}\,@\,T_{cell\,max} \leq I_{inv\_max\,DC\,input} \tag{5.56}$$

Table 5.26 summarizes the main technical parameter of different off-grid inverters for use in the sizing and selection procedure. The technical data in the table has been produced following the catalogs of existing commercial products.

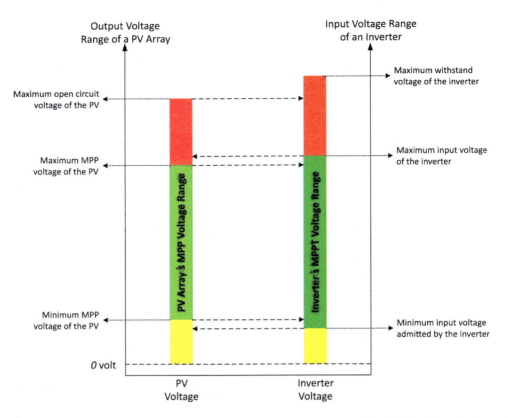

FIGURE 5.85 Comparison of operating voltage range of PV array and designed DC voltage limits at the input of the inverter.

TABLE 5.26
Technical Specifications of Different Off-Grid Inverters for Use in Solar PV Systems

	Off-Grid Inverters—Technical Specifications					
Parameters	Model-1	Model-2	Model-3	Model-4	Model-5	Model-6
Continuous output power (rated power)	1 kW	2 kW	2.4 kW	5 kW	10 kW	20 kW
Surge power (overload capability)	2 kVA	4 kVA	6 kVA	10 kVA	22 kVA	45 kVA
Maximum DC (PV) input current	33 A	20 A	33 A	66 A	50 A	100 A
Maximum DC (PV) input voltage	102 Vdc	75 Vdc	100 Vdc	145 Vdc	600 Vdc	600 Vdc
AC rated output current	4.8 A	9.5 A	11 A	22 A	44 A	88 A
AC nominal output voltage	230 Vac	230 Vac	230 Vac	230 Vac	230 Vac	230 Vac
Output frequency range (auto)	50/60 Hz ±1%	50/60 Hz ±1%	50/60 Hz ±1%	50/60 Hz ±1%	50/60 Hz ±2%	50/60 Hz ±2%
MPPT input voltage range	15–80 Vdc	30–66 Vdc	30–80 Vdc	52–115 Vdc	160–570 Vdc	160–570 Vdc
Battery nominal voltage	12 Vdc	24 Vdc	24 Vdc	48 Vdc	240 Vdc	240 Vdc
Phase type	Single ~	Single ~	Single ~	Single ~	Single ~	Single ~

Example 5.11: Charge-Controller and Off-Grid Inverter Sizing

A 48 V off-grid PV system will be designed by using the 12 V, 200 Ah batteries and eight PV modules, of which parameters are given in Table 5.27.

Assume that a total of 12 batteries will be used for storage needs. If the maximum load current to be drawn at the DC side is 40 A,

Photovoltaic Power Systems

TABLE 5.27
PV Module Parameters

PV Module Specifications: 365Wp Mono-Crystalline Module			
Voltage at P_{max}	Current at P_{max}	Open-Circuit Voltage	Short-Circuit Current
36.7 V	9.95 A	42.8 V	10.8 A

    a. Specify the required charge-controller size and select the possible charger types based on the technical data given in Tables 5.22 and 5.23.
    b. If the continuous power value of the loads is 2200 W and the apparent power of the water pump in the system is 500 VA, specify the required off-grid inverter size and select a suitable inverter from Table 5.26.

Solution

    a. Firstly, let us design a 48 V off-grid system with twelve PV modules. In this design, the PV modules will be connected as $2s \times 4p$ and the battery bank configuration will be $4s \times 3p$ for a 48 V_{DC} system. Since the loads in the system are fed through an inverter, the basic system design can be drawn as below (Figure 5.86).

    We know from the previous explanations that the charge controllers are sized for the voltage and current requirements of the system. As can be seen from the system drawing above, the maximum total PV array current can be as high as:

$$I_T = N_p \times I_{SC} \times f_{safe} = 4 \times 9.95 \times 1.25 = 49.75 \text{ A}$$

As for the maximum PV array input voltage, it is $42.8 \times 2 = 85.6$ V (temperature effect is ignored). Hence, from Table 5.23, the MPPT-based charge controller given by Model-6 with a rated voltage of 48 V and 80 A can be selected for this design. Also, be careful that the battery input range of the selected charger remains within the output voltage interval of the PV array.

    b. Let us follow the three steps described above regarding the inverter sizing and specifying for off-grid PV systems. For this purpose, the following worksheet showing all the steps can be created as below (Table 5.28):

5.5.6.2 Inverter Sizing and Selection for On-Grid Systems

Since the grid-connected PV systems have no battery storage, the generated solar energy is fed directly to the local grid via on-grid inverters. On-grid inverters used in this design can be sized and specified in two ways depending on the project priority. If the inverter type and its size are primarily decided, the PV array is required to be designed to match the selected inverter specifications. As a first consideration, matching the PV array to the specifications of the selected inverter is previously explained in detail under the sub-section of "5.5.1.1.3. PV Array Sizing and Configuration for Grid-Connected Systems." Referring to the relevant part, it will not be repeated here in this section.

The second option regarding the inverter selection is to determine the type and its specifications according to the designed PV array configuration. This second option is explained above under the sub-section of "5.5.6.1. Inverter Sizing and Selection for Off-Grid Systems." The most important difference in sizing on-grid inverters from the off-grid ones is that there is no system load and the generated power is transferred directly to the network. Therefore, the formulas from (5.52) to (5.56) regarding the voltage and current requirements are also valid for specifying the voltage and current limits of on-grid inverters. As for the power requirements, the inverter-rated power must be larger than the maximum PV array power ($P_{max\,PV}$), that is given below.

$$P_{inv\,rated} > P_{max\,PV} \qquad (5.57)$$

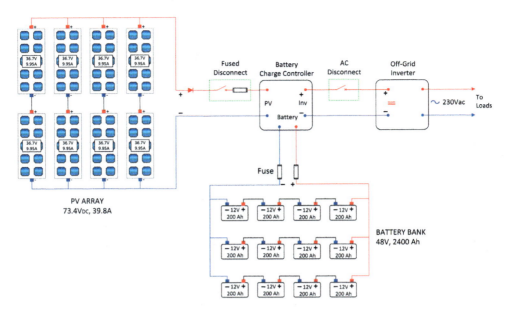

FIGURE 5.86 PV system design.

TABLE 5.28
Calculation Steps for Off-Grid Inverter Sizing

	Explanation	Calculation and Results
Step-1	Identifying load requirements and matching inverter power ratings	Total AC load power is given as $P_{\text{Total AC Load}} = 2200$ W. Hence, the nominal power of the inverter: $P_{\text{inv rated}} = 1.25 \times (P_{\text{Total AC Load}}) = 1.25 \times 2200 = 2750$ W From Table 5.26, we can choose the Model-4 inverter with a rated power of 5 kW as a first step. However, we need to verify the other inequality conditions to reach a final decision. The first condition is about surge loads, which is: $S_{\text{inv surge}}[\text{VA}] \geq S_{\text{Surge Load}} \Rightarrow 6 \text{ kVA} \geq 0.5 \text{ kVA}$ So, the Model-4 inverter is also suitable in terms of surge power requirements.
Step-2	Identifying voltage requirements	It is necessary to verify three voltage conditions: $V_{\text{oc max PV}} \leq V_{\text{max inv}} \Rightarrow 85.6 \text{ V} \leq 145 \text{ V}$ $V_{\text{mp min PV}} \geq V_{\text{MPPT min inv}} \Rightarrow V_{\text{mp min PV}} \geq 52 \text{ V}$ (The MPP output voltage of the PV array will remain above 52 V even under extreme temperature and radiation conditions). $V_{\text{mp max PV}} \leq V_{\text{MPPT max inv}} \Rightarrow 2 \times 36.7 \leq 80 \Rightarrow 73.4 \leq 80$ As can be seen, all three conditions have been met. Note that since the voltage conditions are easily met, the effect of temperature on the PV panel voltage is neglected in this design. Also, be aware that the inverter's battery voltage is 48 Vdc following the given design.
Step-3	Identifying current requirements	As for the current requirements, the following condition must be met. From (a), the highest possible current value was found to be 49.75A. $I_{\text{sc}} @ T_{\text{cell max}} \leq I_{\text{inv max DC input}} \Rightarrow 49.75 \leq 66 \text{ A}$ Satisfying the current requirement, the off-grid inverter labeled **"Model-4"** can be selected from Table 5.26 as a final decision.

Photovoltaic Power Systems

TABLE 5.29
Technical Specifications of Different On-Grid Inverters for Use in Solar PV Systems

Parameters	On-Grid Inverters—Technical Specifications					
	Model-1	Model-2	Model-3	Model-4	Model-5	Model-6
Inverter type	(Multi) String	(Multi) String	String	Central	Central	Central
Grid connection type	3ϕ+Ground	3ϕ+Ground	3ϕ+Ground	3ϕ+Ground	3ϕ+Ground	3ϕ+Ground
Nominal output power	10 kW	20 kW	33 kW	100 kW	500 kW	1000 kW
Nominal grid voltage (AC)	400 V	400 V	400 V	400 V	400 V	400 V
Number of MPPTs	2	2	1	1	1	1
MPPT voltage range (full power)	220–470 V	450–800 V	420–800 V	450–750 V	450–750 V	600–850 V
Max DC input voltage	520 V	1000 V	1000 V	900 V	900 V	1100 V
Max DC input current per MPPT	24 A	25 A	58 A	245 A	1145 A	1710 A
Nominal AC current (three-phase)	14 A	27 A	50.3 A	195 A	965 A	1445 A

Table 5.29 summarizes the main technical parameter of different on-grid inverters for use in the sizing and selection procedure. The technical data in the table has been produced following the catalogs of existing commercial products.

Example 5.12: On-Grid Inverter Sizing

A grid-connected 62 kW$_p$ PV power plant will be designed by using 170 PV modules, of which parameters are given in Table 5.30.

The electrical energy generated from the PV power plant will be transferred to the low-voltage grid through 400 V three-phase system. If the maximum number of series-connected PV modules in a string is 17, select a suitable grid-tie inverter from Table 5.29 and design the final PV array configuration. Finally, draw the system wiring diagram.

Solution

Firstly, let us design the PV array configuration. If the number of series-connected PV modules in a string is 17, then the total number of parallel strings in the array is found to be 170/17 = 10. Hence, the PV array configuration will be $17s \times 10p$. In this case, the PV array power is 170×365 W = 62.05 kW.

From Table 5.29, we can choose the Model-3 inverter with a rated power of 33 kW as a first step. However, we need to use two inverters connected in parallel to meet the 62 kW power requirement. Still, we need to verify the other voltage and current conditions as identified in the above sections.

TABLE 5.30
PV Module Parameters

PV Module Specifications: 365 W$_p$ Mono-Crystalline Module			Module Efficiency: 16.78%		
Voltage at P_{max}	Current at P_{max}	Open-Circuit Voltage	Short-Circuit Current		Power Tolerance
36.7 V	9.95 A	42.8 V	10.8 A		−0 / +3%
Temperature Coefficients					
$\alpha_{I_{sc}}$	$\beta_{V_{mp}}$	$\beta_{V_{oc}}$	P_{max} Coefficient		NOCT
+0.004 A/°C	−0.0757 V/°C	−0.075 V/°C	−0.44 W/°C		45 ± 2°C

By assuming −10°C and +70°C as the minimum and maximum cell temperatures, respectively, the voltage variation of a PV module can be calculated as below:

$$V_{mp} @ T_{cell\_max} = V_{mp,r} + \beta_{V_{mp}} \times (T_{cell\_max} - T_r) = 36.7 - 0.0757 \cdot (70 - 25) = 33.3 \text{ V}$$

$$V_{mp} @ T_{cell\_min} = V_{mp,r} + \beta_{V_{mp}} \times (T_{cell\_min} - T_r) = 36.7 - 0.0757 \cdot (-10 - 25) = 39.35 \text{ V}$$

$$V_{oc\,max} = V_{oc,r} + \beta_{V_{mp}} \times (T_{cell\_min} - T_r) = 42.8 - 0.075 \cdot (-10 - 25) = 45.425 \text{ V}$$

The maximum short-circuit current can be found at $T_{cell\,max}$ (+70°C):

$$I_{sc} @ T_{cell\,max} = I_{scr} + \alpha_{I_{sc}} \times (T_{cell\,max} - T_r) = 10.8 + 0.004 \times (70 - 25) = 10.98 \text{ A}$$

Based on the above calculations, the electrical characteristics of a PV string under extreme temperature conditions can be found as:

$$V_{oc\,max\,PV} = 45.425 \times 17 = 772.2 \text{ V}$$

$$V_{mp\,min\,PV} = 33.3 \times 17 = 566.1 \text{ V}$$

$$V_{mp\,max\,PV} = 39.35 \times 17 = 668.95 \text{ V}$$

From Table 5.29, the voltage specifications of the *Model-3* inverter are:

$$V_{max\,inv} = 1000 \text{ V}$$

$$V_{MPPT\,min\,inv} = 420 \text{ V}$$

$$V_{MPPT\,max\,inv} = 800 \text{ V}$$

Now that we can verify the three voltage conditions:

$$V_{oc\,max\,PV} \leq V_{max\,inv} \Rightarrow 772.2 \text{ V} \leq 1000 \text{ V}$$

$$V_{mp\,min\,PV} \geq V_{MPPT\,min\,inv} \Rightarrow 566.1 \geq 420 \text{ V}$$

$$V_{mp\,max\,PV} \leq V_{MPPT\,max\,inv} \Rightarrow 668.95 \leq 800$$

As can be seen, all three conditions have been met. From the above calculations, the highest possible current of a string is 10.98 A. Due to the 10 parallel-connected strings, the total highest current of the PV array is 10×10.98 A = 109.8 A. Since the maximum current value of the selected 33kW inverter is 58A, the total number of inverters (N_{inv}) can be found by dividing the maximum current of the total PV array by the amperage rating of the selected inverter, $I_{inv-selected}$.

$$N_{inv} = \frac{I_{Total\,PV}}{I_{inv-selected}} = \frac{109.8 \text{ A}}{58 \text{ A}} = 1.89 \xrightarrow{\text{Rounded Up}} \approx 2$$

As a result of calculations, the final configuration is as follows. A total of two string inverters will be used in parallel connection and each inverter will be supplied from $17s \times 5p$ PV array. So, the final system design can be drawn as below (Figure 5.87):

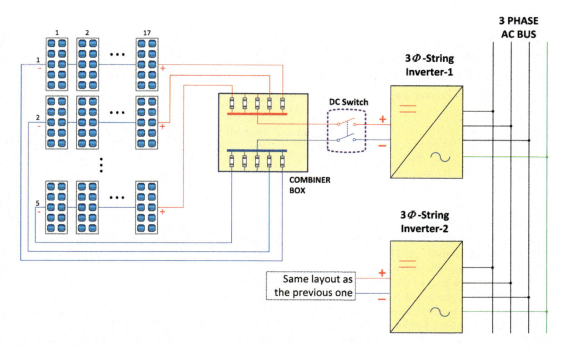

FIGURE 5.87 PV system design.

5.5.6.3 Inverter Sizing and Selection for Hybrid Systems

Due to the operating mode requirements, hybrid inverters must have both on-grid and off-grid inverter features. Therefore, hybrid inverters should meet all the inequality conditions used in both on-grid and off-grid PV inverter sizing. In other words, all the equations from (5.51) to (5.57) can be used in the same way for the sizing and specifying of hybrid inverters. In this context, Table 5.31 summarizes the main technical parameter of different hybrid inverters for use in the sizing and selection procedure. The technical data in the table has been produced following the catalogs of existing commercial products.

TABLE 5.31

Technical Specifications of Different Hybrid Inverters (with Energy Storage) for Use in Solar PV Systems

Parameters	Hybrid Inverters—Technical Specifications				
	Model-1	Model-2	Model-3	Model-4	Model-5
Grid connection type	Single ~	Single ~	3ϕ + Ground	3ϕ + Ground	3ϕ + Ground
Rated output power	3 kW	5 kW	9 kW	10 kW	20 kW
Surge (overload) power	5 kVA	8 kVA	13 kVA	15 kVA	30 kVA
Number of MPPTs	1	2	2	2	1
Max DC (PV) input current (per MPPT)	18 A	14 A per MPPT	23 A per MPPT	23 A per MPPT	92 A
Max DC (PV) input voltage	500 V	600 V	850 V	1000 V	1000 V
MPPT voltage range	80–450 V	125–550 V	380–750 A	330–800 V	330–750 V
AC-rated output voltage	230 V	230 V	400 Vac	400 Vac	400 Vac
AC (max)-rated output current	13.5 A	21.7 A	13 A per Phase	15 A per Phase	30 A per Phase
Battery rated voltage	48 Vdc	96 Vdc	120 Vdc	240 Vdc	240/360 Vdc

TABLE 5.32
PV Module Parameters

PV Module Specifications: 350 W_p Mono-Crystalline Module			Module Efficiency: 16.78 %	
Voltage at P_{max}	Current at P_{max}	Open-Circuit Voltage	Short-Circuit Current	Power Tolerance
36.2 V	9.65 A	42.1 V	10.5 A	−0 / +3%
		Temperature Coefficients		
$\alpha_{I_{sc}}$	$\beta_{V_{mp}}$	$\beta_{V_{oc}}$	P_{max} coefficient	NOCT
+0.004 A/°C	−0.0757 V/°C	−0.075 V/°C	−0.44 W/°C	45 ± 2°C

Example 5.13: Hybrid Inverter Sizing and Specifying

A 10.5 kW_p gird-connected hybrid PV system will be designed by using 12 V, 200 Ah batteries, and 30 PV modules, of which parameters are given in Table 5.32.

The primary goal of the hybrid PV system is to feed local loads. The excess electrical energy from the PV system will be transferred to the low-voltage grid via 400 V, three-phase system. Assume that the continuous power value of the loads is 6000 W, the surge power of the system is negligible and a total of 40 batteries will be used for storage needs.

If the maximum number of series-connected PV modules in a string is 15, design the system DC voltage and battery storage configuration based on the specified design parameters and technical data given in Table 5.31. Also, select a suitable hybrid inverter from Table 5.31. Draw the system wiring diagram considering the final design parameters.

Solution

In this design, the PV modules will be connected as $15s \times 2p$ due to the usage of 30 PV panels in total. Considering the possible use of a 10 kW hybrid inverter from Table 5.31, the battery bank configuration can initially be designed as $20s \times 2p$ for a 240 Vdc system.

Let us follow the steps regarding the sizing and specifying of on-grid and off-grid inverters. For this purpose, the following table showing all the sizing steps can be created below (Table 5.33):

As a result of calculations, the final system configuration can be drawn as below (Figure 5.88):

5.5.7 Protection System Requirements, Sizing, and Selection

Proper sizing and specifying of protection system components for photovoltaic systems are crucial for safe, reliable, and long-term operation. The protection devices for PV systems may need different circuit protection requirements based on the application ranging from residential-scale to grid-connected scales. The protection devices are needed at numerous locations of the PV system both on the DC side and AC side. Wherever they are used, it is necessary to analyze the PV system protection locations in terms of overcurrent, fault current (short-circuit current and leakage current), and overvoltage. Thus, appropriate PV system protection devices can be selected against overvoltage and overcurrent damages. Fuses, disconnect switches, surge protection devices, contactors, circuit breakers (miniature or molded case circuit breakers), and residual current devices are the most commonly used protection devices used in photovoltaic applications. An important note herein is that grounding and lightning protection issues are beyond the scope of this chapter and they will be examined in detail in Chapter 6.

5.5.7.1 Fuse Sizing and Selection for Photovoltaic Systems

The available short-circuit currents are limited in photovoltaic systems. Hence, PV overcurrent protection devices will operate on low levels of fault current. Depending on the installed PV capacity, there may be several parallel-connected PV strings to obtain higher currents and accordingly

TABLE 5.33
Calculation Steps for Hybrid Inverter Sizing

	Explanation	Calculation and Results
Step-1	Identifying load requirements and matching inverter power ratings	Total AC load power is given as $P_{\text{Total AC Load}} = 6000$ W. Hence, the nominal power of the inverter:

$$P_{\text{inv rated}} \geq 1.25 \times \left(P_{\text{Total AC Load}}\right) = 1.25 \times 6000 = 7500 \text{ W}$$

From Table 5.31, we can choose the Model-4 inverter with a rated power of 10 kW as a first step. However, we need to verify the other inequality conditions to reach a final decision. The first condition is surge loads, which are negligible.

Note-1: 9 kW inverter may not be suitable due to the 10.5 kW$_p$ installed capacity of the PV array.

Note-2: The installed power of the PV system is slightly larger than the inverter power (10.5 kW$_p$ > 10 kW). However, the inverter power and PV array power will match each other due to the PV derating factors such as temperature, mismatch, and soiling. Besides, inverters are also designed to carry larger powers than their nominal power. This power is usually denoted as the maximum inverter power in their catalogs.

Step-2	Identifying voltage requirements	By assuming −10°C and +70°C as the minimum and maximum cell temperatures, respectively, the voltage variation of a PV module can be calculated as below:

$$V_{\text{mp}} @ T_{\text{cell\_max}} = V_{\text{mp},r} + \beta_{V_{\text{mp}}} \times \left(T_{\text{cell\_max}} - T_r\right) = 36.2 - 0.0757 \cdot (70 - 25) = 32.8 \text{ V}$$

$$V_{\text{mp}} @ T_{\text{cell\_min}} = V_{\text{mp},r} + \beta_{V_{\text{mp}}} \times \left(T_{\text{cell\_min}} - T_r\right) = 36.2 - 0.0757 \cdot (-10 - 25) = 38.85 \text{ V}$$

$$V_{\text{oc max}} = V_{\text{oc},r} + \beta_{V_{\text{mp}}} \times \left(T_{\text{cell\_min}} - T_r\right) = 42.1 - 0.075 \cdot (-10 - 25) = 44.725 \text{ V}$$

Based on the above calculations, the electrical characteristics of a PV string consisting of 15 serial modules under extreme temperature conditions can be found as:

$$V_{\text{oc max PV}} = 44.725 \times 15 = 670.875 \text{ V}$$

$$V_{\text{mp min PV}} = 32.8 \times 15 = 392 \text{ V}$$

$$V_{\text{mp max PV}} = 38.85 \times 15 = 582.75 \text{ V}$$

Now that we can verify the three voltage conditions:

$$V_{\text{oc max PV}} \leq V_{\text{max inv}} \Rightarrow 670.875 \text{ V} \leq 1000 \text{ V}$$

$$V_{\text{mp min PV}} \geq V_{\text{MPPT min inv}} \Rightarrow 392 \geq 330 \text{ V}$$

$$V_{\text{mp max PV}} \leq V_{\text{MPPT max inv}} \Rightarrow 582.75 \leq 800$$

As can be seen, all three conditions have been met.

Step-3	Identifying current requirements	The maximum possible current value of a PV module can be found at $T_{\text{cell max}}$ (+70°C):

$$I_{\text{sc}} @ T_{\text{cell max}} = I_{\text{scr}} + \alpha_{I_{\text{sc}}} \times \left(T_{\text{cell max}} - T_r\right) = 10.5 + 0.004 \times (70 - 25) = 10.68 \text{ A}$$

Since the inverter input has 2 MPPTs, each PV string will be connected to a separate MPPT unit. So, we need to verify that the maximum input current of each MPPT should be larger than that of the PV string.

$$I_{\text{sc}} @ T_{\text{cell max}} \leq I_{\text{inv max DC input}} \Rightarrow 10.68 \leq 23 \text{ A}$$

Satisfying the current requirement, the hybrid inverter labeled "**Model-4**" can be selected from Table 5.31 as a final decision.

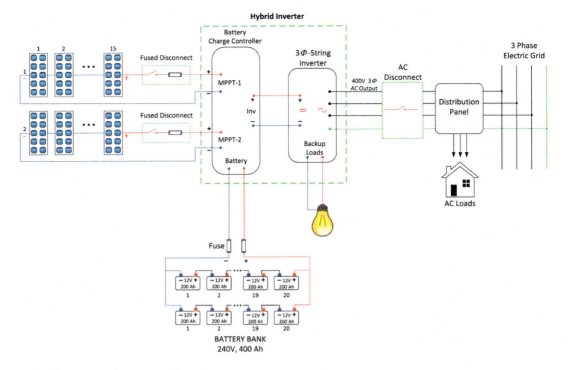

FIGURE 5.88 PV system configuration.

more power. PV systems with single or two strings in parallel will not produce sufficient fault current to damage conductors/equipment as long as they are sized properly (in this case, the cable should be rated at $1.56 \times I_{sc}$). If the cable is not rated for $1.56 \times I_{sc}$, a fuse must be selected to protect the cable (in this case, the cable should be rated as: "Fuse Current Rating < Cable Ampacity"). As a result, the fusing requirement depends on the above-mentioned conditions for PV systems that have less than three strings connected in parallel. For PV systems with three or more strings, the following steps can be used for fuse sizing in photovoltaic circuits.

Step 1 – Determine Maximum PV Circuit Current: According to national electric codes, the maximum circuit current is defined as 125% of the short-circuit current of the PV module (I_{sc}). If there are two or more parallel PV sources in the system, then the maximum circuit current (I_{max}) can be calculated as 1.25 multiplied by the sum of all short-circuit currents coming from ($N_p - 1$) strings of a PV array with N_p parallel branch. It is expressed mathematically as (5.57).

$$I_{max} = 1.25 \times \left[I_{sc1} + I_{sc2} \cdots I_{sc,\,(N_p-1)} \right] = 1.25 \times \sum_{i=1}^{N_p-1} I_{sc,\,i} \tag{5.57}$$

where $I_{sc,\,i}$ is the short-circuit current of i th PV string and N_p represents the total number of parallel-connected PV strings in an array. It is useful to examine Figure 5.89, which identifies the analysis of the PV array system to find necessary data for overcurrent protection.

It is clear from the figure that the total short-circuit current value on a faulted string can be found by multiplying the short-circuit current of a PV string by ($N_p - 1$). More specifically, total short-circuit currents for the points of Fault-1 and Fault-2 locations can be calculated as (5.58) and (5.59), respectively. Note that the short-circuit currents of $I_{sc,\,string}$ and $I_{sc,\,group}$ are specified in Figure 5.89.

$$I_{sc,\,total} @ \text{Fault1} = (N_p - 1) \times I_{sc,\,string} + (N_c - 1) \times I_{sc,\,group} \tag{5.58}$$

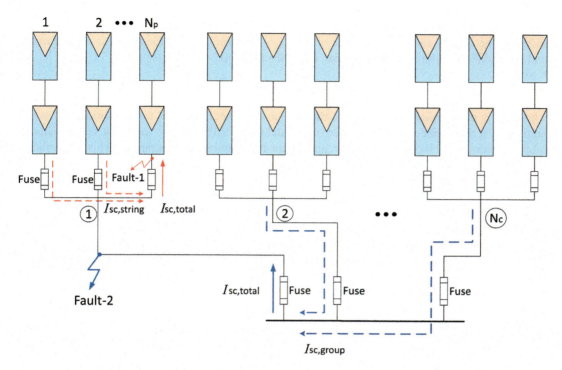

FIGURE 5.89 Analysis of PV array system to find necessary data for overcurrent protection.

$$I_{sc,\,total} @ \text{Fault} 2 = (N_c - 1) \times I_{sc,\,group} \tag{5.59}$$

Step 2 – Determine the Fuse Ampere Rating: According to national electric codes, overcurrent protection devices are needed to be sized to carry current not less than 125% of the maximum circuit current (I_{max}) due to safety reasons. Therefore, nominal fuse ampere rating (I_n) can be calculated by multiplying I_{max} (found in Step-1) by a safety factor of 1.25 as below:

$$I_n = 1.25 \times (I_{max}) = 1.56 \times \left[I_{sc1} + I_{sc2} \cdots I_{sc,\,(N_p-1)} \right] = 1.56 \times (I_{sc\,Total}) \tag{5.60}$$

The coefficient of 1.56 in (5.60) is taken as 1.4 in the IEC 60269-6 fuse standard. If the "IEC fuse standard" is considered as the reference code, then equation (5.60) is rearranged as (5.61).

$$I_n \geq 1.4 \times (I_{sc\,Total}) \tag{5.61}$$

As a result, equation (5.62) can be written by arranging (5.60) and (5.61) together.

$$I_n \geq \begin{cases} 1.56 \times (I_{sc\,Total}), & \text{for NEC standard} \\ 1.4 \times (I_{sc\,Total}), & \text{for IEC standard} \end{cases} \tag{5.62}$$

Step 3 – Determine the Fuse Voltage Rating: It is necessary to verify that the nominal fuse voltage (V_n) should be equal to or greater than the maximum open-circuit voltage of the PV array ($V_{oc\,max\,PV}$), that is given below.

$$V_n \geq V_{oc\,max\,PV} \tag{5.63}$$

The minimum cell temperature is typically taken as −10°C for calculating $V_{oc\,max\,PV}$ and equation (5.21) can be used to calculate the value of $V_{oc\,max\,PV}$. However, a safety factor of 1.2 is satisfactory and commonly used for most applications, as shown in (5.64).

$$V_n \geq 1.2 \times V_{oc} \times N_s \tag{5.64}$$

where V_{oc} is the open-circuit voltage of the PV module under standard test conditions, N_s is the number of series-connected modules per string, and the coefficient 1.2 is used as a safety factor to take into account the increased open-circuit voltage at low ambient temperatures. This safety factor may further be increased for cold climate conditions.

Step 4 – Modify Fuse Rating for Abnormal Temperature Conditions (If required): Fuse ampere ratings are determined under 25°C standard test conditions. As known PV fuses are thermal devices. Hence, if they are exposed to temperatures above 40°C for a long time (more than 2 hours), fuse trip time (or fuse element melting time) may be shortened. Conversely, if the fuses are exposed to ambient temperatures less than 0°C may delay the trip time. Therefore, a temperature correction factor shall be used under prolonged abnormal temperature conditions. Note that if the fuses are in shade and covered against direct sunlight, a temperature correction coefficient for the fuse (f_{temp}) may not need to be applied. In this case, the corrected fuse current ($I_{corrected}$) is calculated as (5.65).

$$I_{corrected} = \left(\frac{1}{f_{temp}}\right) \times I_n \tag{5.65}$$

A typical temperature correction coefficient for PV fuses is given by (5.66). Note that for more accurate heat correction coefficients, one should look for the temperature correction chart provided by the fuse manufacturers.

$$f_{temp} = -0.004 \times T + 1.08 \tag{5.66}$$

where T is the ambient temperature in Celsius degree for prolonged abnormal temperature conditions (typically more than two hours under 0°C or above 40°C).

Step 5 – Determine Final Fuse Ampere Rating: If the corrected ampere rating of the fuse ($I_{corrected}$) is not exactly available among the standard fuse ratings, the next available rating is needed to be selected. For example, if the calculated fuse rating is 18 A, then a 20 A fuse should be used as the next higher ampere rating. Note that the standard fuse ampere ratings for PV applications are as follows: 1 – 2 – 3 – 4 – 5 – 6 – 7 – 8 – 10 – 12 – 15 – 20 – 25 – 30 – 35 – 40 – 45 – 50 – 60 – 70 – 80 – 90 – 100 – 110 – 125 – 150 – 175 – 200 – 225 – 250 – 300 – 350 – 400 – 450 – 500 and 600 A.

Step 6 – Verify if the Fuse Rating Match to Conductor Ampacity: As identified in (5.67), the nominal fuse current (I_n) found in Step-5 must not exceed the ampacity of the selected conductor (I_z). If so, a larger conductor size must be selected to ensure an acceptable safety limit.

$$I_n \leq I_z \tag{5.67}$$

Example 5.14: Fuse Sizing for a Grid-Connected System

A 54 kW$_p$ grid-connected PV system will be designed by using 270 PV modules. The PV module and array parameters are given in Table 5.34. Determine the fuse and conductor ratings for PV strings and PV sub-arrays. Consider Table 5.5 for the fuse/conductor selection.

Solution

Since the PV array design configuration is specified, let us follow the steps regarding the fuse sizing explained above. For this purpose, the following table showing all fuse sizing steps can be created as below (Table 5.35):

TABLE 5.34
PV Module/Array Parameters

200 W_p PV Module Specifications

Voltage at P_{max}	Current at P_{max}	Open-Circuit Voltage	Short-Circuit Current
36 V	5.56 A	41.4 V	6.05 A

10.8 kW_p Sub-Array Configuration

$N_{s,sub}$ = 18 (18 PV modules are in series per string) $N_{p,sub}$ = 3 (3 PV strings are connected in parallel)

54 kW_p Array Configuration

$N_{p,array}$ = 5 (Five 10.8 kW_p Sub-Arrays are connected in parallel)

TABLE 5.35
Calculation Steps for Fuse Sizing

	Explanation	Calculation and Results
Step-1	Determine maximum PV circuit current	Maximum short-circuit current for PV string: $I_{max,\,string} = 1.25 \times (3-1) \times 6.05 = 15.125$ A String Conductor Ampacity: $I_z \geq 1.56 \times (3-1) \times 6.05 = 18.876$ A From Table 5.5, the string conductor size is 1.5 mm^2 = 30 A Maximum short-circuit current for PV sub-arrays: $I_{max,\,sub-array} = 1.25 \times (5-1) \times (3 \times 6.05) = 90.75$ A Sub-array Conductor Ampacity: $I_z \geq 1.56 \times (5-1) \times (3 \times 6.05) = 113.44$ A From Table 5.5, the sub-array conductor size is 16 mm^2 = 132 A
Step-2	Determine the fuse ampere rating	String fuse rating: $I_n = 1.25 \times (I_{max,\,string}) = 1.25 \times 15.125 = 18.90$ A Sub-array fuse rating: $I_n = 1.25 \times (I_{max,\,sub-array}) = 1.25 \times 90.75 = 113.43$ A
Step-3	Determine the fuse voltage rating	Fuse voltage rating: $V_n \geq 1.2 \times V_{oc} \times N_s \Rightarrow V_n \geq 1.2 \times 41.4 \times 18 \Rightarrow V_n \geq 895$ V. Thus, fuses with a nominal voltage of 900 V or 1000 V can be selected.
Step-4	Modify fuse rating for abnormal conditions	No abnormal temperature condition is defined. So, let's skip this step.
Step-5	Determine final fuse ampere rating	The standard fuse ampere ratings for PV string: 20 A The standard fuse ampere ratings for PV sub-array: 125 A
Step-6	Verify the fuse if it protects the conductor	The nominal fuse current (I_n) must be less than the ampacity of the conductor: For string conductor: $I_n \leq I_z \Rightarrow 20$ A ≤ 30 A. It is suitable. For sub-array conductor: $I_n \leq I_z \Rightarrow 125$ A ≤ 132 A. It is suitable

5.5.7.2 Switch Disconnector Sizing and Selection for Photovoltaic Systems

A switch disconnector combines the properties of a load (or power) switch and a disconnector switch. Hence, switch disconnectors in photovoltaic applications can be used as power (load) switches, isolating switches, emergency switches, and for overcurrent protection. In this case, a variety of different designs are available in the market. The most common ones are normal switch disconnectors, fuse disconnectors, and circuit breakers. Normally, sizing and selection procedures for any overcurrent protection devices in PV systems are the same or very similar to each other. Therefore, all steps explained under the section "5.5.7.1. Fuse Sizing and Selection for Photovoltaic Systems" can also be followed in the same way for other overcurrent protection devices. Referring to the relevant section above, it will not be repeated here in this section. The sizing and selection methodology for different designs of switch disconnectors will be explained below through worked examples.

Example 5.15: DC Switch Disconnector Sizing

Assume that a DC switch-disconnect will be sized for disconnecting a central inverter from a PV generator. The PV generator consists of 32 parallel-connected PV strings and each string has 19 series-connected PV modules. The PV module parameters are as follows (Table 5.36):
Determine the current and voltage ratings of switch-disconnect, which will be installed between the PV array and the inverter.

Solution

We can follow the same steps used in fuse sizing. For this purpose, the following calculation table can be created as below (Table 5.37):

TABLE 5.36
PV Module Parameters

200 W_p PV Module Specifications			
Voltage at P_{max}	Current at P_{max}	Open-Circuit Voltage	Short-Circuit Current
36 V	5.56 A	51.5 V	6.05 A

TABLE 5.37
Calculation Steps for DC Switch Sizing

	Explanation	Calculation and Results
Step-1	Determine PV circuit short-current at STC	Maximum short-circuit current for the PV array with 32 parallel strings: $$I_{SC,PV} = 32 \times 6.05 = 193.6 \text{ A}$$
Step-2	Determine the ampere rating	Switch-disconnect ampere rating according to IEC: $$I_n = 1.4 \times (I_{SC,PV}) = 1.4 \times 193.6 = 271.04 \text{ A}$$
Step-3	Determine the voltage rating	Switch-disconnect voltage rating: $$V_n \geq 1.2 \times V_{oc} \times N_s \Rightarrow V_n \geq 1.2 \times 41.4 \times 19 \Rightarrow V_n \geq 943 \text{ V}.$$ Thus, switch-disconnect with a nominal voltage of 1000 V can be selected.
Step-4	Modify the ratings for abnormal conditions	No abnormal temperature condition is defined. So, let's skip this step.
Step-5	Determine final ampere rating	The typical ampere rating from the commercially available products can be selected as 300 A. Note that the typical ampere ratings of commercial switch-disconnect devices are as follows: 25 –32 – 40 – 50 – 70 – 80 – 100 – 125 – 150 – 175 – 200 – 225 – 250 – 300 – 350 – 400 – 450 – 500 A. **Note:** Amperage ratings of a commercial product are not limited to these values only.
Step-6	Verify the final ampere and voltage ratings to see if they match to ratings of the inverter and conductor.	Conductor sizing is not determined in this example. However, the current-carrying capacity of the selected conductor must be greater than or equal to 300 A. Hence, from Table 5.5, the conductor size can be selected as 50 mm² (= 274 A). Note that the suitability of inverter input ratings should also be verified. In this example, inverter parameters are not provided. However, details about inverter sizing and selection are already explained above in the relevant section. The final ratings for the inverter disconnect can be selected as 300 A and 1000 V from the commercially available products.

Photovoltaic Power Systems

TABLE 5.38
PV Module Parameters

350 W_p PV Module Specifications			
Voltage at P_{max}	Current at P_{max}	Open-Circuit Voltage	Short-Circuit Current
36.2 V	9.65 A	44.6 V	10.5 A

Example 5.16: DC and AC Switch Disconnector Sizing

Assume that a DC and an AC switch-disconnect device will be sized for a 5.6 kW_p gird-connected PV system. The DC switch-disconnect will be used for disconnecting the inverter from the PV array, while the AC switch-disconnect will be used for disconnection purposes at the AC side. The PV array consists of two parallel-connected PV strings and each string has eight series-connected PV modules. The PV module parameters are as follows (Table 5.38):

Determine the current and voltage ratings of DC and AC switch disconnects and draw the system wiring diagram.

Solution

We can follow the same steps used in fuse sizing. For this purpose, the following calculation table can be created as below (Table 5.39):

The system configuration can be drawn as below (Figure 5.90):

Example 5.17: DC and AC Switch Disconnector Sizing in Different Options

Assume that a DC and an AC switch-disconnect devices will be sized for a 14 kW_p gird-connected PV system. The typical data for installation and system are given as below (Table 5.40):

DC switch-disconnect options for this configuration are given below:

- Option-1: Each string is to be switched individually.
- Option-2: Strings per MPPT are to be switched as pairs.
- Option-3: All strings on the DC side are to be isolated via one disconnect switch.

As for the AC side, there will be two switch-disconnect as identified below:

- A local switch disconnector to be installed between the inverter and G59 grid protection panel (It monitors the grid parameters such as voltage, phase angle, and frequency. If any of these go outside the grid connection limits, a relay triggers the protective device such as a circuit breaker to open the circuit).
- Main AC switch disconnector to be installed between protection panel and AC distribution board.

Determine the current and voltage ratings of DC and AC switch disconnects and draw the system wiring diagram.

Solution

We can follow the similar steps used in fuse sizing. For this purpose, the following worksheet can be created as below (Table 5.41):

The system configuration (based on option-2) can be drawn as below (Figure 5.91):

TABLE 5.39
Calculation Steps for DC/AC Switch Disconnector Sizing

	Explanation	Calculation and Results
Step-1	Determine PV circuit short-current at STC	Maximum short-circuit current for the PV array with two parallel strings: $$I_{SC,PV} = 2 \times 10.5 = 21 \text{ A}$$
Step-2	Determine the ampere rating	Switch-disconnect ampere rating according to IEC: $$I_n = 1.4 \times (I_{SC,PV}) = 1.4 \times 21 = 29.4 \text{ A}$$
Step-3	Determine the voltage rating	Switch-disconnect voltage rating: $$V_n \geq 1.2 \times V_{oc} \times N_s \Rightarrow V_n \geq 1.2 \times 44.6 \times 8 \Rightarrow V_n \geq 428.16 \text{ V}.$$ Thus, switch-disconnect with a nominal voltage of 450 or 500 V can be selected.
Step-4	Modify the ratings for abnormal conditions	No abnormal temperature condition is defined. So, let's skip this step.
Step-5	Determine final ampere rating	The typical ampere rating from the commercially available products can be selected as 32 A. Note that the typical ampere ratings of commercial switch-disconnect devices are as follows: 25 – 32 – 40 – 50 – 60 – 70 – 80 – 100 – 125 – 150 – 175 – 200 – 225 – 250 – 300 – 350 – 400 – 450 – 500 A. **Note:** Amperage ratings of the commercial products are not limited to these values only.
Step-6	Verify the final ampere and voltage ratings to see if they match to ratings of the inverter and conductor.	Conductor sizing is not determined in this example. However, the current-carrying capacity of the selected conductor must be greater than or equal to 32 A. Hence, from Table 5.5, the conductor size can be selected as 2.5 mm² (–41 A). For this configuration, any switch-disconnect with a minimum of 30 A and 430 V ratings can be selected for the DC side of the inverter. **In the case of an AC disconnect switch**, voltage and current ratings are based on the inverter output ratings, either in a single-phase or three-phase. According to NEC, the multiplier 1.56 in (5.60) and (5.62) changes on the AC side of the circuit. More specifically, the multiplier of 1.25 is used Instead of 1.56 when choosing the AC disconnect switch. Hence, AC switch-disconnector ampacity ($I_{n\ AC}$) shall need to provide the following condition: $I_{n\ AC} \geq 1.25 \times (I_{max,\ inv-output})$. For instance, if the single-phase inverter output specifications are 230 VAC and 25A ($I_{max,\ inv-output}$), then an AC disconnect switch rated at 32 A ($I_{n\ AC} \geq 1.25 \times 25 \text{ A} \Rightarrow I_{n\ AC} \geq 31.25 \text{ A}$) for 230 V can be selected for the AC side.

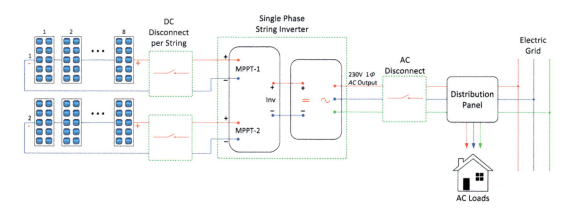

FIGURE 5.90 PV system configuration.

TABLE 5.40
PV Module/Array and Inverter Specifications

350 W_p PV Module Specifications			
Voltage at P_{max}	Current at P_{max}	Open-Circuit Voltage	Short-Circuit Current
36.2 V	9.65 A	44.6 V	10.5 A
Multi-String Inverter Data (Three-Phase, 400 Vac)			
# of MPPT	Max Input Voltage	Max Input Current	Max Output Current
2	600 Vdc	28 A per String	25 A per phase
14 kW_p PV Array Specifications			
10 PV modules per string		2 Parallel string per MPPT	

TABLE 5.41
Calculation Steps for DC/AC Disconnector Sizing

	Explanation	Calculation and Results
Step-1	Determine PV circuit short-current at STC	Maximum short-circuit current per string: $I_{SC,\,String} = 10.5$ A per string Maximum short-circuit current per MPPT (PV sub-array with two parallel strings): $$I_{SC,\,MPPT} = 2 \times 10.5 = 21 \text{ A per MPPT}$$ Maximum short-circuit current for 2-MPPT (PV inverter input current): $$I_{SC,\,PV} = 2 \times 21 = 42 \text{ A}$$
Step-2	Determine the ampere rating	DC switch-disconnector ampere ratings (according to IEC): Option-1: $I_n = 1.4 \times (I_{SC,\,String}) = 1.4 \times 10.5 = 14.7$ A Option-2: $I_n = 1.4 \times (I_{SC,\,MPPT}) = 1.4 \times 21 = 29.4$ A Option-3: $I_n = 1.4 \times (I_{SC,\,PV}) = 1.4 \times 42 = 58.8$ A
Step-3	Determine the voltage rating	Switch-disconnect voltage rating: $$V_n \geq 1.2 \times V_{oc} \times N_s \Rightarrow V_n \geq 1.2 \times 44.6 \times 10 \Rightarrow V_n \geq 535.2 \text{ V}.$$ Thus, switch-disconnect with a nominal voltage of 600 V can be selected.
Step-4	Modify the ratings for abnormal conditions	No abnormal temperature condition is defined. So, let's skip this step.
Step-5	Determine final ampere rating	The typical ampere ratings from the commercially available products can be selected as follow. Option-1: 18 A; Option-2: 32 A; Option-3: 60 A **Note:** Amperage ratings may differ depending on the manufacturers.
Step-6	Verify the final ampere and voltage ratings to see if they match to ratings of the inverter and conductor.	Conductor sizing is not determined in this example. Be sure that the ampacity of the selected conductor must be greater than the calculated current value. For the given configuration options: Option-1: Any switch disconnect with a minimum of 15 A and 550 V can be selected. Option-2: Any switch disconnect with a minimum of 30 A and 550 V can be selected. Option-3: Any switch disconnect with a minimum of 60 A and 550 V can be selected. **As for the AC disconnect switch**, voltage and current ratings are based on the inverter output ratings, either in a single-phase or three-phase. According to NEC, the multiplier 1.56 in (5.60) and (5.62) changes on the AC side of the circuit. More specifically, the multiplier 1.25 is used Instead of 1.56 when choosing the AC disconnect switch. Hence, AC switch-disconnector ampacity ($I_{n\,AC}$) shall need to provide the following condition: $I_{n\,AC} \geq 1.25 \times (I_{max,\,inv-output})$. For this configuration, $I_{n\,AC} \geq 1.25 \times 25 \text{ A} \Rightarrow I_{n\,AC} \geq 31.25$ A. In this context, identified AC switch disconnectors can be selected as follows. Local AC switch-disconnector: 415 V_{AC} (Phase to Phase) and 32 A per phase, Main AC switch-disconnector: 415 V_{AC} (Phase to Phase) and 32 A per phase.

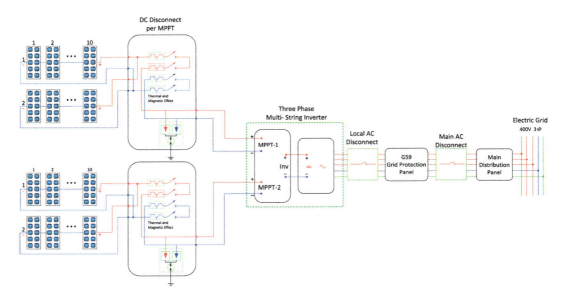

FIGURE 5.91 PV system configuration based on Option-2.

5.5.7.3 Circuit Breaker Sizing and Selection for Photovoltaic Systems

Circuit breakers can be installed in the photovoltaic system to provide overcurrent protection on the DC and AC sides. On the DC sides of the system, the circuit breakers can be used between PV array and inverter, and also between inverter and battery bank. As for the AC sides of the system, they can be used at the output of the inverter. As explained above, sizing and selection procedures for overcurrent protection devices in PV systems are almost similar to each other. For this reason, circuit breakers on DC and AC sides also use the sizing methodology described above in the same way. However, the multiplier 1.56 in (5.60) and (5.62) changes on the AC side of the circuit according to NEC. More specifically, the multiplier 1.25 is used Instead of 1.56 when choosing an AC circuit breaker. Besides, the NEC ampacity formula also changes for the overcurrent protection between the inverter and battery bank on the DC side. Hence, circuit breaker ampacity (I_n) on DC and AC sides can be rearranged as (5.68).

$$I_n \geq \begin{cases} 1.56 \times I_{sc\,Total}, & \text{for DC side (between inv and array)} \\ 1.25 \times I_{max}, & \text{for DC side (between inv and battery)} \\ 1.25 \times I_{max,\,inv-output}, & \text{for AC side (at inverter output)} \end{cases} \quad (5.68)$$

Remember that the analysis and calculation of $I_{sc\,Total}$ in (5.68) was previously explained in the above sections. It is also clear from (5.68) that the protection requirement between inverter and battery bank is equal to 125% of maximum circuit current at full load condition. Hence, the I_{max} (maximum current from the battery) in (5.68) under full load can be calculated by (5.69).

$$I_{max} = \frac{P_{inv}\,[\text{Watts}]}{\eta_{inv} \times V_{system(DC)}} \quad (5.69)$$

where P_{inv} is the nominal inverter output power, η_{inv} is the inverter efficiency under full load, and $V_{system(DC)}$ is the nominal system voltage on the DC side. The third current in (5.68) is the inverter maximum AC current ($I_{max,\,inv-output}$), which is written on the nameplate. The worked examples for the circuit breakers sizing with different needs are studied below.

Photovoltaic Power Systems

TABLE 5.42
PV Module/Array and Inverter Specifications

200 W_p PV Module Specifications			
Voltage at P_{max}	**Current at P_{max}**	**Open-Circuit Voltage**	**Short-Circuit Current**
36 V	5.56 A	41.4 V	6.05 A
115 kW Central Inverter Data (Three-Phase, 400 Vac)			
Full-Load Inverter Efficiency	**Max Input Voltage**	**Max Input Current**	**Max Output Current**
95%	1000 Vdc	550 A	150 A per phase
100 kW_p PV Array Specifications			
16 PV modules per string		32 Parallel string	
1600 Ah and 240 Vdc Battery Bank Specifications			
20 Batteries (each: 12 V and 200 Ah) per string		32 Parallel string	

Example 5.18: Circuit Breaker Sizing on DC and AC Sides

Assume that three circuit breakers on DC and AC sides will be sized and selected for a 100 kW_p gird-connected PV system with battery backup. The corresponding three locations for circuit breakers are as follows.

- Location-1: The first circuit breaker will be installed between the PV array and the inverter (DC side).
- Location-2: The second circuit breaker will be installed between the inverter and the battery bank (DC side).
- Location-3: The third circuit breaker will be installed at the output of the inverter (AC side).

The typical data for installation and system are given as in Table 5.42.
Determine the current and voltage ratings of circuit breakers specified above.

Solution

We can follow the similar steps described above. Hence, the following calculation table can be created for sizing and specifying the circuit breakers as below (Table 5.43).

Important Note: Circuit breakers must be sized not only to protect the system against overcurrents but also against the highest possible short-circuit currents. For this reason, all circuit breakers on the AC sides should also protect against short-circuit currents originating from the grid. In this respect, short-circuit analysis should be done for all grid-connected systems and circuit breaker ratings shall be specified based on the obtained results. In the following section, required short-circuit analyses are explained to determine the breaking capacity of circuit breakers used in grid-connected PV systems.

5.5.7.4 Sizing and Selection of Circuit Breakers against Grid Short Circuits

Circuit breakers are normally rated for five basic parameters. These parameters are rated current (A), rated operating voltage (V), rated insulation voltage (V), rated impulse (lightning) withstand voltage (kV), and rated breaking capacity (kA). Rated breaking capacity also involves two currents, which are maximum short-circuit breaking capacity and nominal operating short-circuit breaking capacity. For this reason, circuit breaker sizing calculation requires a short-circuit analysis for the point of common coupling (PCC) as well as other circuit breaker locations in the PV system. The following diagram shown in Figure 5.92 illustrates the single-line wiring of a typical grid-connected PV power plant with the circuit breakers used on the AC side.

TABLE 5.43
Calculation Steps for Circuit Breaker Sizing

	Explanation	Calculation and Results
Step-1	Determine PV circuit short-current at STC	**Location-1 Calculations (between PV array and Inverter):** Maximum short-circuit current per string: $$I_{SC,\,String} = 6.05 \text{ A per string}$$ Maximum short-circuit current of the entire array (includes 32 parallel strings): $$I_{SC,\,PV} = 32 \times 6.05 = 193.6 \text{ A}$$ **Location-2 Calculations (between Inverter and battery bank):** Maximum circuit current from (5.69): $$I_{max} = \frac{P_{inv}\,[\text{Watts}]}{\eta_{inv} \times V_{system(DC)}} = \frac{115000 \text{ W}}{0.95 \times 240 \text{ Vdc}} = 504 \text{ A}$$ **Location-3 Calculations (output of Inverter):** It is based on the inverter's maximum output current: $$I_{max,\,inv-output} = 150 \text{ A}.$$
Step-2	Determine the ampere rating	Circuit breaker ampere ratings: **Location-1:** $I_n = 1.56 \times (I_{SC,\,String}) = 1.56 \times 6.05 = 9.44$ A per string per string $$I_n = 1.56 \times (I_{SC,\,PV}) = 1.56 \times 193.6 = 302.01 \text{ A}$$ **Location -2:** $I_n = 1.25 \times (I_{max}) = 1.25 \times 504 = 630$ A **Location -3:** $I_n = 1.25 \times (I_{max,\,inv-out}) = 1.25 \times 150 = 187.5$ A per phase
Step-3	Determine the voltage rating	Voltage ratings for circuit breakers: **Location-1:** $V_n \geq 1.2 \times V_{oc} \times N_s \Rightarrow V_n \geq 1.2 \times 41.4 \times 16 \Rightarrow V_n \geq 794.8$ V. **Location-2:** $V_n \geq 1.2 \times V_{System\,DC} \Rightarrow V_n \geq 1.2 \times 240 \Rightarrow V_n \geq 288$ V. **Location-3:** $V_n \geq 1.2 \times V_{inv\,out} \Rightarrow V_n \geq 1.2 \times 400 \Rightarrow V_n \geq 480$ V Thus, circuit-breaker with nominal voltages of 800 V, 300 V, and 500 V can respectively be selected for the locations of 1 to 3.
Step-4	Modify the ratings for abnormal conditions	No abnormal temperature condition is defined. So, let's skip this step.
Step-5	Determine final ampere rating	The typical ampere ratings from the commercially available products can be selected as follow. Location-1: 320 A; Location-2: 650 A; Location-3: 200 A per phase. **Note:** Amperage ratings may differ depending on the manufacturers.
Step-6	Verify the final ampere and voltage ratings to see if they match to ratings of the inverter and conductor.	Conductor sizing is not determined in this example. Be sure that the ampacity of the selected conductor must be greater than the calculated current value. **As a result**, the final current and voltage ratings are: Location-1: Any circuit-breaker with a minimum of 10 A and 795 V can be selected. Location-2: Any circuit-breaker with a minimum of 630 A and 290 V can be selected. Location-3: Any circuit-breaker with a minimum of 188 A (per phase) and 480 V can be selected.

It is clear from Figure 5.92 that there are three circuit breaker locations for large-scale grid-connected PV systems. These breaker locations are the output of the transformer (medium voltage side), the busbar at the low-voltage side (AC recombiner box), and the distribution board (AC combiner box) after the inverter output. Subsequently, short-circuit analysis is required for the given locations to size the breaker ratings suitably.

5.5.7.4.1 Short-Circuit Analysis for Circuit Breaker Selection

All types of the electrical network require circuit breakers to protect the system against short circuits, which are expressed either as short-circuit power (MVA) or short-circuit current (kA).

Photovoltaic Power Systems

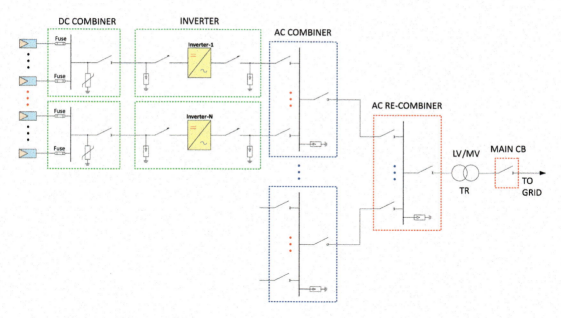

FIGURE 5.92 Typical single-line diagram of a grid-connected PV power plant with the circuit breakers used on the AC side.

To select and adjust the protective devices correctly, short-circuit analysis in the worst possible case must be performed for the circuit breaker locations at each level, which often corresponds to network points where there is a change in conductor cross section. The short-circuit power at any given point not only depends on the network configuration but also on components on the network such as generators, transformers, lines, cables, large motors, etc. As a result of corresponding short-circuit analysis, the breaking and closing current capacities for circuit breakers are determined. The short-circuit analysis required for circuit breaker selection is described below in steps.

Step 1 – Determine the System Single-Line Model with Component Impedances: The network impedances ($Z = X + jR$) shall be calculated with their respective voltages in the system. The impedance method or symmetrical components method can be used to calculate short-circuit currents. Here in this section, the impedance method is considered for the calculation of short-circuit currents (I_{sc}). As a first step, the single-line model of the system, including the PCC (point of common coupling), shall be obtained with the impedances of all system elements such as upstream network, transformers, lines, cables, etc. For this purpose, let us consider the sample network model given in Figure 5.93.

The calculation of impedances for each component in the given network model is explained below.

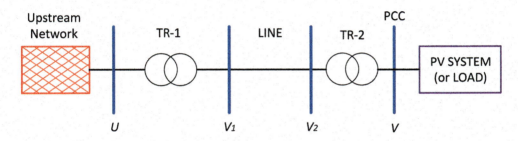

FIGURE 5.93 Sample network model for short circuit analysis

The impedance of the upstream network: The impedance of the upstream network, Z_k [Ω] can be calculated from the known short-circuit power of the upstream network, S_k [MVA] as (5.70).

$$Z_k = \frac{U^2}{S_k} \qquad (5.70)$$

where U is the phase-to-phase voltage of the upstream network in [kVA] and Z_k is the network impedance, which is equal to $R_k + jX_k$. In high and extra-high-voltage networks, the resistive component R_k is small enough to be neglected when compared to network reactance X_k. Hence, the R_k value in the impedance of the upstream network can be taken as zero ($R_k \approx 0$). In this case, the grid reactance is approximately equal to the grid impedance ($Z_k \cong X_k$). Note that impedance components R_k and X_k can be obtained from the network impedance angle (φ_k) as below.

$$R_k = Z_k \cdot \cos\varphi_k$$

$$X_k = Z_k \cdot \sin\varphi_k \qquad (5.71)$$

Normally, the impedance angle (φ) in networks with known (X/R) the ratio is calculated according to the following equation.

$$\varphi = \tan^{-1}\left(\frac{X}{R}\right) \qquad (5.72)$$

The impedance of two-winding transformer: Transformer impedance consists of a resistor (R_T) and a reactance (X_T). The components R_T [Ω] and X_T [Ω], relative to the low-voltage side, can be obtained from the known transformer data as (5.73) and (5.74).

$$R_T = \frac{U_T^2}{S_T} \cdot \frac{u_r}{100} \qquad (5.73)$$

$$X_T = \frac{U_T^2}{S_T} \cdot \frac{u_x}{100} \qquad (5.74)$$

where U_T is rated voltage of transformer at low-voltage side [kV], S_T is rated apparent power of transformer [MVA], u_r is the active component of short-circuit voltage u_k [%], and u_x is a reactive component of short-circuit voltage u_k [%].

Reminder: Short-circuit voltage of a transformer U_k is a primary side voltage that creates nominal current flow at the secondary side when the high-voltage winding is short-circuited. The ratio of this short-circuit voltage (U_k) to the rated primary voltage (U_T) is also called the relative short-circuit voltage and usually given in percent of primary voltage (U_k[%] or u_k). Figure 5.94 shows the circuit diagram for the short-circuit test.

$$u_k = \frac{U_k}{U_T} \times 100 \qquad (5.75)$$

Active and reactive components of relative short-circuit voltage (u_r and u_x) can then be determined by the following equations:

$$u_k = u_r + ju_x \Rightarrow u_k^2 = u_r^2 + u_x^2 \qquad (5.76)$$

$$u_r = \frac{\Delta P_k}{S_T} \times 100 \qquad (5.77)$$

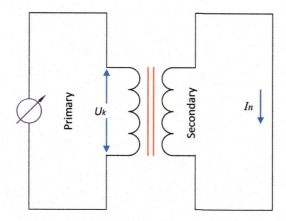

FIGURE 5.94 Measurement of short circuit voltage in a transformer.

$$u_x = \sqrt{u_k^2 - u_r^2} \tag{5.78}$$

where ΔP_k is the short-circuit losses of the transformer [MW] and u_k is the relative short-circuit voltage of the transformer in [%].

Line impedance: The impedance of a line consists of a resistor (R_L) and a reactance (X_L). The components R_L [Ω] and X_L [Ω] can be determined from the known per-unit length data as (5.79).

$$R_L = R'_L \cdot \ell$$

$$X_L = X'_L \cdot \ell \tag{5.79}$$

where ℓ [km] is the line length, R'_L [Ω/km] is line resistance per unit length and X'_L [Ω/km] is the line reactance per unit length. Note that line specifications per unit length can be obtained from the manufacturer datasheets. Otherwise, tables given in national standards can be used for the respective resistance and reactance values of the overhead lines/cables. Normally, the line resistance per unit length R'_L [Ω/km] can be found from the following equation as (5.80).

$$R'_L = \frac{\rho}{S} = \frac{1}{k \cdot S} \tag{5.80}$$

where S [mm^2] is the effective cross-sectional area of a line, ρ [Ω·mm^2/km] is line-specific resistivity, and k [km/Ω·mm^2] is the line-specific conductivity, relative to 20°C. The line resistance values for other temperatures can be calculated from the following equation.

$$R_L(T) = R_L \times [1 + \alpha_{20}(T - 20)] \tag{5.81}$$

where $R_L(T)$ is the line resistance in ohms at conductor temperature T(°C), R_L is the line resistance in ohms at conductor temperature ($T = 20$°C), T is the conductor temperature in °C, and α_{20} is the conductor temperature coefficient. The coefficients of ρ, k, and α_{20} in equations (5.80) and (5.81) are given in Table 5.44 for copper and aluminum conductors.

Impedances of other components: In most cases, the component impedances described above are generally sufficient for short-circuit analysis of grid-connected photovoltaic systems. However, there may be other network components affecting the short-circuit currents in the system. These components can be listed as three-winding transformers, short-circuit limiting reactors, synchronous

TABLE 5.44
ρ, k and α_{20} Values for Copper and Aluminum Conductors

	Copper (Cu)	Aluminum (A1)
ρ [$\Omega \cdot mm^2/km$]	19	29
k [$km/\Omega \cdot mm^2$]	0.0526	0.0345
α_{20} [$1/°C$]	0.00392	0.00403

generators, asynchronous motors, capacitors, and non-rotating loads. Referring to power system books and/or IEC-60909-0 standard, impedance calculations of these components are not given here in this section.

Step 2 – Transforming the System Impedances to the Voltage of PCC: The component impedances determined in Step-1 were calculated based on their own ratings. To be able to calculate the equivalent network impedance ($Z_{k(V)}$) at the point of common coupling (PCC), all previously determined impedances must be converted to PCC voltage, V. Note that the impedances referred to the voltage of PCC are indexed with "(v)." Considering transformer conversion rate "a," the value of impedance components (R_m and X_m) of a network element "m" after n-transformer can be calculated as (5.82) and (5.83).

$$R_{m(V)} = R_m \cdot \frac{1}{a_{T1}^2} \cdot \frac{1}{a_{T2}^2} \cdots \frac{1}{a_{Tn}^2} \tag{5.82}$$

$$X_{m(V)} = X_m \cdot \frac{1}{a_{T1}^2} \cdot \frac{1}{a_{T2}^2} \cdots \frac{1}{a_{Tn}^2} \tag{5.83}$$

where $R_{m(V)}$ in ohms is the converted resistance of a component m to the voltage level of PCC, $X_{m(V)}$ in ohms is the converted reactance of a component m to the voltage level of PCC, and a_{Ti} is the voltage ratio of the transformer i ($i = 1 \cdots n$). Note that the voltage ratio of a transformer is expressed as the ratio of the transformer's secondary voltage v_2 (high-voltage side) to the primary voltage v_1 (low-voltage side) ($a = v_2/v_1$).

Reminder: Considering Figure 5.95, the secondary side resistance of a transformer seen at the primary side (R_{in}) can be calculated as below.

$$v_1 = R_{in} \cdot i_1 \Rightarrow R_{in} = \frac{v_1}{i_1} = \frac{(N_1/N_2) \cdot v_2}{(N_2/N_1) \cdot i_2}$$

$$R_{in} = \left(\frac{N_1}{N_2}\right)^2 \cdot R \Rightarrow R_{in} = \frac{R}{(N_2/N_1)^2} = R \cdot \frac{1}{a^2}$$

where a is the voltage ratio of a transformer and it is expressed as (5.84).

$$a = \frac{\text{Secondary Voltage (High-Voltage Side) [kV]}}{\text{Primary Voltage (Low-Voltage Side) [kV]}} \tag{5.84}$$

Hence, considering Figure 5.93, the impedance components of the upstream network can be transformed to the voltage of PCC as (5.85) and (5.86).

$$R_{k(V)} = R_k \cdot \left(\frac{U_V}{U}\right)^2 \tag{5.85}$$

Photovoltaic Power Systems

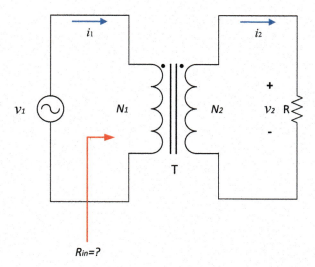

FIGURE 5.95 Transforming secondary side resistance to the primary side (N_1: number of turns on primary and N_2: the number of turns on secondary).

$$X_{k(V)} = X_k \cdot \left(\frac{U_V}{U}\right)^2 \tag{5.86}$$

where U is the phase-to-phase voltage of the upstream network, R_k is the resistance of upstream network, and $R_{k(V)}$ is the transformed resistance of the upstream network, relative to the voltage of PCC (U_V: phase-to-phase voltage). Similarly, X_k is the reactance of the upstream network and $X_{k(V)}$ is the transformed reactance, relative to the voltage of PCC (U_V: phase-to-phase voltage).

The impedance of a transformer is always related to its low-voltage side. Besides, the voltage ratio of a given transformer is ignored when its impedance is transformed to other network points. Similar to the impedance transformations described above, the impedance components of a transformer (T) can be converted to the voltage of PCC as (5.87) and (5.88).

$$R_{T(V)} = R_T \cdot \left(\frac{U_V}{U_T}\right)^2 \tag{5.87}$$

$$X_{T(V)} = X_T \cdot \left(\frac{U_V}{U_T}\right)^2 \tag{5.88}$$

where U_T is the phase-to-phase voltage at the low-voltage side of the transformer, R_T is the resistance of the transformer relative to its low-voltage side, and $R_{T(V)}$ is the converted resistance of the transformer according to the voltage of PCC (U_V: phase-to-phase voltage). Similarly, X_T is the reactance of the transformer relative to its low-voltage side and $X_{T(V)}$ is the converted reactance of the transformer to the voltage of PCC (U_V).

Similarly, the conversion of line impedances can be performed via the following equations.

$$R_{L(V)} = R_T \cdot \left(\frac{U_V}{U_L}\right)^2 \tag{5.89}$$

$$X_{L(V)} = X_T \cdot \left(\frac{U_V}{U_L}\right)^2 \tag{5.90}$$

where U_L is the phase-to-phase voltage of the line, R_L is the resistance of the line and $R_{TL(V)}$ is the transformed resistance of the line relative to the voltage of PCC (U_V). Similarly, X_L is the reactance of the line and $X_{L(V)}$ is the converted reactance of the line to the voltage of PCC (U_V). Note that all resistance and reactance values in equations (5.85)–(5.90) are given in ohms and the voltages are given in kVs.

Step 3 – Calculating the Resulting Network Impedances at PCC: As a result, all the resistances and reactances transformed to the voltage of PCC are summed up separately to calculate the total impedance value seen from the "V" voltage. The following equations are used to calculate the resulting network resistance, network reactance, network impedance, and impedance angle at the point of common coupling.

$$R_{kV} = \sum_{m=1}^{M} R_{m(V)} \tag{5.91}$$

$$X_{kV} = \sum_{m=1}^{M} X_{m(V)} \tag{5.92}$$

$$|Z_{kV}| = \sqrt{R_{kV}^2 + X_{kV}^2} \tag{5.93}$$

$$\varphi_{kV} = \tan^{-1}\left(\frac{X_{kV}}{R_{kV}}\right) \tag{5.94}$$

$$Z_{kV} = R_{kV} + jX_{kV} = |Z_{kV}| \cdot e^{j\varphi_{kV}} = |Z_{kV}| \cdot \cos\varphi_{kV} + j|Z_{kV}| \cdot \sin\varphi_{kV} \tag{5.95}$$

Step 4 – Calculate Short-Circuit Powers and Currents: Maximum possible fault currents for the network points of interest are considered to choose the right circuit breaker and set the protection functions accurately. The maximum fault current occurs due to three-phase short-circuit faults. The waveform of a short-circuit current varies depending on the fault location in the network. PV systems are typically connected to the local networks at low or medium voltage levels. Therefore, short circuits concerning the grid-connected PV systems are considered failures far from the generators. Figure 5.96 shows the typical waveform of short-circuit currents far from the generator. Total short-circuit current has two main components, which are the AC component and transient DC component.

As known, a three-phase short-circuit is a type of symmetrical fault. For this reason, three-phase short-circuits faults can be analyzed based on a single-phase equivalent, which involves only positive-sequence impedances per phase. From the result of fault analysis (far from the generator case), the following short-circuit values are needed to be known for sizing and specifying the protection switchgear. The respective parameters are breaking capacity, closing capacity (or peak short-circuit current), and thermal withstand capacity. The general equation for short-circuit breaking current is expressed by (5.96).

$$I_b = \mu \cdot I_{sc}'' \tag{5.96}$$

where I_b is the breaking capacity in kA, I_{sc}'' is the initial symmetrical short-circuit current, and μ is a factor used in near-to-generator short circuits. In far-from-generator faults, the factor μ is equal to 1 and I_{sc}'' is equal to steady-state short-circuit current I_{sc} ($I_{sc} = I_{sc}''$). As a result, breaking capacity for network faults far from generators will equal equation (5.97).

$$I_b = I_{sc} = \frac{(c \cdot U_n)/\sqrt{3}}{|Z_{kV}|} \tag{5.97}$$

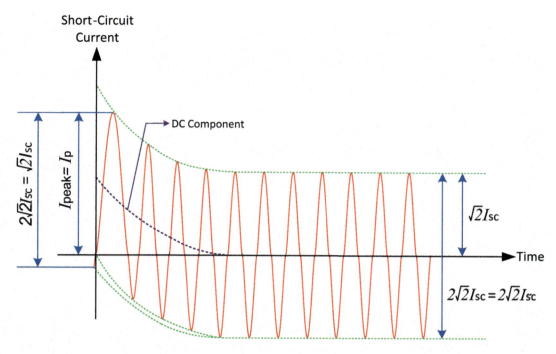

FIGURE 5.96 Short-circuit current of a network far from generator (I_{sc}: steady-state short-circuit current).

TABLE 5.45
Voltage Factor "c" according to Grid Rated Voltage (U_n)

U_n Rated Grid Voltage	Voltage Factor "c"	
	c_{max} for Max Short-Circuit Currents	c_{min} for Min Short-Circuit Currents
Low-voltage systems (100 V–1000 V) (with a tolerance of +6%)a	1.05	0.95
Low-voltage systems (100 V–1000 V) (with a tolerance of +10%)	1.10	0.95
Medium- and high-voltage systems (>1 kV)	1.10	1.00

[a] For example, systems renamed from 380 V to 400 V

where U_n is the phase-to-phase rated voltage at the short-circuit location, $|Z_{kV}|$ is the magnitude of equivalent network impedance seen from the short-circuit location, the expression "$(c \cdot U_n)/\sqrt{3}$" is called equivalent voltage source at the short-circuit location and c is a voltage factor given according to Table 5.45. Since the highest possible short-circuit currents are essential in specifying the ratings of circuit breakers, $c = c_{max}$ value is used for breaking capacity given by equation (5.97).

Considering equation (5.97) given above, the circuit breaker power (S_b in kVA) can be obtained as equation (5.98).

$$\begin{cases} S_{k(V)} = \dfrac{(c_{max} \cdot U_n)^2}{|Z_{kV}|} = \sqrt{3} \cdot (c_{max} \cdot U_n) \cdot I_{sc} \\ S_b \geq S_{k(V)} \end{cases} \quad (5.98)$$

Peak short-circuit current due to far-from-generator faults can be expressed for radial networks as in (5.99).

$$I_p = K\sqrt{2} \cdot I_{sc}'' = K\sqrt{2} \cdot I_{sc} \tag{5.99}$$

where I_p is the peak short-circuit current in kA and K is a factor defined as a function of $\left(\dfrac{R}{X}\right)$ ratio as in (5.100).

$$K = 1.02 + 0.98 \times e^{-3\left(\frac{R}{X}\right)} \tag{5.100}$$

According to IEC 60909-0 standard, thermal equivalent short-circuits current can be expressed as below. Remember that I_{sc}'' is equal to steady-state short-circuit current I_{sc} in three-phase short circuits ($I_{sc} = I_{sc}''$).

$$I_{th} = I_{sc}''\sqrt{m+n} = I_{sc}\sqrt{m+n} \tag{5.101}$$

where I_{th} is the thermal equivalent short-circuit in kA, m is a factor regarding the heat effect of the DC component, and n is a factor regarding the heat effect of the AC component for short-circuit currents. The heat factor n is approximated to 1 for network faults far from generators. Besides, the heat factor m is approximated to zero for far-from-generator faults with the rated short-circuit duration of 0.5 s or more ($T_k \geq 0.5$ s). In this case, it is allowable to take ($m + n = 1$). The exact value for the parameter m can be calculated from the expression given by (5.102).

$$m = \dfrac{1}{2 \cdot f \cdot T_k \cdot \ln(K-1)}\left[e^{4 \cdot f \cdot T_k \cdot \ln(K-1)} - 1\right] \tag{5.102}$$

where f is the grid frequency (50 Hz or 60 Hz), T_k is the (allowable) short-circuit duration in seconds, and the factor K is derived from (5.100).

An important reminder: Equivalent circuits and their respective short-circuit currents by impedance method are also summarized for other fault types in Table 5.46.

Step 5 – Specify the Circuit Breaker Ratings: Various short-circuit values are commonly required for specifying the ratings of switchgear components. These short-circuit values (three-phase steady-state short-circuit current, peak short-circuit current, thermal equivalent short-circuit current), and their common usages in grid-connected PV systems are briefly summarized in Table 5.47.

Example 5.19: Determination of Short-Circuit Current and Power

Consider a photovoltaic power plant that is connected to a 31.5 kV network via LV/MV substation and 0.75 km overhead line. The network configuration and system parameters are presented in Figure 5.97 and Table 5.48.

Calculate the following three-phase short-circuit values for the fault locations (at points: A, B, C, and D) indicated in the network diagram. The required parameters to be calculated are short-circuit power (S_k), short-circuit current (I_{sc3}), peak short-circuit current (I_p), and thermal equivalent short-circuit current (I_{th}).

Solution:

Let us follow the short-circuit calculation steps described above. Therefore, the following worksheet can be created for the short-circuit analysis as below (Table 5.49):

TABLE 5.46
Various Fault Types and Calculation of Short-Circuit Currents by Impedance Method

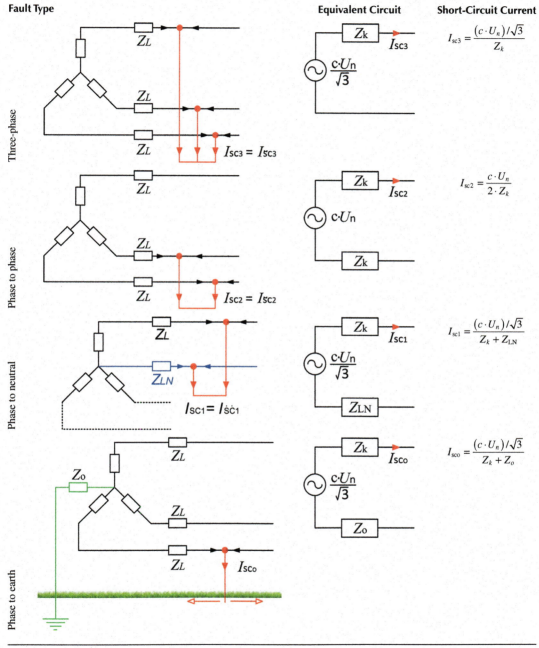

TABLE 5.47
Various Short-Circuit Values and Their Usage Purposes in Grid-Connected PV Systems

Short-Circuit Values	Purpose of Usage and Explanation
Three-phase steady-state short-circuit current, I_{sc} in $(kA)_{rms}$	It is used to determine the breaking capacity of the circuit breaker. The I_{sc} value calculated by (5.97) should be rounded up to the nearest standard circuit breaker rating.
Short-Circuit Power, S_k [kVA] @ fault location	The circuit breaker power (S_b in kVA) shall be larger than or equal to short circuit power at fault location: $S_b \geq (S_k$ @ fault location).
Peak short-circuit current, I_p in $(kA)_{rms}$	It is used to determine the closing (or making) capacity of the circuit breaker. The I_p value calculated by (5.99) should be rounded up to the next standard rating. Another use of I_p is for determining the electro-dynamic withstand capacity of the wiring systems and switchgear components.
Thermal equivalent short-circuit current, I_{th} in $(kA)_{rms}$	I_{th} value is used for the selection of measurement transformers in the electrical station and used for verifying the minimum cross-sectional area of conductors and cables. The use of I_{th} is studied in detail under the relevant sections below.

TABLE 5.48
System Parameters

Network Components			Component Parameters	
31.5 kV network	$S_{k(31.5\ kV)}$ = 160 MVA		Power Factor $(\cos\varphi_k)$=0.15	
MV line	31.5 kV	ℓ_1 = 0.75 km	R'_{L1} = 0.58 Ω/km	X'_{L1} = 0.14 Ω/km
Busbar-1 (B_1)	R and X values of busbar B_1 are neglected.			
MV/LV transformer	31.5/0.4 kV	S_T = 630 kVA	ΔP_k = 8 kW	u_k = 4.5%
Busbar-2 (B_2)	ℓ = 2 m	k = 52.6 m/Ω·mm²	$S = (50 \times 10)$ mm²	X'_{bus} = 0.00024 Ω/m
LV line	0.4 kV	ℓ_1 = 0.03 km	R'_{L1} = 0.025 Ω/km	X'_{L1} = 0.075 Ω/km
Busbar-3 (B_3)	R and X values of busbar B_3 are neglected.			

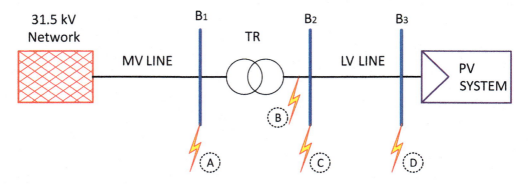

FIGURE 5.97 Network configuration having PV power system.

TABLE 5.49
Calculation Steps for Short-Circuit Analysis

Explanation	Calculation and Results

Step-1 — Determine the single line model and component impedances

The system model is given in the example. As for the component impedances, reactances (X) and resistances (R) are calculated according to their respective voltages in the system.

Network Impedance (Relative to 31.5 kV): $Z_k = \dfrac{U^2}{S_k} = \dfrac{(31.5\,\text{kV})^2}{160\,\text{MVA}} = 6.20\,\Omega$

$R_k = Z_k \cdot \cos\varphi_k = 6.20 \times 0.15 = 0.93\,\Omega;\ X_k = \sqrt{6.20^2 - 0.93^2} = 6.13\,\Omega$

MV Line Impedance (Relative to 31.5 kV):

$R_{L1} = R'_{L1} \cdot \ell_1 = 0.58 \times 0.75 = 0.435\,\Omega;\ X_{L1} = X'_{L1} \cdot \ell_1 = 0.14 \times 0.75 = 0.105\,\Omega$

Transformer Impedance (Relative to LV Side 0.4 kV):

$u_r = \dfrac{\Delta P_k}{S_T} \times 100 = \dfrac{8\,\text{kW}}{630\,\text{kVA}} \times 100 = 1.27\%;\ u_x = \sqrt{4.5^2 - 1.27^2} = 4.32\%$

$R_T = \dfrac{U_T^2}{S_T} \cdot \dfrac{u_r}{100} = \dfrac{(0.4\,\text{kV})^2}{0.63\,\text{MVA}} \cdot \dfrac{1.27}{100} = 0.0032\,\Omega;\ X_T = \dfrac{(0.4\,\text{kV})^2}{0.63\,\text{MVA}} \cdot \dfrac{4.32}{100} = 0.011\,\Omega$

LV Busbar Impedance (Relative to 0.4 kV):

$R_{\text{bus}} = \dfrac{\ell}{k \cdot S} = \dfrac{2}{52.6 \times 500} = 0.000076\,\Omega;\ X_{\text{bus}} = X'_{\text{bus}} \cdot \ell = 0.00024 \times 2 = 0.00048\,\Omega$

LV Line Impedance (Relative to 0.4 kV): $R_{L2} = R'_{L2} \cdot \ell_2 = 0.025 \times 0.03 = 0.00075\,\Omega$;

$X_{L2} = X'_{L2} \cdot \ell_2 = 0.075 \times 0.03 = 0.00225\,\Omega$

Step-2 — Transforming the system impedances to fault location voltages

The component impedances from Step-1 is the converted to the 0.4 kV level as below.

Network Impedance (Relative to 0.4 kV):

$R_{k(0.4\,\text{kV})} = 0.93 \times \left(\dfrac{0.4}{31.5}\right)^2 = 0.00015\,\Omega;\ X_{k(0.4\,\text{kV})} = 6.13 \times \left(\dfrac{0.4}{31.5}\right)^2 = 0.00099\,\Omega$

MV Line Impedance (Relative to 0.4 kV):

$R_{L1(0.4\,\text{kV})} = 0.435 \times \left(\dfrac{0.4}{31.5}\right)^2 = 0.00007\,\Omega;\ X_{L1(0.4\,\text{kV})} = 0.105 \times \left(\dfrac{0.4}{31.5}\right)^2 = 0.000017\,\Omega$

Transformer Impedance (Relative to 0.4 kV): $R_{T(0.4\,\text{kV})} = 0.0032\,\Omega;\ X_{T(0.4\,\text{kV})} = 0.011\,\Omega$

LV Bus Impedance (Relative to 0.4 kV): $R_{\text{bus}(0.4\,\text{kV})} = 0.000076\,\Omega;\ X_{\text{bus}(0.4\,\text{kV})} = 0.00048\,\Omega$

LV Line Impedance (Relative to 0.4 kV): $R_{L2(0.4\,\text{kV})} = 0.00075\,\Omega,\ X_{L2(0.4\,\text{kV})} = 0.00225\,\Omega$

Step-3 — Calculating the equivalent network impedances at fault locations

Equivalent Network Impedances at Point A (Relative to 31.5 kV):

$Z_{kA(31.5\,\text{kV})} = R_{kA(31.5\,\text{kV})} + jX_{kA(31.5\,\text{kV})} = (0.93 + 0.435) + j(6.13 + 0.105) = 1.365 + 6.235\,\Omega$

$\left|Z_{kA(31.5\,\text{kV})}\right| = \sqrt{R_{kA(31.5\,\text{kV})}^2 + X_{kA(31.5\,\text{kV})}^2} = \sqrt{1.365^2 + 6.235^2} = 6.3826\,\Omega$

Equivalent Network Impedances at Point B (Relative to 0.4 kV):

$R_{kB} = 0.00015 + 0.00007 + 0.0032 = 0.00342\,\Omega;\ X_{kB} = 0.00099 + 0.000017 + 0.011 = 0.012007\,\Omega$

$Z_{kB} = R_{kB} + jX_{kB} = 0.00342 + 0.012007\,\Omega;\ |Z_{kB}| = \sqrt{R_{kB}^2 + X_{kB}^2} = \sqrt{0.00342^2 + 0.012007^2} = 0.01248\,\Omega$

Equivalent Network Impedances at Point C (Relative to 0.4 kV):

$R_{kC} = 0.00015 + 0.00007 + 0.0032 + 0.000076 = 0.003496\,\Omega$;

$X_{kC} = 0.00099 + 0.000017 + 0.011 + 0.00048 = 0.012487\,\Omega$

$Z_{kC} = R_{kC} + jX_{kC} = 0.003496 + 0.012487\,\Omega \Rightarrow |Z_{kC}| = \sqrt{0.003496^2 + 0.012487^2} = 0.01297\,\Omega$

Equivalent Network Impedances at Point D (Relative to 0.4 kV):

$R_{kD} = R_{kC} + 0.00075 = 0.004246\,\Omega;\ X_{kD} = X_{kC} + 0.00225 = 0.014737\,\Omega$

$Z_{kD} = R_{kD} + jX_{kD} = 0.004246 + 0.014737\,\Omega \Rightarrow |Z_{kD}| = \sqrt{0.004246^2 + 0.014737^2} = 0.01534\,\Omega$

(*Continued*)

TABLE 5.49 (*Continued*)
Calculation Steps for Short-Circuit Analysis

Explanation	Calculation and Results

The following equations will be used to calculate respective short-circuit values:

$$I_{sc} = \frac{(c \cdot U_n)/\sqrt{3}}{|Z_{kV}|} \text{ (Steady-state short-circuit current)}$$

$$S_{k(V)} = \frac{(c_{max} \cdot U_n)^2}{|Z_{kV}|} = \sqrt{3} \cdot (c_{max} \cdot U_n) \cdot I_{sc} \text{ (Three-phase short circuit power)}$$

$$K = 1.02 + 0.98 \times e^{-3\left(\frac{R}{X}\right)}; I_p = K\sqrt{2} \cdot I_{sc} \text{ (Peak short circuit current)}$$

$$m = \frac{\left[e^{4 \cdot f \cdot T_k \cdot \ln(K-1)} - 1\right]}{2 \cdot f \cdot T_k \cdot \ln(K-1)}; I_{th} = I_{sc}\sqrt{m+n}; \text{ (Thermal equivalent short-circuit current. Use 1 second for } T_k)$$

Note that it is allowable to take $(m + n = 1)$ for $T_k \geq 0.5$ s. In this case, $I_{th} \cong I_{sc}$.

Short-Circuit Calculations at Fault Location A:

$$I_{sc} = \frac{(1.1 \times 31.5 \text{ kV})/\sqrt{3}}{6.382 \, \Omega} = 3.135 \text{ kA}; S_{k(V)} = \frac{(1.1 \times 31.5 \text{ kV})^2}{6.382 \, \Omega} = 188.126 \text{ MVA}$$

$$K = 1.02 + 0.98 \times e^{-3\left(\frac{1.365}{6.235}\right)} = 1.53 \Rightarrow I_p = 1.53 \times \sqrt{2} \times 3.135 = 6.78 \text{ kA}$$

Since $T_k = 1$ s, $I_{th} \cong I_{sc} = 3.135$ kA

Short-Circuit Calculations at Fault Location B (*with voltage tolerance of +10%*):

$$I_{sc} = \frac{(1.1 \times 0.4 \text{ kV})/\sqrt{3}}{0.01248 \, \Omega} = 20.35 \text{ kA}; S_{k(V)} = \frac{(1.1 \times 0.4 \text{ kV})^2}{0.01248 \, \Omega} = 15.51 \text{ MVA}$$

$$K = 1.02 + 0.98 \times e^{-3\left(\frac{0.00342}{0.012007}\right)} = 1.44 \Rightarrow I_p = 1.44 \times \sqrt{2} \times 20.35 = 41.44 \text{ kA}$$

Since $T_k = 1$ s, $I_{th} \cong I_{sc} = 20.35$ kA

Short-Circuit Calculations at Fault Location C (*with voltage tolerance of +10%*):

$$I_{sc} = \frac{(1.1 \times 0.4 \text{ kV})/\sqrt{3}}{0.01297 \, \Omega} = 19.58 \text{ kA}; S_{k(V)} = \frac{(1.1 \times 0.4 \text{ kV})^2}{0.01297 \, \Omega} = 14.92 \text{ MVA}$$

$$K = 1.02 + 0.98 \times e^{-3\left(\frac{0.003496}{0.012487}\right)} = 1.44 \Rightarrow I_p = 1.44 \times \sqrt{2} \times 19.58 = 39.87 \text{ kA}$$

Since $T_k = 1$ s, $I_{th} \cong I_{sc} = 19.58$ kA

Short-Circuit Calculations at Fault Location D (*with voltage tolerance of +10%*):

$$I_{sc} = \frac{(1.1 \times 0.4 \text{ kV})/\sqrt{3}}{0.01534 \, \Omega} = 16.56 \text{ kA}; S_{k(V)} = \frac{(1.1 \times 0.4 \text{ kV})^2}{0.01534 \, \Omega} = 12.62 \text{ MVA}$$

$$K = 1.02 + 0.98 \times e^{-3\left(\frac{0.004246}{0.014737}\right)} = 1.44 \Rightarrow I_p = 1.44 \times \sqrt{2} \times 16.56 = 33.72 \text{ kA}$$

Since $T_k = 1$ s, $I_{th} \cong I_{sc} = 16.56$ kA

Step-4: Calculate short circuit powers and currents

Example 5.20: Determination of Circuit Breaker Ratings in a Grid-Connected PV Power Plant

Consider a 500 kW$_p$ photovoltaic power plant that is connected to a 66 kV network via 0.4/31.5 kV substation, 500 m underground cable, 15 km overhead line, and 31.5/66 kV transformer station. The single-line network diagram including the PV system structure is presented in Figure 5.98.

The system parameters for the above-given network configuration are listed in the following tables (Tables 5.50 and 5.51).

a. Calculate the three-phase short-circuit values (short-circuit power, steady-state short-circuit current and peak short-circuit current) for the circuit breaker locations at points C, D, E, and F as specified in the network diagram.
b. Specify the ratings of current, voltage, power, breaking capacity, and closing capacity for the circuit breakers located at the points C, D, E, and F.

Photovoltaic Power Systems

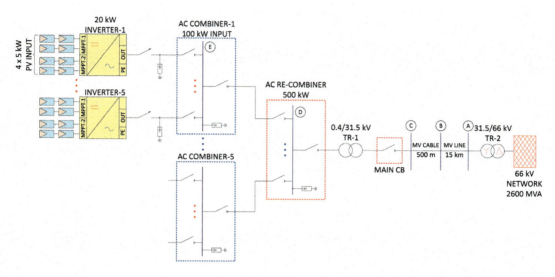

FIGURE 5.98 Single-line diagram of the PV system with grid connection.

TABLE 5.50
System Parameters

Network Components			Component Parameters		
66 kV network	$S_{k(66\,kV)}$ = 2600 MVA		$\cos\varphi_k = 0.1$		
Transformer-2 (TR-2)	31.5/66 kV	S_{Tr2} = 14 MVA	$\Delta P_{k(Tr2)}$ = 75 kW		$u_k = 9.9\%$
MV line	31.5 kV	ℓ_1 = 15 km	$R'_{L1} = 0.337\,\Omega/km$	$X'_{L1} = 0.385\,\Omega/km$	
MV cable	31.5 kV	ℓ_2 = 0.5 km	$R'_{L2} = 0.124\,\Omega/km$	$X'_{L2} = 0.204\,\Omega/km$	
Transformer-1 (TR-1)	0.4/31.5 kV	S_{Tr1} = 1.26 MVA	$\Delta P_{k(Tr1)}$ = 15 kW		$u_k = 5.8\%$
LV cable (between AC combiner and recombiner)	0.4 kV	ℓ_3 = 0.2 km	$R'_{L3} = 0.012\,\Omega/km$	$X'_{L3} = 0.071\,\Omega/km$	
LV cable (between AC combiner and inverter)	0.4 kV	ℓ_4 = 0.05 km	$R'_{L4} = 1.12\,\Omega/km$	$X'_{L4} = 0.08\,\Omega/km$	

TABLE 5.51
Inverter Data

Multi-String Inverter Data (3-Phase, 400 Vac)				
# of MPPT	Max Input Voltage	Max Input Current	Max Output Current	Nominal Power
2	950 Vdc	32 A per String	45 A per phase	28600 W

Solution:

a. Let us follow the short-circuit analysis steps given in Example 5.19.

Step 1 – Determine the Single-Line Model and Component Impedances: The system's single-line model for the required short-circuit analysis can be depicted as below (Figure 5.99):

As for the component impedances, reactances (X) and resistances (R) are calculated relative to their respective voltages as below.

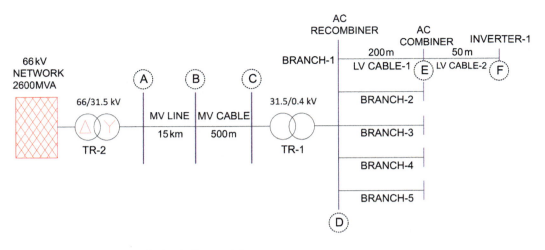

FIGURE 5.99 The system's single-line model.

Network Impedance (Relative to 66 kV):

$$Z_k = \frac{U^2}{S_k} = \frac{(66\text{ kV})^2}{2600\text{ MVA}} = 1.675\ \Omega;\ R_k = Z_k \cdot \cos\varphi_k = 1.675 \times 0.1 = 0.1675\ \Omega$$

$$X_k = \sqrt{1.675^2 - 0.1675^2} = 1.67\ \Omega$$

Transformer-2 Impedance (Relative to LV Side 31.5 kV):

$$u_r = \frac{\Delta P_k}{S_T} \times 100 = \frac{75\text{ kW}}{14000\text{ kVA}} \times 100 = 0.54\%;\ u_x = \sqrt{9.9^2 - 0.54^2} = 9.88\%$$

$$R_{T2} = \frac{U_T^2}{S_T} \cdot \frac{u_r}{100} = \frac{(31.5\text{ kV})^2}{14\text{ MVA}} \cdot \frac{0.54}{100} = 0.382\ \Omega;\ X_{T2} = \frac{(31.5\text{ kV})^2}{14\text{ MVA}} \cdot \frac{9.88}{100} = 7.002\ \Omega$$

MV Line Impedance (Relative to 31.5 kV):

$$R_{L1} = R'_{L1} \cdot \ell_1 = 0.337 \times 15 = 5.055\ \Omega;\ X_{L1} = X'_{L1} \cdot \ell_1 = 0.385 \times 15 = 5.775\ \Omega$$

MV Cable Impedance (Relative to 31.5 kV):

$$R_{L2} = R'_{L2} \cdot \ell_2 = 0.124 \times 0.5 = 0.062\ \Omega;\ X_{L2} = X'_{L2} \cdot \ell_2 = 0.204 \times 0.5 = 0.102\ \Omega$$

Transformer-1 Impedance (Relative to LV Side 0.4 kV):

$$u_r = \frac{\Delta P_k}{S_T} \times 100 = \frac{15\text{ kW}}{1260\text{ kVA}} \times 100 = 1.19\%;\ u_x = \sqrt{5.8^2 - 1.19^2} = 5.67\%$$

$$R_{T1} = \frac{U_T^2}{S_T} \cdot \frac{u_r}{100} = \frac{(0.4\text{ kV})^2}{1.26\text{ MVA}} \cdot \frac{1.19}{100} = 0.00151\ \Omega;\ X_{T1} = \frac{(0.4\text{ kV})^2}{1.26\text{ MVA}} \cdot \frac{5.67}{100} = 0.0072\ \Omega$$

LV Cable-1 Impedance (Relative to 0.4 kV):

$$R_{L3} = R'_{L3} \cdot \ell_3 = 0.0012 \times 0.2 = 0.00024\ \Omega;\ X_{L3} = X'_{L3} \cdot \ell_3 = 0.0071 \times 0.2 = 0.00142\ \Omega$$

LV Cable-2 Impedance (Relative to 0.4 kV):

$$R_{L4} = R'_{L4} \cdot \ell_4 = 1.12 \times 0.05 = 0.056 \, \Omega; \, X_{L4} = X'_{L4} \cdot \ell_4 = 0.08 \times 0.05 = 0.004 \, \Omega$$

Step 2 – Transforming the System Impedances to Fault Location: The component impedances from Step-1 is then converted to the 31.5 kV and 0.4 kV levels as below.

Network Impedance (Relative to 31.5 kV):

$$R_{k(31.5 \, kV)} = 0.1675 \times \left(\frac{31.5}{66}\right)^2 = 0.03815 \, \Omega; \, X_{k(31.5 \, kV)} = 1.67 \times \left(\frac{31.5}{66}\right)^2 = 0.3804 \, \Omega$$

Network Impedance (Relative to 0.4 kV):

$$R_{k(0.4 \, kV)} = 0.1675 \times \left(\frac{0.4}{66}\right)^2 = 0.000006152 \, \Omega; \, X_{k(0.4 \, kV)} = 1.67 \times \left(\frac{0.4}{66}\right)^2 = 0.00006134 \, \Omega$$

Transformer-2 Impedance (Relative to 0.4 kV):

$$R_{T2(0.4 \, kV)} = 0.382 \times \left(\frac{0.4}{31.5}\right)^2 = 0.00005159 \, \Omega; \, X_{T2(0.4 \, kV)} = 7.002 \times \left(\frac{0.4}{31.5}\right)^2 = 0.00113 \, \Omega$$

MV Line Impedance (Relative to 0.4 kV):

$$R_{L1(0.4 \, kV)} = 5.055 \times \left(\frac{0.4}{31.5}\right)^2 = 0.0008151 \, \Omega; \, X_{L1(0.4 \, kV)} = 5.775 \times \left(\frac{0.4}{31.5}\right)^2 = 0.0009312 \, \Omega$$

MV Cable Impedance (Relative to 0.4 kV):

$$R_{L2(0.4 \, kV)} = 0.062 \times \left(\frac{0.4}{31.5}\right)^2 = 0.00001 \, \Omega; \, X_{L2(0.4 \, kV)} = 0.102 \times \left(\frac{0.4}{31.5}\right)^2 = 0.00001644 \, \Omega$$

Transformer-1 Impedance (Relative to 0.4 kV):

$$R_{T1(0.4 \, kV)} = 0.00151 \, \Omega; \, X_{T1(0.4 \, kV)} = 0.0072 \, \Omega$$

LV Cable-1 Impedance (Relative to 0.4 kV):

$$R_{L3(0.4 \, kV)} = 0.00024 \, \Omega; \, X_{L3(0.4 \, kV)} = 0.00142 \, \Omega$$

LV Cable-2 Impedance (Relative to 0.4 kV):

$$R_{L4(0.4 \, kV)} = 0.056 \, \Omega; \, X_{L4(0.4 \, kV)} = 0.004 \, \Omega$$

Step 3 – Calculating the Equivalent Network Impedances at Fault Locations: Equivalent network impedances for points C, D, E, and F are calculated as below.

Equivalent Network Impedances at Point C (Relative to 31.5 kV):

$$R_{kC(31.5 \, kV)} = 0.03815 + 0.382 + 5.055 + 0.062 = 5.53715 \, \Omega$$

$$X_{kC(31.5 \, kV)} = 0.3804 + 7.002 + 5.775 + 0.102 = 13.2594 \, \Omega$$

$$Z_{kC(31.5 \, kV)} = R_{kC(31.5 \, kV)} + jX_{kC(31.5 \, kV)} = 5.53715 + j13.2594 \, \Omega$$

$$\left|Z_{kC(31.5 \, kV)}\right| = \sqrt{R_{kC(31.5 \, kV)}^2 + X_{kC(31.5 \, kV)}^2} = \sqrt{5.53715^2 + 13.2594^2} = 14.3691 \, \Omega$$

Equivalent Network Impedances at Point D (Relative to 0.4 kV):

$$R_{kD} = 0.000006152 + 0.00005159 + 0.0008151 + 0.00001 + 0.00151 = 0.002392 \, \Omega$$

$$X_{kD} = 0.00006134 + 0.00113 + 0.0009312 + 0.00001644 + 0.0072 = 0.009338 \, \Omega$$

$$Z_{kD} = R_{kD} + jX_{kD} = 0.002392 + j0.009338 \, \Omega$$

$$|Z_{kD}| = \sqrt{R_{kD}^2 + X_{kD}^2} = \sqrt{0.002392^2 + 0.009338^2} = 0.00964 \, \Omega$$

Equivalent Network Impedances at Point E (Relative to 0.4 kV):

$$R_{kE} = R_{kD} + 0.00024 = 0.002392 + 0.00024 = 0.002632 \, \Omega$$

$$X_{kE} = X_{kD} + 0.00142 = 0.009338 + 0.00142 = 0.010758 \, \Omega$$

$$Z_{kE} = R_{kE} + jX_{kE} = 0.002632 + j0.010758 \, \Omega$$

$$|Z_{kE}| = \sqrt{0.002632^2 + 0.010758^2} = 0.01107 \, \Omega$$

Equivalent Network Impedances at Point F (Relative to 0.4 kV):

$$R_{kF} = R_{kE} + 0.056 = 0.002632 + 0.056 = 0.058632 \, \Omega$$

$$X_{kF} = X_{kE} + 0.004 = 0.010758 + 0.004 = 0.014758 \, \Omega$$

$$Z_{kF} = R_{kF} + jX_{kF} = 0.058632 + j0.014758 \, \Omega$$

$$|Z_{kF}| = \sqrt{0.058632^2 + 0.014758^2} = 0.0604 \, \Omega$$

Step 4 – Calculate Short-Circuit Power and Currents at Fault Locations: Following equations will be used to calculate respective short-circuit values.

- *Steady-state short-circuit current:* $I_{sc} = \dfrac{(c \cdot U_n)/\sqrt{3}}{|Z_{kV}|}$

- *Three-phase short-circuit power:* $S_{k(V)} = \dfrac{(c_{max} \cdot U_n)^2}{|Z_{kV}|} = \sqrt{3} \cdot (c_{max} \cdot U_n) \cdot I_{sc}$

- *Peak short-circuit current:* $K = 1.02 + 0.98 \times e^{-3\left(\frac{R}{X}\right)}$; $I_p = K\sqrt{2} \cdot I_{sc}$

Short-Circuit Calculations at Fault Location C:

$$I_{sc} = \frac{(1.1 \times 31.5 \, \text{kV})/\sqrt{3}}{14.3691 \, \Omega} = 1392.27 \, \text{A}; \quad S_{k(V)} = \frac{(1.1 \times 31.5 \, \text{kV})^2}{14.3691 \, \Omega} = 83.55 \, \text{MVA}$$

$$K = 1.02 + 0.98 \times e^{-3\left(\frac{5.53715}{13.2594}\right)} = 1.30 \Rightarrow I_p = 1.30 \times \sqrt{2} \times 1392.27 = 2552.03 \, \text{A}$$

Short-Circuit Calculations at Fault Location D (*with voltage tolerance of +10%*)**:**

$$I_{sc} = \frac{(1.1 \times 0.4 \, \text{kV})/\sqrt{3}}{0.00964 \, \Omega} = 26.35 \, \text{kA}; \quad S_{k(V)} = \frac{(1.1 \times 0.4 \, \text{kV})^2}{0.00964 \, \Omega} = 20.08 \, \text{MVA}$$

$$K = 1.02 + 0.98 \times e^{-3\left(\frac{0.002392}{0.009338}\right)} = 1.47 \Rightarrow I_p = 1.47 \times \sqrt{2} \times 26.35 = 54.77 \, \text{kA}$$

Short-Circuit Calculations at Fault Location E (*with voltage tolerance of +10%*):

$$I_{sc} = \frac{(1.1 \times 0.4 \text{ kV})/\sqrt{3}}{0.01107 \, \Omega} = 22.94 \text{ kA}; \; S_{k(V)} = \frac{(1.1 \times 0.4 \text{ kV})^2}{0.01107 \, \Omega} = 17.48 \text{ MVA}$$

$$K = 1.02 + 0.98 \times e^{-3\left(\frac{0.002632}{0.010758}\right)} = 1.49 \Rightarrow I_p = 1.49 \times \sqrt{2} \times 22.94 = 48.33 \text{ kA}$$

Short-Circuit Calculations at Fault Location F (*with voltage tolerance of +10%*):

$$I_{sc} = \frac{(1.1 \times 0.4 \text{ kV})/\sqrt{3}}{0.0604 \, \Omega} = 4.205 \text{ kA}; \; S_{k(V)} = \frac{(1.1 \times 0.4 \text{ kV})^2}{0.0604 \, \Omega} = 3.20 \text{ MVA}$$

$$K = 1.02 + 0.98 \times e^{-3\left(\frac{0.058632}{0.014758}\right)} = 1.02 \Rightarrow I_p = 1.02 \times \sqrt{2} \times 4.205 = 6.064 \text{ kA}$$

b. Remember that the busbar D represents the AC recombiner box and E represents the busbar of the AC combiner box. The protection elements at the input and output sides of each combiner box may have different ratings. Let's assume that the combiner/recombiner box configurations considered in this example are designed as shown below (Figure 5.100).

Considering the above calculations/designs, the following table can be created to determine the ratings of current, voltage, breaking capacity, power, and closing capacity for the circuit breakers located at points C, D, E, and F (Table 5.52).

5.5.7.5 Surge Protection Device Sizing and Specifying for Photovoltaic Systems

A surge protection device (SPD) is used in electrical power networks to protect the system equipment from the current, voltage, and power surges. To ensure a safe operation during the entire life cycle, SPDs are needed to be appropriately implemented throughout the PV installation. For this

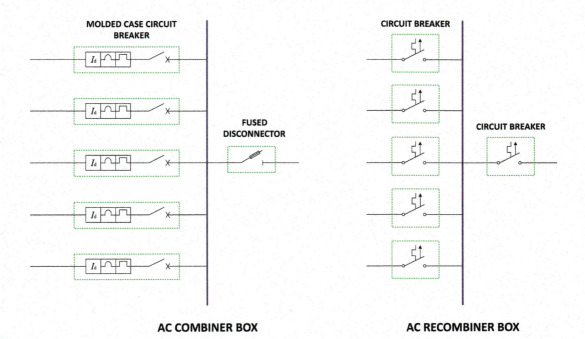

FIGURE 5.100 AC side combiner/recombiner box configurations.

TABLE 5.52
Determination of Circuit Breaker Ratings for the Given Locations

Ratings	Location E (AC Combiner Input)	Location E (AC Combiner Output)	Location F (Inverter Output)
Nominal current	$I_n = 1.25 \times (I_{\max, \text{inv-out}}) = 1.25 \times 45 = 56.25$ A per phase. It is rounded up to the next available value. For instance, $I_n = 63$ A per phase can be selected.	$I_n = 5 \times [1.25 \times (I_{\max, \text{inv-out}})] = 5 \times (1.25 \times 45) = 281.25$ A per phase. It is rounded up to the next available value. For instance, $I_n = 300$ A can be selected.	$I_n = 1.25 \times (I_{\max, \text{inv-out}}) = 1.25 \times 45 = 56.25$ A per phase. It is rounded up to the next available value. For instance, $I_n = 63$ A per phase can be selected. The protection device could be either a circuit breaker or a fuse.
Nominal voltage/ Insulation voltage	Nominal operating voltage V_n is 415 V. As for the insulation voltage V_{ins}; $V_{\text{ins}} \geq 1.2 \times V_n \Rightarrow V_{\text{ins}} \geq 1.2 \times 400 \Rightarrow V_{\text{ins}} \geq 480$ V. Hence, a circuit-breaker with a minimum insulation voltage of 500 V can be selected.	Nominal operating voltage V_n is 415 V. As for the insulation voltage V_{ins}; $V_{\text{ins}} \geq 1.2 \times V_n \Rightarrow V_{\text{ins}} \geq 1.2 \times 400 \Rightarrow V_{\text{ins}} \geq 480$ V. Hence, a circuit-breaker with a minimum insulation voltage of 500 V can be selected.	Nominal operating voltage V_n is 415 V. As for the insulation voltage V_{ins}; $V_{\text{ins}} \geq 1.2 \times V_{\text{inv out}} \Rightarrow V_{\text{ins}} \geq 1.2 \times 400 \Rightarrow V_{\text{ins}} \geq 480$ V. Hence, a circuit-breaker with a minimum insulation voltage of 500 V can be selected.
Breaking capacity ($I_b \geq I_{\text{sc}}$)	$I_b \geq 22.94$ kA. It is rounded up to the next available value.	$I_b \geq 22.94$ kA. It is rounded up to the next available value.	$I_b \geq 4.205$ kA. It is rounded up to the next available value.
CB power ($S_b \geq S_{k(V)}$)	$S_b \geq 17.48$ MVA. It is rounded up to the next available value.	$S_b \geq 17.48$ MVA. It is rounded up to the next available value.	$S_b \geq 3.20$ MVA. It is rounded up to the next available value.
Closing capacity ($\geq I_p$)	Closing capacity is selected to be $I_p \geq 48.33$ kA.	Closing capacity is selected to be $I_p \geq 48.33$ kA.	Closing capacity should be selected to be $I_p \geq 6.064$ kA.

Ratings	Location C	Location D (AC Recombiner Input)	Location D (AC Recombiner Output)
Nominal current	The transformer output current at the HV side can be determined as: $1260/(31.5 \times \sqrt{3}) = 23.09$ A. It is rounded up to the next available value. Hence, a 30 A circuit breaker will be sufficient.	$I_n = 5 \times [1.25 \times (I_{\max, \text{inv-out}})] = 5 \times (1.25 \times 45) = 281.25$ A per phase. It is rounded up to the next available value. For instance, $I_n = 300$ A can be selected.	$I_n = 5 \times \{5 \times (1.25 \times 45)\} = 1406.25$ A per phase. It is rounded up to the next available value. For instance, $I_n = 1450$ A or 1500 A can be selected.
Nominal voltage/ Insulation voltage	Nominal operating voltage V_n is 31.5 kV. As for the insulation voltage V_{ins}; $V_{\text{ins}} \geq 1.2 \times 31.5 \Rightarrow V_{\text{ins}} \geq 37.8$ kV. Hence, a circuit-breaker with a minimum insulation voltage of 38 kV can be selected.	Nominal operating voltage V_n is 415 V. As for the insulation voltage V_{ins}; $V_{\text{ins}} \geq 1.2 \times 400 \Rightarrow V_{\text{ins}} \geq 480$ V. Hence, a circuit-breaker with a minimum insulation voltage of 500 V can be selected.	Nominal operating voltage V_n is 415 V. As for the insulation voltage V_{ins}; $V_{\text{ins}} \geq 1.2 \times 400 \Rightarrow V_{\text{ins}} \geq 480$ V. Hence, a circuit-breaker with a minimum insulation voltage of 500 V can be selected.
Breaking capacity ($I_b \geq I_{\text{sc}}$)	$I_b \geq 1.392$ kA. It is rounded up to the next available value.	$I_b \geq 26.35$ kA. It is rounded up to the next available value.	$I_b \geq 26.35$ kA. It is rounded up to the next available value.
CB power ($S_b \geq S_{k(V)}$)	$S_b \geq 83.55$ MVA. It is rounded up to the next available value.	$S_b \geq 20.08$ MVA. It is rounded up to the next available value.	$S_b \geq 22.08$ MVA. It is rounded up to the next available value.
Closing capacity ($\geq I_p$)	Closing capacity is selected to be $I_p \geq 2552$ A.	Closing capacity is selected to be $I_p \geq 54.77$ kA.	Closing capacity is selected to be $I_p \geq 54.77$ kA.

purpose, three important decisions should be made regarding the photovoltaic applications of SPDs, which are listed as follows.

- Firstly, the locations where SPDs shall be applied are needed to be decided.
- Secondly, the type of SPD for each location should be decided.
- Finally, the ratings of SPD for each location should be decided.

Detailed explanations regarding the above-mentioned SPD application criteria are given in the steps below.

Step 1 – Determining the Locations of SPDs: SPDs in PV systems are mainly used to protect the system equipment such as PV arrays, inverters, monitoring devices, and other sensitive devices powered by grid voltage, 230 or 380 Vac. To reach a good suppression level throughout the PV system, SPDs in cascading structures should be installed at all voltage levels of PV systems. For this purpose, the following assessment should be made for different scales of photovoltaic systems as below.

If the length between the two sensitive equipment (either on the DC side or AC side) is less than the critical length ($L_{\text{Between 2 Equipment}} < L_{\text{crt}}$), a single SPD will be sufficient. If this length is greater than or equal to the critical length ($L_{\text{Between 2 Equipment}} \geq L_{\text{crt}}$), then it will be safer to use two SPDs at both ends between two pieces of equipment. This critical length for solar PV systems on the DC side is defined as a function of lightning ground flash density (N_g in flash/km²/year) in the IEC *60364-7-712* standard ($L_{\text{crt}} = 115/N_g$ for the building-integrated PV system, $L_{\text{crt}} = 200/N_g$ for free-standing PV systems). However, the critical length in practice is generally considered to be 10 meters, which provides a good safety margin. As a result, we can give the following simple formulation regarding the number of SPD required between two sensitive equipment.

$$\# \text{of SPD} = \begin{cases} 1 & \text{for } L_{\text{Between 2 Equipment}} < 10 \text{ m} \\ 2 & \text{for } L_{\text{Between 2 Equipment}} \geq 10 \text{ m} \end{cases} \quad (5.103)$$

As explained above, the formula (5.103) is a simple approach for determining the SPD installation points. However, the specific SPD locations should be decided separately according to the different architectures of a photovoltaic system. In this context, deciding about SPD application locations for different PV system architectures can be assessed as below.

First layout – PV system including a single-string (or multi-string) array and an inverter: The configurations illustrated in Figure 5.101 are the simplest PV system, which is generally used for residential/building applications. Importantly note that the element S_i in Figure 5.101 represents the overcurrent protection devices, while the SPD_i symbolizes the surge protection devices. By considering equation (5.103) and Figure 5.101 together, the number and location of the SPDs can easily be determined as follows:

- If L_{DC} on the DC side is greater than or equal to 10 m, two SPDs between PV array and inverter are used as indicated in Figure 5.101. In this case, the first one is located in the array box and the second one is located in the DC enclosure close to the inverter. Some manufacturers incorporate surge protection equipment on the DC side of the inverter. In this case, the inverter's SPD ratings (nominal discharge current, maximum discharge current, impulse discharge current, short-circuit withstand level, etc.) provided by the manufacturer should be checked according to the criteria in Step-3 given below. If the manufacturer-provided SPD values are satisfied, then there is no need to need to apply an external SPD to the inverter DC side. Otherwise, an external DC SPD is required on the inverter DC side. Keep in mind that an SPD application is required for each MPPT input of the inverter.

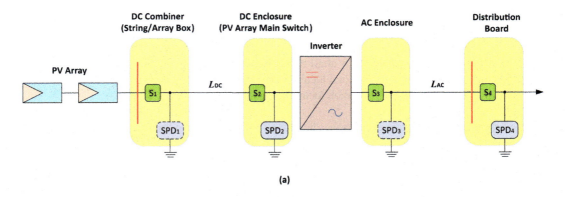

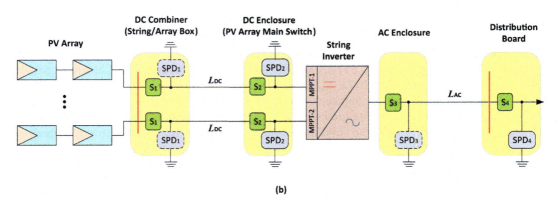

FIGURE 5.101 Photovoltaic system configurations: (a) with a single-string array and an inverter and (b) with a multi-string array and an inverter. Note that S_i in the diagrams represents the overcurrent protection elements and SPD_i represents the surge protection devices.

- If L_{DC} on the DC side is less than 10 m, and only a single SPD between PV array and inverter is sufficient. In this case, the SPD should be installed in the PV generator box, which is at the DC connection location before the inverter.
- If L_{AC} on the AC side is greater than or equal to 10 m, two SPDs between inverter and grid are used as indicated in Figure 5.101. In this case, the first one is located in the AC enclosure and the second one is located in the distribution board.
- If L_{AC} on the AC side is less than 10 m, and only a single SPD between PV array and inverter is sufficient. In this case, a single SPD (as illustrated in Figure 5.101) may be installed in the distribution board in the vicinity of the grid side.

The SPD connections in the PV systems should be as short as possible to be more effective in protecting sensitive equipment. As indicated in Figure 5.102, the total connection length (between SPD terminal blocks and network busbars) should not exceed 50 cm ($\ell_1 + \ell_2 < 50$ cm; $\ell_1 + \ell_3 < 50$ cm; $\ell_2 + \ell_3 < 50$ cm).

As known, an SPD connection may also include a disconnecting circuit breaker. Hence, there is another connection option for SPDs with an integrated or separate disconnecting circuit breaker. In this case, the 50 cm rule can be applied as indicated in Figure 5.103 ($\ell_0 + \ell_1 + \ell_2 < 50$ cm).

Second layout – PV system including multi-array with several single-phase inverters: The PV system configuration (see Figure 5.104) including multi-string array and several single-phase inverters is generally used for residential buildings and can be duplicated as needed. Important to note that single-phase inverters in the design are installed together in a three-phase arrangement.

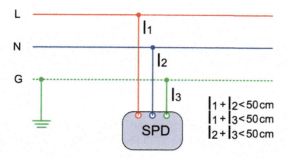

FIGURE 5.102 50 cm rule in the network connection of SPDs.

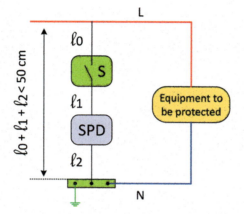

FIGURE 5.103 50 cm rule for the SPDs with integrated or separate disconnecting circuit breaker.

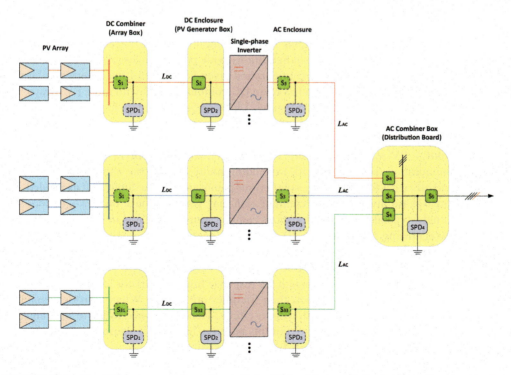

FIGURE 5.104 The PV system configuration with multiple strings and single-phase inverters to form a three-phase system.

By considering the criteria given by equation (5.103) and Figure 5.104 together, the number and location of the SPDs can be determined as below:

- If L_{DC} on the DC side is greater than or equal to 10 m, two SPDs between PV array and inverter are used as indicated in Figure 5.104. In this case, the first one is located in the array box (DC combiner) and the second one is located in the DC enclosure close to the inverter.
- If L_{DC} on the DC side is less than 10 m, and only a single SPD between PV array and inverter is sufficient. In this case, the SPD should be installed in the PV generator box, which is at the DC connection location after the PV array.
- If L_{AC} on the AC side is greater than or equal to 10 m, two SPDs between inverter and grid are used as indicated in Figure 5.104. In this case, the first one is located in the AC enclosure and the second one is located in the distribution board.
- If L_{AC} on the AC side is less than 10 m, and only a single SPD between PV array and inverter is sufficient. In this case, the element SPD_4 should be installed in the distribution board as indicated in Figure 5.104.

Third layout – PV system including multi-arrays and central inverter: The photovoltaic arrays are divided into subgroups as the power levels are increased to about 40 kW or greater. In these types of designs, PV strings are usually paralleled in two steps (in DC combiner and DC recombiner). The typical configurations for the respective PV system designs can be categorized according to their power ranges. PV system configurations and installation capacity levels are normally determined by the types and power limits of available inverter technologies. In this context, it is possible to categorize the central inverter-based power ranges as (30 – 60 kW), (60 – 100 kW), (100 – 500 kW), and (500 kW – MW scales). Hence in this framework, typical PV system configurations and possible SPD usages can briefly be given in the following figures.

In Figure 5.105, the photovoltaic systems indicating the location of overcurrent and surge protection devices are illustrated for the configurations with (30 – 60 kW) and (60 – 100 kW) power ranges.

In Figure 5.106, the photovoltaic systems indicating the location of overcurrent and surge protection devices are illustrated for the configurations with power ranges of (100 – 500 kW) and (500 kW – MW scales). The PV system configuration given in Figure 5.106a is very similar to the one given in Figure 5.105b, except that it has more DC combiner boxes with more PV arrays.

The distances L_{DC} and L_{AC} indicated in Figures 5.105 and 5.106 often go higher than 10 m as the PV system scale increases. Based on this condition, SPDs are required to be used in each connection box as specified in Figures 5.105 and 5.106.

Fourth layout – PV system including multi-arrays and multi-three-phase inverter: The PV system configuration (see Figure 5.107) including multi-string array and several three-phase inverters is generally used for large-scale solar power plants and can be duplicated as needed. Each multi-string three-phase inverter in the configuration may generally have one, two, or four-string inputs. Accordingly, typical SPD usage locations can be given as indicated in Figure 5.107.

Step 2 – Determining the Type of SPD for Each Location: It is crucial to know SPD types to utilize them correctly in PV systems. SPDs ideal for PV systems generally use spark-gap and/or varistor-based technologies. Spark-gap-based technologies are mainly used for Type-1 SPDs, while varistors are used for Type-2 SPDs. Although the response times of varistor technologies are quick when compared to spark-gap technologies, their lifetimes are shorter. When varistor technologies get older, their lives usually end up with a short circuit. Hence, it is necessary to disconnect the SPDs when a short circuit occurs on them. For this purpose, an internal thermal disconnect switch should be included in varistor-based SPDs because of possible short-circuit conditions. However, due to low current and high DC voltages, it is much more difficult to disconnect the varistor-based

Photovoltaic Power Systems

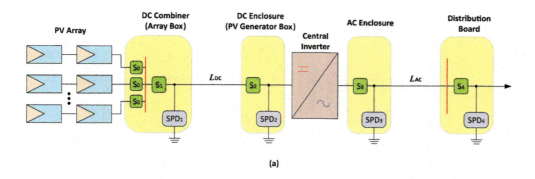

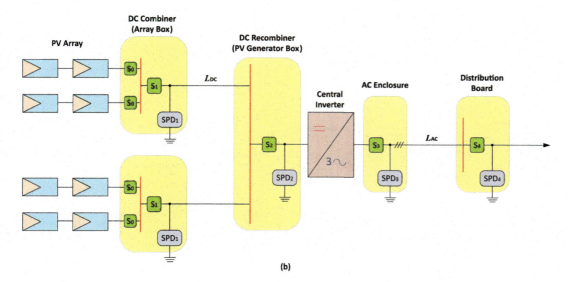

FIGURE 5.105 The PV system configurations include multiple arrays and a central inverter: the typical designs (a) from 30 kW to 60 kW and (b) from 60 kW to 100 kW.

SPDs in photovoltaic DC networks in the event of short circuits. Accordingly, specific backup protection (fuse or MCB: molded case circuit breaker) is also advised for most cases.

SPDs can be classified in several ways such as:

- **By Location:** Type-1 (Class-I), type-2 (Class-II), type-III (Class-III), and/or a hybrid combination such as Type-1+2.
- **By AC/DC Application:** AC power SPD or DC power SPD
- **By Working Principle:** Voltage switching (spark-gap- or gas discharge tube-based technology) and voltage limiting [suppression diodes or varistors such as metal-oxide varistor (MOV), for example, zinc-oxide varistors].
- **By Installation:** DIN-rail type and panel (box) type.

The method we follow here is based on determining the SPD type according to their application locations. For this purpose, the basic characteristics and intended uses of Type-1, Type-2, and Type-3 SPDs should be known. While Table 5.53 gives the basic SPD types, their typical characteristics, and intended usages.

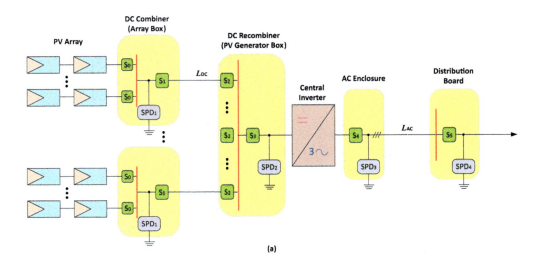

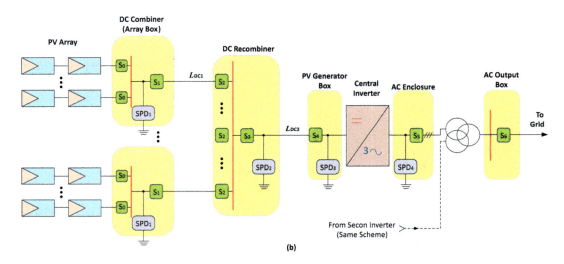

FIGURE 5.106 The PV system configurations include multiple arrays and a central inverter: the typical designs (a) from 100 kW to 500 kW and (b) from 500 kW to MW scales.

It will be more appropriate to determine the usage criteria of SPDs in PV systems according to roof-mounted and ground-mounted structures separately. In the case of an external lightning protection system exists in the installation, the SPD requirement can be determined based on LPS class and separation distance adequacy between the LPS and the PV installation (i.e., isolated or non-isolated LPS system). Figures 5.108–5.110 describe the use of SPDs in roof-mounted PV systems. Figure 5.108 illustrates the lightning zones and the SPD usages in a roof-mounted PV system with no external lightning protection system. As shown in Figure 5.108, the roof-mounted PV system with no external lightning protection can be used in areas where the risk of lightning is low ($N_g < 0.5$ strikes/m² · year). In this case, it is appropriate to use Type-2 SPDs in the PV system (Lightning Protection Zone-1, LPZ-1) and Type-3 SPDs in the vicinity of sensitive equipment (Lightning Protection Zone-2, LPZ-2). Additional Type-2 protection at the DC box is needed if the distance between PV generator and inverter is greater than or equal to 10 m ($L_{DC} \geq 10$ m). Note that the SPDs before the inverter is DC type and the ones after the inverter is AC-type SPDs.

Photovoltaic Power Systems

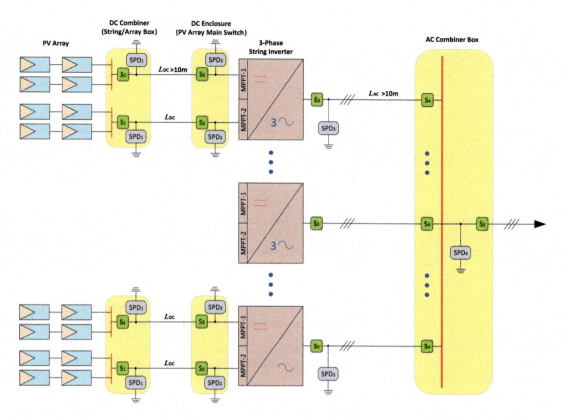

FIGURE 5.107 The PV system configurations include multiple arrays and several three-phase string inverters.

TABLE 5.53
SPD Types and Typical Characteristics

Basic Characteristics	SPD Types		
	Type-1[a]	Type-2[a]	Type-3[a]
Purpose of usage	• Against transient overvoltages due to direct lightning • Lightning current arrester	• Against transient overvoltages due to indirect lightning and switching	• Against overvoltages for sensitive equipment • Device protection (local protection)
Technology used	Spark-gap (SG), gas discharge tube (GDT), or a combination of GDT and varistor technology (VT)	GDT and/or VT	VT and/or suppression diode (SD)
Typical symbols and response characteristics	SG, GDT (Type-1 waveform)	GDT, VT (Type-2 waveform)	VT, SD (Type-3 waveform)
Wave characterization	10/350 μs current wave	8/20 μs current wave	Combination of 1.2/50 μs voltage and 8/20 μs current waves

(Continued)

TABLE 5.53 (Continued)
SPD Types and Typical Characteristics

Basic Characteristics	SPD Types		
	Type-1[a]	Type-2[a]	Type-3[a]
Main applications and SPD usage locations	• Feed point of electrical systems • Service entrance box and • Main switchboard • Central Inverters	• Near the feed point of the electrical system • Distribution/sub-distribution boards[b] • Array/PV generator junction boxes[b] • MPPTs of string inverters[c]	• Near the equipment • Local protection (for sensitive loads)
SPD usage according to Lightning Zone Transition concept	LPZ 0_A → LPZ 0_B	LPZ 0_B → LPZ 1 LPZ 0_B → LPZ 2	LPZ 2 → LPZ 3 Inside (LPZ 3)
	LPZ 0_A → LPZ 1 (Type 1 +Type 2, or Type1+2); LPZ 0_B → LPZ 3 (Type 2, or Type 2+Type 3)		
Basic ratings	Nominal (System) Voltage (V_n in volts), Maximum Continuous Operating Voltage (V_c in volts), Voltage Protection Level (kV), Impulse Discharge Current (kA, only for Type-1 technology), Nominal Discharge Current (kA, only for Type-2 technology), Maximum Discharge Current (kA, only for Type-2 technology), Temporary Overvoltage (V_T in volts), Short-Circuit Withstand Level (I_{scw} in kA), Follow Current Interrupt Rating (kA, only for spark gap technology).		

SG: Spark-Gap, GDT: Gas-Discharge-Tube, VT: Varistor Technology, SD: Suppression Diode.

[a] Both DC and AC types are available. It is possible to use different types of SPDs in a single hybrid structure such as (Type1+Type2) and (Type2+Type3).
[b] If the box shape and/or its material tends to attract lightning strikes, then a type-1 SPD or a lightning rod should be used.
[c] It can be either installed within string inverters or can be connected upstream of the inverters in PV junction boxes.

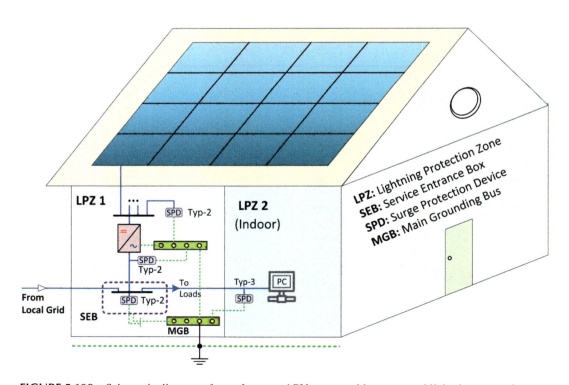

FIGURE 5.108 Schematic diagram of a roof-mounted PV system without external lightning protection.

In the second layout, the roof-mounted PV structure is located within the protection zone of an external lightning system. In this configuration, the separation distance is assumed to be kept and the PV structure is not directly connected to the lightning system (see Figure 5.109). Hence, it is appropriate to use Type-2 SPDs on the DC and AC sides of the inverter and Type-1 or Type-1/2 SPD in the service entrance box (or, in the main distribution board). Additional Type-2 protection at the DC box is needed if the distance between PV generator and inverter is greater than or equal to 10 m ($L_{DC} \geq 10$ m). As in the first configuration, Type-3 SPDs are also used in the load zone (see Figure 5.109). Note that IEC 62305-3 standard gives the details of the separation distance requirements for external lightning protection systems.

In the final layout, the roof-mounted PV structure is directly bonded to the external lightning system due to insufficient separation distance (see Figure 5.110). In this configuration, it is appropriate to use Type-1 or Type-1+2 SPDs on the DC and AC sides of the inverter as well as in the service entrance box (or, in the main distribution board). Additional Type-1/2 combined protection at the DC box is needed if the distance between PV generator and inverter is greater than or equal to 10 m ($L_{DC} \geq 10$ m). As in the other configurations, Type-3 SPDs can also be in the load zone as shown in Figure 5.110.

As shown in Figure 5.111, lightning protection zone transitions in ground-mounted PV systems are similar to those of the building-integrated systems. However, large-scale ground-mounted PV systems may have high-energy equipotential bonding currents due to the long-length cabling (DC, AC, and data cables). Since the ground-mounted large-scale PV systems require the use of different types of inverters (string or central inverters) and widespread grounding systems with meshed equipotential bonding, further aspects are needed when designing lightning and surge protection systems. For example, if a flash of lightning hits the external protection system, a great portion of

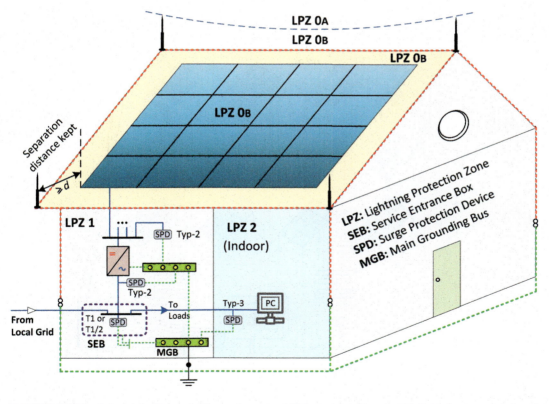

FIGURE 5.109 Schematic diagram of a roof-mounted PV system with external lightning protection and sufficient separation distance.

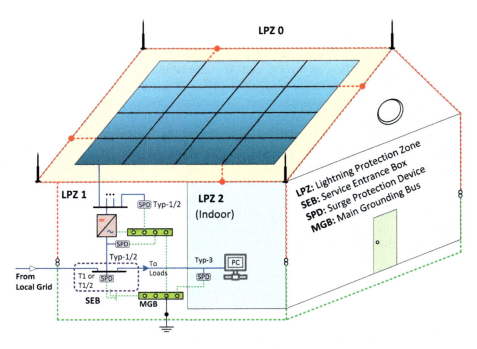

FIGURE 5.110 Schematic diagram of a roof-mounted PV system bonded to the external lightning protection system.

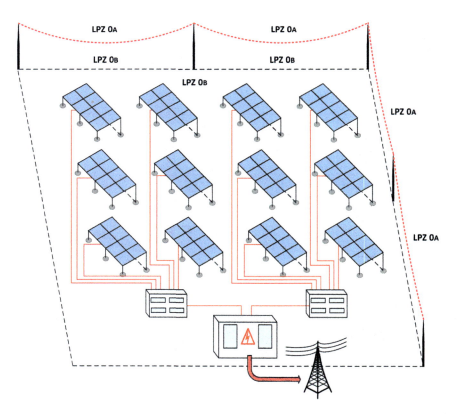

FIGURE 5.111 Ground-mounted PV system with external lightning protection [**Note**: Another lightning protection method is to use protection rods designed according to the rolling sphere method. These protection rods can be coupled to the panel frame (non-isolated) or can be used as isolated].

Photovoltaic Power Systems

the lightning currents enter the grounding systems and are coupled into the equipotential bonding structure. This partial lightning current may also pursue a path to the earth via DC cabling, which may affect the central inverter and string inverter cases at different levels. Therefore, it would be appropriate to consider the use of SPD in ground-mounted PV systems separately for the central inverter and string inverter cases.

Figure 5.112 shows the central inverter-based ground-mounted PV system and SPD usage in the installation. As shown in Figure 5.112, the DC input of the central inverter, as well as the output of each PV generator on the DC side, should be protected using Type-1 or Type-1/2 SPDs due to possible lightning current flow into the meshed equipotential bonding system. Based on the system requirements, a Type-1, Type-1/2, or high-performance Type-2 surge arrester can be installed at the AC output of the central inverter.

Figure 5.113 shows the string inverter-based ground-mounted PV system and SPD usage in the installation. In case of a lightning strike on the protection system of a ground-mounted PV structure,

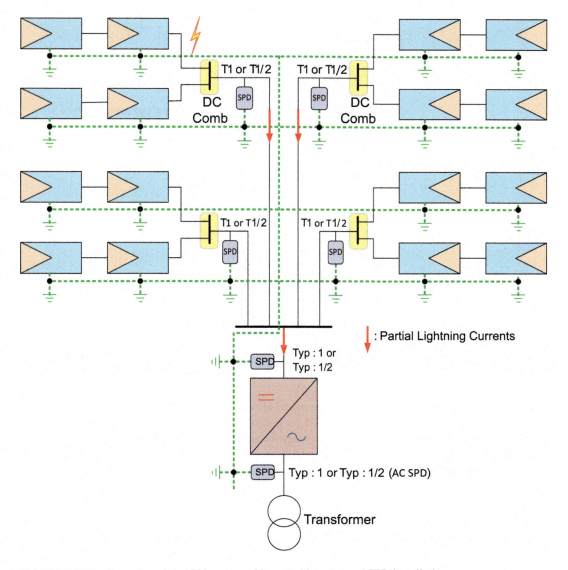

FIGURE 5.112 Ground-mounted PV system with central inverter and SPD installation.

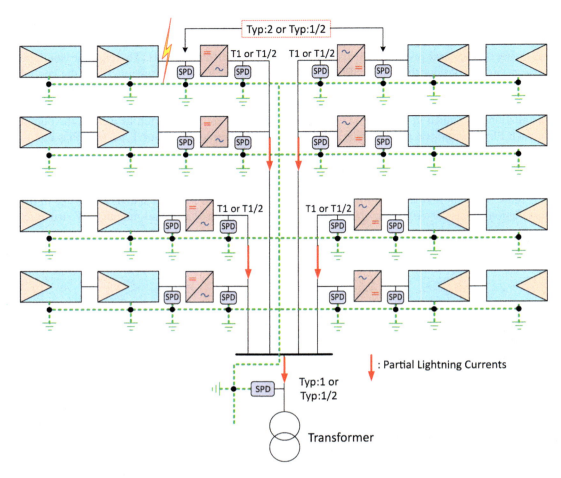

FIGURE 5.113 Ground-mounted PV system with a string inverter and SPD installation.

the partial lightning currents can flow on the AC lines. Hence in the string inverter case, the following rules should be applied for the selection of SPDs:

- To protect the AC side of the string inverters, choose Type-1 or Type:1/2 SPDs with a suitable discharge capacity.
- To protect the DC side of the string inverter, Type-1/2 or only Type-2 SPDs with a suitable discharge capacity can be used as per the mode of protection.

In addition to the SPD selection criteria identified above, the required protection modes of SPDs are needed to be determined, which depend on the grounding system design. As known, earthing systems specify the connection configurations of an electrical system and its equipment to the earth. Table 5.54 summarizes the general earthing system types (TT, TN-C, TN-S, and IT), their basic characteristics, and SPD usages in common and differential modes. Note that common-mode overvoltages occur between active conductors and earth (phase-earth or neutral-earth), while differential mode overvoltages occur between active conductors (e.g., phase-phase or phase-neutral)

Referring to Table 5.54, the most common SPD configurations and electrical circuit diagrams are given in Table 5.55 based on the protection modes. Protection from overvoltages in common and differential modes as well as the combination of both can be made by different nonlinear components such as varistors and spark gaps.

TABLE 5.54
Earthing System Types, Their Typical Characteristics, and SPD Usage

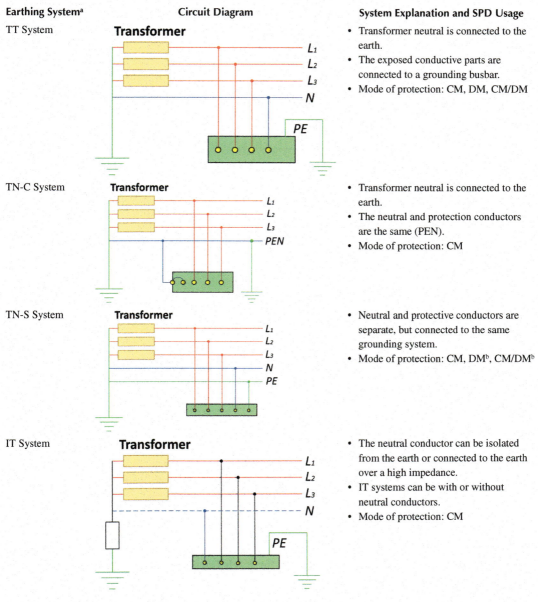

Earthing System[a]	Circuit Diagram	System Explanation and SPD Usage
TT System		• Transformer neutral is connected to the earth. • The exposed conductive parts are connected to a grounding busbar. • Mode of protection: CM, DM, CM/DM
TN-C System		• Transformer neutral is connected to the earth. • The neutral and protection conductors are the same (PEN). • Mode of protection: CM
TN-S System		• Neutral and protective conductors are separate, but connected to the same grounding system. • Mode of protection: CM, DM[b], CM/DM[b]
IT System		• The neutral conductor can be isolated from the earth or connected to the earth over a high impedance. • IT systems can be with or without neutral conductors. • Mode of protection: CM

CM: Common Mode, DM: Differential Mode, CM/DM: Combination of CM and DM.

[a] **T** (Terre in French) as a first letter denotes a direct connection to the earth, and **T** as a second letter denotes a direction to the earth by a local connection system such as a grounding rod. **I** (isolé in French), as a first letter, denotes no connection to the earth (isolated-earth) or an earth connection via a high impedance. **N** (Neutre in French) as a second letter denotes the earth connection via the neutral conductor of the electrical supply system, either separately (denoted by S) to the neutral conductor (TN-S), combined (denoted by C) with the neutral conductor (TN-C), or both (TN-C-S).

[b] It is used in case of a significant difference between neutral cable and protective cable lengths.

TABLE 5.55

Protection Modes in SPDs, Circuit Diagrams, and Their Usage in Earthing Systems

Protection Mode	SPD Circuit Diagram	Explanation and SPD Usage
Common Mode (CM)	[diagram: L_1, L_2, L_3, N with four identical varistors to ground]	• Can be used in all earthing systems (TT, TN-C, TN-S, IT) • Connection Type: 4P (4+0). It consists of four identical SPDs.
Differential Mode (DM)	[diagram: L_1, L_2, L_3, N with three varistors between lines and N]	• Can be used in TT and TN-S systems. • Connection Type: 3P (3+0). It consists of three identical SPDs.
Combination of CM and DM (CM/DM)	Circuit-1 / Circuit-2 [diagrams with L_1, L_2, L_3, N]	• **Circuit-1:** Protection in CM/DM with varistor technology. • **Circuit-2:** Protection in CM/DM with varistor and spark-gap (gas discharge tube) technology. • Circuit-2 is more common than circuit-1. • They can be used in TT and TN-S systems. • Connection Type: 3P+N (3+1). It consists of three identical SPDs and one sum spark gap.

Step 3 – Determining the Ratings of SPD for Each Location: In the steps given above, the requirements for determining the SPD types and locations are identified in detail for the PV systems. However, to finalize the selection of SPDs, their ratings must be determined correctly. For this purpose, the parameters such as lightning ground flash density (risk of lightning is low for $N_g < 0.5$, medium for $0.5 < N_g < 2$, and high for $N_g > 2$, where N_g is given in strikes/m^2/year), operating voltage, protection level, short-circuit current rating, nominal discharge current, and the protection level of waveform against direct or indirect lightning must be known. Specifying the required ratings for SPD selection is described below as follows.

Identifying voltage requirements: SPD voltage criteria should be evaluated separately for DC and AC side applications. Two voltage values must be known to characterize a surge protection device, which are voltage protection level (kV) and maximum continuous operating voltage (V_c in volts).

The maximum continuous operating voltage of an SPD is the DC or AC voltage above which SPD becomes active. This value is evaluated for DC and AC systems separately and determined according to the nominal voltage (V_n in volts) and the type of earthing system. Maximum (continuous) operating (DC) voltage of an SPD ($V_{c\ DC}$) should be greater than or equal to the maximum open voltage ($V_{oc\ max}$) of the PV generator. However, this voltage level (where the SPD becomes active) must be less than the maximum input voltage of the inverter ($V_{max\ inv}$). Thus, the SPD to be selected for the DC side can provide overvoltage protection for both the PV array and the inverter DC input.

$$V_{oc\ max\ PV} \leq V_{c\ DC} \leq V_{max\ inv} \tag{5.104}$$

As previously explained, $V_{oc\ max\ PV}$ value is obtained at minimum temperature conditions and this value can be as high as $1.2 \times V_{oc\ PV}$ for photovoltaic modules. Therefore, equation (5.104) can be revised as follows.

$$1.2 \times V_{oc\ PV} \leq V_{c\ DC} \leq V_{max\ inv} \tag{5.105}$$

where $V_{oc\ PV}$ is the open-circuit voltage of the PV generator under standard test conditions.

In the case of AC voltage, the maximum (continuous) operating (RMS) voltage of an SPD ($V_{c\ AC}$) should meet the following requirement.

$$V_{c\ AC} \geq K \cdot V_o \tag{5.106}$$

where V_o is the phase-to-neutral voltage and the K is the voltage coefficient depending on the earthing system of the network. The values for coefficient K are given in Table 5.56 according to the earthing types and SPD protection modes.

The voltage protection level of an SPD is the maximum voltage across its terminals that the SPD can stay as active. This voltage level is reached when the current carried by the SPD is equal to the nominal discharge current (I_n). The voltage protection level (V_p) for an SPD should be lower than the overvoltage withstand capability (V_w) of the equipment to be protected. To obtain a good safety margin, it would be appropriate to select the SPD's voltage protection level (V_p) at least 20% lower than the dielectric strength of the equipment.

$$V_p \leq V_w \tag{5.107}$$

If the distance between SPD and equipment to be protected is greater than 10 m, the selected protection level should comply with the following equation.

TABLE 5.56
The Values of the K Coefficient according to Earthing System Types and SPD Protection Modes

	K Values for SPD Connections Between			
Earthing System	L and N (Phase and Neutral)	L and PE (Phase and PE)	N and PE (Neutral and PE)	L and PEN (Phase and PEN)
TT	1.15	1.15	1	NA
TN-C	NA	NA	NA	1.15
TN-S	1.15	1.15	1	NA
IT with N	1.15	$\sqrt{3}$	1	NA
IT without N	NA	$\sqrt{3}$	NA	NA

NA: Not Applicable

TABLE 5.57
Required Minimum Impulse Withstand Voltage Levels according to IEC 60664 Standard Series

Equipment Category	Nominal System Voltage (V_n)				Equipment Examples	SPD Location	SPD Type
	120/220V	230/400V	400/690V	1000V			
I (Low)	0.8 kV	1.5 kV	2.5 kV	4 kV	Electronic equipment: TV, computer, audio, recorder	Socket outlet	Type-3
II (Normal)	1.5 kV	2.5 kV	4 kV	6 kV	Household devices: Washer, heater, refrigerator, and other power tools	Sub-distribution Board	Type-2
III (High)	2.5 kV	4 kV	6 kV	8 kV	Distribution switchboards, switching devices, conduits, busbars, and junction boxes	Distribution board	Type-2 or Type (1+2)
IV (Very High)	4 kV	6 kV	8 kV	12 kV	Industrial equipment such as motors, transformers, power plugs, etc.	Electricity meters	Type-1 or Type (1+2)

$$V_p \leq \frac{V_w}{2} \tag{5.108}$$

Required minimum impulse withstand voltage levels for equipment (V_w) in 220/380 V distribution system can be given as in Table 5.57 according to IEC 60664 standard series.

Identifying current requirements: To obtain an effective protection performance, two current values for an SPD should be determined depending on the accepted level of overvoltage risk. These current values are nominal discharge current (I_n in kA), maximum discharge current (I_{max} in kA), impulse discharge current (I_{imp} in kA), and short-circuit withstand level (I_{scw} in kA). As it was in the voltage criteria, the current requirements should also be evaluated for the DC and AC systems individually.

Short-circuit withstand current of an SPD (I_{scw}) can be found in the datasheet and this value must be greater than the maximum short-circuit current of the system ($I_{sc\,max}$). In the case of the DC side application, the maximum short-circuit current of the PV generator ($I_{sc\,max\,PV}$) must be greater than or equal to short-circuit withstand current of the SPD.

$$I_{scw} \geq I_{sc\,max\,PV} \tag{5.109}$$

As previously explained, $I_{sc\,max\,PV}$ value is obtained at maximum temperature conditions and this value can be as high as $1.25 \times I_{sc\,PV}$ for photovoltaic modules. Therefore, equation (5.109) can be revised as follows.

$$I_{scw} \geq 1.25 \times I_{sc\,PV} \tag{5.110}$$

where $I_{sc\,PV}$ is the total short-circuit current of the PV generator under standard test conditions.

For the AC side SPD applications, the maximum short current of the system must be considered when selecting I_{scw}. Maximum short current in a three-phase system occurs under three-phase short-circuit faults. Hence, short-circuit withstand current (I_{scw}) of the SPD must be greater than the three-phase short-circuit current of the AC system (I_{sc3}). However, in single-phase connected systems, it will be sufficient to take the phase-neutral short-circuit current (I_{sc1}) as a reference in I_{scw} selection.

$$I_{scw} > \begin{cases} I_{sc3} & \text{for three - phase system} \\ I_{sc1} & \text{for single - phase system} \end{cases} \quad (5.111)$$

Please remind that the calculations for the three-phase short-circuit current (I_{sc3}) and phase-neutral short-circuit current (I_{sc1}) were explained in detail in the previous sections.

Important note: If the SPD itself is not sized to be able to carry the system short-circuit current, then a serially connected fuse or an overcurrent protection device (backup protection) needs to be added to protect the SPD from the short-circuit faults.

The second current value that we need to consider is the impulse discharge current (I_{imp}). It is the peak value of current through the SPD with a waveform of (10/350 μs). This current, which represents the direct lightning discharges, is used to test Type-1 (or Class-1) SPDs. Therefore, the impulse discharge current is the parameter that is used to characterize the Type-1 SPDs. The following simple approach can be used to determine the minimum required I_{imp} for networks with the neutral line:

- Firstly, the highest possible lightning current value in the application area shall be determined. Possible lightning currents and protection levels according to IEC 62305-2 standard are given in Table 5.58.
- Secondly, the distribution of lightning current between earth and network conductors is found. According to International Standard IEC 62305-2, 50% of the lightning current passes to the earth and 50% to the network conductors. By assuming equal distribution of the current between network conductors, following I_{imp} value per pole can be found as:

$$I_{imp}(\text{kA/pole}) = \frac{[I_{Lightning}(\text{kA})]/2}{N_c} \quad (5.112)$$

where $I_{Lightning}$ is the highest possible lightning current for the application zone and N_c is the number of the conductor in a network. For example, if we consider (3L+N) network and 200 kA lightning current, I_{imp} value is found to be $\frac{200 \text{ kA}/2}{4} = 25$ kA/pole. Similarly, the minimum required I_{imp} values for the (3L+N) network are calculated as 18.75 kA/pole and 12.5 kA/pole respectively for 150 kA and 100 kA lightning currents.

The other current value that we need to consider for Type-2 SPD is the nominal discharge current (I_n). The nominal discharge current is the peak value of current through the SPD with a waveform of (8/20 μs). This current represents the switching and/or indirect lightning discharges, and shows the durability of a Type-2 SPD to operate correctly at least 20 times under nominal discharge current I_n having (8/20 μs) waveform surges. To be more specific about the duration of an SPD, let us consider the following case:

TABLE 5.58
Protection Levels and Lightning Currents according to IEC 62305-2 Standard

Protection Level	I	II	III/IV
Lightning current	200 kA	150 kA	100 kA

- If the PV site under lightning threat is 0.5 km² and the expected number of lightning strikes per km² in a year is 2 (N_g = 4 strikes/(km²·year)), what will be the expected duration (t_D) of the SPD with I_n = 5 kA. Assuming the maximum expected discharge current on the PV system is 5 kA, the estimated duration t_D (years) of the SPD can approximately be calculated as:

$$t_D = \frac{\text{Designated discharge number [strikes]}}{\text{Area}\left[\text{km}^2\right] \times N_g \left[\text{strikes}/\left(\text{km}^2 \cdot \text{year}\right)\right]} \quad (5.113)$$

$$t_D = \frac{20 \text{ strikes}}{0.5 \text{ km}^2 \times 4 \text{ strikes}/\left(\text{km}^2 \cdot \text{year}\right)} = 10 \text{ years}$$

The number of SPD discharges in a healthy state is dependent upon the expected overcurrent level in the system. To be more specific, if the expected discharge current on the SPD is around or equal to the designated nominal discharge current value (I_n), the SPD can discharge the overcurrent 20 times. If the discharge current becomes greater than the designed I_n value, then the discharge capacity of the SPD will decrease, or vice-versa. Hence, it is advisable to choose the SPDs with a higher-rated discharge current, which ensures a longer period of protection and less maintenance. For example, the number of SPD discharges (N_{SPD}) can empirically be written below as a function of discharge current level I_d [kA] for an SPD having I_n value of 20 kA.

$$N_{SPD} = 4037.6 \times (I_d)^{-2.016} \quad (5.114)$$

From (5.114), we can conclude that the corresponding SPD can discharge the 20 kA current (I_d = 20 kA) 20 times, while the current of 10 kA can be discharged 40 times and the current of 5 kA 200 times. Importantly note that equation (5.103) is given as an example for a specific product and it is not a general equation. The relationship between SPD life and discharge current may vary from product to product and from company to company. Therefore, the relevant datasheets should be checked for the correct data.

The last current for Type-2 SPD is the maximum discharge current (I_{max}). It is the peak value of current with a waveform of (8/20 μs) and shows the maximum discharge current limit that the SPD can handle. For this reason, I_{max} is always greater than I_n ($I_{max} > I_n$). If the I_{max} value is chosen larger, then the durability of the SPD will be longer. In a conclusion, it will be more appropriate to choose the larger I_{max} value between two SPDs having the same properties.

Example 5.21: SPD Sizing and Selection for a Roof-Mounted Single-Phase PV System

Consider a 10 kW$_p$ roof-mounted single-phase photovoltaic system that feeds the local loads over a 230 V single-phase electrical system. The solar PV system will be assembled based on the following PV modules and central inverter, which are given in Table 5.59.

The system impedances required to calculate phase-to-neutral short-circuit current are Z_{LN} = 50 mΩ (neutral line impedance relative to service entrance box) and Z_k = 70 mΩ (network impedance relative to service entrance box). Design and size the overvoltage protection system for the following cases:

- **Case-1:** There is no external lightning protection and the distance between PV array and inverter (L_{DC} on the DC side), and the distance between the PV inverter and service entrance box (L_{AC} on the AC side) is less than 10 m (L_{DC} < 10 m and L_{AC} < 10 m).

Photovoltaic Power Systems   361

TABLE 5.59
PV Module and Inverter Specifications

PV Module Specifications: 280 Wp Mono-Crystalline Module			
Voltage at P_{max}	Current at P_{max}	Open Circuit Voltage	Short Circuit Current
31.75 V	8.55 A	39.6 V	9.60 A

Single-Phase and 10.5 kW Central Inverter Specifications			
Input Side (DC)		Output Side (AC)	
Max Short-Circuit Current	Max Input Voltage	Max Output Current	Rated Output Voltage
59 A	630 V	52.5 A	230 V

- **Case-2:** There is no external lightning protection and the distance between PV array and inverter (L_{DC} on the DC side), and the distance between the PV inverter and service entrance box (L_{AC} on the AC side) is greater than or equal to 10 m ($L_{DC} \geq 10$ m and $L_{AC} \geq 10$ m).
- **Case-3:** Assume that an external lightning protection system is installed on the roof and a separation distance between lightning rod and PV array is kept. Also, assume with this case that the distances L_{DC} and L_{AC} are less than 10 m or greater than 10 m.
- **Case-4:** Assume that an external lightning protection system is installed on the roof and the lightning rod and PV array are bonded together due to insufficient separation distance. Also, assume with this case that the distances L_{DC} and L_{AC} are less than 10 m or greater than 10 m.

Solution:

The above steps regarding SPD application/sizing include briefly SPD localization, type determination, and rating specifications. Hence as a first step, let's draw a general PV system scheme including all given cases to determine SPD application points. In the design, PV modules are assembled as $12s \times 3p$ to obtain 10 kW$_p$ (≈ 10.08 kW$_p$) array (Figure 5.114).

Let us follow the SPD sizing/specifying steps for the given cases described in Example 5.21.

Steps 1 and 2 – Locating SPDs and Determining SPD Types: Based on the system drawn in Figure 5.114, we can determine the SPD locations and types for all given cases as shown in Table 5.60.

Steps 3 – Specifying SPD Ratings: The parameters required for the SPD ratings can be calculated in Table 5.61 for each SPD location.

From Table 5.61, the specifications for SPD ratings can be decided as in Table 5.62.

From the above results, voltage and current specification summary for DC and AC SPDs can be given in Table 5.63.

Example 5.22: SPD Sizing and Selection for a Ground-Mounted Photovoltaic Power System

Consider the 500 kW$_p$ PV power plant worked in Example 5.20. This PV power plant was created using 250 W$_p$ PV modules ($V_{oc} = 37.13$, $V_{mp} = 30.2$, $I_{sc} = 8.96$, $I_{mp} = 8.28$). Assume that all given data in Example 5.20 is also valid for this example. Under these conditions, determine the requirements of all SPDs and size them for the DC and AC sides of the PV system.

Solution:

The SPD sizing steps, applied in Example 5.21, can also be used for this example in the same way.

Steps 1 and 2 – Locating SPDs and Determining SPD Types: The wiring diagram regarding the AC side of the PV system is sketched in detail in Example 5.20, but no more detail for the DC

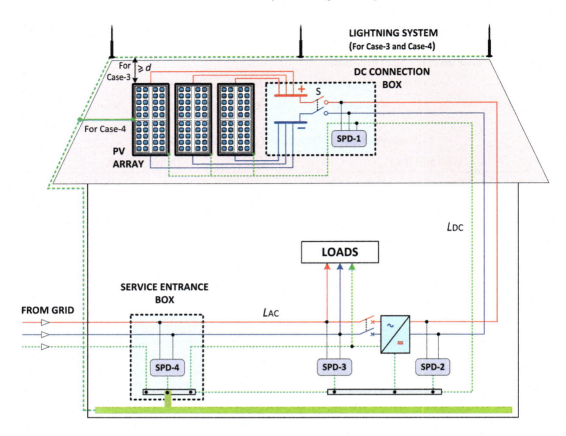

FIGURE 5.114 Roof-mounted PV system showing all cases.

TABLE 5.60
SPD Locations and Types for all Cases

		SPD Locations and Types		
		DC SPD		AC SPD	
PV System Cases		DC Box (Combiner)	Inverter DC Side	Inverter AC Side	Service Entrance Box
Case-1	$L_{DC} < 10$ m, $L_{DC} < 10$ m, No LPS	Type-2	NR/O	NR/O	Type-2
Case-2	$L_{DC} \geq 10$ m, $L_{DC} \geq 10$ m, No LPS	Type-2	Type-2	Type-2	Type-2
Case-3	$L_{DC} < 10$ m, $L_{AC} < 10$ m, LPS and $\geq d$ [a]	Type-2	NR/O	NR/O	Type-1 or Type-1/2 [b]
Case-3	$L_{DC} \geq 10$ m, $L_{AC} \geq 10$ m, LPS and $\geq d$ [a]	Type-2	Type-2	Type-2	Type-1 or Type-1/2 [b]
Case-4	$L_{DC} < 10$ m, $L_{AC} < 10$ m, LPS bonded to PV	Type-1/2	NR/O	NR/O	Type-1 or Type-1/2 [b]
Case-4	$L_{DC} \geq 10$ m, $L_{AC} \geq 10$ m, LPS bonded to PV	Type-1/2	Type-1/2	Type-1/2	Type-1 or Type-1/2 [b]

LPS: Lightning Protection System, NR/O: Not Required or Optional.
[a] Separation distance must be greater than 0.5 m.
[b] Type1/2 combined SPD is recommended.

---

side is given. Hence as a first step, let's draw the PV system scheme including DC and AC sides to determine SPD application points (Figures 5.115 and 5.116).

Based on the system drawn in Figures 5.115 and 5.116, we can determine the SPD locations and types as shown in Table 5.64.

TABLE 5.61
The Required Parameters for Determining the Current and Voltage Ratings of SPDs

	Voltage Ratings			Without Lightning Protection System	Current Ratings		
	$V_{oc\,max\,PV}$	$K \cdot V_o$	$V_w{}^a$	$I_{sc\,max\,PV}$	$I_{sc1}{}^b$	$I_{imp}{}^c$	$I_{max}{}^d$
SPD Locations							
DC box	$1.2 \times V_{oc\,PV} = 1.2 \cdot (39.6\,V \times 12)$ $= 570.24\,V$	NA	4000 V	$1.25 \times I_{sc\,PV} = 1.25 \cdot (9.6 \times 3)$ $= 36\,A$	NA	$I_{imp} = \dfrac{I_L}{2N_c}$ $= \dfrac{100\,kA}{2 \cdot 2}$ $= 25\,kA/\text{pole}$	40 kA/pole
Inverter DC side	570.24 V	NA	4000 V	36 A	NA	25 kA/pole	40 kA/pole
Inverter AC side	NA	$1.15 \times 230\,V = 264.5\,V$	2500 V	NA	$I_{sc1} = \dfrac{(c \cdot U_n)/\sqrt{3}}{Z_k + Z_{LN}}$ $= \dfrac{(1.05 \cdot 400)/\sqrt{3}}{(75+60) \times 10^{-3}}$ $= 2020\,A$	25 kA/pole	40 kA/pole
Service entrance	NA	264.5 V	2500 V	NA	2020 A	$I_{imp} = \dfrac{I_L}{2N_c}$ $= \dfrac{150\,kA}{2 \cdot 2}$ $= 37.5\,kA/\text{pole}$	80 kA/pole

(Continued)

TABLE 5.61 (Continued)
The Required Parameters for Determining the Current and Voltage Ratings of SPDs

	Voltage Ratings				Current Ratings		
	$V_{oc\,max\,PV}$	$K \cdot V_o$	$V_w{}^a$	$I_{sc\,max\,PV}$	$I_{sc1}{}^b$	$I_{imp}{}^c$	$I_{max}{}^d$
			With Lightning Protection System				
SPD Locations							
DC box	570.24 V	NA	4000 V	36 A	NA	$I_{imp} = \dfrac{I_L}{2N_c}$ $= \dfrac{150\,kA}{2 \cdot 2}$ $= 37.5\,kA/pole$	40 kA/pole
Inverter DC side	570.24 V	NA	4000 V	36 A	NA	37.5 kA/pole	40 kA/pole
Inverter AC side	NA	264.5 V	2500 V	NA	2020 A	37.5 kA/pole	40 kA/pole
Service entrance	NA	264.5 V	2500 V	NA	2020 A	$I_{imp} = \dfrac{I_L}{2N_c}$ $= \dfrac{200\,kA}{2 \cdot 2}$ $= 50\,kA/pole$	100 kA/pole

NA: Not Applicable.
[a] From Table 5.57
[b] According to Table 5.46; Note that $U_n = 400\,V$ (grid side, phase to phase voltage)
[c] Based on Table 5.58 and equation (5.112)
[d] As explained above, it will be more appropriate to choose the larger I_{max} value for longer durability. Therefore, the values for I_{max} maybe selected as given.

TABLE 5.62
Current and Voltage Ratings of SPDs

SPD Locations	Voltage Ratings			Current Ratings		
	$V_{c\,DC}$	$V_{c\,AC}$	V_p [a]	I_{scw}	I_{imp}	I_{max}
Without Lightning Protection System						
DC box	$V_{oc\,max\,PV} \leq V_{c\,DC} \leq V_{max\,inv}$ 570.24 V $\leq V_{c\,DC} \leq$ 630 V	NA	$V_p \leq V_w$ $V_p \approx 0.8 \times 4000$ V $V_p \approx 3200$ V	$I_{scw} \geq I_{sc\,max\,PV}$ $I_{scw} \geq 36$ A	25 kA/pole	40 kA/pole
Inverter DC side	570.24 V $\leq V_{c\,DC} \leq$ 630 V	NA	$V_p \approx 3200$ V	$I_{scw} \geq 36$ A	25 kA/pole	40 kA/pole
Inverter AC side	NA	$V_{c\,AC} \geq K \cdot V_o$ $V_{c\,AC} \geq 264.5$ V	$V_p \leq V_w$ $V_p \approx 0.8 \times 2500$ V $V_p \approx 2000$ V	$I_{scw} \geq I_{sc1}$ $I_{scw} \geq 2020$ A	25 kA/pole	40 kA/pole
Service entrance	NA	$V_{c\,AC} \geq 264.5$ V	$V_p \approx 2000$ V	$I_{scw} \geq 2020$ A	37.5 kA/pole	80 kA/pole
With Lightning Protection System						
DC box	$V_{oc\,max\,PV} \leq V_{c\,DC} \leq V_{max\,inv}$ 570.24 V $\leq V_{c\,DC} \leq$ 630 V	NA	$V_p \leq V_w$ $V_p \approx 0.8 \times 4000$ V $V_p \approx 3200$ V	$I_{scw} \geq I_{sc\,max\,PV}$ $I_{scw} \geq 36$ A	37.5 kA/pole	40 kA/pole
Inverter DC side	570.24 V $\leq V_{c\,DC} \leq$ 630 V	NA	$V_p \approx 3200$ V	$I_{scw} \geq 36$ A	37.5 kA/pole	40 kA/pole
Inverter AC side	NA	$V_{c\,AC} \geq K \cdot V_o$ $V_{c\,AC} \geq 264.5$ V	$V_p \leq V_w$ $V_p \approx 0.8 \times 2500$ V $V_p \approx 2000$ V	$I_{scw} \geq I_{sc1}$ $I_{scw} \geq 2020$ A	37.5 kA/pole	40 kA/pole
Service entrance	NA	$V_{c\,AC} \geq 264.5$ V	$V_p \approx 2000$ V	$I_{scw} \geq 2020$ A	50 kA/pole	100 kA/pole

NA: Not Applicable.
[a] Typically, about 80% of the withstand voltage is selected.

Steps 3 – Specifying SPD Ratings: The parameters required for the SPD ratings can be calculated as in Table 5.65 for each SPD location.

From Table 5.65, the specifications for SPD ratings can be decided as in Table 5.66.

From the above results, voltage and current specification summary for DC and AC SPDs can be given in Table 5.67.

5.5.7.6 SPD Selection and Installation Requirements for Communication Lines

Lightning discharges may cause high overvoltages on photovoltaic installations. Hence, data cables, monitoring equipment, and other external communication systems in a solar site are also exposed to the risk of transient overvoltages. It is therefore vital to install SPDs on the data and communication lines to get remote and reliable data, which is essential in operating the PV system effectively and safely. As understood from the explanations, SPDs are required not only for power lines but also for wired communication lines such as Ethernet and RS485. Since commercially available solar PV inverters mostly include Ethernet and RS485 ports, it is essential to select suitable SPDs for these types of interfaces. Figure 5.117 shows a sample SPD application diagram for a typical PV system including RS485 and Ethernet protection.

The details of selection requirements for the SPDs connected to data/communication networks are covered by IEC 61643-22. This section focuses specifically on SPD selection for ethernet and RS485 lines used in PV systems. The general characteristics of SPDs for data lines are essentially

TABLE 5.63
Voltage and Current Specification Summary for DC and AC SPDs

	DC SPD		AC SPD	
SPD Ratings	DC Box (Combiner)	Inverter DC Side	Inverter AC Side	Service Entrance Box
Max continuous operating voltage (V_c)	600 V	600 V	275 V	275 V
Voltage protection level (V_p)	< 3200 V	< 3200 V	< 2000 V	< 2000 V
Short-circuit withstand current (I_{scw})	> 36 A	> 36 A	> 2.5 kA[b]	> 2.5 kA[b]
Impulse discharge current (I_{imp})	≥ 25 kA/pole (without LPS)	≥ 25 kA/pole (without LPS)	≥ 25 kA/pole (without LPS)	≥ 37.5 kA/pole (without LPS)
	≥ 37.5 kA/pole (with LPS)	≥ 37.5 kA/pole (with LPS)	≥ 37.5 kA/pole (with LPS)	≥ 50 kA/pole (with LPS)
Maximum discharge current (I_{max})	40 kA/pole	40 kA/pole	40 kA/pole	80 kA/pole (without LPS) 100 kA/pole (with LPS)
Nominal discharge current (I_n)[a]	20 kA/pole	20 kA/pole	20 kA/pole	40 kA/pole

LPS: Lightning Protection System.
[a] It is recommended to select I_n value as high as possible ($I_{max} > I_n$) for longer durability.
[b] Since network short circuit currents can often reach much higher values, commercially available AC SPDs are provided with very high short circuit withstand levels such as 25 kA and 50 kA.

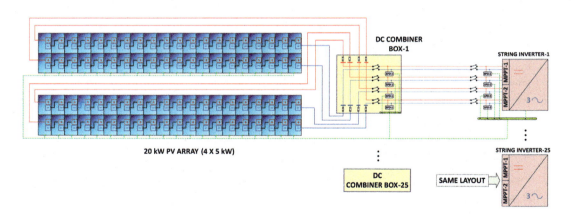

FIGURE 5.115 Wiring diagram of DC side for 500 kW PV power plant (see Example 5.20).

similar to those used in power lines. For example, like other SPDs, data line SPDs are characterized by the same parameters such as maximum continuous operating voltage (V_c), voltage protection level (V_p), nominal discharge current (I_n), impulse current (I_{imp}), and maximum discharge current (I_{max}). Based on the application type, nominal voltages in telecom and signaling networks range from 5 V to 150 V (5 V, 12 V, 24 V, 48 V, 110 V, and 150 V). The characteristic parameters of data line SPDs for PV systems are not as versatile as in classical SPDs. For instance, I_{max} values are generally sized from 10 kA to 20 kA, while I_n values are rated from 5 kA to 10 kA. As a result, SPD selection and installation requirements for photovoltaic data lines can be listed briefly in Table 5.68.

The use of SPD in wired data lines is described above for a single photovoltaic device. However, in large-scale solar PV systems, multi-string inverters are used commonly. In the case

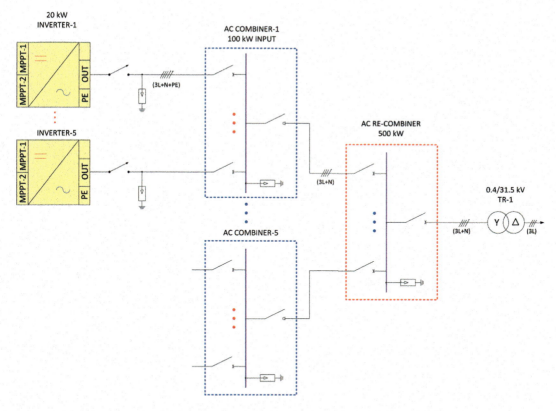

FIGURE 5.116 Single-line diagram of AC side for 500 kW PV power plant (see Example 5.20).

TABLE 5.64
SPD Locations and Types

SPD Locations		Types	Poles/Connection	Earthing System	Protection Mode
DC	DC combiner	T2 or T1/2	(L+,L−); (L+,PE); (L-PE)	PV array earthed	CM (V Circuit), CM/DM (Y Circuit)
SPD	Inverter DC side	T2 or T1/2	(L+,L−); (L+,PE); (L-PE)	PV array earthed	CM (V Circuit), CM/DM (Y Circuit)
AC	Inverter AC side	T1 or T1/2	4P/(4+0); (3P+N)/(3+1)	TN-S	CM/DM
SPD	AC combiner	T1 or T1/2	4P/(4+0); (3P+N)/(3+1)	TN-S	CM/DM
	AC recombiner	T1 or T1/2	(3P+N)/(3+1)	TT	CM/DM

CM: Common Mode, DM: Differential Mode

of using multiple inverters, there is a variety of communication architectures. In the below figures (Figure 5.118 and 5.119), the most commonly used communication options (i.e., ethernet-based and RS485-based communications) and SPD applications for the wired data lines are illustrated in the event of multiple inverter usages.

According to the configuration given in Figure 5.118, inverters are connected via CAT5 or CAT6-type ethernet cables. System data received from the inverters are transferred to the monitoring server through an ethernet router. In this configuration, the maximum distance per device connection is 100 m.

According to the configuration given in Figure 5.119, multiple inverters are connected in a master/slave arrangement through the same RS485 bus system. In this configuration, only the master

TABLE 5.65
The Required Parameters for Determining the Current and Voltage Ratings of SPDs

	Voltage Ratings				Current Ratings		
SPD Locations	$I_{oc\,max\,PV}$	$K \cdot V_o$	V_w [a]	$I_{sc\,max\,PV}$	I_{sc3} [b]	I_{imp} [c]	I_{max} [d]
DC box	$1.2 \times V_{oc\,PV} = 1.2 \cdot (37.13\,V \times 20)$ $= 891.12\,V$	NA	4000 V	$1.25 \times I_{sc\,PV} = 1.25 \times (8.96)$ $= 11.2\,A/string$	NA		50 kA/pole
Inverter DC side	891.12 V	NA	4000 V	11.2 A/string	NA	$I_{imp} = \dfrac{I_L}{2N_c}$ $= \dfrac{150\,kA}{2 \cdot 2}$ $= 37.5\,kA/pole$ 37.5 kA/pole	50 kA/pole
Inverter AC side	NA	$1.15 \times 230\,V$ $= 264.5\,V$ (L-N)	2500 V	NA	4.205 kA	$I_{imp} = \dfrac{I_L}{2N_c}$ $= \dfrac{200\,kA}{2 \cdot 4}$ $= 25\,kA/pole$	40 kA/pole
AC combiner	NA	264.5 V (L-N)	2500 V	NA	22.94 kA	25 kA/pole	40 kA/pole
AC recombiner	NA	264.5 V (L-N)	2500 V	NA	26.35 kA	25 kA/pole	40 kA/pole

NA: Not Applicable.

[a] From Table 5.57
[b] From Example 5.20; Note that $U_n = 400\,V$ (grid side, phase to phase voltage)
[c] Based on Table 5.58 and equation (5.112)
[d] As explained above, it will be more appropriate to choose the larger I_{max} value for longer durability. Therefore, the values for I_{max} maybe selected as given.

TABLE 5.66
Current and Voltage Ratings of SPDs

SPD Locations	Voltage Ratings			Current Ratings	
	$V_{c\,DC}$[a]	$V_{c\,AC}$	V_p[b]	I_{scw}	I_{imp}	I_{max}
DC box	$V_{oc\,max\,PV} \leq V_{c\,DC} \leq V_{max\,inv}$ $891.12\,V \leq V_{c\,DC} \leq 950\,V$	NA	$V_p \leq V_w$ $V_p \approx 0.8 \times 4000\,V$ $V_p \approx 3200\,V$	$I_{scw} \geq I_{sc\,max\,PV}$ $I_{scw} \geq 11.2\,A$	37.5 kA/pole	50 kA/pole
Inverter DC side	$891.12V \leq V_{c\,DC} \leq 950V$	NA	$V_p \approx 3200\,V$	$I_{scw} \geq 11.2\,A$	37.5 kA/pole	50 kA/pole
Inverter AC side	NA	$V_{c\,AC} \geq K \cdot V_o$ $V_{c\,AC} \geq 264.5\,V$ (L-N)	$V_p \leq V_w$ $V_p \approx 0.8 \times 2500\,V$ $V_p \approx 2000\,V$	$I_{scw} \geq I_{sc3}$ $I_{scw} \geq 4.205\,kA$	25 kA/pole	40 kA/pole
AC combiner	NA	$V_{c\,AC} \geq 264.5\,V$ (L-N)	$V_p \approx 2000\,V$	$I_{scw} \geq 22.94\,kA$	25 kA/pole	40 kA/pole
AC recombiner	NA	$V_{c\,AC} \geq 264.5\,V$ (L-N)	$V_p \approx 2000\,V$	$I_{scw} \geq 26.35\,kA$	25 kA/pole	40 kA/pole

NA: Not Applicable.
[a] See Example 5.20 for the inverter specifications.
[b] Typically, about 80% of the withstand voltage is selected.

TABLE 5.67
Voltage and Current Specification Summary for DC and AC SPDs

	DC SPD		AC SPD		
SPD Ratings	DC Box(Combiner)	Inverter DC Side	Inverter AC Side	AC Combiner	AC Recombiner
Max continuous operating voltage (V_c)	920 V	920 V	275 V (L-N)	275 V (L-N)	275 V (L-N)
Voltage protection level (V_p)	< 3200 V	< 3200 V	< 2000 V	< 2000 V	< 2000 V
Short-circuit withstand current (I_{scw})	> 11.2 A	> 11.2 A	> 4.205 kA	> 22.94 kA	> 26.35 kA
Impulse discharge current (I_{imp})	≥ 37.5 kA/pole	≥ 37.5 kA/pole	≥ 25 kA/pole	≥ 25 kA/pole	≥ 25 kA/pole
Maximum discharge current (I_{max})	50 kA/pole	50 kA/pole	40 kA/pole	40 kA/pole	40 kA/pole
Nominal discharge current (I_n)[a]	20 kA/pole	20 kA/pole	20 kA/pole	20 kA/pole	20 kA/pole

[a] It is recommended to select I_n value as high as possible ($I_{max} > I_n$) for longer durability.

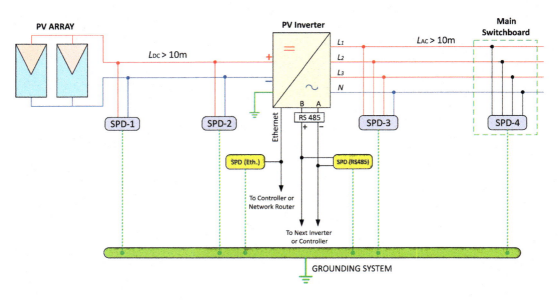

FIGURE 5.117 A typical PV system includes RS485 and Ethernet protection.

TABLE 5.68
SPD Selection and Installation Requirements for Photovoltaic Data Lines

Surge Protection for Data Lines	Wiring/Installation Requirements	Typical SPD Characteristics
Ethernet surge protection	CAT5 or CAT6 type cable can be used. The distance between the inverter and the router must be less than 100 m.	V_n = 5 V (nominal line voltage) I_n = 10 kA (Type-2: 8/20 μs) I_{max} = 20 kA (Type-2: 8/20 μs)
RS485 surge protection	Minimum 3-wire shielded twisted cable should be used (4-wire shielded twisted cable may also be used). The distance between the first and last devices must be less than 1000 m. The maximum number of nodes in the entire data line is 32.	V_n = 12 V/24 V (nominal line voltage) V_c = 16 V/30 V I_n = 5 kA (Type-2: 8/20 μs) I_{max} = 15 kA (Type-2: 8/20 μs) V_p = up to 70 V

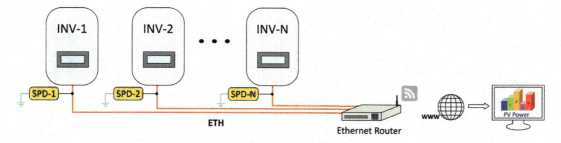

FIGURE 5.118 SPD usage in protections of ethernet lines for multiple inverter cases.

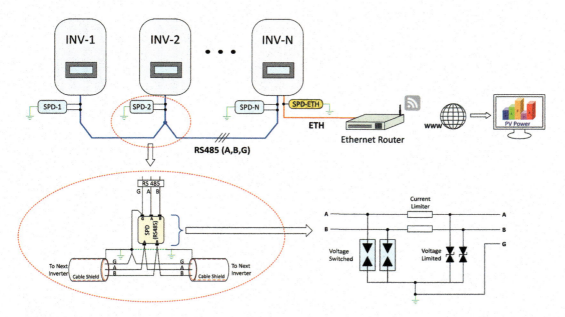

FIGURE 5.119 SPD usage in the protection of RS485 line for multiple inverter cases [G: Ground, A: Data A (+), B: Data B (−)].

inverter is physically connected to the internet through its ethernet port. The length of the CAT5 or CAT6 cable connecting the master device to the ethernet router can be a maximum of 100 m. The total length of RS485 cables from the first inverter to the last inverter should be a maximum of 1000m.

5.5.7.7 Residual Current Device Sizing and Selection for Photovoltaic Systems

Residual current devices (RCDs), also known as ground-fault interrupters (GFIs), are the protection devices specifically designed to provide additional safety for personnel (against electric shock) and devices/installations (against fault currents and fire hazards). RCDs disconnect the circuit when a leakage current above a predefined threshold value (tripping current value) is detected between the live conductor and return conductor (neutral conductor).

RCDs are normally designed to detect different types of current waveforms such as AC, DC, and combination of AC/DC. Hence, RCDs in general can be classified according to the waveform of the earth leakage currents as shown in Table 5.69.

To identify RCD requirements in photovoltaic installations, it is essential to examine the residual current flows in PV systems (also see Section *5.2.7.2. Residual Current Devices* for more details).

TABLE 5.69
RCD Classification Summary

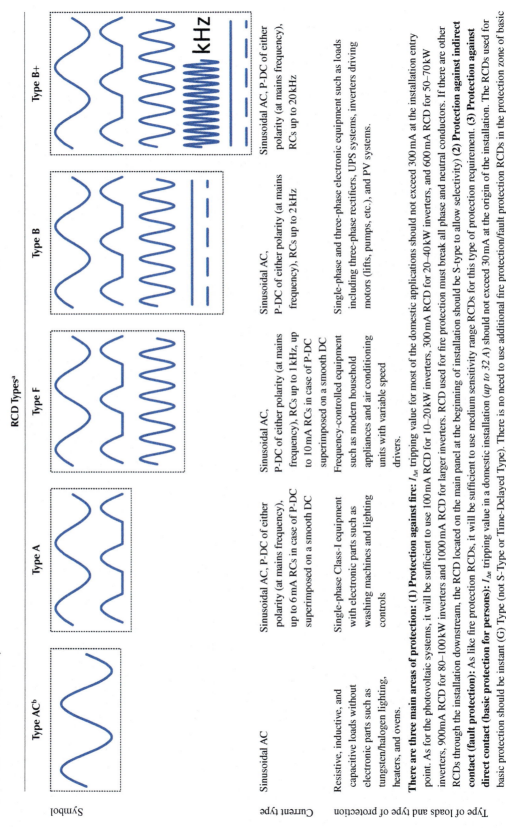

RCD Types[a]

	Type AC[b]	Type A	Type F	Type B	Type B+
Symbol	(sinusoidal waveform)	(sinusoidal waveform with P-DC)	(sinusoidal with P-DC and superimposed smooth DC)	(sinusoidal with P-DC, RCs up to 2 kHz, DC line)	(sinusoidal with P-DC, RCs up to 20 kHz, DC line)
Current type	Sinusoidal AC	Sinusoidal AC, P-DC of either polarity (at mains frequency), up to 6 mA RCs in case of P-DC superimposed on a smooth DC	Sinusoidal AC, P-DC of either polarity (at mains frequency), RCs up to 1 kHz, up to 10 mA RCs in case of P-DC superimposed on a smooth DC	Sinusoidal AC, P-DC of either polarity (at mains frequency), RCs up to 2 kHz.	Sinusoidal AC, P-DC of either polarity (at mains frequency), RCs up to 20 kHz.
Type of loads and type of protection	Resistive, inductive, and capacitive loads without electronic parts such as tungsten/halogen lighting, heaters, and ovens.	Single-phase Class-I equipment with electronic parts such as washing machines and lighting controls	Frequency-controlled equipment such as modern household appliances and air conditioning units with variable speed drivers.	Single-phase and three-phase electronic equipment such as loads including three-phase rectifiers, UPS systems, inverters driving motors (lifts, pumps, etc.), and PV systems.	

There are three main areas of protection: (1) Protection against fire: $I_{\Delta n}$ tripping value for most of the domestic applications should not exceed 300 mA at the installation entry point. As for the photovoltaic systems, it will be sufficient to use 100 mA RCD for 10–20 kW inverters, 300 mA RCD for 20–40 kW inverters, and 600 mA RCD for 50–70 kW inverters, 900 mA RCD for 80–100 kW inverters and 1000 mA RCD for larger inverters. RCD used for fire protection must break all phase and neutral conductors. If there are other RCDs through the installation downstream, the RCD located on the main panel at the beginning of installation should be S-type to allow selectivity **(2) Protection against indirect contact (fault protection):** As like fire protection RCDs, it will be sufficient to use medium sensitivity range RCDs for this type of protection requirement. **(3) Protection against direct contact (basic protection for persons):** $I_{\Delta n}$ tripping value in a domestic installation (*up to 32 A*) should not exceed 30 mA at the origin of the installation. The RCDs used for basic protection should be instant (G) Type (not S-Type or Time-Delayed Type). There is no need to use additional fire protection/fault protection RCDs in the protection zone of basic protection RCDs.

(*Continued*)

TABLE 5.69 (Continued)
RCD Classification Summary

	RCD Types[a]				
	Type AC[b]	Type A	Type F	Type B	Type B+
Typical specifications	**Number of poles:** 1P+N, 2P, 4P **Tripping current values ($I_{\Delta n}$):** 10 mA, 30 mA, 300 mA, 500 mA **Nominal voltages:** 230 Vac, 230/400 Vac **Nominal currents:** 16 A, 25 A, 40 A, 63 A, 80 A, 100 A; 125 A **Respond time:** 10 ms for Type G RCDs, 40 ms for Type S RCDs **Surge current proof (peak withstand current):** 3 kA or 5 kA **NOTE:** In addition, back-up fuse and rated short-circuit capacity values are provided for RCBO devices.				
Tripping sensitivity range	**High Sensitivity:** 5 mA – 10 mA – 30 mA (for direct contact and personnel safety) **Medium Sensitivity:** 100 mA, 300 mA, 500 mA, 1000 mA (for fire protection) **Low Sensitivity:** 3 A, 10 A, 30 A (for protection of machines)				

RC: Residual Current, P-DC: Pulsating DC

[a] According to IEC 60755, only three types of RCD are defined. RCDs can further be divided into two modes, which are General Use (G type: instantaneous) and Selective (S Type: time-delayed) RCDs.

[b] Not authorized in every country.

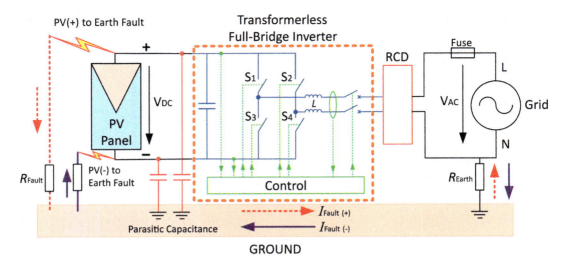

FIGURE 5.120 Residual current flows in a PV installation with faults of "PV(+) to Earth" and "PV(−) to Earth" at the DC side.

As shown in Figure 5.120, possible ground faults on the DC side (PV+ to Earth or PV− to Earth faults) may cause residual currents on the AC part of the PV system having a non-isolated inverter.

Normally, residual current waveform (DC component, AC component, and its frequency content) in a PV system depends on the inverter design/topology, inverter control strategy/switching frequency, PV fault type/location, and the DC voltage of the solar generator. Therefore, possible residual current patterns in PV systems can be quite diverse. As an example, let us consider the topology given in Figure 5.120. Assuming different switching strategies, residual current waveforms and RCD type selections can be given for the possible earth faults (PV+ to Earth, PV− to Earth) as outlined in Table 5.70.

It is clear from Table 5.70 that a transformerless inverter can pass DC residual currents to the AC side unless this transition is somehow prevented by the inverter design. For example, DC residual current resulting from a ground fault on the DC side can be transmitted to the AC side through the AC neutral-to-earth connection, and back to the DC side through the non-isolated inverter. However, an inverter with galvanic isolation between AC and DC sides does not transmit DC residual currents to the AC side.

Most of the currently available photovoltaic inverters include RCD or RCMU (Residual Current Monitoring Unit) functions. Therefore, in practice, it should be checked if the RCD inside the inverter meets the system requirements. Otherwise, an external RCD set at 300 mA (or higher for larger systems) is required at the AC side of the inverter. Whether it is integrated into an inverter or not, an RCMU does not have any disconnection function. Instead, it only activates an alarm if a residual current is detected.

Considering the aforementioned explanations, the RCD selection and specifying steps for photovoltaic systems can be outlined below:

Step 1 – RCD Requirement Analysis and Type Determination: Photovoltaic systems are at risk of fire due to ground faults. The fuses used in the system may not be sufficient to clear the earth faults. Therefore, it is extremely important to detect residual currents and accordingly, to determine the RCD requirements in PV systems. Firstly, it should be analyzed whether the system protection devices can provide adequate safety for installations, fires, and individuals. For this purpose, the following issues should be examined to determine the RCD requirement for the given PV system. For instance:

TABLE 5.70
Residual Current Waveforms and RCD Selection for Different Switching Techniques and Fault Conditions

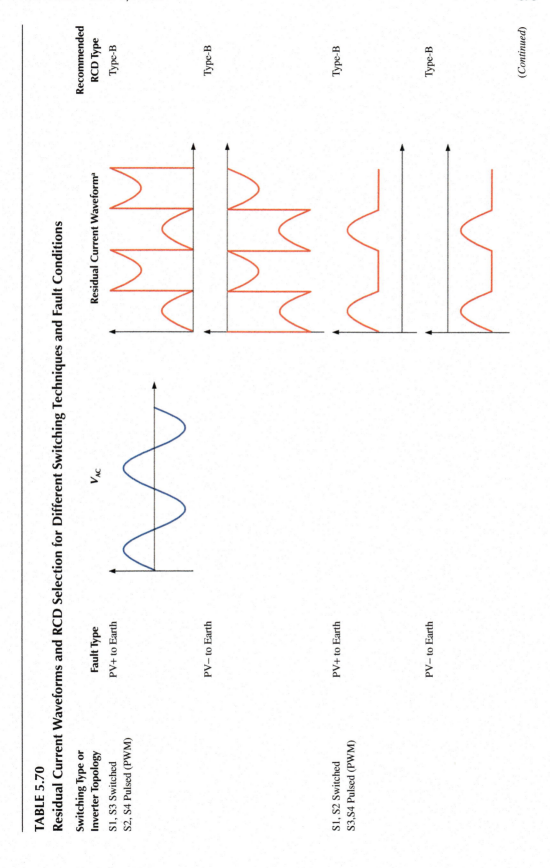

Switching Type or Inverter Topology	Fault Type	Residual Current Waveform[a]	Recommended RCD Type
S1, S3 Switched S2, S4 Pulsed (PWM)	PV+ to Earth		Type-B
	PV− to Earth		Type-B
S1, S2 Switched S3, S4 Pulsed (PWM)	PV+ to Earth		Type-B
	PV− to Earth		Type-B

(*Continued*)

TABLE 5.70 (Continued)
Residual Current Waveforms and RCD Selection for Different Switching Techniques and Fault Conditions

Switching Type or Inverter Topology	Fault Type	V_{AC}	Residual Current Waveform[a]	Recommended RCD Type
S1, S2 Pulsed (PWM) S3, S4 Switched	PV+ to Earth			Type-B
	PV− to Earth			Type-B
All Switches Pulsed (PWM)	PV+ to Earth			Type-B
	PV− to Earth			Type-B

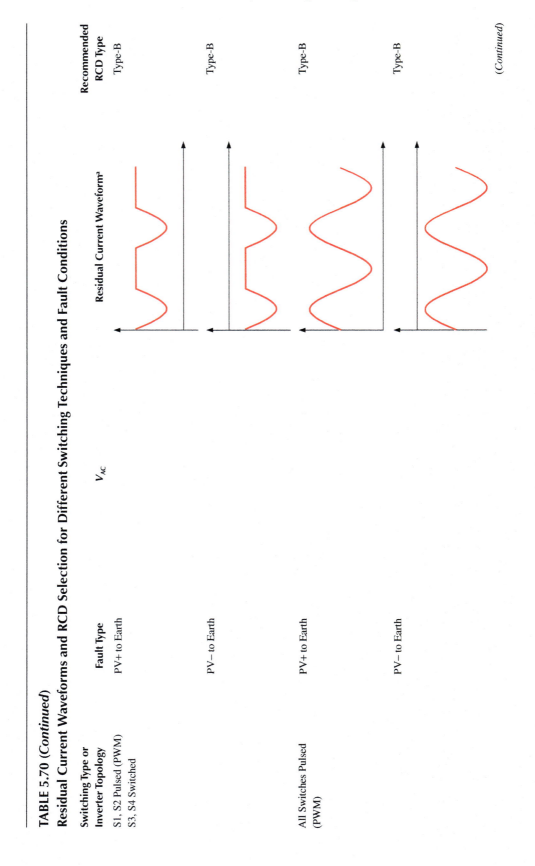

(Continued)

Photovoltaic Power Systems

TABLE 5.70 (Continued)
Residual Current Waveforms and RCD Selection for Different Switching Techniques and Fault Conditions

Switching Type or Inverter Topology	Fault Type	V_{AC}	Residual Current Waveform[a]	Recommended RCD Type
With Separated DC Link and Step-Up Converter	PV+ to Earth		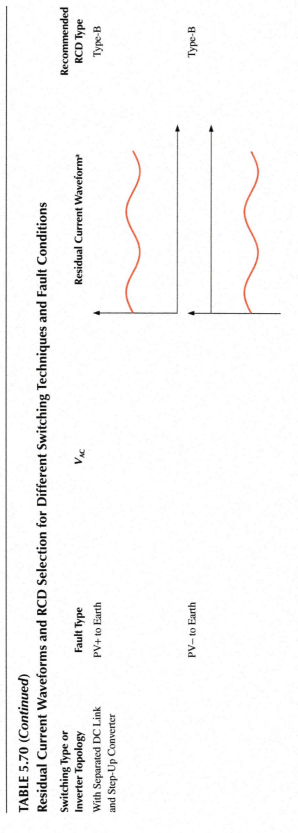	Type-B
	PV− to Earth			Type-B

[a] Residual current waveforms are given without high-frequency content.

- Does the inverter have an isolation transformer? If not, is there any prevention against DC residual currents in the PV inverter?
- Does the inverter contain an RCD function? If so, does the RCD function meet the system protection needs?

Considering the above-mentioned criteria, the following simple algorithm (Figure 5.121) can be created to determine the RCD requirement and type determination for a photovoltaic system.

Step 2 – Determination of RCD Location: According to international standards, RCDs should be installed on the AC side of a PV installation (inverter AC side) to obtain protection for individuals and fires. If the PV system includes more than one inverter, a separate RCD should be added to the AC side of each inverter. The location and the number of RCDs to be used in photovoltaics can be different depending on the system configuration. For example, let's examine the PV system with a load branch in Figure 5.32 (see Section *5.2.7.2. Residual Current Devices*). If a transformerless inverter is selected for the given system, then it is necessary to use two RCDs as shown in Figure 5.32b, the first RCD before the load point and the second one after the load point. A single RCD after load point may be sufficient if an isolated inverter is selected instead of a transformerless one.

The location of the SPDs in the system should also be taken into account when determining the location of RCDs. Because every earth leakage current as a result of SPD activation is evaluated as residual current by the RCDs, which may result in false tripping in the system. For this reason, the coordination of an RCD with SPD in the system is very important and explained in detail below.

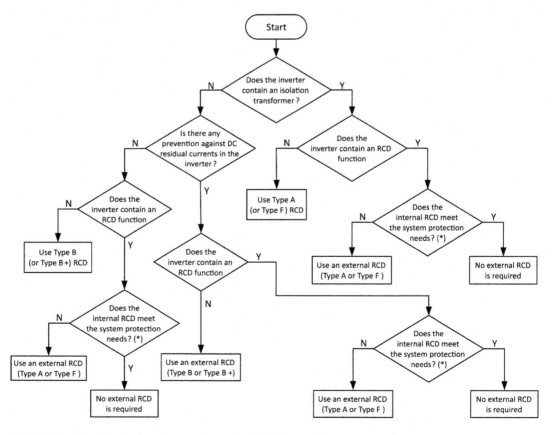

FIGURE 5.121 A basic flowchart to determine the RCD requirement and type determination for photovoltaic systems. (*) RCDs shall be evaluated according to the protection purposes and their specification criteria given in Step-3.

Photovoltaic Power Systems

Coordination with SPD: From the perspective of an RCD, all ground-leakage currents are perceived as residual currents. For this reason, RCDs should only be installed after an SPD (or SPDs before RCD). In other words, SPDs should be installed upstream of RCDs to avoid false tripping due to lightning discharges and transient overvoltages. The correct utilization of an RCD in coordination with an SPD is shown in Figure 5.122.

Note that if an SPD has to be installed downstream of an RCD, the RCD should be of the time-delayed (S Type) to prevent unwanted tripping. In this case, the S-Type RCD should have a minimum surge current immunity of 3 kA with an 8/20 μs waveform. In other words, the expected surge currents to earth should not exceed the immunity level of the RCD. Proper selection of RCD operating times minimizes the unwanted trips in the system. Hence, maximum tripping (operating) times for G- and S-Type RCDs given in Table 5.71 should be considered to design a robust system. According to international standard IEC 61008-1, RCDs should trip if the residual current level is higher than the rated residual current ($I_{\Delta n}$:100%) and the RCDs should not trip if the residual current level is smaller than $0.5 I_{\Delta n}$ (50%).

Step 3 – Specification of RCD Ratings: The RCDs are designed at values of current, voltage, and break times which should properly be related to the circuit characteristics, and the risk level of

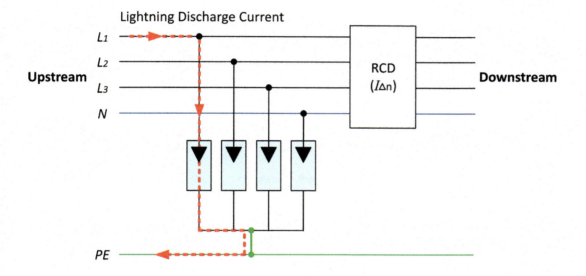

FIGURE 5.122 RCD properly installed on the downstream side of the SPD.

TABLE 5.71
Maximum Break Times for G- and S-Type RCDs

Residual Currents	$I_{\Delta n}$	$2 \times I_{\Delta n}$	$5 \times I_{\Delta n}$
Maximum break time for G-type RCD (for any I_n and $I_{\Delta n}$ values)	0.3 s	0.15 s	0.04 s
Maximum break time for S-type RCD (for $I_n > 25$ A and $I_{\Delta n} > 30$ mA)	0.5 s	0.3 s	0.15 s
Minimum non-actuating time for S-type RCD (for upstream RCD)a	0.13 s	0.06 s	0.05 s
Residual Current Examplesb	**30 mA**	**60 mA**	**150 mA**
Maximum break times	0.3 s	0.15 s	0.04 s

[a] Minimum non-actuating time of the upstream RCD should be higher than the maximum trip time of the RCDs installed downstream.
[b] 30 mA value can be selected as the protection level for general-purpose RCDs.

the system. In this sense, the following additional rules regarding RCD ratings must also be specified before application.

- Since the working range of an RCD is normally from $0.5I_{\Delta n}$ to $1I_{\Delta n}$, the operational leakage current of the protected equipment ($I_{\Delta i}$) should not exceed 30%–40% of the RCD sensitivity to prevent unnecessary tripping. The load factor approach can be used in calculating the total leakage current. Hence, the calculation of the total leakage current ($I_{\Delta total}$) of an electrical installation can be found by multiplying the arithmetic sum of the individual leakage currents by the load factor (L_f) of the system (typically around 70%–80%).

$$\left. \begin{array}{l} I_{\Delta total} = L_f \left[\sum_{i=1}^{N} I_{\Delta i} \right] \\ I_{\Delta n} \geq (I_{\Delta total})/0.3 \end{array} \right\} \quad (5.115)$$

where index "i" represents the individual loads and N represents the total number of loads in the system.
- The nominal tripping current of the upstream RCD should be rated at least three times the nominal tripping current of the RCDs installed downstream.

$$I_{\Delta n(\text{upstream})} \geq 3 \times I_{\Delta n(\text{downstream})} \quad (5.116)$$

- The current rating of the RCD ($I_{n\ RCD}$) should not be less than the maximum load demand ($I_{L\ max}$) of the circuit being protected.

$$I_{n\ RCD} \geq I_{L\ max} \quad (5.117)$$

- The current rating of the RCD ($I_{n\ RCD}$) should not be less than the highest current rating of any overload protective device in the circuit part being protected.

$$I_{n\ RCD} \geq \max\left(I_{pd1}, I_{pd2}, \ldots, I_{pdN}\right) \quad (5.118)$$

- where I_{pdi} represents the current rating of ith a protective device and N represents the current rating of the Nth protective device in the circuit part being protected. Importantly note that meeting any of equation (5.117) or (5.118) is sufficient for RCD selection.

Combination of RCDs with other protective devices: A pure residual current device can detect the current imbalance between supply conductors and return conductors, but they cannot provide protection against overloads and short circuits. Hence, RCDs in practice can be used as a pure leakage current detection device or can be integrated into circuit breakers (CBs) such as residual current breakers without overcurrent protection (RCCB) and residual current breakers with overcurrent protection (RCBO). In other words, an RCCB acts as an RCD and a circuit breaker, while an RCBO combines three protection devices, that is, RCD, fuse, and circuit breaker. However, although the characteristics of RCDs and CBs used as pure devices are quite diverse, the RCCB/RCBO characteristics provided by the manufacturers are limited in practice. For this reason, it should be ensured

TABLE 5.72
Typical Characteristic Values for RCCB and RCBO

Typical Parameters	RCCB	RCBO
Rated voltage	230/400 V or 240/415 V	230/400 V or 240/415 V
Rated currents	10 A-16 A-20 A-25 A-32 A-40 A-63 A-80 A-100 A-125 A	10 A-16 A-20 A-25 A-32 A-40 A-63 A-80 A-100 A-125 A
Rated frequency	50/60 Hz	50/60 Hz
Number of poles and current paths	1P with two current paths; 2P; 3P; 3P with four current paths; 4P	1P with one OPP and UN; 2P with one OPP; 2P with two OPPs; 3P with three OPPs; 3P with three OPPs and UN; 4P with three OPPs; 4P with four OPPs
Rated residual currents	30 mA, 300 mA	30 mA, 300 mA
Rated breaking capacities	3 kA, 4.5 kA, 6 kA, 10 kA	3 kA, 4.5 kA, 6 kA, 10 kA
Time delay	With (S Type) and without (G Type)	With (S Type) and without (G Type)

OPP: Overcurrent Protected Pole, UN: Uninterrupted Neutral

to see if the available RCCB/RCBO provides suitable protection for the system. Otherwise, CBs and RCDs must be used separately in the system.

When specifying the RCCB/RCBO, the characteristic values given in Table 5.72 are needed to be selected. The manufacturer's data should be checked to ensure that appropriate parameters have been selected for the application of interest.

Options for disconnection and their selection (CB, RCD, or RCBO?): According to international standard IEC 60364-4-41, automatic disconnection can be achieved through a circuit breaker or a residual current device. Hence, it should initially be examined if the circuit breaker given is sufficient for automatic line protection or not. In this sense, ground faults for different grid configurations and respective disconnection options are summarized in Table 5.73.

TABLE 5.73
Ground Faults and Disconnection Options for Different Grid Configurations

Grid Types	Ground Faults and Automatic Disconnection	Explanations and Disconnection Options
TT System	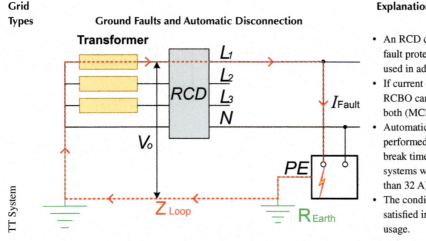	• An RCD can be used as primary fault protection, or RCDs can be used in addition to circuit breakers. • If current conditions are suitable, RCBO can also be used instead of both (MCB and RCD). • Automatic disconnection should be performed within the permissible break time (0.2 s for 230/400 V TT systems with a current value less than 32 A). • The condition $I_{Fault} \geq I_a$ must be satisfied in case of circuit breaker usage.

(Continued)

TABLE 5.73 (*Continued*)
Ground Faults and Disconnection Options for Different Grid Configurations

Grid Types	Ground Faults and Automatic Disconnection	Explanations and Disconnection Options
TN-S System / TN-C System / TN-CS System	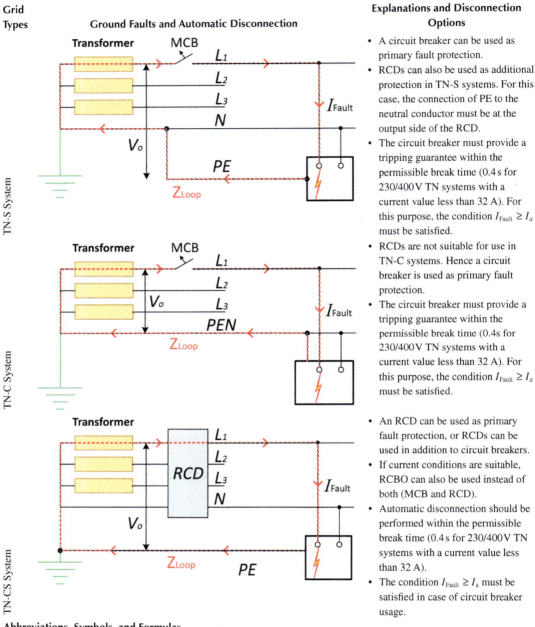	• A circuit breaker can be used as primary fault protection. • RCDs can also be used as additional protection in TN-S systems. For this case, the connection of PE to the neutral conductor must be at the output side of the RCD. • The circuit breaker must provide a tripping guarantee within the permissible break time (0.4 s for 230/400 V TN systems with a current value less than 32 A). For this purpose, the condition $I_{Fault} \geq I_a$ must be satisfied. • RCDs are not suitable for use in TN-C systems. Hence a circuit breaker is used as primary fault protection. • The circuit breaker must provide a tripping guarantee within the permissible break time (0.4 s for 230/400 V TN systems with a current value less than 32 A). For this purpose, the condition $I_{Fault} \geq I_a$ must be satisfied. • An RCD can be used as primary fault protection, or RCDs can be used in addition to circuit breakers. • If current conditions are suitable, RCBO can also be used instead of both (MCB and RCD). • Automatic disconnection should be performed within the permissible break time (0.4 s for 230/400 V TN systems with a current value less than 32 A). • The condition $I_{Fault} \geq I_a$ must be satisfied in case of circuit breaker usage.

Abbreviations, Symbols, and Formulas

MCM: Miniature circuit breaker (traditionally up to 125 A). **RCBO:** Residual current breaker with overcurrent protection (traditionally up to 40 A). V_o: Line-to-ground AC voltage. R_{Earth}: Total resistance of grounding conductor and electrode (typically up to 3 Ω). Z_{Loop}: Total impedance of the fault loop (typically up to 3 Ω for TN systems, and up to 100 Ω for TT systems). $I_a = k \cdot I_{nom}$: Short-circuit operating current of the circuit breaker, where the coefficient k is 3–5 for the trip curve B (suitable for resistive loads and fault loop impedances), 5–10 for the trip curve C (suitable for common loads such as lighting, sockets, and small motors), and 10–20 for the trip curve D (suitable for high inrush loads). I_{Fault}: Fault current flowing through the fault loop [equal to V_o/Z_{Loop} for TN systems and $V_o/(Z_{Loop} + R_{Earth})$ for TT systems].

Example 5.23: RCD Sizing and Selection for a Roof-Mounted Photovoltaic System

Consider the 10 kW roof-mounted PV system worked in Example 5.21. This grid-connected PV system also feeds the local residential loads with a maximum power of 6 kW. Assume that all given data in Example 5.21 is also valid for this example. Work out the following examples.

a. Determine the RCD requirements of the PV system and size them for the required locations (the RCD requirement for the in-house distribution board will also be considered). If the average leakage current value for residential loads is 1.5 mA/kW, perform the sensitivity analysis for the selected indoor RCD (Load Factor = 1).
b. Evaluate the alternative solution options for the potential earth faults and size the relevant protective devices (such as MCB and RCBO). Assume that the Z_{Loop} impedance due to ground fault is 1.5 Ω in evaluating the use of MCB or RCBO.

Solution:

a. The above steps regarding RCD application/sizing include RCD requirement assessment, type determination, RCD localization, and rating specifications. Hence as a first step, let's draw the PV system's single-line diagram with its technical data to be able to work out the RCD sizing steps described above. According to the design in Example 5.21, the grid-connected PV system includes a 10 kW PV array and a load branch as shown in Figure 5.123.

Let us now follow the RCD sizing/specifying steps as below.

Steps 1 and 2 – Requirement Analysis, Type Determination, and RCD Localization: Based on the system in Figure 5.123, we can determine the RCD locations and types as shown in Table 5.74.

Step 3 – Specifying RCD Ratings: The ratings can be specified for each RCD as in Table 5.75.

b. To provide a tripping guarantee in case of circuit breaker usage, the condition $I_{Fault} \geq I_a$ must be satisfied. It is given that Z_{Loop} value is 1.5 ohms. Hence, the fault current I_{Fault} is calculated as $I_{Fault} = 230 / 1.5 = 153.3$ A. Since short-circuit operating current of the

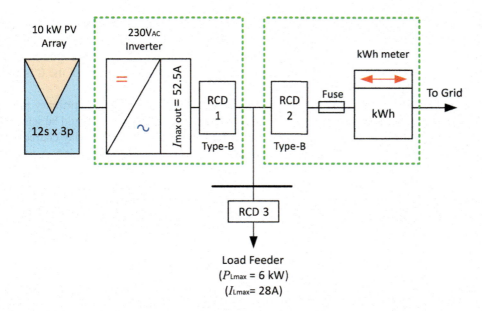

FIGURE 5.123 10kWp grid-connected PV system with RCD application.

TABLE 5.74
RCD Locations and Types for the PV System in Example

	RCD Locations and Types		
Inverter Configuration[a]	Inverter AC Side (RCD-1)	Service Entrance Box (RCD-2)	In-house Distribution Panel (RCD-3)
Inverter without transformer[b]	Type-B	Type-B	Type-B
Inverter with transformer	Not required	Type-A	Type-B

[a] The inverter does not contain an RCD unit.
[b] To provide personnel safety, the RCD must have a residual current rating of 30 mA. However, attention should also be paid to the potential high leakage currents. For example, the inverter may have a larger standing leakage current, or the insulation of the equipment may be of poor quality and wear out faster.

circuit breaker $I_a = k \cdot I_{nom}$, the condition to be met is $I_{Fault} \geq k \cdot I_{nom}$. By considering the tripping curve B ($k = 5$), we can choose the MCB with a rated current of less than 30.66 A ($I_{nom} \leq I_{Fault}/k$). For example, we can choose an MCB with a rated current of 20 A. In this case, it has a short-circuit operating current I_a of 100 A (5×20 A), which is less than 153.3 A ($I_a \leq I_{Fault}$).

As explained above, minimum breaking capacity is another parameter needed when selecting MCB or RCBO. From the short-circuit calculations in Example 5.21, the breaking capacity must be greater than 2020 A. For this reason, the rated breaking capacity can be selected as 3 kA (see Table 5.72).

5.5.7.8 Specification and Selection of Instrument Transformers

Instrument transformers (i.e., current and voltage transformers) in power systems are generally used for monitoring or protection purposes. This section particularly focuses on the application of instrument transformers to the protective relays of solar PV systems. To design a cost-effective and reliable measurement/protection system, the following steps can be created for the current transformer (CT) and voltage transformer (VT) separately.

Current Transformer (CT): Normally, there are several criteria needed to be considered when selecting a CT. The considerations include the type of use, ambient conditions, primary and secondary rated currents (I_{pn} and I_{sn}), accuracy class, withstand voltage, overcurrent strength, overcurrent constant, and rated burden criteria. In the following steps, the selection criteria of the CTs particularly used in the relay circuits of solar PV systems are explained.

Step 1 – Determine the Current Transformation Ratio (I_{pn}/I_{sn}): For the primary rated current of the CT, the next standard value above the highest current in the circuit should be selected. As for the secondary current, choose 1 A or 5 A depending on the instrument or relay characteristics. If the distance between CT and relay is less than 10 m (close measurement), use 5 A secondary. In cases where the distance is greater than 10 m (remote measurement), 1 A secondary current is preferably selected to reduce the burden on the CTs. Note that Joule losses on the lead wires with 5 A are 25 times greater than with 1 A (due to $I^2 R$ losses). Considering the protective relays on the high and low-voltage sides of a power transformer, the CT primary current can be calculated as in equation (5.119).

$$I_p = \begin{cases} \dfrac{S_T}{\sqrt{3} \times U_2}, & \text{for HV side} \\ \dfrac{S_T}{\sqrt{3} \times U_1}, & \text{for LV side} \end{cases} \quad (5.119)$$

TABLE 5.75
Current and Voltage Ratings of RCDs

RCD Locations	Specifications
Inverter AC side (RCD-1)	**Number of poles:** 1P+N or 2P (suitable for single-phase circuits) **Tripping current value** ($I_{\Delta n}$): 30 mA (suitable for personnel protection and general-purpose protection). However, RCDs with lower trip sensitivity such as 100 mA (or above), can also be used to avoid unwanted tripping. In this case, 100 mA trip sensitivity shall be suitable for 10 kW inverter protection or protection against fire hazards due to earth fault currents. **Nominal voltage:** 230 Vac or 240 Vac (It must be compatible with the inverter output voltage) **Nominal current:** 63 A [The inverter output current is 10500 W/230 V = 45.65 A. Therefore, the inverter output can be protected with an overcurrent protection device of 57 A (1.25 × 45.65 = 57 A). In this case, it is sufficient to have $I_{n\,RCD} \geq 57$ A. As a result, the next standard value 63 A can be selected. $I_{n\,RCD} = 63$ A] **Respond time:** There is no need for a hierarchy between RCDs in terms of response time. Therefore, Type G RCDs can be selected (its response time is 10 ms). **Surge current proof (peak withstand current):** 3 kA or 5 kA.
Service entrance box (RCD-2)	Same as RCD-1 specifications.
In-house distribution panel (RCD-3)	**Number of poles:** 1P+N or 2P (suitable for single-phase circuits) **Tripping current value** ($I_{\Delta n}$): 30 mA (suitable for personnel protection and general-purpose protection). **NOTE:** $$I_{\Delta total} = L_f \left[\sum_{i=1}^{N} I_{\Delta i}\right] \Rightarrow I_{\Delta total} = 1 \times \left[(1.5\text{ mA / kW}) \times 6\text{ kW}\right]$$ $= 9\text{ mA} \Rightarrow I_{\Delta n} \geq (I_{\Delta total})/0.3 \Rightarrow I_{\Delta n} \geq \dfrac{9}{0.3} \Rightarrow I_{\Delta n} \geq 30\text{ mA} \Rightarrow I_{\Delta n} = 30\text{ mA}$ can be selected. **Impact of Standing Leakage Current on Tripping Sensitivity:** Two cases are considered. In the first case, let's assume that the average leakage current value is 2.5 mA/kW. What is the tripping sensitivity of the RCD that should be used for a 6 kW load? The answer is: $I_{\Delta total} = 1 \times \left[(2.5\text{ mA/kW}) \times 6\text{kW}\right] = 15\text{mA} \Rightarrow I_{\Delta n} \geq \dfrac{I_{\Delta total}}{0.3} \Rightarrow I_{\Delta n} \geq \dfrac{15}{0.3} \Rightarrow I_{\Delta n} \geq 50\text{mA}$. Hence, the next available value 60 mA can be selected ($I_{\Delta n} = 60$ mA). In the second case, let's assume that the average leakage current value is 1 mA/kW. What is the maximum load level that can be connected to a 30 mA RCD? So, maximum allowable leakage current = 30 mA × 30% = 9 mA $\Rightarrow$ Max load = 9 mA/(1 mA/kW) = 9 kW. **Nominal voltage:** 230 Vac or 240 Vac (It must be compatible with the network voltage). **Nominal current:** 32 A [The max load current is 6000 W/230 V = 26 A. In this case, it is sufficient to have $I_{n\,RCD} \geq 26$ A $(I_{n\,RCD} \geq I_{L\,max})$. As a result, the next standard value 32 A can be selected. $I_{n\,RCD} = 32$ A]. **Respond time:** There is no need for a hierarchy between RCDs in terms of response time. Therefore, Type G RCDs can be selected (its response time is 10 ms). **Surge current proof (peak withstand current):** 3 kA or 5 kA.

where S_T is the transformer apparent power in [VA], U_1 is the phase-to-phase voltage (V) on the low-voltage side of the power transformer, and U_2 is the phase-to-phase voltage (V) on the high-voltage side of the power transformer. As explained above, the nominal primary current of the CT (I_{pn}) is selected as the next available value greater than the highest current I_p in the circuit.

$$I_{pn} > I_p \tag{5.120}$$

To select suitable values of I_{pn} (defined by equation 5.120), standard primary currents of CTs are listed in Table 5.76.

TABLE 5.76
Available Primary Current (A) Values for CTs

5	10	12.5	15	20	25	30	40	50	60	75	100
120	150	160	180	200	240	250	300	400	500	600	750
800	1000	1250	1500	1600	2000	2500	3000	4000	5000	6000	7500

Considering equations (5.119), (5.120), and Table 5.76, the rated transformation ratio is determined as I_{pn}/I_{sn} (Irated primary/Irated secondary).

Step 2 – Determine the Thermal and Dynamic Current Values: CTs should be designed to withstand continuous and short-time thermal currents as well as dynamic currents. Therefore, withstand capability of CTs to these currents must be checked before the final decision.

The continuous rated thermal current (I_{thc}) is the current which can be allowed continuously in the primary without causing a temperature increase beyond the specified values in the standards. Normally, I_{thc} equals to the rating factor (RF) times rated primary current ($I_{thc} = RF \times I_{pn}$). However, in most cases, RF is equal to 1 unless otherwise specified. In this case, I_{thc} will equal to the nominal primary current ($I_{thc} = I_{pn}$). Note that typical RF values in standards are given as 1, 1.2, 1.5, and 2.

Short-time thermal current (I_{th}) is the maximum current value that the CT can withstand for 1 second. The maximum current value in power systems is the short-circuit current. Therefore, the maximum short-circuit current at the point of CT application is equal to the short-time thermal current value. Accordingly, I_{th} value is equal to a three-phase short-circuit current (I_{sc3}) value in three-phase systems and phase-neutral short-circuit current (I_{sc1}) value in single-phase systems.

$$I_{th} = \begin{cases} I_{sc3} & \text{for three} - \text{phase system} \\ I_{sc1} & \text{for single} - \text{phase system} \end{cases} \quad (5.121)$$

If the short-circuit power at the installation point of CT is known, I_{th} value can be calculated as below.

$$I_{th} = \frac{S_k}{\sqrt{3} \times U_n} \quad (5.122)$$

where S_k is the short-circuit power at the installation point of CT in MVA and U_n is rated system voltage (line-to-line voltage) in kV. Note that if I_{th} is given for values other than 1 second, calculate the value for 1 second using equation (5.123).

$$I_{th}\left(\text{for 1 second}\right) = I_{th}\left(\text{for } t \text{ second}\right) \times \sqrt{t} \quad (5.123)$$

The current I_{th} has a thermal effect on the primary windings of the CT. In case of a short circuit, the peak value can go up to approximately $2.5 \times I_{th}$ for a 50Hz system. This current causes electromagnetic forces on the CT's primary windings and its connections. Hence, withstand capability of CTs to these dynamic currents must also be checked. Note that according to IEC standard, the minimum dynamic current of a CT should be greater or equal to $2.5 \times I_{th}$ for 50 Hz systems and $2.6 \times I_{th}$ for 60 Hz systems. According to ANSI standard, this value should equal $2.7 \times I_{th}$.

Note that withstand limits of CTs against short-time thermal currents are given in multiples of nominal primary current, as listed in Table 5.77.

Step 3 – Determine the CT Burden and Apparent Power Values: To precisely determine the output of a CT in VA, a burden to be connected across the transformer should be identified.

TABLE 5.77
Typical Short-Time Thermal Current Withstands Values for CTs

$60 \times I_{pn}$	$80 \times I_{pn}$	$100 \times I_{pn}$	$120 \times I_{pn}$	$150 \times I_{pn}$	$200 \times I_{pn}$	$300 \times I_{pn}$	$400 \times I_{pn}$	$500 \times I_{pn}$	$600 \times I_{pn}$

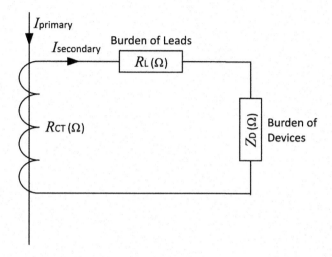

FIGURE 5.124 Representation of a CT and its burdens.

To be able to calculate the burden of a current transformer, the CT and its burdens are illustrated in Figure 5.124.

It is clear from Figure 5.124 that the total burden of a CT (Z_B) consists of lead wire burden (R_L), CT's secondary winding resistance (R_{CT}), and device burden (Z_D). Hence, the total burden of a CT in VA can be calculated as in (5.124).

$$\text{CT Burden [VA]} = I_{sn}^2 \times Z_B \qquad (5.124)$$

where I_{sn} is the nominal secondary current in [A] and Z_B [Ω] is the total burden of the CT ($Z_B = R_{CT} + R_L + Z_D$).

The most suitable sources for obtaining R_{CT}, R_L, and Z_D values are the manufacturer's catalogs. Typical burden values and output powers [VA] of current transformers are given in Table 5.78.

Step 4 – Determine the Accuracy Classes (Metering and Protection Classes): Possible faults or measurement errors in current transformers can cause the protection devices to malfunction, equipment damage, and metering error. For this reason, it is extremely important to select metering and protection classes properly to ensure the safe and reliable operation of CTs.

The metering and protection class definitions of CTs and their characteristic behaviors are given in Table 5.79.

Voltage Transformer (VT): The VT selection criteria include the type of use, ambient conditions, primary and secondary rated voltages (V_{pn} and V_{sn}), accuracy class, withstand voltage, and rated burden parameters. In the following steps, the selection criteria of the VTs particularly used in the relay/metering circuits of solar PV systems are explained.

Step 1 – Determine the Voltage Ratio (V_{pn}/V_{sn}): For the primary rated voltage of the VT, determine the voltage relative to the nominal service voltage of the network. As for the secondary, choose the rated voltage according to the input requirements of the instrument or relay to be fed. Note that

TABLE 5.78
Typical Burdens and Apparent Powers of Current Transformers

R_{CT} Values for Typical CTs [ohms @ 75°C]			
50 A:5 A	75 A:5 A	100 A:5 A	150 A:5 A
0.014 Ω	0.042 Ω	0.056 Ω	0.121 Ω
	R_L Values for Different Lead Wires[a]		
1.5 mm² Cu Cable	2.5 mm² Cu Cable	4 mm² Cu Cable	6 mm² Cu Cable
0.0133 [Ω/m]	0.00798 [Ω/m]	0.00495 [Ω/m]	0.00365 [Ω/m]
	Instrument Burdens in VA		
Protection Relay	Electricity Meters	AC VAR Meters	Amp/Volt Meters
0.2 to 30 VA	0.5 to 5 VA	1 to 5 VA	1 to 5 VA
	Typical Apparent Powers of CTs [VA]		
	1.5 – 3 – 5 – 10 – 15 – 20 – 30 – 45 – 60 – 100		

[a] Round-trip length should be considered.

there are two types of VTs, that is, earthed and unearthed types. Since the solar PV power plants have a grounded circuit assembly, it would be appropriate to choose earthed type VT for the relay/metering circuits of the PV systems. Accordingly, it would be appropriate to select a single-phase earthed VT with a voltage ratio of $\left[V_{pn}/V_{sn} = \left(\frac{33000}{\sqrt{3}}\right) / \left(\frac{110}{\sqrt{3}}\right) \right]$ for the high-voltage side of an MV transformer station (31.5 kV). If two devices are to be connected to the voltage transformer, the VT must have a double secondary. For example, if two meters are used on the MV side, then a VT with double secondary $\left[\left(\frac{33000}{\sqrt{3}}\right) / \left(\frac{110}{\sqrt{3}}\right), \left(\frac{110}{\sqrt{3}}\right) \right]$ can be selected for a 31.5 kV transformer.

Step 2 – Determine the Voltage Factor: Unbalanced fault conditions on unearthed or impedance earthed systems may result in voltage rises on the healthy phase lines. For this reason, VTs may encounter overvoltages under unbalanced fault conditions, which may cause a malfunction in relay operation. Hence a quantity called voltage factor (VF) is defined to determine the upper limits of VT operating voltage. The VF of a VT should be determined according to the system earthing conditions and primary winding connections. Permissible durations of VTs' maximum voltage are given in Table 5.80 for different earthing systems and VT connections.

Step 3 – Determine the VT Burden and Apparent Power Values: As in the current transformer burden calculation, the burden of a VT in VA can be calculated as in (5.125).

$$\text{VT Burden [VA]} = \frac{V_{sn}^2}{Z_B} \quad (5.125)$$

where V_{sn} is the nominal secondary voltage in [V] and Z_B [Ω] is the total burden of the VT connected across the transformer. The impedance Z_B ($Z_B = R_{CT} + R_L + Z_D$) consists of lead wire burden (R_L), VT's secondary winding resistance (R_{CT}), and device burden (Z_D). Typical apparent powers [VA] of voltage transformers are given in Table 5.81. Please see Table 5.78 for instrument and lead wire burden values.

Step 4 – Determine the Accuracy Classes (Metering and Protection Classes): Table 5.82 gives the metering and protection class requirements of VTs according to IEC 60044-2.

CT and VT in Protective Relaying: An example of a protective relay system for a circuit breaker is given in Figure 5.125. Multiple protective functions can be performed by a single protective relay system similar to the one in Figure 5.125. For example, the most commonly used protection

Photovoltaic Power Systems

TABLE 5.79
Accuracy Classes, Their Definitions, and Characteristic Behaviors of CTs

	Definition	Characteristic Curves	Class/Error (IEC 60044)a	Explanation
Metering class	The metering class guarantees the accuracy of the secondary current from 5% to 125% of the rated primary current. Beyond this level, the CT saturates and the secondary current is clipped.	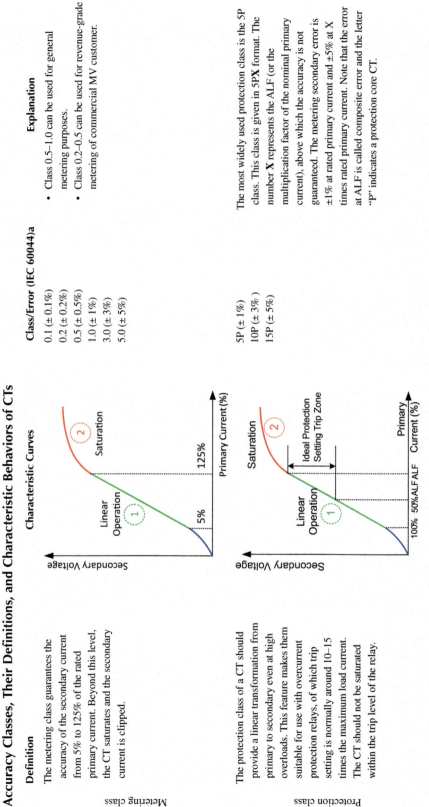	0.1 (± 0.1%) 0.2 (± 0.2%) 0.5 (± 0.5%) 1.0 (± 1%) 3.0 (± 3%) 5.0 (± 5%)	• Class 0.5–1.0 can be used for general metering purposes. • Class 0.2–0.5 can be used for revenue-grade metering of commercial MV customer.
Protection class	The protection class of a CT should provide a linear transformation from primary to secondary even at high overloads. This feature makes them suitable for use with overcurrent protection relays, of which trip setting is normally around 10–15 times the maximum load current. The CT should not be saturated within the trip level of the relay.		5P (± 1%) 10P (± 3%) 15P (± 5%)	The most widely used protection class is the 5P class. This class is given in 5PX format. The number X represents the ALF (or the multiplication factor of the nominal primary current), above which the accuracy is not guaranteed. The metering secondary error is ±1% at rated primary current and ±5% at X times rated primary current. Note that the error at ALF is called composite error and the letter "P" indicates a protection core CT.

ALF: Accuracy Limit Factor.

a According to IEEE C57.13 or ANSI, the metering accuracy class is defined as 0.15, 0.15S (Special High Accuracy Metering), 0.3, 0.6, and 1.2. As for the protection applications, ANSI defines the accuracy classes as C100 (or T100), C200 (or T200), C400 (or T400), C800 (or T800), where the numbers (100 to 800) indicate the secondary terminal volts at ALF=20 [Terminal Volts = (20 times rated secondary current) × (Total external burden (lead wire plus instrument resistances) of 1 Ω, Terminal Voltage = (20×5 A)×(1 Ω) = 100 V. In this case, C100 (or T100) class CT can be selected. If the total external burden was 2 Ω, then C200 class CT would be appropriate [(20×5 A)×(2 Ω) = 200 V].

TABLE 5.80
Permissible Durations of Voltage Transformers' Maximum Voltage for Different Earthing Systems and VT Connections

Primary Winding Connection and System Earthing	Voltage Factors and Durations	VT Connection
Between two phases in any network	VF = 1.2 (continuous)	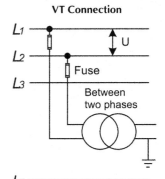
Between phase and earth in an effectively/earthed/directly earthed network	VF = 1.2 (continuous) VF = 1.5 (for 30 seconds)	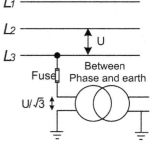
Between phase and earth in a non-effectively earthed neutral system (earthed via limiting resistor) with automatic earth-fault tripping	VF = 1.2 (continuous) VF = 1.9 (for 30 seconds)	
Between phase and earth in an isolated neutral system, or a resonant earthed (Peterson Coil Grounding[a]) system without earth-fault tripping.	VF = 1.2 (continuous) VF = 1.9 (for 8 hours)	

[a] Peterson coil (also known as a ground-fault neutralizer or arc suppression coil) is an iron core-based reactor connected between transformer neutral and earth. It is used for limiting the capacitive earth fault currents due to the phase-to-ground faults. Due to its tapping feature, the reactor can be adjusted according to system capacitance.

TABLE 5.81
Typical Apparent Powers of Voltage Transformers (at 0.8pf Inductive)

10 VA	15 VA	25 VA	30 VA	60 VA	75 VA
100 VA	150 VA	200 VA	300 VA	400 VA	500 VA

functions are under-voltage protection (27), instantaneous overcurrent protection (50), inverse-time overcurrent protection (51), and overvoltage protection (59), where the numbers given in parentheses are the ANSI codes corresponding to the respective protection functions. These codes are often used with specific suffixes, which indicate the zone of protection. For example, the letter "G" represents the ground conductor, while the letter "P" represents the phase conductor. As an example, it is understood from these explanations that 50P/50G functions correspond to instantaneous overcurrent protection and 51P/51G functions correspond to inverse-time overcurrent protections for phase and ground conductors, respectively.

If the signals from the CT and/or VT detect any of the above-mentioned abnormal conditions, the protective relay will send a trip command to the circuit breaker to clear the fault. Note that the circuit breaker in Figure 5.125 is indicated by the ANSI code 52.

TABLE 5.82
Accuracy Classes of Voltage Transformers (according to IEC60044-2a)

	Explanation	Class/Error/(IEC 60044)[a]
Metering class	The metering class guarantees the accuracy of the secondary voltage from 80% to 120% of the rated primary voltage. It is valid for all loads with an inductive power factor of 0.8 at nominal powers from 25% to 100%.	0.1 (± 0.1%) 0.2 (± 0.2%) 0.5 (± 0.5%) 1.0 (± 1%) 3.0 (± 3%)
Protection class	The metering class guarantees the accuracy of the secondary voltage from 25% to 120% of the rated primary voltage. It is valid for all loads with an inductive power factor of 0.8 at nominal powers from 25% to 100%.	3P (± 3%) 6P (± 6%)

[a] According to IEEE C57.13 or ANSI, the accuracy classes are defined as 0.3, 0.6, and 1.2. The standard has letter codes for different burdens of VTs. The letter codes are W (for 12.5 VA at 0.1pf), X (25 VA at 0.7pf), M (35 VA at 0.8pf), Y (75 VA at 0.85pf), Z (200 VA at 0.85pf) and ZZ (400 VA at 0.85pf). These accuracy classes and burden ratings are typically combined in a single label to express accuracy class and burden together. For example, 0.6M indicates a VT with an accuracy of ±6% under the burden of 35 VA.

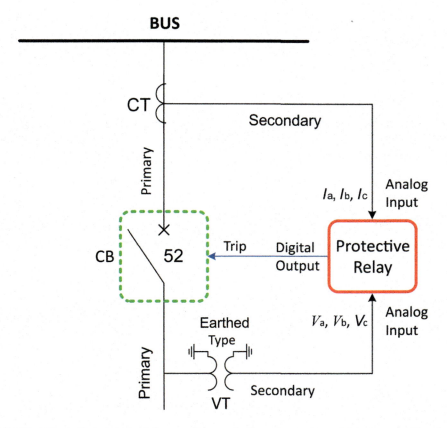

FIGURE 5.125 Single-line electrical diagram for a protective relay system with instrument transformers.

Example 5.24: CT and VT Sizing and Selection for a Photovoltaic Power System

Consider the 500 kW$_p$ PV power plant worked in Example 5.20. Size and specify the CT and VT required for the protection circuit of the main breaker located at point C (high-voltage side of the 0.4/31.5 kV transformer). Please see Figures 5.98 and 5.99 for the system detail.

Note that the system parameters required for CT and VT sizing and selection are as follows:

- The apparent power of the 0.4/31.5 kV transformer is 1250 kVA,
- The distance between the relay and measuring equipment/current transformers is approximately $\ell = 2.5$ m,
- Conductor resistance = 0.087 Ω/m,
- The power drawn by the relay is 0.4 VA per phase,
- Power drawn by the measuring circuit is 0.1 VA per phase.
- The power drawn by the electricity meter is 4 VA (to be connected to VT)
- The secondary coil resistance of the VT is 403 Ω.

Solution:

The specification/selection steps for CT and VT include the determinations of transformation ratio, instrument burden/apparent power values, accuracy class, thermal/dynamic current values for CTs, and voltage factor for VTs. So let's apply these steps for current and voltage transformers.

CT Specification and Selection: Depending on the steps described above, the following specification/selection table can be created for CTs (Table 5.83).

By considering the calculations in Table 5.83, final CT specifications can be decided as in Table 5.84.

VT Specification and Selection: Depending on the steps described above, the following specification/selection table can be created for VTs (Table 5.85).

By considering the calculations in Table 5.85, final VT specifications can be decided as in Table 5.86.

TABLE 5.83
CT Specification and Selection Table

Specification and Selection Steps		Calculation and Results
Step-1	Determine the current transformation ratio (I_{pn} / I_{sn})	From equation (5.119), CT primary current can be calculated as: $I_p = \dfrac{S_T}{\sqrt{3} \times U_2} = \dfrac{1250}{\sqrt{3} \times 31.5} = 22.9$ A. So, the next standard value for CT primary current is 25 A. Since the secondary current is typically 5 A, the CT transformation ratio can be selected as 25/5 A.
Step-2	Determine the thermal and dynamic current values	• The continuous rated thermal current: $I_{thc} = I_{pn} = 25$ A. • Short-time thermal current for 1 second: $I_{th} = I_{sc3}$. From Example 5.20, three-phase short-circuit current at point C, $I_{sc3} = 1428.47$ A. So, $I_{th} = I_{sc3} = 1428.47$ A. • According to IEC, the dynamic current withstand level should at least equal to $2.5 \times I_{th}$. So, $I_{dyn} = 2.5 \times 1428.47 = 3571.175$ A.
Step-3	Determine the CT burden and apparent power values	Considering the given information and equation (5.124), the CT burden value (as per winding) can be calculated as: CT Burden $[\text{VA}] = I_{sn}^2 \times R_B + S_{\text{Relay\&Circuit}} [\text{VA}]$, where $R_B = 2.5 \text{ m} \times (0.087 \, \Omega/\text{m}) = 0.2175 \, \Omega$. Hence, CT Burden $[\text{VA}] = 5^2 \times 0.2175 + (0.4 + 0.1)$ VA / phase = 5.9375 VA per phase. Therefore, the CT apparent power should be selected to be greater than the calculated CT burden value.
Step-4	Determine the accuracy classes	It is appropriate to choose metering class 0.5 and protection class 5P20.

TABLE 5.84
Selected CT Specifications

Type of Use	Indoor Type CT
Current transformation ratio	25 A/5 A
Apparent power as per secondary winding	10 VA or 15 VA can be selected
Short-time thermal current	Short-time thermal current withstand value can be selected as $60 \times I_{pn}$ ($60 \times 25 = 1500$ A). In this case, the calculated I_{th} value (1428.47 A) is less than 1500 A. So, it is suitable. • Minimum dynamic current withstand level of the CT is equal to $2.5 \times I_{th}$. So, I_{dyn} (of the CT) $\geq 2.5 \times 1428.47 \geq 3571.175$ A.
Rated voltage	36 kV (It should be selected according to the network voltage).
Accuracy classes	Metering class is 0.5 and protection class is 5P20.

TABLE 5.85
VT Specification and Selection Table

Specification and Selection Steps		Calculation and Results
Step-1	Determine the voltage ratio (V_{pn} / V_{sn})	A single-phase earthed VT with a voltage ratio of $\left(\dfrac{33000}{\sqrt{3}}\right) / \left(\dfrac{110}{\sqrt{3}}\right)$ can be selected. A VT with double secondary should be selected in case of using two measuring circuits, which then a voltage transformation ratio $\left(\dfrac{33000}{\sqrt{3}}\right) / \left(\dfrac{110}{\sqrt{3}}\right), \left(\dfrac{110}{\sqrt{3}}\right)$ can be selected.
Step-2	Determine the voltage factor	Considering Table 5.80, the voltage factor for the VT can be selected as VF = 1.2 (continuous) and VF = 1.9 (for 30 seconds). In other words, the VT should be designed to withstand 1.2 times the rated voltage continuously and 1.9 times the nominal voltage for 30 seconds. Note that the system supply side is assumed to be non-effectively earthed or earthed through a resistor.
Step-3	Determine the VT burden and apparent power values	Considering the given information and equation (5.125), VT burden value (as per secondary winding) can be calculated as: VT Burden $[VA] = \dfrac{V_{sn}^2}{Z_B}[VA]$, Hence, VT Burden $[VA] = \dfrac{\left(\dfrac{110}{\sqrt{3}}\right)^2}{403} = 10$ VA per winding. With the inclusion of an external 5 VA meter power of 5 VA, the VT burden value should be chosen greater than 15 VA. Therefore, we can choose the VT burden value as 30 VA.
Step-4	Determine the accuracy classes	Metering class is 0.5 and protection class is 3P ($\pm$ 3%).

5.5.7.9 Specification and Selection of DC Auxiliary Supply System

DC auxiliary supply systems are used for the operation of switch tripping, protection tripping, and other ancillary equipment within substations. Transformer stations may have one or several DC systems due to the need for more than one voltage level or the need for system duplication. Nowadays, 110 V or 220 V DC systems are commonly used in most power stations. However, DC auxiliary supply systems at lower voltage levels such as 24 V, 30 V, or 48 V also exist.

The basic elements of a DC auxiliary system include a battery bank, rectifier/charger unit, and a switchboard generally having a DC system monitoring relay. Conditions for monitoring and battery requirements in DC auxiliary supply systems are summarized in Table 5.87.

TABLE 5.86
Selected VT Specifications

Type of Use	Indoor Type VT (Earthed Type, Single Phase)
Voltage transformation ratio	$V_{pn}/V_{sn} = \left(\dfrac{33000}{\sqrt{3}}\right) \bigg/ \left(\dfrac{110}{\sqrt{3}}\right), \left(\dfrac{110}{\sqrt{3}}\right)$
Apparent power as per secondary winding	30 VA as per secondary winding
Accuracy classes	Metering class is 0.5 and protection class is 3P (± 3%).

TABLE 5.87
Battery and Monitoring System Requirements in Auxiliary DC Supply Systems

Battery Requirement Options	Monitoring Requirement
To perform switch tripping only (no standing load)	Not required
To perform switch tripping and to support standing load	Required
To support standing load only (no switch tripping)	Required

TABLE 5.88
Charger/Rectifier Requirements for Auxiliary DC Supply Systems

Input and Output Requirements	Environmental Conditions
• Input Voltage: 230 V (AC) • Output Voltage Levels: 110 V-48 V-30 V-24 V (DC) • Input Supply Variation: ±10% • Galvanic isolation between input and output should be provided. • It is suggested that DC output ripple content should not exceed 3%.	The equipment to be provided should be resistant to ambient temperature (−10°C to +55°C) and humidity level (5%–93%). The system should be installed in a dust-free space.

TABLE 5.89
Monitoring and Alarm Requirements for Auxiliary DC Supply Systems

Monitoring Requirements	Alarm Requirements
• Voltage across each battery set shall be monitored and displayed locally and transmitted remotely. • Battery voltage asymmetry should be monitored periodically. • Battery impedance can be monitored periodically (recommended, but not essential). • High and low voltage detection • State of charge monitoring	• Voltage asymmetry alarm • Battery impedance alarm (not essential) • High voltage alarm and high volts shutdown • Low voltage alarm • Earth fault alarm for switch tripping and load-supporting batteries (not necessary for telecom and SCADA battery systems) • Supply fail or charger failure alarm • Fuse alarm for telecom and SCADA battery systems (not necessary for tripping and load-supporting batteries)

As for the charger/rectifier requirements, the general specifications in Table 5.88 are needed to be considered when designing an auxiliary DC supply system.

Finally, monitoring and alarm requirements should be specified, which are summarized in Table 5.89.

Figure 5.126 shows a typical switch-tripping and load-supporting battery system, which includes a single-battery set and charger/rectifier unit.

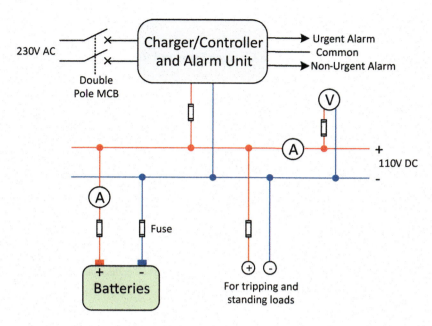

FIGURE 5.126 A typical DC auxiliary supply system for switch tripping and load-supporting.

Once the general requirements identified above are determined, the DC auxiliary supply system can be dimensioned. The sizing of a DC auxiliary supply system for primary and secondary substations should include the following steps.

Step 1 – Determination of Battery-Rated Capacity: The capacity of a battery is determined by the load and the time that the battery is expected to supply the load. The battery capacity for a DC auxiliary supply system depends on the amount of load and the time expected to feed this load. To estimate the load demands and their durations, we need to set the load profile as accurately as possible. It is possible to specify three different load behaviors for the load profile. These loads can be defined as continuous loads (continuously active throughout the entire duty cycle such as signal/alarm lamps and overcurrent relay), non-continuous loads (may become active at any time such as circuit breaker motor, breaker trip coil, and indoor lighting lamps), and momentary loads (short-duration loads in less than one minute). In the literature, several methods have been developed to determine the minimum battery capacity depending on the specified load profile. The corresponding methods in the literature can be considered to obtain more precise results and to determine the correct cell type, depending on the parameters such as load profile, discharge time, minimum/maximum battery voltages, and ambient conditions. However, as a simple approach, the battery capacity can be calculated by considering the average load during the whole duty cycle. It is generally assumed that this average load (L_{avg} in watts) is expected to be met for 24 hours in practice. Accordingly, the required storage capacity in watt-hours is calculated as $L_{avg} \times t$, where t is the discharge time and can be taken as 24 hours. Subsequently, net battery capacity can be calculated using the equations from (5.39) to (5.42).

Step 2 – Determination of Charger/Rectifier Specifications: The battery charger used in a DC auxiliary supply system has several tasks to be performed. It is possible to determine the correct charger size and specifications by determining these tasks and desired charger features. The sizing and selection criteria for the battery charger of a DC supply system are summarized in Table 5.90.

Step 3 – Determination of DC Switchgear Size and Its Components: The DC switchboard of an auxiliary supply system is recommended to be a free-standing metal enclosure. It would be appropriate to choose the nominal current of the switchgear larger than the highest available load

TABLE 5.90
Basic Considerations for Selecting the Charger Type and Rating

Sizing Criteria

- The charger should always keep the battery at full capacity. For this reason, the battery must have the capability to compensate for self-discharges and discharge currents due to load supply. In this context, the required battery charging time should be determined.
- DC system normal load current should be considered when determining the charger-rated output power.
- The charger/rectifier rated output power should be capable of supplying the sum of the continuous and non-continuous loads during a whole duty cycle (identified in step-1). In this case, the charger output power can be determined as follows:

$$P_{\text{charger}}[\text{Watts}] \geq (P_{\text{cont load1}} + P_{\text{cont load2}} + \cdots)$$
$$+ (P_{\text{non-cont load1}} + P_{\text{non-cont load2}} + \cdots)$$

Selection Criteria

- For a good charger performance, the output ripple content should be low, and the dynamic response of the charger should be strong.
- If a large enough charger is not available for a single-phase supply, then the parallel operation of two or more chargers can be considered.
- The monitoring unit of a charger system should monitor the battery voltage, battery cell voltages, charger output voltage, system insulation resistance, battery operating temperature, charger AC supply (voltage, frequency, and phase numbers such as one phase or three-phase), and SOC of the battery.

---

current in the system, with a reasonable safety margin of possible future load increase. This safety margin is usually obtained by multiplying the maximum load current by a factor of 1.25.

The short-circuit strength sizing of a DC switchgear, unlike the AC systems, depends on the DC short-circuit current of the battery group. Short-circuit currents caused by batteries generally depend on the battery's internal resistance and its Ah capacity. Hence, actual values of short-circuit currents of battery groups should be obtained from the manufacturer's catalogs. From the manufacturer's manuals, it is understood that battery short-circuit currents can go up to several kA levels easily. As a result, the short-circuit strength level of a DC switchgear should be chosen so that it must be greater than the battery short-circuit current level in the system.

Step 4 – Cable Size Determination (between switchboard and load): Voltage drop control should be made according to the selected cross-sectional area of the DC cable. If the percentage voltage drop is less than 1%, then the selected cross section is considered to be appropriate.

An Important Reminder: Since the details of DC and AC cable sizing are studied in detail in the following sections, no further details are given here.

Example 5.25: Auxiliary DC System Sizing and Selection for a Photovoltaic Power System

Consider the 500 kWp PV power plant worked in Example 5.20. The load profiles of the auxiliary DC power supply system for the primary and secondary sides of the 0.4/31.5 kV transformer are given in Table 5.91.

Assume that the circuit breaker motor (spring charging motor) in the distribution board of the low-voltage side has been selected as 230V AC. The motor of the circuit breaker at the high-voltage side has been selected as 24V DC. Hence, the DC supply voltage for the primary side is 110V, while it is designed as 24V DC for the secondary side. The cross-sectional area of the cable between distribution switchgear and load is 2.5 mm² for the primary side and 6 mm² for the secondary side. The cable lengths are 10 m and 2 m for the low and high-voltage sides, respectively.

Based on the given data, the size of the DC auxiliary supply system for primary and secondary sides.

TABLE 5.91
Load Profile of Auxiliary DC Supply System

Continuous			Non-continuous			
Load (DC)	**Power**	**Number**	**Load (DC)**	**Power**	**Number**	**Duration**
Low-Voltage Side (0.4 kV)						
Overcurrent relay	30 W	7	Trip coil	150 W	2	Negligible
Signal lamp	2 W	7	Signal lamp	2 W	14	Negligible
Alarm lamp	2 W	4	Lighting lamp	25 W	6	8 hours
			Alarm lamp	2 W	24	Negligible
High-Voltage Side (31.5 kV)						
Overcurrent relay	30 W	2	Trip coil	150 W	2	Negligible
Signal lamp	2 W	2	Circuit breaker motor	200 W	2	Negligible
Alarm lamp	2 W	4	Signal lamp	2 W	4	Negligible
			Lighting lamp	25 W	5	8 hours
			Alarm lamp	2 W	4	Negligible

Solution

Sizing of Primary Side DC Auxiliary Supply System: Since the breaker motor voltage on the primary side is 230V AC, it cannot be considered as a DC system load. According to Table 5.91, the sum of continuous loads:

$$\sum_i (P_{\text{cont load}})_i = 7 \times 30 \text{ W} + 7 \times 2 \text{ W} + 4 \times 2 \text{ W} = 232 \text{ W}$$

And, the sum of non-continuous loads:

$$\sum_i (P_{\text{non-cont load}})_i = 2 \times 150 \text{ W} + 14 \times 2 \text{ W} + 6 \times 25 \text{ W} + 24 \times 2 \text{ W} = 528 \text{ W}$$

Accordingly, the total DC system load is 232 W + 528 W = 760 W. Taking into account a safety margin, the rated power of the charge/rectifier unit can be selected from around 900 W to 1000 W. Among the commercially available products, 220V AC/110V DC-11A unit can be selected. In this case, the charger power is 110 V × 11 A = 1210 W. Since 1210 W > 760 W, the selected product is suitable.

Let us calculate the average load amount to be able to determine the required battery capacity.

$$L_{\text{avg}} = \frac{232 \text{ W} \times 24 \text{ h} + 150 \text{ W} \times 8 \text{ h}}{24 \text{ h}} = 282 \text{ W}$$

Based on equation (5.42), net battery capacity in Ah can be calculated as below (DoD$_{\max}$ = 50%; $V_{\text{DC}} = 110$ V; $\eta_b = 0.90$; $f_{\text{Temp-b}} \approx 1$):

$$C_{\text{Battery-net}} [\text{Ah}] = \frac{282 \text{ W} \times 24 \text{ h}}{0.5 \times 110 \times 0.9 \times 1} = 136 \text{ Ah} @ 110 \text{ V}_{\text{DC}}$$

Note that lead-acid battery parameters are used in the battery capacity calculation given above.

The constant short-circuit current of a lead-acid battery with a capacity of 150 Ah is determined as approximately 5000 A in the manufacturer's catalogs. Hence, the short-circuit withstands of the DC switchgear must be larger than 5 kA.

Finally, the maximum percent voltage drop on DC cables shall be checked. This voltage drop should be less than 1% on the dc cables [$\ell = 10$m; $s = 6$ mm²; $P = 194$ W; $k = 56$ km/(Ω·mm²); $V = 110$ V].

$$\varepsilon\% = \frac{200 \times \ell \times P}{k \times s \times V^2} = \frac{200 \times 10 \times 194}{56 \times 2.5 \times 110^2} = 0.229\% < 1\%$$

As a result, the selected dc cable is suitable for voltage drop.

Reminder: To find the highest voltage drop in the DC auxiliary supply system, the modular cell load at the farthest point from the rectifier DC terminal should be considered. The loads at the farthest point include one trip coil, one overcurrent relay, and alarm/signal lamps in the modular cell. So, the sum of load powers at the farthest point is equal to 150 W + 30 W + 7 × 2 W = 194 W.

Sizing of Secondary Side DC Auxiliary Supply System: According to Table 5.91, the sum of continuous loads:

$$\sum_i (P_{\text{cont load}})_i = 2 \times 30 \text{ W} + 2 \times 2 \text{ W} + 4 \times 2 \text{ W} = 72 \text{ W}$$

And, the sum of non-continuous loads:

$$\sum_i (P_{\text{non-cont load}})_i = 2 \times 150 \text{ W} + 2 \times 200 \text{ W} + 4 \times 2 \text{ W} + 5 \times 25 \text{ W} + 4 \times 2 \text{ W} = 841 \text{ W}$$

Accordingly, the total DC system load is 72 W + 841 W = 913 W. Taking into account a safety margin, the rated power of the charge/rectifier unit can be selected as approximately 1000 W or above. Among the commercially available products, 220V AC/24V DC-42A unit can be selected. In this case, the charger power is 24 V × 42 A = 1008 W. Since 1008 W > 913 W, the selected product is suitable.

Let us calculate the average load amount to be able to determine the required battery capacity.

$$L_{\text{avg}} = \frac{72 \text{ W} \times 24 \text{ h} + 125 \text{ W} \times 8 \text{ h}}{24 \text{ h}} = 114 \text{ W}$$

Based on equation (5.42), net battery capacity in Ah can be calculated as below ($\text{DoD}_{\max} = 50\%$; $V_{\text{DC}} = 24$ V; $\eta_b = 0.90$; $f_{\text{Temp-b}} \approx 1$):

$$C_{\text{Battery-net}} [\text{Ah}] = \frac{114 \text{ W} \times 24 \text{ h}}{0.5 \times 24 \times 0.9 \times 1} = 253 \text{ Ah} @ 24 \text{ V}_{\text{DC}}$$

Note that lead-acid battery parameters are used in the battery capacity calculation given above.

The constant short-circuit current of a lead-acid battery with a capacity of 250 Ah is determined as approximately 5900 A in the manufacturer's catalogs. Hence, the short-circuit withstands of the DC switchgear must be larger than 5.9 kA.

Finally, the maximum percent voltage drop on DC cables shall be checked. This voltage drop should be less than 1% on the dc cables [$\ell = 2$m; $s = 6$ mm²; $P = 396$ W; $k = 56$ m/(Ω·mm²); $V = 24$ V].

$$\varepsilon\% = \frac{200 \times \ell \times P}{k \times s \times V^2} = \frac{200 \times 2 \times 396}{56 \times 6 \times 24^2} = 0.818\% < 1\%$$

As a result, the selected dc cable is suitable for voltage drop.

Reminder: To find the highest voltage drop in the DC auxiliary supply system, the modular cell load at the farthest point from the rectifier DC terminal should be considered. The loads at the farthest point include one circuit breaker motor, one trip coil, one overcurrent relay, and alarm/signal lamps in the modular cell. So, the sum of load powers at the farthest point is equal to 200 W + 150 W + 30 W + 8 × 2 W = 396 W.

5.5.7.10 Overcurrent Protection of Grid-Connected PV Systems

Accurate settings of protection relays in the substations are vital for the safe operation of a power system. Hence, protective relay coordination for a grid-connected PV system must be performed to operate the relevant PV system safely. In this context, to carry out the relay coordination and setting, it is necessary to have basic knowledge about the types of overcurrent protective relays and their trip characteristics.

5.5.7.10.1 Overcurrent Protection Relays

Power system components can be protected most simply and cheaply with the help of overcurrent relays. Overcurrent relays can be divided into three different types in terms of their operating characteristics (time versus current curves). Overcurrent relay types and their characteristic behaviors are summarized in Table 5.92.

The main purpose of the relay setting is to enable the relay to trip the relevant circuit in an acceptable time limit in the cases of possible short-circuit failures. To achieve this goal, two basic parameters, that is, PSM (Plug Setting Multiplier) and TSM (Time Setting Multiplier), must be set in the relays. These parameters are also known as PMS (Plug Multiplier Setting) and TMS (Time Multiplier Setting), respectively.

The PSM of a relay is referred to as the ratio of the actual fault current in the relay to its pickup current and can be expressed mathematically as (5.126).

$$\begin{aligned} \text{PSM} &= \frac{\text{Actual Fault Current on Secondary Coil}}{\text{Relay Pickup Current}} \\ &= \frac{\text{Actual Fault Current on Secondary Coil}}{(\% \text{ Current Setting}) \cdot (\text{Rated Secondary Current of CT})} \\ &= \frac{\text{Actual Fault Current on Primary Coil}}{\text{Relay Pickup Current} \times \text{CTR}} \end{aligned} \quad (5.126)$$

where relay pickup current is equal to % current setting × rated secondary current of CT (current transformer) and CTR is the current transformer ratio, which can be expressed as $\left(\dfrac{I_{\text{rated primary}}}{I_{\text{rated secondary}}}\right)$.

The relay pickup current can be set by adjusting the taps of the relay coil as shown in Figure 5.127.

% Current settings in Figure 5.127 are expressed in the percentage ratio of relay pickup current to the rated secondary current of CT [% Current Setting = (Relay Pickup Current)/(Rated Secondary Current) × 100]. The %Current-Setting for an overcurrent relay usually ranges from 25% to 200%, in steps of 25%. As for earth fault relay, it generally ranges from 10% to 80% in steps of 10%. Finally, earth leakage relays are generally available 10%–40% in steps of 5%. Note that the current setting values in numerical relays can be adjusted almost in continuous steps.

Similar to the current setting, the operation time of a relay can be achieved by adjusting the relay's TSM parameter. The ratio of actual operating time to the operating time obtained from relay characteristic at the corresponding PSM value (for TSM = 1) is called the time setting multiplier (TSM) and can be expressed as (5.127).

$$\text{TSM} = \frac{\text{Actual (Required) Operation Time}}{\text{Time Obtained from Relay Charcteristic at Correspnding PSM (for TSM = 1)}} \quad (5.127)$$

Hence, the actual operation time of a relay (t_{trip}) can be found by (5.128).

$$\text{Actual Time of Operation}\left(t_{\text{trip}}\right) = \text{TSM} \times \left(\text{corresponding operating time at TSM} = 1\right) \quad (5.128)$$

TABLE 5.92
Overcurrent Protective Relays and Their Characteristic Behaviors

Types	Definition, Basic Features, and Characteristic Curves
Instantaneous	<ul><li>These relays operate without time delay and trip in 0.1 seconds.</li><li>The pickup (threshold) current is adjustable and can be set by the user.</li><li>Pickup current setting can be adjusted in different ways depending on the structural type of relay. For example, the pickup current can be tuned by spring adjustment, air-gap adjustment, or by taps of the relay current coil.</li><li>ANSI Code: 50</li></ul> 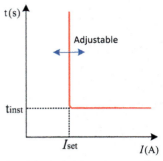
Time-dependent	<ul><li>These relays (time overcurrent relays) operate with a time delay. Both time delay and pick-up current values can be adjusted. Characteristic curves of time overcurrent relays are generally divided into two groups. These are definite time and inverse time curves. Inverse time curves are also divided into subgroups such as moderately inverse (MI), normal inverse (NI), very inverse (VI), and extremely inverse (EI) curves.</li><li>While the delay time and threshold current value can be adjusted in definite-time overcurrent relays, the pickup current, as well as curve level (MI, NI, VI, or EI), can be adjusted in inverse-time overcurrent relays.</li><li>ANSI Code: 50 (for definite time curve)</li><li>ANSI Code: 51 (for inverse time curves)</li></ul> 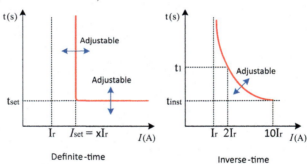
Mixed curve	Mixed curve overcurrent relays are obtained by combining the properties of different types of overcurrent relays in a single relay. Therefore, these relays have all the advantages of other overcurrent relays. Since the overcurrent elements of the relays are manufactured as separate units, it is possible to create different combinations in mixed curve overcurrent relays. Possible combinations are: (i) combination of instantaneous and definite-time curves, (ii) combination of instantaneous and inverse-time curves, (iii) combination of instantaneous, definite-time, and inverse-time curves, and (iv) combination of definite-time and inverse-time curves (called IDMT: Inverse Definite Minimum Time). The relay curve in the fourth combination (definite-time + inverse-time) is one of the most preferred characteristic curves in practice. In this type of relay, the operating time is almost inversely proportional to the overcurrent near the threshold value. And this value becomes largely constant just above the pickup value of the relay. This characteristic curve is given in the third figure below. 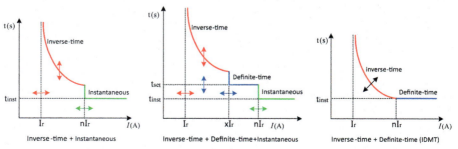

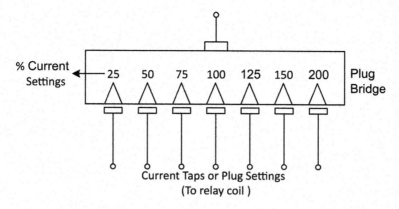

FIGURE 5.127 Tappings for % current setting (or plug settings).

The relay characteristics specified in (5.127) can be derived based on different mathematical equations. The generic form of relay characteristics is represented by the following equations. Equation (5.129) is the IDMT curve format in the IEC standard, while equation (5.130) represents the general IDMT curve format in the ANSI/IEEE standard.

$$t_{trip} = TSM \times \left[\frac{C_{slope}}{(PSM)^{\alpha} - 1} \right] \quad (5.129)$$

$$t_{trip} = \frac{TSM}{5} \times \left[\frac{C_{slope}}{(PSM)^{\alpha} - 1} + B \right] \quad (5.130)$$

In other words, any of the appropriate PSM equations given by (5.126) can be inserted into general IDMT equations (5.129) and (5.130). In this context, the above equations can be written as (5.130) and (5.131) equations.

$$t_{trip} = TSM \times \left[\frac{C_{slope}}{\left(\frac{\text{Actual Fault Current on Secondary Coil}}{\text{Relay Pickup Current (or Current Set Value)}} \right)^{\alpha} - 1} \right] \quad (5.131)$$

$$t_{trip} = TSM \times \left[\frac{C_{slope}}{\left(\frac{\text{Actual Fault Current on Secondary Coil}}{\text{Relay Pickup Current (or Current Set Value)}} \right)^{\alpha} - 1} + B \right] \quad (5.132)$$

where t_{trip} is the actual relay operation (trip) time in seconds, TSM (also known as TDS: time dial setting) is the time setting multiplier (which controls the relay tripping time by setting the pickup current value), C_{slope} is the slope constant (which controls the slope of relay characteristic), α is a constant which represents the degree of the inverse-time curve ($\alpha > 0$), and B is the curve constant.

Note that the constants defined in equations from (5.129) to (5.132) are given in Table 5.93 for each IDMT curve type according to IEC and ANSI/IEEE standards.

The trip curves of an IDMT relay for different TMS settings using the standard inverse (SI) curve are shown in Figure 5.128. The relay curves in Figure 5.128 are given only for discrete TMS values. However, the TMS settings can be adjusted in continuous values in electromechanical relays. In digital relays, continuous tuning can be achieved by keeping the TMS setting steps very small. Note that it is sufficient to use the standard SI curve in most cases of protection. However, in cases where satisfactory grading cannot be achieved, using very inverse (VI) or extremely inverse (EI) curves may help to solve the problem.

TABLE 5.93
IDMT Curve Types and Their Constants according to IEC and ANSI/IEEE Standards

IDMT Curve Types	Standard	C_{slope}	α	B
Moderately inverse	ANSI/IEEE	0.0515	0.02	0.114
Very inverse	ANSI/IEEE	19.61	2	0.491
Extremely inverse	ANSI/IEEE	28.2	2	0.1217
US CO8 inverse	ANSI/IEEE	5.95	2	0.18
US CO2 short-time inverse	ANSI/IEEE	0.16758	0.02	0.11858
Standard inverse	IEC	0.14	0.02	-
Very inverse	IEC	13.5	1	-
Extremely inverse	IEC	80.0	2	-
Long time inverse	IEC	120	1	-

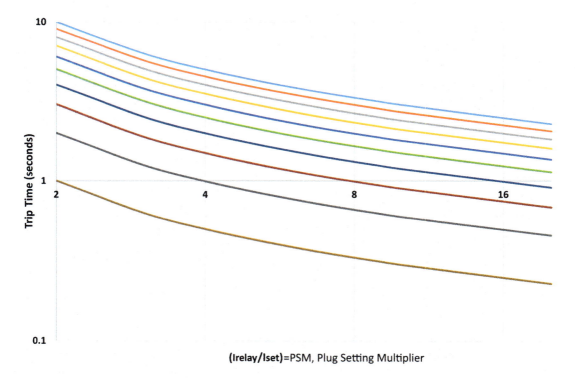

FIGURE 5.128 Typical trip-time/current characteristics of standard IDMT relay (as a multiple of plug settings on the logarithmic axis).

For the relay setting calculations, we need to know the parameters/operations such as fault current level, current transformer ratio (CTR), current settings, the characteristic curve of the relay (trip time versus PSM curves), and time settings. For example, let us assume that we have a 1500 A fault current on the primary side of CT and its CTR is 100 A/1 A. For the case given, what would be the actual operation times if the percent current settings were to be 100%, 125%, and 150%, respectively? So, let us explain the calculation steps of the actual relay operating times at the corresponding current settings.

Step-1 (Pickup Current Calculation): From the current transformer ratio, the secondary current of CT is known as 1A. Therefore, the pickup current of the relay can be calculated for the given %current settings as below:

$$\text{Pickup Currents} = \begin{cases} 1 \times 100\% = 1 \text{ A} \\ 1 \times 125\% = 1.25 \text{ A} \\ 1 \times 150\% = 1.5 \text{ A} \end{cases}$$

Step-2 (PSM Calculation): The PSM values for the given fault current can then be calculated as below:

$$\text{PSM Values} \left(\frac{\text{Actual Fault Current on Secondary}}{\text{Relay Pickup Current}} \right) = \begin{cases} 15/1 = 15 \\ 15/1.25 = 12 \\ 15/1.5 = 10 \end{cases}$$

Note that the actual fault current on the secondary coil is equal to $1500 \times (1/100) = 15$ A.

Step-3 (Operating Time Calculation of the Relay): Once the PSM value is calculated, we need to find out the time of operation from the relay's characteristics. It is possible to use different approaches to find the required operating time at an adjusted TSM value. For example, the time of operation value can directly be read from the available characteristic curves (if it matches) at the corresponding TSM. Secondly, the time of operation can be directly calculated from the mathematical equation of the relay characteristics. For this purpose, equations from (5.129) to (5.132) or equations provided by the manufacturer can be used. Thirdly, trip time reading can be derived from the tabulated values, which represent the relay characteristics. Now, let us use equation (5.129) to calculate the time of operations at TSM = 1. In this case, (5.129) can be written as below ($C_{slope} = 0.14$, $\alpha = 0.02$).

$$t_{trip} = 1 \times \left[\frac{0.14}{(\text{PSM})^{0.02} - 1} \right]$$

From the above-given equation, t_{trip} times for the PSM values found in Step-2 can be calculated as below.

$$t_{trip} = \begin{cases} 2.97s & \text{for PSM} = 10 \\ 2.75s & \text{for PSM} = 12 \\ 2.52s & \text{for PSM} = 15 \end{cases}$$

Step-4 (Actual Operating Time Calculation): After obtaining the relay operating time at TSM = 1, t_{trip} times are needed to be multiplied with an adjusted time setting multiplier to get the actual trip time. Assume that the TSM value is set as 0.2. Hence, actual trip times can be found below:

$$\text{Actual trip times} = \begin{cases} 2.97 \text{ s} \times 0.2 = 0.594 \text{ s} & \text{for PSM} = 10 \\ 2.75 \text{ s} \times 0.2 = 0.550 \text{ s} & \text{for PSM} = 12 \\ 2.52 \text{ s} \times 0.2 = 0.504 \text{ s} & \text{for PSM} = 15 \end{cases}$$

Example 5.26: Calculation of Setting Values for IDMT Relays

Assume that the current rating of an IDMT curve type overcurrent relay is 5 A and it is set to 150% current and 0.3 TSM value. If the CTR is 300 A/5 A, determine the actual operation time of the relay for a fault current of 4500 A.

Solution:

For the relay setting calculations, let us follow the calculation steps explained above:

Step-1 (Pickup Current Calculation): Secondary current of CT is 5 A. Therefore, the pickup current of the relay can be calculated for the given plug setting (150%) as below:

$$\text{Pickup Currents} = 5 \times 150\% = 7.5 \text{ A}$$

Step-2 (PSM Calculation): Firstly, let us calculate the fault current on the secondary coil, which is equal to $4500 \times (5/300) = 75$ A. The PSM value for the given fault current can then be calculated as:

$$\text{PSM} = \frac{\text{Actual Fault Current on Secondary}}{\text{Relay Pickup Current}} = \frac{75 \text{ A}}{7.5 \text{ A}} = 10$$

Step-3 (Operating Time Calculation of the Relay): Now, let us use equation (5.129) to calculate the time of operations at TSM = 1.

$$t_{trip} = 1 \times \left[\frac{0.14}{(\text{PSM})^{0.02} - 1} \right] = \left[\frac{0.14}{(10)^{0.02} - 1} \right] = 2.97 \text{ s}$$

Step-4 (Actual Operating Time Calculation): Since the TSM value is set as 0.3, actual trip times can be found as:

$$\text{Actual trip time} = 2.97 \text{ s} \times 0.3 = 0.891 \text{ s}$$

Note that the actual time of operation can also be obtained directly from the IDMT curve equation by substituting the adjusted TSM value into the formula, which is given below:

$$t_{trip} = \text{TSM} \times \left[\frac{0.14}{(\text{PSM})^{0.02} - 1} \right] = 0.3 \times \left[\frac{0.14}{(10)^{0.02} - 1} \right] = 0.891 \text{ s}$$

5.5.7.10.2 Overcurrent Relay Coordination in Radial Networks

Relay coordination is the process of enabling protective relays to trip in a pre-specified sequence in the electrical power system. For example, protection devices in a downstream network should activate before the upstream protection devices. In general, the protection scheme in a coordinated network should include the following requirements: (i) Under normal operating conditions, circuit breakers should not trip, (ii) during the fault, only the circuit breaker closest to the fault location on the source side should open, and (iii) if the first circuit breaker fails, then the second circuit breaker closer to the source should clear the fault. The process of this relay coordination can be achieved by selecting the proper plug and time setting multiplications for the possible highest fault current levels at the relay location. To verify the suitability of the desired selectivity, the coordination curves should also be checked graphically by considering the selected plug settings and time multiplier settings.

For the correct overcurrent relay applications, the following points should be addressed:

- Precise fault current flow in each part of the network should be known.
- Relay settings should be adjusted to trip in the shortest time at maximum short-circuit current and also it should provide satisfactory protection at the minimum fault current expected.
- All protection elements connected in series on the same line should be in coordination with each other. Hence, it is recommended that the relays in series with each other should have the same operating characteristic.
- The current setting of a relay must be adjusted so as not to operate for the maximum load current. However, it must be operated for a current greater than or equal to the minimum expected fault current.

There are three relay grading methods used commonly to provide the desired selectivity level, which are: (i) time grading (discrimination by time), (ii) current-grading (discrimination by current), and (iii) time-current-grading (discrimination by both time and current). Let us consider a radial network given in Figure 5.129 to demonstrate these grading principles.

Time-Grading Method: Definite-time overcurrent relays are used in this method. According to the time-grading approach, sequential time settings are assigned to these relays to ensure that the relay closest to the fault should trip first, if it fails, then the next one closer to the fault sends the trip signal. The time setting of the relays gets higher as they approach the source. This principle is demonstrated in Figure 5.130 for the radial system given in Figure 5.129.

Protection units with definite-time overcurrent relays are placed at each of locations 1, 2, and 3, of which time delays provide protection selectivity. The relay integrated with CB-1 is set to the shortest possible time and these times are gradually increased in the next relays. These values typically take 0.25, 0.50, 1, and 1.5 seconds, respectively. The main drawback of this method is that in case of a fault occurrence at a point closer to the source, for example, at point F, the large fault current will be cleared in a longer time, which may be destructive to the system.

Current-Grading Method: As known, fault currents vary with respect to the fault location due to the impedance difference between the source and the fault. Hence, only the relay closest to the fault is typically set to operate at an appropriate current value to clear the respective fault. The current setting of the relays increases as they get closer to the source. This principle is demonstrated in Figure 5.131 for the radial system given in Figure 5.129.

The main drawback of this method is that it is only suitable for an LV system that has a substantial impedance difference between two consecutive CB zones. Important to note that short-circuit analysis with respect to different fault locations is essential to determine the required set values.

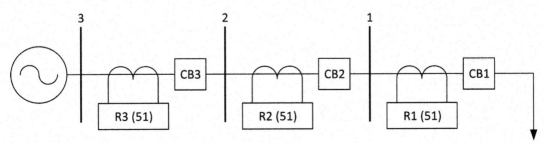

FIGURE 5.129 A typical radial network with an overcurrent protection scheme.

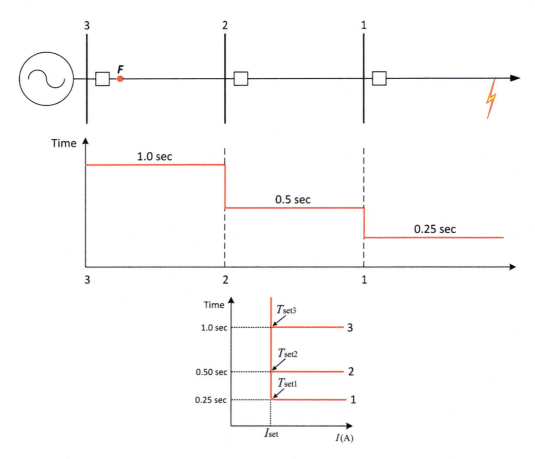

FIGURE 5.130 A typical radial network and the principles of time grading (radial network, time grading, and definite-time relay setting).

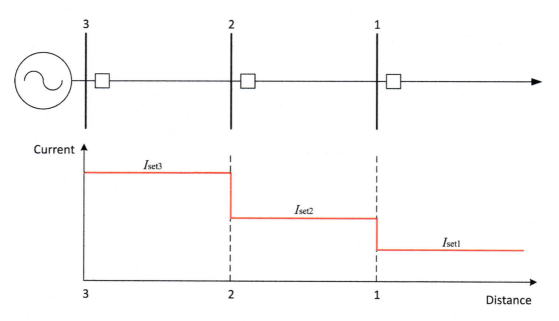

FIGURE 5.131 A typical radial network and the principles of current grading.

Time-Current-Grading Method: Inverse-time overcurrent relays, explained in the above sections in detail, is used in this method. As known, the operating time of these types of relays is inversely proportional to the current level and their characteristics are dependent on both time and current settings (TSM and PSM). The principle of the time-current-grading method is demonstrated in Figure 5.132 with the help of inverse-time overcurrent relaying.

It is clear from Figure 5.132 that discrimination by time-current-grading works as follows for the fault location at point **F**:

Relay1 operates immediately at the time T_1. If Relay1 fails, then Relay2 sends a trip signal for CB2 at the time T_2. Similarly, if Relay2 fails as well, Relay3 activates the CB3 at the time T_3 to clear the fault. In this context, inverse-time relay grading can be done with respect to the first relay as follows:

Step-1 (Setting Calculations for Relay1): First of all, all the steps to determine the actual operating time are needed to be followed for Relay1 as explained in the above sections (from step1 to step4). Let's assume that the actual trip time of Relay1 is calculated as T_1.

Step-2 (Setting Calculations for Relay2): Once the time T_1 is determined, the actual operation time T_2 for Relay2 is calculated as follows.

$$T_2 = T_1 + DT \tag{5.133}$$

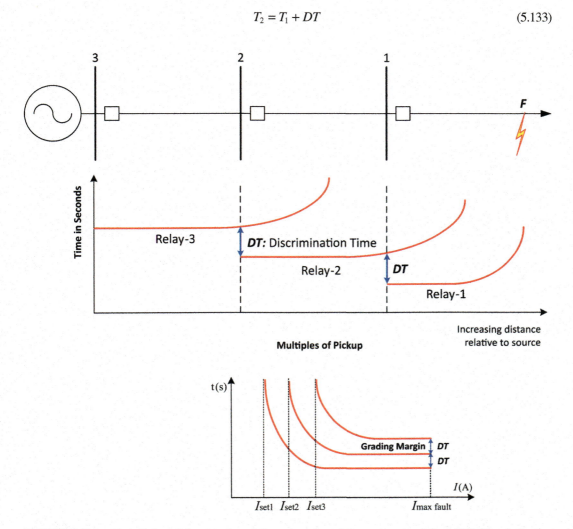

FIGURE 5.132 A typical radial network and the principles of time-current grading (*DT*: Discrimination Time).

where *DT* is the discrimination time in seconds and it is the smallest possible time interval at which the relay and CB can clear the fault. This discrimination time is taken as 0.4 s in IEEE/ANSI and 0.5 s in IEC.

Then, regardless of Relay1, PSM and t_{trip} values for Relay2 (PSM and t_{trip-2}) are calculated based on the relay's characteristics and respective fault current value. As a result of this, the TMS value for Relay2 is calculated as follows.

$$\text{TMS (for Relay2)} = \frac{T_2}{t_{trip-2}} \tag{5.134}$$

Step-3 (Setting Calculations for Relay3): Similar to Step-2, setting values for Relay3 can also be calculated as follows.

$$T_3 = T_2 + DT$$

$$t_{trip-3} = 1 \times \left[\frac{C_{slope}}{\left(PSM_{Relay3}\right)^\alpha - 1} \right]$$

$$\text{TMS (for Relay3)} = \frac{T_3}{t_{trip-3}} \tag{5.135}$$

Example 5.27: IDMT Relay Coordination on a Radial Network

Consider the radial network shown in Figure 5.133. According to the system data given in the figure, determine the relay settings assuming the standard IDMT curve type.

Solution:

Let us follow the calculation steps to grade Relay2 with respect to Relay1.

Step-1 (Setting Calculations for Relay1): Actual trip time of Relay1 ($T_1 = t_{trip-1}$ for Relay 1) is calculated as below:

Pickup current calculation: Secondary current of CT is 5 A. Therefore, the pickup current of the relay can be calculated for the given plug setting (100%) as below:

$$\text{Pickup Currents} = 5 \times 100\% = 5 \text{ A}$$

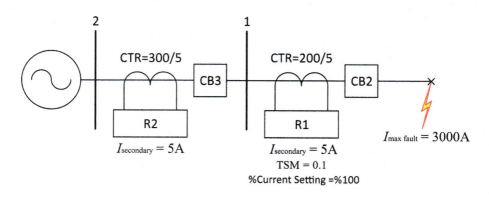

FIGURE 5.133 Considered radial system.

PSM calculation: Fault current on the secondary coil, which is equal to $3000 \times (5/200) = 75$ A. The PSM value for the given fault current can then be calculated as:

$$PSM = \frac{\text{Actual Fault Current on Secondary}}{\text{Relay Pickup Current}} = \frac{75 \text{ A}}{5 \text{ A}} = 15$$

Operating time calculation of the relay: Now, let us use equation (5.129) to calculate the time of operations at TSM = 1.

$$t_{trip} = 1 \times \left[\frac{0.14}{(PSM)^{0.02} - 1} \right] = \left[\frac{0.14}{(15)^{0.02} - 1} \right] = 2.51 \text{ s}$$

Actual operating time calculation: Since the TSM value is set as 0.1, actual trip times can be found as:

$$\text{Actual trip time } (T_1) = 2.51 \text{ s} \times 0.1 = 0.251 \text{ s}$$

Step-2 (Setting Calculations for Relay2): Actual trip time of Relay2 (T_2) is calculated as below:

$$T_2 = T_1 + DT$$

$$T_2 = 0.251 + 0.4 = 0.651 \text{ s}$$

For the PSM calculation, we need to find the fault current on the secondary coil as $3000 \times (5/300) = 50$ A, which leads to PSM = $50/5 = 10$. Hence, t_{trip-2} can be found from the standard IDMT relay characteristic curve at TSM = 1.

$$t_{trip-2} = 1 \times \left[\frac{C_{slope}}{(PSM_{Relay2})^{\alpha} - 1} \right] = \left[\frac{0.14}{(10)^{0.02} - 1} \right] = 2.97 \text{ s}$$

As a result, the TMS value for Relay2 is calculated as follows:

$$TMS \text{ (for Relay2)} = \frac{T_2}{t_{trip-2}} = \frac{0.651}{2.97} = 0.22$$

Example 5.28: IDMT Relay Coordination for Radial Network Having Different Fault Levels

For the 11 kV system shown in Figure 5.134, determine the relay setting based on the provided data (take discrimination time as 0.5 seconds).

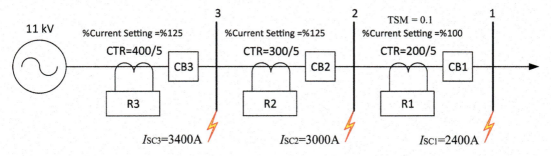

FIGURE 5.134 Considered 11 kV radial system.

Solution:

Let us follow the calculation steps to grade Relay2 and Relay3 with respect to Relay1.

Step-1 (Setting Calculations for Relay1): Actual trip time of Relay1 ($T_1 = t_{trip-1}$ for Relay1) is calculated as below:

Pickup current calculation: Secondary current of CT is 5 A. Therefore, the pickup current of the relay can be calculated for the given plug setting (100%) as below:

$$\text{Pickup Currents} = 5 \times 100\% = 5 \text{ A}$$

PSM calculation: Fault current on the secondary coil, which is equal to $2400 \times (5/200) = 60$ A. The PSM value for the given fault current can then be calculated as:

$$\text{PSM} = \frac{\text{Actual Fault Current on Secondary}}{\text{Relay Pickup Current}} = \frac{60 \text{ A}}{5 \text{ A}} = 12$$

Operating time calculation of the relay: From (5.129), we have the trip time operations at TSM = 1.

$$t_{trip} = 1 \times \left[\frac{0.14}{(\text{PSM})^{0.02} - 1} \right] = \left[\frac{0.14}{(12)^{0.02} - 1} \right] = 2.75 \text{ s}$$

Actual operating time calculation: Since the TSM value is set as 0.1, actual trip times can be found as:

$$\text{Actual trip time } (T_1) = 2.75 \text{s} \times 0.1 = 0.275 \text{s}$$

Step-2 (Setting Calculations for Relay2): Actual trip time of Relay2 (T_2) is calculated as below:

$$T_2 = T_1 + DT$$

$$T_2 = 0.275 + 0.5 = 0.775 \text{ s}$$

Since the secondary current of CT is 5 A. Therefore, the pickup current can be calculated for the given plug setting (125%) as below:

$$\text{Pickup Currents} = 5 \times 125\% = 6.25 \text{ A}$$

For the PSM calculation, we need to find the fault current on the secondary coil as $3000 \times (5/300) = 50$ A, which leads to PSM = 50/6.25 = 8. Hence, t_{trip-2} can be found from the standard IDMT relay characteristic curve at TSM = 1.

$$t_{trip-2} = 1 \times \left[\frac{C_{slope}}{(\text{PSM}_{Relay2})^{\alpha} - 1} \right] = \left[\frac{0.14}{(8)^{0.02} - 1} \right] = 3.3 \text{ s}$$

As a result, the TMS value for Relay2 is calculated as follows:

$$\text{TMS (for Relay2)} = \frac{T_2}{t_{trip-2}} = \frac{0.775}{3.3} = 0.24$$

Step-3 (Setting Calculations for Relay3): Actual trip time of Relay2 (T_3) is calculated as below:

$$T_3 = T_2 + DT$$

$$T_3 = 0.775 + 0.5 = 1.275 \text{ s}$$

Since the secondary current of CT is 5 A. Therefore, the pickup current can be calculated for the given plug setting (125%) as below:

$$\text{Pickup Currents} = 5 \times 125\% = 6.25 \text{ A}$$

For the PSM calculation, we need to find the fault current on the secondary coil as $3400 \times (5/400) = 42.5$ A, which leads to PSM $= 42.5/6.25 = 6.8$. Hence, t_{trip-3} can be found from the standard IDMT relay characteristic curve at TSM $= 1$.

$$t_{trip-3} = 1 \times \left[\frac{C_{slope}}{(\text{PSM}_{Relay2})^{\alpha} - 1} \right] = \left[\frac{0.14}{(6.8)^{0.02} - 1} \right] = 3.58 \text{ s}$$

As a result, the TMS value for Relay3 is calculated as follows:

$$\text{TMS (for Relay3)} = \frac{T_3}{t_{trip-3}} = \frac{1.275}{3.58} = 0.36$$

5.5.7.10.3 Specifying and Setting of Protection Relays for PV Power Plants

Protective relaying is a vital application for the grid connection and safe operation of PV systems. It is possible to divide the PV system protective relaying system into two levels, which are interconnection protection at PCC (point of common coupling) and protection within the facility of PV generator units. The protective relaying at the interconnection point may further be assessed according to the voltage level of the grid. Table 5.94 shows the requirements for PV system protective relaying within the facility and PCC at the LV, MV, and HV levels.

TABLE 5.94

Requirements for PV System Protective Relaying within the Facility and at LV, MV, and HV PCCs

Protection Requirement	ANSI Code	Explanation/Recommended Relay Setting
PV System Protective Relaying within the Installation		
Phase Fault (Overcurrent)/ Ground Fault (Overcurrent) Protection for Feeders	50/51 50N/51N	Instantaneous overcurrent (**50**)/time overcurrent (**51**); Neutral instantaneous overcurrent (**50N**)/ neutral time overcurrent (**51N**).
Phase Fault/Overcurrent Protection for Transformers	87T	Transformer differential relay (**87T**)
Undervoltage Detection*	27	$V < 80\%$; Disconnection time is 1.5 seconds. (*For PV systems connected at HV PCC)
Overvoltage Detection*	59	$V > 120\%$; Disconnection time is 0.1 seconds. (*For PV systems connected at HV PCC)
Under-frequency Detection*	81U	$f < 95\%$; Disconnection time is 0.1 seconds. (*For PV systems connected at HV PCC)
Over-frequency Detection*	81O	$f > 105\%$; Disconnection time is 0.1 seconds. (*For PV systems connected at HV PCC)
Protective Relaying at LV PCC		
Undervoltage Detection	27	$V < 85\%$; Disconnection time is 0.3 seconds.
Overvoltage Detection	59	$V > 110\%$; Disconnection time is 0.3 seconds.
Under-frequency Detection	81U	$f < 99\%$; Disconnection time is 0.3 seconds.
Over-frequency Detection	81O	$f > 101\%$; Disconnection time is 0.3 seconds.

(Continued)

TABLE 5.94 (*Continued*)
Requirements for PV System Protective Relaying within the Facility and at LV, MV, and HV PCCs

	PV System Protective Relaying within the Installation	
Protection Requirement	**ANSI Code**	**Explanation/Recommended Relay Setting**
Overcurrent /Earth Fault (or overcurrent) Protection	50/51 50/51N	Instantaneous (**50**)/time overcurrent (**51**) protection; Instantaneous (**50**)/ neutral time overcurrent (**51N**) protection.
Residual Current Protection	50N/51N	Independent high-sensitivity RCD is also recommended.
Anti-Islanding Protection (AIP) and Reclosing	81R 78 27X 25	AIP in case of grid outage can be achieved by employing different methods such as rate-of-frequency change (**81R**) and vector shift methods. AIP relays can also monitor or detects other protection requirements such as Undervoltage, overvoltage, under frequency, and over frequency. Hence network disconnection can also happen if one of the protection functions is triggered. Phase-angle measuring (**78**), synchronizing (**25**), and auxiliary Undervoltage relays (**27X** < 1 min delayed closing) may also be required for reclosing strategy.
Other Protection and Control Functions (power factor, phase and neutral unbalancing, PV connection at PCC, and supervision functions)	55 60N 46 89 74CT, 74TCS BF	The limitations for power factor relaying (**55**), phase current unbalancing (**46**), and neutral current unbalancing (**60N**) can also be adapted to the protection system in need of using such additional protections. PV connection at PCC can be implemented over a visible and lockable switch (**89**). Supervision functions such as BF (supervision for breaker failure), 74CT (current transformer supervision), and 74TCS (trip circuit supervision) are also included in relays' control functions. Note that **negative sequence relays** can also be used to protect the system from unbalanced loading and/or unbalanced faults (such as phase-to-phase faults).
	Protective Relaying at MV PCC	
Undervoltage Detection	27	$V < 85\%$; Disconnection time is 0.3 seconds.
Overvoltage Detection	59	$V > 110\%$; Disconnection time is 0.3 seconds.
Under-frequency Detection	81U	$f < 99\%$; Disconnection time is 0.3 seconds.
Over-frequency Detection	81O	$f > 101\%$; Disconnection time is 0.3 seconds.
Overcurrent/Earth Fault (or overcurrent) Protection	50/51 50/51N 67/67N 59G	Instantaneous (**50**)/time overcurrent (**51**) protection; Instantaneous (**50**)/time earth overcurrent (**51N**) protection. AC directional overcurrent (**67**)/neutral directional overcurrent (**67N**) relay in case of the parallel feeder circuit. Distance (**21**)/neutral distance (**21N**) relays are also used in the case of a parallel feeder circuit. 59G relay is used for unearthed primary interconnection transformers. Note that **50/51N** or **67N** relays are suitable for earthed primary interconnection transformers.
Anti-Islanding Protection (AIP) and Reclosing	79, 27X	Although some DNOs demand, a specific AIP is not necessary for MV PCC (unlike LV). AC reclosing (**79**) in coordination with auxiliary Undervoltage relays (**27X** < 1 min delayed closing) may also be required for auto-reconnection.
PV Connection Switch Lock-out Relay	89 86	PV connection at PCC is implemented over a visible and lockable switch (**89**). Lock-out (or master trip) relay (**86**) holds the equipment out of service during abnormal conditions.
	Protective Relaying at HV PCC	
Undervoltage Detection	27	$V < 80\%$; Disconnection time is 2.5 seconds.
Overvoltage Detection	59	$V > 115\%$; Disconnection time is 0.1 seconds.
Overcurrent Protection	50/51	Instantaneous/time overcurrent protection

(*Continued*)

Photovoltaic Power Systems 413

TABLE 5.94 (*Continued*)
Requirements for PV System Protective Relaying within the Facility and at LV, MV, and HV PCCs

PV System Protective Relaying within the Installation

Protection Requirement	ANSI Code	Explanation/Recommended Relay Setting
Earth Fault Protection	51N	Earth faults can be controlled by utilizing **51N** (time overcurrent) protection
Phase/Earth Fault Protection by Distance Relay	21/21N	Distance (**21**)/neutral distance (**21N**) relay (recommended if PV output level is greater than a specific threshold value such as 100MVA)
Phase Fault/Overcurrent Protection for Transformers	87T	Transformer differential relay (**87T**)
Protection Against Reverse Power Flow	32	A directional power relay (**32**) is used to separate the PV generator unit from the HV network in case of power loss on the primary side of the transformer (PV side).

As known, large PV power plants have several protection zones both on DC and AC sides. For this purpose, the required relay functions and their settings should be done within the facility and at all connection voltage levels (LV, MV, and HV). Typical protection schemes for PV power plants and their connection to the local distribution grid at the LV level can be demonstrated in Figure 5.135. Similarly, other protection schemes at medium and high-voltage PCCs are given in Figures 5.136 and 5.137.

The elements 27, 59, and 81U/O for the respective disturbances provide basic protection functions to isolate the PV system from the utility. Supplementary protective functions, as shown in Figures 5.135–5.137, can also be added to these basic protection schemes to increase the systems' protective performance. Note that the protection elements given in the respective Figures and their usage purposes are explained in detail in Table 5.94.

As a general rule, the graded overcurrent system with discriminative protection times can also be used for the relay coordination of the photovoltaic protection systems. A sample numerical study of relay coordination is explained in Example 5.29 for a grid-connected large PV system.

As explained above, the protective relay should not generate a trip signal under nominal operating conditions but should be able to trip within the predefined tripping time under short circuits. Hence, the relay's current setting in the substation zone should be determined with respect to the nominal current and maximum short-circuit current values. If the system short-circuits currents in the protection zone are already known, they can be used directly for the relay settings. Alternatively, the short-circuit current of the transformer can also be simplified by neglecting the system impedance. For example, the short-circuit current on the secondary side of an MV/LV distribution transformer (if the system impedance is neglected) can be found by equation (5.136).

$$I_{sc} = \frac{\left(\frac{S_T \cdot 10^3}{\sqrt{3} \cdot U_T}\right) \cdot 100}{u_k} = \frac{I_n \cdot 100}{u_k} \quad (5.136)$$

where
  U_T: Low-voltage side nominal voltage of the transformer (no-load voltage) [V]
  S_T: Nominal apparent power of the transformer [kVA]
  u_k: Relative short-circuit voltage of the transformer [%]
  I_n: Nominal current of the transformer on the LV side [A]

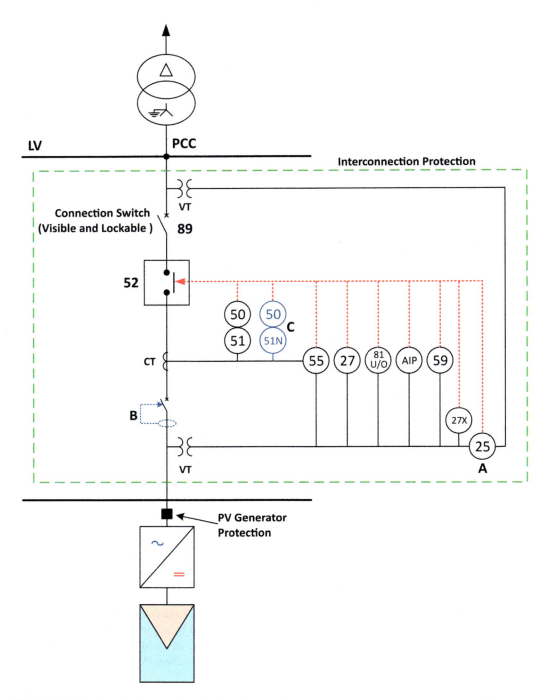

FIGURE 5.135 A typical protective relaying diagram for a grid-connected PV system at LV PCC (A: Only for self-commutated voltage source-based inverters, B: Option-1 for earth fault protection, C: Option-2 for earth fault protection).

IMPORTANT NOTE ABOUT EARTH FAULT RELAY SETTINGS: The methods used for setting the phase fault/overcurrent relays and their coordination procedures are explained in detail in the above sections. As for the earth fault relay settings and their selectivity can also be carried out in the same way as phase overcurrent relays. Pickup currents for earth fault relays typically range

Photovoltaic Power Systems

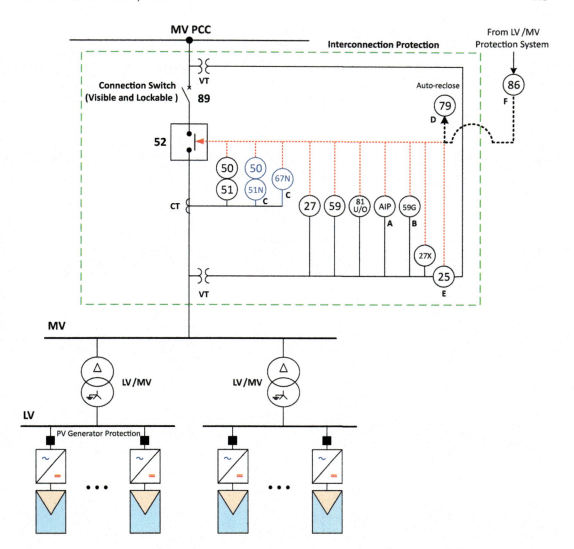

FIGURE 5.136 A typical protective relaying diagram for a grid-connected PV system at MV PCC (A: Anti-islanding protection (not required by every DNO), B: For unearthed primary interconnection transformer, C: For earthed primary interconnection transformer, D: Instantaneous reclosing (reclose interval is 1 second or higher), E: Only for self-commutated voltage source based inverters, F: Required if there is only single LV/MV transformer).

from 1 A to 4 A. For example, the pickup currents can be selected as 1 A for a 0.4 kV network, 3.6 A for a 15 kV network, and 4 A for a 30 kV network. The trip time of an earth fault relay and the phase overcurrent relay, both of which are placed in the same protection zone, can also be set to be the same trip value. In other words, the coordination settings applied to phase overcurrent relays are also suitable for earth fault relays.

Example 5.29: Relay Coordination for a PV System Connected to Local Grid

Consider the 500 kWp PV power plant worked in Example 5.20. A simplified single-line diagram of the PV power plant and its connection to the local grid can be re-drawn as Figure 5.138. The system data and short-circuit currents at different locations of the network are listed in Table 5.95.

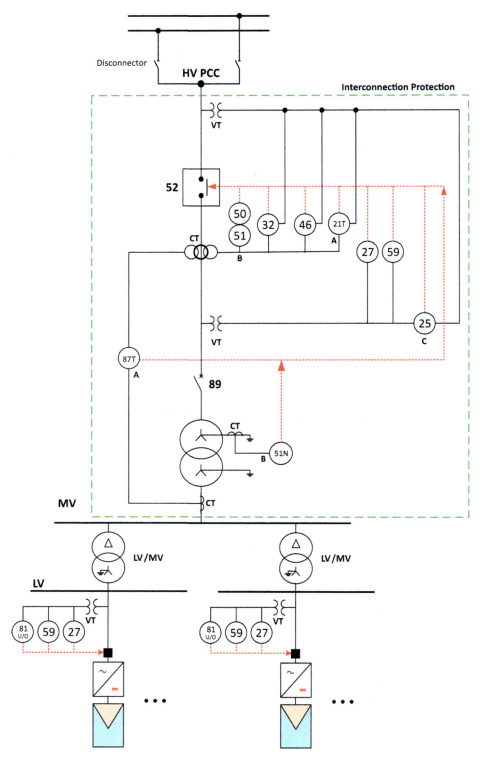

FIGURE 5.137 A typical protective relaying diagram for a grid-connected PV system at HV PCC (A: For large PV power plants (≥100 MVA), B: For PV power plants up to 100 MVA, C: Only for self-commutated voltage source based inverters). Note that the elements 27, 59, and 81U/O at PV generator level are required only for a PV power plant connecting to the HV network.

Photovoltaic Power Systems

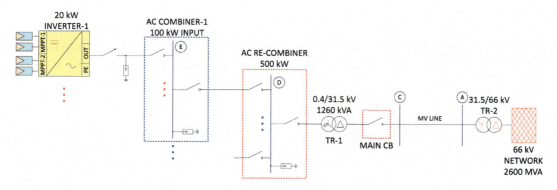

FIGURE 5.138 The simplified single-line diagram of the 500 kW PV power plant.

TABLE 5.95
The System Data and Short-Circuit Currents

Location	Three-Phase Short Circuit Current I_{sc3}	Recommended Current Transformer	Three Phase Transformer Data (0.4/31.5 kV)
C	1428 A	150/5 A	1260 kVA (Y/Δ: star point grounded), u_k = %9.9,
D	24350 A	750/5 A	Transformer maximum short circuit withstand
E	22940 A	600/5 A	duration = 2 sn.

Assume that the maximum allowable trip time at 66 kV connection point A is determined by TSO (Transmission System Operator) to be 1 second.

Based on the provided data, determine the protection requirements at points E, D, and C. Perform the discriminative relay coordination throughout the system to reach the protection objectives and do the relay settings accordingly. Finally, draw the protective relaying diagram for both LV and MV PCCs.

Solution:

From Table 5.94, the fundamental protection objectives and required relay functions can be listed in Table 5.96 for the respective locations.

As explained above, protection devices in a downstream network should activate before the upstream protection devices. In this example, the lowermost downstream location is the PV generator, while the network upstream side is toward the 66 kV network. Therefore, if the downstream point is taken as a reference for determining the relay trip time, the discrimination time is gradually increased toward the upstream direction. However, in this example, the maximum trip time at the upstream point was already determined by the TSO $(t_{max\ trip\ at\ A} = 1\ second)$. So in this case, the relay trip times toward the downstream direction can then be determined by decreasing the trip times gradually from 1 second.

Before calculating the trip times of Relay C, let's calculate the transformer's nominal current and short-circuit current at point C. The nominal current of the transformer on the 31.5 kV side is as follows:

$$I_n(31.5\ \text{kV side}) = \frac{S_T}{\sqrt{3} \cdot U_T} = \frac{1260\ \text{kVA}}{\sqrt{3} \times 31.5\ \text{kV}} = 23.09\ \text{A}$$

Since the current transformer is 25/5 A, the relay nominal input is also selected as 5 A. Remember that the current transformer with a primary current more than the load current (or nominal current) can be selected. So 25/5 A CT is acceptable. Also, make sure that the current setting should never

TABLE 5.96
The Basic Protection Objectives and Required Relay Functions for the Locations of C, D, and E

Protection Location	Protection Objectives	Required Relays
Protection at LV (Point E)	Fault Protection	Instantaneous (**50**), time (**51**) overcurrent, and neutral time overcurrent (**51N**) relays may be required.
Interconnection protection at LV (Point D)	Fault Protection	Instantaneous (**50**), time (**51**) overcurrent, and neutral time overcurrent (**51N**) relays may be required.
	Anti-islanding Protection (AIP)	AIP disconnects the network as soon as the grid power outage occurs. The operating range for undervoltage, overvoltage, under-frequency, and over-frequency should also be detected by the elements of (**27**), (**59**), and (**81U/O**) in coordination with AIP.
	Reclose	Reclosing can be done with the help of synchronizing relay (**25**).
Interconnection protection at MV (Point C)	Fault Protection	Instantaneous overcurrent (**50**)/time overcurrent (**51**)
	Voltage and Frequency Protection	Undervoltage (**27**), overvoltage (**59**), and under/over-frequency relays (**81U/O**) can be used to detect respective disturbances.
	Reclose	Reclosing can be realized with the help of synchronizing (**25**) relay.

exceed the rated current value of the protected equipment. Besides, the short-circuit current of the transformer on the primary and secondary sides can simply be found as below:

$$I_{sc}(31.5 \text{ kV side}) = \frac{I_n \cdot 100}{u_k} = \frac{23.09 \times 100}{9.9} = 233.23 \text{ A}$$

$$I_{sc}(0.4 \text{ kV side}) = \frac{\left(\frac{S_T}{\sqrt{3} \cdot U_T}\right) \cdot 100}{u_k} = \frac{\left(\frac{1260 \text{ kVA}}{\sqrt{3} \times 0.4 \text{ kV}}\right) \cdot 100}{9.9} = 18370 \text{ A}$$

The short-circuit current results obtained from the above-simplified method are lower than the actual short-circuit currents given in Table 5.95, which were calculated according to the worst-case scenario. For this reason, instead of the values given in Table 5.95, it is better to take the newly calculated fault currents as the basis for relay settings.

Now, let us follow the calculation steps to grade Relay C, Relay D, and Relay E with respect to maximum trip time at point A. Since the time T_A was given as 1 second, the actual operation time T_C for Relay C can be calculated as follows.

Step-1 (Setting Calculations for Relay C): Actual trip time of Relay C ($T_C = t_{\text{trip-C}}$ for Relay C) is calculated as below:

Pickup current calculation: Secondary current of CT is 5 A. Therefore, the pickup current of the relay can be calculated for the given plug setting (100%) as below:

$$\text{Pickup Currents} = 5 \times 100\% = 5 \text{ A}$$

PSM calculation: Fault current on the secondary coil, which is equal to $233.23 \times (5/25) = 46.64$ A. The PSM value for the given fault current can then be calculated as:

$$\text{PSM} = \frac{\text{Actual Fault Current on Secondary}}{\text{Relay Pickup Current}} = \frac{46.64 \text{ A}}{5 \text{ A}} = 9.4$$

Photovoltaic Power Systems

Operating time calculation of the relay: From (5.129), we have the trip time operations at TSM = 1.

$$t_{trip} = 1 \times \left[\frac{0.14}{(PSM)^{0.02} - 1} \right] = \left[\frac{0.14}{(9.4)^{0.02} - 1} \right] = 3.05 \text{ s}$$

Actual operating time calculation: Since we need to coordinate Relay C according to the trip time on feeder-A, the actual trip time of Relay C will equal (1 − 0.4 = 0.6 s). So, the TSM value is required to be set as 0.2 to get an actual trip time of 0.6 seconds. Note that the discrimination time DT is taken as 0.4 seconds.

$$0.6 \text{ s} = 3.05 \text{ s} \times TSM \Rightarrow TSM = \frac{0.6 \text{ s}}{3.05 \text{ s}} \approx 0.2$$

Step-2 (Setting Calculations for Relay D): Two relays need to be coordinated downstream. The remaining tripping time for these two relays is 0.6 seconds in total. Therefore, let's set the relay at point D for 0.3 seconds, which means that DT is taken as 0.3 seconds. Actual trip time of Relay D (T_D) is calculated as below:

$$T_D = T_C - DT$$

$$T_D = 0.6 - 0.3 = 0.3 \text{ s}$$

Since the nominal load current (PV current) on the primary side is $\frac{500 \text{ kW}}{\sqrt{3} \times 0.4 \text{ kV}} = 721 \text{ A}$, the selected. So 750/5 A CT is acceptable. This time, let's assume that the PS value is set to 150%. Hence, the pickup current can be calculated for the given plug setting (150%) as below:

$$\text{Pickup Currents} = 5 \times 150\% = 7.5 \text{ A}$$

For the PSM calculation, we need to find the fault current on the secondary coil as $18370 \times (5/750) = 122.46 \text{ A}$, which leads to PSM = 122.46/7.5 ≈ 16.32. Hence, t_{trip-D} can be found from the standard IDMT relay characteristic curve at TSM = 1.

$$t_{trip-2} = 1 \times \left[\frac{C_{slope}}{(PSM_{Relay2})^{\alpha} - 1} \right] = \left[\frac{0.14}{(16.32)^{0.02} - 1} \right] = 2.43 \text{ s}$$

As a result, the TMS value for Relay D is calculated as follows:

$$TMS \text{ (for Relay2)} = \frac{T_D}{t_{trip-D}} = \frac{0.3}{2.43} \approx 0.125$$

Step-3 (Setting Calculations for Relay E): Actual trip time of Relay E (T_E) is calculated as below (DT is taken as 0.2 seconds):

$$T_E = T_D - DT$$

$$T_3 = 0.3 - 0.2 = 0.1 \text{ s}$$

Since the nominal load current (PV current) on the 100 kW branch is $\frac{100 \text{ kW}}{\sqrt{3} \times 0.4 \text{ kV}} = 144.2 \text{ A}$, So the selected 600/5 A CT is acceptable. Let's set the PS value as 200%. Hence, the pickup current can be calculated for the given plug setting (200%) as below:

$$\text{Pickup Currents} = 5 \times 200\% = 10 \text{ A}$$

TABLE 5.97
System Parameters and Standard IDMT Relay Settings for the Locations of C, D, and E

Relay Locations	Max Load (PV) Current	CT Ratio Selected	Fault Currents		Relay Current Settings		Coordination Parameters		
			Short Circuit Result	Simplified Method	Percent	Primary/Secondary Currents	PSM	TSM	Actual Trip Time
C	23.09 A	150/5 A	1428 A	233.23 A	100%	150 A/5 A	9.4	0.2	0.6 second[a]
D	721 A	750/5 A	24350 A	18370 A	150%	1125 A/7.5 A	16.32	0.125	0.3 second
E	144.2 A	600/5 A	22940 A	N.A.	200%	1200 A/10 A	14.32	0.04	0.1 second

[a] The maximum expected clearance time of the fault is 0.6 seconds. This value should never exceed the transformer maximum short-circuit withstand duration (0.2 seconds). Since (0.6<2), the selected trip time is suitable.

Let's multiply the system short-circuit current at Point E by a factor of 75% to obtain a safety tripping margin. Also for the PSM calculation, we need to find out the fault current on the secondary coil as $(0.75 \times 22940) \cdot (5/600) = 143.375$ A, which leads to PSM $= 143.275/10 = 14.32$. Hence, t_{trip-E} can be found from the standard IDMT relay characteristic curve at TSM $= 1$.

$$t_{trip-E} = 1 \times \left[\frac{C_{slope}}{(PSM_{Relay2})^\alpha - 1} \right] = \left[\frac{0.14}{(14.32)^{0.02} - 1} \right] = 2.56 \text{ s}$$

As a result, the TMS value for Relay E is calculated as follows:

$$\text{TMS (for Relay} - \text{E)} = \frac{T_E}{t_{trip-E}} = \frac{0.1}{2.56} \approx 0.04$$

Importantly note that in some cases, the standard inverse-time curve may not be sufficient for the desired time grading. In such cases, this time grading difficulty can be resolved by using other relay characteristics such as very inverse and extremely inverse.

System parameters used in relay coordination and relay settings are summarized for the respective C, D, and E locations in Table 5.97.

From the above results, fundamental protection schemes for the considered PV power plant can be demonstrated in Figure 5.139.

5.5.7.11 Cable Sizing and Selection for Photovoltaic Systems

Proper selection of cables as well as their correct sizing is one of the most important factors that affect the economic and safe operation of PV systems. There are five key factors for cable sizing. These are the ampacity (current-carrying capacity), short-circuit withstand capability, voltage withstand capability, the maximum allowable voltage drop, and the maximum allowable power loss. More specifically, (i) the cables are required to be sized to withstand the current flowing through them, (ii) the cables must be able to withstand the highest short-circuit currents for short periods until the protection elements break the circuit, (iii) the cable insulations must be able to withstand the highest possible voltage level in the system, (iv) the cables are needed to ensure the minimum permissible voltage drop, (v) the cables are needed to ensure the minimum power loss requirement, and (vi) finally, they should be able to handle the variation of temperature and moisture.

Considering the aforementioned explanations, the cable sizing steps for photovoltaic systems can be outlined below:

Step 1 – Maintaining Cable Ampacity: The current-carrying capacity of a power cable is mainly dependent upon the installation conditions and material properties used. The minimum ampacity value required for a PV cabling system can be decided based on the knowledge of the

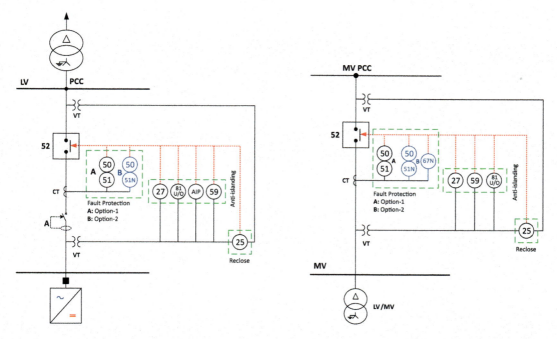

FIGURE 5.139 Fundamental protective relaying for the considered PV power plant via LV and MV PCCs.

maximum current-carrying capacity of the respective cables. As is known, the current-carrying capacity of a cable under standard conditions (I_Z) will vary depending on the cable grouping, insulation type, and operating conditions (such as temperature and moisture), which are taken into account by applying all the cable correction factors ($c_f = k_1 \times k_2 \cdots$)

$$I_{Zc} = c_f \cdot I_Z \tag{5.137}$$

The maximum operating current (I_{max}) in the system shall be lower or equal to the corrected ampacity (I_{Zc}) of the cable.

$$I_{Zc} \geq I_{max} \Rightarrow c_f \cdot I_Z \geq I_{max} \Rightarrow I_Z \geq \frac{I_{max}}{c_f} \tag{5.137a}$$

Detailed tables about cable correction factors (based on different methods of cable installation/grouping and ambient temperatures) are available in the IEC 60364-5-52 standard. Herein, the frequently used correction factors according to ground/air temperatures are given in Table 5.98. Correction factors according to the cable grouping are also shown in Table 5.99.

When installing a roof-mounted PV system, an additional ambient temperature factor is required to be considered for particular raceways, conduits, or cables. Cables or conductors mounted on the rooftop can be exposed to higher temperatures than conductors in other types of installation. This is because the roof exposed to solar radiation turns into a thermal mass, and the heat stored by the roof is transferred to the conduit. Therefore, the NEC specified various temperature adders to the ambient temperature based on the conductor mounting distance from the roof surface. Table 5.100 addresses some temperature add-ons for solar cables mounted on the rooftop to be able to modify a safe ampacity level for a particular installation. It is clear from Table 5.100 that to reduce the heat effect caused by the roof, it is necessary to create as much distance as possible between the cable trays and the roof surface.

TABLE 5.98
Correction Factors according to Ambient Temperatures

Air Temperature[a]/ Ground Temperature[g]	Insulation Type[a]			
	PVC[a]	PVC[g]	XLPE/EPR[a]	XLPE/EPR[g]
10°C	1.22	1.10	1.15	1.07
15°C	1.17	1.05	1.12	1.04
20°C	1.12	1	1.08	1
25°C	1.06	0.95	1.04	0.96
30°C	1	0.89	1	0.93
35°C	0.94	0.84	0.96	0.89
40°C	0.87	0.77	0.91	0.85
45°C	0.79	0.71	0.87	0.80
50°C	0.71	0.63	0.82	0.76
55°C	0.61	0.55	0.76	0.71
60°C	0.50	0.45	0.71	0.65
65°C	-	-	0.65	0.60
70°C	-	-	0.58	0.53
75°C	-	-	0.50	0.46
80°C	-	-	0.41	0.38

PVC: Polyvinyl-chloride, XLPE: Cross-linked polyethylene, EPR: ethylene propylene rubber.

[a] Temperature limits at the conductor are 70°C for PVC insulation and 90°C for XLPE and EPR insulation. Temperature limits at the sheath are 70°C for PVC material and 105°C for XLPE and EPR materials. As known, the ampacity values of grouped cables will be less than the same cable installed in isolation. This decrease happens due to the mutual heating mechanism.

Cable grouping factors in NEC standards are given in a simpler way versus the number of live conductors. Therefore, as an alternative approach, the DC cable grouping factor for rooftop PV applications can be applied as in Table 5.101, based on the NEC standard. Note that unlike Table 5.99, the grouping factor in Table 5.101 considers only the number of live conductors, not the other installation properties.

Note to refer that the universal wire sizes and typical ampacity values are given in Table 5.5. However, detailed ampacity values can also be obtained from manufacturer catalogs or the IEC-60512 standard.

The following essential points should also be addressed when sizing the photovoltaic DC cables.

1. If a string cable is protected by an overcurrent protection device, the string cable must be sized to have a rating equal to or greater than the overcurrent current protection device. For instance, if the overcurrent current protective device is rated at 10 A, then the string cable shall be required to be rated at an ampacity level of 10 A at least.
2. If no overcurrent device is located in the string cable, the ampacity level of the string cable shall be rated to have an ampacity level equal to or greater than $1.25 \times \sum_{i=1}^{N_p-1} I_{sc,i}$ (see equation 5.57 and Figure 5.86).

Step 2 – Ensuring the Short-Circuit Temperature Rating of the Cable: Each cable has a withstand capability against short-circuit currents. The effective short-circuit current in cable sizing is the thermally equivalent short-circuit current, which was given by equation (5.101). The minimum

TABLE 5.99
Correction Factors for Grouped Cables/Circuits

Cable Arrangement Type	Number of Circuit or Multi-Core Cables												Explanation
	1	2	3	4	5	6	7	8	9	12	16	20	
Bunched in air, on a surface, embedded or enclosed	1	0.8	0.7	0.65	0.6	0.57	0.54	0.52	0.5	0.45	0.41	0.38	• Surrounded by thermal insulation or Fixed on a surface • On unperforated/perforated trays (horizontal or vertical) • Buried direct in the ground[a]/Laid in underground ducts, pipes, or conduits
Single layer on wall, floor, or unperforated tray	1	0.85	0.79	0.75	0.73	0.72	0.72	0.71	0.7	No further correction factor			• Fixed on a surface • On unperforated trays (horizontal or vertical)
Single layer fixed directly under a wooden ceiling	0.95	0.81	0.72	0.68	0.66	0.64	0.63	0.62	0.61				
Single layer on a perforated horizontal or vertical tray	1	0.88	0.82	0.77	0.75	0.73	0.73	0.72	0.72				• On perforated trays (horizontal or vertical)
Single-layer on ladder support or cleats etc.	1	0.87	0.82	0.8	0.8	0.79	0.79	0.78	0.78				

[a] Another correction factor shall apply depending on the nature of the soil (or soil thermal resistivity). The k_3 values herein are 1.21 for very wet (saturated) soil, 1.13 for wet soil, 1.05 for damp soil, 1 for dry soil, and 0.86 for very dry (sunbaked) soil.

TABLE 5.100
Temperature Add-Ons for the Roof-Mounted Conduits/Cables

The Distance between the Roof Surface and the Conduit Base (or Cable)	Ambient Temperature Add-Ons	
	°C	°F
On the roof (up to 13 mm)	33	60
Above roof (from 13 mm to 90 mm)	22	40
Above roof (from 90 mm to 300 mm)	17	30
Above roof (from 300 mm to 900 mm)	14	25

TABLE 5.101
Cable Grouping Factors as a Function of Live Conductors in a Raceway or Cable

# of conductors	1–3	4–6	7–9	10–20	21–30	31–40	≥41
Grouping factor	1	0.80	0.70	0.50	0.45	0.40	0.35

conductor cross-sectional area of a cable S_{min} (in mm²) must be able to withstand the thermal short-circuit currents (I_{th} in amps) for short-circuit durations until the protection system clears the fault (t_{trip} in seconds). Hence, the minimum required cable size (S_{min}) considering the shirt-circuit temperature increase can typically be found by equation (5.138).

$$S_{min} = \frac{I_{th} \cdot \sqrt{t_{trip}}}{J_{thr}} \leq S_{cable} \tag{5.138}$$

where S_{cable} [mm²] is the nominal cross-sectional area of the conductor, and J_{thr} is the short-circuit current density of the cable in [A/mm²], which is determined according to the insulation type and published in the cable catalogs by the manufacturers. Typical short-time rated current density values for PVC and XLPE cables are given in Table 5.102.

Step 3 – Maintaining Voltage Drop Requirements: The power cables shall be designed to minimize voltage drop. Ideally, the voltage drop in a solar PV system should be kept to less than 2%. However, it is recommended to keep the voltage drop to be less than 3% to ensure that the cost is down. In terms of system safety, the voltage drop should always be kept below 5%. According to the regulations of most countries, a maximum voltage drop of 3% is permissible for both AC and DC sides of PV systems.

In general, the percent voltage drop is calculated with the formula $\%u = (\Delta U/U) \cdot 100$ for 3-phase systems, and $\%u = (\Delta U/V) \cdot 100$ for single-phase systems, where U is phase-phase voltage, and V is the phase-neutral voltage. The voltage drop can be calculated as $\Delta U = 2I(R \cdot \cos\varphi + X \cdot \sin\varphi)$ for single-phase systems, and $\Delta U = \sqrt{3}I(R \cdot \cos\varphi + X \cdot \sin\varphi)$ for balanced 3-phase systems. Since the reactance is negligible for the low and medium voltage cables, specifically in lower lengths, the percent voltage drop calculations for AC and DC cables can be given below.

The percent voltage drop for a photovoltaic **DC cabling system** ($\%v$) can easily be calculated by equation (5.139).

$$\%v = \frac{2 \cdot 100 \cdot \ell_{dc} \cdot P_{max}}{k \cdot S_{dc\,cable} \cdot (V_{MPP})^2} \tag{5.139}$$

where ℓ_{dc} is the length of DC cable in meters, P_{max} is the maximum PV power [Watts] to be carried by the respective conductor, k is the electrical conductivity $\left(k_{Cu} = 56 \frac{m}{\Omega \cdot mm^2}, k_{A\ell} = 34 \frac{m}{\Omega \cdot mm^2}\right)$,

TABLE 5.102
Rated Short-Time Current Density Values (J_{thr} in A/mm² for Short-Circuit of 1 Second)

Conductor Temperature	Insulation Type (J_{thr} in A/mm² for Short-Circuiting of 1 second)		
	PVCa (≤ 300 mm²)	PVCa (> 300 mm²)	(XLPE/EPR)[a]
20°C	150	140	181
30°C	143	133	176
40°C	136	126	170
50°C	129	118	165
60°C	122	111	159
65°C	119	107	157
70°C	115	103	154
80°C	-	-	148
90°C	-	-	143

[a] Permissible operating temperature is 70°C for PVC insulation and 90°C for XLPE. The permissible operating temperature is 160°C for PVC (≤ 300 mm²), 140°C for PVC (> 300 mm²), and 250°C for XLPE.

Photovoltaic Power Systems

and V_{MPP} is the string voltage [volts] at MPP. By inserting $P_{max} = I_{max} \cdot V_{MPP}$ into (5.139), equation (5.140) is obtained as below.

$$\%v = \frac{2 \cdot 100 \cdot \ell_{dc} \cdot I_{max}}{k \cdot S_{cable} \cdot V_{MPP}} \tag{5.140}$$

If the allowable percent voltage drop value %3 is substituted into the above equations, the minimum cross section of the string cable ($S_{dc\ cable(min)}$) can be obtained as 5.141.

$$S_{dc\ cable(min)} = \frac{200 \cdot \ell_{dc} \cdot P_{max}}{k \cdot (3) \cdot (V_{MPP})^2} = \frac{200 \cdot \ell_{dc} \cdot I_{max}}{k \cdot (3) \cdot V_{MPP}} \tag{5.141}$$

The percent voltage drop for a photovoltaic AC cabling system (%u) can be obtained similarly. Subsequently, the percent voltage drop for single-phase (%$u_{1\Phi}$) and balanced three-phase AC cables (%$u_{3\Phi}$) are obtained as (5.142) and (5.143), respectively.

$$\%u_{1\Phi} = \frac{2 \cdot 100 \cdot \ell_{ac} \cdot P_n}{k \cdot S_{cable} \cdot (V_n)^2} \tag{5.142}$$

$$\%u_{3\Phi} = \frac{100 \cdot \ell_{ac} \cdot P_n}{k \cdot S_{cable} \cdot (U_n)^2} \tag{5.143}$$

where ℓ_{ac} is the length of AC cable in meters, P_n is the nominal system power [Watts] to be carried by the respective conductor, V_n is the rated phase-to-neutral voltage [volts], and U_n is the rated phase-to-phase voltage [volts]. If the allowable percent voltage drop value %3 is substituted into the above equations, the minimum cross section of the AC cable ($S_{ac\ cable(min)}$) can be obtained for single-phase and three-phase circuits as (5.144) and (5.145).

$$S_{ac\ cable(min)} = \frac{2 \cdot 100 \cdot \ell_{ac} \cdot P_n}{k \cdot (3) \cdot (V_n)^2} \tag{5.144}$$

$$S_{ac\ cable(min)} = \frac{100 \cdot \ell_{ac} \cdot P_n}{k \cdot (3) \cdot (U_n)^2} \tag{5.145}$$

Step 4 – Meeting the Power Loss Conditions: Both DC and AC power cables shall be designed to minimize power losses on the conductors. Ideally, the percent power loss on each DC and AC cable segment of PV power systems should always be kept to less than 1%. Following a similar method to that of voltage drop calculations, the percent power loss formula for DC, single-phase AC, and three-phase AC cables can be obtained as below.

$$\%p_{dc} = \frac{\Delta P_{dc}}{P_{string}} \cdot 100 = \frac{200 R_{dc} I_{string}^2}{V_{MPP} I_{string}} = \frac{200 \left(\frac{\ell_{dc}}{k \cdot S_{dc}} \right) I_{string}}{V_{MPP}} = \frac{200 \cdot \ell_{dc} \cdot I_{string}}{k \cdot S_{dc} \cdot V_{MPP}} \tag{5.146}$$

$$\%p_{ac\ (1\Phi)} = \frac{\Delta P_{ac}}{P_{inv}} \cdot 100 = \frac{200 R_{ac} I_{ivn}^2}{V_n I_{inv} \cos\varphi} = \frac{200 \left(\frac{\ell_{ac}}{k \cdot S_{ac}} \right) I_{inv}}{V_n \cos\varphi} = \frac{200 \cdot \ell_{ac} \cdot I_{inv}}{k \cdot S_{ac} \cdot V_n \cdot \cos\varphi} \tag{5.147}$$

$$\%p_{ac\ (3\Phi)} = \frac{\Delta P_{ac}}{P_{inv}} \cdot 100 = \frac{300 R_{ac} I_{ivn}^2}{\sqrt{3} U_n I_{inv} \cos\varphi} = \frac{\sqrt{3} \cdot 100 \left(\frac{\ell_{ac}}{k \cdot S_{ac}} \right) I_{inv}}{U_n \cos\varphi} = \frac{\sqrt{3} \cdot 100 \cdot \ell_{ac} \cdot I_{inv}}{k \cdot S_{ac} \cdot U_n \cdot \cos\varphi} \tag{5.148}$$

where S_{dc} is the cross section of DC cable, S_{ac} is the cross section of the AC cable, and $\cos\varphi$ is the power factor of power flowing through the AC cables. Other parameters in the above formulas are the same as defined previously. Remember that the percent power loss for the selected cable section should meet the condition $\%p < 1\%$. If this condition is not met, then the cable cross section should be increased.

Step 5 – Ensuring the Withstand Voltage Capability: The power cables shall be designed to withstand the voltage stresses such as maximum system voltage and impulse (lightning) voltage. Therefore, the operating voltage and impulse withstand voltage levels of DC and AC cables should also be checked to ensure safe and reliable photovoltaic system operation. Since the voltage verification issues have been discussed sufficiently in the previous sections, it will not go into further detail here in this section.

Example 5.30: Sizing of DC and AC Cables in a Roof-Mounted Photovoltaic System

Consider the 5 kW$_p$ PV system shown in Figure 5.140. The PV array consists of four parallel strings, while the strings consist of five modules connected in series, each having a power of 250 W. The PV module and cabling data are given in Table 5.103.

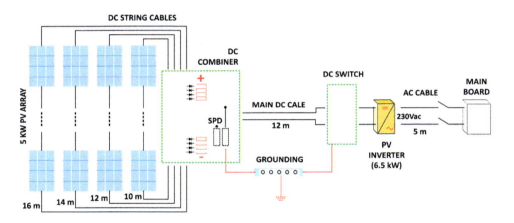

FIGURE 5.140 Circuit diagram of 5 kW rooftop PV system with a central inverter.

TABLE 5.103
The Parameters for PV Module and Cabling

PV Module Specifications: 250 W$_p$ Poly Crystalline Module			Module Efficiency: 15.58 %	
Voltage at P_{max}	**Current at P_{max}**	**Open-Circuit Voltage**	**Short-Circuit Current**	**Power Tolerance**
30.7 V	8.15 A	37.4 V	8.63 A	−0 / +3%
Cabling Parameters				
DC string cables	String cables are respectively 16 m, 14 m, 12 m, and 10 m. Cables are mounted directly on the roof surface in groups of four in a raceway.			
Main DC cable	The main DC cable length is 12 m. These cables are grouped in pairs in a raceway.			
AC cable	AC cable length is 5 m. The AC cables are mounted directly on the wall in groups of two in a conduit. Assume that the **maximum short-circuit current** (due to grid failures) on the AC side of the inverter is 2 kA.			
Other parameters	All cables will be XLPE insulated. The ambient air temperature at the site is 35°C. The conduit is mounted directly on the roof surface. Four conductors are laying in the conduit.			

Considering the given data, size the DC and AC cables in the PV system, and select the appropriate cables from the manufacturers' catalogs.

Solution:

Let us follow the cable sizing steps to select appropriate DC and AC cables of the respective rooftop PV system.

Step 1 – Maintaining Cable Ampacity (DC String Cables): As previously explained, the minimum ampacity should be decided based on the maximum current of our particular PV system. It is clear from the circuit diagram and system parameters that the current of the PV module at MPP is 8.15A under STC. Hence, the maximum current of the PV string by considering the over-irradiance factor (1.25) can be calculated as $8.15 \times 1.25 = 10.1875$ A. To calculate the corrected ampacity of the cable, it is necessary to include the factors of temperature and grouping. Firstly, let's consider the temperature correction factor (k_1). Assume that XLPE-based solar cable (rated for 90°C) will be used for the string and main DC cable. Taking into account that the cable conduit is in direct contact with the roof surface, the factor k_1 is then determined as below:

From Table 5.100, the add-on temperature (for on-roof installation) is selected as 33°C. Therefore, the equivalent ambient temperature for the particular PV installation will be 35°C + 33°C = 68°C. Now, from Table 5.98, the temperature factor for XLPE cable can be selected as $k_1 = 0.58$.

From Table 5.101, the grouping factor (for four to six cables) can be selected as $k_2 = 0.80$. Considering all these correction factors, we can now calculate the required minimum ampacity value of the string cable as below:

$$I_{Z\,min} = \frac{I_{max}}{k_1 \times k_2} = \frac{10.1875\,A}{0.58 \times 0.80} = 21.95\,A$$

As a result, the ampacity value of the DC string cable to be selected should be greater than 21.95 A ($I_Z > 21.95$ A).

Step 2 – Ensuring the Short-Circuit Temperature Rating (DC String Cables): The thermal short-circuit current passing through the DC cable is equal to the possible maximum short-circuit current of the PV system. From equation (5.57),

$$I_{max} = 1.25 \times \sum_{i=1}^{N_p-1} I_{sc,\,i} = 1.25 \times \sum_{i=1}^{4-1} I_{sc,\,i} = 1.25 \times (3 \times 8.63) = 32.36\,\text{Amps}$$

From Table 5.102, the parameter J_{thr} is selected 143 A/mm² for short-circuit 1 second ($t_{trip} = 1$ s, $J_{thr} = 143$ A/mm²). Hence from (5.138), the minimum conductor cross-sectional area of the DC cable can be calculated as below.

$$S_{min} = \frac{I_{th} \cdot \sqrt{t_{trip}}}{J_{thr}} = \frac{32.36 \times \sqrt{1}}{143} = 0.227\,\text{mm}^2 \leq S_{cable}$$

Step 3 – Maintaining Voltage Drop Requirements (DC String Cables): The DC power cables shall meet the voltage drop condition $\%v < 3\%$. Hence from (5.141), the minimum cross section of the DC cable for String-1 can be obtained as below for the worst-case scenario (i.e., longest string cable at $I_{max} = 1.25 \times I_{nom}$, or $P_{max} = 1.25 \times P_{nom}$).

$$S_{dc\,cable(min)} = \frac{200 \cdot \ell_{dc} \cdot I_{max}}{k \cdot (3) \cdot V_{MPP}} = \frac{200 \cdot 16 \cdot 10.1875}{56 \cdot (3) \cdot (30.7 \times 5)} = 1.26\,\text{mm}^2$$

Similarly, the minimum cable cross sections for String-2, String-3, and String-4 should also be selected to be greater than 1.26 mm², even if the cable lengths differ.

Step 4 – Meeting the Power Loss Conditions: The DC power cables shall meet the percent power loss condition $\%p_{dc} < 1\%$. Hence from (5.146), the minimum cross section of the DC cable for String-1 can be obtained as below for the worst-case scenario (i.e., longest string cable at $I_{max} = 1.25 \times I_{nom}$, or $P_{max} = 1.25 \times P_{nom}$).

TABLE 5.104
Sizing Steps for the Main DC Cable of the Given PV System

Step 1—Cable ampacity selection	Since the main DC cable carries the currents of four strings, its maximum current is equal to $10.1875 \times 4 = 40.75$ A. From Tables 5.98 and 5.99, the temperature and grouping factors can be selected as $k_1 = 0.96$ (ambient temperature is 35°C), and $k_2 = 1$ (# of the circuit is 1). Hence, the required minimum ampacity value of the main DC cable: $$I_{Z\,min} = \frac{40.75 \text{ A}}{0.96 \times 1} = 42.45 \text{ A} \therefore (I_Z > 42.45 \text{ A}).$$
Step 2—Short-circuit temperature rating	The thermal short-circuit current passing through the main DC cable is equal to the sum of the maximum short-circuit currents of the PV strings. Hence, $I_{max} = 4 \times (1.25 \times 8.63) = 43.15$ A. From (5.138), the minimum conductor cross-section of the main DC cable can be calculated as $$S_{min\,(dc)} = \frac{I_{th} \cdot \sqrt{t_{trip}}}{J_{thr}} = \frac{43.15 \times \sqrt{1}}{143} = 0.302 \text{ mm}^2 \therefore S_{cable\,(dc)} > 0.302 \text{ mm}^2.$$
Step 3—Meeting the voltage drop condition	The main DC power cables shall meet the voltage drop condition $\%v < 3\%$. Hence from (5.141), the minimum cross-section of the main DC cable ($I_{max} = 1.25 \times \sum I_{string}$, or $P_{max} = 1.25 \times \sum P_{string}$) is: $$S_{dc\,main\,cable(min)} = \frac{200 \cdot \ell_{dc} \cdot I_{max}}{k \cdot (3) \cdot V_{MPP}} = \frac{200 \cdot 12 \cdot (4 \cdot 10.1875)}{56 \cdot (3) \cdot (30.7 \times 5)} = 3.8 \text{ mm}^2$$
Step 4—Meeting the power loss condition	The main DC power cables shall meet the percent power loss condition $\%p_{dc} < 1\%$. Hence from (5.146), the minimum cross-section of the main DC cable ($I_{max} = 1.25 \times \sum I_{string}$, or $P_{max} = 1.25 \times \sum P_{string}$) is: $S_{dc\,main\,cable(min)} = \dfrac{200 \cdot \ell_{dc} \cdot I_{max(main\,DC)}}{k \cdot (1) \cdot V_{Array}} = \dfrac{200 \cdot 12 \cdot (4 \cdot 10.1875)}{56 \cdot (1) \cdot (30.7 \times 5)} = 11.38 \text{ mm}^2$
Step 5—Final cable size	The main DC cable to be selected should meet all the conditions calculated in the above steps. Hence, $S_{dc\,main\,cable} \geq \max [0.302, 3.8, 11.38] \Rightarrow S_{dc\,main\,cable} \geq 11.38$ mm². As a result, $S_{dc\,main} = 16$ mm² can be selected as the next standard cross section.

$$S_{dc\,cable(min)} = \frac{200 \cdot \ell_{dc} \cdot I_{string}}{k \cdot (1) \cdot V_{MPP}} = \frac{200 \cdot 16 \cdot 10.1875}{56 \cdot (1) \cdot (30.7 \times 5)} = 3.79 \text{ mm}^2$$

Similarly, the minimum cable cross sections for String-2, String-3, and String-4 should also be selected to be greater than 1.26 mm² to meet the percent power loss conditions.

The DC cable specifications should be determined to meet all the conditions calculated in the above steps. Hence, the DC cable cross section should be selected such that $S_{dc} \geq \max [S_{dc\,cable(min)}, S_{dc\,cable(min)}, S_{dc\,cable(min)}]$. From this equation, $S_{dc} \geq \max [0.227, 1.26, 3.79] \Rightarrow S_{dc} \geq 3.79$ mm². As a result, $S_{dc} = 4$ mm² can be selected as the next standard cross section. We should also check that the ampacity of the cable must be greater than 21.95 A and the cable insulation must meet the voltage conditions defined in Step-5.

The main DC cable sizing can be done in the same way. Hence, the sizing steps for the main DC cable can be summarized in Table 5.104.

Lastly, the sizing steps for the AC cable are given in Table 5.105.

As a result of the above calculations, sample cable properties that can be selected from the manufacturer's catalogs are given in Table 5.106.

Example 5.31: Sizing of DC and AC Cables in a Ground-Mounted Photovoltaic Power System

Consider the 500 kW$_p$ PV power plant worked in Example 5.20. A simplified single-line diagram of the PV power plant and its connection to the local grid can be re-drawn as Figure 5.141.

The system data and short-circuit currents at different locations of the network can be found from the worked Example 5.20. The PV array consists of four parallel strings, while the strings consist of 20 modules connected in series, each has a power of 250 W. The PV module specifications

TABLE 5.105
AC Cable Sizing Steps for the Given PV System

Step 1—Cable ampacity selection	Since the main AC cable carries the total power of the PV array, its maximum power is equal to $5\,kW \times 1.25 = 6.25\,kW$. This power (6.25 kW) can be taken as a reference in the calculation of the ≈maximum AC current. However, since the maximum PV array power is already taken into account when determining the inverter-rated power, it is sufficient to consider the inverter-rated power (6.5 kW) at this stage. Hence, the maximum output current of the inverter will be 28.27 A $\left(I_{max\,AC} = \dfrac{6500}{230}\right)$. From Tables 5.98 and 5.99, the temperature and grouping factors can be selected as $k_1 = 0.96$ (ambient temperature is 35°C), and $k_2 = 1$ (# of the circuit is 1). Hence, the required minimum ampacity value of the main DC cable: $$I_{Z\,min} = \dfrac{28.27\,A}{0.96 \times 1} = 29.5\,A \therefore (I_Z > 29.5\,A).$$
Step 2—Short-circuit temperature rating	The thermal short-circuit current that is effective on the AC cable happens due to the network short-circuit currents. The short-circuit current is given as 2.5 kA. From (5.101), thermal equivalent short-circuit current can be expressed as equal to the system's short-circuit current, that is, $I_{th} = I_{sc} = 2\,kA$. From (5.138), the minimum conductor cross section of the AC cable can be calculated as $S_{min\,(ac)} = \dfrac{I_{th} \cdot \sqrt{t_{trip}}}{J_{thr}} = \dfrac{2000 \times \sqrt{1}}{143} = 13.98\,mm^2 \therefore S_{cable\,(ac)} > 13.98\,mm^2$.
Step 3—Meeting the voltage drop condition	The AC power cable shall also meet the voltage drop condition $\%u < 3\%$. Hence from (5.144), the minimum cross-section of the single-phase AC cable is: $$S_{ac\,cable(min)} = \dfrac{200 \cdot \ell_{ac} \cdot P_n}{k \cdot (3) \cdot (V_n)^2} = \dfrac{200 \cdot 5 \cdot 6500}{56 \cdot (3) \cdot (230)^2} = 0.73\,mm^2$$
Step 4—Meeting the power loss condition	The AC power cables shall meet the percent power loss condition $\%p_{ac} < 1\%$. Hence from (5.147), the minimum cross-section of the AC cable is ($I_{inv} = 28.27\,A$, and $\cos\varphi \approx 1$): $$S_{ac\,cable(min)} = \dfrac{200 \cdot \ell_{ac} \cdot I_{inv}}{k \cdot (1) \cdot V_n \cdot \cos\varphi} = \dfrac{200 \cdot 5 \cdot 28.27}{56 \cdot (1) \cdot 230 \cdot 1} = 2.2\,mm^2$$
Step 5—Final cable size	The single-phase AC cable to be selected should meet all the conditions calculated in the above steps. Hence, $S_{ac\,cable} \geq \max[13.98, 0.73, 2.2] \Rightarrow S_{ac\,cable} \geq 13.98\,mm^2$. As a result, $S_{ac\,cable} = 16\,mm^2$ can be selected as the next standard cross section.

TABLE 5.106
Sample DC and AC Cable Specifications for the Given PV System

Cable Type	Cable Length	Cable Cross section	Maximum Power	Maximum System Current	Ampacity (in Air/on Surface) at 60°C	Maximum System Voltage	Cable Voltage	% Voltage Drop	% Power Loss
String cable	16 m	$2 \times 4\,mm^2$	1.5625 kW	10.1875 A	55 A/52 A	233.75 V[a]	Up to 0.6 kV	%0.948	%0.948
Main DC cable	12 m	$2 \times 16\,mm^2$	6.25 kW	40.75 A	132 A/125 A	233.75 V	Up to 0.6 kV	%0.711	%0.711
AC cable	5 m	$2 \times 16\,mm^2$	6.5 kW	28.27 A	132 A/125 A	400 V	Up to 0.6 kV	%0.132	%0.137

[a] Temperature correction coefficient for the PV voltage is taken as $1.25\,(37.4 \times 5 \times 1.25 = 233.75\,V)$.

are the same as in Example 5.30. The data for cable grouping, laying, and ambient conditions for the considered PV system are given in Table 5.107.

Cable K-factors regarding the depth of cable laying are obtained from the manufacturer data sheets as in Table 5.108.

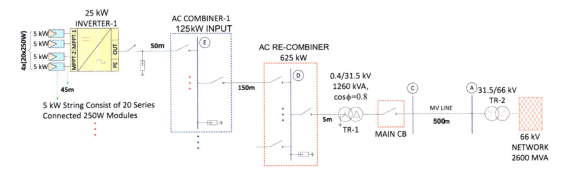

FIGURE 5.141 The simplified single-line diagram of the 500 kW PV power plant (the powers are updated considering the over-irradiance conditions).

TABLE 5.107
The Parameters for the PV Cabling System

Cable Type and Cabling Section	Cable Length[a]	Cable/Conductor Grouping[b]	Cable Depth[c]
DC string cables	40 m	Conductors are laid in groups of 8 (# of the circuit is 8/2 = 4). Note that earthing conductor is considered as separately laying.	The DC cables are laid 75 cm below the ground.
LV AC cable between inverters and Point E	50 m	Conductors are laid in groups of 25 (# of the circuit is 25/5 = 5).	The AC cables are laid 90 cm below the ground.
LV AC cable between points E and D	200 m	Conductors are laid in groups of 25 (# of the circuit is 25/5 = 5).	The AC cables are laid 90 cm below the ground.
LV AC cable between point D and transformer station	5 m	Conductors are laid in groups of 5 (single circuit). Note that there are sufficient distances between the single-core cables in order not to derate because of mutual heating.	The AC cables are laid 1 m below the ground.
MV AC cable between points C and A	500 m	Conductors are laid in groups of 4 (single circuit). Note that there are sufficient distances between the single-core cables in order not to derate because of mutual heating.	The AC cables are laid 1 m below the ground.
Ambient conditions and other parameters	All cables will be XLPE insulated. The ground temperature at the site is 30°C, and assume the type of soil as damp soil.		

[a] The longest cable distance in the respective cable section.
[b] The maximum number of conductors grouped in a cable trench.
[c] There is an inversely proportional relationship between soil thermal resistivity and the ampacity of the cable buried in that ground. In other words, as the soil thermal resistivity increases, the ampacity of the cable buried in the same soil decreases. Besides, the deeper you go, the lower the thermal resistivity you will get. As a result, as the cable depth increases, the cable ampacity value increases.

Considering the given data, size the DC and AC cables in the PV system, and select the appropriate cables from the manufacturers' catalogs.

Solution:

Let us follow the cable sizing steps to select appropriate DC and AC cables for the ground-mounted PV system.

Step 1 – Cable Ampacity Selection: The ampacity selection for each cable type is carried out in Table 5.109.

TABLE 5.108
Cable Correction Factors for the Depth of Cable Laying

Conductor Size	Depth of Laying						
	50 cm	70 cm	90 cm	100 cm	120 cm	150 cm	≥180 cm
Up to 25 mm²	1.1	1.0	0.99	0.98	0.97	0.96	0.95
Above 25 mm²–300 mm²	1.1	1.0	0.98	0.97	0.96	0.94	0.92
Above 300 mm²	1.1	1.0	0.97	0.96	0.95	0.92	0.91

TABLE 5.109
The Ampacity Selection for Each Cable Type of the PV System

Cable Type and Cabling Section	Maximum Current[a]	Cable Correction (K) Factors[b]	Minimum Ampacity[c]
DC string cables [Ground T: 30°C, Damp soil, Depth: 75 cm, # of the circuit: 4]	$I_{max} = 8.15 \times 1.25 = 10.1875$ A	$k_1 = 0.93; k_2 = 1.05$ $k_3 = 1.0; k_4 = 0.65$	16.05 A
LV AC cable (from inverters to point E) [Ground T: 30°C, Damp soil, Depth: 90 cm, # of the circuit: 5]	$I_{max\,AC} = \dfrac{25000}{\sqrt{3} \cdot 400 \cdot 1} = 36.09$ A	$k_1 = 0.93; k_2 = 1.05$ $k_3 = 0.99; k_4 = 0.60$	62.22 A
LV AC cable (from points E to D) [Ground T: 30°C, Damp soil, Depth: 90 cm, # of the circuit: 5]	$I_{max\,AC} = 36.09 \times 5 = 180.45$ A	$k_1 = 0.93; k_2 = 1.05$ $k_3 = 0.99; k_4 = 0.60$	311.1 A
LV AC cable (from D to power station) [Ground T: 30°C, Damp soil, Depth: 100 cm, # of the circuit:1]	$I_{max\,AC} = 180.45 \times 5 = 902.25$ A	$k_1 = 0.93; k_2 = 1.05$ $k_3 = 0.97; k_4 = 1.0$	952.8 A
MV AC cable (from points C to A) [Ground T: 30°C, Damp soil, Depth: 100 cm, # of the circuit: 1]	$I_{max\,AC} = \dfrac{1260}{\sqrt{3} \cdot 31.5 \cdot 0.8} = 28.87$ A	$k_1 = 0.93; k_2 = 1.05$ $k_3 = 0.97; k_4 = 1.0$	30.48 A

[a] Each 5 kW PV string is connected to the inverter via a separate DC cable.
[b] k_1 (ground temperature) is obtained from Table 5.98, k_2 (soil correction) from the footnote of Table 5.99, k_3 (cable depth) from Table 5.108, and k_4 (cable grouping) from Table 5.99.
[c] The minimum ampacity value is calculated from the formula $I_{Z\,min} = I_{max} / (k_1 \times k_2 \times k_3 \times k_4)$.

Step 2 – Short-Circuit Temperature Rating: Thermal equivalent short-circuit currents for each cable type are calculated as in Table 5.110.

Step 3 – Meeting the Voltage Drop Conditions: Minimum cross section calculations based on maximum permissible percent voltage drop are given for each cable section as in Table 5.111.

Step 4 – Meeting the Power Loss Conditions: Minimum cross section calculations based on maximum permissible percent power loss are given for each cable section as in Table 5.112.

Step 5 – Final Cable Size: The cables in the PV system should meet all the conditions calculated in the above steps. Therefore, final cable sizes can be selected as explained in Table 5.113.

As a result of the above calculations, sample cable properties that can be selected from the manufacturer's catalogs are given in Table 5.114.

5.5.7.12 Busbar Sizing and Selection for Photovoltaic Systems

The selection of a busbar inside a combiner box depends on the highest short-circuit current level at the point of use. These busbars should also be able to handle the rated system current continuously. Like all other conductors, each busbar used in distribution boxes has a specific withstand capability

TABLE 5.110
Thermal Equivalent Short-Circuit Currents for Each Cable Type of the PV System

Cable Type and Cabling Section	Short-Circuit Current[a]	Short-Time Current Density (J_{thr})[b]	Minimum Cross-Section[c]
DC string cables	$I_{max} = 4 \times (1.25 \times 8.63) = 43.15$ A	$J_{thr} = 143$ A/mm²	0.302 mm²
LV AC cable (inverters to E)	$I_{sc} = 4.205$ kA	$J_{thr} = 143$ A/mm²	29.4 mm²
LV AC cable (from E to D)	$I_{sc} = 22.94$ kA	$J_{thr} = 143$ A/mm²	160 mm²
LV AC cable (from D to Tr.)	$I_{sc} = 26.35$ kA	$J_{thr} = 143$ A/mm²	184.2 mm²
MV AC cable (from C to A)	$I_{sc} = 1428.47$ A	$J_{thr} = 143$ A/mm²	9.908 mm²

[a] The highest possible short circuit currents at the respective cable locations were calculated in Example 5.20.
[b] Short-time current density values (J_{thr}) are obtained from Table 5.102.
[c] The minimum cross section of the conductor is calculated from the formula $S_{min} = I_{th} \cdot \sqrt{t_{trip}} / J_{thr}$. Note that $I_{th} = I_{sc}$ for the given network.

TABLE 5.111
Minimum Cross-Section Calculation to Meet the Percent Voltage Drop Conditions

Cable Type and Cabling Section	Maximum Current[a]	System Voltage	Cable Length	Min Cross Section[b]
DC string cables	$I_{max} = 10.1875$ A	$V_{MPP} = 30.7 \times 20 = 614$ V	45 m	0.89 mm²
LV AC cable (inverters to E)	$I_{inv\,AC} = 36.09$ A	$U_n = 400$ V	50 m	2.69 mm²
LV AC cable (from E to D)	$I_{max\,AC} = 180.45$ A	$U_n = 400$ V	150 m	40.28 mm²
LV AC cable (from D to Tr.)	$I_{max\,AC} = 902.25$ A	$U_n = 400$ V	5 m	6.72 mm²
MV AC cable (from C to A)	$I_{max\,AC} = 28.87$ A	$U_n = 31500$ V	500 m	21.5 mm²

[a] Maximum system currents are calculated in Table 5.109.
[b] The minimum cross-section is calculated from the formula $S_{dc\,cable(min)} = 200 \cdot \ell_{dc} \cdot I_{max} / (k \cdot 3 \cdot V_{MPP})$ for DC string cable, and from the formula $S_{ac\,cable(min)} = 100 \cdot \ell_{dc} \cdot I_n / (k \cdot 3 \cdot U_n)$ for three-phase AC cables.

TABLE 5.112
Minimum Cross-Section Calculation to Meet the Percent Power Loss Conditions

Cable Type and Cabling Section	Maximum Current[a]	System Voltage	Cable Length	Min Cross Section[b]
DC string cables	$I_{max} = 10.1875$ A	$V_{MPP} = 614$ V	45 m	2.67 mm²
LV AC cable (inverters to E)	$I_{inv\,AC} = 36.09$ A	$U_n = 400$ V	50 m	13.95 mm²
LV AC cable (from E to D)	$I_{max\,AC} = 180.45$ A	$U_n = 400$ V	150 m	209.2 mm²
LV AC cable (from D to Tr.)	$I_{max\,AC} = 902.25$ A	$U_n = 400$ V	5 m	26.6 mm²
MV AC cable (from C to A)[c]	$I_{max\,AC} = 28.87$ A	$U_n = 31500$ V	500 m	27.90 mm²

[a] Maximum system currents are calculated in Table 5.109.
[b] The minimum cross section is calculated from the formula $S_{dc\,cable(min)} = 200 \cdot \ell_{dc} \cdot I_{string} / (k \cdot 1 \cdot V_{MPP})$ for DC string cable, and from the formula $S_{ac\,cable(min)} = \sqrt{3} \cdot 100 \cdot \ell_{ac} \cdot I_{inv} / (k \cdot 1 \cdot U_n \cdot \cos\varphi)$ for three-phase AC cables.
[c] $\cos\varphi = 0.8$, and $\%p_{ac(3\Phi)} < 5\%$.

TABLE 5.113
Final Cable Size Selection to Meet All the Considered Conditions

Cable Type and Cabling Section	Min Cross-Section according to Thermal Short-Circuit	Min Cross Section according to Voltage Drop	Min Cross-Section Based on Percent Power Loss	Min Cross Section to Meet All Conditions	Selected Cross Section
DC string cables	0.302 mm²	0.89 mm²	2.67 mm²	≥ max [0.302, 0.89, 2.67]	4 mm²
LV AC cable (inverters to E)	29.4 mm²	2.69 mm²	13.95 mm²	≥ max [29.4, 2.69, 13.95]	35 mm²
LV AC cable (from E to D)	160 mm²	40.28 mm²	209.2 mm²	≥ max [160, 40.28, 209.2]	240 mm²
LV AC cable (from D to Tr.)	184.2 mm²	6.72 mm²	26.6 mm²	≥ max [184.2, 6.72, 26.6]	240 mm²
MV AC cable (from C to A)	9.908 mm²	21.5 mm²	27.90 mm²	≥ max [9.908, 21.5, 27.90]	35 mm²

TABLE 5.114
Sample DC and AC Cable Specifications for the Given PV System

Cable Type	Length	Cross Section	Max Power	Max Current	Ampacity (in Ground)	Max System Voltage	Cable Voltage	% Voltage Drop	% Power Loss
DC string cables	45 m	2×4 mm²	6.25 kW	10.18 A	32 A (>16.05 A)	935 V[a]	>1 kV	0.67	0.67
LV AC cable (inverters to E)	50 m	5×35 mm²	25 kW	36.09 A	130 A (>62.22 A)	400 V	0.6 kV	0.23	0.23
LV AC cable (from E to D)	150 m	5×240 mm²	125 kW	180.45 A	415 A (>311.1 A)	400 V	0.6 kV	0.28	0.28
LV AC cable (from D to Tr.)	5 m	3× (5×240 mm²)	625 kW	902.25 A	3 × 415 A (>952.8 A)	400 V	0.6 kV	0.028	0.028
MV AC cable (from C to A)	500 m	4×35 mm²	1260 kVA	28.87 A	175 A (>30.48 A)	31.5 kV	36 kV	0.032	0.063

[a] Temperature correction coefficient for the PV voltage is taken as 1.25 (37.4 × 20 × 1.25 = 935 V).

Note: Percent voltage drop $\%u = (\Delta U/U) \times 100$, where $\Delta U = \sqrt{3} IR \cos\varphi$, and the cable reactance X is neglected. Similarly, percent power loss is obtained from $\%p = (\Delta P/P) \times 100$, where $\Delta P = 3RI^2$, and $P = S \times \cos\varphi$.

against short-circuit currents. The minimum busbar cross-sectional area S_{min} (in mm²) can also be found by equation (5.138), which is $S_{min} = I_{th} \cdot \sqrt{t_{trip}}/J_{thr}$ where J_{thr} is the short-circuit current density of the busbar in [A/mm²]. The typical short-time rated current density value for a copper busbar is 159 A/mm².

Example 5.32: Busbar Sizing at the Point of Combiner Boxes in a Photovoltaic Power System

Consider the 500 kW$_p$ PV power plant worked in Example 5.31. Calculate the required minimum cross section of the copper busbars at the E (AC combiner) and D (AC recombiner) points. Note that short-circuit current values at points E and D are given as 4.205 kA and 22.94 kA, respectively, from example 5.31.

Solution:

The parameter J_{thr} for Cu busbar is provided as 159 A/mm². These bus conductors are expected to withstand short circuits for a duration of 1 second ($t_{trip} = 1$ s, $J_{thr} = 159$ A/mm²). Hence from (5.138), the minimum conductor cross-sectional area of the busbar can be calculated as below.

$$S_{min} \text{ (for point E)} = \frac{I_{th} \cdot \sqrt{t_{trip}}}{J_{thr}} = \frac{4205 \times \sqrt{1}}{159} = 26.45 \text{ mm}^2 \leq S_{bus}$$

$$S_{min} \text{ (for point D)} = \frac{I_{th} \cdot \sqrt{t_{trip}}}{J_{thr}} = \frac{22940 \times \sqrt{1}}{159} = 144.28 \text{ mm}^2 \leq S_{bus}$$

5.5.7.13 AC Combiner and Recombiner Sizing and Specifying

AC combiner boxes are the first-level combiners located on the AC side. Hence, a typical AC combiner contains the first level of protective devices for inverters on the AC side. Then an AC recombiner box, generally placed inside the LV/MV transformer kiosk (or outside on a concrete base), re-combines all the AC combiner boxes on a single busbar system and transfers the incoming powers to the LV/MV transformer station over a single circuit system. Since the AC combiners and the recombiner are located on the grid side, short-circuit currents because of the grid faults are the effective currents in their sizing and selection. The sizing and selection of a combiner box also depend on the highest short-circuit current level at the point of use. Hence, the AC combiner/recombiner boxes must have the capability to handle the short-circuit currents for short periods until the protection devices clear the fault. The short-circuit current on the AC combiner busbar is calculated using equation (5.97), which can be rewritten as (5.149).

$$I_{SC} = \frac{(c \cdot U_n)/\sqrt{3}}{\sqrt{(R_{kV})^2 + (X_{kV})^2}} \tag{5.149}$$

where U_n is the phase-to-phase rated voltage at the short-circuit location, R_{kV} is the equivalent system resistance seen from the short-circuit location, X_{kV} is the equivalent system reactance seen from the short-circuit location, and the expression "$(c \cdot U_n)/\sqrt{3}$" is called equivalent voltage source at the short-circuit location. Typically, the R_{kV} value is equal to the total resistances of the cable (R_{Cable}: AC cable resistance), transformer ($R_{LV, Tr}$: Transformer resistance on LV side), and grid ($R_{LV, Grid}$: LV grid resistance), while the X_{kV} value is the sum of the reactances of the same elements ($X_{LV, Grid}$; $X_{LV, Tr}$; X_{Cable}). Hence, the R_{kV} and X_{kV} values can be expressed as (5.150).

$$R_{kV} = R_{LV, Grid} + R_{LV, Tr} + R_{Cable}$$
$$X_{kV} = X_{LV, Grid} + X_{LV, Tr} + X_{Cable} \tag{5.150}$$

Importantly note that the cable impedance on the downstream side of the AC recombiner is not taken into account when calculating the short-circuit current on the AC recombiner. In this case, the short-circuit current on the AC recombiner can be modified as (5.151).

$$I_{SC} = \frac{(c \cdot U_n)/\sqrt{3}}{\sqrt{(R_{LV, Grid} + R_{LV, Tr})^2 + (X_{LV, Grid} + X_{LV, Tr})^2}} \tag{5.149}$$

Example 5.33: AC Combiner and Recombiner Box Sizing in a Photovoltaic Power System

Consider a photovoltaic power plant that is connected to a 31.5 kV network via LV/MV substation and underground LV cable. The network configuration with the system parameters is presented in Figure 5.142.

Photovoltaic Power Systems   435

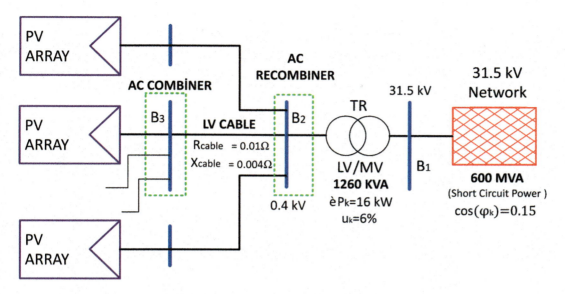

FIGURE 5.142 The considered network configuration and system parameters.

Considering the network configuration, size of the AC combiner, and recombiner boxes located at B_3 and B_2, respectively.

Solution:

According to the system model given in Figure 5.142, component impedances, reactances (X), and resistances (R) are calculated according to their respective voltages in the system.

Network Impedance (Relative to 31.5 kV):

$$Z_k = \frac{U^2}{S_k} = \frac{(31.5 \text{ kV})^2}{600 \text{ MVA}} = 1.65 \, \Omega$$

$$R_k = Z_k \cdot \cos\varphi_k = 1.65 \times 0.15 = 0.2475 \, \Omega; \, X_k = \sqrt{1.65^2 - 0.93^2} = 1.6313 \, \Omega$$

Network Impedance (Relative to 0.4 kV):

$$R_{k(0.4 \text{ kV})} = 0.2475 \times \left(\frac{0.4}{31.5}\right)^2 = 0.00004 \, \Omega; \, X_{k(0.4 \text{ kV})} = 1.6313 \times \left(\frac{0.4}{31.5}\right)^2 = 0.00026 \, \Omega$$

Transformer Impedance (Relative to LV Side 0.4 kV):

$$u_r = \frac{\Delta P_k}{S_T} \times 100 = \frac{16 \text{ kW}}{1260 \text{ kVA}} \times 100 = 1.27\%; \, u_x = \sqrt{6^2 - 1.27^2} = 5.86\%$$

$$R_T = \frac{U_T^2}{S_T} \cdot \frac{u_r}{100} = \frac{(0.4 \text{ kV})^2}{1.26 \text{ MVA}} \cdot \frac{1.27}{100} = 0.0016 \, \Omega; \, X_T = \frac{(0.4 \text{ kV})^2}{1.26 \text{ MVA}} \cdot \frac{5.86}{100} = 0.075 \, \Omega$$

LV Cable Impedance (Relative to 0.4 kV):

$$R_{\text{Cable}} = 0.01 \, \Omega; \, X_{\text{Cable}} = 0.004 \, \Omega$$

Equivalent Network Impedances at Point B_2 (Relative to 0.4 kV):

$$R_{B2} = 0.00004 + 0.0016 = 0.00164 \, \Omega;$$

$$X_{B2} = 0.00026 + 0.075 = 0.07526\ \Omega$$

$$Z_{B2} = R_{B2} + jX_{B2} = 0.00164 + 0.07526\ \Omega \Rightarrow |Z_{B2}| = \sqrt{0.00164^2 + 0.07526^2} = 0.07528\ \Omega$$

Equivalent Network Impedances at Point B_3 (Relative to 0.4 kV):

$$R_{B3} = R_{B2} + 0.01 = 0.01164\ \Omega;\ X_{B3} = X_{B2} + 0.004 = 0.07926\ \Omega$$

$$Z_{B3} = R_{B3} + jX_{B3} = 0.01164 + 0.07926\ \Omega \Rightarrow |Z_{B3}| = \sqrt{0.01164^2 + 0.07926^2} = 0.0801\ \Omega$$

Short-Circuit Calculations at Fault Location B_2 (*with voltage tolerance of +10%*):

$$I_{sc} = \frac{(1.1 \times 0.4\ \text{kV})/\sqrt{3}}{0.07528\ \Omega} = 3.375\ \text{kA};\ I_{th} \cong I_{sc} = 3.375\ \text{kA}$$

Short-Circuit Calculations at Fault Location B_3 (*with voltage tolerance of +10%*):

$$I_{sc} = \frac{(1.1 \times 0.4\ \text{kV})/\sqrt{3}}{0.0801\ \Omega} = 3.171\ \text{kA};\ I_{th} \cong I_{sc} = 3.171\ \text{kA}$$

Hence from the above calculations, minimum short-circuit current ratings for the AC combiner and recombiner boxes shall respectively be selected as (≥ 3.171 kA) and (≥ 3.375 kA).

5.5.7.14 Bonding and Grounding Methods in Photovoltaic Systems

A properly designed bonding and grounding system increases safety for both the user and equipment, improves facility performance, and even reduces operating costs. The grounding concept for PV systems is not only related to exposed conductive parts but also to the live parts in the system. Various aspects should be considered when determining the most appropriate grounding strategy for photovoltaic structures. The criteria for determining the PV system grounding strategy can be listed as (i) grid configuration, (ii) PV system configuration, (iii) inverter structure, (iv) insulation level, (v) PV system size, (vi) PV system mounting base, and (vii) lightning risk level. These grounding aspects are briefly explained in Table 5.115.

As shown in Table 5.115, various types of grounding configurations exist for photovoltaic systems in terms of different aspects such as grid types, system configuration, lightning hazards, etc. For example, buildings are typically powered by the TN-S network. For this reason, the grounding for building-integrated PV systems can be done by earthing the DC neutral conductor in accordance with the TN-S network structure. As another example of solar PV systems used for telecommunication powering, typically 48VDC PV systems, the exposed conductive parts are required to be earthed, whereas the live parts are insulated from the earth per the isolation concept in IT networks.

To better comprehend the earthing concepts, PV system grounding configurations are illustrated in the figures below. Figure 5.143 shows the grounding systems of negative, positive, and floating DC concepts for solar PV applications.

Considering Figure 5.143 and Table 5.115 together with grounding standards, the general guidelines for PV system grounding can be deduced as follows:

- Regardless of system configuration and network types, exposed conductive parts in a PV system such as PV frames, charge-controller chassis, monitoring system chassis, enclosures, inverter housing, distribution panels, PV mounting structures, conduits, and the chassis of end-use appliances are required to be grounded.
- The neutral conductor of a PV system can be earthed only if the PV system is galvanically isolated from the power grid.

TABLE 5.115
PV System Grounding Requirements for Various Considerations

Considered Aspects	Sub-Details	Exposed Conductive Parts[a]	Neutral Conductor (N)	Ground Conductor (PE)	Combined Ground and Neutral (PEN)	PV System Ground Terminals[b]	Positive Conductor Grounding	Earthing Via a High Impedance	No Earth Connection	Supply (Tr) Star Point or Neutral	Loads/Equipment Locally Grounded	Household Equipment/Loads	Grounding Meshes / (Rods + Strips)	Power Station Grounding Mesh	Technical Room/ Transformer Building	Fences of PV Farm/ Park	Lightning Protection System
Grid configuration[c]	TN-S	✓	✓			✓				✓							
	TN-C	✓	✓		✓	✓				✓							
	TN-C-S	✓		✓		✓				✓							
	IT	✓		✓				✓			✓						
	TT	✓								✓							
PV system configuration	Floating (ungrounded) PV system(4*)	✓						✓	✓	✓							
	Negative DC system grounding	✓				✓				✓							
	Positive DC system grounding(5*)	✓					✓			✓							
Inverter structure	Transformerless inverter	✓							✓								
	Transformer included inverter	✓		✓		✓											
Insulation level	Class-I insulated	✓				✓											
	Class-II (double) insulated								✓								
PV system size	Small-scale	✓				✓						✓					
	Large-scale	✓				✓							✓	✓	✓	✓	
PV system mounting base	Roof-mounted	✓				✓						✓					
	Ground-mounted	✓				✓							✓		✓	✓	
Lightning risk level(6*)	Low lightning risk	✓				✓											
	High lightning risk	✓				✓									✓		✓

(4*) A floating (or ungrounded) PV system has neither positive nor negative DC conductors connected to the earth. It is generally applied in mobile/moving systems such as vehicles.
(5*) If the PV system includes components having a positive ground requirement (such as positive ground solar panels, chargers, and battery storage), then a positive grounded PV system will be an option. It is available on some older model vehicle systems and is not preferred in today's modern applications.
(6*) Normally, Lightning protection and grounding system requirements should be considered together. However, since the lightning protection and its design are explained in a separate section, it is not studied in detail herein.

[a] PV frames, charge controller chassis, monitoring system chassis, enclosures, inverter housing, distribution panels, conduits, PV mounting structures, and the chassis of end-use appliances.
[b] SPD ground terminals, negative terminals of the battery system, inverter ground terminal, and AC system ground points.
[c] TN-S (earth and neutral conductors are separated), TN-C (earth and neutral conductors are combined), TN-C-S (includes both TN-S and TN-C), IT (isolated earth: no earth connection, or earth connection over a high impedance), TT (grid neutral conductor earthed at transformer starpoint, and load/equipment earthed locally).

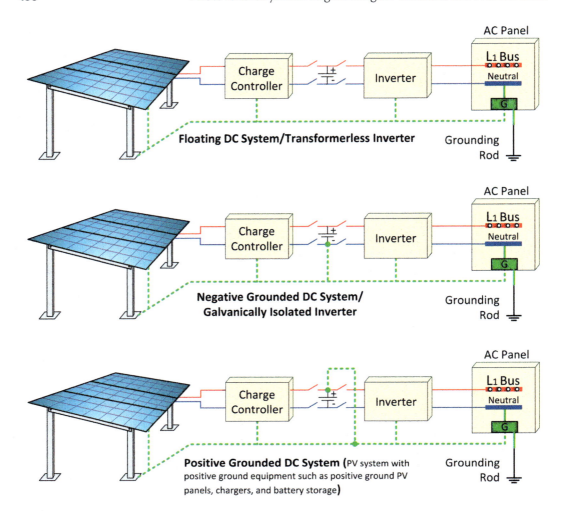

FIGURE 5.143 PV grounding concepts for negative, positive, and floating DC systems.

- In a floating or double-insulated PV system (Class II insulation), none of the current-carrying conductors are grounded.
- Due to their structural properties, some PV module technologies require positive conductor grounding to operate. In such systems, equipment such as charge controllers and battery storage should also have positive grounding features.
- To increase the earthing system performance, the grounding system must also be supported with equipotential bonding, if the PV site includes non-homogeneous soil type and/or If the PV system goes to larger sizes.
- A grounding system (or a grounding chain) contains components GEC (Grounding Electrode Conductor), grounding connections, GE (Ground Electrode), electrode-to-soil resistance, and the soil. Hence, the performance of the grounding system simply depends on the quality of these components.
- The earthing conductor must also be sized to carry the highest possible fault current in the system, like any other conductor.
- System grounding in large-scale PV structures should be carried out in three stages, individual assembly grounding, group assembly grounding, and interconnection of grounded subgroups. These subgroups also include the grounding of transformer building, star-point grounding, and the fences grounding of PV farm/park.

- The fences of PV farms are also required to be grounded properly, depending on the specific lightning risks and the identified hazards.
- If the spread of the PV system in the field is vast, the grounding system can be designed to form a loop around the whole PV array block. Alternatively, the PV array block may just have an earthing at one end, and all earthed blocks can then be tied to each other by the grounding leads to reach a common ground.
- To check the safety performance of the grounding system, touch and step voltage must be calculated in all PV facilities.

As summarized above, equipotential bonding in PV systems increases the safety of the grounding system and makes the protection against lightning and overvoltage more effective. Generally, the requirement for equipotential bonding increases due to the increase in system size or non-homogeneous soil structure. Therefore the metal components of the PV power plant should be interconnected together to reach common ground, even if they are located on different buildings and spread over a large area. Good or bad examples for grounding/bonding designs at different levels are illustrated in Figure 5.144.

One of the crucial problems in grounding is the corrosion of earthing components over time. To minimize corrosion in the PV metal frame and ground conductors (earth rods, bonding conductors, and clamps), the material differences between all the metal parts should be minimized as well. Therefore, the foundation type applied to the PV frame is an important criterion in the selection of materials in the PV grounding system. The most commonly used foundation types and the grounding system materials recommended for the respective foundations are shown in Figure 5.145 for the ground-mounted PV applications.

As shown in Figure 5.145, the photovoltaic ground mounting system can be categorized into five types as ramming method, screw method, cast-in method, concrete block method, and concrete foundation method. The PV frame piles are driven directly into the ground in the ramming method. Screw-tipped piles are rotated and fixed into the ground in the screw method. In the cast-in method, the holes are drilled in the ground and wet concrete is poured to fix the piles. In the concrete methods (concrete block and concrete foundation), the piles are fixed on a concrete base using an anchor bolt.

When designing earthing systems for a large-scale PV power plant, sections of the site should be carefully planned to get an optimized grounding system. As well-known, large-scale PV power plants consist of hundreds of PV panels with support frames, steel piles, and footings, to get a strong ground-mounted system. All these PV structures constitute large PV arrays, connected together through DC connections to the inverters to feed the local grid over step-up transformers. Then in some cases, certain numbers of PV array sections, inverter groups, and step-up transformers are also combined into larger blocks to power an HV substation. As can be understood, several sub-grounding systems, far from each other, shall be interconnected in a large-scale PV farm. This work requires a design optimization for an economical and effective grounding solution. For this reason, performing soil resistivity tests throughout the whole solar site should be an essential requirement before starting the PV projects. In other words, soil resistivity tests throughout the entire PV field should be performed beforehand to be able to reach an optimized PV grounding solution.

Soil resistivity (Ω-meters) surveying in large solar fields is a difficult and complex task. A standardized guide on the soil resistivity survey of large-scale earthing systems is not readily available due to the consideration need for a large number of variables such as variations in land structures, non-homogeneous soil structures, and seasonal effects on the soil resistivity. The Wenner 4-Probe method in general is used for the field soil resistivity survey. However, the parameters such as the number/directions of test traverses, minimum/maximum probe spacing, and the required number of measurements should carefully be determined and the test results should be interpreted by experienced engineers. Several test traverses in different directions are necessary for the accurate earthing analysis of large solar farms. For this purpose, suitably powered test equipment is necessary for

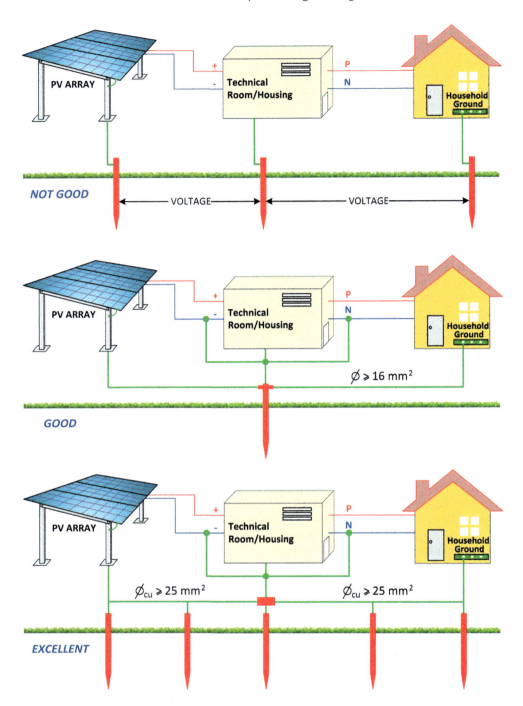

FIGURE 5.144 Good or bad examples of ground bonding for a simple PV system.

longer-spaced measurements. Note that low-power off-the-shelf testers are not often sufficient for the soil resistivity testing of large solar farms. Figure 5.146 shows a sample solar field, and determination of test traverses for the soil resistivity survey.

Different soil resistivity values are usually obtained from the measurement results in large-scale solar fields. When interpreting these soil resistivity test data, field engineers should also consider

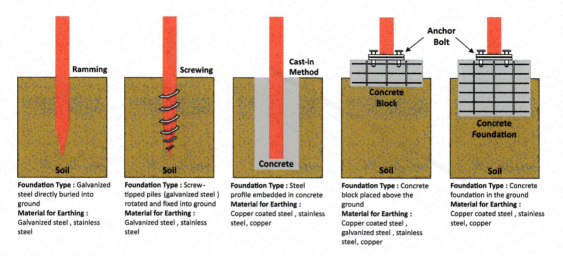

FIGURE 5.145 Foundation types and recommended earthing system materials for ground-mounted PV systems (Note that aluminum is not allowed to be buried in the ground).

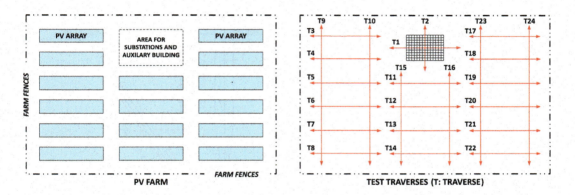

FIGURE 5.146 A sample solar field and determination of test traverses.

various scenarios such as atmospheric conditions, ambient temperature, soil moisture content, and seasonal effects.

From the soil resistivity survey results, it is expected to learn the variation of soil resistivity values with respect to depth and spacing. A worst-case scenario should then be considered for designing a safe and effective earthing system. In other words, earthing systems and earthing protection schemes should be designed according to the worst-case scenario. As well known, the soil resistivity value varies depending not only on the soil type but also on the soil moisture content. In general, wet soil has a lower resistivity value than dry soil. From another perspective, the required number of grounding electrodes will be less for wet soil, while the required number of grounding rods will be higher for dry soil. Studies show that the number of grounding electrodes required for dry grounds can be three to four times higher than for wet grounds. For example, if 10 electrodes are required for wet soil, this number can go up to 40 electrodes for dry soil.

The grounding rules used for a typical electrical installation are also valid for the grounding system of a solar PV facility. Therefore the grounding requirements outlined in the international standards shall also apply to any of the PV power plants. The first step of this effort should start with the correct grounding of a PV array unit mounted on a metallic frame system. Figure 5.147 shows a sample grounding system for a foundation type where the PV frame piles are directly buried in the ground.

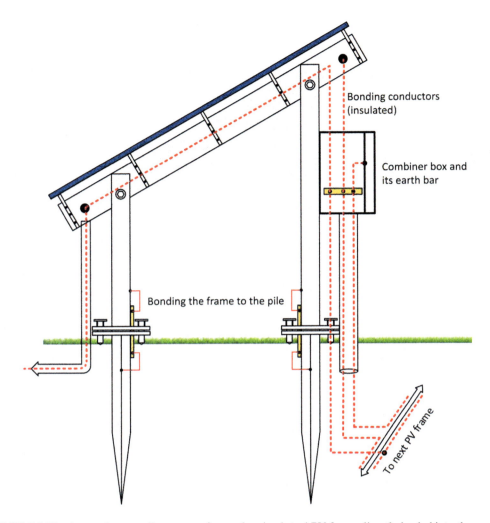

FIGURE 5.147 A sample grounding system for a galvanized steel PV frame directly buried into the ground.

Since the PV frame is a natural path for earth fault current and it is required to be bonded to the grounding system, as shown in Figure 5.147. Another earthing example is given in Figure 5.148 for a concrete block-based foundation type.

Earthing bonds of PV frames are exposed to corrosion and failures. Dissimilar metal contacts will accelerate the corrosion effect under outdoor atmospheric conditions. Therefore, it is important to choose corrosion-resistant materials to eliminate this effect on the contact surfaces.

The next step for grounding a large-scale PV power plant is to ground the PV array groups in blocks and then interconnect them, of which structures mainly depend on the PV frame mounting types. Figure 5.149 shows the grounding system schematic of a sample PV farm assembled according to the ramming or screw method.

The hot-dip galvanized steel piles of the PV frames in Figure 5.149 were considered as grounding electrodes and no additional auxiliary ground electrodes were assumed to be used. The circles shown on the PV arrays represent the piles of the PV frames and each array has 20 piles in the given sample system. The 20 piles of a PV array are all linked to each other above the ground by the metal structures of the photovoltaic panels or by a suitably selected ground conductor. Also, as seen in Figure 5.149, the interconnections between the PV arrays and between PV array groups were made by employing ground strip conductors, horizontally buried underground at a depth of at least 0.5 m.

Photovoltaic Power Systems

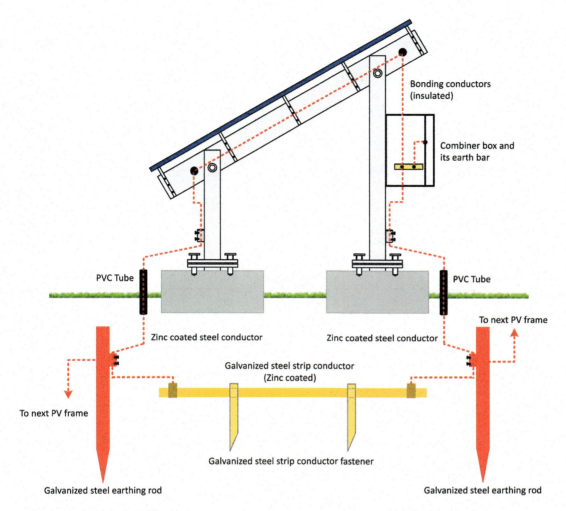

FIGURE 5.148 PV frame earthing example for a concrete block-based foundation type (earthing rod length should be 1.5 m at least, and the distance between two rods should be twice the length of a rod).

Figure 5.150 shows the grounding schematic of the PV farm constructed according to the concrete foundation method. As in Figure 5.149, the metal parts of the PV array in Figure 5.150 are also interconnected above the ground by the metal structures of the PV panels. Similarly, interconnections between the PV arrays and between PV array groups in Figure 5.150 can be fulfilled by using ground strip conductors buried at a depth of 0.5m. The entire PV farm grounding system is completed by interconnecting the grounds of the remaining system components to the grounding network of the PV array groups. The remaining system components herein are usually the power station building, the lightning protection system, and the metallic fences of the PV farm. Figure 5.151 shows the integration of earthing networks with each other on a sample PV farm. Grounding details of PV farm fences are also shown in Figure 5.152.

The requirements of general grounding design for a solar site have now been outlined above. Afterward, the suitability of the grounding system designs, as well as the limits of the ground resistance values, should then be checked for the entire PV field. For this purpose, the following steps can be followed in general.

Step 1 – Determination of Grounding System Configuration: Generally, four types of design configurations exist for grounding a PV field. The designs often include the grounding configurations

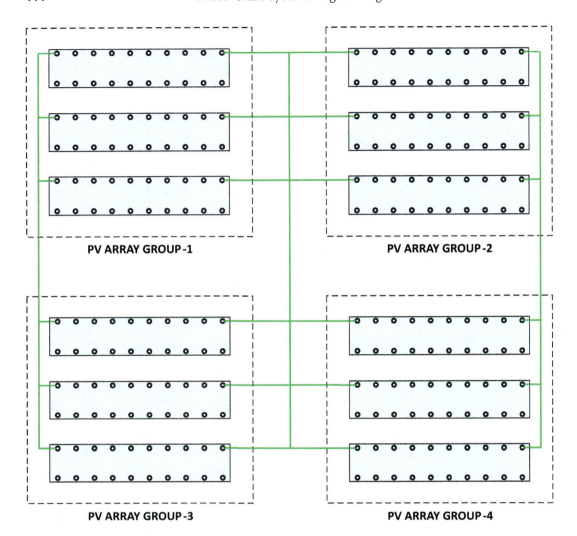

FIGURE 5.149 Schematic drawing of the grounding system of PV array groups for a sample large-scale PV farm, constructed by ramming or screw method.

such as a single ground electrode, multiple ground electrodes, mesh network, and ground plate as shown in Figure 5.153. As the most common earthing method, basic grounding systems consist of a single ground rod. Generally, power generation facilities and power stations contain more complex grounding configurations such as multiple ground electrodes, ground loops, mesh grids, and ground plates.

Step 2 – Calculation of Ground Resistance for the Configurations: First of all, ground resistance values are required to be calculated for each earthing configuration. Different methods such as Laurent-Niemann, Sverak, Schwarz, Dwight, and Thapar-Gerez have been defined in the literature and standards. In the following equations, commonly used formulas used in ground resistance calculations are given. According to the Schwarz Method, the ground resistance of a single vertical rod (R_r) is calculated by (5.150). As for straight-ground conductors laid horizontally on the surface, the grounding resistance (R_c) is given by equation (5.151). Equation (5.152) can be used for the ground resistance of round or strip conductor buried at a depth of h, that is, R_{ch}.

$$R_r = \frac{\rho}{2\pi L_r}\left[\ln\left(\frac{8L_r}{D}\right) - 1\right] \quad (5.150)$$

Photovoltaic Power Systems

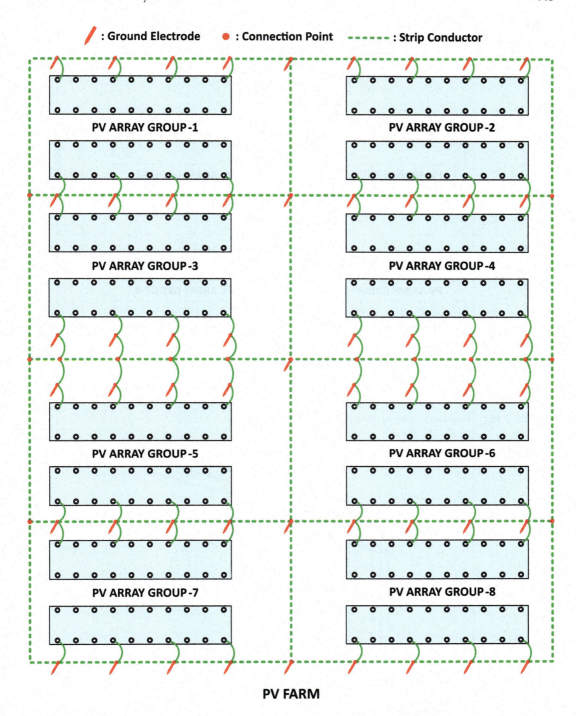

FIGURE 5.150 Schematic drawing of the grounding system for a large-scale PV farm, constructed according to the concrete foundation method.

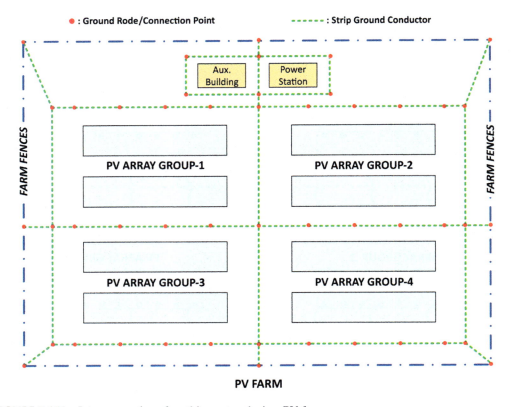

FIGURE 5.151 Interconnection of earthing networks in a PV farm.

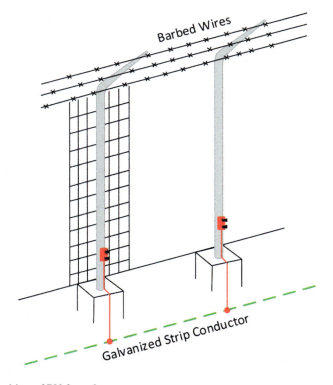

FIGURE 5.152 Earthing of PV farm fences.

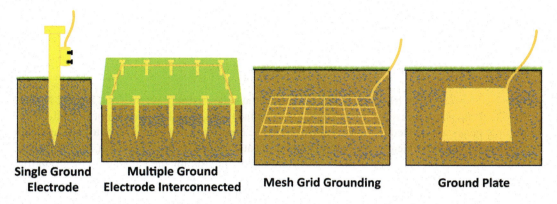

Single Ground Electrode **Multiple Ground Electrode Interconnected** **Mesh Grid Grounding** **Ground Plate**

FIGURE 5.153 Earthing system configurations.

$$R_c = \frac{\rho}{\pi L_c}\left[\ln\left(\frac{4L_c}{D}\right)-1\right] \tag{5.151}$$

$$R_{ch} = \frac{\rho}{2\pi L_{ch}}\ln\left(\frac{L_{ch}^2}{1.85hD}\right) \tag{5.152}$$

where R_r, R_c, and R_{ch} are the ground resistances [Ω] for the respective electrodes, ρ is the soil resistivity [Ω·m], L_r is the buried length of the ground rod [m], L_c is the horizontal length of the conductor [m], L_{ch} is the horizontal conductor length [m] buried at a depth of h [m], and D is the diameter of the ground rod [m]. Note that the diameter of non-cylindrical electrodes, such as strip ground conductors, can be obtained from the equivalent cross section. Alternatively, the diameter of a thin strip conductor can approximately be assumed as the half-width of the strip.

The ground resistance formed by multiple earthing rods, placed far from each other (the spacing between rods ≥2 times the length of the rod), can simply be obtained from the parallel equivalent of resistances, that is, $\frac{1}{R} = \frac{1}{R1} + \frac{1}{R2} + \cdots + \frac{1}{Rn}$. If the spacing between grounding electrodes is less than twice the length of the rod, the combined resistance of aligned multiple electrodes is approximated by Heppe R.J (1998) as equation (5.153).

$$R_e = \frac{\rho}{2\pi L_r \cdot n}\left[\ln\left(\frac{8L_r}{D}\right)-1+\frac{L_r}{d}\ln\left(\frac{1.78n}{e}\right)\right] \tag{5.153}$$

where R_e is the equivalent ground resistance [Ω] of n vertically driven rods, ρ is the soil resistivity [Ω·m], n is the number of ground rods, L_r is the buried length of the ground rod [m], D is the diameter of the ground rod [m], d is the spacing between rods [m], and e is the Euler's number 2.71828.

The ground resistance of a mesh grid (R_m) can simply be obtained from equation (5.154).

$$R_m = 0.443\frac{\rho}{\sqrt{S}} + \frac{\rho}{L} \tag{5.154}$$

where R_m is the ground resistance [Ω] of a mesh, ρ is the soil resistivity [Ω·m], S is the area [m²] covered by the mesh grid, and L is the total length [m] of the conductor used in the mesh.

Finally, the approximate resistance of a ground plate (R_p) can be calculated by Equation 5.155.

$$R_p = \frac{\rho}{4}\sqrt{\frac{\pi}{S}} \qquad (5.155)$$

where R_p is the resistance [Ω] to earth of a plate, ρ is the soil resistivity [$\Omega \cdot m$], and S is the surface area [m^2] of the plate. It is recommended that the one face area of a plate should not exceed $1.5\,m^2$. If multiple plates are needed to be used, they should be connected in parallel vertically, at least 2 m apart from one another. Ideally, they should be buried at a minimum depth of 0.6 m into damp soil.

There are different approaches in the literature and standards for the calculation of grounding resistance. Besides, there are also numerous formulas for different grounding arrangements. Since other formulas and approaches are available in the literature, no further details are given here.

Step 3 – Calculating and Checking Equivalent Ground Resistances: After finding the ground resistance of each system component in the field, the equivalent ground resistance of the entire PV farm should then be calculated to see if it meets the design goal. The equivalent resistance (R_e) of the PV field can be calculated by $\frac{1}{R_e} = \sum_{i=1}^{n}\left(\frac{1}{R_i}\right)$, where R_i represents the component's earth resistances and n counts the number of grounded systems in the field. The grounding resistance values should then be tested to see if the design goals of the entire PV farm meet the maximum grounding resistance limits as per Table 5.116 (ANSI/IEEE 142 Grounding of Industrial and Commercial Power Systems).

The values required for grounding resistance should be less than the maximum reference values. Otherwise, the earthing system is required to be improved until we reach the design goals. For this purpose, the main improvements in the grounding design could be listed as follow:

- Increase the electrode cross sections and rod lengths. In addition, bury them deeper into the ground. If it is difficult to drive the rods deeper in the ground, a horizontal crow's foot ground termination kit (‡) can also be used.
- Increase the number of ground rods (multiple-rod approach).
- The final approach would be to treat the soil by using ground-enhancing materials (GEMs) such as salt and chemical rods.

Step 4 – Check the Touch and Step Voltages: After meeting the resistance goals, step and touch voltage values should also be calculated to check the safety level of the PV farm's grounding system. A touch voltage normally occurs due to a hand-to-foot or a hand-to-hand contact, while a step voltage is created through the legs from one foot to the other. These voltage limits are derived from tolerable body currents (I_B), which depend on the ground resistance and the duration of the shock

TABLE 5.116
Maximum Acceptable Grounding Resistance Values

Substation	Maximum Grounding Resistance	
	≤ 4.16 kV	> 4.16 kV
≤ 1000 kVA	5 Ω	10 Ω
From 1000 to 5000 kVA	2 Ω	5 Ω
> 5000 kVA	1 Ω	2 Ω
Lightning protection	10 Ω	
For hazardous and non-hazardous areas		

current that a person may expose to. According to IEEE 80-2015, the allowable limits for touch and step voltages are calculated by equations (5.156) and (5.157).

$$E_{touch} = (R_B + 1.5\rho_s C_s) I_B = (R_B + 1.5\rho_s C_s)\frac{k}{\sqrt{t_s}} \tag{5.156}$$

$$E_{step} = (R_B + 6\rho_s C_s) I_B = (R_B + 6\rho_s C_s)\frac{k}{\sqrt{t_s}} \tag{5.157}$$

where I_B is the tolerable body current [A], ρ_s is the resistivity of the surface material [$\Omega \cdot$ m], t_s is the duration of (shock/fault) current [s], and k is a constant equal to 0.116 or 0.157 for people with a body weight of 50 kg and 70 kg, respectively. If a surface material is not covered on the ground surface, the surface resistivity ρ_s is taken as equal soil resistivity ρ. The C_s, given by equation (5.158), is a correction factor that is considered if the ground surface is covered by a thin layer of high resistivity surface material (such as pebbles and small pieces of rocks). This is usually done in substations to increase the contact resistance between the ground and persons' feet. If no surface material (protective surface layer) is used, ρ_s will then equal to ρ, and C_s will be 1.

$$C_s = 1 - \frac{0.09\left(1 - \frac{\rho}{\rho_s}\right)}{2h_s + 0.09} \tag{5.158}$$

where ρ is the soil resistivity, and ρ_s is the surface material resistivity in [$\Omega \cdot$ m]. The thickness of the surface material is represented by h_s [m]. The duration of shock current, often 0.5 seconds for photovoltaic systems, is limited by the designed clearing times of the protection system. Hence, the tolerable body current value can approximately be taken as 150 mA for solar PV farms. It is important to note that using the body resistance value of a 50 kg person ($R_B = 1000\,\Omega$) in the calculations of step and touch voltages keep the design on the safe side. However, grounding designs made for 70 kg people ($R_B = 1500\,\Omega$) are also sufficient for most cases.

Subsequently, the tolerable touch and step voltage values obtained from equations (5.156) and (5.157) should then be compared with the maximum touch and step voltages (V_m and V_s), that occur within a mesh of a ground grid. For personal safety, the conditions $E_{touch} \geq V_m$ and $E_{step} \geq V_s$ are required to be satisfied. The mesh touch and step voltage values are obtained from (5.159) and (5.160).

$$V_m = \frac{\rho K_m K_i I_g}{L_m} \tag{5.159}$$

$$V_s = \frac{\rho K_s K_i I_g}{L_s} \tag{5.160}$$

where V_m and V_s are the maximum touch and step voltages [Volts] within a mesh of a ground grid, respectively. K_i, defined as $K_i = 0.644 + 0.148 \times n$, is the correction factor for current irregularity, where "n," expressed in detail below, is the shape factor of the grounding grid. K_m and K_s are the geometrical mesh and step factors, which are defined in (5.162) and (5.163). Similarly, L_m and L_s are the effective lengths for mesh and step voltages, respectively. Finally, I_g is the RMS ground return current flowing between the ground grid and earth [Amps]. An earth fault current is split into two main paths. A portion of this fault current returns to the source through the system-neutral conductor. The remaining second part of the fault current returns to the source through the ground (I_g). This amount of current that flows into the ground, also known as the ground return current (I_g), can simply be calculated by a split factor (S_f) as per IEEE 80 (equation 5.161).

$$I_g = S_f \cdot I_f \tag{5.161}$$

where I_g is the total fault current, and S_f is the fault current split factor. In most cases, considering the split factor as 60% provides an adequate level of safety. As for the details of the S_f value, the standard of IEEE-80 can be referenced when designing the earthing system of the complex substations.

After calculating the I_g value, one further condition (GPR $\leq E_{\text{touch}}$) regarding the ground potential rise (GPR = $I_g \times R_e$, where R_e is equivalent earthing resistance) should also be checked to ensure human safety in the vicinity of the installation. Importantly note that if the GPR $\leq E_{\text{touch}}$ condition is fulfilled, the conditions $E_{\text{touch}} \geq V_m$ and $E_{\text{step}} \geq V_s$ will also be met. Therefore, the condition GPR $\leq E_{\text{touch}}$ should be checked initially. If the condition GPR $\leq E_{\text{touch}}$ was already satisfied, there would be no need to check the conditions $E_{\text{touch}} \geq V_m$ and $E_{\text{step}} \geq V_s$.

The above-mentioned geometrical factors K_m and K_s can be obtained respectively from equations (5.162) and (5.163).

$$K_m = \frac{1}{2\pi}\left[\ln\left(\frac{d^2}{16hD} + \frac{(d+2h)^2}{8dD} - \frac{h}{4D}\right) + \frac{K_{ii}}{\sqrt{1+h}} \cdot \ln\left(\frac{8}{\pi(2n-1)}\right)\right] \tag{5.162}$$

$$K_s = \frac{1}{\pi}\left[\frac{1}{2h} + \frac{1}{d+h} + \frac{(1-2^{2-n})}{d}\right] \tag{5.163}$$

where K_{ii} is a weighting factor for earth rods. If the mesh grid does not include any earth electrodes or includes only a few ground rods (none located on the corners or the perimeter), K_{ii} is expressed as $K_{ii} = 1/(2n)^{\frac{2}{n}}$. Otherwise, K_{ii} is equal to 1 for grids with ground rods along the grid perimeter and/or corners. The "d" is the average spacing between parallel conductors, D is the diameter of the grid conductor, and h is the burial depth, which is defined in the ranges 0.25 m $\leq h \leq 2.5$ m, $D < 0.25h$, and $d > 2.5$ m. The average mesh spacing (d) for rectangular grids can be calculated as (5.164).

$$d = \frac{1}{2}\left(\frac{W_g}{n_r - 1} + \frac{L_g}{n_c - 1}\right) \tag{5.164}$$

where W_g and L_g [m] are the width and length of the mesh grid, respectively. Similarly, n_r and n_c are the number of parallel conductors in rows and columns, respectively.

The shape factor "n" is defined as a function of the dimensions of the mesh grid and the total length of the conductor used in that grid. For square mesh grids, $n = 2L_T/L_P$, where L_T is the total length of the conductor in the horizontal grid while L_P is the peripheral length of the grid in [meters]. For rectangular mesh grids, $n = (2L_T/L_P) \times \left[\sqrt{(L_P/4\sqrt{S})}\right]$, where S is the area of the grid in [m^2].

For the other shapes of grounding grids, the general form of the "n" is expressed as below equations (from 5.165 to 5.169).

$$n = n_A \cdot n_B \cdot n_C \cdot n_D \tag{5.165}$$

$$n_A = \frac{2L_T}{L_P} \tag{5.166}$$

$$n_B = \sqrt{(L_P/4\sqrt{S})} \tag{5.167}$$

$$n_C = (L_x L_y/S)^{(0.7S/L_x L_y)} \tag{5.168}$$

$$n_D = \frac{d_{\max}}{\sqrt{(L_x^2 + L_y^2)}} \tag{5.169}$$

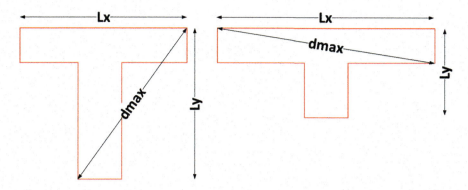

FIGURE 5.154 Illustration of L_x, L_y, and d_{max} dimensions on T-shape grids.

where L_x and L_y are the maximum length of the grid in meters in the directions of x and y, respectively. The maximum distance between any two points on the mesh grid is represented by d_{max} in meters. For example, the parameters L_x, L_y, and d_{max} can be illustrated on the two different T-shape grids as in Figure 5.154.

The parameter L_m is the effective length for mesh voltage. For grounding grids without earthing electrodes (or containing very few ground rods, but none of them located at the corner and along the periphery line), the effective mesh length (L_m) is calculated by (5.170).

$$L_m = L_T + L_R \tag{5.170}$$

where L_T is the total length of grid conductor [m], and L_R is the total length of all ground rods [m]. For other grounding grids having earthing electrodes throughout the periphery line, the effective buried length of the mesh is calculated by (5.171).

$$L_m = L_T + \left[1.55 + 1.22 \times \frac{L_r}{\sqrt{\left(L_x^2 + L_y^2\right)}}\right] L_R \tag{5.171}$$

where L_r [m] is the length of a single ground rod used in the grid. The other parameters in (5.170) are already explained in the above sections.

The parameter L_s [m] is the effective buried length for step voltage, and it can simply be calculated for all cases by equation (5.172). The parameters L_T and L_R in (5.172) are the same as above.

$$L_s = 0.75 L_T + 0.85 L_R \tag{5.172}$$

Step 5 – Size and Check the Cross Section to Finalize the Grounding Design: It is vital to obtain the minimum cross section of the grounding conductors to ensure that the grounding grid shall withstand the maximum earth fault current. For fault currents that can be interrupted within 5 seconds, the minimum cross-sectional area of the grounding conductor, derived from the general adiabatic temperature rise equation ($S^2 k^2 = I^2 t$, where $k = K\sqrt{\ln\frac{T_f + \beta}{T_i + \beta}}$), can be calculated by equation (5.173) as per IEC 60364.

$$S_{min} = \frac{I}{K} \sqrt{\frac{t}{\ln\left(\frac{T_f + \beta}{T_i + \beta}\right)}} \tag{5.173}$$

TABLE 5.117
Required Coefficients for the Cross-Section Calculations of Different Earthing Conductors

Material	β[°C]	K[Amp(sec)$^{1/2}$/mm^2]	T_i[°C]	T_f[°C]
Copper	234.5	226	20	300 (bare or galvanized)
				150 (tin or lead plated)
Aluminum	228	148	20	300
Steel	202	78	20	300 (galvanized)

where S_{min} is the minimum cross-sectional area of the grounding conductor [mm^2], I is the RMS earth fault current [A], t is the fault current duration [s], and K is a factor depending on the conductor material, T_f is the conductor's final (maximum) short-circuit temperature [°C], T_i is the initial operation temperature (maximum) short-circuit temperature [°C] of the conductor, and β is the temperature coefficient inverse of the resistance [°C] for the conductor material. The coefficients in equation (5.173) are given in Table 5.117 for different conductor materials.

Example 5.34: Earthing Calculation of a PV Farm

A solar PV power plant is planned to be built on 2600 m^2 of land. The photovoltaic solar farm consists of different grounded subsystems such as metallic wire fences, substations, distribution panels, and PV arrays. The initial earthing design of a PV farm and required system parameters for earthing calculation are respectively given in Figure 5.155 and Table 5.118.

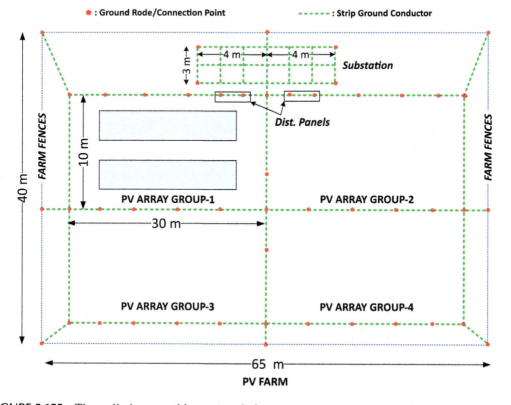

FIGURE 5.155 The preliminary earthing system design.

TABLE 5.118
The Preliminary Earthing System Parameters

Subcomponents of Earthing System	Earthing Configuration	Quantity (and/or) Units
Wire fence	Ground electrode	8 Rods
Substation	Mesh grid	The mesh grid area, which includes 4 rods in the mesh, is 24 m² (circumference: 22 m and total conductor length: 48 m)
Star point	Solidly grounded	Tr: 31,5/0,4 kV and 800 kVA, I_{sc} at LV side = 1900 A
Distribution panel	Ground electrode	2 Rods
PV array group	Multiple ground rods connected	6 Rods (total conductor length: 80 m)
PV site	All earthing systems (4 PV array group+2 distribution panel+1 substation + wire fence + additional 9 rods are interconnected via 320 m strip conductor) are interconnected.	
Other System Parameters		
Phase to neutral short-circuit current (at the low voltage side of the transformer)		$I_{sc} = I_{fault} = 1900$ A
Leakage current protection		A leakage current protection device of 300 mA is used at each inverter output.
Short circuit duration		0.5 second
Soil resistivity		400 Ω·m
Grounding rod length		1 m
Conductor burial depth		0.8 m
Earthing rod diameter		0.02 m
Strip conductor type		Round grounding conductor
Grounding strip diameter		0.034 m

a. Check the suitability of the initial grounding design of the given PV farm in terms of grounding resistance, touch voltage, and conductor cross sections. If necessary, determine the improvements that can be made in the grounding design, and decide the final configuration.

b. If the soil resistivity was 100 Ω·m, what modifications would have to be made in the grounding system?

Solution:

a. Let's follow the grounding steps of a solar field described above to check the suitability of initial grounding designs.

Step 1 – Grounding System Configuration: Design configurations are already described in Table 5.118.

Step 2 – Ground Resistance Calculations: All required data to calculate the grounding resistances are given in Table 5.118. Firstly, let us calculate the resistance of a single ground rod from equation (5.150).

$$R_r = \frac{\rho}{2\pi L_r}\left[\ln\left(\frac{8L_r}{D}\right) - 1\right] = \frac{400}{2\pi \times 1}\left[\ln\left(\frac{8 \times 1}{0.02}\right) - 1\right] = 317.7\ \Omega$$

From equation (5.152), ground resistance of round (or strip) conductor, buried at a depth of h, can be found as below.

$$R_{ch1}(80 \text{ m Loop}) = \frac{\rho}{2\pi L_{ch}} \ln\left(\frac{L_{ch}^2}{1.85hD}\right) = \frac{400}{2\pi \times 80} \ln\left(\frac{80^2}{1.85 \times 0.8 \times 0.034}\right) = 9.35 \, \Omega$$

$$R_{ch2}(320 \text{ m Interconnection}) = \frac{\rho}{2\pi L_{ch}} \ln\left(\frac{L_{ch}^2}{1.85hD}\right) = \frac{400}{2\pi \times 320} \ln\left(\frac{320^2}{1.85 \times 0.8 \times 0.034}\right) = 2.89 \, \Omega$$

From equation (5.154), the ground resistance of a mesh grid can be calculated as below.

$$R_m(\text{substation}) = 0.443 \frac{\rho}{\sqrt{S}} + \frac{\rho}{L} = 0.443 \frac{400}{\sqrt{24}} + \frac{400}{48} = 44.5 \, \Omega$$

Step 3 – Calculation of Equivalent Ground Resistances: After finding the ground resistance of each system component in the field, the equivalent ground resistance of each grounding configuration is calculated. Finally, the equivalent ground resistance of the entire PV farm should be then calculated.

- *Ground Resistance Calculation of Substation Building (Mesh Grid+4 Rods):*

  Equivalent grounding resistance of 4 rods, $R_1 = \frac{R_r}{4} = \frac{317.7}{4} = 79.45 \, \Omega$

  From Step-2, mesh grid resistance, $R_m = 44.5 \, \Omega$

  The total grounding resistance of the substation building (R_{e1}) is equal to the parallel equivalent of the ground rod resistance (R_1) and mesh grid grounding resistance (R_m).

  $$R_{e1} = \frac{R_1 \times R_m}{R_1 + R_m} = \frac{79.45 \times 44.5}{79.45 + 44.5} = 28.52 \, \Omega$$

- *Ground Resistance Calculation of Distribution Panel* (**2 Rods**):

  Equivalent grounding resistance of 2 rods, $R_{e2} = \frac{R_r}{2} = \frac{317.7}{2} = 158.89 \, \Omega$

- *Ground Resistance Calculation of a PV Array Group* (**6 Rods + strip conductor**):

  Equivalent grounding resistance of 6 rods, $R_2 = \frac{R_r}{6} = \frac{317.7}{6} = 52.96 \, \Omega$

  From Step-2, strip conductor, $R_{ch1} = 9.35 \, \Omega$

  The total grounding resistance of the substation building (R_{e3}) is equal to the parallel equivalent of the ground rod resistance (R_2) and strip conductor grounding resistance (R_{ch1}).

  $$R_{e3} = \frac{R_2 \times R_{ch1}}{R_2 + R_{ch1}} = \frac{52.96 \times 9.35}{52.96 + 9.35} = 7.94 \, \Omega$$

- *Ground Resistance Calculations of Wire Fence (8 Rods) and Additional 9 Rods on the Grid* (**8+9 =17 Rods**):

  Equivalent grounding resistance of 8 rods, $R_{e4} = \frac{R_r}{17} = \frac{317.7}{17} = 18.68 \, \Omega$

- *Equivalent Ground Resistance of the Entire PV Farm* (**4 PV array group + 2 distribution panel + 1 substation + wire fence + additional 9 rods are interconnected via 320 m strip conductor**):

$$\frac{1}{R_e} = \frac{1}{R_{e1}} + \frac{1}{\left(\frac{R_{e2}}{2}\right)} + \frac{1}{\left(\frac{R_{e3}}{4}\right)} + \frac{1}{R_{e4}} + \frac{1}{R_{ch2}} \Rightarrow$$

$$\frac{1}{R_e} = \frac{1}{28.52} + \frac{1}{\left(\frac{158.89}{2}\right)} + \frac{1}{\left(\frac{7.94}{4}\right)} + \frac{1}{18.68} + \frac{1}{2.89} \Rightarrow$$

$$\frac{1}{R_e} = 0.9504 \Rightarrow R_e = 1.052\,\Omega > 1\,\Omega$$

According to Table 5.116, the maximum acceptable grounding resistance value for a 0.4 kV system should be less than 1 Ω. Therefore, it is necessary to make improvements in the initial grounding design. To keep the grounding design on the safe side, let us try to reduce the equivalent ground resistance of the PV farm by at least 30%. Considered modifications in the grounding system and the resultant ground resistance values are respectively given in Figure 5.156 and Table 5.119. After Modification-1, the equivalent ground resistance of the whole PV farm decreased from 1.084 Ω to 0.82 Ω. In addition, if we increase the electrode length from 1 m to 1.5 m, the equivalent ground resistance value decreases from 0.82 Ω to 0.77 Ω.

Step 4 – Touch and Step Voltage Calculations: Since the resistance goals are reached, the step and touch voltage values can now be calculated. From (5.156) and (5.157), the allowable limits of touch and step voltages for a person of 70 kg are calculated as below ($R_B = 1500\,\Omega, k = 0.157, C_s = 1$).

$$E_{touch} = (R_B + 1.5\rho_s C_s)\frac{k}{\sqrt{t_s}} = (1500 + 1.5 \times 400)\frac{0.157}{\sqrt{0.5}} = 466.27\,\text{V}$$

$$E_{step} = (R_B + 6\rho_s C_s)I_B = (1500 + 6 \times 400)\frac{0.157}{\sqrt{0.5}} = 865.92\,\text{V}$$

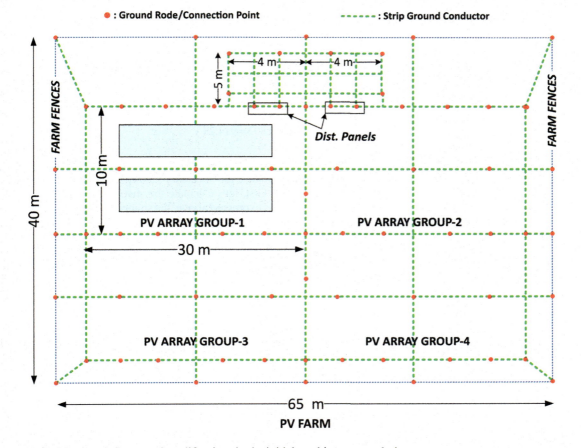

FIGURE 5.156 Suggested modifications in the initial earthing system design.

TABLE 5.119
The Initial Grounding Design and Suggested Modifications Parameters

Earthing System Components	Initial Design	Suggested Modification-1 (Case-1)	Additional Modification-2 (Case-2)
Wire fence	8 Rods	16 Rods	In addition to Modification-1, the length of the grounding rods is increased from 1 meter to 1.5 meters.
Substation	Mesh grid area: 24 m² # of Rods: 4 Conductor length: 48 m	Mesh grid area: 24 m² # of Rods: 6 Conductor length: 60 m	
Distribution panel	2 Rods	2 Rods	
PV array group	# of Rods: 6 Conductor length: 80 m	# of Rods: 9 Conductor length: 80 m	
PV site	4 PV array group 2 Distribution panel 1 Substation Additional 9 rods on the grid Strip Length: 320 m	4 PV array group 2 Distribution panel 1 Substation Additional 17 rods on the grid Strip Length: 610 m	
Ground resistance of the PV farm's grounding grid	$R_e = 1.052\ \Omega$	$R_e = 0.766\ \Omega$	$R_e = 0.713\ \Omega$

Initially, the condition GPR $\leq E_{touch}$ should be checked. From (5.161), the ground return current (I_g) is calculated as below ($S_f = 60\%$).

$$I_g = S_f \cdot I_f = 0.60 \times 1900\ \text{A} = 1140\ \text{A} = 1.14\ \text{kA}$$

The ground potential rise can then be calculated for the suggested modification (Case-1).

$$\text{GPR} = I_g \times R_e = 1140 \times 0.802 = 914.28\ \text{V}$$

Since GPR > E_{touch} (914.28 V > 466.27 V), we need to check the conditions $E_{touch} \geq V_m$ and $E_{step} \geq V_s$. To be able to calculate the V_m and V_s values, the necessary parameters in equations (5.159) and (5.160) should be computed. First, let's calculate the shape factor "n" for the corresponding mesh grid architecture. As expressed above, the shape factor "n" is calculated for rectangular grids as follows (from Figure 5.156 & Table 5.119; $L_T \cong 610 + 60$, $L_p = 40 + 120 + 10$, $S = 120 \times 20 = 2400\ \text{m}^2$).

$$n = (2L_T / L_P) \times \sqrt{\frac{L_P}{4\sqrt{S}}} = (2 \times 670/170) \times \sqrt{\frac{170}{4\sqrt{2400}}} = 7.34$$

Again as given above, the parameter K_i is calculated as follows.

$$K_i = 0.644 + 0.148 \times n = K_i = 0.644 + 0.148 \times 7.34 = 1.73$$

Average mesh grid spacing (d) can be calculated from Figure 5.156 and equation (5.164).

$$d = \frac{1}{2}\left(\frac{W_g}{n_r - 1} + \frac{L_g}{n_c - 1}\right) = \frac{1}{2}\left(\frac{20}{5-1} + \frac{60}{5-1}\right) = 10\ \text{m}$$

The parameters K_m and K_s are calculated from equations (5.162) and (5.163) as follows ($D = 0.034$ m, $d = 10$ m, $h = 0.8$ m, $n = 7.34$, $K_{ii} = 1$).

Photovoltaic Power Systems

$$K_m = \frac{1}{2\pi}\left[\ln\left(\frac{10^2}{16\times 0.8\times 0.034}+\frac{(10+2\times 0.8)^2}{8\times 10\times 0.034}-\frac{0.8}{4\times 0.034}\right)+\frac{1}{\sqrt{1+0.8}}\cdot \ln\left(\frac{8}{\pi(2\times 7.34-1)}\right)\right]=0.892$$

$$K_s = \frac{1}{\pi}\left[\frac{1}{2\times 0.8}+\frac{1}{10+0.8}+\frac{1}{10}\left(1-2^{2-7.34}\right)\right]=0.259$$

The lengths L_m and L_s are calculated from equations (5.171) and (5.172) as follows [$L_r = 1$ m, $L_T = 670$ m, $L_x = 120$, $L_y = 20$, $L_R = $ (Total # of Rods)$\times L_r = 89 \times 1$]

$$L_m = L_T + \left[1.55+1.22\times \frac{L_r}{\sqrt{(L_x^2+L_y^2)}}\right]L_R = 670+\left[1.55+1.22\times \frac{1}{\sqrt{(120^2+20^2)}}\right]\times 89 = 808.84 \text{ m}$$

$$L_s = 0.75 L_T + 0.85 L_R = 0.75\times 670 + 0.85\times 89 = 578.15 \text{ m}$$

Now, the parameters V_m and V_s can be calculated from equations (5.159) and (5.160) as follows.

$$V_m = \frac{\rho K_m K_i I_g}{L_m} = \frac{400\times 0.892\times 1.73\times 1140}{808.84} = 870.55 \text{ V}$$

$$V_s = \frac{\rho K_s K_i I_g}{L_s} = \frac{400\times 0.259\times 1.73\times 1140}{578.15} = 354.16 \text{ V}$$

As can be seen from the above results, although the condition $E_{step} \geq V_s \left(865.92 \text{ V} \geq 354.16 \text{ V}\right)$ is satisfied, the condition $E_{touch} \geq V_m$ is not met yet (466.27 V $\not\geq$ 870.55 V). Therefore, the tolerable voltage value E_{touch} should be increased to ensure the condition ($E_{touch} \geq V_m$). To do this, the crushed stones at a thickness of 0.1m can be used as a protection layer in the substation area. The crushed rock resistivity ρ_s is approximately 3000 $\Omega\cdot$m ($\rho_s = 3000$ $\Omega\cdot$m). Let us now recalculate the E_{touch} value based on the $\rho_s = 3000$ $\Omega\cdot$m.

$$C_s = 1 - \frac{0.09\left(1-\frac{\rho}{\rho_s}\right)}{2h_s + 0.09} = 1 - \frac{0.09\left(1-\frac{400}{3000}\right)}{2\times 0.1 + 0.09} = 0.731$$

$$E_{touch} = (R_B + 1.5\rho_s C_s)\frac{k}{\sqrt{t_s}} = (1500+1.5\times 3000\times 0.731)\frac{0.157}{\sqrt{0.5}} = 1063.45 \text{ V}$$

Now that the condition $E_{touch} \geq V_m \left(1063.45 \text{ V} \geq 870.55 \text{ V}\right)$ is also met. Hence all the conditions of the safe grounding design are met (except for the cross-section check).

Step 5 – Cross Section Calculation of the Grounding Conductor: The cross section of the grounding conductor can be determined based on the phase-to-earth fault current that may occur at the LV ends of the transformer. Since the fault current was given as 1900 A, the required minimum cross section of the copper conductor is calculated from (5.173) as below (from Table 5.117, $K = 226$ A(s)$^{1/2}$/mm^2, $T_f = 300°$C, $T_i = 20°$C, $\beta = 234.5°$C).

$$S_{min} = \frac{I}{K}\sqrt{\frac{t}{\ln\left(\frac{T_f+\beta}{T_i+\beta}\right)}} = \frac{1900 \text{ A}}{226\left[\text{A}(s)^{1/2}/\text{mm}^2\right]}\sqrt{\frac{0.5 \text{ s}}{\ln\left(\frac{300°C+234.5°C}{20°C+234.5°C}\right)}} = 4.20 \text{ mm}^2$$

TABLE 5.120

The Comparative Grounding Results for the Soil Resistivity of 100 Ω · m and 400 Ω · m

Soil Resistivity	Results	Initial Design	Modified Design (Case-1)	Suitability Check $E_{touch} \geq V_m$ & $E_{step} \geq V_s$
$\rho = 400$ Ωm	R_e	1.052 Ω	0.766 Ω	Initial Design→Not Suitable
	GPR	1235.7 V	914.28 V	Modified Design (for $\rho_s = 400$ Ω · m)
	E_{touch}	466.27 V	466.27 V for $\rho_s = 400$ Ω · m	→ Not Suitable
			1063.45 V for $\rho_s = 3000$ Ω · m	Modified Design (for $\rho_s = 3000$ Ω · m)
	E_{step}	865.92 V	865.92 V for $\rho_s = 400$ Ω · m	→ Suitable
			3254.68 V for $\rho_s = 3000$ Ω · m	
	V_m	1399.3 V	870.55 V	
	V_s	403.05 V	354.16 V	
$\rho = 100$ Ωm	R_e	0.263 Ω	0.192 Ω	Initial Design→ Suitable
	GPR	299.82 V	218.88 V	Modified Design (for $\rho_s = 400$ Ω · m)
	E_{touch}	366.35 V	366.35 V for $\rho_s = 100$ Ω · m	→ Suitable
			1032.45 V for $\rho_s = 3000$ Ω · m	Modified Design (for $\rho_s = 3000$ Ω · m)
	E_{step}	466.27 V	466.27 V for $\rho_s = 100$ Ω · m	→ Suitable
			3130.64 V for $\rho_s = 3000$ Ω · m	
	V_m	349.82 V	217.63 V	
	V_s	100.76 V	88.54 V	

As an alternative approach, the minimum conductor cross section can be calculated from (5.138) as follows [$I_{th} = 1900$ A, J_{thr} (for bare copper conductor) = 67 A/mm² for short-circuit duration of 1 second].

$$S_{min} = \frac{I_{th} \cdot \sqrt{t_{trip}}}{J_{thr}} = \frac{1900 \cdot \sqrt{1}}{67} = 28.35 \text{ mm}^2$$

It is understood from the above S_{min} values, the selected conductor cross section at the beginning is suitable.

b. The soil resistivity was assumed to be 400 Ω · m in option (a). In option (b), the suitability of the initial grounding design is being asked for the assumption that the soil resistivity was 100 Ω · m. Since the grounding calculations are given in detail in option (a), the grounding results are given only in Table 5.120 for the respective soil resistivity values.

5.5.7.14.1 Potential Induced Degradation Effect on Photovoltaic Modules

The potential induced degradation (PID) effect may take place on PV modules due to the high negative voltage occurrence between solar cells and module frames. The PID phenomenon may have a negative impact not only on crystalline PV technologies but also on thin-film PV technologies.

Figure 5.157 shows the negative voltage occurrence between PV frames and cells in a PV system with a transformerless inverter (or in a bipolar PV system). Under this negative high voltage, the positive mobile ions inside the PV modules are pushed toward the solar cells, while the negative ions are migrated away from the cells. For instance, the positive sodium ions (Na+) in the glass cover can migrate toward the cells under this voltage. Figure 5.158 illustrates the PID mechanisms in a PV module under negative high DC voltage. This ion mobility causes leakage currents in different paths (such as from the module's front side or backside to the solar cells through the glass or encapsulating material), which results in degradation in the module power output.

In simple terms, PID affects the ions of a solar cell and consequently reduces the power output of that cell. In practice, this reduction in the output power may reach up to 30% or even higher in

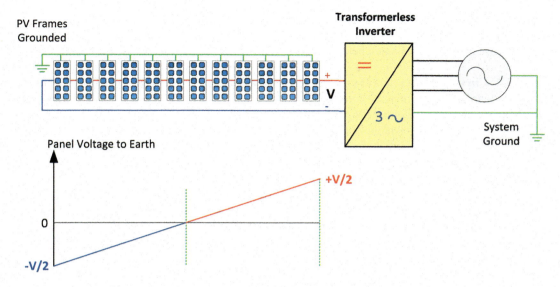

FIGURE 5.157 Ungrounded PV system and array voltage with respect to earth.

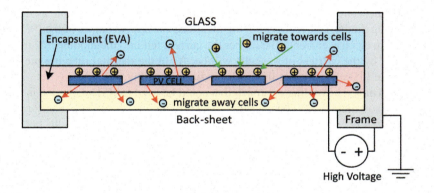

FIGURE 5.158 Illustration of PID effect in PV panels.

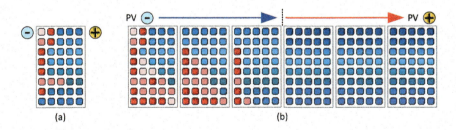

FIGURE 5.159 PID formation in an individual panel (a) and a string (b).

some cases. The occurrence of PID mainly depends on the electrical configuration of the system and the cell technology. However, factors such as system voltage, temperature, humidity, and solar radiation also contribute to PID. Once a PID phenomenon starts to occur in a PV string, a few weeks or months later, PID occurs at the entire negative side of the string. Figure 5.159 shows the PID formation in an individual panel and a string.

The PID occurrence in a PV module cannot be distinguishable by the naked eye. Therefore, several methods are used to determine if a PV module is affected by a PID or not. The PID detection methods are listed and explained below.

- **I–V Curve:** The presence of possible defects in a PV module can easily be detected under standard test conditions by comparing its IV curve with the reference curve. PID reduces the shunt resistance of the PV cells and thus causes the performance of the PV modules to degrade. As it is known, the decrease in the shunt resistance causes the leakage currents to increase, which eventually results in the module output current decreasing. In brief, with the PID effect, the MPP of a PV module moves downward on the IV curve.
- **Open-Circuit Voltage:** If a PV panel is affected by PID, the open-circuit voltage will likely be lower than expected. Some of the solar cells in a PV module are damaged by the PID effect. Therefore, the open-circuit voltage of the module will decrease due to PID. It is possible to measure the open-circuit voltage of the module with the help of a sensitive voltmeter. This method works well when the PID is high. Otherwise, voltage reduction maybe not be distinguishable easily.
- **Thermal Imaging or Infrared Inspection:** Solar PV cells that are affected by PID have higher temperatures than other neighboring cells. Imaging with an infrared camera (or thermographic camera) allows the detection of unhealthy cells, which are appeared darker than healthy cells.
- **Electroluminescence (EL) Imaging:** EL imaging uses the same principle as that of LED (light-emitting diode). Hence, EL imaging measures the optical output of the PV module in response to an electric current. The higher the current fed into the module, the brighter the light emitted by the carriers of the cells will be. As in the thermal imaging method, dark cells show the PID effect, while light color indicates healthy cells.
- **Field Data Analysis:** Another approach that could be used for PID detection is to analyze the field measurement data. The measured data provided by the online monitoring system can be analyzed per inverter, string, and panel. And the comparative results between their operating voltages, powers, or efficiencies can then be used to detect the presence of PID.

As can be understood, the PID occurrence has serious economic and technical impacts on the solar PV systems. Therefore, the PID effect in a photovoltaic system should be mitigated at all levels. There are many ways to prevent or mitigate the PID occurrence in a solar PV system. The general PID mitigation techniques are listed and explained below.

- **Cell/Module Level:** The sheet resistivity, chemical composition of ARC, thickness, and homogeneity of each layer in a PV cell, resistivity/water absorptivity of encapsulation, and Na (sodium) content of the cover glass are the major factors that affect the quality of a PV module. Manufacturers should pay attention to these factors to avoid PID damages on PV modules.
- **System Level:** The PID effect at the PV system level can also be mitigated by applying several specific technologies and taking measures. These PID mitigation ways are briefly described below.
    - **PV System Configuration/Earthing:** Inverter topologies are of great importance in the development of PID. The inverters with galvanic isolation allow the negative potential of the module to be grounded, keeping the negative pole at the ground potential to eliminate the PID effects. Therefore, a PV system configuration that allows negative grounding of the PV array should be preferred wherever possible. Figure 5.160 shows a grounded PV system and PV array voltage relative to earth.
    - **System DC Voltage Level:** Negative high DC voltages might trigger PID formation. In PV arrays, do not occur for systems with a voltage of less than 500 Vdc.

Photovoltaic Power Systems

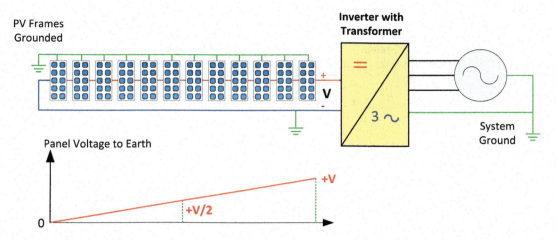

FIGURE 5.160 Grounded PV system and array voltage with respect to earth.

- **Charge Equalizers/PV Offset Box/Anti-PID Box:** There are two types of power degradation in PV systems, reversible (surface polarization) and non-reversible (electro-corrosion observed mostly on thin films). The reversible surface polarization in a PV system with a transformerless inverter can be controlled and reversed by using charge equalizers, so-called anti-PID or PV offset boxes. When the transformerless inverters are not active at night, the charge equalizers apply a reverse bias voltage to the PV array during the night to cancel out the reversible type PID effects. Hence, the recovery of PV modules is accelerated, and accordingly, the PV efficiency lost during the daytime can be regenerated by applying charge equalizers. Charge equalizers, connected between the PV panels and the inverter, typically apply a voltage between 400 V and 1000 V relative to the earth after sunset.
- **High Impedance Grounding:** A transformerless system can be grounded at the negative pole of the inverter with a high-value resistor, for example, 22 kΩ. In this case, additional equipment is required to be installed to detect the earth fault. This technique can stop or prevent PID by reducing the negative voltage potential in the PV array.
- **Site Level:** Sites with cool weather conditions and low humidity levels are ideal for PV systems. Also, locating the PV site in a windy field helps prevent the PID effect as it keeps the PV modules cooler.

5.5.7.14.2 Bonding and Interconnection Optimization of PV Arrays

The primary and cost-effective approach to protecting an electrical installation against lightning, fire, and short-circuit hazards can be achieved by a proper grounding system and avoiding unnecessary cable lengths in the installation. The electromagnetic interference of the lightning discharge current may result in inducing overvoltages in the wiring loops of the PV installation. Therefore, the internal loop of each PV panel/string should be designed carefully. More specifically, the total surface area covered by the interconnection loops should be as small as possible to minimize the voltages induced by lightning strikes. Figure 5.161 illustrates the possible induction loops formed in a PV installation and shows the sample reduction techniques in the surface loop areas.

It is necessary to draw attention to an important point about the wiring methods in Figure 5.161. Some PV module manufacturers may have specific restrictions on the PV module installation to prevent water storage or to allow water drainage around the junction box. For example, it may not be allowed to install the junction box upside down. If this is the case, connecting modules in a daisy-chain structure provides an alternative solution. In this connection technique, excess module cable lengths can be coiled up and arranged with the help of tape or clamp. If the module connection

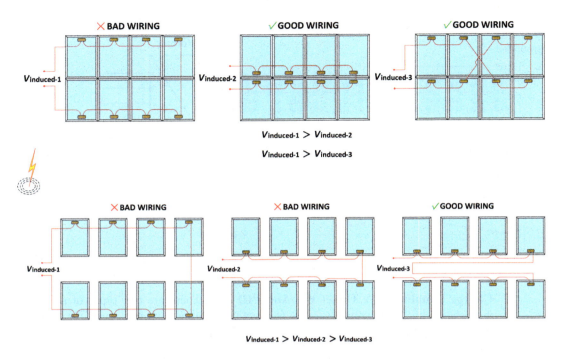

FIGURE 5.161 Induction loops and sample loop area reduction techniques in a PV installation.

cable length is sufficient, the leapfrog wiring method can also be used as an alternate for the serial connection of the PV modules to avoid coiled-up wiring. Daisy-chain and leapfrog wiring methods are shown in Figure 5.162.

Importantly note that excess cable lengths should never be fixed inside the PV module metal frames or around metal legs.

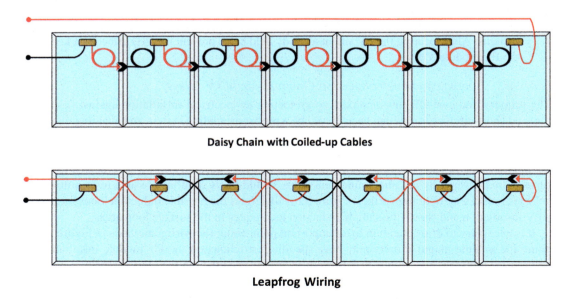

FIGURE 5.162 PV string wiring in a daisy chain (with coiled-up cables) and leapfrog wiring.

Photovoltaic Power Systems

5.5.7.15 Lightning System Design for Photovoltaic Power Systems

Solar PV sites are frequently exposed to direct and indirect flashes of lightning, and PV equipment at the site is very susceptible to these lightning strikes. Experience worldwide has shown the high economic loss and devastating impact of not installing adequate lightning/surge protection on PV systems. The cost of a lightning protection system is negligibly low compared to the total PV plant cost. For this reason, a properly designed lightning protection system is an essential component for solar PV sites. Protection angle, rolling sphere, and mesh methods are specified in the literature to determine the position and size of the air-termination system. Although all these three methods are applicable for rooftop and small-scale PV systems, the rolling sphere application is technically more suitable for large-scale PV sites.

5.5.7.15.1 Protection Angle Method (PAM)

If the structure/PV system to be protected is completely contained by the protection volume of the air-termination system, it is considered that the position and selected size of the respective air-termination system are sufficient. The PAM is commonly used to protect the simple-shaped structures and buildings. The protection zone of the PAM can be illustrated as a conical volume with a semi-apex angle α (protection angle), as shown in Figure 5.163.

Air-termination rod application in PV systems can be divided into two categories as insulated and non-insulated LPS. It would be more appropriate to use an isolated lightning protection system in areas with high lightning risk. An insulated LPS simply consists of a free-standing air-termination rod with down conductors. As shown in Figure 5.164, the air termination should be placed at a distance "d" from the metallic frame of the PV structure. The distance d given by (5.174) should be greater than or equal to the minimum distance "s" ($d \geq s$).

$$s = \frac{k_i \cdot k_c}{k_m} \times L \tag{5.174}$$

where the distance L is taken as the distance from the highest point of the PV frame, where the separation distance is considered, to the nearest equipotential bonding point (along with the down conductor). As for the other parameters, k_i is 0.08 for protection class-I, 0.06 for protection class-II, and 0.04 for protection classes III and IV. The parameter k_m is a variable depending on the isolation material between the air-termination rod and the PV frame, and it is 1 for air, 0.5 for concrete or brick, and 0.7 for other isolation materials. Lastly, k_c is defined as equal to 1/n (n: number of down conductors).

A sample PAM application of an isolated LPS is shown in Figure 5.164 for PV panels placed on a flat roof plane. In the case of using more than one air-termination rod, the distance between two

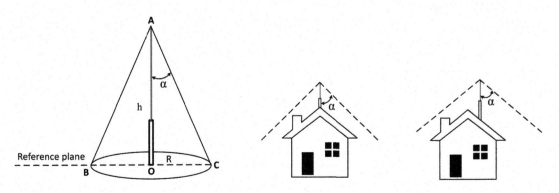

FIGURE 5.163 A simple shaped structure with protection angle method (h: height of air-termination rod above the reference plane, α: protection angle, R: protection radius at reference plane).

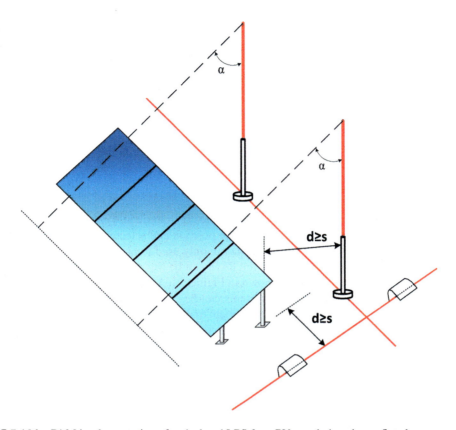

FIGURE 5.164 PAM implementation of an isolated LPS for a PV panel placed on a flat plane.

consecutive rods (in the horizontal and vertical directions) must also meet the safe distance condition of the rolling sphere method (explained below).

An air terminal-based LPS protection for PV systems can be isolated from the module frames (see Figure 5.164) or mounted directly on them (non-isolated system). Figure 5.165 shows non-isolated and isolated simple LPS applications for a ground-mounted PV system. It should be noted that a non-isolated air terminal approach is often preferred in practice, as it is a more economical solution than the isolated system.

The protection angle method is a mathematical simplification of the rolling sphere method for a given range of h [m], the height of air-termination rods above the reference plane of the structure to be protected. The PAM also depends on the four classes of lightning protection systems (LPS). From IEC 62305-3, the protective angles for different classes of LPS can be expressed by equations (5.175)–(5.178) as below.

$$\alpha_I = \begin{cases} 70 & \text{for } h \leq 2 \\ 80.908 \times e^{-0.06h} & \text{for } 2 < h \leq 20 \\ \text{Apply Rolling Sphere \& Mesh Methods} & \text{for } h > 20 \end{cases} \quad (5.175)$$

$$\alpha_{II} = \begin{cases} 73.2 & \text{for } h \leq 2 \\ 79.937 \times e^{-0.039h} & \text{for } 2 < h \leq 30 \\ \text{Apply Rolling Sphere \& Mesh Methods} & \text{for } h > 30 \end{cases} \quad (5.176)$$

Photovoltaic Power Systems

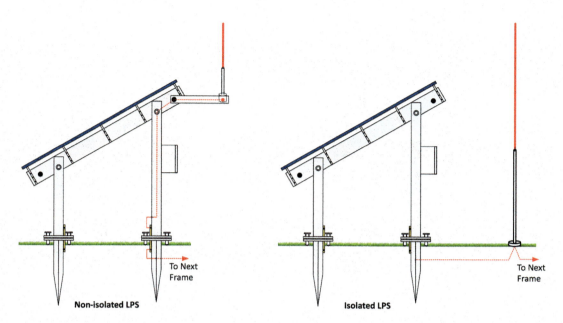

FIGURE 5.165 An isolated and non-isolated air terminal-based LPS protection for ground-mounted PV panels.

$$\alpha_{III} = \begin{cases} 76.3 & \text{for } h \leq 2 \\ 81.234 \times e^{-0.027h} & \text{for } 2 < h \leq 45 \\ \text{Apply Rolling Sphere \& Mesh Methods} & \text{for } h > 45 \end{cases} \quad (5.177)$$

$$\alpha_{IV} = \begin{cases} 78.7 & \text{for } h \leq 2 \\ 80.304 \times e^{-0.02h} & \text{for } 2 < h \leq 60 \\ \text{Apply Rolling Sphere \& Mesh Methods} & \text{for } h > 60 \end{cases} \quad (5.178)$$

where α_I, α_{II}, α_{III}, and α_{IV} are the protective angles of air-termination rods for the four classes of LPS, respectively. Note that Franklin rods in the PAM methods should be placed on the higher and most vulnerable places such as edges and overhangs of the structure to be protected.

Example 5.35: Application of Protection Angle/Mesh Methods for a Roof-Mounted PV System

A roof-mounted solar PV system is planned to be installed as shown in Figure 5.166. Calculate the required protection angles for the four levels of LPS and the respective air-termination rod lengths. Determine the minimum separation distance between the PV Panel and the air-termination rod for an isolated lightning protection system.

Solution:

To keep the building within the protection volume of the rod, the protection distance for the given case must be greater than 4.165 m, as shown in Figure 5.167.

Let us now consider a Franklin rod with an overall height of 10 m (building height+protection rod height). From equations (5.175) to (5.178), we can find the protective angles for all levels of LPS.

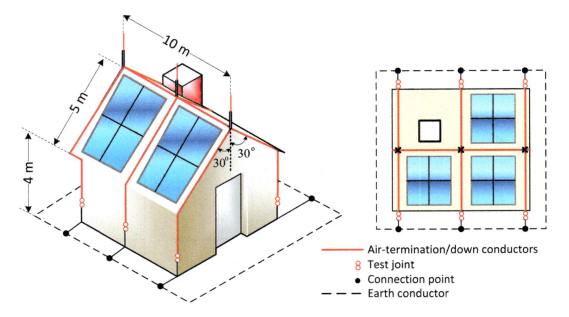

FIGURE 5.166 Application of protection angle and mesh methods for a roof-mounted PV system.

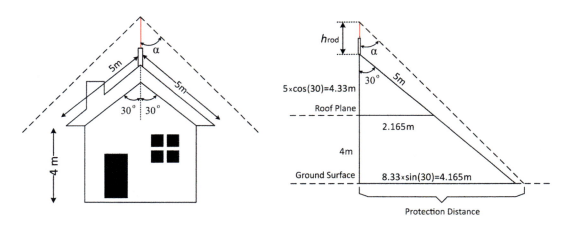

FIGURE 5.167 Application of protection angle and mesh methods for a roof-mounted PV system.

From (5.175), α_I is calculated as below.

$$\alpha_I = 80.908 \times e^{-0.06h} = 80.908 \times e^{-0.06 \times 10} = 44.4°$$

Similarly, the protective angles for other LPS classes are calculated as follows.

$$\alpha_{II} = 79.937 \times e^{-0.039h} = 79.937 \times e^{-0.039 \times 10} = 54.12°$$

$$\alpha_{III} = 81.234 \times e^{-0.027h} = 81.234 \times e^{-0.027 \times 10} = 62.01°$$

$$\alpha_{IV} = 80.304 \times e^{-0.02h} = 80.304 \times e^{-0.02 \times 10} = 65.74°$$

TABLE 5.121
Protective Angles, Protection Distances, and Separation Distances for All Levels of LPS

LPS	$h_{overall}$	$h_{building}$	h_{rod}	Protection Angle (α)	Protection Distance ($h_{overall} \times \tan \alpha$)	L (5+4+1.67)	k_i	k_c (n=6)	k_m	Min Separation Distance $s_{min} = \dfrac{k_i \cdot k_c}{k_m} \times L$
I	10 m	8.33 m	1.67 m	44.4°	9.79 m	10.67 m	0.08	1/6	1	0.142 m
II	10 m	8.33 m	1.67 m	54.12°	13.82 m	10.67 m	0.06	1/6	1	0.106 m
III	10 m	8.33 m	1.67 m	62.01°	18.81 m	10.67 m	0.04	1/6	1	0.071 m
IV	10 m	8.33 m	1.67 m	65.74°	22.18 m	10.67 m	0.04	1/6	1	0.071 m

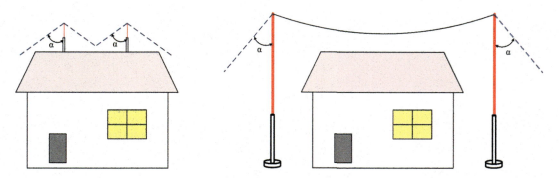

FIGURE 5.168 PAM application examples (left: two air-termination rods, right: wired air termination system).

From the above results, the protection distance values can easily be calculated as shown in Table 5.121. In addition, the minimum separation distances required for the four classes of LPS are also given in Table 5.121.

It is clear from the above results that the selected air-termination rod creates sufficient protection volume for the given building. It should be noted that if a structure cannot be covered using a single rod, we need to increase the number of terminals as per the coverage volume. In this case, all the air terminals should be interconnected. Two application samples are given in Figure 5.168.

5.5.7.15.2 Rolling Sphere Method (RSM)

PV farms are highly susceptible to lightning strikes as they are often installed on large and open fields. PV systems installed in large areas are generally protected against direct lightning strikes by the rolling sphere method (RSM). As a requirement of an effective LPS, the RSM should include air terminals, down conductors, equipotential bonding, separation distance applications, and low impedance grounding. As explained above in the PAM method, a single air terminal maintains a conical protection volume depending on the height of the arrester rod. However, several air terminals create continuous protection zones in the outer sub-region of the rolling spheres. Figure 5.169 shows the RSM implementation and protected volume for isolated and non-isolated PV arrays.

The main goal in designing the RSM-based LPS is to ensure that the rolling sphere does not touch the surface to be protected. To achieve this goal, the parameters of sphere radius (R), distance between two sequential air-termination rods (d), the rod height above the reference plane (h_{rod}) and sphere penetration depth (p) should be determined. The value of the lightning strike current is the most important factor in determining the design parameters. Generally, LPS designs are performed according to four classes of lightning strikes (Class-I, Class-II, Class-III, and Class-IV). If the lightning current values are not known in the area, air-termination design according to LPS-I keeps the system on the safe side. Be careful that whatever the protection class is, the cross section of the LPS

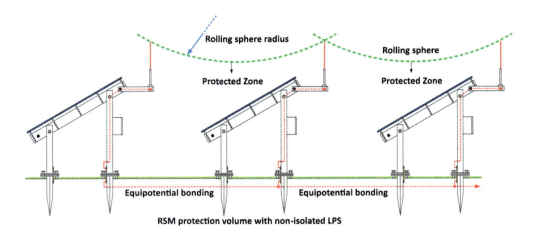
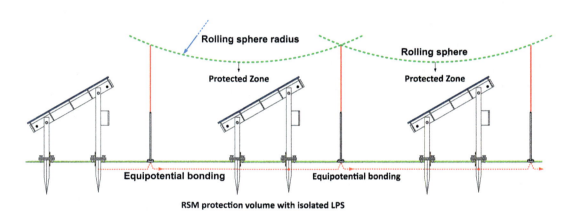

FIGURE 5.169 RSM-based LPS protection for ground-mounted PV systems having isolated and non-isolated air-terminal applications.

conductors should be sized for the highest possible lightning current in the area. In this context, it is recommended to work with the local meteorological office to determine the lightning probability and the required level of protection for the given solar site. Table 5.122 shows the rolling sphere radiuses and expected lightning currents for the four levels of protection.

TABLE 5.122
Rolling Sphere Radiuses and Expected Lightning Currents for the Four Levels of Protection

Protection Level	Sphere Radius (R)[a]	Expected Lightning Current (I_p)
I	20 m	2.9 kA
II	30 m	5.4 kA
III	45 m	10.1 kA
IV	60 m	15.7 kA

[a] Rolling sphere radius (or striking distance), $R = 10 \cdot I_p^{0.65}$

Photovoltaic Power Systems

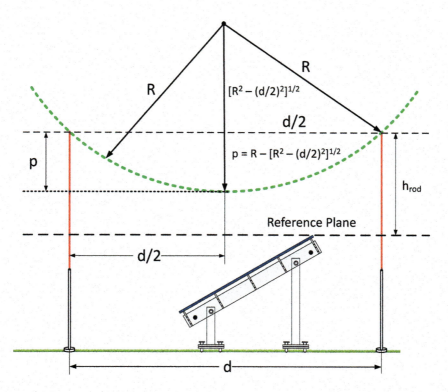

FIGURE 5.170 RSM-based LPS for PV arrays having two masts and calculation of the penetration depth.

Figure 5.170 shows RSM-based LPS having two or more masts, which are located on the horizontal reference plane for a PV frame structure.

It is clear from Figure 5.170 that the penetration depth (p) of the rolling sphere between the two masts can be calculated by equation (5.179). Here, the height of the air-termination rod above the reference plane (h_{rod}) must be greater than the penetration depth ($h_{rod} > p$).

$$p = R - \sqrt{R^2 - (d/2)^2} \tag{5.179}$$

It should be noted that the initial value of the distance between two air-termination rods (d) can be selected as a certain percentage of the radius of the rolling sphere. For example, this value can be set to 70% of the sphere radius for the initial design.

Example 5.36: Application of Rolling Sphere Method to a Ground-Mounted PV System

A PV farm is planned to be installed as shown in Figure 5.171. Considering the RSM method, calculate the LPS design parameters (i.e., distance between two sequential rods, the rod height above the reference plane, and sphere penetration depth) for the protection level-I.

Solution:

In this example, only protection level-I is considered for the lightning risk assessment. To design the air termination-based LPS, RSM-based approach shall be considered. From Figure 5.171, the PV field to be protected is a rectangular area, of which dimension is 62 m by 62.5 m.

From Table 5.122, the radius of the rolling sphere (R) is 20 m. The distance between two air-termination rods can be selected as $d = 20 \times 0.70 = 14$ m for the initial design. Accordingly, the sphere penetration depth, p is calculated from (5.179) as follows.

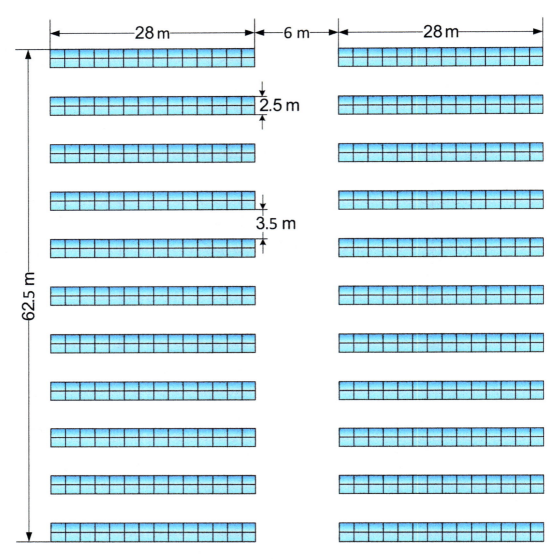

FIGURE 5.171 The layout of the given PV farm and its basic dimensions.

$$p = R - \sqrt{R^2 - (d/2)^2} \Rightarrow p = 20 - \sqrt{20^2 - \left(\frac{14}{2}\right)^2} = 1.26 \text{ m}$$

According to this result, the height of the air-termination rod above the PV frame plane can be selected as 1.5 m or higher ($h_{rod} > p$) for a safe LPS design. Once these initial design parameters are determined as above, the actual LPS sizes to be applied should be determined on the given PV field. Two approaches can be proposed for this purpose. In the first approach, air-termination rods can be spread evenly over the entire PV field. In the second approach, air-termination rods can be positioned to create a protection volume, which will cover each PV array group individually. In this example, the second approach is employed.

Since the distance between the two rods is initially determined as 14 m, the distance between the rods in the first row can be selected as 28/2 = 14 m. Accordingly, the rod distance between two rows can be set to 12 m. (2.5 + 3.5 + 2.5 + 3.5) = 12 m. It is assumed that the first air-termination rod in the first row is placed at the front side of the PV frame. Then no air-termination rod was placed in the next row (the second row was skipped). Similarly, air-termination rods were again

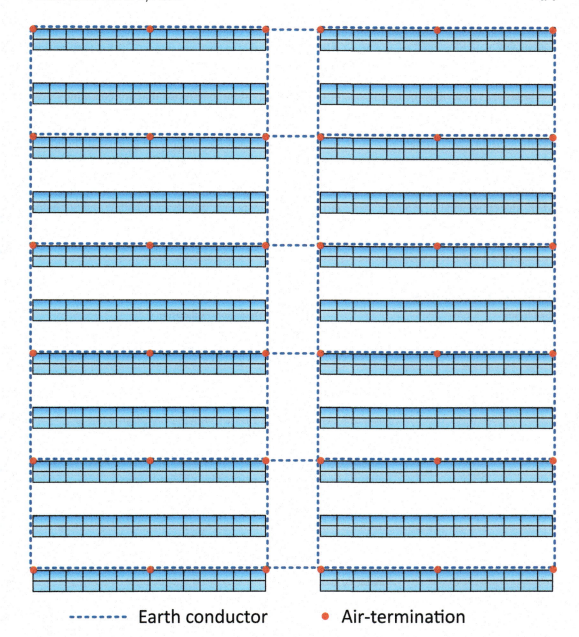

FIGURE 5.172 Final lightning protection design for the given PV field.

placed in the third row. In summary, only PV rows 1, 3, 5, 7, 9, and 11 are having Franklin rods. From the above results, the final lightning protection design for the given PV field can be depicted in Figure 5.172.

5.5.8 Identifying Load Requirements: Estimation and Calculations

As in all electrical power systems, the correct determination of load demands is also a fundamental requirement for the sizing and designing of solar PV systems. Depending on the design objectives, the load requirements to be calculated for PV systems may differ. Therefore, the electrical design professionals should analyze the load requirements in detail at the beginning of the project.

In this framework, a relationship between the total loads connected to the system and the actual load demand on the system should be established. To be able to perform an accurate load estimation, the frequently confused terms about electrical load demands and their characteristic behaviors are explained in Table 5.123.

Table 5.123 describes the definitions of the most important load terms that are commonly used in electrical load estimation. In addition to these, there are other preliminary (initial) load estimation approaches such as building area and space-by-space methods. However, these methods are mostly used for load estimation of large-scale buildings. Solar PV systems for large-scale buildings are often used for supplementary generation because of insufficient space. Therefore, building load is not important when sizing PV arrays applied to large-scale buildings. In other words, only the available area used for the PV array is essential for large-scale building applications. Applications, where the entire load is required to be fed by PV systems, are usually small-scale remote house applications or various off-grid PV applications.

TABLE 5.123
Important Terms and Definitions about Electrical Loads Connected to the System

Load Terms	Definition/Explanation
Total Connected (Installed) Load	It is the sum of all loads connected to the system. It can be expressed in Watts or kW.
Demand Load	It is the average electrical load (at the receiving terminal) over a given period, often expressed in amps, kA, kW, kVAR, or kVA. Generally, 10-minute, 30-minute, or 1-hour averages can be taken depending on the time interval of the particular electrical utility.
Demand Interval	It is the range in which the load is averaged for a specific period. There may be short-term averages such as 10 minutes, 30 minutes, or 1 hour, or long-term averages such as daily or monthly.
Max Demand (Peak Demand)	It is the highest of all power demands that occur in a certain period such as 10 minutes, 30 minutes, or an hour. Weekly, monthly, or annual peak power demands may also be taken into account for particular evaluations or billing services.
Demand Factor (k_u) (in IEC Max Utilization Factor)	Under normal operating conditions, certain loads, such as motors and building loads, often operate at less than their rated (or installed) powers. In this context, the demand factor is the ratio of the maximum demand in a system to the total connected load of the system. It is often expressed as a percentage (%) or in a ratio (≤ 1). Demand Factor (k_u) = Maximum demand/Total connected load Note that demand factors for buildings usually range from 45% to 85% of the total connected load.
Coincidence Factor (k_s)	Under normal operating conditions, each group of loads (fed from a particular distribution or sub-distribution board) does not operate simultaneously and there is always some degree of diversity. Therefore, this fact must be into account for the load estimating purposes (by the use of a factor k_s). In this context, the coincidence factor is the ratio of the maximum demand of the complete system to the sum of the individual maximum loads. It is often expressed as a percentage (%) or in a ratio (≤ 1). $$\text{Coincidence Factor } (k_s) = \frac{\sum_{i=1}^{n} \max(\text{Aggreated Load}_i)}{\sum_{i=1}^{n} \text{Individual Peak Load}_i}$$ Note that coincidence factors for buildings typically range between 40% and 95%.
Diversity Factor ($1/k_s$)	The diversity factor is the inverse of the coincidence factor ($1/k_s$). Hence, it is simply defined as the ratio of sum of individual maximum demands to maximum system demand. Diversity factor = Sum of individual max demands / Max system demand

DC and AC load demand to be taken as a basis for PV array sizing for remote houses can be calculated as equations (5.180) and (5.181).

$$\text{Energy Demand}\left(\frac{\text{Wh}}{\text{week}}\right)^{DC} = \sum_{i=1}^{n}\left(P_i^{DC}\right) \times \Delta t_i\left(\frac{\text{hour}}{\text{day}}\right) \times \Delta t_i\left(\frac{\text{day}}{\text{week}}\right) \quad (5.180)$$

$$\text{Energy Demand}\left(\frac{\text{Wh}}{\text{week}}\right)^{AC} = \sum_{i=1}^{n}\left(P_i^{AC}\right) \times \Delta t_i\left(\frac{\text{hour}}{\text{day}}\right) \times \Delta t_i\left(\frac{\text{day}}{\text{week}}\right) \quad (5.181)$$

where the index "i" counts the electrical devices connected to the system and the parameter "n" shows the total number of electrical devices in the system. The P_i represents the value of the i th load in Watts. The parameter $\Delta t_i\left(\frac{\text{hour}}{\text{day}}\right)$ shows the expected working time of the related load in a day, while the $\Delta t_i\left(\frac{\text{day}}{\text{week}}\right)$ shows the number of working days in a week for that load. If loads are operating differently during weekdays and weekends, then the load demand shall be calculated separately for weekdays and weekends. Later, the results found shall be aggregated together.

If there are both DC loads and AC loads in the system, the weekly total electrical energy demand (Wh/week) can be calculated by equation (5.182).

$$\text{Total Energy Demand}\left(\frac{\text{Wh}}{\text{week}}\right) = \frac{\text{Energy Demand}\left(\frac{\text{Wh}}{\text{week}}\right)^{DC}}{\eta_{\text{conv}}} + \frac{\text{Energy Demand}\left(\frac{\text{Wh}}{\text{week}}\right)^{AC}}{\eta_{\text{inv}}} \quad (5.182)$$

where η_{conv} and η_{inv} are the DC/DC converter and DC/AC inverter efficiencies, respectively.

If the system has battery storage, the weekly total electrical energy demand is then converted to (Ah/day) by considering the battery voltage to be used in the system as below.

$$\text{Daily Energy Demand}\left[\frac{\text{Ah}}{\text{day}}\right] = \frac{\sum\left(\frac{\text{Wh}}{\text{week}}\right)}{\text{Battery Voltage}} \times \frac{1}{7}\left[\frac{\text{week}}{\text{day}}\right] \quad (5.183)$$

Example 5.37: Determination of Load Demand and System Requirements for a PV-Powered Remote House

A photovoltaic energy system will be designed to feed a small chalet, which will consist of PV modules with the following parameters (Table 5.124).

TABLE 5.124
PV Module Data

Rated Power	150 Watts
Number of cells	72
Rated voltage	34 volts
Rated current	4.45 Amps
Open circuit voltage	42.8 volts
Short circuit current	4.75 Amps
NOCT	50°C

TABLE 5.125
Household Appliances and Their Daily Working Hours

Electrical Loads	Power	Hours/day	Days/week
Lighting Devices (AC)	60 Watts	5	7
Television (AC)	100 Watts	4	5
Laptop (AC)	50 Watts	3	4
Electric Cooker (AC)	300 Watts	1	7
Refrigerator (AC)	200 Watts $\left[\dfrac{ON}{OFF} = \dfrac{1}{2}\right]$	24	7
Microwave Oven (AC)	600 Watts	1/4	3

Household appliances in this chalet and their daily working hours are given in Table 5.125. 100 Ah and 12 V batteries will be used as energy storage in the PV energy system to be designed. The system will only be used during the ski season and in the winter months when the air temperature is around (−5°C). According to meteorological data, the days of autonomy (DoA) in this ski season are determined as 3, and the PSH value is given as 5 hours/day. Note that the charge-discharge efficiency of the batteries is given as 80% and the inverter efficiency as 90%. Also, assume that V_{DC} = 24 volts, f_{Temp-b} is 0.65, and DoD_{max} is 50%.

a. Calculate the daily load demand in (Ah/day).
b. Determine the required number of batteries and their connection type.
c. Determine the total number of PV modules and their connection type required.
d. Determine the inverter-rated power required.

Solution:

a. There are only AC loads in the system. The weekly energy demand can be calculated from equation (5.181) as follows.

$$\sum [\text{Wh/week}] = (60 \times 5 \times 7) + (100 \times 4 \times 5) + (50 \times 3 \times 4) + (300 \times 1 \times 7)$$
$$+ \left(200 \times 24 \times 7 \times \frac{1}{3}\right) + \left(600 \times \frac{1}{4} \times 3\right) = 18450 [\text{Wh/week}]$$

Hence from (5.183), the energy demand in Ah/day can be calculated as below.

$$\frac{\sum [\text{Wh/week}]}{V_b} = \frac{18450}{12} \cong 1538 \, [\text{Ah/week}]$$

$$\frac{[\text{Ah/week}]}{7} = \frac{1538}{7} \cong 220 \, [\text{Ah/day}]$$

b. The battery requirements can be calculated from equations (5.42) to (5.45) as follows.

$$C_{\text{Battery-net}}[\text{Ah}] = \frac{E_{\text{Daily Avg}}[\text{Wh/day}] \times \text{DoA}[\text{day}]}{\text{DoD}_{\max} \times V_b \times \eta_b \times f_{\text{Temp-b}}} = \frac{\dfrac{18450[\text{Wh/week}]}{7} \times 3}{0.50 \times 12 \times 0.80 \times 0.65} \cong 2535 \, \text{Ah}$$

$$N_{\text{Battery}} = \frac{C_{\text{Battery-net}}[\text{Ah}]}{c_b [\text{Ah}]} = \frac{2535 [\text{Ah}]}{100 [\text{Ah}]} \cong 26$$

$$N_{sb} = \frac{24}{12} = 2$$

$$N_{pb} = \frac{26}{2} = 13$$

As a result, the battery storage configuration is $(2s \times 13p)$, and accordingly, a total of 26 batteries are needed.

c. If we neglect the environmental effect and cabling loss effects on the output power of PV panel ($\Delta P_{PV} = 1$, $\eta_{cab} = 1$), the PV array requirements can be calculated from equations (5.10) to (5.16) as follows. Note that the load growth factor (L_{gf}) is assumed to be 1.2.

$$E_{overall} = \frac{E_{Load} \cdot L_{gf}}{(\Delta P_{PV}) \cdot \eta_{bat} \cdot \eta_{inv} \cdot \eta_{ch} \cdot \eta_{cab}} = \frac{\frac{18450[\text{Wh/week}]}{7} \times 1.2}{0.80 \times 0.90} \cong 4395 \text{ Wh/day}$$

$$P_{Peak(PV)} = \frac{E_{overall}}{PSH_{avg}} = \frac{4395}{5} = 879 \text{ W}$$

$$I_{DC} = \frac{P_{Peak(PV)}}{V_{DC}} = \frac{879}{24} = 36.7 \text{ A}$$

$$N_s = \frac{V_{DC}}{V_{mp}} = \frac{24}{34} \approx 1$$

Note that N_s is rounded up to the nearest integer number. Finally, N_p can be calculated from (5.16) as below.

$$N_p = \frac{I_{DC}}{I_{mp}} = \frac{36.7}{4.45} \approx 9$$

Again, N_p is rounded up to the nearest integer number. As a result, the PV array configuration is $(1s \times 9p)$, and accordingly, a total of 9 PV modules are needed.

d. The inverter power can be estimated as follows.

$$\text{Inverter Power} = \frac{(\text{Total Connected AC Loads}) \times L_{gf}}{(\text{Inverter Efficiency})}$$

$$\text{Inverter Power} = \frac{(60 + 100 + 50 + 300 + 200 + 600) \times 1.2}{0.90} = 1747 \text{ W}$$

As an upper nominal value, a 2000 W inverter will be sufficient.

PROBLEMS

P5.1. A PV module, of which specifications are given in Table 5.8, will be used to supply a resistive load with a rated power of 2200 W ($V_r = 220$ V, $I_r = 10$ A). Assume that the resistive load operates all the day times. The PV system without battery storage will feed the load with a directly coupled stand-alone system and it is assumed to be operated throughout the entire year. Let's use the solar radiation data given in Table 5.9 in the field (please see Example 5.1 for Tables 5.8 and 5.9).
   a. Calculate the PV array size and draw the system design diagram considering the basic protection requirements.

b. Considering the $I-V$ characteristic of the module given by Figure 5.60 (in Example 5.1), calculate the power that can be transferred from the PV array to the load under 1000 W/m², 800 W/m², 600 W/m², and 400 W/m² radiations.

P5.2. The electricity requirement of a small house in a rural area is planned to be met by a stand-alone PV system. The selected PV module is a polycrystalline silicon 200 W_p with the specifications of $V_{oc} = 31.2$ V, $V_{mp} = 25.71$ V, $I_{sc} = 8.15$ A, and $I_{mp} = 7.75$ A. The average daily energy demand of this house is estimated to be 20 kWh/day approximately. Assume that the average daily peak sun hours (PSH_{avg}) through the entire year is 4.9 at the given tilt and azimuth of the site.

a. If the designed DC system voltage is 48 V, calculate the PV array size and draw the system wiring diagram considering the PWM-based solar charger.

b. Calculate the PV array size and configuration in case of MPPT solar charger usage. Draw the PV system wiring diagram and show the basic protection schemes on the design. Note that the basic specification of the MMPT solar charger is given below (Table 5.126). Assume that NOCT=45°C and voltage coefficient of the PV is −0.12%/°C.

P5.3. A 550 kW_p grid-connected PV power plant is planned to be installed in New York. The characteristic values of the selected PV module in this project are given in Table 5.127.

Two different inverter options for the PV array design are considered. These inverters are string-type and central-type inverters. The basic specifications of these inverters are given in Table 5.128.

Assuming −10°C and +70°C as the minimum and maximum cell temperatures of the modules, calculate the following.

a. Size and configure the PV array in case of string inverter usage. Draw the array configuration.

b. Size and configure the PV array in case of central inverter usage. Draw the array configuration.

TABLE 5.126
The Parameters of Solar Chargers

Basic Specifications of the MPPT Solar Charger	
MPP tracking voltage range	60–180 Volt DC
Maximum charging current	32 A
Nominal battery voltage	48 Volt DC
Maximum open circuit voltage of the PV array	205 Volt DC

TABLE 5.127
The PV Module Parameters

PV Module Specifications: 320 W_p Poly Crystalline Module			Module Efficiency: 16.49 %	
Voltage at P_{max}	Current at P_{max}	Open-Circuit Voltage	Short-Circuit Current	Power Tolerance
37.62 V	8.51 A	44.84 V	9.52 A	−0/ + 3%
Temperature Coefficients				
$\alpha_{I_{sc}}$	$\beta_{V_{mp}}$	$\beta_{V_{oc}}$	P_{max} Coefficient	NOCT
+0.005 A/°C	−0.0757 V/°C	−0.034 V/°C	−0.45 W/°C	47 ± 2°C

TABLE 5.128
Inverter Parameters

Option-1: String Inverter (# of Independent MPPT: 6/(# of DC Pairs for Each MPPT: 4)

MPP Voltage Range	Max DC Voltage	Max DC Current	Max Input Power for Each MPPT	Nominal Power
450 V – 880 V	1000 V	38 A	20 kW$_p$	120 kW

Option-2: Central Inverter

MPP Voltage Range	Max DC Voltage	Max DC Input Current for Each MPPT	Max Input Power	Nominal Power
430 V – 850 V	1000 V	1215 A	600 kW$_p$	550 kW

P5.4. Consider the string & central inverter-based 550 kW PV applications given in Problem 5.4 above. Assume that the dimension of the *320 W$_p$* PV module used in Example 5.3 is 960 × 1950 mm. Also assume that the site location is in New York with the coordinate of (40.7128° N, 74.0060° W). Calculate the land area requirements for the following three cases.

  a. Calculate the required land area for the string inverter-based application in option (a) of Problem 5.3. Suppose that the arrays are south facing and the tilt angle is 40°.
  b. Calculate the required land area for the central inverter-based application in option (b) of Problem 5.3. Suppose that the arrays are south facing and the tilt angle is 40°.
  c. Assume that the 550 kW PV system consists of two-axis PV towers with 2.2 m heights and each PV tower has a 2560 W$_p$ array. Calculate the required land area for this case.

P5.5. Consider the string & central inverter-based 500 kW PV applications given in Problem 5.3 above.

  a. Determine the system combiner box requirements for the string inverter-based application in option (a) of Problem 5.3. Select a suitable combiner box from the commercially available products and draw the wiring diagram.
  b. Determine the system combiner and recombiner box requirements for the central inverter-based application in option (b) of Problem 5.3. Select a suitable combiner and recombiner boxes from the commercially available products and draw the wiring diagram.

P5.6. The electricity requirement of a small house in a rural area is planned to be met by a stand-alone PV system. The average daily energy demand of this house is estimated to be 16 kWh/day approximately. Suppose that deep-cycle gel batteries, of which specifications are given below will be used as an energy storage system. Due to the cold winter months, battery storage conditions may drop down to the temperature of −10 °C (Table 5.129).

If the designed system DC voltage is 48 V and the number of autonomy days (DoA) is 2, calculate the battery storage size and draw the PV system diagram with battery bank wiring.

TABLE 5.129
Battery Parameters

Basic Specifications of the VRLA Gel Battery

Rated Ah capacity	50 Ah
Nominal battery voltage	12 Volt DC
Temperature range	−30°C to +50°C
Round trip efficiency	83%

P5.7. A stand-alone PV system will be designed to feed an LED-based lighthouse. The energy needed by the lighthouse is planned to be supplied from the battery system during the night. The lighting system is planned to be consists of 4×18 W LED bulbs. The required battery storage size will be made for the worst-case scenario, that is, winter months when the temperature drops to $-10°C$ and the nighttime is 15 hours. The other design parameters are given as follows:
- 12 V, 120 Ah deep-cycle lead-acid batteries will be used for energy storage.
- The round-trip efficiency is 88% for the selected battery.
- The predicted DoA value is 4.

If the designed system DC voltage is 12 V, calculate the battery storage size and draw the PV system diagram with battery bank wiring.

P5.8. Stand-alone PV systems in different configurations will be designed by using the 12 V, 150 Ah batteries and PV modules, of which parameters are given in Table 5.130.
  a. Design a 12 V off-grid system with four PV modules. Assume that a total of 8 batteries will be used for storage needs. If the maximum load current to be drawn at the DC side is 10 A, specify the required charge-controller size and select the possible charger types based on the technical data given in Tables 5.22 and 5.23.
  b. Design a 24 V off-grid system with six PV modules. Again assume that a total of 8 batteries will be used for storage needs. If the maximum load current to be drawn at the DC side is 20 A, specify the required charge-controller size and select the possible charger types based on the technical data given in Tables 5.22 and 5.23.

P5.9. A 24 V off-grid PV system will be designed by using the 6 V, 180 Ah batteries and six PV modules, of which parameters are given in Table 5.131.

Assume that a total of 12 batteries will be used for storage needs. If the maximum load current to be drawn at the DC side is 30 A, specify the required charge-controller size and select the possible charger types based on the technical data given in Tables 5.22 and 5.23.

P5.10. A 48 V off-grid PV system will be designed by using the 12 V, 300 Ah batteries and eight PV modules of which parameters are given in Table 5.132.

Assume that a total of eight batteries will be used for storage needs. If the maximum load current to be drawn at the DC side is 40 A,
  a. Specify the required charge-controller size and select the possible charger types based on the technical data given in Tables 5.22 and 5.23.

TABLE 5.130
PV Module Parameters

PV Module Specifications: 50 W_p Poly-Crystalline Module			
Voltage at P_{max}	Current at P_{max}	Open Circuit Voltage	Short Circuit Current
17.5 V	2.9 A	21.8 V	3.2 A

TABLE 5.131
PV Module Parameters

PV Module Specifications: 36 W_p Poly-Crystalline Module			
Voltage at P_{max}	Current at P_{max}	Open Circuit Voltage	Short Circuit Current
15.6 V	2.32 A	19.8 V	2.59 A

Photovoltaic Power Systems

TABLE 5.132
PV Module Parameters

PV Module Specifications: 400 W$_p$ Mono-Crystalline Module			
Voltage at P_{max}	Current at P_{max}	Open-Circuit Voltage	Short-Circuit Current
40.6 V	9.86 A	49.3 V	10.47 A

b. If the continuous power value of the loads is 2400 W and the apparent power of the water pump in the system is 600 VA, specify the required off-grid inverter size and select a suitable inverter from Table 5.26.

P5.11. A grid-connected 60 kW$_p$ PV power plant will be designed by using 150 PV modules, of which parameters are given in Table 5.133.

The electrical energy generated from the PV power plant will be transferred to the low-voltage grid through 400 V three-phase system. If the maximum number of series-connected PV modules in a string is 15, select a suitable grid-tie inverter from Table 5.29 and design the final PV array configuration. Finally, draw the system wiring diagram.

P5.12. A 11.2 kW$_p$ gird-connected hybrid PV system will be designed by using 12 V, 120 Ah batteries, and 28 PV modules, of which parameters are given in Table 5.133 (see Problem 5.11). The primary goal of the hybrid PV system is to feed local loads. The excess electrical energy from the PV system will be transferred to the low-voltage grid via 400 V, three-phase system. Assume that the continuous power value of the load is 7000 W, the surge power of the system is negligible, and a total of 60 batteries will be used for storage needs.

If the maximum number of series-connected PV modules in a string is 14, design the system DC voltage and battery storage configuration based on the specified design parameters and technical data given in Table 5.31. In addition, select a suitable hybrid inverter from Table 5.31. Draw the system wiring diagram considering the final design parameters.

P5.13. A 48 kW$_p$ grid-connected PV system will be designed by using 120 PV modules. The PV module and array parameters are given in Table 5.134. Determine the fuse and conductor ratings for PV strings and PV sub-arrays. Consider Table 5.5 for the fuse/conductor selection.

P5.14. Assume that a DC switch-disconnect will be sized for disconnecting a central inverter from a PV generator. The PV generator consists of 16 parallel-connected PV strings and each string has 14 series-connected PV modules. The PV module parameters are as follows (Table 5.135):

Determine the current and voltage ratings of switch-disconnect, which will be installed between the PV array and the inverter.

TABLE 5.133
PV Module Parameters

PV Module Specifications: 400 W$_p$ Mono-Crystalline Module			Module Efficiency: 19.3 %	
Voltage at P_{max}	Current at P_{max}	Open-Circuit Voltage	Short-Circuit Current	Power Tolerance
40.6 V	9.86 A	49.3 V	10.47 A	−0 / +3%
		Temperature Coefficients		
$\alpha_{I_{sc}}$	$\beta_{V_{mp}}$	$\beta_{V_{oc}}$	P_{max} Coefficient	NOCT
+0.002 A/°C	Not Given	−0.026 V/°C	−0.36 W/°C	42 ± 3°C

TABLE 5.134
PV Module/Array Parameters

400 W_p PV Module Specifications			
Voltage at P_{max}	**Current at P_{max}**	**Open-Circuit Voltage**	**Short-Circuit Current**
40.6 V	9.86 A	49.3 V	10.47 A

9.6 kW_p Sub-Array Configuration

$N_{s,sub} = 12$ (12 PV modules are in series per string) $N_{p,sub} = 2$ (2 PV strings are connected in parallel)

54 kW_p Array Configuration

$N_{p,array} = 5$ (five 9.6 kW_p Sub-Arrays are connected in parallel)

TABLE 5.135
PV Module Parameters

500 W_p PV Module Specifications			
Voltage at P_{max}	**Current at P_{max}**	**Open-Circuit Voltage**	**Short-Circuit Current**
43.4 V	11.53 A	41.4 V	12.13 A

TABLE 5.136
PV Module Parameters

475 W_p PV Module Specifications			
Voltage at P_{max}	**Current at P_{max}**	**Open-Circuit Voltage**	**Short-Circuit Current**
41.9 V	11.34 A	50.5 V	11.93 A

P5.15. Assume that a DC and an AC switch-disconnect device will be sized for a 10.5 kW_p grid-connected PV system. The DC switch-disconnect will be used for disconnecting the inverter from the PV array, while the AC switch-disconnect will be used for disconnection purposes at the AC side. The PV array consists of 2 parallel-connected PV strings and each string has 11 series-connected PV modules. The PV module parameters are as follows (Table 5.136):

Determine the current and voltage ratings of DC and AC switch disconnects and draw the system wiring diagram.

P5.16. Assume that three circuit breakers on DC and AC sides will be sized and selected for a 50 kW_p grid-connected PV system with battery backup. The corresponding three locations for circuit breakers are as follows.
- Location-1: The first circuit breaker will be installed between the PV array and the inverter (DC side).
- Location-2: The second circuit breaker will be installed between the inverter and the battery bank (DC side).
- Location-3: The third circuit breaker will be installed at the output of the inverter (AC side).

The typical data for installation and system are given as below (Table 5.137):

Determine the current and voltage ratings of circuit breakers specified above.

TABLE 5.137
PV Module/Array and Inverter Specifications

400W$_p$ PV Module Specifications			
Voltage at P_{max}	**Current at P_{max}**	**Open-Circuit Voltage**	**Short-Circuit Current**
40.6 V	9.86 A	49.3 V	10.47 A
55 kW Central Inverter Data (Three Phase, 400 Vac)			
Full-Load Inverter Efficiency	**Max Input Voltage**	**Max Input Current**	**Max Output Current**
95%	1000 Vdc	123 A	27 A per phase
50 kW$_p$ PV Array Specifications			
14 PV modules per string		9 Parallel string	
240 Vdc Battery Bank Specifications			
20 Batteries (each: 12 V and 100 Ah) per string		16 Parallel string	

P5.17. Consider an 800 kW$_p$ photovoltaic power plant that is connected to a 66 kV network via 0.4/31.5 kV substation, 1000 m underground cable, 5 km overhead line, and 31.5/66 kV transformer station. The single-line network diagram including the PV system structure is presented in Figure 5.173.

The system parameters for the above-given network configuration are listed in the following tables (Tables 5.138 and 5.139).

a. Calculate the three-phase short-circuit values (short-circuit power, steady-state short-circuit current, and peak short-circuit current) for the circuit breaker locations at points C, D, E, and F as specified in the network diagram.

b. Specify the ratings of current, voltage, power, breaking capacity, and closing capacity for the circuit breakers located at the points C, D, E, and F.

P5.18. Consider the 800 kW$_p$ PV power plant given in Problem 5.17. This PV power plant was created using 400 W$_p$ PV modules (V_{oc} = 49.3, V_{mp} = 40.6, I_{sc} = 10.47, I_{mp} = 9.86). Assume that all given data in Problem 5.17 is also valid for this problem. Under these conditions, determine the requirements of all SPDs and size them for the DC and AC sides of the PV system.

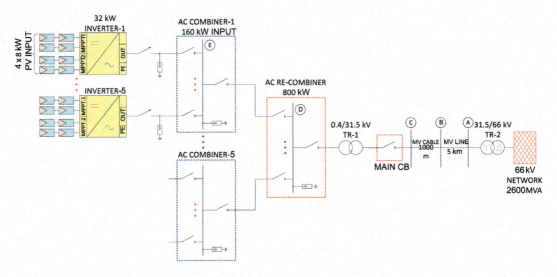

FIGURE 5.173 Single-line diagram of the PV system with grid connection.

TABLE 5.138
System Parameters

Network Components		Component Parameters		
66 kV network	$S_{k(66\ kV)}$ = 2800 MVA		$\cos\varphi_k = 0.1$	
Transformer-2 (TR-2)	31.5/66 kV	S_{Tr2} = 20 MVA	$\Delta P_{k(Tr2)}$ = 100 kW	$u_k = 10.8\%$
MV line	31.5 kV	$\ell_1 = 10$ km	$R'_{L1} = 0.337$ Ω/km	$X'_{L1} = 0.385$ Ω/km
MV cable	31.5 kV	$\ell_2 = 1$ km	$R'_{L2} = 0.124$ Ω/km	$X'_{L2} = 0.204$ Ω/km
Transformer-1 (TR-1)	0.4/31.5 kV	$S_{Tr1} = 1.5$ MVA	$\Delta P_{k(Tr1)} = 20$ kW	$u_k = 6.8\%$
LV cable (between AC combiner and recombiner)	0.4 kV	$\ell_3 = 0.3$ km	$R'_{L3} = 0.012$ Ω/km	$X'_{L3} = 0.071$ Ω/km
LV cable (between AC combiner and inverter)	0.4 kV	$\ell_4 = 0.07$ km	$R'_{L4} = 1.12$ Ω/km	$X'_{L4} = 0.08$ Ω/km

TABLE 5.139
Inverter Data

Multi-String Inverter Data (Three-Phase, 400 Vac)				
# of MPPT	Max Input Voltage	Max Input Current	Max Output Current	Nominal Power
2	1000 Vdc	51 A per String	72 A per phase	40200 W

P5.19. Consider a 15 kW$_p$ roof-mounted single-phase grid-connected PV system that also feeds the local loads over a 230 V single-phase electrical system. The solar PV system will be assembled based on the following PV modules and central inverter, which are given in the Table 5.140.

The system impedances required to calculate phase-to-neutral short circuit current are $Z_{LN} = 50$ mΩ (neutral line impedance relative to service entrance box) and $Z_k = 65$ mΩ (network impedance relative to service entrance box). This grid-connected PV system also feeds the local residential loads with a maximum power of 8 kW. Work out the following RCD requirements.

a. Determine the RCD requirements of the PV system and size them for the required locations (the RCD requirement for the in-house distribution board will also be considered). If the average leakage current value for residential loads is 1.5 mA/kW, perform the sensitivity analysis for the selected indoor RCD (Load Factor = 1 and local load is 6 kW).
b. Evaluate the alternative solution options for the potential earth faults and size the relevant protective devices (such as MCB and RCBO). Assume that the Z_{Loop} impedance due to ground fault is 0.14 Ω in evaluating the use of MCB or RCBO.

TABLE 5.140
PV Module and Inverter Specifications

PV Module Specifications: 400 W$_p$ Mono-Crystalline Module			
Voltage at P_{max}	Current at P_{max}	Open-Circuit Voltage	Short-Circuit Current
40.6 V	9.86 A	49.3 V	10.47 A

Single-Phase and 18 kW Central Inverter Specifications			
Input Side (DC)		Output Side (AC)	
Max Short Circuit Current	Max Input Voltage	Max Output Current	Rated Output Voltage
105 A	660 V	93.6 A	230 V

Photovoltaic Power Systems

P5.20. Consider the 800 kW$_p$ PV power plant given in Problem 5.17. Size and specify the CT and VT required for the protection circuit of the main breaker located at point C (high-voltage side of the 0.4/31.5 kV transformer). Please see Figure 5.173 for the system detail.

Note that the system parameters required for CT and VT sizing and selection are given as follows:
- The apparent power of the 0.4/31.5 kV transformer is 1500 kVA,
- The distance between the relay and measuring equipment/current transformers is approximately $\ell = 3$ m,
- Conductor resistance = 0.087 Ω/m,
- The power drawn by the relay is 0.5 VA per phase,
- Power drawn by the measuring circuit is 0.125 VA per phase.
- The power drawn by the electricity meter is 5 VA (to be connected to VT)
- The secondary coil resistance of the VT is 423 Ω.

P5.21. Consider the 800 kW$_p$ PV power plant given in Problem 5.17. The load profiles of the auxiliary DC power supply system for the primary and secondary sides of the 0.4/31.5 kV transformer are given in Table 5.141.

Assume that the circuit breaker motor (spring charging motor) in the distribution board of the low voltage side has been selected as 230V AC. The motor of the circuit breaker at the high voltage side has been selected as 24 V DC. Hence, the DC supply voltage for the primary side is 110V, while it is designed as 24 V DC for the secondary side. The cross-sectional area of the cable between distribution switchgear and load is 4 mm^2 for the primary side and 6 mm^2 for the secondary side. The cable lengths are 12 m and 2.5 m for the low and high voltage sides, respectively.

Based on the given data, size the DC auxiliary supply system for primary and secondary sides.

P5.22. Consider the 800 kW$_p$ PV power plant given in Problem 5.17. A simplified single-line diagram of the PV power plant and its connection to the local grid can be re-drawn as Figure 5.174. The system data and short circuit currents at different locations of the network are listed in Table 5.142. Assume that the maximum allowable trip time at 66 kV connection point A is determined by TSO (Transmission System Operator) to be 1 second.

TABLE 5.141
Load Profile of Auxiliary DC Supply System

Low Voltage Side (0.4 kV)						
Continuous			Non-continuous			
Load (DC)	Power	Number	Load (DC)	Power	Number	Duration
Overcurrent relay	40 W	7	Trip coil	180 W	2	Negligible
Signal lamp	3 W	7	Signal lamp	3 W	14	Negligible
Alarm lamp	3 W	4	Lighting lamp	20 W	6	8 hours
			Alarm lamp	3 W	24	Negligible
High-Voltage Side (31.5 kV)						
Overcurrent relay	40 W	2	Trip coil	180 W	2	Negligible
Signal lamp	3 W	2	Circuit breaker motor	220 W	2	Negligible
Alarm lamp	3 W	4	Signal lamp	3 W	4	Negligible
			Lighting lamp	20 W	5	8 hours
			Alarm lamp	3 W	4	Negligible

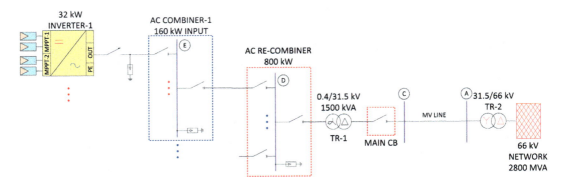

FIGURE 5.174 The simplified single-line diagram of the 800 kW PV power plant.

TABLE 5.142
The System Data and Short-Circuit Currents

Location	Three-Phase Short Circuit Current I_{sc3}	Recommended Current Transformer	Three-Phase Transformer Data (0.4/31.5 kV)
C	2050 A	200/5 A	1500 kVA (Y/Δ: star point grounded), $u_k = \%6.8$,
D	26650 A	750/5 A	transformer maximum short circuit withstand
E	23540 A	600/5 A	duration = 2 sn.

Based on the provided data, determine the protection requirements at points E, D, and C. Perform the discriminative relay coordination throughout the system to reach the protection objectives and do the relay settings accordingly. Finally, draw the protective relaying diagram for both LV and MV PCCs.

P5.23. Consider the 8 kW$_p$ PV system shown in Figure 5.175. The PV array consists of four parallel strings, while the strings consist of five modules connected in series, each having a power of 400 W. The PV module and cabling data are given in Table 5.143.

Considering the given data, size the DC and AC cables in the PV system, and select the appropriate cables from the manufacturers' catalogs.

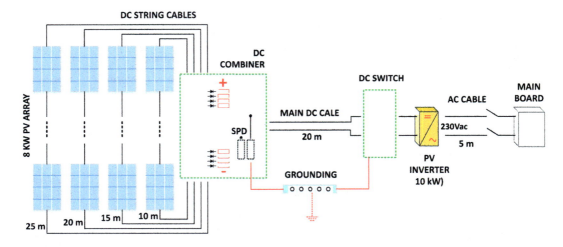

FIGURE 5.175 Circuit diagram of 8 kW rooftop PV system with a central inverter.

TABLE 5.143
The Parameters for PV Module and Cabling

PV Module Specifications: 400 W$_p$ Poly-Crystalline Module			Module Efficiency: 17.73 %	
Voltage at P_{max}	Current at P_{max}	Open-Circuit Voltage	Short-Circuit Current	Power Tolerance
40.6 V	9.86 A	49.3 V	10.47 A	−0 / +3%

	Cabling Parameters
DC string cables	String cables are respectively 25 m, 20 m, 15 m, and 10 m. Cables are mounted directly on the roof surface in groups of four in a raceway.
Main DC cable	Main DC cable length is 20 m. These cables are grouped in pairs in a raceway.
AC cable	AC cable length is 5 m. The AC cables are mounted directly on the wall in groups of two in a conduit. Assume that the **maximum short-circuit current** (due to grid failures) on the AC side of the inverter is 2.5 kA.
Other parameters	All cables will be XLPE insulated. The ambient air temperature at the site is 35°C. The conduit is mounted directly on the roof surface. Four conductors are laying in the conduit.

P5.24. Consider the 800 kW$_p$ PV power plant given in Problem 5.17. Calculate the required minimum cross section of the copper busbars at the E (AC combiner) and D (AC recombiner) points. Assume that short-circuit current values at points E and D are given as 6.25 kA and 26.65 kA, respectively.

P5.25. Consider a PV power plant that is connected to a 31.5 kV network via LV/MV substation and underground LV cable. The network configuration with the system parameters is presented in Figure 5.176.

Considering the network configuration, and size the AC combiner and recombiner boxes located at B_3 and B_2, respectively.

P5.26. A solar PV power plant is planned to be built on 5400 m^2 of land. The photovoltaic solar farm consists of different grounded subsystems such as metallic wire fences, substations, distribution panels, and PV arrays. The initial earthing design of a PV farm and required system parameters for earthing calculation are respectively given in Figure 5.177 and Table 5.144.

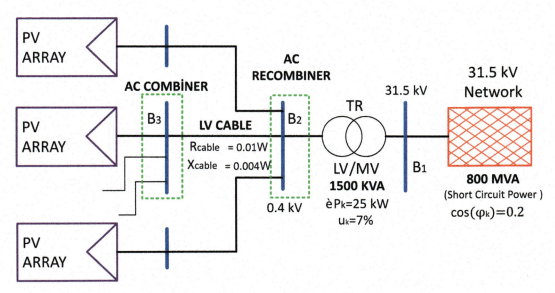

FIGURE 5.176 The considered network configuration and system parameters.

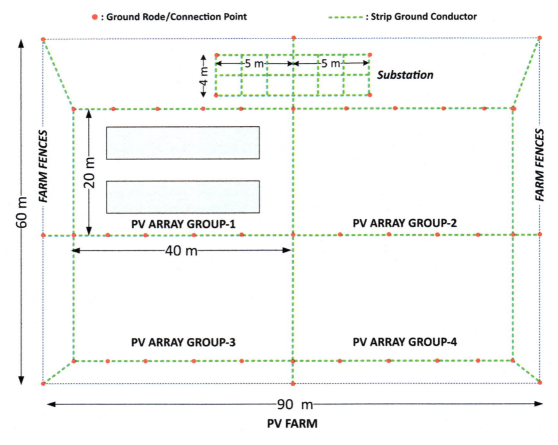

FIGURE 5.177 The preliminary earthing system design.

TABLE 5.144
The Preliminary Earthing System Parameters

Subcomponents of Earthing System	Earthing Configuration	Quantity (and/or) Units
Star point	Solidly grounded	Tr: 31,5/0,4 kV and 900 kVA, I_{sc} at LV side = 2100 A
PV site		All earthing systems are interconnected as shown in Figure 5.177. The red dots in the figure represent the earthing electrodes. The system consists of 4 PV array groups, 1 substation, and a wire fence around the solar farm.
Other System Parameters		
Phase to neutral short-circuit current (at the low voltage side of the transformer)		$I_{sc} = I_{fault} = 2100$ A
Leakage current protection		A leakage current protection device of 300 mA is used at each inverter output.
Short circuit duration		0.5 second
Soil resistivity		200 Ω·m
Grounding rod Length		1.5 m
Conductor burial depth		0.8 m
Earthing rod diameter		0.02 m
Strip conductor type		Round grounding conductor
Grounding strip diameter		0.034 m

Check the suitability of the initial grounding design of the given PV farm in terms of grounding resistance, touch voltage, and conductor cross-sections. If necessary, determine the improvements that can be made in the grounding design, and decide the final configuration.

P5.27. A PV farm is planned to be installed as shown in Figure 5.178. Considering the rolling sphere method, calculate the LPS design parameters (i.e., distance between two sequential rods, the rod height above the reference plane, and sphere penetration depth) for all the protection levels of I, II, III, and IV.

P5.28. A photovoltaic energy system will be designed to feed a small remote house, which will consist of PV modules with the following parameters (Table 5.145).

Household appliances in this house and their daily working hours are given in Table 5.146.

200 Ah and 12 V batteries will be used as energy storage in the PV energy system to be designed. The system will be used during the entire year. The coldest air temperature in the region is about −5°C. According to meteorological data, the days of autonomy

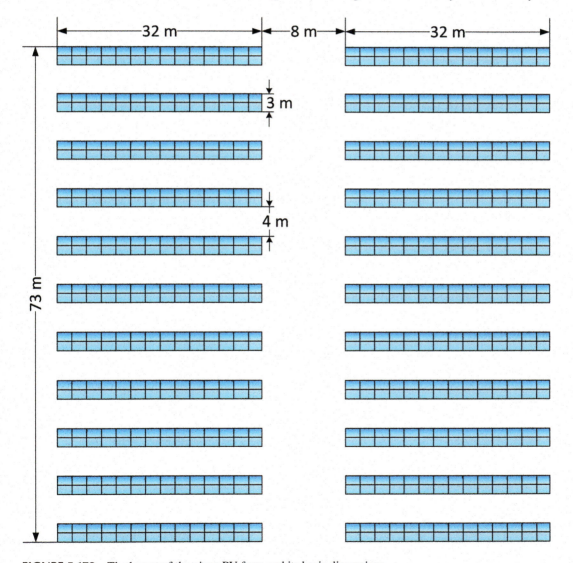

FIGURE 5.178 The layout of the given PV farm and its basic dimensions.

TABLE 5.145
PV Module Data

Rated power	400 Watts
Rated voltage	40.6 V
Rated current	9.86 A
Open circuit voltage	49.3 V
Short circuit current	10.47 A
NOCT	47°C

TABLE 5.146
Household Appliances and Their Daily Working Hours

Electrical Loads	Power	Hours/day	Days/week
Lighting Devices (AC)	100 Watts	5	7
Television (AC)	500 Watts	4	5
Laptop (AC)	50 Watts	3	4
Electric Cooker (AC)	500 Watts	1	7
Refrigerator (AC)	1200 Watts $\left[\dfrac{ON}{OFF} = \dfrac{1}{2}\right]$	24	7
Washing Machine (AC)	2000 Watts	1	1
Microwave Oven (AC)	800 Watts	1/4	3
Other Equipment (AC)	1000 Watts	1/3	3

(DoA) in the winter months are determined as 4, and the PSH value is given as 5.5 hours/day. Note that the charge-discharge efficiency of the batteries is given as 80% and the inverter efficiency as 90%. In addition, assume that $V_{DC} = 48$ volts, f_{Temp-b} is 0.65, and DoD_{max} is 50%.

a. Calculate the daily load demand in (Ah/day).
b. Determine the required number of batteries and their connection type.
c. Determine the total number of PV modules and their connection type required.
d. Determine the inverter-rated power required.

BIBLIOGRAPHY

1. R. Solar and M. Development, "Solar PV Training & Referral Manual" SNV-Stichting Nederlandse Vrijwilligers, Zimbabwe, 2015.
2. SIEMENS, "Standard-Compliant Components for Photovoltaic Systems", Siemens Catalog, Germany, 2012.
3. Eaton, "Solar circuit protection application guide: Complete and reliable solar circuit protection", *Eat. Prod. Cat.*, Dublin, Ireland, 2016.
4. A. Al-khazzar, "The required land area for installing a photovoltaic power plant", *Iran. J. Energy Environ.*, vol. 8, no.1, pp. 11–17, 2017.
5. A. A. Khamisani, "Design Methodology of Off-Grid PV Solar Powered System (A Case Study of Solar Powered Bus Shelter)", pp. 1–23, 2018.
6. D. F. Al Riza and S. I. U. H. Gilani, "Standalone photovoltaic system sizing using peak sun hour method and evaluation by TRNSYS simulation", *Int. J. Renew. Energy Res.*, vol. 4, no. 1, pp. 109–114, 2014.
7. M. S. S. Basyoni, M. S. S. Basyoni, and K. Al-Dhlan, "Design, sizing and implementation of a PV system for powering a living room", *Int. J. Eng. Res. Sci.*, vol. 3, no. 5, pp. 64–68, 2017.
8. A. N. Al-shamani et al., "Design & sizing of stand-alone solar power systems a house Iraq", *Recent Adv. Renew. Energy Sources*, pp. 145–150, 2013.

9. A. Phayomhom, S. Sirisumrannukul, T. Kasirawat, and A. Puttarach, "Safety design planning of ground grid for outdoor substations in MEA's power distribution system", *ECTI Trans. Electr. Eng. Electron. Commun.*, vol. 9, no. 1, pp. 102–112, 2011.
10. B. Teitelbaum, "Off-grid system design", *Solar*, vol. 20, pp. 7–17, 2017.
11. J. Gwamuri and S. Mhlanga, "Design of PV solar home system for use in urban Zimbabwe", *Plenary Pape. Sess. I Water*, 2012.
12. A. Abu-Jasser, "A stand-alone photovoltaic system, case study: A residence in Gaza", *J. Appl. Sci. Environ. Sanit.*, vol. 5, no. 1, pp. 81–91, 2010.
13. J. Gravelle and P. E. E. Ramirez-Bettoni, "Substation Grounding Tutorial", Minnesota Power Systems Conference, Minnesota, 2017.
14. A. N. Madkor, W. R. Anis, and I. Hafez, "The effect of numbers of inverters in photovoltaic grid connected system on efficiency reliability and cost", *Int. J. Sci. Technol. Res.*, vol. 4, no. 9, pp. 99–107, 2015.
15. H. N. Amadi, "Design of grounding system for A. C. substations with critical consideration of the mesh, touch and step potentials", *Eur. J. Eng. Technol.*, vol. 5, no. 4, pp. 44–57, 2017.
16. E. R. Sverko, "Ground measuring techniques: electrode resistance to remote earth & soil resistivity", Erico Inc., USA, 1999.
17. S. Myint and Y. Aung Oo, "Performance analysis of actual step and mesh voltage of substation grounding system with the variation of length and number of ground rod", *Int. J. Sci. Eng. Appl.*, vol. 3, no. 2, pp. 10–17, 2014.
18. V. N. Katsanou and G. K. Papagiannis, "Substation grounding system resistance calculations using a FEM approach", in *2009 IEEE Bucharest Power Tech Innovative Ideas Toward the Electrical Grid of the Future*, Bucharest, Romania, August 2009.
19. L. H. Chen, J. F. Chen, T. J. Liang, and W. I. Wang, "Calculation of ground resistance and step voltage for buried ground rod with insulation lead", *Electr. Power Syst. Res.*, vol. 78, no. 6, pp. 995–1007, 2008.
20. J. C. Wiles, "Photovoltaic System Grounding Prepared by: Solar America Board for Codes and Standards", New Mexico State University, 2012.
21. C. A. Charalambous, N. D. Kokkinos, and N. Christofides, "External lightning protection and grounding in large-scale photovoltaic applications", *IEEE Trans. Electromagn. Compat.*, vol. 56, no. 2, pp. 427–434, 2014.
22. ABB solutions for photovoltaic applications Group, "Technical application papers no. 10. photovoltaic plants", *Tech. Appl. Pap.*, vol. 10, no. 10, p. 107, 2010.
23. T. Izhar, "Techniques to obtain the permissible grounding resistance", *Science International,* vol. 27, no. 6, pp. 6157–6164, 2015.
24. A. Demetriou, D. Buxton, and C. A. Charalambous, "Stray current DC corrosion blind spots inherent to large PV systems fault detection mechanisms: elaboration of a novel concept", *IEEE Trans. Power Deliv.*, vol. 33, no. 1, pp. 3–11, 2018.
25. M. C. Falvo and S. Capparella, "Safety issues in PV systems: Design choices for a secure fault detection and for preventing fire risk", *Case Stud. Fire Saf.*, vol. 3, pp. 1–16, 2015.
26. M. Kazanbas, L. Menezes, and P. Zacharias, "Considerations on grounding possibilities of Transformerless grid-connected photovoltaic inverters", in *2012 IEEE International Energy Conference and Exhibition (ENERGYCON 2012)*, pp. 1–6, Florence, Italy, 2012.
27. R. Schaerer and D. Lewis, "Large utility-scale photovoltaic solar power plant grounding system safety design - General practices and guidance", in *IEEE Power & Energy Society General Meeting*, vol. 2015-September, Denver, Colorado, 2015.
28. C. A. Christodoulou, V. T. Kontargyri, A. C. Kyritsis, K. Damianaki, N. P. Papanikolaou, and I. F. Gonos, "Lightning Performance Study for Photovoltaic Systems", 2015.
29. D. S. Pillai and N. Rajasekar, "A comprehensive review on protection challenges and fault diagnosis in PV systems", *Renew. Sustain. Energy Rev.*, vol. 91, no. March, pp. 18–40, 2018.
30. S. C. Malanda, I. E. Davidson, E. Singh, and E. Buraimoh, "Analysis of Soil Resistivity and its Impact on Grounding Systems Design", in *2018 IEEE PES/IAS Power Africa, 2018*, pp. 324–329, Cape Town, South Africa, 2018.
31. B. Brooks, "Getting Grounded in Correct Grounding", National Electric Code, 2017.
32. M. Nassereddine, K. Ali, and C. Nohra, "Photovoltaic solar farm: Earthing system design for cost reduction and system compliance", *Int. J. Electr. Comput. Eng.*, vol. 10, no. 3, pp. 2884–2893, 2020.
33. J. Huang, H. Li, Y. Sun, H. Wang, and H. Yang, "Investigation on potential-induced degradation in a 50MWp crystalline silicon photovoltaic power plant", *Int. J. Photoenergy*, vol. 2018, 2018.

34. W. I. Bower and J. C. Wiles, "Analysis of grounded and ungrounded photovoltaic systems", *Proceedings of 1994 IEEE 1st World Conference on Photovoltaic Energy Conversion-WCPEC (A Joint Conference of PVSC, PVSEC and PSEC)*, vol. 1, pp. 809–812, Hawaii, USA, 1994.
35. A. S. Ayub, W. H. Siew, and F. Peer, "Grounding strategies for solar PV panels", in *2018 IEEE International Symposium on Electromagnetic Compatibility and 2018 IEEE Asia-Pacific Symposium on Electromagnetic Compatibility (EMC/APEMC)*, pp. 418–422, Singapore, 2018.
36. J. Šlamberger, M. Schwark, B. B. Van Aken, and P. Virtič, "Comparison of potential-induced degradation (PID) of n-type and p-type silicon solar cells", *Energy*, vol. 161, pp. 266–276, 2018.
37. S. Pingel et al., "Potential induced degradation of solar cells and panels", *2010 35th IEEE Photovoltaic Specialists Conference 2010*, November 2014, pp. 2817–2822, Piscataway, NJ, 2010.
38. W. Luo et al., "Potential-induced degradation in photovoltaic modules: A critical review", *Energy Environ. Sci.*, vol. 10, no. 1, pp. 43–68, 2017.
39. F. P. Mohamed, W. H. Siew, and S. Mahmud, "Effect of group grounding on the potential rise across solar PV panels during lightning strike", in *2019 11th Asia-Pacific International Conference on Lightning (APL 2019)*, pp. 5–9, Hong Kong, China, 2019.
40. W. Luo et al., "Potential-induced degradation in photovoltaic modules: A critical review", *Energy Environ. Sci.*, vol. 10, no. 1, pp. 43–68, 2017.
41. T. C. Chen, T. W. Kuo, Y. L. Lin, C. H. Ku, Z. P. Yang, and I. S. Yu, "Enhancement for potential-induced degradation resistance of crystalline silicon solar cells via anti-reflection coating by industrial PECVD methods", *Coatings*, vol. 8, no. 12, pp. 1–12, 2018.
42. SMA Solar Technology AG, "Potential Induced Degradation", pp. 1–4, 2011.
43. Z. G. Datsios and P. N. Mikropoulos, "Safe grounding system design for a photovoltaic power station", in *8th Mediterranean Conference on Power Generation, Transmission, Distribution and Energy Conversion (MEDPOWER 2012)*, vol. 2012, no. 613 CP, Cagliari, Italy, 2012.
44. N. Kokkinos, N. Christofides, and C. Charalambous, "Lightning protection practice for large-extended photovoltaic installations", in *2012 31st International Conference on Lightning Protection (ICLP 2012)*, Vienna, Austria, 2012.
45. M. Nassereddine and A. Hellany, "Designing a lightning protection system using the rolling sphere method", in *2009 Second International Conference on Computer and Electrical Engineering (ICCEE 2009)*, vol. 1, pp. 502–506, Dubai, UAE, 2009.
46. M. Zuercher, R. Schlesinger, "Protecting Electrical PV Systems from the Effects of Lightning", Solectria Renewables, pp. 1–6, USA, 2012.
47. N. Fallah, C. Gomes, M. Z. A. Ab Kadir, G. Nourirad, M. Baojahmadi, and R. J. Ahmed, "Lightning protection techniques for roof-top PV systems", in *2013 IEEE 7th International Power Engineering and Optimization Conference (PEOCO 2013)*, pp. 417–421, Langkawi, Malaysia, June 2013
48. F. Q. Alenezi, J. K. Sykulski, and M. Rotaru, "Grid-connected photovoltaic module and array sizing based on an iterative approach", *Int. J. Smart Grid Clean Energy*, vol. 3, no. 2, pp. 247–254, 2014.
49. E. M. Zaw, "External lightning protection and earthing system design of PV plant in Thilawa special economic zone", *Peer-Reviewed, Ref. Index. J. with IC Value*, vol. 4, no. 12, pp. 56–63, 2018.
50. P. Y. Okyere and G. Eduful, "Evaluation of rolling sphere method using leader potential concept: A case study", *IJME-Intertech*, pp. 1–20, 2006.
51. J. Crispino, "Rolling spheres method for lightning protection", *Complexity*, pp. 1–25, 2007.
52. I. Balouktsis and T. D. Karapantsios, "Methodology for load matching and optimization of directly coupled PV pumping systems", *7th WSEAS International Conference on Electric Power Systems, High Voltages, Electric Machines (POWER '08)*, pp. 227–232, Venice, Italy, November 2007
53. C. E. Council, "International conference power generation and sustainable development", *Sol. Pv Syst.*, vol. 2013, pp. 233–238, 2001.
54. A. Andersson, "Battery Energy Storage Systems in Sweden: A National Market Analysis and a Case Study of Behrn Sport Arena", p. 9, 2018.
55. J. C. Sivertsen, "Design and Installation of a Grid-Connected PV System", Master's Thesis, University of Agder, 2014.
56. L. Khouzam, "A New Approach to a Analyze the Matching of", pp. 733–738, 1991.
57. A. Balouktsis, T. D. Karapantsios, K. Anastasiou, A. Antoniadis, and I. Balouktsis, "Load matching in a direct-coupled photovoltaic system-application to Thevenin's equivalent loads", *Int. J. Photoenergy*, pp. 1–8, vol. 2006, 2006.
58. J. A. Azzolini and M. Tao, "A control strategy for improved efficiency in direct-coupled photovoltaic systems through load management", *Appl. Energy*, vol. 231, pp. 926–936, 2018.

59. O. H. O. F. Direct-coupled, E. L. To, P. Generator, K. Khouzam, and L. Khouzam, "Optimum matching of direct-coupled electronechanical loads to a photovolataic generator", Solar Energy, vol. 8, no. 3, pp. 343–349, 1993.
60. M. Abu-Aligah, "Design of photovoltaic water pumping system and compare it with diesel powered pump", *Jordan J. Mech. Ind. Eng.*, vol. 5, no. 3, pp. 273–280, 2011.
61. K. Khouzam, L. Khouzam, and P. Groumpos, "Optimum matching of ohmic loads to the photovoltaic array", *Sol. Energy*, vol. 46, no. 2, pp. 101–108, 1991.
62. K. Y. Khouzam, "The load matching approach to sizing photovoltaic systems with short-term energy storage", *Sol. Energy*, vol. 53, no. 5, pp. 403–409, 1994.
63. D. Johnson, "Cables and connectors", *Control Eng.*, vol. 55, no. 10, pp. 85–88, 2008.
64. Legrand, "Sizing Conductors and Selecting Protection Devices", Power Guide Book, Netherlands, 2009.
65. I. Photovoltaic, "DC - Distribution and Protection Components", pp. 11–56, 2011.
66. C. A. D. Mosheer, "Optimal solar cable selection for photovoltaic systems", *Int. J. Renew. Energy Resour. (formerly Int. J. Renew. Energy Res.)*, vol. 5, no. 2, pp. 28–37, 2016.
67. S. H. Alwan, J. Jasni, M. Z. A. Ab Kadir, and N. Aziz, "Factors affecting current ratings for underground and air cables", *World Acad. Sci. Eng. Technol. Int. J. Electr. Comput. Energ. Electron. Commun. Eng.*, vol. 10, no. 11, pp. 1422–1428, 2016.
68. F. O. Electrical, "Cable Sizing Calculation", pp. 1–10, 2016.
69. B. Dana, "Cable selection in small off-grid solar energy installations", *Quetsol*, vol. 5, pp. 1–9, 2012.
70. L. R. Maillo, "Application Note Economic Cable Sizing in PV Systems: Case Study", Euripian Copper Institute and Leonarda Energy, no. cu0167, Italy, 2017.
71. Eland Cables, "Photovoltaic solar cable PV1-F", Eland Cables, vol. 44, no. 1, pp. 1–2, 2008.
72. E. T. El Shenawy, A. H. Hegazy, and M. Abdellatef, "Design and optimization of stand-alone PV system for Egyptian rural communities", *Int. J. Appl. Eng. Res.*, vol. 12, no. 20, pp. 10433–10446, 2017.
73. H. A. Hamza, Y. M. Auwal, and M. I. Sharpson, "Standalone PV system design and sizing for a household in Gombe, Nigeria", *Int. J. Interdiscip. Res. Innov.*, vol. 6, no. 1, pp. 96–101, 2018.
74. P. Mohanty, K. R. Sharma, M. Gujar, M. Kolhe, and A. N. Azmi, "PV system design for off-grid applications", Springer International Publishing, pp. 49–83, vol. 196, 2016.
75. J. F. Wansah, A. Appl, and S. Res, "Sizing a stand-alone solar photovoltaic system for remote homes at Bakassi Peninsula", *Advance in Applied Science Research*, vol. 6, no. 2, pp. 20–28, 2015.
76. W. Feng and Z. M. Slameh, "Off-grid photovoltaic system design for Haiti school project", *J. Power Energy Eng.*, vol. 02, no. 11, pp. 24–32, 2014.
77. M. Leach and R. Oduro, "Preliminary Design and Analysis of a Proposed Solar and Battery Electric Cooking Concept: Costs and Pricing", United Kingdom Government, pp. 26–37, UK, 2016.
78. S. Report, "Photovoltaic Power Systems and the National Electrical Code: Suggested Practices", New Mexico State University, New Mexico, 1996.
79. Y. M. Irwan, Z. Syafiqah, A. R. Amelia, M. Irwanto, W. Z. Leow, and S. Ibrahim, "Design the balance of system of photovoltaic for low load application", *Indones. J. Electr. Eng. Comput. Sci.*, vol. 4, no. 2, pp. 279–285, 2016.
80. M. S. Jadin, I. Z. M. Nasiri, S. E. Sabri, and R. Ishak, "A sizing tool for PV standalone system", *ARPN J. Eng. Appl. Sci.*, vol. 10, no. 22, pp. 10727–10732, 2015.
81. P. Manimekalai, R. Harikumar, and S. Raghavan, "An overview of batteries for photovoltaic (PV) systems", *Int. J. Comput. Appl.*, vol. 82, no. 12, pp. 28–32, 2013.
82. T. M. Ndagijimana and B. Kunjithapathan, "Design and implementation PV energy system for electrification rural areas", *Int. J. Eng. Adv. Technol.*, vol. 8, no. 5, pp. 2340–2352, 2019.
83. J. C. Hernández, P. G. Vidal, and F. Jurado, "Lightning and surge protection in photovoltaic installations", *IEEE Trans. Power Deliv.*, vol. 23, no. 4, pp. 1961–1971, 2008.
84. J. M. Doyle, "Surge protection", *Aviat. Week Sp. Technol. (New York)*, vol. 164, no. 9, p. 29, 2006.
85. CITEL company, "Surge Protection for Photovoltaic Systems", pp. 1–20, 2013.
86. O. Surge and P. Technical, "Overvoltage Surge Protection - Technical Note, Europe and APAC", vol. 2, pp. 1–7, 2017.
87. I. Note, E. Networks, A. Code, T. Rules, and E. R. Authority, "Technical Rules for the South West Interconnected Network", Western Power, Australia, 2016.
88. ABB, "MV/LV Transformer Substations: Theory and Examples of Short-Circuit MV / LV Transformer Substations: Theory and Examples of Short-Circuit Calculation", Technical Application Paper, Germany, 2005
89. K. Bataineh and D. Dalalah, "Optimal configuration for design of stand-alone PV system", *Smart Grid Renew. Energy*, vol. 03, no. 02, pp. 139–147, 2012.

90. P. CONTACT, "Lightning and Surge Protection for Photovoltaic Systems", pp. 1–20, 2001.
91. V. Lackovic, "Introduction to short circuit current calculations", *CED Engineering Course Notes*, vol. 490, no. 877, pp. 1–46, 2015.
92. ABB, "Technical guide The MV/LV Transformer Substations (Passive Users)", Technical Guide. MV/LV Transformer Substations (Passive Users), p. 85, 2015.
93. Schneider Electric, "Short-circuit power", MT Parten., vol. B12, 2004.
94. International Electro Technical Commission, "Technical Report IEC 60909-1 Second Edition-Short-Circuit Currents in Three-Phase A.C. Systems", vol. 2, 2002.
95. DIgSilent, "Short-Circuit Calculation with Full-size Converters according to IEC 60909", Comparison Report, DIgSILENT GmbH, Germany, 2013.
96. J. M. Gers and E. J. Holmes, "Calculation of short-circuit currents", *Prot. Electr. Distrib. Networks*, Schneider-Electric, no. 158, pp. 11–30, Germany, 2011.
97. D. Sweeting, "Applying IEC 60909, short-circuit current calculations", in *2011 Record of Conference Papers Industry Applications Society 58th Annual IEEE Petroleum and Chemical Industry Conference (PCIC) 2011*, pp. 1–6, Pennsylvania, 2011.
98. D-A-CH-CZ_Kompendium, "Technical Rules for the Assessment of Network Disturbances", 2012.
99. ABB, *"Distribution Automation Handbook"*, ABB Technical Papers, p. 120, Germany, 2013.
100. M. J. Thompson and D. Wilson, "Auxiliary DC control power system design for substations", in *2007 60th Annual Conference for Protective Relay Engineers*, pp. 522–533, Atlanta, Georgia, March 2007
101. P. Dolan, "Auxilary D.C. Power Supplies for Primary & Secondary Substations", no. 9, pp. 1–40, 2017.
102. I. E. C. Short and C. Analysis, "Short Circuit Current Calculation in Three-Phase AC Systems", pp. 1–18, 2019.
103. L. G. Brazier and A. L. Williams, "Transmission and distribution lines and cables", *J. Inst. Electr. Eng.*, vol. 8, no. 88, pp. 213–214, 1962.
104. S. Nikolovski, V. Papuga, and G. Knežević, "Relay protection coordination for photovoltaic power plant connected on distribution network", *Int. J. Electr. Comput. Eng. Syst.*, vol. 5, no. 1, pp. 15–20, 2014.
105. U. U. Uma, "Overcurrent relay setting model for effective substation relay coordination", *IOSR J. Eng.*, vol. 4, no. 5, pp. 26–31, 2014.
106. Schneider Electric Network Protection & Automation Guide, "Overcurrent Protection for Phase and Earth faults Network Protection & Automation Guide", 2014.
107. R. Leelaruji and L. Vanfretti, "Power System Protective Relaying: basic concepts, industrial-grade devices, and communication mechanisms." *Intern. Rep.*, p. 35, 2011.
108. L. Namangolwa and E. Begumisa, "Impacts of Solar Photovoltaic on the Protection System of Distribution Networks: A case of the CIGRE low voltage network and a typical medium voltage distribution network in Sweden", p. 103, 2016.
109. S. Gooding, K. Harley, and G. Antonova, "Considerations for connecting photovoltaic solar plants to distribution feeders", in *2016 69th Annual Conference for Protective Relay Engineers (CPRE 2016), Texas, USA*, 2017.
110. M. V. ABB Products, "Technical guide Protection criteria for medium voltage networks", ABB S.p.A. Unità Operativa Sace-MV, p. 52, 2016.
111. K. A. Kumar and J. Eichner, "Guidance on Proper Residual Current Device Selection for Solar Inverters", pp. 1–5, 2013.
112. Siemens, "Residual Current Protective Devices", 2010.
113. M. Guillot, "Safe and Reliable Photovoltaic Energy Generation", pp. 1–44, 2012.
114. J. Schonek, "Residual current devices in LV", *Cah. Tech. Schneider Electr.* no. 114, Germany, 2006.
115. S. Kebaïli et al., "Sizing a stand-alone solar photovoltaic system for remote homes at Bakassi Peninsula", *Indones. J. Electr. Eng. Comput. Sci.*, vol. 6, no. 2, pp. 1–8, 2016.
116. A. Kingsley, G. Quansah, and A. Bernard, "Stand-alone solar PV system sizing for critical medical equipment to meet an emergency energy demands in Ghana: Tanoso Community Hospital in Kumasi", vol. 13, no. 5, pp. 53–61, 2018.
117. S. Kebaïli and H. Benalla, "Optimal sizing of a stand-alone photovoltaic systems under various weather conditions in Algeria", *Rev. des Energies Renouvelables*, vol. 18, no. 2, pp. 179–191, 2015.
118. M. Ishaq and U. H. Ibrahim, "Design of an off grid photovoltaic system: A case study of Government Technical College, Wudil, Kano State", *Int. J. Sci. Technol. Res.*, vol. 2, no. 12, pp. 175–181, 2013.
119. S. Tenaga, "Residual Current Devices (RCD) in electrical installation", in *National Conference on Electrical Safety*, Kuala Lumpur, Malaysia, July 2011, pp. 21–22.

6 Photovoltaic System Applications

6.1 INTRODUCTION

Photovoltaic panels have a wide range of applications, particularly for stand-alone and mobile systems. Common application areas of PV systems are given below:

- **Warning Signals:** Warning signals are frequently used for property, life & public safety. For example, the main usage areas are traffic warning signs, fog signals, railroad signals, aircraft warning lights, navigation indicators, etc.
- **Lighting Systems:** Due to their high reliability and decreasing costs, the use of PV-powered lighting systems has increased rapidly in recent years. For instance, the main usage areas are billboards, security lighting, emergency warning, area lighting, road/street lighting, residential use, etc.
- **Telecommunication Systems:** Telecommunication systems have to work with almost 100% availability, especially in our globalized world. Therefore, high-reliability power system applications for telecommunication systems are critical. In addition, most communication systems are located in remote areas where they are often exposed to adverse weather conditions. For these reasons, PV-powered telecommunication systems have increased rapidly during the last few decades. The main uses of photovoltaics in telecommunication systems are microwave repeaters, base transmission stations, radio stations, telephone exchanges, satellite earth stations, etc.
- **Residential:** Especially people living in remote areas often prefer PV systems for their domestic use as PV systems are becoming more reliable and economically viable. Typical residential applications of PVs are residences, cottages, and recreational vehicles.
- **Water Pumping**: Especially in rural areas, PV-powered water pumping applications are often preferred as they offer simplicity, reliability, and low maintenance. Therefore, PV systems are applied on a wide scale, from hand pumps to large-scale irrigation pumps. Typical water pumping applications of photovoltaics are agricultural irrigation, remote area water supply, water stocking, and domestic water supply.
- **Other Applications**: The application areas of photovoltaics are indeed not limited to those given above. Some of the other implementation areas are refrigeration, remote-sensing/monitoring, direct-drive supply systems, cathodic protection, vehicles/aircraft/satellites, and military applications.

Systematic methods and approaches for designing and sizing PV power systems are described in detail in Chapter 5. As a general design approach, we can determine the following steps to design a PV system.

- Step 1: Determine the load requirements
- Step 2: Determine the design current and tilt angle of the PV array
- Step 3: Determine the storage requirements
- Step 4: Determine the PV array requirements
- Step 5: Determine the power conditioning unit requirements
- Step 6: Determine the protection requirements and select the components
- Step 7: Determine the system wire sizes

This chapter only focuses on specific application areas listed above and shows how to adapt the previously described methodologies to these areas. Given that, the aforementioned applications are examined in detail in the following sections.

6.2 WARNING SIGNALS

The warning signals and the use of photovoltaics in warning signaling are becoming very diverse. For example, PV-powered warning systems are often preferred in many areas such as maritime, navy & coast guard, oil industry, highways, railways, communication towers, high buildings, and piers. The electrical loads of warning systems are usually lamps or sirens and are typically powered by 12 V_{DC}.

Today's modern warning systems include the battery & load energy management system (system control unit). However, it is also possible to design a PV-powered direct-drive warning system without having a smart controller unit. For the applications without a control unit, the battery storage capacity should be chosen to be large enough relative to the PV array size so that the probability of overcharging the battery can be kept low. A low-voltage disconnect switch should also be used to protect the battery from over-discharging. In addition, a blocking diode system is required to protect the PV modules from the battery discharge currents during the night times. Finally, it must also be ensured that the PV module has sufficient voltage to fully charge the battery unit. The criteria explained above must be taken into account when designing a controller-free direct-drive warning system. The design and sizing of the warning system are explained step by step with worked examples given below.

Example 6.1: PV-Powered 24-Hour Road Warning System

A 24-hour road warning system will be designed to alert motor vehicle drivers. The design criteria of the warning system are listed below.

- The road warning system will be a PV-powered stand-alone system.
- The system will run continuously 24 hours a day, all year round.
- The PSH (peak sun hours) value of the region is given as 5 h/day at the given tilt and azimuth of the site.
- The light source consists of a LED bulb operating 0.5 seconds on and 0.5 seconds off.
- The LED bulb draws 1.5 amps while it is on. The designed system voltage is 12 V_{DC}.
- The highest DoA (Days of Autonomy) value is 10 for the given region. Assume that the average ambient temperature during DoA days is −10°C.
- 12 V and 100 Ah lead-acid batteries will be used for energy storage needs. The provided temperature correction curve for the lead-acid battery is given in Figure 6.1. The maximum DoD (Depth of Discharge) value of the battery is 50%.
- The system will include a charge controller and the combined round-trip efficiency for the selected battery & charger is 90%.
- The PV module specifications to be used are given in Table 6.1.

Calculate the load demand, battery storage & array size, as well as DC cable size. In addition, determine the protection requirements and draw the complete PV system diagram.

Solution

The design of any electrical power system always starts with determining the load requirements. So, let's determine the load requirements in the first step, the battery storage requirements in the second step, and the PV array requirements in the third step. Afterward, let's specify the protection & system components and size of the DC cable. We can draw the wiring diagram of the final design as the last step.

Photovoltaic System Applications

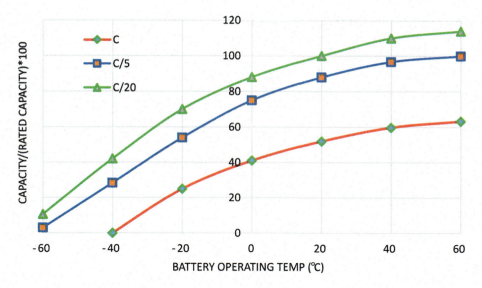

FIGURE 6.1 Variation of the lead-acid battery capacity versus temperature.

TABLE 6.1
PV Module Data

Rated Power	Rated Voltage	Rated Current	Open Circuit Voltage	Short-Circuit Current	NOCT
63 Watts	14.15 volts	4.45 Amps	17.85 volts	4.75 Amps	50°C

Step 1 – Determination of Load Requirements: There are only DC loads in the system. The weekly and daily energy demands can be calculated from equation (5.180) as follows (Note that the load power is $12 \times 1.5 = 18$ W, and the duty cycle is $\dfrac{0.5}{0.5+0.5} = 1/2$).

$$\sum [\text{Wh/week}] = \left(18 \times 24 \times 7 \times \frac{1}{2}\right) = 1512\,[\text{Wh/week}]$$

Hence from (5.183), the energy demand in Ah/day can be calculated as below.

$$\frac{\sum [\text{Wh/week}]}{V_b} = \frac{1512}{12} = 126\,[\text{Ah/week}]$$

$$\frac{[\text{Ah/week}]}{7} = \frac{126}{7} = 18\,[\text{Ah/day}]$$

Since the round-trip efficiency is given as 0.90, the daily total electrical energy demand (Ah/day) can be calculated with the help of equation (5.182) as below:

$$\text{Total (Corrected) Energy Demand}\left(\frac{\text{Ah}}{\text{day}}\right) = \frac{\text{Energy Demand}\left(\dfrac{\text{Ah}}{\text{day}}\right)^{\text{DC}}}{\eta_{\text{conv}}} = \frac{18\left(\dfrac{\text{Ah}}{\text{day}}\right)^{\text{DC}}}{0.90} = 20\,\text{Ah/day}$$

Step 2 – Determination of Storage Requirements: The battery requirements can be calculated using the equations from (5.42) to (5.45) as follows (from Figure 6.1, the $f_{\text{Temp-b}}$ value for the C/20 curve is approximately 70% at −10°C).

$$C_{\text{Battery-net}}\,[\text{Ah}] = \frac{E_{\text{Daily Avg}}\,[\text{Wh/day}] \times \text{DoA}\,[\text{day}]}{\text{DoD}_{\max} \times V_b \times \eta_b \times f_{\text{Temp-b}}} = \frac{\dfrac{1512\,[\text{Wh/week}]}{7} \times 10}{0.50 \times 12 \times 0.90 \times 0.70} \cong 572\;\text{Ah}$$

$$N_{\text{Battery}} = \frac{C_{\text{Battery-net}}\,[\text{Ah}]}{c_b\,[\text{Ah}]} = \frac{572\,[\text{Ah}]}{100\,[\text{Ah}]} \cong 6$$

$$N_{\text{sb}} = \frac{12}{12} = 1$$

$$N_{\text{pb}} = \frac{6}{1} = 6$$

As a result, the battery storage configuration is $(1s \times 6p)$, and accordingly, a total of six batteries are needed.

Step 3 – Determination of PV Array Requirements: Let us follow the steps regarding the PV array sizing for stand-alone PV systems (described in Chapter 5). From equations (5.8) to (5.17), PV array sizing can be performed as follows. Note that PV system losses are not given. Hence, typical values are used in this example ($f_{\text{Manufacturing}}$: 0.98, f_{Temp} : 0.94, f_{Soiling} : 0.95 $\Rightarrow$ PV Array Losses $\Delta P_{\text{PV}} = 0.98 \times 0.94 \times 0.95 = 0.875$).

$$E_{\text{overall}} = \frac{E_{\text{Load}}\,[\text{Wh/day}]}{(\Delta P_{\text{PV}}) \cdot \eta_{\text{overall}}} = \frac{\dfrac{1512\,[\text{Wh/week}]}{7}}{(0.875) \times (0.90)} = \frac{216\,[\text{Wh/day}]}{0.7875} \cong 275\;\text{Wh/day}$$

$$P_{\text{Peak(PV)}} = \frac{E_{\text{overall}}}{\text{PSH}_{\text{avg}}} = \frac{275\,\text{Wh/day}}{5\,\text{h/day}} = 55\,\text{W}_p$$

$$I_{\text{DC}} = \frac{P_{\text{Peak(PV)}}}{V_{\text{DC}}} = \frac{55\,\text{W}}{12\,\text{V}} \approx 4.58\;\text{A}$$

$$N_s = \frac{V_{\text{DC}}}{V_{\text{mp}}} = \frac{12}{14.15} = 0.85 \approx 1$$

$$N_p = \frac{I_{\text{DC}}}{I_{\text{mp}}} = \frac{4.58}{4.45} = 1.02 \approx 1$$

N_p value can be taken as "1" since the calculated design current and PV module's I_{mp} current is very close to each other. Consequently, only a single 63-watt PV module of which specifications are given in Table 6.1, is sufficient to feed the system.

$$N_T = N_s \times N_p = 1s \times 1p$$

Step 4 – Specification of Protection Components and Charger Sizing: According to the design so far, the PV modules will be connected as $1s \times 1p$ and the battery bank configuration will be $1s \times 6p$ for a 12 V_{DC} system. As explained in Chapter 5, the charge controllers are sized for the voltage and current requirements of the system. From the design, the maximum total PV array current can be as high as:

Photovoltaic System Applications

$$I_T = N_p \times I_{SC} \times f_{safe} = 1 \times 4.45 \times 1.25 = 5.5625 \text{ A}$$

Accordingly, a 12 V/75 W_p solar charger with a rated charging current of 5 A and a maximum charging current of 6.25 A will be suitable for the given example. It should also be ensured that the input voltage range of the charger matches the voltage range of the PV module.

Protection components are required to be specified for each circuit in the system. Generally, it is recommended to put protection devices at the array output, between charger and load, at the charger input, and on the battery circuit (i.e. battery input and output). Accordingly, the protection components specification for the given example is summarized in Table 6.2.

Step 5 – DC Wire Sizing: Cable sizing and selection criteria have already been explained in Chapter 5. Therefore, in this example, only the DC wire sizing results are given, as summarized in Table 6.3.

Considering the design criteria determined above, the system wiring diagram can be drawn as in Figure 6.2.

6.3 LIGHTING SYSTEMS

Recent improvements in the efficiency of lamps and a decreasing trend in the cost of photovoltaics have led to the application of PV-powered lighting systems in many locations. In these applications, lighting control can be accomplished with different control elements such as photocells, timers, switches, motion sensors, or infrared sensors. Today, pre-packaged off-the-shelf systems containing PV power supply, batteries, ballasts, lamps, and controllers are now commercially available. Hence,

---

TABLE 6.2
Protection Components Specification

General Protection Requirements

Modern chargers usually have the following protection functions. These are:

Reverse Polarity Protection: It is protected against reverse polarity of both PV and battery, so the charger will not be damaged until the polarities are corrected. Note that it is protected against the reverse polarity of both PV and battery. In other words, the charger will not be damaged if the PV output voltage matches the charger voltage range. Otherwise, the charger may be damaged.

Night Times Reverse Charging Protection: It prevents the battery from being discharged through the solar PV module at night hours.

Battery protection: Provides protection against overvoltage, over-discharge, and overheating.

Load Protection: Disconnects the load from the circuit in case of short-circuit at the load terminals. Additionally, when the load power exceeds its rated value, the charge controller automatically cuts off the output after a time delay.

Protection Against Overcharging: If the charging current of the PV array exceeds the controller-rated current, it will limit the charging current at its rated current.

ATTENTION: If the solar charger does not have the protection functions explained above, the system can be protected with the following protection devices.

Protected Circuit	Protection Device, Specification, & Description[a]
Array Output	• Blocking diode for reverse charging protection (>6 A, >20 V). Alternatively, a normally closed relay (>5.56 A, 20 V) can be used for reverse charging protection
	• Fused disconnect switch for overcurrent and short-circuit protection (>5.56 A, >20 V)
	• Surge protector between array and charger (6 A, >20 V)
Charger to Load	• Fuse or fused switch for load protection (1.8 A, ≥15 V)
Charger to Battery	• Fuse or fused switch for battery protection (6 A, ≥15 V)

[a] See Chapter 5 for details on sizing and selection of protection elements.

TABLE 6.3
DC Wire Sizing

Wire Section	System Voltage	Max. Current	Cable Length	Max. Voltage Drop	Wire Size
Array to Charger	12 V_{DC}	1.25 × 4.75 = 5.94 A	1.5 m	2%	1.5 mm²
Charger to Battery	12 V_{DC}	~ 6 A	0.5 m	2%	1.5 mm²
Charger to Load	12 V_{DC}	~ 1.8 A	0.5 m	1%	1.5 mm²

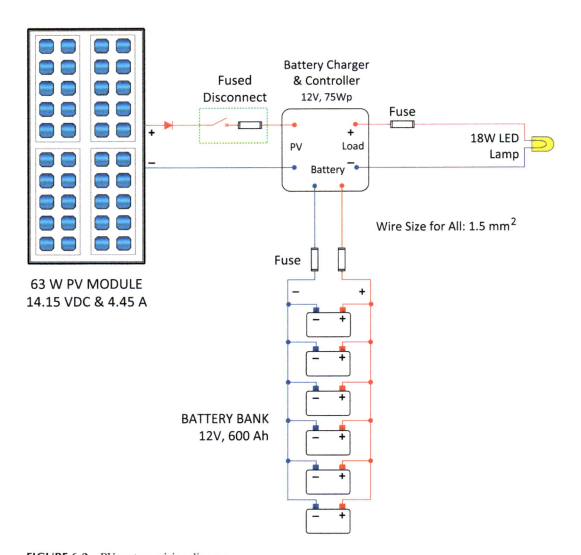

FIGURE 6.2 PV system wiring diagram.

it is possible to find the desired system with quick market research. Most stand-alone PV lighting systems use LED-based luminaires, operating at 12 or 24 V_{DC}. It should be noted that LED-based lighting systems need a well-designed reflector and/or diffuser to focus the available light on the area to be illuminated and reduce the total PV power. For system security, the array frame should be grounded strictly. The PV panel azimuth should be faced to true south in the northern hemisphere and true north in the southern hemisphere with a tilt angle as discussed in Chapter 4.

Photovoltaic System Applications

Example 6.2: PV-Powered Garden Lighting System

A PV-powered stand-alone system will be designed to feed the garden lighting of a boutique hotel in Austin, Texas. The design criteria for the lighting system are listed below.

- The boutique hotel is located at the coordinates 30.26N and 97.73W.
- The boutique hotel is open 7 months a year, only during the tourist season from April 1 to October 31.
- The garden lighting system will run continuously at night times only, from sunset to sunrise.
- The monthly PSH (peak sun hours) values of the site are listed in Table 6.4.
- A total of 20 lighting poles will be used for garden lighting. Each lighting pole has 2 × 20 W LED bulbs.
- The designed system voltage is 24 V_{DC}.
- PV panels will be placed on the roof of the hotel and the roof surface is suitable for placing the panels at an optimum tilt angle.
- The distance between the PV array and the farthest light pole is 30 m.
- The highest DoA (Days of Autonomy) value is assumed to be 3 days for the given region. Assume that the average ambient temperature during DoA days is 20°C.
- 12 V and 200 Ah lead-acid batteries will be used for energy storage needs. The provided temperature correction curve for the lead-acid battery is given in Figure 6.1. The maximum DoD (Depth of Discharge) value of the battery is 50%.
- The system will include a charge controller and the combined round-trip efficiency for the selected battery & charger pair is 90%.
- The PV module specifications are given in Table 6.5.

Design a PV-powered stand-alone system for the garden lighting of this boutique hotel and determine its specifications. Afterward, draw the PV system diagram of the final design.

Solution

It is possible to adapt the previously described design steps to this example as follows.

Step 1 – Determination of Load Requirements: The worst-case scenario should be considered in the design process of a stand-alone PV system. The system is expected to operate at high performance between April 1 and October 31. Therefore, the month with the highest load demand but the lowest insolation should be considered to be the design month. The highest load demand occurs during the longest nights in the given period. As known, the months with the longest nights are April and October in the period of interest. Therefore, it would be appropriate to select the month having the lowest ratio of "solar insolation to load demand" as the design month.

TABLE 6.4
Monthly PSH Values

Months	Jan.	Feb.	March	April	May	June	July	Aug.	Sep.	Oct.	Nov.	Dec.
PSH (h/day)	3.92	4.69	5.88	6.70	6.84	6.90	6.93	7.00	6.01	5.91	4.55	3.98

TABLE 6.5
PV Module Data

Rated Power	Rated Voltage	Rated Current	Open Circuit Voltage	Short-Circuit Current	NOCT
100 Watts	18 volts	5.56 Amps	20.08 volts	6.06 Amps	47°C

The easiest way to find the design month is to compare these ratios on April 15 and October 15. Because October 15 and April 15 reflect the average characteristics of the month they are in, in terms of load demand and solar insolation. The nighttime can be found by subtracting the daylight duration (DD) from 24 hours (ND = 24 − DD). However, for this particular calculation, the declination angles must be determined first. Julian day number, $N = 105$ on April 15, and from equation (2.9), we get:

$$\delta = 23.45 \cdot \sin\left[\frac{360}{365}(105 + 284)\right] = 9.415°$$

Julian day number, $N = 288$ on October 15, and from equation (2.9), we have:

$$\delta = 23.45 \cdot \sin\left[\frac{360}{365}(288 + 284)\right] = -9.6°$$

Since $L = 30.26°$ for the region, DD can be found from 2.38 as below:

$$DD(\text{on April 15}) = \frac{1}{7.5}\cos^{-1}(-\tan 9.415 \times \tan 30.26) = 12.74 \text{ hours}$$

$$DD(\text{on Oct 15}) = \frac{1}{7.5}\cos^{-1}(-\tan(-9.6) \times \tan 30.26) = 11.24 \text{ hours}$$

From the results above, the nighttime durations on the relevant days are found as follows:

$$ND(\text{on April 15}) = 24 - 12.74 = 11.53 \text{ hours}$$

$$ND(\text{on Oct 15}) = 24 - 11.24 = 12.76 \text{ hours}$$

To determine the design month, the worst-case scenario can be found by comparing the "insolation to load" ratios on the respective days.

$$\left[\text{Insolation to Load Ratio (Oct)} = \frac{5.91}{12.76 \times 20 \times 40}\right] < \left[\text{Insolation to Load Ratio (April)} = \frac{6.70}{11.53 \times 20 \times 40}\right]$$

As can be seen, the worst-case scenario occurs in October. For this reason, October is considered as the design month for PV array sizing.

We have DC loads only on 20 lighting poles, each has $2 \times 20\,W$ LED bulbs. The weekly and daily energy demands can be calculated from equation (5.180) as follows.

$$\sum[\text{Wh/week}] = (40 \times 20 \times 12.76 \times 7) = 71456\,[\text{Wh/week}]$$

Hence from (5.183), the energy demand in Ah/day can be calculated as below.

$$\frac{\sum[\text{Wh/week}]}{V_b} = \frac{71456}{12} = 5955\,[\text{Ah/week}]$$

$$\frac{[\text{Ah/week}]}{7} = \frac{5995}{7} = 850.7\,[\text{Ah/day}]$$

Photovoltaic System Applications

Since the round-trip efficiency is given as 0.90, the daily total electrical energy demand (Ah/day) can be calculated with the help of equation (5.182) as below:

$$\text{Total (Corrected) Energy Demand}\left(\frac{Ah}{day}\right) = \frac{\text{Energy Demand}\left(\frac{Ah}{day}\right)^{DC}}{\eta_{conv}} = \frac{850.7\left(\frac{Ah}{day}\right)^{DC}}{0.90} = 945.2 \text{ Ah/day}$$

Step 2 – Determination of Tilt Angle: Although the design month for the PV array is October, the entire period of interest should be taken into account when determining the optimum tilt angle. By following the solution approach in *Example 4.14*, the optimum tilt angles for the given months can be determined as in Table 6.6. Note that tilt angle calculation is not repeated here as it has already been explained in detail in Chapter 4.

Note that periodically tilt angle ($b_{opt(period)} = 10.226°$) is calculated by taking the average of the monthly tilt angles in the relevant period (from April 1 to October 31). In other words, the tilt angle of the stand-alone PV system will be set to 10.226 degrees.

Step 3 – Determination of Storage Requirements: The battery requirements can be calculated using the equations from (5.42) to (5.45) as follows (from Figure 6.1, the f_{Temp-b} value for the C/20 curve is 100% at 20°C).

$$C_{Battery-net}[Ah] = \frac{E_{Daily\,Avg}[Wh/day] \times DoA[day]}{DoD_{max} \times V_b \times \eta_b \times f_{Temp-b}} = \frac{\frac{71456[Wh/week]}{7} \times 3}{0.50 \times 12 \times 0.90 \times 1} \cong 5671 \text{ Ah}$$

$$N_{Battery} = \frac{C_{Battery-net}[Ah]}{c_b[Ah]} = \frac{5671[Ah]}{200[Ah]} \cong 30$$

$$N_{sb} = \frac{V_{DC}}{V_b} = \frac{24}{12} = 2$$

$$N_{pb} = \frac{N_{Battery}}{N_{sb}} = \frac{30}{2} = 15$$

As a result, the battery storage configuration is ($2s \times 15p$), and accordingly, a total of 30 batteries are needed.

Step 4 – Determination of PV Array Requirements: Before going into the *PV array sizing* phase, the given PSH value for the horizontal surface must be converted to its equivalent for a surface tilted at 10.226 degrees. We already know from Step-1 that the design month for PV array

TABLE 6.6
Monthly Tilt Angles for the Latitude 30.26

Months	L = 30.26 $b_{opt(m)}$	L = 30.26 $b_{opt(period)}$
April (30 days)	15.251	10.226
May (31 days)	−1.578	
June (30 days)	−12.834	
July (31 days)	−7.711	
August (31 days)	7.852	
September (30 days)	26.775	
October (31 days)	43.83	

sizing is October. From equation (4.68), let's calculate the insolation conversion factor on October 15, representing the month of October ($L = 30.26$, $\delta = -9.6$, $\omega_{ss} = 84.34$, $b = 10.226$).

$$C_{b(sf)} = \frac{\frac{\pi 84.34}{180}\sin(-9.6)\sin(30.26-10.226)+\cos(-9.6)\cos(30.26-10.226)\sin(84.34)}{\frac{\pi 84.34}{180}\sin(-9.6)\sin(30.26)+\cos(-9.6)\cos(30.26)\sin(84.34)} = 1.00007$$

So PSH at tilt b, $\text{PSH}_b = 1.00007 \times 5.91 = 5.9104$ h/day. Now let us follow the steps regarding the PV array sizing for stand-alone PV systems as described earlier in Chapter 5.

From equations (5.8) to (5.17), PV array sizing can be performed as follows. Note that PV system losses are not given. Hence, typical values are used in this example ($f_{\text{Manufacturing}} : 0.98$, $f_{\text{Temp}} : 0.94$, $f_{\text{Soiling}} : 0.95 \Rightarrow$ PV Array Losses $\Delta P_{\text{PV}} = 098 \times 0.94 \times 0.95 = 0.875$).

$$E_{\text{overall}} = \frac{E_{\text{Load}}[\text{Wh/day}]}{(\Delta P_{\text{PV}}) \cdot \eta_{\text{overall}}} = \frac{\frac{71456[\text{Wh/week}]}{7}}{(0.875) \times (0.90)} = \frac{10208[\text{Wh/day}]}{0.7875} \cong 12963 \text{ Wh/day}$$

$$P_{\text{Peak(PV)}} = \frac{E_{\text{overall}}}{\text{PSH}_{\text{avg}}} = \frac{12963 \text{ Wh/day}}{5.9104 \text{ h/day}} \cong 2193 \text{ W}_p$$

$$I_{\text{DC}} = \frac{P_{\text{Peak(PV)}}}{V_{\text{DC}}} = \frac{2193 \text{ W}}{24 \text{ V}} = 91.375 \text{ A}$$

$$N_s = \frac{V_{\text{DC}}}{V_{\text{mp}}} = \frac{24}{18} = 1.34 \approx 2$$

$$N_p = \frac{I_{\text{DC}}}{I_{\text{mp}}} = \frac{91.375}{5.56} = 16.43 \approx 17$$

Consequently, $2 \times 17 \times 100 \text{ W} = 3400 \text{ W}_p$ PV array is required to feed the system.

$$N_T = N_s \times N_p = 2s \times 17p$$

Step 5 – Specification of Protection Components and Charger Sizing: According to the design so far, the PV modules will be connected as $2s \times 17p$ and the battery bank configuration will be $2s \times 15p$ for a 24 V_{DC} system. The charge controllers should be sized based on the voltage and current requirements of the system. From the design, the maximum total PV array current can be as high as:

$$I_{\max} = N_p \times I_{\text{SC}} \times f_{\text{safe}} = 17 \times 6.06 \times 1.25 = 128.775 \text{ A}$$

The nominal PV system current is $17 \times 5.56 = 94.52$ Amps. Therefore, the rated solar charger current can be selected as 100 A. In addition, maximum PV array voltage can be as high as:

$$V_{\max} = N_s \times V_{\text{OC}} \times f_{\text{safe}} = 2 \times 20.8 \times 1.2 = 48.192 \text{ V}$$

Accordingly, a 24 V/3500 W$_p$ solar charger with a rated charging current of 100 A and a maximum charging current of (>130 A) will be suitable for the given example. It should also be ensured that the input voltage range of the charger matches the voltage range of the PV module. For this example, it is sufficient to select a solar charger with a maximum PV voltage of 50 V. The specification of protection components can be given in Table 6.7.

TABLE 6.7
Protection Components Specification[a]

Protected Circuit	Protection Device, Specification, & Description[b]
Array Output	• Fused disconnect switch for overcurrent and short-circuit protection (≥160 A, ≥50 V) • Surge protector between array and charger (>160 A, >50 V) Note that $I_n \geq 1.56 \times I_{sc}$ (1.56×17×6.06=160.72 A) & $V_n \geq 1.2 \times V_{OC} \times N_s$ (1.2×20.8×2=48.192 V)
Charger to Load	• Fuse or fused switch for load protection (40 A, ≥30 V) Note that $I_{Load} = 800$ W/24 $V_{DC} = 33.34$ A
Charger to Battery	• Fuse or fused switch for battery protection (>177 A, ≥30 V) Note that $I_{Charge} = 3400$ W × 1.25/24 $V_{DC} = 177$ A

[a] If the solar charger does not have the general protection functions specified in Table 6.2, the system can be protected by the indicated protection devices.
[b] See Chapter 5 for details on sizing and selection of protection elements.

TABLE 6.8
DC Wire Sizing[a]

Wire Section	System Voltage	Max. Current	Cable Length	Max. Voltage Drop	Wire Size
Array to Charger	24 V_{DC}	128.275 A	2 m	3%	25 mm² (~12 AWG)
Charger to Battery	24 V_{DC}	~177 A	2 m	2%	35 mm² (~12 AWG)
Charger to Load	24 V_{DC}	~34 A	30 m	3%	4 mm² (~12 AWG)

[a] See Chapter 5 for details on sizing and selection of DC cables.

Step 6 – DC Wire Sizing: The DC wire sizing results can be given in Table 6.8. Considering the design criteria determined above, the system wiring diagram can be drawn as in Figure 6.3.

6.4 TELECOMMUNICATION SYSTEMS

Power system reliability is of great importance for telecommunication systems, which have become an integral part of today's modern technologies. Most telecommunications applications are located in remote areas with limited access and are often subject to adverse weather conditions. Therefore, the applications of PV-powered telecommunication systems are being progressively attractive, specifically for remote applications. The need for energy storage is critical in telecommunications systems where the peak load demand is much greater than the average load demand. For the energy storage needs of these applications, deep-cycle lead-acid batteries and NiCad batteries are often preferred.

To design a PV-powered telecommunication system, their load profiles and electrical supply details are required to be known. Some of the technical details & design criteria are listed below.

- DC bus options for the receivers and transmitters of UHF, VHF, AM, and FM telecommunication systems can be 12, 24, and 48 V_{DC}.
- Load profile depends on the duty cycle and operating mode. Transmitting loads are the greatest loads, and can typically go up to tens of amps, depending on the transmitters' power range. The load currents typically range from a few amps to 10 amps in the receiver modes, while it is less than 3 amps in the standby modes.

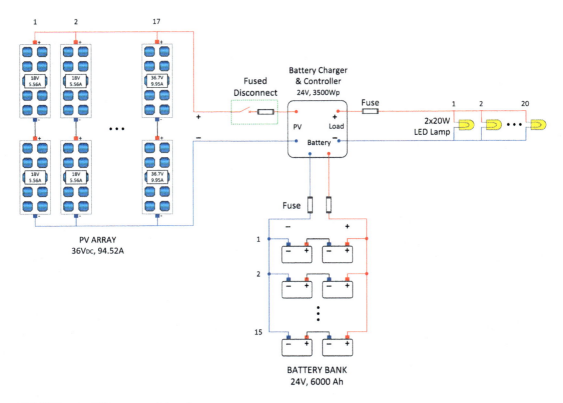

FIGURE 6.3 PV system wiring diagram.

- Other on-site loads such as alarms, ventilation equipment, and lights should also be taken into account when specifying the entire load profile of the telecommunication stations.
- PV arrays feeding telecommunication systems can be mounted on towers, on the roof of communication buildings, or the ground. Adverse weather conditions on the site, such as wind loads, snow accumulation, and icing should also be taken into account when designing the mechanical system.
- In general, batteries with minimum maintenance, low self-consumption rate, low-temperature derating, and high energy density should be selected for telecommunication applications. If non-sealed batteries are used, the battery rooms (or cabins) must be equipped with a ventilation system to evacuate the explosive gases released during charging.
- For the battery packs, including a large number of batteries, all battery strings must be protected against over-currents, for example with a fuse at the output terminal of each parallel string.
- The operating temperature limits of the charge controllers must be compatible with the field temperature conditions. To keep system reliability high, multiple charge controllers connected in parallel can be used instead of a single, high-current charger.

The end-users in cellular wireless access networks connect to base stations via wireless channels. And each base station in the network is also connected to other network elements via wired or wireless connections. There are different generations of wireless access technology, as commonly called 3G, 4G, and 5G. However, components of a base station in the network are similar to each other. A base station includes a baseband unit (BBU), which is fed via an AC-DC or a DC-DC converter. And each BBU consists of one or multiple transceivers containing a radio frequency (RF) unit, a power amplifier (PA), and an antenna. Moreover, the base stations should include a cooling system

Photovoltaic System Applications

as well. Figure 6.4 shows the basic components of a PV-powered base station, and Figure 6.5 shows a simple power flow in the base station.

The energy management & control system monitors the power flow in the system and feeds the required power to the cellular base station according to the following two cases.

Case-1: The PV array feeds the base station during daylight hours when solar energy is sufficient. At the same time, stored energy in the battery bank is used to power the base station during nighttime hours and daylight when solar energy is insufficient.

Case-2: The battery group should be able to supply the shortage of energy during the days of autonomy (DoA) without reaching the maximum depth of discharge (DoD).

Note that an electrical grid or diesel generator can also be used as a backup to power the base station in case the battery bank loses its ability to meet the load demand.

The percentage power consumption of components in a 4G base station is typically in the range of 8%–10% for the cooling system, 12%–14% for the BBU, 5%–7% for the RF unit, 5%–7% for the DC-DC converter, 7%–9% for the main power supply, 55%–60% for the PA unit. Base stations have different modes of operation and their power consumption change over time. For example, a base station can be either on-state or off-state. On the other hand, there are different traffic modes of operation (such as low traffic and heavy traffic mode) under the on-state mode. Besides, there may be transition modes between different operating states. For instance, a base station needs a certain amount of time and energy to be fully powered up. Hence, the typical power consumption modes of a base station can be illustrated as in Figure 6.6.

Accurate power consumption estimation for a base station is essential to design a stand-alone PV-powered station. As can be seen from Figure 6.2, the total power consumption of a base station (P_{BS}) is equal to the sum of the powers consumed by its components and can be given as below.

$$P_{BS} = P_{BBU} + P_{CS} + P_{PS} + N_{Sector} \times (P_{RF} + P_{PA}) + P_{AUX} \quad (6.1)$$

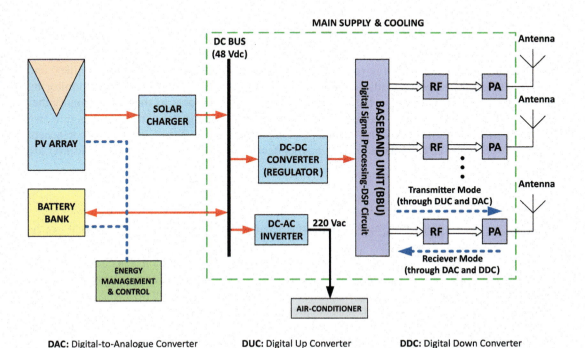

FIGURE 6.4 Components of a PV-powered base station.

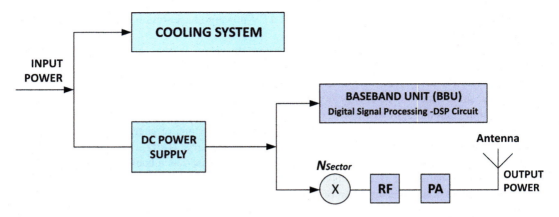

FIGURE 6.5 Power flow diagram of a base station.

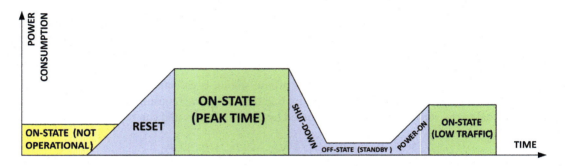

FIGURE 6.6 Typical power consumption modes of a base station.

where P_{BBU} is the power consumption of the baseband unit, P_{CS} is the cooling system power, P_{PS} is the total power of the supply system, N_{Sector} is the number of radio units embedded in the BBU, P_{RF} is the power of the radio frequency unit, P_{PA} is the power of the amplifier, and P_{AUX} is the power of other loads remaining in the system, such as lighting and monitoring needs. While some loads on the base station are almost independent of the traffic, some other loads are traffic-dependent. Therefore, it is possible to divide the base station loads into two groups as traffic-dependent and traffic-independent (equation 6.2). While it is possible to accept BBU, CS, PS, and AUX loads as traffic-independent, radio unit (RU) loads (RF and PA) are traffic-dependent.

$$P_{BS} = P_{t-dependent} + P_{t-independent} \qquad (6.2)$$

where $P_{t-dependent}$ is equal to $P_{BBU} + P_{CS} + P_{PS} + P_{AUX}$, and $P_{t-independent}$ is equal to P_{RU}, which can be expressed as $N_{Sector} \times (P_{RF} + P_{PA})$.

As for PV array sizing, the amount of energy consumed by the base station is needed rather than its power. The load profile of a base station is usually repeated weekly. However, it is possible to size the PV array based on the daily energy consumption if there is no significant difference between the load profiles on weekdays and weekends. A calculation approach for weekly energy consumption demand is given below. Since energy consumption is defined as the product of power and the working times, weekly energy demand can be expressed by (6.3).

$$\text{Energy Demand}\left(\frac{\text{Wh}}{\text{week}}\right)^{BS} = \left(P_{\text{t-independent}}\right) \times \Delta t_i \left(\frac{\text{hour}}{\text{day}}\right) \times \Delta t_j \left(\frac{\text{day}}{\text{week}}\right)$$
$$+ \left(\overline{P_{\text{t-dependent}}}\right) \times \Delta t_k \left(\frac{\text{hour}}{\text{day}}\right) \times \Delta t_m \left(\frac{\text{day}}{\text{week}}\right) \tag{6.3}$$

If multiple base stations are mounted on the telecommunication tower, the same energy demand calculation is repeated for each base station to find the total energy requirement. Note that the traffic-dependent energy demand can be expressed based on the daily average load ($\overline{P_{\text{t-dependent}}}$) if a time-varying load profile is available for the base station. However, to keep the energy supply availability higher, peak (or rated) load demand can also be taken into account when sizing the PV array.

Example 6.3: PV-Powered Telecommunication System

A stand-alone PV system will be designed to power a remote GSM base station located at the coordinates 33.75 N, 84.39 W in Atlanta, Georgia. The criteria for designing the remote telecommunication system are listed below.

- The GSM base station is mounted on a tower and runs continuously, day and night, all year round.
- The average monthly PSH (peak sun hours) values on the horizontal surface of the site are listed in Table 6.9.
- The designed system voltage (DC bus Voltage) is 48 V_{DC}.
- PV panels will be ground-mounted and shade-free.
- The electrical loads of the base station and their typical operating times are given in Table 6.10. Assume that the effect of seasonal variations on the load demand of the base station is negligible.
- The distance between the PV array and the base station is 40 m.
- The highest DoA (Days of Autonomy) value is assumed to be 2 days for the given region. Assume that the average ambient temperature during DoA days is −10°C.
- 12 V and 500 Ah lead-acid batteries will be used for energy storage needs. The provided temperature correction curve for the lead-acid battery is given in Figure 6.1. The maximum DoD (Depth of Discharge) value of the battery is 70%.
- The system will include a charge controller and the combined round-trip efficiency for the selected battery & charger pair is 90%.
- The efficiency of the converter and inverter feeding the DC and AC loads is approximately 95%.
- The PV module specifications are given in Table 6.11.

Design a PV-powered stand-alone system for the given GSM base station and determine its specifications. Afterward, draw the PV system diagram of the final design.

TABLE 6.9
The Electrical Loads of the Base Station

Months	Jan.	Feb.	Mar.	Apr.	May	Jun.	Jul.	Aug.	Sep.	Oct.	Nov.	Dec.
PSH (h/day)	2.45	2.86	4.26	4.90	5.92	6.95	6.63	6.15	5.21	4.41	3.55	2.95

TABLE 6.10
The Electrical Loads of the Base Station

Electrical Load	DC or AC	Power (W)	Operating Time (h/day)
Radio Unit & RF Feeder	DC	4200	24
Baseband Unit	DC	2100	24
Signal Transmitting Loads	DC	130	24
Power Supply Unit	DC	1200	24
Remote Monitoring	DC	20	24
Lighting & Security	AC	200	12
Fan	DC	120	24
Air-Conditioner	AC	2500	6.5

TABLE 6.11
PV Module Data

Rated Power	Rated Voltage	Rated Current	Open Circuit Voltage	Short-Circuit Current	NOCT
200 Watts	7.833 pt	8.14 Amps	30 volts	8.52 Amps	50°C

Solution

Let's follow the design steps similar to the examples worked above.

Step 1 – Determination of Load Requirements: The energy demand of the base station is largely independent of seasonal variations. Therefore, the month with the lowest insolation, which is January, can be considered as the design month (the worst-case scenario) for PV array sizing.

The base station contains both DC and AC loads. We can calculate DC and AC load demands separately. From Table 6.9, the maximum daily energy demand for DC and AC loads can be calculated as below.

$$\sum [\text{Wh/day}]^{DC} = (4200 + 2100 + 130 + 1200 + 20 + 120) \times 24 = 186480 \ [\text{Wh/day}]$$

$$\sum [\text{Wh/day}]^{AC} = (200 \times 12 + 2500 \times 6.5) = 18650 \ [\text{Wh/day}]$$

Accordingly, energy demand in Ah/day (with respect to 12 V battery voltage) can be calculated as below.

$$\frac{\sum [\text{Wh/day}]^{DC}}{V_b} = \frac{186480}{12} = 15540 \ [\text{Ah/day}]^{DC}$$

$$\frac{\sum [\text{Wh/day}]^{AC}}{V_b} = \frac{18650}{12} = 1554 \ [\text{Ah/day}]^{AC}$$

By considering the efficiency of the rectifier and inverter, the daily total electrical energy demand (Ah/day) can be calculated on the DC bus as below.

$$\text{Total Energy Demand}\left(\frac{Ah}{day}\right) = \frac{\text{Energy Demand}\left(\frac{Ah}{day}\right)^{DC}}{\eta_{conv}} + \frac{\text{Energy Demand}\left(\frac{Ah}{day}\right)^{AC}}{\eta_{inv}}$$

Photovoltaic System Applications

$$\text{Total Energy Demand}\left(\frac{\text{Ah}}{\text{day}}\right) = \frac{15540\left(\frac{\text{Ah}}{\text{day}}\right)^{DC}}{0.95} + \frac{1554\left(\frac{\text{Ah}}{\text{day}}\right)^{AC}}{0.95} \cong 17994 \text{ Ah/day}$$

Since the round-trip efficiency is given as 0.90, the daily total electrical energy demand (Ah/day) at the PV level can be calculated as follows:

$$\text{Total (Corrected) Energy Demand}\left(\frac{\text{Ah}}{\text{day}}\right) = \frac{\text{Total Energy Demand}\left(\frac{\text{Ah}}{\text{day}}\right)^{DC} \text{ on the DC bus}}{\eta_{\text{round-trip}}}$$

$$\text{Total (Corrected) Energy Demand}\left(\frac{\text{Ah}}{\text{day}}\right) = \frac{17994\left(\frac{\text{Ah}}{\text{day}}\right)}{0.90} \cong 19994 \text{ Ah/day}$$

Step 2 – Determination of Tilt Angle: Although the design month for determining the photovoltaic needs is January, the yearly optimum tilt angle should be considered when positioning the PV array. By following the solution approach in *Example 4.14*, the optimum tilt angles for the given months can be determined as in Table 6.12 (see Chapter 4 for details).

It is clear from Table 6.11 that the yearly tilt angle of the stand-alone PV system shall be set to 30.17 degrees. Considering the seasonal basis, the tilt angle will be set to 56.95 degrees for winter, 19.57 degrees for spring, 0 degrees for summer (negative-value tilt angle is impractical), and 44.87 degrees for autumn. Monthly tilt angles are also given in Table 6.11. Once again, negative-value tilt angles can practically be considered zero degrees.

Step 3 – Determination of Storage Requirements: The battery requirements can be calculated using the equations from (5.42) to (5.45) as follows (from Figure 6.1, the $f_{\text{Temp-b}}$ value for the C/20 curve is 80% at −10°C).

$$C_{\text{Battery-net}}[\text{Ah}] = \frac{E_{\text{Daily Avg}}[\text{Ah/day}] \times \text{DoA}[\text{day}]}{\text{DoD}_{\max} \times \eta_b \times f_{\text{Temp-b}}} = \frac{19994 \times 2}{0.70 \times 0.90 \times 0.8} \cong 79341 \text{ Ah}$$

$$N_{\text{Battery}} = \frac{C_{\text{Battery-net}}[\text{Ah}]}{c_b[\text{Ah}]} = \frac{79341[\text{Ah}]}{500[\text{Ah}]} = 158.68 \approx 160$$

TABLE 6.12
Monthly Tilt Angles for the Latitude 33.75

Months	$b_{\text{opt}(m)}$	$b_{\text{opt}(s)}$	$b_{\text{opt}(y)}$
December (31 days)	60.49	56.95	30.17
January (31 days)	58.91		
February (28 days)	51.44		
March (31 days)	38.05	19.57	
April (30 days)	18.74		
May (31 days)	1.91		
June (30 days)	−9.32	−0.73	
July (31 days)	−4.22		
August (31 days)	11.34		
September (30 days)	30.26	44.87	
October (31 days)	47.20		
November (30 days)	57.14		

$$N_{sb} = \frac{V_{DC}}{V_b} = \frac{48}{12} = 4$$

$$N_{pb} = \frac{N_{Battery}}{N_{sb}} = \frac{160}{4} = 40$$

As a result, the battery storage configuration is $(4s \times 40p)$, and accordingly, a total of 160 batteries are needed.

Step 4 – Determination of PV Array Requirements: Before going into the *PV array sizing* phase, the given PSH value for the horizontal surface must be converted to its equivalent for a surface tilted at 30.17 degrees. We already know from Step-1 that the design month for PV array sizing is January. From equation (4.68), let's calculate the insolation conversion factor on January 15, representing the month of January ($L = 33.75$, $\delta = -21.27$, $\omega_{ss} = 74.92$, $b = 30.17$).

$$C_{b(sf)} = \frac{\frac{\pi 74.92}{180}\sin(-21.27)\sin(33.75-30.17) + \cos(-21.27)\cos(33.75-30.17)\sin(74.92)}{\frac{\pi 74.92}{180}\sin(-21.27)\sin(33.75) + \cos(-21.27)\cos(33.75)\sin(74.92)} = 1.77$$

So PSH at tilt b, $PSH_b = 1.77 \times 2.45 = 4.336$ h / day. Now let us follow the steps regarding the PV array sizing for stand-alone PV systems as described earlier in Chapter 5.

From equations (5.8) to (5.17), PV array sizing can be performed as follows. Note that PV system losses are not given. Hence, typical values are used in this example ($f_{Manufacturing} : 0.98$, $f_{Temp} : 0.94$, $f_{Soiling} : 0.95 \Rightarrow$ PV Array Losses $\Delta P_{PV} = 098 \times 0.94 \times 0.95 = 0.875$).

$$E_{overall} = \frac{E_{Load}[Wh/day]}{(\Delta P_{PV}) \cdot \eta_{overall}} = \frac{12V \times 19994 \left[\frac{Ah}{day}\right]}{(0.875) \times (0.90)} = \frac{239928 \, [Wh/day]}{0.7875} \cong 304670 \text{ Wh / day}$$

$$P_{Peak(PV)} = \frac{E_{overall}}{PSH_{avg}} = \frac{304670 \text{ Wh / day}}{4.336 \text{ h / day}} \cong 70265 \, W_p$$

$$I_{DC} = \frac{P_{Peak(PV)}}{V_{DC}} = \frac{64962 \text{ W}}{48 \text{ V}} \cong 1464 \text{ A}$$

$$N_s = \frac{V_{DC}}{V_{mp}} = \frac{48}{24.6} = 1.95 \approx 2$$

$$N_p = \frac{I_{DC}}{I_{mp}} = \frac{1464}{8.14} = 179.85 \approx 180$$

Consequently, $2 \times 180 \times 200 \text{ W} = 72000 \, W_p$ PV array is required to feed the system.

$$N_T = N_s \times N_p = 2s \times 180p$$

Step 5 – Specification of Protection Components and Charger Sizing: According to the design so far, the PV modules will be connected as $2s \times 167p$ and the battery bank configuration will be $4s \times 42p$ for a 48 V_{DC} system. The charge controllers should be sized based on the voltage and current requirements of the system. From the design, the maximum total PV array current can be as high as:

$$I_{T\max} = N_p \times I_{SC} \times f_{safe} = 180 \times 8.52 \times 1.25 \cong 1917 \text{ A}$$

Photovoltaic System Applications

For this application, it was decided to use 48 V/120 A MPPT solar chargers. In this case, the total number of charge controllers ($N_{Charger}$) can be found by dividing the maximum current of the total PV array by the amperage rating of the selected charger, $I_{Charger\text{-}selected}$ (equation 5.49).

$$N_{Charger} = \frac{I_{T\max}}{I_{Charger\text{-}selected}} = \frac{1917}{120\,A} = 15.975 \approx 16$$

So, a minimum of 16 charge controllers must be connected in parallel to feed the system safely.

Also, maximum PV array voltage can be as high as:

$$V_{\max} = N_s \times V_{OC} \times f_{safe} = 2 \times 30 \times 1.2 = 72\,V$$

Accordingly, a 48 V/6500 W_p solar charger with a rated charging current of 120 A and a maximum charging current of (>130 A) will be suitable for the given example. It should also be ensured that the input voltage range of the charger matches the voltage range of the PV module. For this example, it is sufficient to select a solar charger with a maximum PV voltage of 80 V.

Attention: In the fourth and fifth steps above, PV array and charge controller sizes are determined. However, different options are possible for their connection configurations.

Option-1: A single PV array with a $2s \times 180p$ connection feeds the base station over 16 parallel-connected solar chargers. The main disadvantage of this structure is that a very high DC current may flow through a single path. In other words, the current of 1917 A, generated by the 180 parallel-connected strings, flows through a single DC wire path. In this case, a very high cross-section DC cable (or multiple DC cable) is needed. However, it is quite difficult to create PV array and DC box connections using these high cross-section cables. Another disadvantage of this structure is that the PV array & DC cabling system is more prone to failures, which results in relatively low system reliability.

Option-2: In this option, the $2s \times 180p$ PV array structure is divided into sub-array groups. Here, each sub-array has its charge controller. Thus, the parallel-connected sub-array groups feed the base station over a common DC bus system. In this configuration, the need for a very high cross-section DC cable will be eliminated. Thus, the system protection against overcurrent will become easier.

In this example, Option-2 was chosen due to its ease of implementation and high system reliability. We know from the above calculations that a minimum of 16 chargers are needed. However, it is possible to obtain equally-sized sub-array & charger pairs by dividing the main PV array into 18 sub-array groups. Thus, the final PV array configuration will be $18p \times (10p \times 2s)$. Here, each $10p \times 2s$ sub-array structure has its charge controller. In this structure, 18 charge controllers charge the battery storage system altogether.

Considering the designs determined above, the system wiring diagram can be drawn as in Figure 6.7.

Based on the system configuration in Figure 6.7, the specification of protection components can be given in Table 6.13.

Step 6 – DC Wire Sizing: The DC wire sizing results can be given in Table 6.14.

It is important to note that the PV system can be designed in a hybrid structure with a diesel generator or local grid to reduce the battery size and increase the system reliability.

6.5 WATER PUMPING SYSTEMS

Accessing water resources is a universal need. That is mostly met by water pumping systems. The energy requirement for water pumping can be met by a wide variety of sources. In this regard, PV-powered stand-alone systems for small and medium-sized water pumping systems have been used for many years due to their flexibility, simplicity, and high reliability. To be able to design a PV-powered water pumping system, first of all, it is necessary to know the structure and working principle of the water pumping system. A typical PV-powered (submersible) water pumping system, which consists of a PV array, a controller, and an electrical pump, is shown in Figure 6.8.

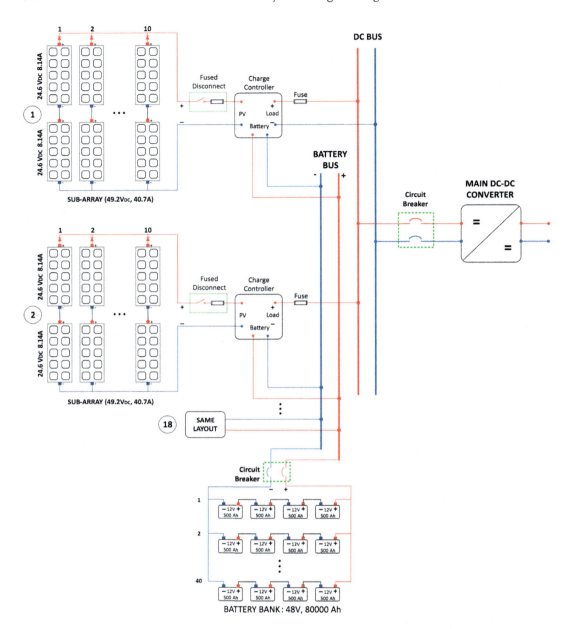

FIGURE 6.7 PV system wiring diagram.

In addition to the submersible configuration, the surface-mounted water pumping configuration is also frequently used, specifically for intakes from tanks, rivers, dams, and lakes. The pump of a surface configuration system is mounted at ground level, just outside or near the water source. The surface-mounted pumps are connected to the water sources through an inlet suction pipe and the output of this pump is connected to the delivery pipe.

The configuration of the electrical system to feed the water pumping depends on the type of electric motor used in the pump system. In general, both DC and AC motors are available for use in water pumping systems having displacement and centrifugal pumps. The required voltage is typically 120 or 240 volts for AC pumping systems, while voltages for DC pumps can be rated as 6 V, 12 V, 24 V, 32 V, or 48 V depending on the power needed.

TABLE 6.13
Protection Components Specification

Protected Circuit	Protection Device, Specification, & Description
Sub-Array Output	• Blocking diodes at the output of each of 10 strings [>97 A ($9 \times 8.56 \times 1.25 \cong 97$ A), >72 V ($2 \times 30 \times 1.2 = 72$ V)]. Note that multiple blocking diodes connected in parallel can also be used to meet the protection level of 97 Amps • Fused disconnect switch for overcurrent and short-circuit protection (≥134 A, >72 V) • Surge protector between sub-array and charger (>134 A, >72 V) Note that $I_n \geq \mathbf{1.56} \times I_{SC}$ ($1.56 \times 10 \times 8.56 \cong 134$ A) & $V_n \geq \mathbf{1.2} \times V_{OC} \times N_s$ ($1.2 \times 30 \times 2 = 72$ V)
Array Output	• Molded case circuit breaker for overcurrent and short-circuit protection (≥2404 A, >72 V) • Surge protector between array and charger (>2404 A, >72 V) Note that $I_n \geq \mathbf{1.56} \times I_{SC}$ ($1.56 \times 180 \times 8.56 = 2404$ A) & $V_n \geq \mathbf{1.2} \times V_{OC} \times N_s$ ($1.2 \times 30 \times 2 = 72$ V)
Charger to DC Bus	• Fuse or fused switch (170 A, >50 V) Note that $I = 6500$ W $\times 1.25/48$ $V_{DC} = 170$ A
Battery Bus to Battery Group	• Circuit breaker for battery protection (>3046 A, >50 V) Note that $I_{Charge} = 18 \times (6500$ W $\times 1.25/48$ $V_{DC} = 3046$ A)

TABLE 6.14
DC Wire Sizing[a]

Wire Section	System Voltage	Max. Current	Cable Length	Max. Voltage Drop	(Single Wire) Size
String Cable	48 V_{DC}	$8.56 \times 1.25 = 10.7$ A	2 m	2%	1.5 mm² (~16 AWG)
Sub-Array to Charger	48 V_{DC}	$10 \times 8.56 \times 1.25 \cong 107$ A	2 m	3%	25 mm² (~4 AWG)
Charger to Battery Bus	48 V_{DC}	~170 A	2 m	2%	35 mm² (~2 AWG)
Battery Bus to Battery Group	48 V_{DC}	$18 \times (6500$ W$/48$ V$) = 2438$ A	10 m	3%	5 × 120 mm² (5 × 4/0 AWG)
DC Bus to DC-DC Converter	48 V_{DC}	$1.25 \times 180 \times 8.56 = 1926$ A	5 m	3%	4 × 120 mm² (4 × 4/0 AWG)

[a] See Chapter 5 for details on sizing and selection of DC cables.

Several different factors affect the efficiency of the water pumping system. Our goal in this book is not to design an optimally efficient pump system from scratch. Rather, it is to design a stand-alone PV system for a decided configuration to feed the energy demand of the pumping system. The basic principle of energy demand calculation is the same in both submersible and surface-mounted configurations. A surface-mounted system is a special case of a submersible configuration in terms of total head calculations. For this reason, it is sufficient to explain the energy requirement and sizing calculations only with the help of a submersible system. Hence, the necessary steps for designing a PV-powered water pumping system can be described below.

Step 1 – Calculation of Total Head and Design Yield Requirements: The total head in (m) and the design yield in (m³/h), defining the pump's operating point, are the two main parameters

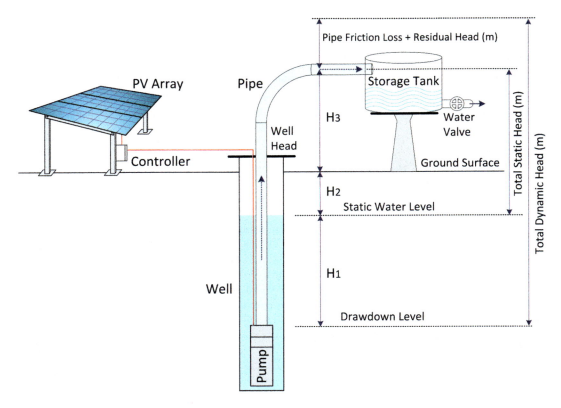

FIGURE 6.8 Typical PV-powered pumping system (submersible configuration).

used in pump selection and sizing. As illustrated in Figure 6.8, total dynamic head (TDH) in meters can be expressed by (6.4).

$$\text{TDH [m]} = H_1 + H_2 + H_3 + H_4 \tag{6.4}$$

where

H_1 = Drawdown Height in [m]
H_2 = Static Water Level in [m], that is, the depth between the top surface of the water in the well and the ground surface
H_3 = Vertical height from the ground surface to the tank inlet in [m]
H_4 = Equivalent height in [m] corresponds to friction loss (pressure drop in the pipe due to friction losses) and residual head requirements (additional pressure needs at the water inlet). In general terms, H_4 can be expressed by the following equation, where F_{Loss} is the friction loss coefficient in [1/100 m] and RH is the residual head in meters.

$$H_4 = \left[L_T + \sum (n_f \cdot f_e) \right] \times F_{\text{Loss}} \times 100^{-1} + \text{RH} \tag{6.5}$$

where n_f is the number of fittings (such as elbows, return bends, tees, gate valves, and non-return valves) in the pipeline, and f_e is the equivalent length of fittings inside the pipeline in meters. If the pipeline includes multiple pipe types, such as steel and PVC pipes, the friction loss coefficient is calculated for each of the pipe segments separately. Then, the equivalent friction losses found for

Photovoltaic System Applications

each pipe segment are summed to obtain H_4. Table 6.15 gives the typical friction loss coefficients for PVC and GI (galvanized iron or steel) pipes, while Table 6.16 gives the equivalent length of different fittings contained in the pipeline.

As for the surface pumps, the term TH (Total Head) is used instead of TDH, where $H_1 = 0$ and H_2 is suction depth (vertical distance between the water surface and pump inlet).

$$\text{TH } [m] = H_2 + H_3 + H_4 \tag{6.6}$$

The second parameter needed in this step is the design yield in [m³/h]. The design yield can simply be obtained by dividing the daily water demand [m³/day] by the pump's daily operating hours [h/day].

$$\text{Design Yield } \left[m^3/h \right] = \frac{\text{Daily Water Demand } \left[m^3/\text{day} \right]}{\text{Hours of Operation } \left[h/\text{day} \right]} \tag{6.7}$$

Note that the design yield should never exceed the safe yield level of the water well (maximum extraction potential of the borehole). Specifically, the design yield must always satisfy the following condition for battery-free off-grid PV systems.

$$\text{Design Yield } \left[m^3/h \right] \leq \text{Safe Yield } \left[m^3/h \right] \tag{6.8}$$

Note that if the design yield is greater than the safe yield (Design Yield > Safe Yield), then it will be necessary to design a PV system with a battery backup to be able to meet the daily water demand. Besides, a battery backup is also needed if the pumping is wanted at night times or if the operating hours are higher than the PSH.

Step 2 – Pump and Motor Selection Based on the Duty Point: The total head and design yield values found in Step-1 are used for a suitable pump selection. For this purpose, performance curves provided by the manufacturers can be used to select the pump with a specified duty point. A typical pump performance curve is given in Figure 6.9. On the pump's performance curve, the horizontal axis represents the pump flow rate [m³/h], while the vertical axis shows the head [m]. The second curve below the performance curve is the pump efficiency curve. To be able to keep the overall efficiency higher, the design yield and the flow rate at the maximum efficiency point should be matched as much as possible. Literally, all the curves above the duty point can meet the system design requirements. However, for a cost-effective solution, the closest curve covering the duty point should be chosen (Curve-2).

After selecting the required pump, the pump and motor powers in kW can easily be obtained from the power flow of the motor-pump assembly, as shown in Figure 6.10.

The theoretical hydraulic power P_h (kW) can be obtained from equation (6.8) for a total head H (m), where Q = flow rate (m³/h), ρ = water density (1000 kg/m³), and g = gravitational acceleration (9.81 m/s²).

$$P_h [\text{kW}] = \frac{Q[m^3/h] \cdot \rho[kg/m^3] \cdot g[m/s^2] \cdot H[m]}{3.6 \times 10^6} \tag{6.9}$$

After calculating the hydraulic power P_h, the minimum rated power of the electric motor P_e can be calculated via equation (6.10).

$$P_e [\text{kW}] \geq \frac{P_h [\text{kW}]}{\eta_{\text{pump}} \cdot \eta_{\text{motor}}} \tag{6.10}$$

TABLE 6.15
Friction Loss Coefficients per 100 m of PVC and GI Pipes

Flow (m³/h)	3/4" PVC D	3/4" PVC E	3/4" GI	1" PVC D	1" PVC E	1" GI	1¼" PVC D	1¼" PVC E	1¼" GI	1½" PVC C	1½" PVC D	1½" PVC E	1½" GI	2" PVC C	2" PVC D	2" PVC E	2" GI	2½" PVC B	2½" PVC C	2½" PVC D	2½" PVC E	2½" GI
Class[a]	D	E	GI	D	E	GI	D	E	GI	C	D	E	GI	C	D	E	GI	B	C	D	E	GI
1.5	2.4	3.2	8			1.9																
2	5	6.8	17.8			4.2																
2.5	8.6	11	31.9	1.6	2.2	7.5			1.5													
3	12.9	17.4	49.8	2.8	3.8	11.6	1.3	1.8	2.6													
3.5	19	25.8	71.7	4.2	5.7	17.7	2	2.7	3.9				1.5									
4				6.2	8.4	25.2	2.7	3.7	5.7				2.3				1					
5				8.5	11.6	32.7	3.5	4.7	8.0	1.2	1.4	1.9	3.3				1.5					
6				10.8	14.8	50.5	4.9	6.7	10.3	1.5	1.8	2.4	4.3	1	1.2	1.6	2.1					
7				15.5	21.3	72.7	6.9	9.3	15.5	2.2	2.6	3.5	6.3	1.4	1.6	2.2	2.9					
8				21.6	29.6	98.9	9.2	12.5	21.7	3.1	3.6	4.9	9	1.7	2	2.7	3.7					
9				28.8	41.8		11.5	15.6	29.3	4.1	4.7	6.5	12.3	2.1	2.5	3.4	4.7					
10							14.4	19.6	38.8	5.1	5.9	8.1	15.6	2.6	3	4.1	5.8					
12							17.5	23.8	49.1	6.4	7.5	10.2	20	3.6	4.2	5.8	8.4					
14								33.3	60.7	7.8	9	12.4	24.6	4.8	5.6	7.6	11.5					1.1
16									87.4	10.8	12.6	17.3	35.3	6.1	7.2	9.8	16.2					1.5
18										14.3	13.9	22.9	48.3	7.6	8.9	14.4	20.9		1	1.3	1.8	2.5
20										18.3	25.9	29.3	63	9.2	10.8	18.7	25.6	1.2	1.4	1.6	2.2	3.1

Note: F_{Loss} for pipes with larger diameters and higher flow rates can be found in the manufacturer's catalogs or the literature.

[a] Class B: 6 Bar, Class C: 9 Bar, Class C: 12 Bar, Class D: 15 Bar.

TABLE 6.16
The Equivalent Length of Straight Pipeline for Fittings and Valves in (meters)

Fitting Element		Screwed Fittings						Flanged Fittings				
		\multicolumn{12}{c}{Pipe Size in Inches}
		3/4"	1"	$\left(1\frac{1}{4}\right)''$	$\left(1\frac{1}{2}\right)''$	2"	$\left(2\frac{1}{2}\right)''$	3/4"	1"	$\left(1\frac{1}{4}\right)''$	$\left(1\frac{1}{2}\right)''$	2"	$\left(2\frac{1}{2}\right)'$
		\multicolumn{12}{c}{Equivalent Friction Losses in meters}
Elbows	Regular 90°	1.3	1.6	2	2.3	2.6	2.8	0.4	0.5	0.6	0.7	0.9	1.1
	Long Radius 90°	0.7	0.8	1	1	1.1	1.1	0.4	0.5	0.6	0.7	0.8	0.9
	Regular 45°	0.3	0.4	0.5	0.6	0.8	1	0.2	0.2	0.3	0.4	0.5	0.6
Tees	Line Flow	0.7	1	1.4	1.7	2.3	2.8	0.3	0.3	0.4	0.5	0.5	0.6
	Branch Flow	1.6	2	2.7	3	3.7	4	0.8	1	1.3	1.6	2	2.3
Return Bends	Regular 180°	1.3	1.6	2	2.3	2.6	2.8	0.4	0.5	0.6	0.7	0.9	1.1
Valves	Globe	7.3	8.8	11.3	12.8	16.5	18.9	12.2	13.7	16.5	18	21.4	23.5
	Gate	0.2	0.3	0.3	0.4	0.5	0.5	0	0	0	0	0.8	0.8
	Angle	4.6	5.2	5.5	5.5	5.5	5.5	4.6	5.2	5.5	5.5	6.4	6.7

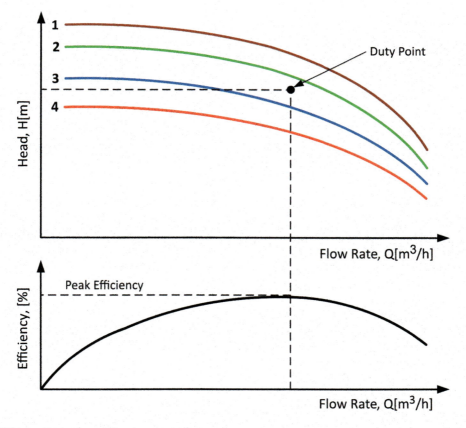

FIGURE 6.9 Typical centrifugal pump performance curve.

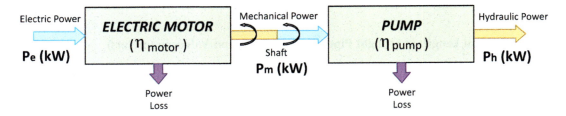

FIGURE 6.10 Energy conversion in a motor & pump assembly.

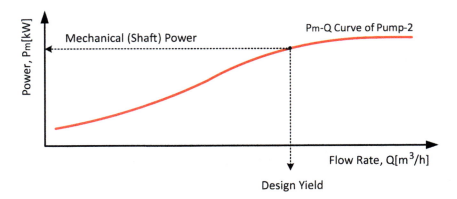

FIGURE 6.11 Typical P_m–Q curve of a pump.

Alternatively, the electric motor power can be found with the help of the $P_m - Q$ (Mechanical Power vs. Flow Rate) curves of the selected pump. For instance, in the graphical analysis given above, Pump-2, which represents Curve-2, was selected. Typically, the $P_m - Q$ curve of a pump is shown in Figure 6.11. Then, the mechanical power corresponding to the design yield is determined from the $P_m - Q$ curve of the pump, as shown in Figure 6.11. After determining the mechanical power P_m, the minimum rated power of the electric motor (P_e) can be calculated with equation (6.11).

$$P_e[\text{kW}] \geq \frac{P_m[\text{kW}]}{\eta_{\text{motor}}} \qquad (6.11)$$

Step 3 – Determination of Load Requirements: Once the motor and pump requirements are obtained from the above steps, it is now possible to calculate the energy demand of the load. For a stand-alone PV system without energy storage, the daily maximum working hours are limited by the peak solar hours PSH (h/day). In other words, the total daily energy demand of the pump motor is equal to the motor's energy demand at its nominal power for the PSH period. In this context, the electrical energy demand of the pump system is given by (6.12).

$$E_{\text{pump}}[\text{kWh}] = P_{er}[\text{kW}] \times t_{\text{pump}}[\text{h}] \qquad (6.12)$$

where t_{pump} is the operating hours of the pump and P_{er} is the rated electric power of the electric motor.

As for the stand-alone PV application with an energy storage system, the daily operating hours are not limited by the PSH. Rather, the designed working hours should be taken into account when calculating the energy demand of the pump.

Step 4 – Determination of PV Array Requirements: The energy output from the PV array should meet both the system losses and the pump demand for the PSH period. Here, the energy generated by the PV array E_{PV} can be given by (6.13)

Photovoltaic System Applications

$$E_{PV}[\text{kWh}] = P_{PV}[\text{kW}] \times \text{PSH}[\text{h}] \times \text{PR} \tag{6.13}$$

where P_{PV} is the peak power of the PV array, PR is the system performance ratio, and given by $(\Delta PV) \times \eta_{\text{overall}}$. Herein, ΔPV is the overall PV array efficiency, and η_{overall} is the overall efficiency of the power electronics equipment in the system, which is studied in detail in the previous sections. By equalizing the energies given by 6.10 and 6.9 ($E_{PV} = E_{\text{pump}}$), the required PV array size can be easily calculated by 6.14.

$$p_{PV}[\text{kW}] = \frac{P_{er}[\text{kW}] \times t_{\text{pump}}[\text{h}]}{\text{PSH}[\text{h}] \times \text{PR}} = \frac{P_{er}[\text{kW}] \times t_{\text{pump}}[\text{h}]}{\text{PSH}[\text{h}] \times \Delta PV \times \eta_{\text{overall}}} \tag{6.14}$$

Calculations regarding PV array configuration have been studied in detail in the previous sections with worked examples. Therefore, it is not repeated here again. The remaining steps (controller selection, protection system requirements, and conductor sizing) are the same as in the other applications. In the examples below, different types of PV-powered water pumping applications are worked out and explained.

Example 6.4: PV-Powered Water Pumping System (Surface-Mounted System)

A stand-alone PV-powered water pumping system will be designed to feed the water demand of animals on a farm, located at the coordinates 33.75 N, 84.39 W in Atlanta, Georgia. The design considerations for the remote area water pumping system are listed below.

- The daily water requirement of the farm animals is approximately 40 m³/day.
- The total capacity of the water storage tank is about 150 m³.
- Farm animals can go without water for a maximum of one day.
- The grassland where the water source is located is only used during the winter months.
- The average monthly PSH (peak sun hours) values on the horizontal surface of the farm are the same as in Example 6.3.
- The voltage of the pump motor is 24 V_{DC}.
- The pump takes the water from a depth of 20 m to the tank placed at about 3 m height above the ground (static lift is 3 m).
- The safe yield limit of the water source is 10 m³/h.
- The pump's suction pipe is connected to a 2-inch galvanized iron pipe, and the pump outlet is connected to a 2-inch PVC pipe (Class E) with three elbows (90 degrees), one gate valve, and one non-return valve.
- Assume that the total length of the suction pipe is 20 m, the total length of the delivery pipe is 18 m, and the residual head is zero.
- PV panels will be ground-mounted and shade-free.
- The distance between the PV array and the pump controller is 3 m.
- The distance between the pump controller and the motor is 30 m.
- The PV system does not contain any battery storage unit.
- The system includes a pump controller (linear current booster for use on direct-drive solar systems), which keeps the voltage at a constant level around the panel's MPP, and in turn, matches the electrical requirements of the panel to maximize the daily water.
- A float switch is used to control the water level of the tank.
- The PV module specifications are given in Table 6.17.

Design a PV-powered stand-alone solar pumping system for the given specifications and draw the PV system diagram of the final design.

Solution

Let's design the water pumping system by following the steps explained above.

TABLE 6.17
PV Module Data

Rated Power	Rated Voltage	Rated Current	Open Circuit Voltage	Short-Circuit Current	NOCT
200 Watts	24.6 volts	8.14 Amps	30 volts	8.52 Amps	50°C

Step 1 – Calculation of Total Head and Design Yield: From expressions (6.5) and (6.6), the total head is calculated as below. Note that F_{Loss} coefficients for GI and PVC pipes are considered separately. F_{Loss} coefficients and equivalent lengths of pipe fittings are respectively taken from Tables 6.15 and 6.16 ($H_1 = 0$, $H_2 = 20$ m, $H_3 = 3$ m, RH = 0, $F_{Loss(GI)} = 5.8$ m, $F_{Loss(PVC)} = 4.1$ m, $L_{T(GI)} = 20$ m, $L_{T(PVC)} = 18$ m, 3 elbows: $3 \times 2.6 = 7.8$ m, 1 gate valve = 0.5 m, 1 non-return valve = 5.5 m).

$$\text{TH}[m] = H_2 + H_3 + \left[L_{T(GI)}\right] \times F_{Loss(GI)} \times 100^{-1} + \left[L_{T(PVC)} + \sum(n_f \cdot f_e)\right] \times F_{Loss(PVC)} \times 100^{-1} + RH$$

$$\text{TH}[m] = 20 + 3 + 20 \times 5.8 \times 100^{-1} + [18 + (3 \times 2.6 + 0.5 + 5.5)] \times 4.1 \times 100^{-1} + 0 \cong 25.5 \text{ m}$$

To find the design yield, it is necessary to know the equivalent daily operating hours of the pump. As explained above, in PV systems without batteries, the daily operating hours of the pump are limited by the PSH value. Therefore, the optimum tilt angle at the solar site must be found first. Then the PSH value at that tilt angle should be calculated. Since the solar data and the site are the same as in Example 6.3, the solar data calculated for the winter months can be directly taken from this example, as in Table 6.18.

From Table 6.18, the average PSH value is 4.706 for the winter months. In other words, the daily working hours of the pump is 4.706 h. Hence, the minimum design yield can be obtained from (6.7) as below:

$$\text{Design Yield}\left[m^3/h\right] = \frac{\text{Daily Water Demand}\left[m^3/day\right]}{\text{Hours of Operation}\left[h/day\right]} = \frac{40 \text{ m}^3/\text{day}}{4.706 \text{ h/day}} = 8.49 \text{ m}^3/\text{h}$$

As can be seen, the calculated design yield is smaller than the safe yield of the water well (8.49 m³/h ≤ 10 m³/h). Therefore, it is possible to design a stand-alone PV system with no battery storage. Note that if the design yield was greater than the safe yield, then it would be necessary to design a PV system with a battery backup.

Step 2 – Pump and Motor Selection: Since pump and motor performance curves are not given in this example, their requirements can be calculated theoretically. From equation (6.8), the hydraulic power P_h (kW) is obtained as below.

$$P_h[kW] = \frac{Q[m^3/h] \cdot \rho[kg/m^3] \cdot g[m/s^2] \cdot H[m]}{3.6 \times 10^6} = \frac{9 \times 1000 \times 9.81 \times 25.5}{3.6 \times 10^6} = 0.625 \text{ kW}$$

Note that the minimum design yield was 8.49 m³/h. Hence, it is sufficient to design the system with a flow rate between 8.49 and a safe yield of 10 m³/h. So, the flow rate Q is taken as 9 m³/h. By considering the typical pump efficiency of 70% and the motor efficiency of 90%, the minimum rated power of the motor can be found from equation (6.8) as follows:

$$P_e[kW] \geq \frac{0.625 \text{ kW}}{0.70 \times 0.90} = 0.992 \text{ kW}$$

Accordingly, it will be sufficient to choose a 1 kW DC pump ($P_{er} = 1$ kW) to feed the water tank. The duty point for the pump system is 9 m³/h at 25.5 m head.

TABLE 6.18
Monthly Tilt Angles and Average PSH Calculation for Winter at the Latitude 33.75

Months	$b_{opt(m)}$	$b_{opt(s)}$
December (31 days)	60.49	56.95
January (31 days)	58.91	
February (28 days)	51.44	

	Insolation Conversion Factor and PSH	
Months	$C_{b(st)}$ [a]	PSH_b (PSH at tilt 56.95)
December (31 days)	1.85	1.85 × 2.95 = 5.451 h/day
January (31 days)	1.77	1.77 × 2.45 = 4.336 h/day
February (28 days)	1.51	1.51 × 2.86 = 4.330 h/day
Average PSH for Winter		4.706 h/day

[a] A sample calculation is given in Example 6.3.

Step 3 – Determination of Load Requirements: From the above step, we know the motor and pump ratings. Now that the electrical energy demand of the pump system is calculated by 6.12. Note that $t_{pump} = PSH_b$.

$$E_{pump}[\text{kWh}] = P_{er}[\text{kW}] \times t_{pump}[\text{h}] = (1\,\text{kW}) \times (4.706\,\text{h}) = 4.706\,\text{kWh}$$

Step 4 – Determination of PV Array Requirements: The PV array size can be obtained from 6.14 as below. Since the site-specific details are not given, the overall performance ratio of the system is assumed to be 70% (PR = ΔPV × $\eta_{overall}$ for a typical PV array is around 70%).

$$p_{PV}[\text{kW}] = \frac{E_{Pump}[\text{kWh}]}{PSH[\text{h}] \times PR} = \frac{4.706\,\text{kWh}}{4.706 \times 0.70} \cong 1.45\,\text{kW}$$

Since the peak power of the PV module we have is 200 W_p, the total number of PV modules required for the system (N_{min}) is:

$$N_T = \frac{p_{PV}[\text{kW}]}{p_{Peak\,(module)}[\text{kW}]} = \frac{1.45\,\text{kW}}{0.2\,\text{kW}} = 7.25 \cong 8$$

$$N_s = \frac{V_{DC}}{V_{mp}} = \frac{48}{24.6} = 1.95 \approx 2$$

$$N_p = \frac{N_T}{N_s} = \frac{8}{2} = 4$$

Consequently, $2 \times 4 \times 200\,W = 1600\,W_p$ PV array is required to feed the solar pumping system.

$$N_T = N_s \times N_p = 2s \times 4p$$

Step 5 – Specification of Protection Components and Controller Sizing: According to the design so far, the PV modules will be connected as $2s \times 4p$ for a 48 V_{DC} system. The pump controllers should be sized based on the voltage and current requirements of the system. From the design, the maximum total PV array current can be as high as:

$$I_{T\max} = N_p \times I_{SC} \times f_{safe} = 4 \times 8.52 \times 1.25 \cong 43\,\text{A}$$

Also, maximum PV array voltage can be as high as:

$$V_{\max} = N_s \times V_{OC} \times f_{safe} = 2 \times 30 \times 1.2 = 72\,\text{V}$$

Accordingly, a 48 V/2000 W pump controller with a maximum input current of (>43 A) and maximum withstand voltage (>72 V) will be suitable for the given example. Considering the designs determined above, the system wiring can be drawn as in Figure 6.12.

Based on the system configuration above, the specification of protection components can be given in Table 6.19.

Step 6 – DC Wire Sizing: The DC wire sizing results can be given in Table 6.20.

Example 6.5: PV-Powered Water Pumping System (AC Submersible System)

The water needs of a small residential area in a rural region of an island will be met by a stand-alone PV-powered water pumping system. An average of 30 m³/day of water is required for

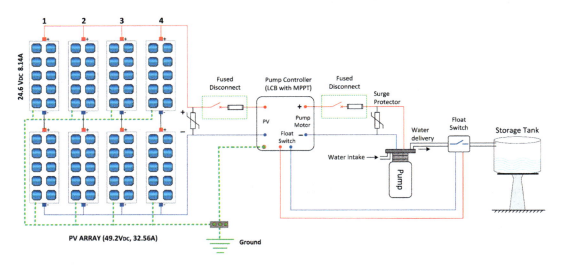

FIGURE 6.12 PV pumping system wiring.

TABLE 6.19
Protection Components Specification

Protected Circuit	Protection Device, Specification, & Description
Array Output	• Fused disconnect switch for overcurrent and short-circuit protection (≥54 A, >72 V) • Surge protector between array and controller (>54 A, >72 V) Note that $I_n \geq 1.56 \times I_{SC}$ (1.56×4×8.56 ≅ 54A) & $V_n \geq 1.2 \times V_{OC} \times N_s$ (1.2×30×2=72V)
Controller Output	• Fused disconnect switch for overcurrent and short-circuit protection (≥54 A, >72 V) • Surge protector between array and controller (>134 A, >72 V) Note that $I_n \geq 1.56 \times I_{SC}$ (1.56×4×8.56 ≅ 54A) & $V_n \geq 1.2 \times V_{OC} \times N_s$ (1.2×30×2=72V)

TABLE 6.20
DC Wire Sizing[a]

Wire Section	System Voltage	Max. Current	Cable Length	Max. Voltage Drop	(Single Wire) Size
String Cable	48 V_{DC}	8.56×(4−1)×1.25=32.1 A	2 m	2%	4 mm² (~12 AWG)
Array to Controller	48 V_{DC}	8.56 × 4 ×1.25 = 42.8 A	3 m	3%	6 mm² (~10 AWG)
Controller to DC Motor	48 V_{DC}	8.56 × 4 ×1.25 = 42.8 A	30 m	3%	6 mm² (~10 AWG)

[a] See Chapter 5 for details on sizing and selection of DC cables.

---

domestic use in the given residential site, which is located at the coordinates 33.75 N, 74.5 W in the North Athletic Ocean. The design considerations for the remote area water pumping system are listed below.

- The water pumping system will be used all year round.
- The total capacity of the water storage tank is about 100 m³.
- The average monthly PSH (peak sun hours) values on the horizontal surface of the remote residential site are considered the same as in Example 6.3 since the latitudes are the same.
- It is decided to use a multistage submersible AC pump system with a three-phase induction motor.
- The power cables are fixed in a plastic pipe in the well. The power cables outside the well are installed in a conduit from the junction box at the array level to the wellhead.
- The voltage of the pump motor is 110 V_{AC}. A 48 V_{DC}/110 V_{AC} inverter will be used to feed the motor.
- The pump, placed in the 6-inch borehole, is connected to the 2-inch galvanized steel pipe.
- The pump takes the water from a depth of 8 m (drawdown height is 4 m, and static level is 4 m) to the tank placed on a hill about 30 m high above the wellhead (Static lift is: 30+2=32 m). The water tank is mounted at about 2 m in height from the ground surface. Due to the shape of the land, the total length of the pipe used between the wellhead and the tank inlet is 60 m. Assume that the residual head is zero.
- The pump is attached to a 2-inch galvanized steel pipe in the well, and the steel pipe is connected to a 2-inch PVC pipe (Class E) at the wellhead. The pipe system in total includes four elbows (2×90 degrees and 2×45 degrees), one gate valve, and one non-return valve.
- The safe yield limit of the borehole is 9 m³/h.
- PV panels will be ground-mounted and shade-free.
- The distance between the PV array and the inverter is 3 m.
- The distance between the inverter and the motor is 18 m.
- The PV system does not contain any battery storage unit.
- The PV module specifications are the same as given in Table 6.17.

Design a PV-powered stand-alone solar pumping system for the given specifications and draw the PV system diagram of the final design.

Solution

Step 1 – Calculation of Total Head and Design Yield: From expressions (6.4) and (6.5), the total head is calculated as below. Note that F_{Loss} coefficients for GI and PVC pipes are considered separately. F_{Loss} coefficients and equivalent lengths of pipe fittings are respectively taken from Tables 6.15 and 6.16 ($H_1 = 4$ m, $H_2 = 4$ m, $H_3 = 32$ m, RH = 0, $F_{Loss(GI)} = 3.7$ m, $F_{Loss(PVC)} = 2.7$ m, $L_{T(GI)} = 8$ m, $L_{T(PVC)} = 60$ m, 2 elbows (90 degrees): $2 \times 2.6 = 5.2$ m, 2 elbows (45 degrees): $2 \times 0.8 = 1.6$ m, 1 gate valve = 0.5 m, 1 non-return valve = 5.5 m).

$$TH\,[m] = H_1 + H_2 + H_3 + \left[L_{T(GI)}\right] \times F_{Loss(GI)} \times 100^{-1} + \left[L_{T(PVC)} + \sum\left(n_f \cdot f_e\right)\right] \times F_{Loss(PVC)} \times 100^{-1} + RH$$

$$TH\,[m] = 4 + 4 + 32 + 8 \times 3.7 \times 100^{-1} + \left[60 + (2 \times 2.6 + 2 \times 0.8 + 0.5 + 5.5)\right] \times 2.7 \times 100^{-1} + 0 \cong 42.25\text{ m}$$

As explained above, the PSH value should be found at the optimum tilt angle to calculate the design yield. Since the solar data is the same as in Example 6.3, the yearly optimal tilt angle value ($b_{opt(y)} = 30.17°$) can directly be taken from Example 6.3. We already know that the design month for PV array sizing is January. Note that the load is constant all year round and the solar insolation in January is the lowest. Therefore, the corrected PSH value can also be taken from Example 6.3 (PSH = 4.336 h / day). Hence, the minimum design yield can be obtained from 6.7 as below:

$$\text{Design Yield}\,[m^3/h] = \frac{\text{Daily Water Demand}\,[m^3/\text{day}]}{\text{Hours of Operation}\,[h/\text{day}]} = \frac{30\text{ m}^3/\text{day}}{4.336\text{ h/day}} = 6.91\text{ m}^3/\text{h}$$

As can be seen, the calculated design yield is smaller than the safe yield of the borehole (6.91 m³/h ≤ 9 m³/h). Therefore, it is possible to design a stand-alone PV system without battery storage.

Step 2 – Pump and Motor Selection: From equation (6.8), the hydraulic power P_h (kW) is obtained as below.

$$P_h\,[kW] = \frac{Q\,[m^3/h] \cdot \rho\,[kg/m^3] \cdot g\,[m/s^2] \cdot H\,[m]}{3.6 \times 10^6} = \frac{8 \times 1000 \times 9.81 \times 42.25}{3.6 \times 10^6} = 0.921\text{ kW}$$

Note that the minimum design yield was 6.91 m³/h. Hence, it is sufficient to design the system with a flow rate between 6.91 and a safe yield of 9 m³/h. So, the flow rate Q is taken as 8 m³/h. By considering the typical pump efficiency of 70% and the motor efficiency of 90%, the minimum rated power of the motor can be found from equation (6.8) as follows:

$$P_e\,[kW] \geq \frac{0.921\text{ kW}}{0.70 \times 0.90} = 1.462\text{ kW}$$

Accordingly, it will be sufficient to choose a 1.5 kW AC pump ($P_{er} = 1.5$ kW) to feed the water tank. The duty point for the pump system is 8 m³/h at 42.25 m head.

Step 3 – Determination of Load Requirements: Now, the electrical energy demand of the pump system can be calculated by 6.12. Note that $t_{pump} = PSH_b$.

$$E_{pump}\,[kWh] = P_{er}\,[kW] \times t_{pump}\,[h] = (1.5\text{ kW}) \times (4.336\text{ h}) = 6.504\text{ kWh}$$

Step 4 – Determination of PV Array Requirements: By assuming the overall performance ratio of the system is 70%, the PV array size can be obtained from (6.14) as below:

$$p_{PV}\,[kW] = \frac{E_{Pump}\,[kWh]}{PSH\,[h] \times PR} = \frac{6.504\text{ kWh}}{4.336 \times 0.70} \cong 2.15\text{ kW}$$

Since the peak power of the PV module we have is 200 W_p, the total number of PV modules required for the system (N_{min}) is:

$$N_T = \frac{p_{PV}[\text{kW}]}{p_{\text{Peak (module)}}[\text{kW}]} = \frac{2.15 \text{ kW}}{0.2 \text{ kW}} = 10.75 \cong 12 \text{ (Rounded to next even number)}$$

$$N_s = \frac{V_{DC}}{V_{mp}} = \frac{48}{24.6} = 1.95 \approx 2$$

$$N_p = \frac{N_T}{N_s} = \frac{12}{2} = 6$$

Consequently, $2 \times 6 \times 200$ W = 2400 W_p PV array is required to feed the solar pumping system.

$$N_T = N_s \times N_p = 2s \times 6p$$

Step 5 – Specification of Protection Components and Inverter Sizing: According to the design so far, the PV modules will be connected as $2s \times 6p$ for a 48 V_{DC} system. The inverter should be sized based on the voltage and current requirements of the system. Total AC load power is given as $P_{\text{Total AC Load}} = 1.5$ kW. Hence, the nominal power of the inverter:

$$P_{\text{inv rated}} \geq 1.25 \times (P_{\text{Total AC Load}}) \Rightarrow P_{\text{inv rated}} \geq 1.25 \times 1500 = 1.875 \text{ kW}$$

Accordingly, an inverter with a rated power of 2000 W can be used. The inverter to be selected should also meet the following voltage and current conditions. From the design, the maximum PV array voltage can be as high as: $V_{max} = N_s \cdot V_{OC} \cdot f_{safe} = 2 \times 30 \times 1.2 = 72$ V, and the maximum total PV array current can be as high as: $I_{T\,max} = N_p \cdot I_{SC} \cdot f_{safe} = 6 \times 8.52 \times 1.25 \cong 64$ A.

$$V_{OC\,max\,PV} \leq V_{max\,inv} \Rightarrow 72 \text{ V} \leq V_{max\,inv}$$

$$(I_{SC} @ T_{cell\,max} = I_{T\,max}) \leq I_{inv_{max\,DC\,input}} \Rightarrow I_{T\,max} \leq 64 \text{ A}$$

As a result, a 48 V_{DC}/110 V_{AC} inverter with a maximum input current of (>64 A) and maximum withstand voltage of (>72 V) shall be suitable for the given example. Considering the designs determined above, the system wiring can be drawn as in Figure 6.13.

Based on the system configuration above, the specification of protection components can be given in Table 6.21.

Step 6 – DC Wire Sizing: The DC wire sizing results can be given in Table 6.22.

6.6 REMOTE MONITORING AND CATHODIC PROTECTION

Low-powered remote applications such as antenna, data accusation (SCADA), and cathodic protection require reliable power to maintain high safety in system operation. For example, measuring the parameters such as flow rate, pressure, and temperature along the oil & gas pipelines/stations is an important health indicator. With these data, important warnings regarding the upcoming potential problems are provided. In addition, metallic underground systems containing toxicants or petrochemicals (such as storage tanks or pipelines) often have cathodic protection as it provides an effective solution against corrosion. Because of the reliability and simplicity, PV systems are ideal for these kinds of applications. In this context, PV systems are often preferred, especially in such applications where no grid access is provided.

The load profiles and electrical supply details are required to be known to be able to design PV-powered remote monitoring and cathodic protection systems. Some of the technical details & design criteria are listed below.

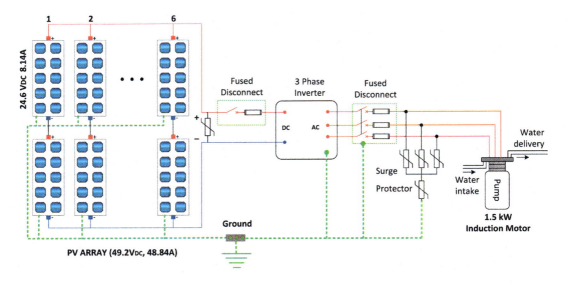

FIGURE 6.13 PV pumping system wiring.

TABLE 6.21
Protection Components Specification[a]

Protected Circuit	Protection Device, Specification, & Description
Array Output	• Fused disconnect switch for overcurrent and short-circuit protection (≥80 A, >72 V) • Surge protector between array and inverter (>80 A, >72 V) Note that $I_n \geq 1.56 \times I_{SC}$ (1.56×6×8.56 ≅ 80A) & $V_n \geq 1.2 \times V_{OC} \times N_s$ (1.2×30×2=72V)
Inverter Output	• Fused disconnect switch for overcurrent and short-circuit protection (≥15 A, >76 V) Note that: $I_n \geq 1.56 \times I_{max\ inv}$ ($1.56 \times \dfrac{1500}{\sqrt{3} \times 110 \times 0.85} \cong 14.45\ A \xrightarrow{\text{next rated value}} 15\ A$) $V_n \geq V_{max} \Longrightarrow V_{max\ inv} = 1.2 \times \dfrac{110}{\sqrt{3}} \cong 76\ V$ • Surge protector between inverter and pump (≥15 A). The SPD ratings should be ranged to be activated before the withstand voltage of the motor and inverter

[a] If the input and output switches are included in the switch box of the pump/inverter, then there is no need to use external protection elements.

TABLE 6.22
DC Wire Sizing[a]

Wire Section	System Voltage	Max. Current	Cable Length	Max. Voltage Drop	(Single Wire) Size
String Cable	48 V_{DC}	8.56×(6−1)×1.25=53.5 A	2 m	2%	6 mm² (~10 AWG)
Array to İnverter	48 V_{DC}	8.56 × 6×1.25 ≈ 64 A	3 m	3%	6 mm² (~10 AWG)
Inverter to AC Motor	110/$\sqrt{3}$ V_{AC}	15 A / phase	18 m	3%	3×2.5 mm² (~ 3× 10 AWG)

[a] See Chapter 5 for details on sizing and selection of DC cables.

Photovoltaic System Applications

- **Remote Monitoring:** Most of the solar-powered off-grid monitoring systems operate at 12 V_{DC}. The load demand generally varies depending on the number of sensors, sampling rate, and the requirements for data logging & transmission.
- Due to their low power demands, only a single PV module is mostly sufficient for remote monitoring applications. Some of the data loggers may already contain battery storage (such as rechargeable nickel-cadmium or lead-acid) that can be charged with a PV module. In this case, the allowable charging currents and operating voltage must be considered in the PV system design. Battery charge control may not be necessary if the load demand and PV output current are less than 1 amp. If an external energy storage system is needed for data loggers, an external charge controller may also be used. Usually, all equipment and wiring systems are located on the PV frame.
- **Cathodic Protection:** Metallic structures corrode because of ion losses to electrolytic surroundings. For example, water and acidic structures inside a soil volume act as electrolytes for the buried metal, thereby providing a suitable medium for electron flow. This process, which occurs naturally, can be reversed, and corrosion of the metal can be prevented. This method, also called cathodic protection, can be fulfilled by either burying a sacrificial anode or by using an external power source to inject a DC current (**impressed current**) into the metallic structure. The actual current amount to protect a metal is indeed a complex task. Since solar-powered cathodic protection is considered only, a simplified current injection technique for small-sized cathodic protection systems is explained herein. The amount of load current here is the current required to overcome the potential between the metal (anode) and the surrounding electrolyte. Once the necessary current amount is determined, the PV sizing can be done similarly as the other applications. To keep the wiring voltage drop to a minimum, the PV array should be installed as close as possible to the load (metal structure to be protected). If the cathodic application is in the coastal region or the sea, the PV modules must be resistant to the salty environment.
- Battery storage is used almost for all PV-powered cathodic protection systems. In general, deep-cycle lead-acid or nickel-cadmium batteries in a waterproof enclosure are recommended. Particularly, non-metallic enclosures are strongly advised for marine applications.

The minimum dc voltage value required for a cathodic protection system can be calculated by ohm's law as follows:

$$V_{\min} = R_T \times I_R \tag{6.15}$$

where I_R is the required current for cathodic protection, and the R_T represents the sum of the resistances in the current path between the anode and the cathode. These resistances through the current path consist of anode lead wires resistance (R_W), anode bed resistance (or anode to electrolyte resistance, R_A), and the earth resistance from anode to protected metal (R_E).

$$R_T = R_W + R_A + R_E \tag{6.16}$$

Figure 6.14 illustrates the schematic diagram of the impressed current-based cathodic protection technique, which compromises the rectifier, impressed current anode bed, electrolytic medium, and the metal to be protected. Figure 6.14 also shows the resistances in the current path from the rectifier's positive end to the cathode.

The total resistance R_T can also be found by applying a voltage (ΔV) between the rectifier output and the protected metal and then measuring the current value (I).

$$R_T = \frac{\Delta V}{I} \tag{6.17}$$

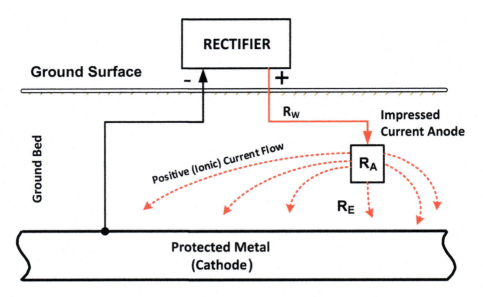

FIGURE 6.14 Schematic representation of cathodic protection based on impressed current.

TABLE 6.23
Typical Current Density Values for Cathodic Protection Systems

Surrounding Electrolyte	Required Current Density (mA/m²) for Bare Pipelines	Special Resistivity of the Medium ($\Omega \cdot$ m)
Sea Water	100	0–1
High Saline Soil	25	1–10
High Corrosive Soil	10	10–50
Corrosive Soil	5	50–100
Low Corrosive Soil	1	>100

The amount of required current to be used for bare and coated metals should be well-studied together with the corrosion engineers. The required current I_R can simply be found by multiplying the total external surface area (A_T) of the protected metal with the required current density J_R(mA / m²).

$$I_R = A_T \cdot J_R \tag{6.18}$$

For example, typical current density values required by the cathodic protection are given for pipelines in Table 6.23.

Now that V_{min} value can be found by using equations from (6.15) to (6.18). The design voltage of the PV system (in case of no battery storage) can then be set to 150% of the V_{min} to compensate for the aging and other derating factors.

$$V_{PV} \text{ (without battery storage)} = V_{min} \times 1.5 \tag{6.19}$$

For PV systems having batteries, it is sufficient to design the system DC voltage to be slightly higher than V_{min} value. Because required current shall be provided by the battery pack, which is almost stable during the useful life.

$$V_{PV} \text{ (with battery storage)} \geq V_{min} \tag{6.20}$$

Photovoltaic System Applications

Example 6.6: PV-Powered Pipeline Monitoring and Cathodic Protection for a Metal Tank

Two different PV systems will be designed for a pipeline facility. One of the PV systems will feed the pipeline monitoring system. The second one will power a cathodic protection system for a metal tank in the pipeline facility. Design details are given separately in the options below. For both applications, the PV panels are tilted at about 50 degrees to maximize the solar energy in the winter season. Assume that the winter season PSH value in the region is given as 4.7 h/day at the given tilt of the site. The PV module specifications to be used for both cases are given in Table 6.24.

a. **Pipeline Monitoring:** The 12 V_{DC} monitoring system includes a data logger and a transmitter. While the data logger draws 300 mA for 24 h/day, the transmitter circuit draws 1.25 A current for 1 h/day. 12 V and 100 Ah Ni-Cd batteries will be used for energy storage needs. Assume that the highest DoA (Days of Autonomy) value is 5 for the given region and the maximum DoD (Depth of Discharge) value of the battery is 95%. The system includes a simple controller and the round-trip efficiency for the selected battery & controller is 85%.

b. **Metal Tank Cathodic Protection:** A single 2 m graphite anode with a rated current of 2 A and a diameter of 16 cm is selected for the cathodic protection. The required protection current was determined by the corrosion engineer as 200 mA. A test voltage was applied to the system to measure the total resistance from the lead wire to the tank. As a result of the test, the total resistance (lead wire + anode + ground) was determined as approximately 50 Ω. 12 V and 50 Ah sealed lead-acid batteries will be used for energy storage needs. Assume that the highest DoA (Days of Autonomy) value is 5 for the given region and the maximum DoD (Depth of Discharge) value of the battery is 70%. The system includes a controller and the combined round-trip efficiency for the selected battery & controller is 90%. The battery temperature factor will be taken as 1.

Based on the design criteria given above, design the PV-powered monitoring and cathodic protection systems, and draw the wiring diagrams.

Solution:

a. Firstly, let's carry out the load demand calculations for the remote monitoring system. The system includes DC loads only. The weekly and daily energy demands are:

$$\sum [Wh/week] = (3.6 \times 24 \times 7 + 15 \times 1 \times 7) = 709.8 \ [Wh/week]$$

Note that the power of the data logger and transmitter is: $300 \cdot 10^{-3} \times 12 = 3.6 \ W$; $1.25 \times 12 = 15 \ W$.

From (5.183), the energy demand in Ah/day can be calculated as below.

$$\frac{\sum [Wh/week]}{V_b} = \frac{709.8}{12} = 59.15 \ [Ah/week]$$

TABLE 6.24
PV Module Data

Rated Power	Rated Voltage	Rated Current	Open Circuit Voltage	Short-Circuit Current	NOCT
30 Watts	13.40 volts	2.25 Amps	17.95 volts	2.75 Amps	47°C

$$\frac{[\text{Ah/week}]}{7} = \frac{59.15}{7} = 8.45 \; [\text{Ah/day}]$$

Since the round-trip efficiency is given as 0.85, the daily total electrical energy demand (Ah/day) can be calculated as:

$$\text{Total (Corrected) Energy Demand} \left(\frac{\text{Ah}}{\text{day}} \right) = \frac{8.45 \left(\frac{\text{Ah}}{\text{day}} \right)^{\text{DC}}}{0.85} = 9.94 \; \text{Ah/day}$$

Secondly, the battery requirements can be determined below: be (5.42) to (5.45) as follows (from Figure 6.15, the $f_{\text{Temp-b}}$ value for the C/20 curve is approximately 70% at −10°C).

$$C_{\text{Battery-net}} \; [\text{Ah}] = \frac{E_{\text{Daily Avg}} \; [\text{Wh/day}] \times \text{DoA} \; [\text{day}]}{\text{DoD}_{\max} \times V_b \times \eta_b \times f_{\text{Temp-b}}} = \frac{\frac{709.8 \; [\text{Wh/week}]}{7} \times 5}{0.95 \times 12 \times 0.85 \times 1} \cong 52.3 \; \text{Ah}$$

$$N_{\text{Battery}} = \frac{C_{\text{Battery-net}} \; [\text{Ah}]}{c_b \; [\text{Ah}]} = \frac{52.3 \; [\text{Ah}]}{100 \; [\text{Ah}]} \cong 0.523 \xrightarrow{\text{rounded up}} 1$$

$$N_{\text{sb}} = \frac{12}{12} = 1$$

$$N_{\text{pb}} = \frac{1}{1} = 1$$

As a result, the battery storage configuration is $(1s \times 1p)$, and accordingly, a total of 1 battery is needed.

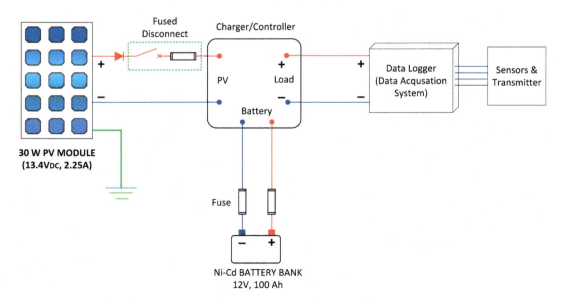

FIGURE 6.15 PV-powered monitoring system.

In the third step, Let us size the PV array based on the typical derating factors ($f_{\text{Manufacturing}} : 0.98, f_{\text{Temp}} : 0.94, f_{\text{Soiling}} : 0.95 \Rightarrow \text{PV Array Losses} \Delta P_{\text{PV}} = 0.98 \times 0.94 \times 0.95 = 0.875$).

$$E_{\text{overall}} = \frac{E_{\text{Load}}[\text{Wh/day}]}{(\Delta P_{\text{PV}}) \cdot \eta_{\text{overall}}} = \frac{\frac{709.8 \,[\text{Wh/week}]}{7}}{(0.875) \times (0.90)} = \frac{101.4 \,[\text{Wh/day}]}{0.7875} \cong 129 \text{ Wh/day}$$

$$P_{\text{Peak(PV)}} = \frac{E_{\text{overall}}}{\text{PSH}_{\text{avg}}} = \frac{129 \text{ Wh/day}}{4.7 \text{ h/day}} = 27.44 \, W_p$$

$$I_{\text{DC}} = \frac{P_{\text{Peak(PV)}}}{V_{\text{DC}}} = \frac{27.44 \text{ W}}{12 \text{ V}} \approx 2.28 \text{ A}$$

$$N_s = \frac{V_{\text{DC}}}{V_{\text{mp}}} = \frac{12}{13.40} = 0.89 \approx 1$$

$$N_p = \frac{I_{\text{DC}}}{I_{\text{mp}}} = \frac{2.28}{2.25} = 1.01 \approx 1$$

N_p value can be taken as "1" since the calculated design current and PV module's I_{mp} current is very close to each other. Consequently, only a single 30-watt PV module of which specifications are given in Table 6.24 is sufficient to feed the system.

$$N_T = N_s \times N_p = 1s \times 1p$$

Since the wire sizing, controller sizing, and the protection components selections are well-studied in the previous examples, it is sufficient to give the wiring diagram only, as shown in Figure 6.15.

b. By following the above steps to design a solar-powered cathodic protection system, the following calculations are performed.

Load Demand Calculation: From (6.15), the minimum required voltage:

$$V_{\text{min}} = R_T \times I_R = 50 \, \Omega \cdot 200 \times 10^{-3} \text{ A} = 10 \text{ V}$$

Therefore, it is sufficient to set the DC system voltage to 12 V [V_{PV} (with battery) $\geq V_{\text{min}}$]. The DC power is 2.4 W ($= 12 \text{ V} \cdot 200 \times 10^{-3}$ A). Note that the cathodic controller unit automatically adjusts the load current of 200 mA. In this case, the weekly and daily energy demands are:

$$\sum [\text{Wh/week}] = (2.4 \times 24 \times 7) = 403.2 \, [\text{Wh/week}]$$

$$\frac{\sum [\text{Wh/week}]}{V_b} = \frac{403.2}{12} = 33.6 \, [\text{Ah/week}]$$

$$\frac{[\text{Ah/week}]}{7} = \frac{33.6}{7} = 4.8 \, [\text{Ah/day}]$$

Since the round-trip efficiency is given as 0.9, the daily total electrical energy demand (Ah/day) can be calculated as:

$$\text{Total (Corrected) Energy Demand} \left(\frac{Ah}{day}\right) = \frac{4.8 \left(\frac{Ah}{day}\right)^{DC}}{0.9} = 5.34 \text{ Ah/day}$$

Battery Calculation: From (5.42) to (5.45), battery requirements can be determined as below:

$$C_{\text{Battery-net}} [Ah] = \frac{E_{\text{Daily Avg}} [Wh/day] \times DoA [day]}{DoD_{\max} \times V_b \times \eta_b \times f_{\text{Temp-b}}} = \frac{\frac{403.2 [Wh/week]}{7} \times 5}{0.7 \times 12 \times 0.9 \times 1} \cong 38.1 \text{ Ah}$$

$$N_{\text{Battery}} = \frac{C_{\text{Battery-net}} [Ah]}{c_b [Ah]} = \frac{38.1 [Ah]}{50 [Ah]} \cong 0.76 \xrightarrow{\text{rounded up}} 1$$

$$N_{sb} = \frac{12}{12} = 1$$

$$N_{pb} = \frac{1}{1} = 1$$

As a result, the battery storage configuration is ($1s \times 1p$), and accordingly, a total of 1 battery is needed.

PV Array Calculation: Based on the typical derating factors ($f_{\text{Manufacturing}} : 0.98$, $f_{\text{Temp}} : 0.94$, $f_{\text{Soiling}} : 0.95 \Rightarrow$ PV Array Losses $\Delta P_{PV} = 098 \times 0.94 \times 0.95 = 0.875$), the PV array sizing can be determined as follows.

$$E_{\text{overall}} = \frac{E_{\text{Load}} [Wh/day]}{(\Delta P_{PV}) \cdot \eta_{\text{overall}}} = \frac{\frac{403.2 [Wh/week]}{7}}{(0.875) \times (0.90)} = \frac{57.6 [Wh/day]}{0.7875} \cong 74 \text{ Wh/day}$$

$$P_{\text{Peak(PV)}} = \frac{E_{\text{overall}}}{PSH_{\text{avg}}} = \frac{74 \text{ Wh/day}}{4.7 \text{ h/day}} = 15.75 W_p$$

$$I_{DC} = \frac{P_{\text{Peak(PV)}}}{V_{DC}} = \frac{15.75 W}{12 V} \approx 1.3125 \text{ A}$$

$$N_s = \frac{V_{DC}}{V_{mp}} = \frac{12}{13.40} = 0.89 \approx 1$$

$$N_p = \frac{I_{DC}}{I_{mp}} = \frac{1.3125}{2.25} = 0.58 \approx 1$$

Consequently, a single 30-watt PV module is sufficient to feed the system.

$$N_T = N_s \times N_p = 1s \times 1p$$

Considering the designs determined above, the wiring diagram for cathodic protection can be drawn as in Figure 6.16.

Photovoltaic System Applications

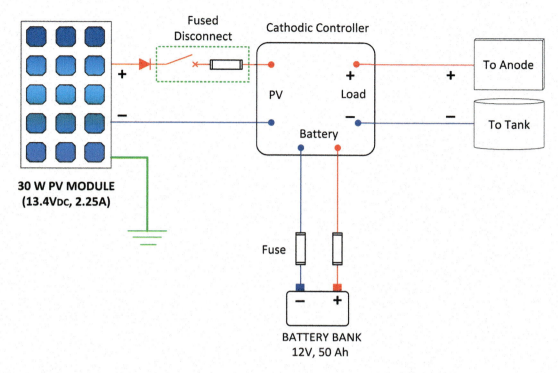

FIGURE 6.16 PV-powered cathodic protection system.

6.7 PROBLEMS

P6.1. A PV-powered lighthouse system will be designed to guide the watercraft. The design criteria are listed below.
- The system will run only at night, and the system voltage is 12 V_{DC}.
- The current drawn by the light source and its control circuit in a single operating cycle is given in Figure 6.17.
- The PV panels are tilted at about 55 degrees to maximize the solar energy in the winter season.
- Assume that the mean PSH value during the winter is 5 h/day at the given tilt of the site.
- The highest DoA (Days of Autonomy) value is 10 for the given region.
- 12 V and 100 Ah lead-acid batteries will be used for energy storage needs. The battery temperature factor will be taken as 1. The maximum DoD (Depth of Discharge) value of the battery is 70%.
- The system will include a charge controller and the combined round-trip efficiency for the selected battery & charger is 90%.
- PV modules will be selected from the commercially available products.
Based on the design criteria given above, design the PV-powered lighthouse system, and draw the wiring diagrams.

P6.2. A PV-powered stand-alone system will be designed to feed square lighting. The design criteria of the lighting system are listed below.
- The square lighting system will run continuously at night times only, from sunset to sunrise.

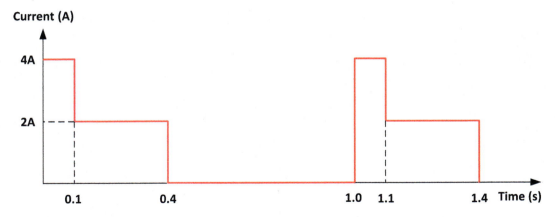

FIGURE 6.17 The current drawn by the lighthouse.

- A total of 10 lighting poles will be used, and each lighting pole has 2×40 W LED bulbs.
- The designed system voltage is 24 V_{DC}.
- Each lighting unit has its PV panel & battery group, which are mounted on a pole.
- The highest DoA (Days of Autonomy) value is assumed to be 4 days for the given region.
- 12 V and 100 Ah lead-acid batteries will be used for energy storage needs. The battery temperature factor will be taken as 1. The maximum DoD (Depth of Discharge) value of the battery is 70%.
- The system will include a charge controller and the combined round-trip efficiency for the selected battery & charger pair is 90%.
- PV modules will be selected from the commercially available products.

Based on the design criteria given above:

a. Design the PV-powered lighting pole and draw the wiring diagram only for a single pole system.
b. If a central PV array and battery pack were designed for the 10 lighting poles (instead of individually designed poles), what would be the system sizes? Compare them for both cases.

P6.3. A stand-alone PV system will be designed to power a remote GSM base station located at the coordinates 37.98 N, 23.72 E in Athens, Greece. The criteria for designing the remote telecommunication system are listed below.

- The GSM base station is mounted on a tower and runs continuously, day and night, all year round.
- The average monthly PSH (peak sun hours) values on the horizontal surface of the site are listed in Table 6.25.
- The designed system voltage (DC bus Voltage) is 48 V_{DC}.
- PV panels will be ground-mounted and shade-free.
- The electrical loads of the base station and their typical operating times are given in Table 6.26. Assume that the effect of seasonal variations on the load demand of the base station is negligible.
- The distance between the PV array and the base station is 30 m.
- The highest DoA (Days of Autonomy) value is assumed to be 2 days for the given region. Assume that the average ambient temperature during DoA days is −10°C.
- 12 V and 500 Ah lead-acid batteries will be used for energy storage needs. The provided temperature correction curve for the lead-acid battery is given in Figure 6.1. The maximum DoD (Depth of Discharge) value of the battery is 70%.

Photovoltaic System Applications

TABLE 6.25
Average Monthly PSH Values on the Horizontal Surface

Months	Jan.	Feb.	Mar.	Apr.	May	Jun.	Jul.	Aug.	Sep.	Oct.	Nov.	Dec.
PSH (h/day)	3.17	4.13	5.32	6.88	7.9	8.09	7.77	6.96	5.79	4.75	3.54	2.93

TABLE 6.26
The Electrical Loads of the Base Station

Electrical Load	DC or AC	Power (W)	Operating Time (h/day)
Radio Unit & RF Feeder	DC	3150	24
Baseband Unit	DC	1575	24
Signal Transmitting Loads	DC	97.5	24
Power Supply Unit	DC	900	24
Remote Monitoring	DC	15	24
Lighting & Security	AC	150	12
Fan	DC	90	24
Air-Conditioner	AC	1875	6.5

- The system will include a charge controller and the combined round-trip efficiency for the selected battery & charger pair is 95%.
- The efficiency of the converter and inverter feeding the DC and AC loads is approximately 95%.
- PV modules will be selected from the commercially available products.

Design a PV-powered stand-alone system for the given GSM base station and determine its specifications. Afterward, draw the PV system diagram of the final design.

P6.4. A Shallow well pump with a DC motor will be designed to provide an average of 20 m³/day of water for the domestic needs of a small village. The design considerations for the remote area water pumping system are listed below.
- The water pumping system will be used all year round.
- The total capacity of the water storage tank is about 80 m³.
- The PV panels are tilted at about 55 degrees to maximize the solar energy in the winter season.
- Assume that the mean PSH value during the winter is 4.6 h/day at the given tilt of the site.
- The highest DoA (Days of Autonomy) value is 4 for the given region.
- The voltage of the pump motor is 24 V_{DC}.
- It is decided to use a battery backup in the system to allow pumping for 24 hours a day. 12 V and 300 Ah lead-acid batteries will be used for energy storage needs. The battery temperature factor will be taken as 1. The maximum DoD (Depth of Discharge) value of the battery is 70%.
- The total dynamic head of the pump system is calculated as 25 m.
- The safe yield limit of the water source is powerful enough.
- PV panels will be ground-mounted and shade-free.
- PV modules will be selected from the commercially available products.

Design a PV-powered stand-alone solar pumping system for the given specifications and draw the PV system diagram of the final design.

P6.5. A PV-powered refrigerator/freezer (RF) system will be designed for the storage needs of medical stockpiles on the island of Cyprus. The design considerations are listed below.
- A dual-compressor RF unit was selected for the cooling system.
- Each compressor operates independently and draws 8 amps while operating.
- PV modules and the required battery pack will be selected from the commercially available products.
- The RF unit is used every day for 10 hours a day for the refrigerator and 6 hours for the freezer in the summer months. Similarly, the RF unit is used every day for 8 hours a day for the refrigerator and 4 hours for the freezer during the winter months. The operating time of refrigerator & freezer compressors are 9 and 5 hours, respectively, for the Spring/Autumn seasons.
- The PV panels are tilted at about 50 degrees to maximize the solar energy in the winter season.
- Assume that the mean PSH value during the winter is 5 h/day at the given tilt of the site.
- The highest DoA (Days of Autonomy) value is 3 for the given region. Assume that the average ambient temperature during DoA days is −10°C.
- The voltage of each compressor motor is 12 V_{DC}.

Design a PV-powered stand-alone refrigeration system for the given specifications and draw the PV system diagram of the final design.

BIBLIOGRAPHY

1. A. Kadri, "Design of PV system for mobile Tele-communication tower", *Int. J. Eng. Res.*, vol. V9, no. 06, pp. 339–345, 2020.
2. G. S. Adly, W. R. Anis, and I. M. Hafez, "Design of photovoltaic powered cathodic protection system", *Int. J. Sci. Technol. Res.*, vol. 06, no. 07, pp. 246–253, 2017.
3. M. Al, "Nauka Technika solar PV powered cathodic protection for a buried pipeline", *Polska Energetyka Sloneczna*, vol. 04, no. I, pp. 5–8, 2018.
4. M. H. Alsharif, "Comparative analysis of solar-powered base stations for green mobile networks", *Energies*, vol. 10, no. 8, p. 1208, 2017.
5. M. H. Alsharif, "Techno-economic evaluation of a stand-alone power system based on solar/battery for a base station of global system mobile communication", *Energies*, vol. 10, no. 3, pp. 1–20, 2017.
6. M. H. Alsharif and J. Kim, "Optimal solar power system for remote telecommunication base stations: A case study based on the characteristics of south Korea's solar radiation exposure", *Sustainability*, vol. 8, no. 9, pp. 1–21, 2016.
7. M. H. Alsharif, R. Nordin, and M. Ismail, "Energy optimization of hybrid off-grid system for remote telecommunication base station deployment in Malaysia", *Eurasip J. Wirel. Commun. Netw.*, vol. 2015, no. 1, pp. 1–15, 2015.
8. A. V. Anayochukwu, "Optimal sizing and application of renewable energy sources at GSM base station site," *Int. J. Renew. Energy Res.*, vol. 3, no. 3, pp. 579–585, 2013.
9. V. A. Ani and N. A. Ndubueze, "Energy optimization at GSM base station sites located in rural areas," *Int. J. Energy Optim. Eng.*, vol. 1, no. 3, pp. 1–31, 2012.
10. O. Arnold, F. Richter, G. Fettweis, and O. Blume, "Power consumption modeling of different base station types in heterogeneous cellular networks," in *2010 Future Network and Mobile Summit*, Florence, Italy, pp. 1–8, 2010.
11. A. A. Atoyebi, "Total dynamic head determination model for submersible pumps installation", *Int. J. Appl. Sci. Technol.*, vol. 5, no. 1, pp. 95–102, 2015.
12. A. Ayang, P.-S. Ngohe-Ekam, B. Videme, and J. Temga, "Power consumption: Base stations of telecommunication in Sahel Zone of Cameroon: Typology based on the power consumption—Model and energy savings", *J. Energy*, vol. 2016, pp. 1–15, 2016.
13. M. Benghanem, K. O. Daffallah, and A. Almohammedi, "Estimation of daily flow rate of photovoltaic water pumping systems using solar radiation data", *Results Phys.*, vol. 8, pp. 949–954, 2018.

14. A. Chatzipapas, S. Alouf, and V. Mancuso, "On the minimization of power consumption in base stations using on/off power amplifiers," *2011 IEEE Online Conference on Green Communications. GreenCom'11*, France, pp. 18–23, 2011.
15. R. J. Chilundo, U. S. Mahanjane, and D. Neves, "Design and performance of photovoltaic water pumping systems: Comprehensive review towards a renewable strategy for Mozambique", *J. Power Energy Eng.*, vol. 06, no. 07, pp. 32–63, 2018.
16. B. Durin, S. Lajqi, and L. Plantak, "'Worst month' and 'critical period' methods for the sizing of solar irrigation systems - A comparison", *Rev. Fac. Ing.*, vol. 88, pp. 100–109, 2018.
17. D. Gadze, S. B. Aboagye, and K. A. P. Agyekum, "Real Time Traffic Base Station Power Consumption," *Int. J. Comput. Sci. Telecommun.*, vol. 07, no. 5, pp. 6–13, November, 2016.
18. J. P. Guyer, F. Asce, and F. Aei, "An introduction to cathodic protection principles", *J. Chem. Eng. Inf.*, vol. 877, no. 5, pp. 1–23, 2014.
19. S. H. Hassmoro, "Cathodic protection power supply implementations using solar photovoltaic system", Final Project Report, Universiti Teknologi PETRONAS, vol. 1, no. 1, pp. 1–55, 2014.
20. M. S. Hossain and M. F. Rahman, "Hybrid solar PV/Biomass powered energy efficient remote cellular base stations", *Int. J. Renew. Energy Res.*, vol. 10, no. 1, pp. 329–342, 2020.
21. K. K. Jasim and M. A. Abdul-Hussain, "Optimization of hybrid PV / diesel power system for remote telecom station", *Int. J. Appl. or Innov. Eng. Manag.*, vol. 5, no. 3, pp. 39–46, 2016.
22. B. Laoun, K. Niboucha, and L. Serir, "Cathodic protection of a buried pipeline by solar energy", *Rev. des Energies Renouvelables*, vol. 12, no. 1, pp. 99–104, 2008.
23. J. Lorincz and I. Bule, "Renewable energy sources for power supply of base station sites", *Int. J. Bus. Data Commun. Netw.*, vol. 9, no. 3, pp. 53–74, 2013.
24. J. Lorincz, T. Garma, and G. Petrovic, "Measurements and modelling of base station power consumption under real traffic loads", *Sensors*, vol. 12, no. 4, pp. 4281–4310, 2012.
25. M. Deruyck, E. Tanghe, W. Joseph, and L. Martens, "Modelling the energy efficiency of microcell base stations", in *1st International Conference on Smart Grids, Green Communications and IT Energy-Aware Technologies (ENERGY) 2011*, p. 6, Ghent University, Belgium, 2011.
26. T. Mohsen, A. Ali, and R. Iman, "Feasibility of using impressed current cathodic protection systems by solar energy for buried oil and gas pips", *Int. J. Eng. Adv. Technol.*, vol. 3, no. 2, pp. 222–225, 2013.
27. T. Morales, "Design of small photovoltaic (PV) solar-powered water pump systems natural resources conservation service" United States Department of Agriculture, Technical Notes, no. 28, Portland, Oregon, 2010.
28. M. A. Al-Refai, "Optimal design and simulation of solar photovoltaic powered cathodic protection for underground pipelines in Libya", *J. Electr. Eng.*, vol. 7, no. 2, pp. 61–73, 2019.
29. O. Navigation, "Impressed current cathodic protection for reinforced", *Afr. J. Electr. Electron. Res.*, vol. 1, no. 1, pp. 2016–2017, 2018.
30. L. J. Olatomiwa, S. Mekhilef, and A. S. N. Huda, "Optimal sizing of hybrid energy system for a remote telecom tower: A case study in Nigeria", in *2014 IEEE Conference on Energy Conversion, CENCON 2014*, no. February 2015, pp. 243–247, Malaysia, 2014.
31. M. Á. Pardo, R. Cobacho, and L. Bañón, "Standalone photovoltaic direct pumping in urban water pressurized networks with energy storage in tanks or batteries", *Sustainability*, vol. 12, no. 2, pp. 1–20, 2020.
32. R. A. Shmais, "Powering of Radio Communication Stations in Remote Areas by Solar PV : Optimal System Design and Economics", Doctoral dissertation, 2015.
33. B. S. V and S. W. S, "Solar photovoltaic water pumping system for irrigation: A review," *African J. Agric. Res.*, vol. 10, no. 22, pp. 2267–2273, 2015.
34. Y. Zhang et al., "An overview of energy-efficient base station management techniques", in *2013 24th Tyrrhenian International Workshop on Digital Communication - Green ICT, TIWDC 2013*, 2013.

7 Grid Codes for Integration of PV Systems and Power Quality Assessment

7.1 INTRODUCTION

The use of renewable energy is constantly expanding all over the world. The applications of solar PV systems are also increasing in many areas, from small stand-alone applications to MW-scale grid-connected systems. As in all power generation systems, the high PV penetration creates several challenges in the operation of power systems. This chapter will describe the grid code requirements for the integration of solar PV systems into electrical grids. In addition, network disturbance limits and their calculation steps are explained with worked examples, specifically for solar PV systems.

Grid code requirements for connecting the PV systems to LV or MV grids are examined below for both normal operation and grid disturbance conditions. Grid code requirements, as given in Figure 7.1, can be split into two main classes, that is, normal (steady-state) operating and grid disturbance (transient) conditions.

Operating ranges of voltage and frequency during normal and faulty conditions are always relevant for all power grids. These operating ranges in general can be specified in Figure 7.2 for a typical power system.

The grid codes in Figure 7.1, indeed, identify the limits of voltage and frequency deviations based on normal operating ranges. In this context, Figure 7.1 shows the basic grid code requirements for a typical PV power plant that must accomplish to be connected to the grid. These grid codes and related power quality assessments are discussed in detail in the following sections.

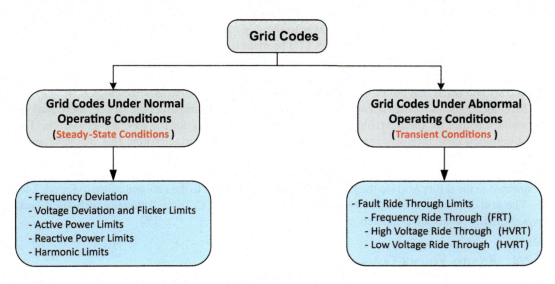

FIGURE 7.1 Schematic representation of cathodic protection based on impressed current.

DOI: 10.1201/9781003415572-7

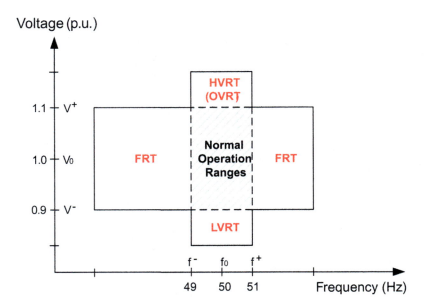

FIGURE 7.2 Typical operating ranges for voltage and frequency during normal and fault ride-through conditions (FRT: frequency ride-through, LVRT: low-voltage ride-through, HVRT: high-voltage ride-through, OVRT: over-voltage ride-through).

7.2 TERMS AND DEFINITIONS

In this section, some specific terms regarding grid codes, power quality assessment, and the respective formulations are defined for ease of explanation.

7.2.1 Point of Common Coupling and Grid Connection Point

PCC (point of common coupling) and GPC (grid connection point) are two terms that are often confused. In the context of a grid-connected PV system, the GPC is the first point at which the PV plant is connected to the grid via the metering and protection system. Usually, after the GCP, there is a transmission line to carry the PV power to the grid. The closest connection point of this line segment to the utility network is called PCC. The main difference between PCC and GCP is that other network users may have already been connected to the PCC, or there may be new users in the future to be connected to the PCC. The PCC and GPC locations can be illustrated in Figure 7.3. It should also be noted that the PCC and GPC locations can be the same in some cases.

PCC locations may differ according to system connection configurations. Hence, there may be different schemes for the interconnection to the grid, particularly for the MV and HV levels. The most common configurations are presented in Figure 7.4a. Importantly note that the schemes in Figure 7.4a are also valid for distributed generations (DGs) connected to the LV or LV/MV grids. DGs in the LV grids have relatively smaller power than those in MV and HV grids. Some will be connected to the utility grid via the LV/MV transformers, similar to the configurations in Figure 7.4a.

The grid disturbances are usually assessed regarding the PCC. In the following sections, related power quality issues are evaluated in detail.

7.2.2 Relative Voltage Change (ε)

As known, every load change causes a current change in the grid and thus a voltage change at the PCC. Herein, the relative voltage changes between the AC side of a PV power plant and PCC will be analyzed using phasor (vector) diagrams. The main purpose of this analysis is to derive a relative

PV Systems and Power Quality Assessment

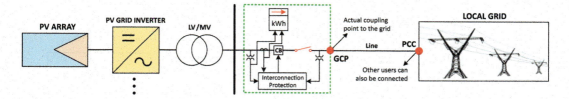

FIGURE 7.3 GCP and PCC locations for a typical network.

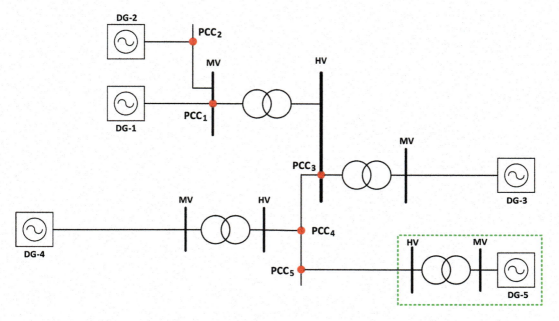

FIGURE 7.4A Various interconnection schemes and associated PCCs for the distributed generations (DGs). The schemes here also apply to DGs connected to the LV or LV/MV networks.

voltage drop formula (ε) based on the short-circuit power at the PCC. Assuming that the system is symmetrical, let us consider the single-line diagram in Figure 7.4b.

Due to a load current change of ΔI, the phasor diagram of the longitudinal and perpendicular voltage drops along the power line is given in Figure 7.4c.

where

- φ: Load angle (degree)
- θ: Impedance angle (degree)
- δ: Angle between the voltages of sending and receiving ends (degree)
- R, X: Resistance and reactance of the line (Ω)
- I: Load current (amps)
- ΔI: Due to the load change ΔS (kVA), the current change in (amps)
- U_1, U_2 Phase-to-phase voltages in (V) at the receiving and sending ends, respectively

Since $\Delta U \approx \Delta U_{\text{Long}}$, the relative voltage drop $\varepsilon = \Delta U / U$ takes into account only the longitudinal voltage drop. So, for the absolute voltage drop ΔU:

$$\Delta U = \Delta I \cdot R \cdot \cos\varphi + \Delta I \cdot X \cdot \sin\varphi \tag{7.1}$$

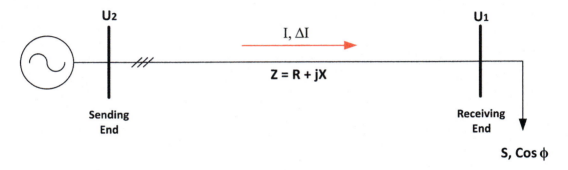

FIGURE 7.4B Typical single-line diagram of a three-phase system under symmetrical loading.

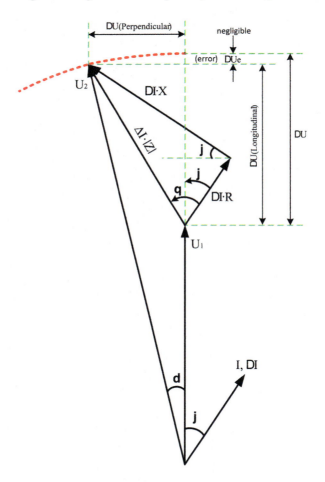

FIGURE 7.4C Phasor diagram of the voltage drops due to a current change of ΔI.

If the current is written in terms of its active and reactive components ($\Delta I_p = \Delta I \cdot \cos\varphi$, and $\Delta I_q = \Delta I \cdot \sin\varphi$), ΔU turns into:

$$\Delta U = \Delta I_p \cdot R + \Delta I_q \cdot X \tag{7.2}$$

The load change $\Delta S = \Delta P + j\Delta Q$, where $\Delta P = \Delta I_p \cdot U_2$, and $\Delta Q = \Delta I_q \cdot U_2$. Hence, let us insert $\Delta I_p = \dfrac{\Delta P}{U_2}$ and $\Delta I_q = \dfrac{\Delta Q}{U_2}$ values into (7.2):

PV Systems and Power Quality Assessment

$$\Delta U = \frac{\Delta P}{U_2} \cdot R + \frac{\Delta Q}{U_2} \cdot X \tag{7.3}$$

Since $\varepsilon = \Delta U / U_2$, the relative voltage change ε can be expressed as below:

$$\varepsilon = \frac{\Delta U}{U_2} = \frac{\Delta P}{U_2^2} \cdot R + \frac{\Delta Q}{U_2^2} \cdot X \tag{7.4}$$

Let's multiply and divide equation (7.4) by the impedance "Z":

$$\varepsilon = \frac{Z}{U_2^2} \cdot \frac{\Delta P}{Z} \cdot R + \frac{Z}{U_2^2} \cdot \frac{\Delta Q}{Z} \cdot X \tag{7.5}$$

Since the short-circuit power at Bus-1 is $S_k = U_2^2 / Z$, we can replace Z / U_2^2 expression by $(1 / S_k)$:

$$\varepsilon = \frac{\Delta P}{S_k} \cdot \frac{R}{Z} + \frac{\Delta Q}{S_k} \cdot \frac{X}{Z} \tag{7.6}$$

Let us substitute the equations $\Delta P = \Delta S \cdot \cos\varphi$, $\Delta Q = \Delta S \cdot \sin\varphi$, $R = Z \cdot \cos\theta$, and $R = Z \cdot \sin\theta$ in (7.6):

$$\varepsilon = \frac{\Delta S}{S_k} \cdot (\cos\varphi \cdot \cos\theta + \sin\varphi \cdot \sin\theta) \tag{7.7}$$

Due to trigonometric expression $\cos(\theta - \varphi) = \cos(\varphi - \theta) = \cos\varphi \cdot \cos\theta + \sin\varphi \cdot \sin\theta$:

$$\varepsilon = \frac{\Delta S}{S_k} \cdot \cos(\theta - \varphi) \tag{7.8}$$

where ε is the relative voltage change at Bus-1, $\theta = \tan^{-1}(X/R)$, and S_k is the short-circuit power at Bus-1. Here, the value of φ is the angle between the current and voltage before the load change. Therefore, changes in load affect the previous angle φ in the (+) or (−) direction. Of course, inductive load changes act as "+" and capacitive load changes as "−". Thus, if equation (7.4) is used instead of (7.8), $\Delta$the Q value takes the "−" value for capacitive changes.

7.3 STEADY-STATE VOLTAGE DEVIATION LIMITS

Voltage is a common characteristic parameter for electrical power systems. Therefore, the voltage magnitude can be used as a fundamental criterion to determine if a power system is operating normally or experiencing challenges during the operation. Typically, all grid operators have imposed voltage limits that should not be exceeded at both LV and MV levels. Although there are slight differences in international standards (IEC 61727 and IEEE 1574) and country standards, commonly accepted voltage limits are summarized in Table 7.1 for the normal operation of PV systems in LV and MV networks.

TABLE 7.1
Steady-State Voltage Tolerance Limits for PV Systems Connected to LV and MV Networks

Voltage Level	Voltage at PCC[a]	Max Trip Time
Low Voltage		Normal operation
120 V-60 Hz System	88% < V < 110% (for 120 V base)	
230/400 V-50 Hz System	85% < V < 110% (for 230/400 V base)	
Medium Voltage	95% < V < 105%	Normal operation

[a] The maximum voltage increase due to PV systems should be less than 3% for LV networks and 2% for MV networks.

According to IEC 61000: 3-3, the voltage change limits that a PV plant may cause, when connecting to or disconnecting from the LV distribution network, must be within the range of $\mp 3.3\%$. As for the LV grid voltage limits, the fluctuation range, due to the activation or deactivation of the PV system, should always remain within the steady-state limits of $\mp 10\%$. The relative voltage change limits due to activation or deactivation of the PV system are illustrated in Figure 7.5.

The voltage rise caused by the PV systems can be found by using the two-bus model, reduced to the Thevenin equivalent. Besides, there are different configurations for the equivalent two-busbar model, depending on whether the load bus exists. These configurations are analyzed separately in the following.

Two-bus Equivalent Model without Load Point: A PV system connected to a distribution feeder can be represented by a Thevenin equivalent model, as shown in Figure 7.6. Since the PV system injects active (P_{PV}) and reactive power (Q_{PV}) into the grid, P_{PV} and Q_{PV} values will be positive.

Referring to equations (7.4) and (7.8), relative voltage rise (ε) at the PCC (PV system busbar) in Figure 7.5 can be calculated as below:

$$\varepsilon = \frac{P_{PV} \cdot R_S + Q_{PV} \cdot X_S}{V_{PV}^2} \tag{7.9}$$

$$\varepsilon = \frac{S_{PV}}{S_{SC}} \cdot \cos(\theta_s - \varphi_{PV}) \tag{7.10}$$

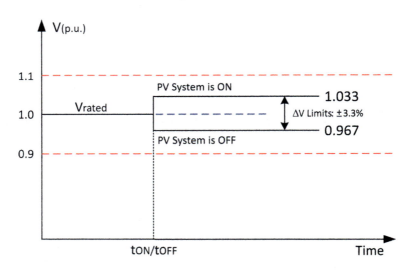

FIGURE 7.5 Allowable voltage rise and drop limits when the PV system is switched on or switched off.

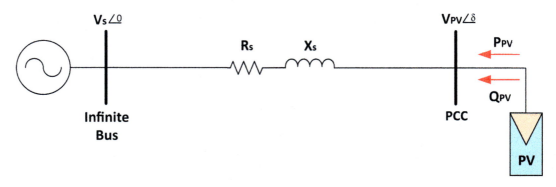

FIGURE 7.6 Thevenin equivalent model of a typical feeder with a PV system.

PV Systems and Power Quality Assessment

where

S_{PV}: Apparent power of PV system at PCC (VA)
P_{PV}: Active power of PV system at PCC (VA)
Q_{PV}: Reactive power of PV system at PCC (VA)
V_{PV}: Phase-to-phase voltage at PCC (V)
S_{SC}: Short-circuit power at PCC (VA)
θ_s: Network impedance angle at PCC (degree)
φ_{PV}: Angle of the apparent power of the PV station at PCC (degree)
R_S: Network resistance at PCC (Ω)
X_S: Network reactance at PCC (Ω)

Note that equations (7.9) and (7.10) are also valid for other DG systems under the following conditions:

- It is valid for DG systems (such as over-excited generators), which supply inductive power to the grid ($P > 0$, $Q > 0$, and $0 \leq \varphi_{DG} \leq 90$).
- It is valid for DG systems (such as under-excited generators), which draws inductive power from the grid ($P > 0$, $Q < 0$, and $0 \leq \varphi_{DG} \leq -90$).
- Therefore, compensation capacitors in the system must be taken into account.

From the above equations, the maximum PV hosting capacity ($P_{PV\,max}$ in watts, or $S_{PV\,max}$ in VA) at PCC can also be found by substituting the maximum allowable relative voltage change (ε_{max}) into equations (7.9) and (7.10). It should be noticed that $Q_{PV} = P_{PV} \cdot \tan(\varphi_{PV})$, and $S_{PV} = P_{PV} / \cos(\varphi_{PV})$.

$$P_{PV\,max} = \frac{\varepsilon_{max} \cdot V_{PV}^2}{R_S + X_S \cdot \tan(\varphi_{PV})} \tag{7.11}$$

$$P_{PV\,max} = \frac{\varepsilon_{max} \cdot S_{SC}}{\cos(\theta_s - \varphi_{PV})} \cdot \cos(\varphi_{PV}) \tag{7.12}$$

For example, the maximum voltage increase due to PV systems should be less than $\varepsilon_{max} = 3\%$ for LV networks, and $\varepsilon_{max} = 2\%$ for MV networks.

Two-Bus Equivalent Model with a Load Point: In this model, the Thevenin equivalent circuit has a load point, as shown in Figure 7.7.

As similar to the previous one, the hosting capacity can again be derived from the voltage rise equations. The voltage rise expression due to the PV plant can easily be written with the help of equation (7.1):

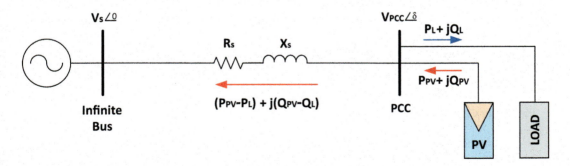

FIGURE 7.7 Thevenin equivalent model of a typical grid-connected PV system with a load point.

$$\Delta U = V_{PV} - V_s \approx \frac{\nabla P}{V_{PCC}} \cdot R_S \cdot \cos\varphi + \frac{\Delta P}{V_{PCC}} \cdot X_S \cdot \sin\varphi \quad (7.13)$$

Let's get ΔP from (7.13) as below:

$$\Delta P = P_{PV} - P_L = \frac{V_{PCC}(V_{PCC} - V_s)}{R_S \left[1 + \frac{X_S}{R_S} \cdot \tan(\varphi)\right]} \quad (7.14)$$

From equation (7.14), the maximum PV hosting capacity ($P_{PV\ max}$) can also be obtained under the minimum load ($P_{L\ min}$) condition as below:

$$P_{PV\ max} = \frac{V_{PCC\ max}(V_{PCC\ max} - V_s)}{R_S \left[1 + \frac{X_S}{R_S} \cdot \tan(\varphi_{PCC})\right]} + P_{L\ min} \quad (7.15)$$

where $V_{PCC\ max}$ is the maximum allowable voltage at PCC. From equation (7.14), the maximum PV hosting capacity under any power factor condition can be found. For example, let's assume that reactive power control is applied at the PCC and the power factor is kept constant at 0.95. Hence, the $P_{PV\ max}$ value for $\cos\varphi_{PCC} = 0.95$ can be calculated by (7.16).

$$P_{PV\ max} = \frac{V_{PCC\ max}(V_{PCC\ max} - V_s)}{R_S \left[1 + \left(\frac{X_S}{R_S}\right) \cdot \tan(\arccos 0.95)\right]} + P_{L\ min} \quad (7.16)$$

If we want to obtain the maximum PV hosting capacity expression depending on the relative voltage rise, it is possible to derive it from equations (7.4) and (7.8) as below. It should be noticed that $S_{PCC} = (P_{PV} - P_{Load})/\cos(\varphi_{PV})$. Also, note that $\varepsilon = \varepsilon_{max}$ is written to get the maximum PV hosting capacity ($P_{PV\ max}$), and the load's active power, in this case, will be $P_{L\ min}$.

$$P_{PV\ max} = \frac{\varepsilon_{max} \cdot V_{PCC}^2 + P_{L\ min} \cdot [R_S + X_S \cdot \tan(\varphi_{PCC})]}{R_S + X_S \cdot \tan(\varphi_{PCC})} \quad (7.17)$$

$$P_{PV\ max} = \frac{\varepsilon_{max} \cdot S_{SC}}{\cos(\theta_s - \varphi_{PCC})} \cdot \cos(\varphi_{PCC}) + P_{L\ min} \quad (7.18)$$

The maximum PV hosting capacity expressions given above can be applied for networks as long as they can be reduced to a two-bus system. With the help of the reduced two-bus model, accurate results can be obtained within a negligible error range. However, for the assessment of complicated or extended grid topologies, it is convenient to obtain the hosting capacity results with the help of load flow analysis.

Example 7.1: Voltage Rise and PV Hosting Capacity Calculation at PCC

Consider the 500 kW$_p$ PV power plant worked in Example 5.20. With the help of obtained results of Example 5.20, the equivalent two-bus diagrams of the PV plant can be drawn with respect to the busbar C, as shown in Figure 7.8.

a. Find the relative voltage rise at PCC due to the 500 kW PV system.
b. Considering the given conditions, calculate the maximum PV capacity that can be connected to the PCC in terms of voltage rise criteria.
c. Assume that the PCC bus also feeds the local loads. If the load profile during a typical daytime is as in Figure 7.9, calculate the maximum PV hosting capacity for PCC.

PV Systems and Power Quality Assessment

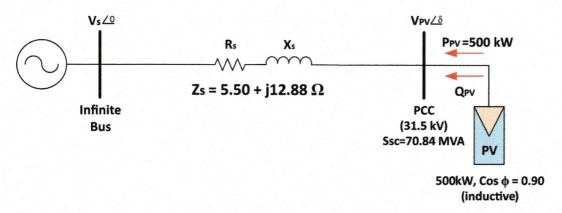

FIGURE 7.8 Thevenin equivalent model of the 500 kW PV power plant with reference to Bus-C (see Example 5.20).

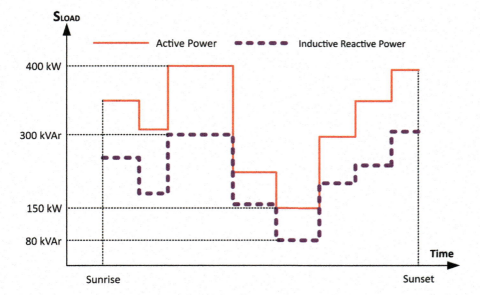

FIGURE 7.9 Typical load profile during the daytime at PCC.

Solution:

a. From (7.9), the relative voltage rise (ε) at the PCC can be calculated as ($\varphi_{PV} = \cos^{-1} 0.90 = 25.84°$):

$$\varepsilon = \frac{P_{PV} \cdot R_S + Q_{PV} \cdot X_S}{V_{PV}^2} = \frac{P_{PV} \cdot R_S + P_{PV} \cdot \tan(\varphi_{PV}) \cdot X_S}{V_{PV}^2} = \frac{500 \cdot 10^3 \times \left[5.50 + \tan(25.84) \times 12.88\right]}{(31500)^2}$$

$$= 0.0059 = 0.59\%$$

$$\Rightarrow \varepsilon(\%) = 0.59\%$$

As can be seen, the relative voltage rise at the PCC is less than the upper limits of 3% and 2%, respectively for LV and MV networks.

b. PV hosting capacity can be calculated using both (7.11) and (7.12) equations. From (7.11), the PV hosting capacity for MV PCC can be calculated as below ($\varepsilon_{max} = 0.02$ for MV distribution system):

$$P_{PV\,max} = \frac{\varepsilon_{max} \cdot V_{PV}^2}{R_S + X_S \cdot \tan(\varphi_{PV})} = \frac{0.02 \times (31500)^2}{5.50 + 12.88 \times \tan(25.84)} = 1.69 \text{ MW}$$

From (7.12), we get ($\theta_s = \tan^{-1}\left(\frac{12.88}{5.50}\right) = 66.87°$, and $S_{SC} = 70.84$ MW):

$$P_{PV\,max} = \frac{\varepsilon_{max} \cdot S_{SC}}{\cos(\theta_s - \varphi_{PV})} \cdot \cos(\varphi_{PV}) = \frac{0.02 \times (70.84 \text{ MW})}{\cos(66.87 - 25.84)} \times \cos(25.84) = 1.69 \text{ MW}$$

As can be seen, the PV hosting capacity at the PCC is found to be 1.69 MW for both equations. Since there is already a 0.5 MW PV system installed, an additional 1.19 MW PV system (1.69 − 0.5 = 1.19 MW) can be installed on the same PCC bus.

c. PV hosting capacity in the case of PCC with a load point can be calculated using both (7.17) and (7.18) equations. From (7.17), the PV hosting capacity for MV PCC can be calculated as below ($\varepsilon_{max} = 0.02$ for MV distribution system, $S_{L\,min} = 150$ kW + $j80$ kVAr, and $\varphi_{PCC} = 25.84°$):

$$P_{PV\,max} = \frac{\varepsilon_{max} \cdot V_{PCC}^2 + P_{L\,min} \cdot [R_S + X_S \cdot \tan(\varphi_{PCC})]}{R_S + X_S \cdot \tan(\varphi_{PCC})}$$

$$= \frac{0.02 \times (31500)^2 + 150000 \cdot [5.50 + 12.88 \times \tan(25.84)]}{5.50 + 12.88 \times \tan(25.84)} = 1.84 \text{ MW}$$

From (7.18), we get ($\theta_s = \tan^{-1}\left(\frac{12.88}{5.50}\right) = 66.87°$ and $S_{SC} = 70.84$ MW):

$$P_{PV\,max} = \frac{\varepsilon_{max} \cdot S_{SC}}{\cos(\theta_s - \varphi_{PCC})} \cdot \cos(\varphi_{PCC}) + P_{L\,min}$$

$$= \frac{0.02 \times (70.84 \text{ MW})}{\cos(66.87 - 25.84)} \times \cos(25.84) + 0.15 = 1.84 \text{ MW}$$

As can be seen, the PV hosting capacity at the PCC is found to be 1.84 MW for both equations. Since there is already a 0.5 MW PV system installed, an additional 1.34 MW PV system (1.84 − 0.5 = 1.34 MW) can be installed on the same PCC bus.

7.4 VOLTAGE FLICKER LIMITS

Unsteadiness in the luminance or spectral distribution of light due to fluctuations in voltage is called flicker. The flickings of the light are perceived by our visual senses and often leave subjective impressions and influence people's brains. The flicker is measured by emulating the characteristic behavior of a 60 W incandescent lamp. The change in the luminance level of the bulb is perceived by the test persons as disturbing after a certain repeat rate (r) is reached. In experiments, at least 50% of the participants found the variation of luminance as disturbing when the flicker intensity P_{st} is greater than 1 pu ($P_{st} > 1$).

According to IEC 64000:4-15, there are two types of flicker intensity measurements as short term (P_{st}) and long term ($P_{\ell t}$). While the P_{st} value is measured over 10 minutes, the $P_{\ell t}$ value is calculated from 12 P_{st} values (equation 7.19), sequentially measured over a 2-hour interval.

$$P_{\ell t} = \sqrt[3]{\frac{1}{12}\sum_{j=1}^{12} P_{st(j)}^3} \tag{7.19}$$

PV Systems and Power Quality Assessment

where "j" is the index to count 12 P_{st} of the 10-minute values within the 2-hour time interval. Here, the $P_{\ell t}$ is a more important indicator than P_{st} in terms of reflecting the severity of the voltage fluctuation. According to EN 61000: 2-2&12, the upper limit value of P_{st} and $P_{\ell t}$ for a low voltage, the network is 1.0 and 0.8, respectively. Considering the planning and compatibility levels, the voltage flicker limits for LV and MV networks are given in Table 7.2.

In general, any voltage flicker caused by a grid-connected inverter must not exceed the allowable limits. Therefore, the voltage flicker due to the connection of a PV plant to the LV and MV networks should not exceed the flicker limit curves in Figure 7.10.

TABLE 7.2
Flicker Intensity Limits for LV and MV Grids (according to IEC 61000 and IEEE 1453 Standards)

Flicker Intensity	LV Networks		MV (11–33 kV)	MV (Above 33 kV)	MV Networks
	Compatibility	Planning	Compatibility	Compatibility	Planning
P_{st} [p.u.]	1	0.5	0.9	0.8	0.7
$P_{\ell t}$ [p.u.]	0.8	0.4	0.7	0.6	0.5

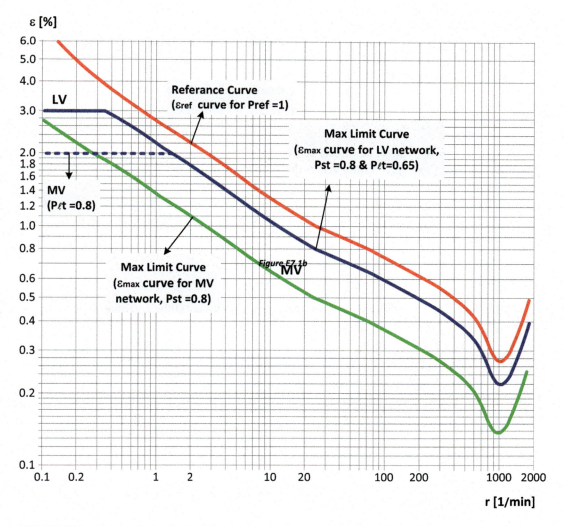

FIGURE 7.10 Voltage flicker limits (flicker repeat rate, r versus relative voltage change, ε).

As can be seen from the flicker limit curves in Figure 7.10, at low repeat rates $(0.1 < r[1/\min] < 1)$, the upper emission limits of the MV and LV networks are restricted to 2% and 3%, respectively. The main reason for this limitation is to ensure that the voltage levels at LV and MV are maintained. It should also be noted that for short-term load changes with a very low repetition rate $(r < 0.01[1/\min])$, which is around 5–10 times per day, the emission limits of the MV and LV network can, respectively, be restricted to 3% and 6%, as an exceptional case (equation 7.20).

$$\varepsilon = \begin{cases} \varepsilon_{\max, i}(\text{LV}) \leq 3\% \text{ for } 0.1 \leq r \leq 1; & \varepsilon_{\max, i}(\text{LV}) \leq 6\% \text{ for } r < 0.01 \\ \varepsilon_{\max, i}(\text{MV}) \leq 2\% \text{ for } 0.1 \leq r \leq 1; & \varepsilon_{\max, i}(\text{MV}) \leq 3\% \text{ for } r < 0.01 \end{cases} \quad (7.20)$$

Voltage flicker limits in an energy system should be assessed in detail for the PV generation, PCC, and load. Because there are contribution and compatibility levels for each party in the total emission value at PCC. For the networks having PV systems, a flicker assessment can be performed based on the flicker emission curves in Figure 7.7. The flicker evaluation can be performed in two steps as given below:

Step 1 – Determination of Relative Voltage Change at PCC: In this step, the relative voltage changes (ε_i) at PCC due to rapidly changing loads ($0.1 \leq$ repetition rate, $r[1/\min] \leq 2000$) can be calculated with the help of equation (7.4) or (7.8). For this purpose, network impedance (Z_s), network impedance angle (θ_s), network short-circuit power (S_{SC}), the load changes at PCC (ΔP & ΔQ), and load angles at PCC (φ_{PCC}) should be calculated (or measured).

Step 2 – Determination of Maximum Relative Voltage Changes and Flicker Intensity Limits at PCC: In this step, the maximum relative voltage changes ($\varepsilon_{\max}$) and flicker intensity limits ($P_{st,i}$) for LV and MV networks are needed to be determined with the help of the emission limit curve of Figure 7.10. Then, these values ($\varepsilon_{\max}$ and $P_{st,i}$) should respectively be compared with the relative voltage change value (ε_i) in Step-1, and the upper limit of P_{st} ($P_{st,\text{limit}}$). This comparison algorithm works as follows.

- First, the repetition rate of the flicker source (r) is determined.
- Then, the "$\varepsilon_{\max}$" and "ε_{ref}" values versus this repeat rate (r) are determined from the flicker emission curve of Figure 7.10.
- Afterward, using the proportionality relationship ($\frac{P_{st,i}}{\varepsilon_i} = \frac{P_{\text{ref}}}{\varepsilon_{\text{ref}}}$), the relevant P_{st} value is calculated using equation (7.21).

$$P_{st} = \frac{\varepsilon_{\max}}{\varepsilon_{\text{ref}}} \cdot P_{\text{ref}} \quad (7.21)$$

- Finally, the following criteria are needed to be checked to see if the conditions are met.

$$\begin{cases} \varepsilon_i(\text{LV}) \leq \varepsilon_{\max}(\text{LV}), & P_{st,i} \leq 0.8 \text{ \& } P_{\text{lt},i} \leq 0.65 \text{ for } r[1/\min] \geq 0.1 \\ \varepsilon_i(\text{MV}) \leq \varepsilon_{\max}(\text{MV}), & P_{st,i} \leq 0.8 \text{ \& } P_{\text{lt},i} \leq 0.65 \text{ for } r[1/\min] \geq 0.1 \end{cases} \quad (7.22)$$

$$\begin{cases} \varepsilon_i(\text{LV}) \leq \varepsilon_{\max}(\text{LV}), & \text{for } 0.01 \leq r[1/\min] \leq 0.1 \\ \varepsilon_i(\text{MV}) \leq \varepsilon_{\max}(\text{MV}), & \text{for } 0.01 \leq r[1/\min] \leq 0.1 \end{cases} \quad (7.23)$$

$$\begin{cases} \varepsilon_i(\text{LV}) \leq 6\%, & \text{for } r[1/\min] \leq 0.01 \\ \varepsilon_i(\text{MV}) \leq 3\%, & \text{for } r[1/\min] \leq 0.01 \end{cases} \quad (7.24)$$

PV Systems and Power Quality Assessment 551

If the above conditions from (7.22) to (7.24) are met, then the relevant energy system is said to be suitable in terms of flicker emissions. Note that if there is only one network user, the long-term limit in (7.22) is taken to be 0.5 ($P_{\ell t,i} \leq 0.5$ for one network user). Additionally, a voltage measurement study or a site-specific simulation may be required in cases where there is insufficient data for flicker calculation, such as in the event of irregular flickers.

In addition to the flicker assessment principles given above, it is useful to explain some of the special cases as below.

Summation Law for Non-synchronous Multiple Flicker Source: The summation law is used to calculate the total flicker effect of the multiple individual sources on the PCC voltage. The flicker summation formula given in (7.25) is included in most standards.

$$P_{st} = \sqrt[3]{P_{st,1}^3 + P_{st,2}^3 + \cdots P_{st,m}^3} = \sqrt[3]{\sum_{i=1}^{m} P_{st,i}^3} \quad (7.25)$$

where P_{st} is the short-term flicker intensity, m is the total number of independent flicker sources connected to the same PCC, and i is the consecutive index from 1 to m. The summation formula in (7.25) was developed with the assumption that there is a low coincidence probability among independent flicker sources. If there is a high probability of coincidence between individual flicker voltage changes, the flicker summation formula can then be modified as in (7.26).

$$P_{st} = P_{st,1} + P_{st,2} + \cdots P_{st,m} = \sum_{i=1}^{m} P_{st,i} \quad (7.27)$$

Flicker Propagation: In radial distribution networks, the flicker effect decreases from the PCC location toward the grid supply. The change in the flicker intensity here is proportional to the ratio of the short-circuit powers at the relevant points. For example, for the grid-connected PV system in Figure 7.11, the $P_{st\,X}$ value at the Bus-X can be calculated by equation (7.27).

$$P_{st(X)} = \frac{S_{SC(PCC)}}{S_{SC(X)}} \cdot P_{st(PCC)} \quad (7.28)$$

Total Relative Voltage Change at PCC: The PCC feeders of the distribution system can often have several loads and more than one generation source. For the flicker evaluation, if measurement is not possible at the PCC, the voltage change of each load that has a dominant effect on the flicker should be taken into account. For this purpose, the superposition method can be used to calculate the total relative voltage change due to multiple independent sources and loads. While the PV system connected to the distribution feeder causes a voltage rise at the corresponding PCC, the loads

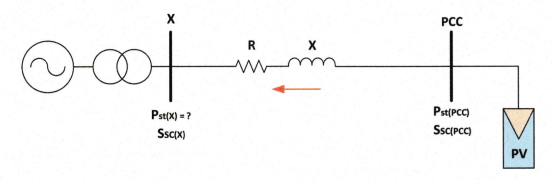

FIGURE 7.11 An example of a radial grid for flicker propagation.

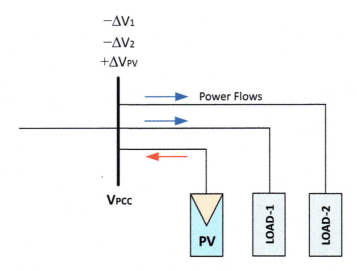

FIGURE 7.12 Total relative voltage change at PCC due to multiple users.

cause a voltage drop. For maximum voltage change, let us consider the situation for the phasors being 100% overlap. For example taking into account the 100% coincidence of the phasors, for the grid-connected PV system in Figure 7.12, the maximum relative voltage change at the PCC can be calculated by equation (7.29).

$$\varepsilon_{\max(\text{PCC})} = \frac{|\Delta V_{\text{PCC}}|}{|V_{\text{PCC}}|} = \frac{|\Delta V_{\text{PV}}| - |\Delta V_1| - |\Delta V_2|}{|V_{\text{PCC}}|} = |\varepsilon_{(\text{PV})}| - |\varepsilon_{(L1)}| - |\varepsilon_{(L2)}| \quad (7.29)$$

The relative voltage change (ε_i) of each user expressed in (7.29) can be calculated with the help of equation (7.4) or (7.8). At this point, it is useful to emphasize an important point. Normally, the impact of a medium-sized PV system on PCC voltage and flicker emission is quite low. However, as the connected PV capacity at the PCC approaches the maximum hosting capacity, the effect of PV generation fluctuations and rapid load changes on the flicker will be more significant. In this case, flicker emission limits may be exceeded. Therefore, a flicker study should be performed carefully for distribution systems with high PV penetration levels. If necessary, flicker mitigation techniques such as TSC (thyristor switched capacitor), TCR (thyristor controlled reactor), active filtering, and dynamic compensation systems should be implemented.

Example 7.2: Voltage Flicker Assessment at PCC with a PV System

Consider the LV distribution feeder given in Figure 7.13 with a 500 kWp PV power plant. As seen in Figure 7.13, there are three dominant sources of flicker connected to the PCC bus. These loads and their characteristic behavior are given below.

Load-1 (Motor Startup): An average of 6 three-phase motors start up in about 10 minutes. The rated voltage of the motors is 400 V and the rated current is 7 A. The power factor of the motors during startup is about 0.6 ($\cos\varphi_a = 0.6$ inductive). The starting current (I_a) of the motor is about six times the rated current (I_r), ($I_a / I_r \approx 6$). The motors run for at least 2 hours after startups and then the operators take a break. Assume that the motors are independent of each other and start up non-synchronously.

Load-2 (Resistive Heaters): There are 12 different three-phase resistance furnaces in an industrial facility. Three of these furnaces (group-1) have 20 kW, five of them (group-2) have 15 kW, and four of them (group-3) have 10 kW heaters. These furnaces are used to test various industrial products and have a thermostat-based temperature control system. Testing of a product often takes more than 2 hours and at least nine of the furnaces work every day (one furnace from each group is kept as a spare). Each furnace is used for testing a different product and there is a gap

PV Systems and Power Quality Assessment

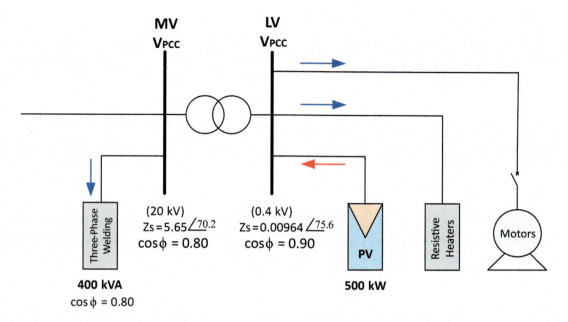

FIGURE 7.13 LV distribution feeder with a 500 kW PV power plant and loads.

of approximately 1 hour between two consecutive tests. During the test process, the furnaces are gradually increased to their maximum power levels. As a result, there are significant fluctuations in the power drawn by the industrial facility due to processes such as thermostat controls, gradual power increases, startups, and shutdowns. In the measurements at the main distribution panel, it was determined that the power fluctuation varies between 100% and 30% of the total installed power of the furnaces. The repeat rate of large magnitude fluctuations was found to be five times in 10 minutes on average.

Load-3 (Three-Phase Welding): In a network user facility, a three-phase spot-welding machine is connected to the medium voltage side of the substation located at the PCC. The repetition rate of a three-phase welding machine is 100 welding pulses per minute.

Required Information: Different approaches are used for the calculation of the relative voltage drop created by the welding machines. One of the accepted methods is to multiply the relative voltage drop formula by a predetermined form factor (F). The characterization of welding machines is not covered in this book since it is out of scope. Just assume the form factor as 1.3 ($F = 1.3$) for the relevant welding machine with 100 pulses per minute. In short, calculate the maximum voltage drop by the $\varepsilon = F \cdot \frac{\Delta S}{S_{SC}} \cdot \cos(\theta - \varphi)$ formula. Note that in general, the form factor should be taken into account for the non-rectangular voltage changes. The details about the form factor are available in the literature.

a. Assess the conformity of the flicker emission limits of the defined loads individually and collectively at LV and MV PCCs for the night times (PV system is not active).
b. Assess the conformity of the flicker emission limits of the defined loads individually and collectively at LV and MV PCCs for the day times (PV system is active).

Solution:

a. **Flicker Assessment for Load-1 (Motor Startup):** Long-term flicker assessment is not required, as there are only 10 minutes of startup time within 2 hours. Since the motors have an average of six startups (N) in 10 minutes (T), the repetition rate r_1:

$$r_1 = \frac{N}{T} = \frac{6}{10} = 0.6\,[1/\min]$$

Since the motors are identical and independent of each other, the startup of each motor causes a separate voltage drop. Therefore, the apparent power change of a single motor is taken into account.

$$\Delta S_1 = \sqrt{3} \cdot U_r \cdot I_a = \sqrt{3} \cdot U_r \cdot (nI_r) = \sqrt{3} \cdot 400\,\text{V} \cdot (6 \cdot 7\,\text{A}) \cong 29.1\,\text{kVA}$$

From (7.8), maximum relative voltage change $\left(S_{SC} = \dfrac{U^2}{Z_s} = \dfrac{(0.4\,\text{kV})^2}{0.00964} = 16.597\,\text{MVA}\right.$; $\varphi_1 = \cos^{-1}(0.6) = 53.13)$:

$$\varepsilon_{\max 1} = \frac{\Delta S_1}{S_{SC}} \cdot \cos(\theta - \varphi_1) = \frac{29.1}{16597} \cdot \cos(75.6 - 53.13) = 0.162\%$$

As shown in Figure 7.14, the $\varepsilon_{\max 1}$ value from the flicker emission limit curve is found to be 2.6% for $r_1 = 0.6[1/\min]$. In addition, from the same curve, the ε_{ref} value is equal to 3.2% ($\varepsilon_{\max 1} = 2.6\%$, $\varepsilon_{ref} = 3.2\%$).

Hence, the short-term flicker intensity (P_{st}) can be calculated from (7.21):

$$P_{st1} = \frac{\varepsilon_{\max 1}}{\varepsilon_{ref}} \cdot P_{ref} = \frac{0.162\%}{3.2\%} \cdot 1 = 0.050625$$

Normally, two types of conditions must be satisfied for conformity [$\varepsilon_{\max} \leq \varepsilon_{\text{limit}}$, and $P_{st} \leq P_{st(\text{limit})}$]. So, in this example:

$$\varepsilon_{\max 1} \leq \varepsilon_{\text{limit}} \Rightarrow 0.162\% \leq 2.6\% \quad (\checkmark)$$

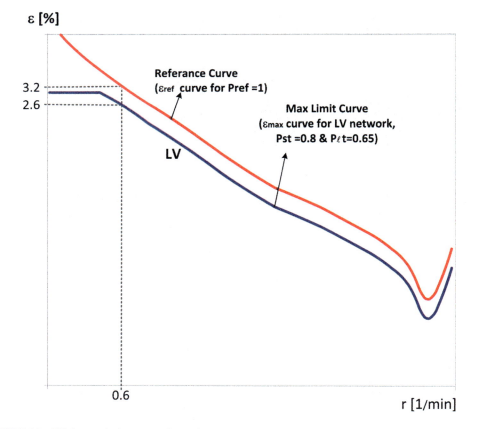

FIGURE 7.14 Flicker emission curve for $r=0.6$ [1/min].

$$P_{st1} \leq P_{st(limit)} \Rightarrow 0.050625 \leq 0.8 \quad (\sqrt{})$$

As a result, the PCC connection of the motors is approved.

Flicker Assessment for Load-2 (Resistive Heaters): Since there is one resistance furnace is kept as redundant from each group, the total installed capacity (P_{total}) is:

$$P_{total} = 2 \cdot 20 \text{ kW} + 4 \cdot 15 \text{ kW} + 3 \cdot 10 \text{ kW} = 130 \text{ kW}$$

Based on the particular load behavior, both short-term and long-term flicker intensity assessments should be performed. Since the fluctuation in the load varies between 30% and 100%, the highest load change of the resistance furnace is 91 kW ($130 - 0.30 \cdot 130 = 91$ kW). Since $\cos\varphi_2 = 1$, $\Delta S_2 = 91$ kW. Therefore, maximum relative voltage change for the resistive heaters:

$$\varepsilon_{max\,2} = \frac{\Delta S_2}{S_{SC}} \cdot \cos(\theta - \varphi_2) = \frac{91}{16597} \cdot \cos(75.6 - 0) = 0.136\%$$

Since the large power fluctuations occur five times (N) per 10 minutes (T), the repeat rate r_2:

$$r_2 = \frac{N}{T} = \frac{5}{10} = 0.5 [1/\min]$$

Again from the flicker emission limit curve, the $\varepsilon_{max\,2}$ value is found to be 2.7% for $r_2 = 0.5[1/\min]$. Also, from the same curve, the ε_{ref} value is equal to 3.3%. Hence, the short-term flicker intensity (P_{st}) can be calculated from (7.21):

$$P_{st\,2} = \frac{\varepsilon_{max\,2}}{\varepsilon_{ref}} \cdot P_{ref} = \frac{0.136\%}{3.3\%} \cdot 1 = 0.0412$$

From (7.29), the long-term flicker intensity (P_{lt}) can be calculated as:

$$P_{lt2} = \sqrt[3]{\frac{1}{12} \sum_{j=1}^{12} P_{st2(j)}^3} = \sqrt[3]{\frac{12 \times 0.0412^3}{12}} = 0.0412 < 0.65 \Rightarrow (\sqrt{})$$

$$\varepsilon_{max\,2} \leq \varepsilon_{limit} \Rightarrow 0.136\% \leq 2.7\% \quad (\sqrt{})$$

$$P_{st1} \leq P_{st(limit)} \Rightarrow 0.0412 \leq 0.8 \quad (\sqrt{})$$

As a result, the PCC connection of the resistance heaters is approved.

Flicker Assessment for Load-3 (Three-Phase Welding Machine): The three-phase welding machine is connected to the medium voltage side of the PCC. As specified in the example, the maximum relative voltage change for the three-phase spot-welding machine ($S_{SC(MV)} = \frac{U^2}{Z_s} = \frac{(20 \text{ kV})^2}{5.65} = 70.78$ MVA, $\Delta S_3 = 0.4$ MVA, $\varphi_3 = \cos^{-1}(0.8) = 36.86$, $F = 1.3$):

$$\varepsilon_{max\,3} = F \cdot \frac{\Delta S_3}{S_{SC(MV)}} \cdot \cos(\theta - \varphi_3) = 1.3 \times \frac{0.4}{70.78} \times \cos(70.2 - 36.86) = 0.472\%$$

Besides, from the flicker emission limit curves (MV and reference curves), the $\varepsilon_{max\,3}$ value is found to be 0.33% for $r_3 = 100[1/\min]$. In addition, from the same curve, the ε_{ref} value is equal to 0.75%. Hence, the short-term and long-term flicker intensities (P_{st3}, P_{ft3}) can be calculated as below. Notice that $P_{ft3} = P_{st3}$ because there is only one flicker source at MV PCC.

$$P_{ft3} = P_{st3} = \frac{\varepsilon_{max3}}{\varepsilon_{ref}} \cdot P_{ref} = \frac{0.33\%}{0.75\%} \cdot 1 = 0.44 \leq 0.5 \quad (\surd)$$

Remember that $P_{ft,i} \leq 0.5$ for one network user. As all emission limits are now met, the connection of the welding machine to the MV PCC is approved.

Aggregated Voltage Flicker at LV PCC: Since the loads are not independent and simultaneous, aggregated voltage flicker at LV PCC can be found from (7.25):

$$P_{st(LV\ PCC)} = \sqrt[3]{P_{st1}^3 + P_{st2}^3} = \sqrt[3]{0.050625^3 + 0.0412^3} = 0.05063$$

As there is only a single source of flickers for the long term, $P_{P_{ft2}(LV\ PCC)} = P_{st2} = 0.0412$. Since both emission values are below the allowable limits, the flicker values are approved for LV PCC. From equation (7.29), the maximum relative voltage drop can be calculated as below for the case of being 100% coincidence in voltage drop phasors.

$$\varepsilon_{max(LV\ PCC)} = |\varepsilon_{(L1)}| + |\varepsilon_{(L2)}| = 0.162\% + 0.136\% = 0.298\%$$

Accordingly, maximum relative voltage drop values are below the allowable limit. Consequently, the voltage drop values are also permitted for LV PCC.

Aggregated Voltage Flicker at MV PCC: There is only a single source of flicker at the MV PCC. However, the effect of LV flickers on the MV side must be taken into account. From the flicker propagation formula of (7.28), MV voltage flickers due to LV flickers can be found as below ($S_{SC(LV\ PCC)} = 16.597$ MVA, $S_{SC(MV\ PCC)} = 70.78$ MVA, $P_{st(LV\ PCC)} = 0.05063$, $P_{ft2}(LV\ PCC) = 0.0412$):

$$P_{st(LV\ to\ MV\ PCC)} = \frac{S_{SC(LV\ PCC)}}{S_{SC(MV\ PCC)}} \cdot P_{st(LV\ PCC)} = \frac{16.597}{70.78} \times 0.05063 = 0.01187$$

$$P_{ft(LV\ to\ MV\ PCC)} = \frac{S_{SC(LV\ PCC)}}{S_{SC(MV\ PCC)}} \cdot P_{ft2}(LV\ PCC) = \frac{16.597}{70.78} \times 0.0412 = 0.00966$$

So, the aggregated voltage flicker at MV PCC can be found from (7.25):

$$P_{st(MV\ PCC)} = \sqrt[3]{P_{st(LV\ to\ MV\ PCC)}^3 + P_{st3}^3} = \sqrt[3]{0.01187^3 + 0.44^3} = 0.440002$$

$$P_{ft(MV\ PCC)} = \sqrt[3]{P_{ft(LV\ to\ MV\ PCC)}^3 + P_{ft3}^3} = \sqrt[3]{0.00966^3 + 0.44^3} = 0.440001$$

As a result, all emission values are also suitable at MV PCC.
b. Many studies in the literature have measured and analyzed the effect of PV systems on the flicker level at PCCs. The main results from these analyses and long-term surveys can be listed as follows:
- In networks with high PV penetration, the long-term flicker level at the PCC of PV systems is usually below the limit of 0.5% ($P_{ft} \leq 0.5$). However, there are exceptional cases for the P_{ft} being violated.
- In the performed analysis, it has been determined that the limit violations in the P_{st} value, observed at certain days and times of the year, are caused by loads connected to PCCs. In other words, no correlation was detected between P_{st} flickers versus global radiation and clear sky indices.

PV Systems and Power Quality Assessment

- Only in distribution feeders with a high level of PV penetration, P_{st} flicker emissions may sometimes exceed the limits if the load is very low (e.g. on Sundays).
- However, the limits for relative voltage change may be exceeded due to load fluctuations in distribution networks with a high level of PV penetration. Therefore, the relative voltage change for such systems should be assessed carefully.
- In short, PV systems are not the source of higher flicker levels in electrical networks.

As explained above, there is no direct analytical method for calculating the P_{st} and P_{lt} flickers caused by PV systems. Reliable values for these flicker intensities can only be determined by measurement. At this point, it is sufficient to perform the calculations only for the relative voltage changes and check the ε_{max} limits accordingly.

First, let's find the voltage rises created by the 500 kW PV system on the LV and MV sides of the PCC. From (7.9), the relative voltage rise at the LV PCC ($\varepsilon_{max(LV\,PCC)}$) can be calculated as below ($\varphi_{PV} = \cos^{-1} 0.90 = 25.84°$, $Z_{s(LV)} = R + jX = 0.002392 + j0.009338\,\Omega$, $Z_{s(MV)} = R + jX = 5.50 + j12.88\,\Omega$):

$$\varepsilon_{max(LV\,PCC)} = \frac{P_{PV} \cdot R_S + Q_{PV} \cdot X_S}{V_{PV}^2} = \frac{P_{PV} \cdot R_S + P_{PV} \cdot \tan(\varphi_{PV}) \cdot X_S}{V_{PV}^2}$$

$$= \frac{500 \cdot 10^3 \times [0.002392 + \tan(25.84) \times 0.009338]}{(400)^2} = 0.0216 = 2.16\%$$

$$\Rightarrow \varepsilon_{max(LV\,PCC)}(\%) = 2.16\% \leq 3\% \quad (\checkmark)$$

$$\varepsilon_{max(MV\,PCC)} = \frac{P_{PV} \cdot R_S + Q_{PV} \cdot X_S}{V_{PV}^2} = \frac{P_{PV} \cdot R_S + P_{PV} \cdot \tan(\varphi_{PV}) \cdot X_S}{V_{PV}^2}$$

$$= \frac{500 \cdot 10^3 \times [5.50 + \tan(25.84) \times 12.88]}{(20000)^2} = 0.0146 = 1.46\%$$

$$\Rightarrow \varepsilon_{max(MV\,PCC)}(\%) = 1.46\% \leq 2\% \quad (\checkmark)$$

Relative voltage drops due to Load-1, Load-2, and Load-3 are calculated above ($\varepsilon_{max\,1} = 0.162\%$, $\varepsilon_{max\,3} = 0.472\%$). Accordingly, the net relative voltage changes at the LV and MV busbars ($\varepsilon_{net(LV\,PCC)}$, $\varepsilon_{net(MV\,PCC)}$) can be calculated from (7.29) as:

$$\varepsilon_{net(LV\,PCC)} = \varepsilon_{max(LV\,PCC)} - \varepsilon_{max\,1} - \varepsilon_{max\,2} \Rightarrow$$

$$\varepsilon_{net(LV\,PCC)} = 1.46\% - 0.162\% - 0.136\% = 1.862\% \leq 3\% \quad (\checkmark)$$

$$\varepsilon_{net(MV\,PCC)} = \varepsilon_{max(MV\,PCC)} - \varepsilon_{max\,3} \Rightarrow$$

$$\varepsilon_{net(MV\,PCC)} = 1.46\% - 0.472\% = 0.988\% \leq 2\% \quad (\checkmark)$$

7.5 STEADY-STATE FREQUENCY LIMITS

The limits for frequency deviations differ slightly between standards. For example, according to IEEE 929-2000, the allowable frequency deviation (Δf) range for a small PV system connected to 60 Hz LV distribution networks is from "plus 0.5 Hz" to "minus 0.7 Hz"

TABLE 7.3
Frequency Tolerance Limits of PV Systems for LV and MV Grids (Normal Operating Conditions)

Standard	Base Frequency	Voltage Level (or PV Size)	Frequency Limits (Hz)	Max Trip Time (in case of exceeding the limits)
IEC 61727	50 Hz	LV ($\leq$ 10 kVA)	$49 < f < 51$	10 cycles (0.2 seconds)
IEEE 929	60 Hz	LV ($\leq$ 10 kVA)	$59.3 < f < 60.5$	6 cycles (0.1 seconds)
IEEE 1547	60 Hz	LV ($\leq$ 30 kVA)	$59.3 < f < 60.5$	10 cycles (0.16 seconds)
	60 Hz	LV (> 30 kVA)	$57 < f < 60.5$	10 cycles (0.16 seconds)
German/VDE 0126:1-1	50 Hz	MV	$47.5 < f < 51.5$	10 cycles (0.2 seconds)

(-0.7 Hz $\leq \Delta f \leq +0.5$ Hz $\Rightarrow 59.3$ Hz $< f < 60.5$ Hz). On the other hand, according to the IEC 61727 standard, the allowable frequency deviation range for a 50 Hz LV grid is from "plus 1 Hz" to "minus 1 Hz" (-1 Hz $\leq \Delta f \leq +1$ Hz $\Rightarrow 49$ Hz $< f < 51$ Hz). For instance, if the system frequency goes outside these limits, a small-scale PV system (≤ 10 kVA) has to trip within 0.1 sec (6 cycles for 60 Hz system) according to IEEE 929-2000, and 0.2 seconds (10 cycles for 50 Hz system) according to IEC 61727. When the PV system is integrated into the MV grid side, the frequency tolerance band becomes wider. The frequency tolerances required for the normal operating conditions of PV systems are listed in Table 7.3 for LV and MV networks.

7.6 ACTIVE POWER CONTROL AND FREQUENCY REGULATION

International standards do not impose any active power control requirement for PV systems connected to the LV grids. However, PV systems integrated into MV grids can be assessed under two categories in terms of their capability to respond to the changes of active power in the grid. While the PV plants in the first category do not participate in the system's active power control, the PV plants in the second group are those planned to contribute to the system's active power control. Generally, the ones in the second group are selected from large-scale PV systems.

If the total generated power is more than the total load, the frequency will increase. If the opposite is true (total generation is less than total load), then the frequency decreases. The initial response to short-term frequency deviations (typically 0–30 seconds) is given by the governor systems of the generator-turbine pairs, which is called primary frequency control. The secondary frequency control system, namely AGC (Area Generation Control), responds to the frequency deviations of longer durations (typically from 30 seconds to 10 minutes). In this case, a set point is sent to the control point of the particular PV plant by the system operators. These set points can range from 100% to 0% of the PV plant's nominal power. At this point, the solar inverters can contribute to the secondary AGC responses by the power settings of their controller.

Figure 7.15 shows the power-frequency output characteristic of a typical solar inverter, where P_{set} is the inverter set point, and f_r is the rated system frequency. According to various national standards (such as Germany and China), the maximum ramp rate for controlling the active power of a PV plant should be less than or equal to 10% per minute of the installed capacity.

Note that PV systems are only capable of responding to frequency increases since their outputs depend on the available solar radiation. However, previously reduced PV generation can relatively be increased again by controlling the MPPs of the solar inverters.

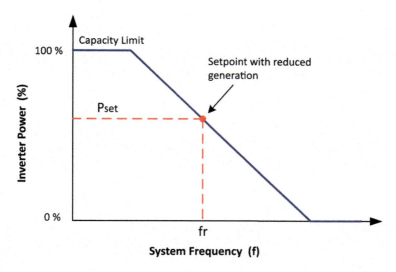

FIGURE 7.15 A typical power-frequency (*P–f*) droop characteristic for PV-based frequency regulation.

7.7 REACTIVE POWER CONTROL AND VOLTAGE REGULATION

As in the active power control requirements, international standards do not impose a primary task on LV PV systems to control the voltage and reactive power flow of the distribution networks. However, most countries have their own technical rules, which regulate the general connection criteria to LV distribution networks. Therefore, LV PV applications must also comply with the reactive power regulations existing in these technical rules. For example, the German "VDE-AR-N 4105" and the European "EN 50438" standards have specified three different power levels for LV photovoltaic applications:

- If $S_{PV} \leq 3.68$ kVA, the PV system should be operated between the power factor range of 0.95 lagging (over-excited) and 0.95 leading (under-excited).
- If $S_{PV} \leq 13.8$ kVA, the PV system should receive any setpoint from the DSO (distribution system operator) between the power factor range of 0.95 lagging (over-excited) and 0.95 leading (under-excited).
- If $S_{PV} > 13.8$ kVA, the PV system should receive any setpoint from the DSO between the power factor range of 0.90 lagging (over-excited) and 0.90 leading (under-excited).

According to several national and international standards, large PV stations connected to the MV power networks should be able to supply reactive power to the grid to reach a power factor between 0.95 lagging and 0.95 leading, thereby supporting the grid voltage stability during the day times of normal operating conditions. The single-line diagram of a large PV system and its plant-level reactive power compensation at the MV level are given in Figure 7.16.

It is possible to use different control approaches for the plant-level reactive power compensation at PCC. These control modes are (i) fixed power factor, (ii) dynamic (variable) power factor (depending on the active power delivered), (iii) fixed reactive power in (VAr), or (iv) dynamic (variable) reactive power (depending on the system voltage). Although reactive power requirements of a PV system may differ according to the various regional grid codes, the general approach is to keep the power factor range within $[-0.95(\text{leading}) \leq \cos\varphi \leq +0.95(\text{lagging})]$ or $[-0.90(\text{leading}) \leq \cos\varphi \leq +0.90(\text{lagging})]$ under the normal operating conditions. Figure 7.17 presents three different grid codes as $(P-Q)$ & $(P-\cos\varphi)$ diagrams for the steady-state reactive power requirements of a PV system.

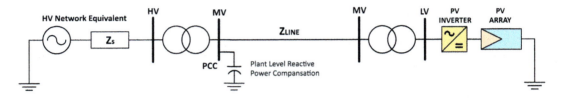

FIGURE 7.16 One-line diagram of a typical PV power plant and its reactive power compensation at MV level.

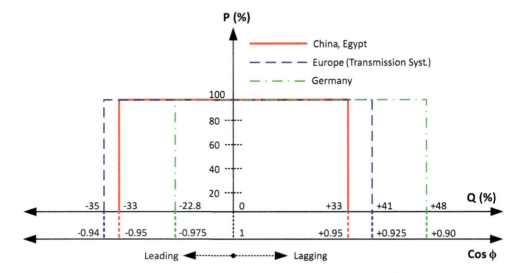

FIGURE 7.17 Steady-state reactive power requirements of an MV PV system in different grid codes.

In cases where the PCC voltage goes out of the determined limits, some standards suggest detailed actions in the reactive power regulation of PV power stations. The reactive power regulation limits of large-scale PV systems, applied in the German and European Transmission Systems, are given in Figure 7.18 for different PCC voltages. In other words, large-scale PV plants connected to MV and HV networks should have the ability to control the voltage at the PCC with the required reactive power compensation and keep it within the specified ranges. Note that the relevant network codes should be investigated for other regulation ranges in different networks.

7.8 VOLTAGE UNBALANCE LIMITS

Voltage unbalance usually occurs as a result of unbalanced loading in 3-phase systems. These loads that cause unbalanced loading are single-phase (phase-to-neutral) and two-phase (phase-to-phase) loads. Phase-to-phase loads are generally connected to medium and high-voltage networks. Two-phase industrial loads, frequently used in these networks, are listed below.

- Resistance spot-welding machines
- Induction furnaces
- Electric arc furnaces

The factor to quantify the voltage unbalance is called unbalance factor (f_u). The approximated "f_u" factor given in (7.30) can be used to assess the unbalance severity at photovoltaic PCCs.

$$f_u \approx \frac{\sum S_{L(\text{single-phase})} + \sum S_{L(\text{two-phase})}}{S_{SC\,(\text{PCC})}} \qquad (7.30)$$

PV Systems and Power Quality Assessment

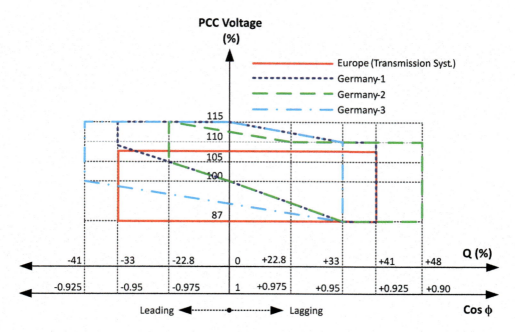

FIGURE 7.18 Reactive power requirements of a large PV system for different PCC voltages.

where

$\sum S_{L(\text{single-phase})}$: Sum of the single-phase loads connected to the PCC (kVA).
$\sum S_{L(\text{two-phase})}$: Sum of the two-phase loads connected to the PCC (kVA).
$S_{SC\,(PCC)}$: Network short-circuit power at the PCC (kVA)

Typical allowable levels for voltage unbalance factor are given as follows for normal operating conditions:

$$f_u \leq 2\% \tag{7.31}$$

$$f_{u,i} \leq 0.7\% \tag{7.32}$$

where f_u is the voltage unbalance factor caused by all grid consumers together, and $f_{u,i}$ is the voltage unbalance factor caused by individual consumers for 10-minute averaged values. If the voltage unbalance limit is exceeded at a photovoltaic PCC, the following corrective measures can be taken to reduce the f_u level.

- Single-phase loads should be equally distributed over three phases.
- Some of the three-phase and single-phase loads should be isolated from each other.
- Reactor and capacitor-based compensation systems should be installed at the PCC.
- Investments should be made to increase the short-circuit power of the network.

Example 7.3: Voltage Unbalance Assessment at PCC with a PV System

Consider the LV distribution feeder given in Example 7.2. If the load-3 was a two-phase spot-welding machine (instead of a three-phase machine), what would be the voltage unbalance factor caused by the load ($S_L = 0.4$ MVA). Considering the network short-circuit power at MV PCC, calculate the maximum value of two-phase and single-phase loads that can be connected to the system.

Solution:

From Example 7.2, we have short-circuit power $S_{SC(MV\ PCC)} = 70.78$ MVA at MV PCC. By using 7.30, the voltage unbalance factor for the spot-welding machine is:

$$f_u = \frac{S_{L(two-phase)}}{S_{SC\ (PCC)}} = \frac{0.4}{70.78} = 0.56\% < 0.7\% \quad (\checkmark)$$

By taking $f_u = 2\%$, the highest level of unbalanced loads that can be connected to the MV PCC:

$$S_{L\ max} = f_u \times S_{SC\ (PCC)} = 2\% \times 70.78\ \text{MVA} = 1.4156\ \text{MVA}$$

7.9 HARMONIC LIMITS

High harmonic content in the system voltage can cause deterioration of both system operation and equipment connected to the PCC. Although harmonic distortion levels are not very high, PV systems may cause harmonic distortion in the system since they are connected to the PCCs via power electronic equipment. Permissible harmonic voltage and current limits for LV & MV networks, accepted by most countries, are defined in EN 61000-2-2 and IEEE 519 standards.

Table 7.4 gives the emission limits for the individual harmonic contents of the voltage in LV distribution networks according to EN 61000-2-2. Alternatively, Table 7.5 defines the voltage harmonic limits based on IEEE 519-1992. Table 7.6 shows the permissible limits for current harmonics according to EN 61000-2-2 and IEEE 519 standards.

The harmonic limits listed above can be summarized for PV inverters as follows:

- THD_I (Total harmonic current distortion) at the PCC should be less than 5% of the fundamental frequency at the rated current output of the PV inverter (up to 50th harmonic).
- Individual harmonic currents of the PV inverter should not exceed the emission limits, as shown in Table 7.6.
- Individual harmonic and total harmonic distortion of the output voltage of the PV inverter should not exceed the emission limits, shown in Tables 7.4 and 7.5. Note that the values in Table 7.5 are valid for the conditions surviving for more than an hour. For harmonics lasting less than 1 hour, these limits may be exceeded by 50%.

TABLE 7.4
Harmonic Voltage Limits in LV Distribution Networks According to EN 61000-2-2 (RMS Values as a Percentage of the Fundamental Component)

Individual Harmonics					
Odd Harmonics				Even Harmonics	
Multiples of 3 (= 3i)		Non-multiples of 3 (≠ 3i)			
Harmonic Number (i)	Harmonic Voltage in [%]	Harmonic Number (i)	Harmonic Voltage in [%]	Harmonic Number (i)	Harmonic Voltage in [%]
3	5	5	6	2	2
9	1.5	7	5	4	1
15	0.4	11	3.5	6	0.5
21	0.3	13	3	8	0.5
$21 < i \leq 45$	0.2	$17 \leq i \leq 49$	$2.27 \times \left(\frac{17}{i}\right) - 0.27$	$10 \leq i \leq 50$	$0.25 \times \left(\frac{10}{i}\right) + 0.25$
Total Harmonic Distortion (THD_V) in [%]					

$THD_V \leq 8\%$ (up to $i = 50$)

TABLE 7.5
Harmonic Voltage Limits in LV & MV Networks according to IEEE 519-1992

Voltage at PCC	Individual Harmonic Distortion [%]	Total Voltage Distortion (THD_V) [%]
$V \leq 69$ kV	3	5
69 kV $< V \leq 160$ kV	1.5	2.5
$V > 160$ kV	1	1.5

TABLE 7.6
Harmonic Current Limits in Distribution Networks according to IEEE 1547 and IEC 61727

	Odd Harmonics			Even Harmonics	
Harmonic Number (i)	Harmonic Voltage in [%] IEEE 1547	Harmonic Voltage in [%] IEC 61727	Harmonic Number (i)	Harmonic Voltage in [%] IEEE 1547	Harmonic Voltage in [%] IEC 61727
$3 \leq i \leq 9$	4	4	$2 \leq i \leq 8$	25% of the odd harmonic limits	1
$11 \leq i \leq 15$	2	2	$10 \leq i \leq 32$	25% of the odd harmonic limits	0.5
$17 \leq i \leq 21$	1.5	1.5	–	–	–
$23 \leq i \leq 33$	0.6	0.6	–	–	–
$i > 33$	0.3	–	–	–	–

Total Harmonic Distortion (THD_V), or Total Demand Distortion (TDD_V) in [%]
IEEE 1547 and *IEC 61727*

$THD_V \leq 5\%$ & $TDD_V \leq 5\%$ (up to $i = 50$)

7.10 VOLTAGE RIDE-THROUGH REQUIREMENTS

Fault ride-through capability of a PV power plant refers to the grid-connected stations being able to remain online during momentary network faults by providing grid support. In the context of voltage ride-through (VRT), there are two capabilities defined for low (LVRT) and high (HVRT) voltage ride-through to recover the voltage at PCC during symmetrical and unsymmetrical network faults. Therefore, the generic form of LVRT and HVRT requirements can be illustrated in Figure 7.19.

It is clear from Figure 7.19 that the PV system should not be disconnected from the grid between two solid lines, but can be disconnected from the grid outside the solid lines. However, each grid code can add different constraints on the connection and the disconnection of the PV plant (depending on the PCC voltage, plant capacity, and grid robustness). For example, Table 7.7 presents the low- and high-voltage ride-through requirements based on the standards IEEE 1574 and IEC 61727. The main purpose of the allowed delays specified in Table 7.7 is to avoid unnecessary tripping due to temporary small distortions in the grid and in this way to contribute to the grid stability.

Details of LVRT and HVRT tolerances required by the Natural Environment Research Council (NERC) are given in Figure 7.20. Note that the zero-voltage ride-through for 150 ms is required according to NERC.

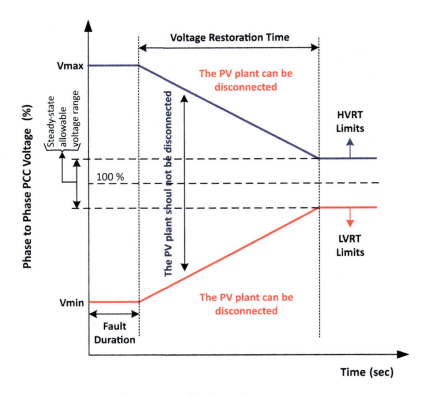

FIGURE 7.19 The general form of LVRT and HVRT requirements.

TABLE 7.7
LVRT and HVRT Voltage Limits for PV Systems Connected to LV Networks

Voltage Level	Voltage at PCC	Max Trip Time
	$V < 50\%$ (for 120 V base)	6 cycles (0.1 seconds for 60 Hz)
Low Voltage	$V < 50\%$ (for 230/400 V base)	5 cycles (0.1 seconds for 50 Hz)
120 V-60 Hz System	$50\% < V < 88\%$ (for 120 V base)	120 cycles (2 seconds for 60 Hz)
(IEEE 1574)	$50\% < V < 85\%$ (for 230/400 V base)	100 cycles (2 seconds for 50 Hz)
---	$88\% < V < 110\%$ (for 120 V base)	Normal operation
230/400 V-50 Hz System	$85\% < V < 110\%$ (for 230/400 V base)	
(IEC 61727)	$110\% < V < 137.5\%$ (for 120 V base)	120 cycles (2 seconds for 60 Hz)
	$110\% < V < 135\%$ (for 230/400 V base)	100 cycles (2 seconds for 50 Hz)
	$V \geq 137.5\%$ (for 120 V base)	2 cycles (0.034 seconds for 60 Hz)
	$V \geq 135\%$ (for 230/400 V base)	2.5 cycles (0.05 seconds for 50 Hz)

7.11 SUPPORTING THE NETWORK WITH REACTIVE CURRENT

Some countries have mandated that PV power stations should support the grid voltage with reactive current injection during the grid faults. Supporting the grid with reactive current is required to be controlled dynamically. For example, when the PCC voltage decreases, the PV system should increase its reactive current to restore the voltage back to the allowable range. On the other hand, when the PCC voltage rises, the PV station should reduce the system voltage by withdrawing

PV Systems and Power Quality Assessment   565

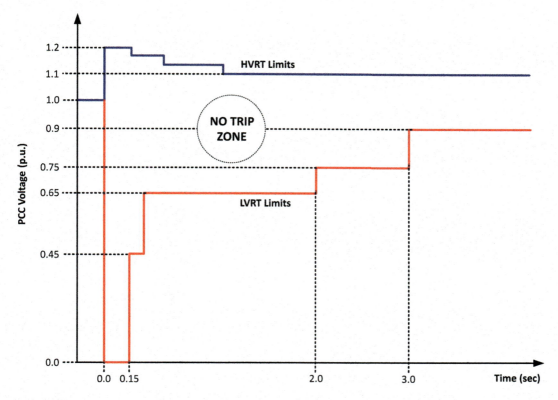

FIGURE 7.20 LVRT and HVRT requirements according to NERC (North America).

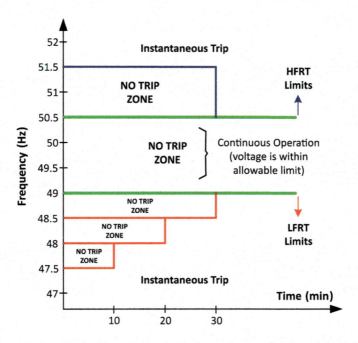

FIGURE 7.21 Reactive current requirements versus voltage changes for the grid codes of Germany and Spain.

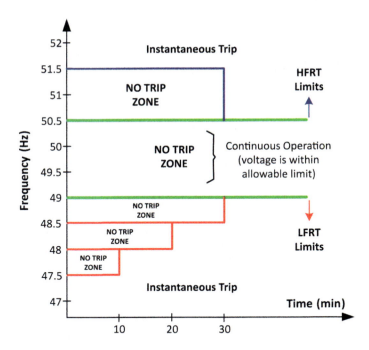

FIGURE 7.22 LFRT and HFRT requirements based on German grid codes.

reactive current from the network. In this context, Figure 7.21 shows the reactive current requirements versus voltage changes for the grid codes of Germany and Spain.

In these grid codes, the zone where the PCC voltage is between 90% and 110% is called the "dead band," and no control action is required within the range of this band. When the voltage goes outside the allowable limits, the response time for the reactive current injection or withdrawal of PV power plants should not exceed about 20–30 ms according to German and Chinese codes.

7.12 FREQUENCY RIDE-THROUGH REQUIREMENTS

The frequency ride-through capability of a PV power plant refers to the grid-connected stations being able to remain in service during the frequency deviations for prearranged periods. In this context, there are two capabilities defined for low (LFRT) and high (HFRT) frequency ride-through capabilities to meet the requirements defined by the grid codes. Allowed frequency-duration profiles differ according to the country's grid codes. For example, LVRT and HVRT requirements based on the German grid codes can be illustrated in Figure 7.22. Another frequency-duration profile for a 60 Hz system is given in Table 7.8 based on WECC (Western Electricity Coordinating Council) in the United States.

TABLE 7.8
LFRT and HFRT Requirements according to the Hydro-Quebec System in Canada

LFRT Limits (Hz)	HFRT Limits (Hz)	Minimum Time
$f > 59.4$	$60 < f < 60.6$	Continuous
$f < 59.4$	$f > 60.6$	3 minutes
$f < 58.4$	$f > 61.6$	30 seconds
$f < 57.8$	–	7.5 seconds
$f < 57.3$	–	45 cycles (0.75 seconds)
$f < 57$	$f > 61.7$	Instantaneous

7.13 PROBLEMS

P7.1. Consider the solar PV farm given in Figure 7.23. The network short-circuit power at PCC is 280 MVA, and the network impedance angle is 70 degrees. All the data required are shown on the network below.

   a. Find the relative voltage rise at the PCC due to a single 2500 kW PV system.
   b. Find the maximum relative voltage rise at the PCC due to the totality of all the PV systems.
   c. Considering the given conditions, calculate the maximum PV capacity that can be connected to the PCC in terms of voltage rise criteria.

P7.2. Consider the energy park given in Figure 7.24. The energy park includes three identical PV systems of 2.5 MW each and three identical wind turbines of 3 MW each. The individual wind turbine generators (WTG) and PV units are equipped with a reactive power compensation system, so that the rated power of a WTG is 3 MW=3 MVA, and the rated power of a PV generator is 2.5 MW=2.5 MVA. Hence, in this case, the power factor at PCC is equal to 1 ($\cos \varphi_{PCC} = 1$). The network data are specified in Figure 7.24.

In the measurements carried out, it was determined that the effects of the PV units on the flicker are small enough to be neglected. On the other hand, WTGs were found to be effective on PCC voltage flicker on high windy days. The flicker emission of a single WTG can be estimated as below:

$$P_{st} = P_{\ell t} = c \cdot \frac{S_{WTG}}{S_{SC\,PCC}}$$

where c is the flicker coefficient of the wind turbine (generally provided by the manufacturer based on a test report), S_{WTG} is the rated apparent power of a WTG in VA, and $S_{SC\,PCC}$ is the short-circuit power of the network at PCC in VA. Note that the long-term permissible emission level at PCC is 0.46 ($P_{\ell t} \leq 0.46$).

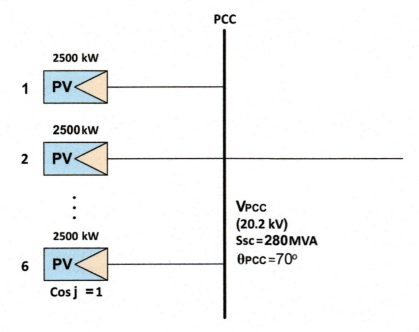

FIGURE 7.23 A solar park consisting of six identical PV systems, each with 2500 kW.

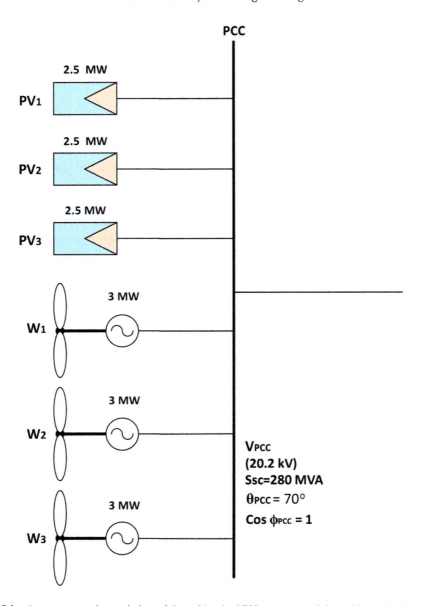

FIGURE 7.24 An energy park consisting of three identical PV systems and three identical wind turbines.

a. If the flicker coefficient for the WTG is given as 4 ($c=4$), calculate the total flicker emission level at PCC.
b. If the long-term flicker level of all three PV units at PCC was 0.12, what would be the total flicker emission at PCC caused by wind and solar units together?

P7.3. Consider the residential low-voltage distribution network consisting of single-phase rooftop PV units in Figure 7.25. The lowest power level allowed for rooftop PV systems in the region is 1 kW, while the highest is 3.5 kW. Perform an assessment of voltage unbalance and for the network according to best, medium, and worst-case scenarios. Note that the following criteria will be considered in the assessment.

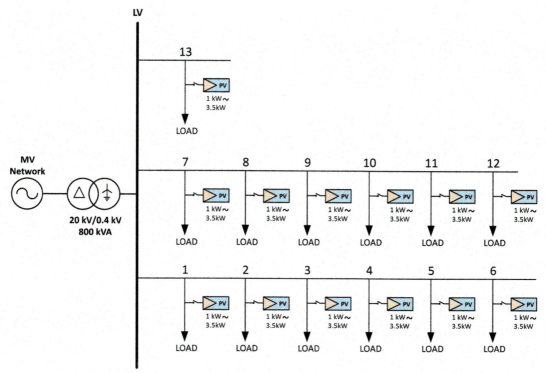

FIGURE 7.25 LV distribution feeder consisting of single-phase PV generation units.

a. Consider the equal distribution of PVs between the three phases. In this case, calculate the relative voltage rise in the distribution feeder. Compare the results with the permissible limits (best scenario).
b. Gradually increase the unbalanced distribution of PVs between the three phases. For each case, calculate the permissible relative voltage rise separately on the phases L_1, L_2, and L_3. Compare the results with the permissible limits (medium and worst scenario).

BIBLIOGRAPHY

1. M. Altin, E. U. Oguz, E. Bizkevelci, and B. Simsek, "Distributed generation hosting capacity calculation of MV distribution feeders in Turkey", in *IEEE PES Innovative Smart Grid Technology Conference on Europe*, vol. 2015-January, no. January, pp. 1–7, 2015.
2. Q. Zheng, J. Li, X. Ai, J. Wen, and J. Fang, "Overview of grid codes for photovoltaic integration", in *IEEE Conference on Energy Internet and Energy System Integration. EI2 2012- Proceedings*, vol. 2018-January, no. March 2019, pp. 1–6, Beijing, China, 2012.
3. K. Doğanşahin, "Dağıtık Üretim Tesislerinin Maksimum Şebeke Entegrasyon Limitinin Belirlenmesi İçin Yeni Bir Metodolojinin Geliştirilmesi," 2018.
4. M. Z. Ul Abideen, O. Ellabban, and L. Al-Fagih, "A review of the tools and methods for distribution networks' hosting capacity calculation", *Energies*, vol. 13, no. 11, pp. 1–25, 2020.
5. K. Doğanşahin, B. Kekezoğlu, R. Yumurtaci, O. Erdinç, and J. P. S. Catalão, "Maximum permissible integration capacity of renewable DG units based on system loads", *Energies*, vol. 11, no. 1, pp. 1–16, 2018.

6. N. C. Yang and T. H. Chen, "A review on evaluation of maximum permissible capacity of distributed generations connected to a smart grid", in *Proceedings of International Conference on Machine Learning and Cybernetics*, vol. 4, pp. 1589–1593, 2012.
7. E. Troester, "New German grid codes for connecting PV systems to the medium voltage power grid", in *2nd International Workshop on Concentrating Photovoltaic Power Plants: Optical Design, Production, Grid Connection*, pp. 9–10, Darmstadt, Germany, 2009.
8. R. Therese, K. Jose, R. T. K. Jose, R. Nisha, P. Ramesh, and K. Jiju, "Active anti-islanding protection for grid connected solar photovoltaic power plant", vol. 5, no. 5, pp. 124–131, 2015.
9. M. Taufiqul Arif, A. M. T Oo, and A. Shawkat Ali, "Integration of renewable energy resources into the distribution network -A review on required power quality", *Int. J. Energy Power*, vol. 1, pp. 37–48, no. 2, 2012.
10. S. Sakar, M. E. Balci, S. H. E. A. Aleem, and A. F. Zobaa, "Hosting capacity assessment and improvement for photovoltaic-based distributed generation in distorted distribution networks", in *EEEIC 2016- International Conference on Environment and Electrical Engineering*, pp. 2–7, 2016.
11. S. Sakar, M. E. Balci, S. H. E. Abdel Aleem, and A. F. Zobaa, "Increasing PV hosting capacity in distorted distribution systems using passive harmonic filtering", *Electr. Power Syst. Res.*, vol. 148, pp. 74–86, 2017.
12. M. Rylander, "Computing Solar PV Hosting Capacity of Distribution Feeders." Technical Notes, Electric Power Research Institute (EPRI), USA, 2017.
13. S. Rahman, M. Moghaddami, A. I. Sarwat, T. Olowu, and M. Jafaritalarposhti, "Flicker estimation associated with PV integrated distribution network", in *Conference Proceedings on IEEE SOUTHEASTCON*, vol. 2018-April, no. 1553494, pp. 1–6, USA, 2018.
14. B. K. Perera, P. Ciufo, and S. Perera, "Point of common coupling (PCC) voltage control of a grid-connected solar photovoltaic (PV) system", *IECON Proceedings (Industrial Electronics Conference)*, pp. 7475–7480, Vienna, Austria, 2013.
15. M. Patsalides, G. Makrides, A. Stavrou, V. Efthymiou, and G. E. Georghiou, "Assessing the photovoltaic (PV) hosting capacity of distribution grids", *IET Conf. Publ.*, vol. 2016, no. CP711, pp. 2–5, 2016.
16. S. A. Papathanassiou and N. D. Hatziargyriou, "Technical requirements for the connection of dispersed generation to the grid", *Proc. IEEE Power Eng. Soc. Transm. Distrib. Conf.*, vol. 2, no. SUMMER, pp. 749–754, 2001.
17. S. A. Papathanassiou, "A technical evaluation framework for the connection of DG to the distribution network", *Electr. Power Syst. Res.*, vol. 77, no. 1, pp. 24–34, 2007.
18. A. Lucas, "Single-phase PV power injection limit due to voltage unbalances applied to an urban reference network using real-time simulation", *Appl. Sci.*, vol. 8, no. 8, p.1333, 2018.
19. C. Lopez and J. Blanes, "Voltage fluctuations produced by the fixed-speed wind turbines during continuous operation. European perspective", *Wind Farm - Impact in Power System and Alternatives to Improve the Integration*, no. July 2011.
20. J. Kondoh and D. Kodaira, "An evaluation of flicker emissions from small wind turbines", *Energies*, vol. 14, no. 21, p. 7263, 2021.
21. H. Khairy, M. EL-Shimy, and G. Hashem, "Overview of grid code and operational requirements of grid-connected solar PV power plants", *Focus*, vol. 1, no. April, p. 2, 2015.
22. H. Karawia, M. Mahmoud, and M. Sami, "Flicker in distribution networks due to photovoltaic systems", *CIRED - Open Access Proc. J.*, vol. 2017, no. 1, pp. 647–649, 2017.
23. R. Hudson and G. Heilscher, "PV grid integration - System management issues and utility concerns", *Energy Procedia*, vol. 25, pp. 82–92, 2012.
24. K. Hossam Abobakr, "Comparison of Egyptian standards for grid-connected photovoltaic power plants with IEC and IEEE standards: A case study in Egypt", *Int. J. Adv. Sci. Technol.*, vol. 28, no. 16, pp. 856–870, 2019.
25. A. F. Hoke, M. Shirazi, S. Chakraborty, E. Muljadi, and D. Maksimovic, "Rapid active power control of photovoltaic systems for grid frequency support", *IEEE J. Emerg. Sel. Top. Power Electron.*, vol. 5, no. 3, pp. 1154–1163, 2017.
26. A. Hoke and D. Maksimović, "Active power control of photovoltaic power systems", *2013 1st IEEE Conference on Technologies for Sustainability (SusTech)*, no. March 2014, pp. 70–77, Portland, Oregon, 2013.
27. J. C. Hernández, M. J. Ortega, J. De La Cruz, and D. Vera, "Guidelines for the technical assessment of harmonic, flicker and unbalance emission limits for PV-distributed generation", *Electr. Power Syst. Res.*, vol. 81, no. 7, pp. 1247–1257, 2011.

28. N. Hatziargyriou, E. Karfopoulos, D. Koukoula, M. Rossi, and V. Giacomo, "On the DER hosting capacity of distribution feeders", in *23rd International Conference on Electricity Distribution*, no. June, pp. 15–18, Lyon, Italy, 2015.
29. S. Hashemi and J. Stergaard, "Methods and strategies for overvoltage prevention in low voltage distribution systems with PV", *IET Renew. Power Gener.*, vol. 11, no. 2, pp. 205–214, 2017.
30. IEEE Standards Coordinating Committee 21 on Fuel Cells Photovoltaics Dispersed Generation and Energy Storage, IEEE Recommended Practice for Utility Interface of Photovoltaic (PV) Systems, vol. 2000. 2000.
31. A. H. Faranadia, A. M. Omar, and S. Z. Noor, "Voltage flicker assessment of 15.3 kWp grid connected photovoltaic systems", in *2017 IEEE 8th Control and System Graduate Research Colloquium, ICSGRC 2017 Proceedings*, no. November, pp. 110–115, Shah Alam, Malaysia, 2017.
32. N. Etherden and M. H. J. Bollen, "Increasing the hosting capacity of distribution networks by curtailment of renewable energy resources", in *2011 IEEE PES Trondheim PowerTech Power Technology for a Sustainable Society, POWERTECH 2011*, pp. 1–7, Trondheim, Norway, 2011.
33. M. Elshahed, "Assessment of sudden voltage changes and flickering for a grid-connected photovoltaic plant", *Int. J. Renew. Energy Res.*, vol. 6, no. 4, pp. 1328–1335, 2016.
34. A. Ellis, B. Karlson, J. Williams, and Sandia National Laboratories, "Utility-scale photovoltaic procedures and interconnection requirements", *Sandia National Laboratories*, no. February 2012.
35. F. Ebe, B. Idlbi, J. Morris, G. Heilscher, and F. Meier, "Evaluation of PV hosting capacities of distribution grids with utilization of solar roof potential analyses", *CIRED - Open Access Proc. J.*, vol. 2017, no. 1, pp. 2265–2269, 2017.
36. A. Dubey, S. Santoso, and A. Maitra, "Understanding photovoltaic hosting capacity of distribution circuits", in *IEEE Power Energy Society General Meeting*, vol. 2015, September, Denver, Colorado, 2015.
37. A. Dubey and S. Santoso, "On estimation and sensitivity analysis of distribution circuit's photovoltaic hosting capacity", *IEEE Trans. Power Syst.*, vol. 32, no. 4, pp. 2779–2789, 2017.
38. K. DoğanŞahin, B. Kekezoğlu, and R. Yumurtacı, "An improved mathematical model for the calculation of maximum permissible DG integration capacity", *J. Fac. Eng. Archit. Gazi Univ.*, vol. 35, no. 1, pp. 275–285, 2020.
39. F. Ding and B. Mather, "On distributed PV hosting capacity estimation, sensitivity study, and improvement", *IEEE Trans. Sustain. Energy*, vol. 8, no. 3, pp. 1010–1020, 2017.
40. Gerhard Bartak, Hansjörg Holenstein, Jan Meyer, "Technical Rules for the Assessment of Network Disturbances," Ceske Sdruzeni Rozvodrycnh Energetickych Spolecnosti (Csres), D-A-CH-CZ_ Kompendium, Aarau, Berlin, Prague, Vienna, 2007.
41. B. I. Crăciun, T. Kerekes, D. Séra, and R. Teodorescu, "Overview of recent grid codes for PV power integration", in *Proceedings of International Conference on Optimization of Electrical and Electronic Equipment, OPTIM*, pp. 959–965, Brasov, Romania, 2012.
42. P. Chirapongsananurak and N. Hoonchareon, "Grid code for PV integration in distribution circuits considering overvoltage and voltage variation", in *IEEE Region 10 Annual International Conference. Proceedings/TENCON*, vol. 2017-December, pp. 1936–1941, 2017.
43. D. S. G. Berriel, G. B. Menoni, T. D. L. Mussi, and D. N. Martín, "Flicker emission analysis of a wind farm", *Renew. Energy Power Qual. J.*, vol. 1, no. 10, pp. 355–360, 2012.
44. A. Spring, G. Wirth, G. Becker, R. Pardatscher, and R. Witzmann, "Effects of flicker in a distribution grid with high PV penetration", 28th European Photovoltaic Solar Energy Conference and Exhibition (28th EU PVSEC), Paris, France, 2013.

8 Economics of Solar Photovoltaic Systems

8.1 INTRODUCTION

Solar PV systems vary widely in size and cost. Therefore, calculating the economics of a solar PV system is the key issue to decide if the investment for that PV project is feasible. There are various methods and economic indicators to assess the profitability or economic aspects of a PV project. Firstly, the basic parameters and calculations required for the economic evaluations of PV systems are studied in the following sections. Then, detailed analysis and assessments were carried out for different PV applications and configurations in the worked examples below.

8.2 BASIC ECONOMIC PARAMETERS AND CALCULATIONS

In a PV project, the main objective from an economic point of view is to maximize the profitability of the project. For this purpose, different decision-making methods can be used. Deciding on the right and appropriate method is very important for the operation of the relevant project as well as for future investments. The most commonly used methods and parameters are explained in the following sections.

8.2.1 Cash Flow Models

There are three basic cash flows used in engineering economics for an investment project with economic life in the range of $[0$ to $N]$.

- Cash flow at present time $t = 0$ (denoted by P)
- Cash flow at a future time $t = N$ (denoted by F)
- Uniform annual cash flows in the range of $[t = 1$ to $N]$ (denoted by A)

If a specific valuation rate (such as a discount rate or interest rate) is set for the range of $[0$ to $N]$, it is possible to convert the **Present**, **Future**, or **Annual** value of the money to their values at another date. For this purpose, one of the cash flow diagrams represented in Figure 8.1 can be selected depending on the preferred economic assessment method.

The economic analysis of photovoltaic systems is usually evaluated annually and therefore, the money flows are also on an annual basis. Assuming the valuation rate is equal to the inflation rate i, the mathematical expressions between PV (present value), FV (future value), and AV (annual value) are given for a single cash flow as below:

$$PV = \frac{FV}{(1+i)^N} \Rightarrow FV = PV \cdot (1+i)^N \tag{8.1}$$

If there are non-uniform cash flows each year (such as $AV_1, AV_2 \ldots AV_N$) in the range of $[0$ to $N]$, the expression between FV and AV is given by equation (8.2).

$$FV = AV \cdot \left[\frac{(1+i)^N - 1}{i}\right] \Rightarrow AV = FV \cdot \left[\frac{i}{(1+i)^N - 1}\right] \tag{8.2}$$

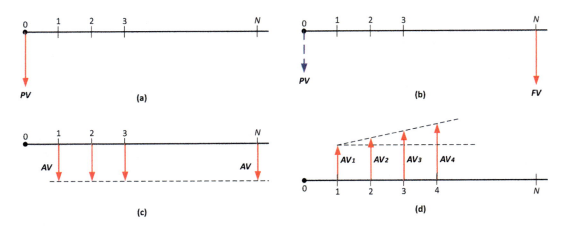

FIGURE 8.1 Cash flow diagrams: (a) present value, (b) future value, (c) annual uniform value, and (d) gradient increment G.

By substituting (8.1) into (8.2), the expression between *PV* and *AV* can be found for non-uniform cash flows by equation (8.3).

$$PV = AV \cdot \left[\frac{(1+i)^N - 1}{i \cdot (1+i)^N}\right] \Rightarrow AV = PV \cdot \left[\frac{i \cdot (1+i)^N}{(1+i)^N - 1}\right] \quad (8.3)$$

8.2.2 Present Value Method

The present value (*PV*) method, also known as the present discounted value, is a method for discounting the cash flows from the future range of [0 to *N*] to the date of the current valuation (*t* = 0). To find the present value (*PV*) of a future amount of money (C_{FV}) for the time of periods *N* (usually in years), it is multiplied by a discount rate. The discount factor is also called the present value coefficient (PVC = $1/(1+d)^t$), where *d* is the discount rate. The discount rate here is usually equal to the interest rate (*i*).

$$PV = \text{PVC} \cdot C_{FV} = \frac{C_{FV}}{(1+d)^t} \quad (8.4)$$

Example 8.1: Present Value of a Battery Replacement

A stand-alone rooftop PV system will be installed in a rural area. The contactor company estimated the economic life of the PV system as 25 years. According to the agreement, the company will replace the battery for $10,000 every 10 years. Assuming the interest rate in that region was 4%, what would be the present cost of the battery replacements?

Solution:

If the batteries have to be replaced every 10 years, their present values can be estimated by equation (8.4) as below (*d* = 0.04, *N* = 10 for the first replacement, and *N* = 20 for the second replacement):

$$PV_{\text{First Replacement}} = \frac{C_{FV}}{(1+d)^N} = \frac{10000}{(1+0.04)^{10}} = \$6755.65$$

$$PV_{\text{Second Replacement}} = \frac{C_{FV}}{(1+d)^N} = \frac{10000}{(1+0.04)^{20}} = \$4563.87$$

So the total present value of battery replacements over 20 years is:

$$PV_{\text{Total Replacement}} = 6755.65 + 4563.87 = \$11319.52$$

8.2.3 Net Present Value Method

Net present value (NPV) is used to find the current total value of all future cash flows in the range [0 to N]. NPV is calculated by subtracting the initial investment amount (C_0) from the converted total present value as shown in (8.5).

$$\text{NPV} = \sum_{t=1}^{N} \frac{C_{AV(t)}}{(1+d)^N} - C_0 \tag{8.5}$$

where

C_0: Initial investment [\$]
$C_{AV(t)}$: Annual net cash flows [\$], found by subtracting the costs in year t from the revenues of that year.
d: Discount rate [%], usually equal to the interest rate (i).

When deciding between several projects, the project with the highest NPV should be preferred. The projects with NPV < 0 are not economically feasible and should be eliminated. If NPV > 0, the investment will economically be viable, which means that the investor gets a profit. If NPV = 0, then the investment is still economically viable. However, the investor does not make any profit, and only recovers the initial investment cost.

Importantly note that equating the discount rate to the interest rate ($d = i$) may cause large deviations in the NPV in inflationary markets, where the inflation rate is not negligible. In this case, the inflation rate (e) should be subtracted from the inflation rate ($d = i - e$). For example, let's consider a solar PV project with an initial investment of 12 M\$ and its annual cash flows for three years are: 4 M\$, 5 M\$, and 6 M\$, respectively. Assuming the interest rate $i = 15\%$/year, then the net present value:

$$\text{NPV} = -12 + \frac{4}{(1+0.15)} + \frac{5}{(1+0.15)^2} + \frac{6}{(1+0.15)^3} = -0.796 \text{ M\$}$$

Assuming the cash flows depreciated by the annual inflation rate (5%), then the new discount rate is equal to $d = i - e = 15\% - 5\% = 10\%$. In this case, the NPV is:

$$\text{NPV} = -12 + \frac{4}{(1+0.1)} + \frac{5}{(1+0.1)^2} + \frac{6}{(1+0.1)^3} = 0.273 \text{ M\$}$$

As can be seen from the above results, NPV is negative (NPV < 0) when inflation is not considered. However, NPV becomes greater than zero if the inflation rate is taken into account (NPV > 0). Note that the numerical example given above is for a specific period. The economic life of a PV system is about 25 years on average. If the NPV is calculated for a longer period, it will be greater than zero for both cases.

If the annual cash flows are uniform ($C_{AV1} = C_{AV2} = \cdots = C_{AV(N)} = C_{AV}$) through the range of [0 to N], then the NPV is calculated as follows:

$$\text{NPV} = C_{AV} \cdot \left[\frac{(1+d)^N - 1}{d \cdot (1+d)^N} \right] - C_0 \qquad (8.6)$$

Example 8.2: Net Present Value of a PV System

Cash flows over the economic life of a utility-scale PV system are shown in Figure 8.2. The initial investment cost of the PV project is 9 M$ and its economic life is estimated to be 28 years. According to the agreement between the company and investor, the contractor will replace the solar invertors for $400,000 every 7 years. The expected annual revenue from the electrical energy sale is $700,000 and the O&M cost of the PV facility is $30,000. Assuming the discount rate is 4%, calculate the net present value of the PV project.

Solution:

Firstly, it's better to have annual net cash flows before calculating NPV. For this purpose, the annual net cash flows can be given as a list, as in Table 8.1.

The cash flows in 28 years can be brought back to the present worth (PV) using equation (8.4). A sample calculation for $N=1$ and $N=7$ is given below. All other calculation results are given in Table 8.2.

$$PV_1 = \frac{C_1}{(1+d)^N} = \frac{670}{(1+0.04)^1} = 644.23 \text{ K\$}$$

$$PV_7 = \frac{C_7}{(1+d)^N} = \frac{270}{(1+0.04)^7} = 205.17 \text{ K\$}$$

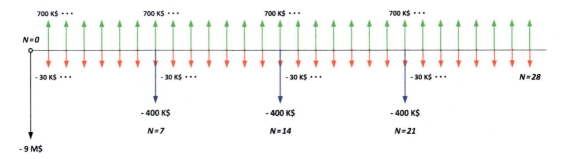

FIGURE 8.2 Cash flow diagram of the PV system over 28 years.

TABLE 8.1
Annual Net Cash Flows Over 28 Years

N (Year)	1	2	3	4	5	6	7	8	9	10	11	12	13	14
Net Cash (K$)	670	670	670	670	670	670	270	670	670	670	670	670	670	270
N (Year)	15	16	17	18	19	20	21	22	23	24	25	26	27	28
Net Cash (K$)	670	670	670	670	670	670	270	670	670	670	670	670	670	670

Economics of Solar Photovoltaic Systems

TABLE 8.2
Present Values of Annual Net Cash Flows

N (Year)	1	2	3	4	5	6	7	8	9	10	11	12	13	14
PV (K$)	644.2	619.5	595.6	572.7	550.7	529.5	205.2	489.6	470.7	452.6	435.2	418.5	402.4	155.9
N (Year)	15	16	17	18	19	20	21	22	23	24	25	26	27	28
PV (K$)	372.0	357.7	344.0	330.7	318.0	305.8	118.5	282.7	271.8	261.4	251.3	241.7	232.4	223.4

So the total present value of the PV project over 28 years is:

$$PV_{\text{Total}} = \sum_{t=1}^{28} \frac{C_{AV(t)}}{(1+0.04)^t} = 10453.8 \text{ K\$}$$

If we subtract the initial investment cost from the total present value (based on equation 8.5), the NPV of the solar project is found as below:

$$\text{NPV} = \sum_{t=1}^{28} \frac{C_{AV(t)}}{(1+0.04)^t} - C_0 \Rightarrow \text{NPV} = 10453.8 \text{ K\$} - 9000 \text{ K\$} = 1453.8 \text{ K\$}$$

Since NPV > 0, the project investment is acceptable.

8.2.4 CAPITAL RECOVERY FACTOR

The capital recovery factor (CRF) is used to find the annual equivalent value (AEV) of an income or cost. If "i" is the interest rate, the CRF [1/year] is given by equation (8.7).

$$\text{CRF} = \left[\frac{i \cdot (1+i)^N}{(1+i)^N - 1} \right] \quad (8.7)$$

In this framework, for economic comparison of several investment projects with different lifetimes ($N_1 \neq N_2 \neq N_3 \cdots$) on a common base, the net present values of these projects are multiplied by the CRF, as shown in (8.8) and (8.9).

$$\text{AEV}_1 \, [\$/\text{year}] = \text{NPV}_1 \cdot \left[\frac{i \cdot (1+i)^{N_1}}{(1+i)^{N_1} - 1} \right] \quad (8.8)$$

$$\text{AEV}_2 \, [\$/\text{year}] = \text{NPV}_2 \cdot \left[\frac{i \cdot (1+i)^{N_2}}{(1+i)^{N_2} - 1} \right] \quad (8.9)$$

The investment project with the highest AEV is preferred among the several projects. In terms of a solar PV project, the CRF factor can be used in the assessment of operation and maintenance (O&M) costs, which are often considered uniform every year throughout the lifetime of the solar PV system. An illustrative study of using CRF in PV system calculations is given in the following examples.

Example 8.3: Use of Capital Recovery Factor in O&M Cost of a PV Project

Assume that the annual O&M cost of a PV system is 1000 $/year. If the interest rate of O&M cost is 5%, compute its present worth over a 28-year period.

Solution:

Substituting the given data into equation (8.7), the CRF value in [1/year] is obtained as below:

$$\text{CRF} = \left[\frac{0.05 \cdot (1+0.05)^{28}}{(1+0.05)^{28} - 1}\right] = 0.0671 \text{ / year}$$

With the help of equation (8.8), the net present value of the total cost of O&M is:

$$\text{Anual Cost} = PV \cdot CRF$$

$$1000\,[\$/\text{year}] = PV \cdot 0.0671\,[1/\text{year}] \Rightarrow PV = \frac{1000}{0.0671} = \$14{,}903$$

Example 8.4: Use of Capital Recovery Factor for Annual Loan Payments

A loan of $10,000 was used to finance a PV project with a payback period of 10 years. If the interest rate is 8%, calculate the annual payments.

Solution:

Substituting the given data into equation (8.7), the CRF value is:

$$\text{CRF} = \left[\frac{0.08 \cdot (1+0.08)^{10}}{(1+0.08)^{10} - 1}\right] = 0.149/\text{year}$$

With the help of equation (8.8), the annual payment will be:

$$\text{Anual Payment}\,[\$/\text{year}] = PV \cdot CRF$$

$$\text{Anual Payment} = 10000\,\$ \cdot 0.149\,[1/\text{year}] = 1490\,[\$/\text{year}]$$

Example 8.5: Use of CRF in Bank Loan Payments

A farmer wants to build a grid-connected solar PV project for use in farm buildings. Two options were offered to the farmer for the PV project.

Option-1: It is a fixed tilt angle PV system, and the farmer himself can completely cover the financing of this project.

Option-2: It is a PV system with a solar tracker. However, the farmer has to use a $9000 bank loan for the tracker unit. On the other hand, the PV system with a sun tracker will generate 3200 [kWh / year] more energy than that of a fixed-angle system. If the farmer chooses Option-1, he will have to buy 3200 [kWh / year] from the grid for 0.40 $ / kWh.

a. The farmer has chosen Option-2. If the interest rate is 8% with a payback period of 7 years, calculate the annual bank payments.

Economics of Solar Photovoltaic Systems

b. The price of purchased energy from the grid will also increase with an inflation rate of 4 [%/year]. If the farmer wants to pay the bank loan with 3200 [kWh/year] solar energy production, what will be the annual payment?
c. If the farmer wants to pay the entire bank loan by selling 3200 [kWh/year] energy, what should be the minimum energy sales price?

Solution:

a. The solar tracker system costs $9000. Its financing will be met by a 7-year bank loan with an interest rate of 8%/year. Thus, from equation (8.8), we get:

$$\text{Annual Payment } [\$/\text{year}] = 9000 \cdot \left[\frac{0.08 \cdot (1+0.08)^7}{(1+0.08)^7 - 1} \right] = 1728.65 \; [\$/\text{year}]$$

b. The farmer will use the 3200 [kWh/year] solar energy amount for loan payments. That is to say, if there was no sun tracker, the farmer would buy 3200 kWh/year of energy from the grid. As the energy purchase price will increase by 4 [%/year], the tariffs and energy savings (E_{saving}) for the next 7 years can be calculated from equation (8.1) as follows:

$$FV_1 = 0.40 \cdot (1+0.04)^1 = 0.416 \; [\$/\text{kWh}] \Rightarrow E_{saving\;1} = 3200 \times 0.416 = \$1331$$

$$FV_2 = 0.40 \cdot (1+0.04)^2 = 0.432 \; [\$/\text{kWh}] \Rightarrow E_{saving\;2} = 3200 \times 0.432 = \$1382$$

$$FV_3 = 0.40 \cdot (1+0.04)^3 = 0.450 \; [\$/\text{kWh}] \Rightarrow E_{saving\;3} = 3200 \times 0.450 = \$1440$$

$$FV_4 = 0.40 \cdot (1+0.04)^4 = 0.468 \; [\$/\text{kWh}] \Rightarrow E_{saving\;4} = 3200 \times 0.468 = \$1498$$

$$FV_5 = 0.40 \cdot (1+0.04)^5 = 0.486 \; [\$/\text{kWh}] \Rightarrow E_{saving\;5} = 3200 \times 0.486 = \$1555$$

$$FV_6 = 0.40 \cdot (1+0.04)^6 = 0.506 \; [\$/\text{kWh}] \Rightarrow E_{saving\;6} = 3200 \times 0.506 = \$1619$$

$$FV_7 = 0.40 \cdot (1+0.04)^7 = 0.526 \; [\$/\text{kWh}] \Rightarrow E_{saving\;7} = 3200 \times 0.526 = \$1683$$

The net annual bank payments are then calculated by subtracting the energy savings from the annual bank payments, as shown in Table 8.3.

TABLE 8.3
Annual Net Bank Payments Over 7 Years

N (Year)	1	2	3	4	5	6	7
Bank Payments ($/year)	1729	1729	1729	1729	1729	1729	1729
Energy Savings ($/year)	1331	1382	1440	1498	1555	1619	1638
Net Payments ($/year)	398	347	289	231	174	110	91

c. If the farmer wants to pay the entire bank loan by selling 3200 [kWh/year] energy all through the loan period of 7 years, the minimum energy cost (EC_{min}) shall be:

$$EC_{min} = \frac{\text{Annual Bank Payment [\$/year]}}{\text{Amount of Energy Sold [kWh/year]}} = \frac{1729}{3200} = 0.540 \, [\$/kWh]$$

8.2.5 Life-Cycle Cost

Performing a life-cycle cost (LCC) analysis for a power system gives the total present value of that project, including all the occurred costs over the economic life. Hence, the LCC of a project can be calculated using equation (8.10).

$$\text{LCC} \, [\$] = C_0 + C_{O\&M(PV)} + C_{E(PV)} + C_{R(PV)} - C_{S(PV)} \qquad (8.10)$$

where

C_0: Capital cost of the project, which includes all the initial expenses such as equipment, design, engineering, and installation [$]. Since this capital cost occurs in the first year of the project, it is the same as its present value. As a simple approach, the capital cost for a PV system [$C_{0(PV)}$] can be calculated in two parts. The first part is the cost depending on the total surface area of the PV panels (C_A), including panels, wiring, racks, etc. The second part is the cost independent of the surface area (C_I), which includes metering, batteries, grid integration cost, etc. Hence, the capital cost is equal to: $C_{0(PV)} = A \times C_A + C_I$, where A is the total surface area of photovoltaic panels.

$C_{O\&M(PV)}$: Sum of all yearly scheduled O&M costs, which is brought back to its present value [$]. Typically, O&M costs include salaries, inspections, insurance, property taxes, and all planned maintenance.

$C_{E(PV)}$: Present value of the energy cost, which is obtained from the sum of yearly fuel costs [$]. This fuel cost is typically zero for PV system applications.

$C_{R(PV)}$: Present value of the replacement cost, which is obtained from the sum of all expected repair and equipment replacement costs over the life of the system [$]. The replacement costs for PV systems usually include the battery and inverter replacement costs.

$C_{S(PV)}$: The present value of the net worth of the system at the end of its lifetime [$]. The present value of this salvage cost must be discounted due to the time value of money.

Example 8.6: Life-Cycle Cost Analysis of a PV System

Let us consider a 30 kW grid-connected solar PV system with a total surface area of 200 m². The unit price of 300W solar panels is $300, each with a surface area of 2 m². The installation cost of metal frameworks is estimated to be 40 $/m². Besides, the initial cost of the inverter is $4000 and will be replaced every 8 years. Furthermore, the cost of the battery pack required for energy storage is $3000 and will be replaced at the same time as the inverter. Finally, the system installation cost is $1200, and the cost of BOS (balance of the system) plus mounting hardware is $2000. Assuming the O&M cost of 150 $/year and the salvage income as 10% of the capital cost, calculate the life-cycle cost of the PV system. Note that the life of the PV system is 24 years and the interest rate is 4%.

Solution:

Firstly, we need to estimate the initial cost to calculate the LCC. Assuming the entire surface is covered with PV panels, a total of 100 panels is needed (200 m² / 2 m² = 100, or 30000 W / 300 W = 100). Hence, the capital cost can be estimated as below with the help of the expression $C_{0(PV)} = A \times C_A + C_I$.

Economics of Solar Photovoltaic Systems

TABLE 8.4
Life-Cycle Cost Analysis of the Solar PV System ($N=24$ years, and $i=4\%$)

LCC	Items		Amount [$]	Present Worth [$]	Percentage of Total LCC [%]
Capital Cost	PV Array		30000	30000	49.00
	Inverter		4000	4000	6.53
	Battery		3000	3000	4.90
	Metal Carcass		8000	8000	13.07
	BOS+Mounting Hardware		2000	2000	3.27
	Installation		1200	1200	1.96
	Sub Total-1		$48200	$48200	78.7%
O&M Cost	Annual O&M (Total value over 24 years) $C_{O\&M} = \sum_{t=1}^{24} \frac{C_{O\&M(t)}}{(1+0.04)^t}$		150	2287	3.7
	Sub Total-2		$150	$2287	3.7%
Replacement Cost		Year			
	Inverter	8	4000	2922.8	4.77
	Inverter	16	4000	2135.6	3.49
	Battery	8	3000	2192.1	3.58
	Battery	16	3000	1601.7	2.62
	Sub Total-3		$14000	$8852.2	14.5%
Salvage Cost		Year			
	Salvage (10% of Capital Cost)	24	4820	1880.4	3.1
	Sub Total-4		$4820	$1880.4	3.1%
LIFE-CYCLE COST (LCC) $= C_0 + C_{O\&M} + C_R - C_S$				$57458.8	100%

$$C_{0(PV)} = 100 \times \$300 + 200\,\text{m}^2 \times 40\frac{\$}{\text{m}^2} + \$4000 + \$3000 + \$1200 + \$2000 = \$48,200$$

It is now useful to create a calculation sheet, as shown in Table 8.4, to calculate the net present value of the PV project.

8.2.6 Levelized Cost of Energy

The levelized cost of energy (LCOE) is a term often used to compare the electrical energy costs (usually in $/kWh) of different types of power plants, regardless of the power plant scale or operating time. Indeed, this concept is an application of the present value method. Firstly, the parameters of the discount rate d and the lifetime (N) in years are determined for each power plant. Then, the present values of the annual costs occurring in the range of [0 to N] are calculated for each power plant. Afterward, considering these present value costs, the LCC is found for each power plant (see Section 8.2.5 for more detail on LCC). In the final step, the computed LCC is divided by the expected total energy production of each plant over the lifetime, as shown in equation (8.11).

$$\text{LCOE}\,[\$/\text{kWh}] = \frac{\text{Life Cycle Cost (LCC)}\,[\$]}{\text{Lifetime Energy Generation}\,[\text{kWh}]} \quad (8.11)$$

The general equation given in (8.11) for LCOE has been calculated by two different methods in the literature. The first method is known as the discounting method, while the second one is known as the annuity method. According to the discounting technique, all future costs are discounted to their total present value (see Section 8.2.5) and then divided by the lifetime energy output. A simplified LCOE equation for a grid-connected PV system can be arranged as equation (8.12).

$$\text{LCOE } [\$/kWh] = \frac{C_0 + \sum_{t=1}^{N} \frac{C_{O\&M(t)}}{(1+d)^t} + \sum_{t=1}^{N} \frac{C_{R(t)}}{(1+d)^t} - \frac{C_S}{(1+d)^N}}{\sum_{t=1}^{N} \frac{E_{(t)} \cdot (1-D)^t}{(1+d)^t}} \quad (8.12)$$

where d is the discount rate, D is the system degradation rate for photovoltaics, and $E_{(t)}$ is the energy output of the system in [kWh] in the year "t." Note that other parameters in equation (8.12) are already explained in Section 8.2.5.

According to the annuity approach in calculating LCOE, all costs throughout the life-cycle are converted into annual equivalent costs. And then, this annual cost is divided by the average annual electrical output of the plant over its lifetime, as shown in equation (8.13).

$$\text{LCOE } [\$/kWh] = \frac{\text{Annual Equivalent Cost}}{\text{Average Energy Output}} = \frac{\left[\sum_{t=0}^{N} \frac{C_{(t)}}{(1+d)^t}\right]\left[\frac{i \cdot (1+d)^N}{(1+d)^N - 1}\right]}{\frac{1}{N}\left[\sum_{t=1}^{N} E_{(t)}\right]} \quad (8.13)$$

The use of the discounting method is more common in the literature when calculating the LCOE for wind and solar power systems. Therefore, the discounting method is preferred here in the LCOE calculation of solar PV systems.

It is worth noting that the concept of levelized cost of energy has been used more frequently in the solar PV industry in recent years. In particular, the manufacturers often refer to the LCOE for marketing purposes. Because the performance and cost comparison of different PV technologies can be fulfilled easily by using LCOE technique. For investors, it provides an opportunity to choose the most suitable PV technology & equipment, depending on the peak solar hours and climatic conditions.

Example 8.7: Levelized Cost of Energy Comparison for Three Different PV Systems

Consider three different 30 kW on-grid solar PV systems as described in Table 8.5. Compare the LCOE (levelized cost of energy) values for each PV system option based on the given data.

Solution:

All data needed for LCOE calculation are available in Table 8.5. Hence from (8.12), we obtain the LCOE values for the three PV system types as follows:

$$\text{LCOE}_1 = \frac{48200 + \sum_{t=1}^{28}\frac{150}{(1+0.05)^t} + \left[\frac{4000}{(1+0.05)^{10}} + \frac{4000}{(1+0.05)^{20}} + \frac{3000}{(1+0.05)^7} + \frac{3000}{(1+0.05)^{14}} + \frac{3000}{(1+0.05)^{21}}\right] - \frac{4820}{(1+0.05)^{28}}}{\sum_{t=1}^{28} \frac{40000 \cdot (1-0.005)^t}{(1+0.05)^t}}$$

Economics of Solar Photovoltaic Systems

TABLE 8.5
Solar PV System Parameters to Calculate LCOE

Parameters	Items		PV System-1 (Fixed Tilt)	PV System-2 (Single-Axis Tracker)	PV System-3 (Two-Axis Tracker)
Financial & Operational	Lifetime (N)		28 years	28 years	28 years
	Discount Rate (d)		5%	5%	5%
	Annual Generation		40000 kWh	54000 kWh	57000 kWh
	Degradation Rate		0.5 %/year	0.5 %/year	0.5 %/year
Capital Cost [$]	PV Array		$30000	$30000	$30000
	Inverter		$4000	$4000	$4000
	Battery		$3000	$3000	$3000
	Metal Carcass/Tracker		$8000	$17000	$20000
	BOS+Mounting Hardware		$2000	$3000	$4000
	Installation		$1200	$2400	$3600
	Sub Total-1		**$48200**	**$59400**	**$65200**
O&M Cost [$]	Annual O & M		$150	$200	$250
	Sub Total-2		**$150**	**$200**	**$250**
Replacement Cost [$]		Year			
	Inverter	10	$4000	$4000	$4000
	Inverter	20	$4000	$4000	$4000
	Battery	7	$3000	$3000	$3000
	Battery	14	$3000	$3000	$3000
	Battery	21	$3000	$3000	$3000
	Sub Total-3		**$17000**	**$17000**	**$17000**
Salvage Cost [$]		Year			
	Salvage (10 % of Capital Cost)	28	$4820	$5940	$6520
	Sub Total-4		**$4820**	**$5940**	**$6520**

$$\text{LCOE}_1 = \frac{48200 + 2234.7 + [2455.7 + 1507.6 + 2132 + 1515.2 + 1076.8] - 1229.6}{563213.5 \text{ kWh}}$$

$$= \frac{57889.4 \text{ \$}}{563213.5 \text{ kWh}} = 0.1027 \text{ \$/kWh}$$

As for the second system:

$$\text{LCOE}_2 = \frac{59400 + \sum_{t=1}^{28} \frac{200}{(1+0.05)^t} + \left[\frac{4000}{(1+0.05)^{10}} + \frac{4000}{(1+0.05)^{20}} + \frac{3000}{(1+0.05)^{7}} + \frac{3000}{(1+0.05)^{14}} + \frac{3000}{(1+0.05)^{21}} \right] - \frac{5940}{(1+0.05)^{28}}}{\sum_{t=1}^{28} \frac{54000 \cdot (1-0.005)^t}{(1+0.05)^t}}$$

$$\text{LCOE}_2 = \frac{59400 + 2979.6 + [2455.7 + 1507.6 + 2132 + 1515.2 + 1076.8] - 1515.3}{760338.2 \text{ kWh}}$$

$$= \frac{69551.6 \text{ \$}}{760338.2 \text{ kWh}} = 0.0914 \text{ \$/kWh}$$

And for the third system:

$$\text{LCOE}_3 = \frac{65200 + \sum_{t=1}^{28}\frac{250}{(1+0.05)^t} + \left[\begin{array}{c}\frac{4000}{(1+0.05)^{10}} + \frac{4000}{(1+0.05)^{20}} + \frac{3000}{(1+0.05)^7} \\ + \frac{3000}{(1+0.05)^{14}} + \frac{3000}{(1+0.05)^{21}}\end{array}\right] - \frac{6520}{(1+0.05)^{28}}}{\sum_{t=1}^{28}\frac{57000\cdot(1-0.005)^t}{(1+0.05)^t}}$$

$$\text{LCOE}_3 = \frac{65200 + 3724.5 + [2455.7 + 1507.6 + 2132 + 1515.2 + 1076.8] - 1663.2}{802579.2 \text{ kWh}}$$

$$= \frac{75948.6 \text{ \$}}{802579.2 \text{ kWh}} = 0.0946 \text{ \$/kWh}$$

According to the results, while the PV system with a two-axis tracker has the highest energy generation, the PV system with a single-axis tracker has the lowest energy cost.

8.2.7 Internal Rate of Return

The internal rate of return (IRR) on an investment or project is a valuation rate that equates the net present value of all cash flows of the investment to zero. The IRR method does not apply to equipment, but only to investment projects. Firstly, the net present value of the investment project is expressed in terms of the discount rate d. The net present value (NPV) expression is then set to zero, as shown in equation (8.14). In the final step, this non-linear equation is solved for d. The solution of d gives the IRR of the project. The projects with IRR < 0 should not be implemented. On the other hand, the investment project with the highest IRR means that it is the most profitable project among the possible options. If there is a minimum acceptable rate of return (MARR) targeted by the business, the condition IRR ≥ MARR must be met for the project to be approved.

$$\sum_{t=1}^{N} \frac{C_t}{(1+d)^N} - C_0 = 0 \tag{8.14}$$

If the annual cash flows are uniform ($C_1 = C_2 = \cdots = C_N = C$) over the lifetime, then the variable d is solved for IRR, as follows:

$$C \cdot \left[\frac{(1+d)^N - 1}{d \cdot (1+d)^N}\right] - C_0 = 0 \tag{8.15}$$

The IRR can also be found graphically. NPV versus d is plotted by giving different d values in the expression "$NPV = f(d)$." By definition, the d value at NPV = 0 is the solution of IRR (Figure 8.3).

Various iterative and regression methods can be used to find d. Analytical and numerical methods for the solution of IRR are available in the literature. Hence, only the use of IRR functions in *MATLAB*® and *Excel* are given here for the solution of IRR. The *MATLAB*® command contains annual cash flows $(-C_0, C_1, C_2, \ldots, C_N)$, where C_0 is the initial investment cost:

$$\text{Return} = \text{irr}\big([-C_0 \ C_1 \ C_2 \ \cdots \ C_N]\big) \tag{8.16}$$

On the other hand, IRR can also be calculated with the help of Excel, as below:

Economics of Solar Photovoltaic Systems

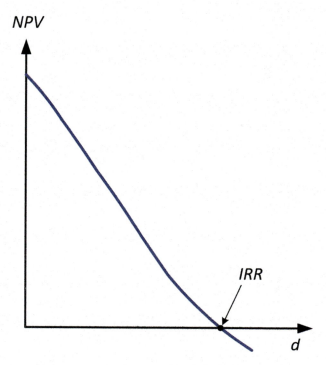

FIGURE 8.3 Graphical representation of the IRR.

$$= \text{IRR}(\text{CELL}_1 : \text{CELL}_N) \qquad (8.17)$$

where the cash flows in Excel are written from Cell-1 to Cell-N. For example, if a series of cash flows from the cells B1 through B7 are entered, it is calculated as "$= \text{IRR}(B1 : B7)$." Note that the commands for Excel functions may vary in different languages.

Example 8.8: Internal Rate of Return of a Solar PV Project

Consider the present values of annual net cash flows given in Example 8.2. Find the IRR of the solar PV project.

Solution:

The present values of annual net cash flows given in Table 8.2 are entered in the MATLAB® command window as follows. Then the IRR value is calculated as follows with the command "**Return = irr(CashFlow).**"
 Since IRR = 0.0131 > 0, the project is suitable for investment.

```
@ CashFlow=[-9000 644.2 619.5 595.6 572.7 550.7 529.5 205.2 489.6
470.7 452.6 435.2 418.5 402.4 155.9 372.0 357.7 344.0 330.7 318.0
305.8 118.5 282.7 271.8 261.4 251.3 241.7 232.4 223.4];

@ Return=irr(CashFlow)

Return =     0.0131
```

8.2.8 SIMPLE PAYBACK PERIOD

The payback period, in general, estimates how long it will take for an investment project to pay for itself. Therefore, the project option with the shortest payback period is preferred. Due to its easiness, the simple payback method is often used for investments although its limitations. For example, the simple payback period does not take into account the time value of money, project risks, and other financing issues (such as lifetime and salvage value). Considering the uniform cash flows ($C_1 = C_2 = \cdots = C_N = C$), the simple payback period (SBP, year) is calculated as follows:

$$\text{SBP [year]} = \frac{C_0\,[\$]}{C\,[\$/\text{year}]} \qquad (8.16)$$

If the annual cash flows are not uniform:

$$\text{SBP[year]} = (T-1) + \frac{C_0 - \sum_{t=1}^{T-1} C_t}{C_T} \qquad (8.17)$$

where T is the first year in which the cumulative cash flow equals or exceeds the initial investment amount (C_0) and C_T is the net cash flow occurring in the year T.

Example 8.9: Simple Payback Period of a Solar PV Project

Consider the annual net cash flows given in Example 8.2. Find the payback period ignoring the time value of money.

Solution:

The annual net cash flows given in Table 8.1 are not uniform. Therefore, the cumulative summation is required, as shown in Table 8.6.

The cumulative sum of annual cash flows exceeds the initial investment cost in the 15th year (9250 K\$ > 9000 K\$). Hence, from equation (8.17), we get:

$$\text{SBP [year]} = (15-1) + \frac{9000 - \sum_{t=1}^{14} C_t}{C_{15}} = 14 + \frac{9000 - 8580}{670} = 14.62 \text{ year}$$

TABLE 8.6
Annual Net Cash Flows over 28 Years and Their Cumulative Summations

N (Year)	1	2	3	4	5	6	7	8	9	10	11	12	13	14
Net Cash (K\$)	670	670	670	670	670	670	**270**	670	670	670	670	670	670	270
N (Year)	15	16	17	18	19	20	21	22	23	24	25	26	27	28
Net Cash (K\$)	670	670	670	670	670	670	**270**	670	670	670	670	670	670	670
Cumulative Summation of Annual Net Cash Flow														
N (Year)	1	2	3	4	5	6	7	8	9	10	11	12	13	14
Cum. Sum	670	1340	2010	2680	3350	4020	4290	4960	5630	6300	6970	7640	8310	8580
N (Year)	15	16	17	18	19	20	21	22	23	24	25	26	27	28
Cum. Sum	9250	9920	10590	11260	11930	12600	12870	13540	14210	14880	15550	16220	16890	17560

Economics of Solar Photovoltaic Systems

8.2.9 Discounted Payback Period

The simple payback period is a kind of static method that does not take into account the time value of money. If the time value of money is taken into account, then the discounted payback period (DPB) is obtained. Namely, the annual cash flows are brought back to $t=0$ by a discounted rate (d). In other words, cash flow for each year is multiplied by $1/(1+d)^t$ and summed by the discounted value of the previous year. That is to say, the cumulative sums of the present values are found. Therefore, DPB can be expressed by equation (8.18), as similar to (8.17).

$$\text{DBP}\,[\text{year}] = (T-1) + \frac{C_0 - \sum_{t=1}^{T-1}\frac{C_t}{(1+d)^t}}{\frac{C_T}{(1+d)^t}} \tag{8.18}$$

where T is the first year in which the cumulative sum of the present value of net cash flows equals or exceeds the initial investment amount (C_0), C_T is the net cash flow occurring in the year T, and d is the discount rate.

Example 8.10: Discounted Payback Period of a Solar PV Project

Consider the cash flows given in Example 8.2. Find the payback period considering the time value of money.

Solution:

We can use the present values of annual net cash flows in Table 8.1 to calculate the payback period. For this purpose, the cumulative summation is required, as shown in Table 8.7.

The cumulative sum of annual cash flows exceeds the initial investment cost in the 23rd year (9244 K\$ > 9000 K\$). Hence, from equation (8.18), we get:

$$\text{SBP}\,[\text{year}] = (23-1) + \frac{9000 - \sum_{t=1}^{22}\frac{C_t}{(1+d)^t}}{\frac{C_{22}}{(1+d)^{22}}} = 22 + \frac{9000 - 8927}{282.7} = 22.25\ \text{year}$$

TABLE 8.7
Present Values of Annual Net Cash Flows Over 28 Years and Their Cumulative Summations

N (Year)	1	2	3	4	5	6	7	8	9	10	11	12	13	14
Net Cash (K\$)	644.2	619.5	595.6	572.7	550.7	529.5	205.2	489.6	470.7	452.6	435.2	418.5	402.4	155.9
N (Year)	15	16	17	18	19	20	21	22	23	24	25	26	27	28
Net Cash (K\$)	372.0	357.7	344.0	330.7	318.0	305.8	118.5	282.7	271.8	261.4	251.3	241.7	232.4	223.4

Cumulative Summation of Annual Net Cash Flow

N (Year)	1	2	3	4	5	6	7	8	9	10	11	12	13	14
Cum. Sum	644	1264	1859	2432	2983	3512	3717	4207	4678	5130	5566	5984	6386	6542
N (Year)	15	16	17	18	19	20	21	22	23	24	25	26	27	28
Cum. Sum	6914	7272	7616	7947	8265	8571	8689	8972	9244	9505	9756	9998	10230	10454

8.3 ECONOMIC ASSESSMENT OF GRID-CONNECTED PV SYSTEMS

For the economic assessment of the grid-connected PV systems, annual net cash flows (total incomes minus total expenses) over the lifetime are brought back to their present values. If a bank loan is used to finance the PV system, it is possible to write the general equation of annual cash flow as follows:

$$C_t = \frac{E_{\text{SOLD}} \cdot (1-D)^t}{(1+i_{\text{FiT}})^t} \cdot \text{FiT} - \text{BP}_t - \left[\frac{C_{\text{OM}(t)}}{(1+i)^t} + \frac{C_{R(t)}}{(1+i)^t} - \frac{C_{S(t)}}{(1+i)^t} \right] \quad (8.19)$$

where C_t is the net cash flow in year t, E_{SOLD} is the annual energy sold to the grid, D is the system degradation rate for photovoltaics, FiT is the price for the feed-in tariff, i_{FiT} is the annual inflation rate for FiT, BP_t is the annual bank payment in case of using the loan, $C_{\text{OM}(t)}$ is the annual O&M cost, $C_{R(t)}$ is the replacement cost in year t, $C_{S(t)}$ is the salvage revenue in the last year of a lifetime, and i is the annual inflation rate used for cost components. Importantly note that replacement cost and salvage income apply only to certain years. For example, if an inverter replacement takes place every 10 years, it is only included in the calculation for $t = 10$ and $t = 20$. Similarly, salvage revenue is considered for the final year only. These costs will be zero for the rest of the other years.

Assuming the loan interest rate i_{LOAN}, annual bank payments (BP_t) of a loan over the respective period (N_{LOAN}) can be calculated based on the capital recovery factor (CRF) as follows:

$$\text{BP}_t = \text{LOAN} \cdot \left[\frac{i_{\text{LOAN}} \cdot (1+i_{\text{LOAN}})^{N_{\text{LOAN}}}}{(1+i_{\text{LOAN}})^{N_{\text{LOAN}}} - 1} \right] \quad (8.20)$$

Depending on the incentives and tax rates of the countries, more economic parameters may also be included in the annual cash flows. For example, net profit for the annual cash flows of investment projects should be taken into consideration. In other words, the net profit is calculated by subtracting the tax amount from the profit before tax:

$$\text{Net Profit} = \text{Profit Before Tax} \cdot (1 - \text{Tax Rate}) \quad (8.21)$$

Accordingly, the Net Present Value (NPV) of the solar PV system is expressed as below in the case of a bank loan:

$$\text{NPV} = \sum_{t=1}^{N} \frac{C_t}{(1+d)^N} - (C_0 - \text{LOAN}) \quad (8.22)$$

Example 8.11: NPV Analysis of a Solar PV Project with Detailed Financial Data

Consider a 50-kW grid-connected PV system, of which financial and operating data are given below.

- Annual energy generation: 58 MWh
- Degradation rate: 0.5% / year
- Feed-in tariff price: 0.46 $ / kWh and has an inflation rate of 2%
- The unit cost of installation: 6500 $ / kW
- VAT (Value Added Tax): 10% of the installation cost
- Operating cost: 100 $ / year and has a valuation rate of 3%
- Maintenance cost: 1% of total installation cost and has a valuation rate of 3%
- Interest rate, i: 6% / year

Economics of Solar Photovoltaic Systems

- The inflation rate, e: 2% / year
- The useful life of the installation: 25 years
  a. If the PV system is built with 100% own capital, perform the NPV analysis. In this case, find the payback period of the system.
  b. Assume that 60% of the installation cost of the PV system was financed with a bank loan. If the loan interest rate is 5% / year, perform the NPV analysis and find the payback period of the system. Note that the bank loan will be paid back with the annual net profit of the PV system.
  c. Assume that the bank loan in option (b) will be paid back in fixed amounts in 15 years. In this case, perform the NPV analysis and find the payback period of the system.

Solution:

a. The following calculations are considered to perform the NPV analysis:
   - Cost of Installation $= 6500 \times 50 = \$325000$
   - VAT included Initial Cost $= 1.1 \times 325000 = \$357,500$
   - O & M Cost $= 0.01 \times 357500 + 100 = \$3675$
   - Minimum acceptable rate of return of the investment $(d) = 4\%$ / year, where d = intereset rate – inflation rate $(6\% - 2\% = 4\%)$

   In the first option, a self-financed PV system is proposed (100% own capital). Details of NPV analysis and relevant cash flows for each year are given in Table 8.8. As can be seen from the NPV results, the payback period, in this case, is approximately 14 years.

b. The following calculations are considered to perform the *NPV* analysis:
   - Cost of Installation $= 6500 \times 50 = \$325000$
   - Loan Amount $= 325000 \times 0.60 = \$195000$
   - VAT included Cost $= 1.1 \times 325000 = \$357500$
   - VAT included Initial Cost $= \$357500 - \$195000 = \$142500$

   In the second option, a 60% financed PV system is proposed (40% own capital). The loan will be paid back by the annual profit of the PV system. Therefore, when calculating the residual debt for each year, a 5% interest rate is applied to the debt carried over from the previous year:

$$(\text{Residual Dept})_{t+1} = (\text{Residual Dept})_t \times (1+i) - (\text{Annual Profit})_{t+1}$$

Details of NPV analysis and relevant cash flows for each year are given in Table 8.9. As can be seen from the NPV results, the payback period in the second case is approximately 16 years.

c. In the third case, the loan will be paid back in equal amounts every year over the 15-year term. Annual loan amounts are then calculated from equation (8.20) as follows:

$$BP = LOAN \cdot \left[\frac{i_{LOAN} \cdot (1+i_{LOAN})^{N_{LOAN}}}{(1+i_{LOAN})^{N_{LOAN}} - 1}\right] = 195000 \cdot \left[\frac{0.01 \cdot (1+0.05)^{15}}{(1+0.05)^{15} - 1}\right] = \$18787.75$$

Therefore, the net present value of earnings is calculated as:

$$(\text{Net NPV of Earning})_t = (\text{NPV Earning})_t - \$18787.75$$

Details of NPV analysis and relevant cash flows for each year are given in Table 8.10. As can be seen from the NPV results, the payback period in the third case is approximately 17 years.

TABLE 8.8
Net Present Values Analysis of the Self-Financed Solar PV Project

Year	Annual Gen. (kWh)	Degradation Rate (%/year)	Degraded Gen. (kWh)	Feed-in Tariff ($/kWh)	Inflation for FiT (%/year)	Fut. Val. of FiT (%/year)	Fut. Val. of Sold Energy ($)	MARR $(i-e)$ (%/year)	Pres. Val. of Sold Energy ($)	O&M Cost ($)	Val. Rate for O&M Cost (%/y)	Pres. Val. of O&M Cost ($)	Pres. Val. of Earnings ($)	Cum. Sums of Pr. Val. Earnings ($)	NPV of the Project ($)
0															−357500
1	58000	0.005	58000	0.46	0.02	0.46	26680	0.004	26574	3675	0.003	3664	22899	22899	−334601
2	58000	0.005	57710	0.46	0.02	0.47	27214	0.004	26997	3675	0.003	3653	23322	46221	−311279
3	58000	0.005	57421	0.46	0.02	0.48	27758	0.004	27427	3675	0.003	3642	23752	69973	−287527
4	58000	0.005	57134	0.46	0.02	0.49	28313	0.004	27865	3675	0.003	3631	24190	94163	−263337
5	58000	0.005	56849	0.46	0.02	0.50	28879	0.004	28309	3675	0.003	3620	24634	118796	−238704
6	58000	0.005	56564	0.46	0.02	0.51	29457	0.004	28760	3675	0.003	3610	25085	143881	−213619
7	58000	0.005	56282	0.46	0.02	0.52	30046	0.004	29218	3675	0.003	3599	25543	169424	−188076
8	58000	0.005	56000	0.46	0.02	0.53	30647	0.004	29684	3675	0.003	3588	26009	195433	−162067
9	58000	0.005	55720	0.46	0.02	0.54	31260	0.004	30157	3675	0.003	3577	26482	221914	−135586
10	58000	0.005	55442	0.46	0.02	0.55	31885	0.004	30637	3675	0.003	3567	26962	248877	−108623
11	58000	0.005	55164	0.46	0.02	0.56	32523	0.004	31126	3675	0.003	3556	27451	276327	−81173
12	58000	0.005	54889	0.46	0.02	0.57	33173	0.004	31622	3675	0.003	3545	27947	304274	−53226
13	**58000**	**0.005**	**54614**	**0.46**	**0.02**	**0.58**	**33837**	**0.004**	**32125**	**3675**	**0.003**	**3535**	**28450**	**332724**	**−24776**
14	**58000**	**0.005**	**54341**	**0.46**	**0.02**	**0.60**	**34513**	**0.004**	**32637**	**3675**	**0.003**	**3524**	**28962**	**361687**	**4187**
15	58000	0.005	54069	0.46	0.02	0.61	35204	0.004	33158	3675	0.003	3514	29483	391169	33669
16	58000	0.005	53799	0.46	0.02	0.62	35908	0.004	33686	3675	0.003	3503	30011	421180	63680
17	58000	0.005	53530	0.46	0.02	0.63	36626	0.004	34223	3675	0.003	3493	30548	451728	94228
18	58000	0.005	53262	0.46	0.02	0.64	37358	0.004	34768	3675	0.003	3482	31093	482821	125321
19	58000	0.005	52996	0.46	0.02	0.66	38106	0.004	35322	3675	0.003	3472	31647	514469	156969
20	58000	0.005	52731	0.46	0.02	0.67	38868	0.004	35885	3675	0.003	3461	32210	546679	189179
21	58000	0.005	52467	0.46	0.02	0.68	39645	0.004	36457	3675	0.003	3451	32782	579461	221961
22	58000	0.005	52205	0.46	0.02	0.70	40438	0.004	37038	3675	0.003	3441	33363	612824	255324
23	58000	0.005	51944	0.46	0.02	0.71	41247	0.004	37628	3675	0.003	3430	33953	646777	289277
24	58000	0.005	51684	0.46	0.02	0.73	42072	0.004	38228	3675	0.003	3420	34553	681330	323830
25	58000	0.005	51426	0.46	0.02	0.74	42913	0.004	38837	3675	0.003	3410	35162	716492	358992

Economics of Solar Photovoltaic Systems

TABLE 8.9
Net Present Values Analysis of the Financed Solar PV Project

Year	Annual Gen. (kWh)	Degradation Rate (%/year)	Degraded Gen. (kWh)	Feed-in Tariff ($/kWh)	Inflation for FiT	Fut. Val. of FiT (%/year)	Fut. Val. of Sold Energy ($)	MARR $(i-e)$ (%/year)	Pres. Val. of Sold Energy ($)	O&M Cost ($)	Pres. Val. of O&M Cost ($)	Pres. Val. of Earnings ($)	Cum. Sums of Pr. Val. Earnings ($)	NPV of the Project ($)	Residual Debt ($)
0															195000
1	58000	0.005	58000	0.46	0.02	0.46	26680	0.004	26574	3675	3664	22899	0	−142500	181851
2	58000	0.005	57710	0.46	0.02	0.47	27214	0.004	26997	3675	3653	23322	0	−142500	167622
3	58000	0.005	57421	0.46	0.02	0.48	27758	0.004	27427	3675	3642	23752	0	−142500	152250
4	58000	0.005	57134	0.46	0.02	0.49	28313	0.004	27865	3675	3631	24190	0	−142500	135673
5	58000	0.005	56849	0.46	0.02	0.50	28879	0.004	28309	3675	3620	24634	0	−142500	117823
6	58000	0.005	56564	0.46	0.02	0.51	29457	0.004	28760	3675	3610	25085	0	−142500	98630
7	58000	0.005	56282	0.46	0.02	0.52	30046	0.004	29218	3675	3599	25543	0	−142500	78018
8	58000	0.005	56000	0.46	0.02	0.53	30647	0.004	29684	3675	3588	26009	0	−142500	55911
9	58000	0.005	55720	0.46	0.02	0.54	31260	0.004	30157	3675	3577	26482	0	−142500	32224
10	58000	0.005	55442	0.46	0.02	0.55	31885	0.004	30637	3675	3567	26962	0	−142500	6873
11	58000	0.005	55164	0.46	0.02	0.56	32523	0.004	31126	3675	3556	27451	20233	−122267	0
12	58000	0.005	54889	0.46	0.02	0.57	33173	0.004	31622	3675	3545	27947	48180	−94320	0
13	58000	0.005	54614	0.46	0.02	0.58	33837	0.004	32125	3675	3535	28450	76630	−65870	0
14	58000	0.005	54341	0.46	0.02	0.60	34513	0.004	32637	3675	3524	28962	105593	−36907	0
15	58000	0.005	54069	0.46	0.02	0.61	35204	0.004	33158	3675	3514	29483	135075	**−7425**	**0**
16	58000	0.005	53799	0.46	0.02	0.62	35908	0.004	33686	3675	3503	30011	165086	**22586**	**0**
17	58000	0.005	53530	0.46	0.02	0.63	36626	0.004	34223	3675	3493	30548	195634	53134	0
18	58000	0.005	53262	0.46	0.02	0.64	37358	0.004	34768	3675	3482	31093	226727	84227	0
19	58000	0.005	52996	0.46	0.02	0.66	38106	0.004	35322	3675	3472	31647	258375	115875	0
20	58000	0.005	52731	0.46	0.02	0.67	38868	0.004	35885	3675	3461	32210	290585	148085	0
21	58000	0.005	52467	0.46	0.02	0.68	39645	0.004	36457	3675	3451	32782	323367	180867	0
22	58000	0.005	52205	0.46	0.02	0.70	40438	0.004	37038	3675	3441	33363	356730	214230	0
23	58000	0.005	51944	0.46	0.02	0.71	41247	0.004	37628	3675	3430	33953	390683	248183	0
24	58000	0.005	51684	0.46	0.02	0.73	42072	0.004	38228	3675	3420	34553	425236	282736	0
25	58000	0.005	51426	0.46	0.02	0.74	42913	0.004	38837	3675	3410	35162	460398	317898	0

TABLE 8.10
Net Present Values Analysis of the Financed Solar PV Project with 15 Years of the Loan

Year	Annual Gen. (kWh)	Degradation Rate (%/year)	Degraded Gen. (kWh)	Feed-in Tariff ($/kWh)	Inflation for FiT (%/year)	Fut. Val. of FiT (%/year)	Fut. Val. of Sold Energy ($)	MARR $(i-e)$ (%/year)	Pres. Val. of Sold Energy ($)	O&M Cost ($)	Pres. Val. of O&M Cost ($)	Net Pres. Val. of Earnings ($)	Cum. Sums of Pr. Val. Earnings ($)	NPV of the Project ($)
0														−142500
1	58000	0.005	58000	0.46	0.02	0.46	26680	0.004	26574	3675	3664	4112	4112	−138388
2	58000	0.005	57710	0.46	0.02	0.47	27214	0.004	26997	3675	3653	4535	8647	−133853
3	58000	0.005	57421	0.46	0.02	0.48	27758	0.004	27427	3675	3642	4966	13613	−128887
4	58000	0.005	57134	0.46	0.02	0.49	28313	0.004	27865	3675	3631	5403	19016	−123484
5	58000	0.005	56849	0.46	0.02	0.50	28879	0.004	28309	3675	3620	5847	24863	−117637
6	58000	0.005	56564	0.46	0.02	0.51	29457	0.004	28760	3675	3610	6298	31161	−111339
7	58000	0.005	56282	0.46	0.02	0.52	30046	0.004	29218	3675	3599	6756	37917	−104583
8	58000	0.005	56000	0.46	0.02	0.53	30647	0.004	29684	3675	3588	7222	45139	−97361
9	58000	0.005	55720	0.46	0.02	0.54	31260	0.004	30157	3675	3577	7695	52834	−89666
10	58000	0.005	55442	0.46	0.02	0.55	31885	0.004	30637	3675	3567	8176	61009	−81491
11	58000	0.005	55164	0.46	0.02	0.56	32523	0.004	31126	3675	3556	8664	69673	−72827
12	58000	0.005	54889	0.46	0.02	0.57	33173	0.004	31622	3675	3545	9160	78833	−63667
13	58000	0.005	54614	0.46	0.02	0.58	33837	0.004	32125	3675	3535	9664	88497	−54003
14	58000	0.005	54341	0.46	0.02	0.60	34513	0.004	32637	3675	3524	10176	98672	−43828
15	58000	0.005	54069	0.46	0.02	0.61	35204	0.004	33158	3675	3514	10696	109368	−33132
16	**58000**	**0.005**	**53799**	**0.46**	**0.02**	**0.62**	**35908**	**0.004**	**33686**	**3675**	**3503**	**30011**	**139379**	**−3121**
17	**58000**	**0.005**	**53530**	**0.46**	**0.02**	**0.63**	**36626**	**0.004**	**34223**	**3675**	**3493**	**30548**	**169927**	**27427**
18	58000	0.005	53262	0.46	0.02	0.64	37358	0.004	34768	3675	3482	31093	201020	58520
19	58000	0.005	52996	0.46	0.02	0.66	38106	0.004	35322	3675	3472	31647	232667	90167
20	58000	0.005	52731	0.46	0.02	0.67	38868	0.004	35885	3675	3461	32210	264877	122377
21	58000	0.005	52467	0.46	0.02	0.68	39645	0.004	36457	3675	3451	32782	297659	155159
22	58000	0.005	52205	0.46	0.02	0.70	40438	0.004	37038	3675	3441	33363	331022	188522
23	58000	0.005	51944	0.46	0.02	0.71	41247	0.004	37628	3675	3430	33953	364976	222476
24	58000	0.005	51684	0.46	0.02	0.73	42072	0.004	38228	3675	3420	34553	399529	257029
25	58000	0.005	51426	0.46	0.02	0.74	42913	0.004	38837	3675	3410	35162	434691	292191

8.4 ECONOMIC ASSESSMENT OF STAND-ALONE PV SYSTEMS

It is possible to perform the economic assessment of stand-alone or self-consuming PV installations using LCC-based approaches, studied in the above sections. For the calculation of annual cash flows, the simple cash flow technique ($C_t = \sum$ inflows $- \sum$ outflows) will be used as usual:

$$C_t = \frac{E_{GEN} \cdot (1-D)^t}{(1+i_{ScT})^t} \cdot ScT - \left[\frac{C_{OM(t)}}{(1+i)^t} + \frac{C_{R(t)}}{(1+i)^t} - \frac{C_{S(t)}}{(1+i)^t} \right] \quad (8.23)$$

where E_{GEN} is the annual generated energy, D is the system degradation rate for photovoltaics, ScT is the self-consumption tariff, and i_{ScT} is the annual inflation rate for ScT. Note that other parameters in equation (8.19) are already explained in Section 8.3.

Herein, there will be savings on the bills due to the self-consumption value of the energy. In general, there are two alternatives to determine the annual economic benefit of the PV system. While the first alternative is to purchase the required energy from the grid, the second one is to use a fuel-fired generator. In other words, the annual economic benefits can be evaluated by comparing the stand-alone PV system with these two alternatives separately. This is explained below with a worked example.

Example 8.12: NPV Analysis of a Stand-Alone PV System

Consider a 4-kW stand-alone PV system, of which financial and operating data are given below.

- Annual energy generation: 4350 kWh
- Degradation rate: 0.5% / year
- Unit cost of installation: 2750 $ / kW
- Operation & maintenance cost: 200 $ / year
- Discount rate, d : 4% / year
- Useful life of the installation: 25 years
- Replacement cost: Included in the initial investment cost.
- Salvage value: Ignorable

Calculate the *NPV* values at the end of 25 years for the following three different alternatives.

- Self-consumption tariff in the case of grid alternative: 0.24 $/kWh.
- 2.5 kW Oil-fired generator – Capital Cost $3500, energy cost 0.11 $/kWh.
- 2.5 kW Gas-fired generator – Capital Cost $4200, energy cost 0.14 $/kWh.

Solution:

NPV analyses are performed below for all three options:
Option 1: In this case, the saving amount due to the photovoltaic generation is 0.24 $/kWh. According to the given data, the NPV equation can be arranged as follows:

$$NPV = \sum_{t=1}^{N} \left[\frac{E_{GEN} \cdot (1-D)^t}{(1+d)^t} \cdot ScT - \frac{C_{OM(t)}}{(1+d)^t} \right] - C_0$$

$$NPV = \sum_{t=1}^{25} \left[\frac{4350 \cdot (1-0.005)^t}{(1+0.04)^t} \cdot 0.24 - \frac{200}{(1+0.04)^t} \right] - (4 \times 2750) = \$1320$$

Option 2: In the second case, the self-consumption tariff will be equal to the levelized cost of energy of the oil-fired generator. Therefore, let's first calculate the LOCE of the oil-fired generator with the help of equation (8.11). Note that the costs for O&M, replacement, and salvage are neglected. For the correct comparison, the generated energy of the oil-fired generator is assumed to be equal to photovoltaic energy generation.

$$\text{LCOE } [\$/\text{kWh}] = \frac{\text{Life Cycle Cost (LCC) } [\$]}{\text{Lifetime Generation } [\text{kWh}]} = \frac{C_0 + \sum_{t=1}^{N} \frac{C_{\text{FUEL}}}{(1+d)^t}}{\sum_{t=1}^{N} \frac{E_{\text{GEN}}}{(1+d)^t}}$$

$$\text{LCOE } [\$/\text{kWh}] = \frac{3500 + \sum_{t=1}^{25} \frac{4350 \times 0.11}{(1+0.04)^t}}{\sum_{t=1}^{N} \frac{4350}{(1+0.04)^t}} = 0.1615 \ \$/\text{kWh}$$

Hence, the self-consumption tariff is 0.1615 $/kWh. Now that the NPV calculation can be performed as follows:

$$\text{NPV} = \sum_{t=1}^{25} \left[\frac{4350 \cdot (1-0.005)^t}{(1+0.04)^t} \cdot 0.1615 - \frac{200}{(1+0.04)^t} \right] - 11000 = -\$3731$$

In the second option, NPV is negative. In other words, the oil-fired generator is more advantageous than the solar PV system. Because the photovoltaic system cannot save enough money due to the low energy cost of the oil-fired generator.

Option 3: Let's follow the same steps as Option-2 for the gas-fired generator system. Therefore, let's first calculate the LOCE of the oil-fired system:

$$\text{LCOE } [\$/\text{kWh}] = \frac{4200 + \sum_{t=1}^{25} \frac{4350 \times 0.14}{(1+0.04)^t}}{\sum_{t=1}^{N} \frac{4350}{(1+0.04)^t}} = 0.2018 \ \$/\text{kWh}$$

So, the self-consumption tariff is 0.2018 $/kWh. Now that the NPV calculation can be performed as follows:

$$\text{NPV} = \sum_{t=1}^{25} \left[\frac{4350 \cdot (1-0.005)^t}{(1+0.04)^t} \cdot 0.2018 - \frac{200}{(1+0.04)^t} \right] - 11000 = -\$1138$$

In the third option, NPV is still negative. In other words, the gas-fired generator is more advantageous than the solar PV system. However, the gas-fired system is more economical than the oil-fired system.

8.5 PROBLEMS

P8.1. Cash flows over the economic life of a utility-scale PV system are shown in Figure 8.2. The initial investment cost of the PV project is 10 M$ and its economic life is estimated to be 28 years. According to the agreement between the company and investor, the contractor will replace the solar invertors for $420,000 every 7 years. The expected annual revenue from the electrical energy sale is $770,000, the O&M cost of the PV facility is $35,000, and salvage revenue is $850,000. Assuming the discount rate of 6%/year, calculate the net present value of the PV project (Figure 8.4).

Economics of Solar Photovoltaic Systems

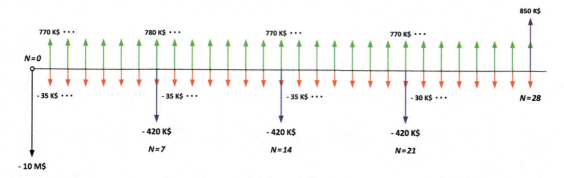

FIGURE 8.4 Cash flow diagram of the PV system over 28 years.

P8.2. Consider a 10 kW stand-alone solar PV system with a total surface area of 72 m². The unit price of 250 W solar panels is $300, each with a surface area of 1.8 m². The installation cost of metal frameworks is estimated to be 40 $/m². Besides, the initial cost of the inverter is $2000 and will be replaced every 10 years. Furthermore, the cost of the battery pack required for energy storage is $3000 and will be replaced at the same time as the inverter. Finally, the system installation cost is $1000, and the cost of BOS (balance of the system) plus mounting hardware is $1000. Assuming the O&M cost of 50 $/year and the salvage income as 10% of the capital cost, calculate the life-cycle cost of the PV system. Note that the life of the PV system is 25 years, and the annual interest rate is 5%.

P8.3. Consider three different 100 kW on-grid solar PV systems as described in Table 8.11. Compare the LCOE (levelized cost of energy) values for each PV system option based on the given data.

P8.4. Consider the present values of annual net cash flows given in Problem 8.2. Find the IRR of the solar PV project.

P8.5. Consider the annual net cash flows given in Problem 8.2. Find the payback period ignoring the time value of money.

P8.6. Consider the cash flows given in Problem 8.2. Find the payback period considering the time value of money.

P8.7. Consider a 100-kW grid-connected PV system, of which financial and operating data are given below.
- Annual energy generation: 120 MWh
- Degradation rate: 0.5%/year
- Feed-in tariff price: 0.45 $/kWh and has an inflation rate of 3%
- The unit cost of installation: 2500 $/kW
- VAT (Value Added Tax): 10% of the installation cost
- Operating cost: 200 $/year and has a valuation rate of 3%
- Maintenance cost: 1% of total installation cost and has a valuation rate of 3%
- Interest rate, i : 7%/year
- The inflation rate, e : 2%/year
- The useful life of the installation: 28 years
  - **a.** If the PV system is built with 100% own capital, perform the NPV analysis. In this case, find the payback period of the system.
  - **b.** Assume that 50% of the installation cost of the PV system was financed with a bank loan. If the loan interest rate is 5% / year, perform the NPV analysis and

TABLE 8.11
Solar PV System Parameters to Calculate LCOE

Parameters	Items		PV System-1 (Fixed Tilt)	PV System-2 (Single-Axis Tracker)	PV System-3 (Two-Axis Tracker)
Financial & Operational	Lifetime (N)		28 years	28 years	28 years
	Discount Rate (d)		6%	6%	6%
	Annual Generation		135 MWh	180 MWh	190 MWh
	Degradation Rate		0.5%/year	0.5%/year	0.5%/year
Capital Cost [$]	PV Array		$100000	$100000	$100000
	Inverter		$15000	$15000	$15000
	Metal Carcass/Tracker		$25000	$55000	$75000
	BOS + Mounting Hardware		$6000	$10000	$15000
	Installation		$4000	$7500	$12000
	Sub Total-1		**$150000**	**$187500**	**$217000**
O&M Cost [$]	Annual O & M		$500	$700	$950
	Sub Total-2		**$500**	**$700**	**$950**
Replacement Cost [$]		Year			
	Inverter	10	$15000	$15000	$15000
	Inverter	20	$15000	$15000	$15000
	Sub Total-3		**$30000**	**$30000**	**$30000**
Salvage Cost [$]		Year			
	Salvage (10 % of Capital Cost)	28	$15000	$18750	$21700
	Sub Total-4		**$15000**	**$18750**	**$21700**

find the payback period of the system. Note that the bank loan will be paid back with the annual net profit of the PV system.

c. Assume that the bank loan in option (b) will be paid back in fixed amounts in 12 years. In this case, perform the NPV analysis and find the payback period of the system.

P8.8. Consider a 12-kW stand-alone PV system, of which financial and operating data are given below.
- Annual energy generation: 13500 kWh
- Degradation rate: 0.5% / year
- Unit cost of installation: 2000 $ / kW
- Operation & Maintenance cost: 500 $ / year
- Discount rate, d : 5% / year
- Useful life of the installation: 25 years
- Replacement cost: Included in the initial investment cost.
- Salvage value: 10% of Capital Cost

Calculate the NPV values at the end of 25 years for the following three different alternatives.
- Self-consumption tariff in the case of grid alternative: 0.24 $/kWh.
- 5 kW Oil-fired generator – Capital Cost $5500, energy cost 0.12$/kWh.
- 5 kW Gas-fired generator - Capital Cost $8000, energy cost 0.15 $/kWh.

BIBLIOGRAPHY

1. S. Ay, "Mühendislikte Ekonomik Analiz", Birsen Yayınevi, Istanbul, 2021.
2. S. Syafii, Z. Zaini, D. Juliandri, and W. Wati, "Design and economic analysis of grid- connected photovoltaic on electrical engineering building in Universitas Andalas", *Int. J. Eng. Technol.*, vol. 10, no. 4, pp. 1093–1101, 2018.
3. ABB solutions for photovoltaic applications Group, "Technical application papers no.10. Photovoltaic plants", *Tech. Appl. Pap.*, vol. 10, no. 10, p. 107, 2010.
4. O. A. Adeaga, A. A. Dare, K. M. Odunfa, and O. S. Ohunakin, "Modeling of solar drying economics using life cycle savings (L.C.S) method", *J. Power Energy Eng.*, vol. 03, no. 08, pp. 55–70, 2015.
5. E. M. Barhoumi, S. Farhani, P. C. Okonkwo, M. Zghaibeh, F. Khadoum Al Housni, and F. Bacha, "Economic analysis and comparison of stand alone and grid connected roof top photovoltaic systems", in 2021 *6th International Conference on Renewable Energy: Generation and Applications (ICREGA 2021)*, pp. 223–228, Al Ain , UAE, 2021.
6. S. Bazyari, R. Keypour, S. Farhangi, A. Ghaedi, and K. Bazyari, "A study on the effects of solar tracking systems on the performance of photovoltaic power plants", *J. Power Energy Eng.*, vol. 02, no. 04, pp. 718–728, 2014.
7. S. B. Darling, F. You, T. Veselka, and A. Velosa, "Assumptions and the levelized cost of energy for photovoltaics", *Energy Environ. Sci.*, vol. 4, no. 9, pp. 3133–3139, 2011.
8. S. H. Han, J. E. Diekmann, Y. Lee, and J. H. Ock, "Multicriteria financial portfolio risk management for international projects", *J. Constr. Eng. Manag.*, vol. 130, no. 3, pp. 346–356, 2004.
9. F. J. Hay, "*Economics of Solar Photovoltaic Systems*", Institute of Agriculture and Natural Resources, University of Nebraska, Linconn, 2013.
10. C. S. Lai and M. D. McCulloch, "Levelized cost of energy for PV and grid scale energy storage systems", pp. 1–11, 2016.
11. J. Lee, B. Chang, C. Aktas, and R. Gorthala, "Economic feasibility of campus-wide photovoltaic systems in New England", *Renew. Energy*, vol. 99, pp. 452–464, 2016.
12. Z. Lu, Y. Chen, and Q. Fan, "Study on feasibility of photovoltaic power to grid parity in china based on LCOE", *Sustainability*, vol. 13, no. 22, p. 12762, 2021.
13. N. A. Ludin et al., "Environmental impact and levelised cost of energy analysis of solar photovoltaic systems in selected Asia pacific region: A cradle-to-grave approach", *Sustainability*, vol. 13, no. 1, pp. 1–21, 2021.
14. T. S. Ong and C. H. Thum, "Net present value and payback period for building integrated photovoltaic projects in Malaysia", *Int. J. Acad. Res. Bus. Soc. Sci.*, vol. 3, no. 2, pp. 2222–6990, 2013.
15. M. T. Patel, R. Asadpour, M. Woodhouse, C. Deline, and M. A. Alam, "LCOE*: Re-thinking LCOE for Photovoltaic Systems", in *2019 IEEE 46th Photovoltaic Specialists Conference (PVSC)*, pp. 1711–1713, Chicago, Illinois, 2019.
16. S. Rodrigues et al., "Economic feasibility analysis of small scale PV systems in different countries", *Sol. Energy*, vol. 131, no. May, pp. 81–95, 2016.
17. D. L. Talavera, E. Muñoz-Cerón, J. de la Casa, D. Lozano-Arjona, M. Theristis, and P. J. Pérez-Higueras, "Complete procedure for the economic, financial and cost-competitiveness of photovoltaic systems with self-consumption", *Energies*, vol. 12, no. 3, pp. 1–23, 2019.
18. D. L. Talavera, G. Nofuentes, and J. Aguilera, "The internal rate of return of photovoltaic grid-connected systems: A comprehensive sensitivity analysis", *Renew. Energy*, vol. 35, no. 1, pp. 101–111, 2010.
19. U.S. Department of Energy Office of Indian Energy Policy and Programs, "Levelized Cost of Energy (LCOE)", US Dep. Energy, p. 9, 2015.

Index

AC cables 239–241, 425–432, 484–485
AC combiner 324, 337, 341–342, 367–369, 433–436, 482, 485
AC-coupled 244–250, 252
AC distribution box 225
AC recombiner 324, 341–342, 367–369, 433–434
active power control 558
air mass 12, 13, 38, 84, 141–150, 192
albedometer 59–60
altitude angle 21–22, 25–27, 33, 39, 48, 49, 51–53, 55–56, 145, 278–181
altitude effect 146
amorphous silicon 108–109, 111–112
approximation method 96
atmospheric transmission 40
auxiliary supply 393–398, 483
auxiliary system 393

backup 199–202, 218, 244–247, 347, 359, 505, 515, 520
balance of system components 234, 580
battery backup 244, 247, 323, 515, 520
battery selection 288, 290
battery sizing 286–296
battery storage 4, 200, 244–253, 293–298, 307, 312, 437, 473–475, 494–496, 501, 510–511, 519–532
bidirectional metering 237, 239
bifacial PV cells 111, 112
bimodal inverters 210, 219
blocking diodes 172–176, 258, 494, 497, 513
bonding 243, 351–353, 436–440, 461, 463
busbar selection 431–433
busbar sizing 431–433
bypass diodes 112–113, 158–172

cable ampacity 239–241, 283, 314–317, 320, 420–422, 428–433
cable selection 396, 420, 422, 427–430, 497
cable sizing 396, 420, 422, 427–430, 497
cable systems 239
cadmium telluride 108, 109, 126
capital recovery factor 577–578, 588
cash flow models 573
cathodic protection 15, 493, 525–532
central inverter 206, 209, 323, 350, 361, 477, 481–482
characteristic curves 73, 164, 173–174, 228, 258, 260, 389, 400, 403, 410–411, 419–420
characteristic resistance 116–117
charge controller 228, 234, 262, 287, 298–303, 437–438, 494–511
charge controller sizing 298–303, 437–438, 494–511
charge controller selection 298–303, 437–438, 494–511
chemical energy 2, 4
CIGS-CIS 109
circuit breaker 231, 317–322, 323–340, 342–345, 380–384, 395–397, 404, 513
circuit breaker selection 231, 317–322, 323–340
circuit breaker sizing 231, 317–322, 323–340
clear-sky 37, 40, 48–49

clear-sky radiation 37
clock time 29, 31, 37
combiner box 15, 201, 208–209, 212, 234, 273, 283–287, 324, 341, 434–437
communication line 224, 225, 365
communication system 236–237, 365, 493, 503–507
component selection 15, 253–253
concentrated PV Cells 110
contactors 232–234, 312
conversion factors 9–10, 42, 190
copper indium gallium selenide 109
correction factors 421–423, 427, 431
cross section calculation 431, 457
crystalline silicon 65, 67, 74, 107–111, 131
current-grading 405–407
current transformer 384–388, 392, 399, 403, 412, 417

daily radiation 41
daily tilt angle 183–185, 188
daylight duration 32–37
DC cables 398, 422, 424, 503
DC combiner box 15, 208–209, 283–285
DC-coupled 244–246, 250–251
DC distribution box 208
DC fuses 208, 223, 227–229, 286
declination angle 16, 20–24, 30, 278
design yield 513, 515, 518, 520, 523–524
direct-coupled 248, 254–258
disconnectors 15, 221, 229–231, 317, 321
discounted payback period 587
distribution box 208–212, 220, 431
double-diode model 106–107
dust deposition 155–156
dust effect 154
dye-sensitized 110

earth's latitude 18–20
earth's longitude 18–20
earth's orbit 15–17, 31, 56
economic assessment 573, 588, 593
economic parameters 573, 588
economics of photovoltaic systems 573, 575
electrical characteristics 70, 72, 241, 310, 313
electrical energy 1, 2, 156, 237, 248, 309, 312, 473, 495, 501, 508–509, 518, 521, 524, 530, 532
electromagnetic energy 2, 5, 68
energy bands 66, 67
energy conversion 1, 2, 5, 10, 65, 108–109, 142
energy conversion factors 10
energy forms 1
energy source 11, 199, 247
energy units 3, 9
extraterrestrial radiation 12, 16, 17, 38, 40, 41, 43–44

fill factor 92–94, 115
forms of energy 1, 2, 7
frequency limits 557
frequency regulation 557–558

frequency ride-through 566
fuse selection 312, 314, 316–318
fuse sizing 312, 314, 316–318

Gallium Arsenide 109, 126
graphene PV Cells 111
grid codes 539, 540, 559–560, 565–566
grid-connected PV systems 177, 200, 244, 247, 399, 588
grid inverters 210, 306, 309
grounding 242–243, 355, 382, 436–460
ground-mounted 348, 351–354, 361, 437, 441, 464, 465, 468–469, 507

harmonic limits 15, 563
hosting capacity 545–548, 552
hotspot phenomenon 159
hourly radiation 40, 41, 42
hourly tilt angle 177–178
hybrid inverters 210, 219
hybrid PV cells 111
hybrid PV systems 250–251
humidity effect 157

icing effect 157
ideal PV model 77–79
IDMT relays 404
incidence angle 26–29, 148–149
insolation 40, 47–48, 190–191, 255, 500, 521
insolation on tracking surfaces 47–48, 190–191
instrument transformers 384
interconnection optimization 461
internal rate of return 584–585
inverter selection 304–311, 525
inverter sizing 304–311, 525
I-V equations 94

junction box 159, 162–163, 193, 350, 358

kinetic energy 2, 4–6, 8

land area requirements 270, 276–278
land requirements 270, 276–278
latitude 18–20
levelized cost of energy 581–582
life-cycle cost 580–581
light energy 1, 3–4
lighting systems 295, 493, 497–498
lightning protection 208, 348–352, 361–366, 463–465, 571
lightning system 351, 463
linear kinetic energy 4–5
load requirements 253–254, 262, 304, 471, 493–495, 499, 508, 518, 521, 524
load switches 231
longitude 18–20

master-slave concept 202–203, 206, 210, 214
mathematical modelling 98, 164
mechanical energy 2, 5–6
microcrystalline 110
mismatch effect 86
module inverter 203
monitoring systems 237, 547
monthly radiation 42
monthly tilt angle 188, 501, 509, 521

monocrystalline 108
multi-junction 110
multi-string inverter 204

net metering 237, 239
net present value method 575–576
network impedance 330–340, 435–436
Newton-Raphson 101–106, 118–122
NOCT 99, 112–113
nominal operating cell temperature 99, 112–113
NPV 575–576, 588–589
nuclear energy 2, 6

off-grid system 188, 252, 300–304, 307
on-grid system 304, 307
operating temperature 112, 129–130, 239–240, 270, 474, 504
operating time 399–400, 403–404, 407–409, 419
optimum daily tilt angle 185
optimum tilt angle 177–189
orbit of earth 15
organic PV Cells 110
overcurrent protection 227, 232, 283, 314, 315, 317, 322, 399, 405, 412, 422
overcurrent protection relays 399

parameter estimation 98–106, 123
partial shading 159, 164–166, 172–174
payback period 586–587
peak solar hours 254
perovskite PV cells 111
photon energy 3, 68–70
photovoltaic applications 493–525
photovoltaic array 77–78
photovoltaic cables 239
photovoltaic equations 94, 98, 164
photovoltaic inverters 210
photovoltaic module 77–78
photovoltaic technologies 107–112
pickup current 399–403, 408–411, 414–415, 418–419
PID 458–461
plug setting multiplier 399
p-n junction 65, 70–75
p-n junction diode 72–75
point of common coupling 199, 323, 325, 411, 540
polycrystalline 108–110
potential energy 2, 7–8
potential induced degradation 458–459
power cables 239–240, 424–428
power quality 539
present value method 574, 575
problems 10, 61, 125, 192, 475, 533, 567, 594
protection angle method 463, 464
protection devices 223–232, 312–317, 343–344, 371, 374, 404
protection relays 399, 411
protection system 223, 312, 463, 525
protection system selection 312, 463
protection system sizing 312, 463
PV array 77–78
PV array sizing 77–78, 252
PV cell 75–76
PV cell efficiency 92–93
PV cell operating temperature 129

Index

PV module 76–77, 129
PV performance 138, 141
PV system sizing 252
pyranometer 57–59
pyrheliometer 59

quantum-dot PV cells 111

radiation calculations 37–43
radiation measurements 56
radiation on horizontal surfaces 39, 43
rain effect 157
rapid shutdown 234
RCD 226, 371
RCD selection 371
reactive current 464, 566
reactive power control 559
recombiner box 283–287, 341, 434–436
relative voltage change 540–545, 549–557
relay coordination 404, 408–409, 415
remote monitoring 525, 527
residual current device 226, 371
residual current device sizing 371
rolling sphere method 464, 467, 469
rotational kinetic energy 2, 5–8
roof-mounted 51, 348–352, 360, 383, 423, 426, 465–466

semiconductors 65–70
seasonally tilt angle 186–190
series-parallel connections 94
series resistance 80
shading analysis 53–54
shading effect 168, 174, 253, 270, 277
shadow analysis 50, 53
shadow geometry 51, 270, 277, 278, 281
shadow types 50–51
short circuit analysis 323–340
shunt resistance 79, 81, 159, 160
silicon lattice 66
simple approximation 96–98
simple payback period 586
single-diode model 94, 98–99
snow effect 157–158
soiling effect 155–157
solar angles 20–27
solar energy 1, 3–4, 254
solar geometry 14
solar incidence angle 26–29, 140
solar radiation calculations 37–43
solar spectrum 11–14, 142
solar time 29–31, 40
solar tracking 47–50
solution methods 94, 96
sound energy 2, 8–9
SPD selection 341–364

stand-alone inverters 216–219
stand-alone PV systems 247, 248, 280, 253–256, 260, 264, 593
steady-state frequency limits 557
steady-state voltage deviation 543–547
step voltage 439, 448–451, 455
string inverter 204, 214, 215, 269, 477
string ribbon 108
substation layout 222
sunlight recorder 59
sunrise time 28, 32–33, 36–37
sunset time 28, 32–33, 36–37
surge protection device 208, 220, 312, 323, 241, 356
surge protection device sizing 341–364
switch disconnector 221, 231, 317–320
switch disconnector selection 317–320
switch disconnector sizing 317–320

team concept inverter 206
technical specifications 112, 214–215, 300, 306, 309, 311
telecommunication systems 493, 503–504
temperature effect 88, 132, 135
thermal energy 2, 3
thin-film technologies 108–109
three-phase connection 219
tilt angle determination 175–190
time-current-grading 405, 407
time-grading 405
time setting multiplier 399, 401, 403
touch voltage 439, 448–451, 455
total head 513, 515, 520, 523
two-diode model 83

utility connection 220–222

voltage deviation limits 543
voltage drop 74, 170, 398, 420, 424–431, 498, 527, 541–542, 553–556
voltage flicker 548–550, 556
voltage flicker limits 548
voltage regulation 559
voltage ride-through 563
voltage rise 544–547, 551, 557, 564
voltage transformer 384, 387, 388, 390–392
voltage unbalance limits 560

warning signals 493–494
water pumping systems 511–512
wavelength 3, 11–14, 68–70
wind effect 154

yearly tilt angle 186–188

zenith angle 13, 21–22, 26–27, 43, 141, 142, 145, 178

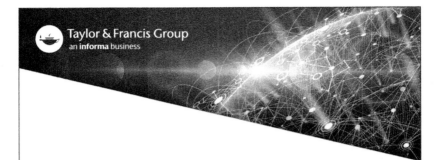